YIELDING FRUIT

The Life and Times
of Royce S. Bringhurst

YIELDING FRUIT

The Life and Times
of Royce S. Bringhurst

by John R. Bringhurst

To my Dad

Introduction

The name of Royce S. Bringhurst has been associated with strawberries for as long as I can remember. As his son, I have long been aware of my father's worldwide prominence as a geneticist and strawberry breeder, but had only the most general idea as to the actual nature of his work and contributions. When, after his death in November of 2005, our family determined to give his personal papers and professional materials to a major library for preservation, it became my duty to prepare those papers for donation. In the process of doing so, I became aware that at some point my father had intended to make a written account of his life, which was likely prevented by his failing health. It was then that I struck upon the idea of writing what I thought at the time would be a brief history of his life and work, both as a keepsake for his descendants and as a supplement to his collection of papers, placing them into their proper context.

I began writing this history during my spare moments in the years following Royce's death, but quickly discovered that I had undertaken a far greater and more complex task than I had ever imagined. Royce had left an immense collection of personal writings in the form of letters and journals, which formed a remarkable chronicle not just of his own life, but of the times in which he lived. I had not gone far into these before a rich tapestry began to emerge, formed by the intertwining threads of individuals, places, and events, only some of which I had even heard of, but which began to come alive for me as they had flowed from the pen of my father while those events were taking place around him. Royce was a thoughtful and expressive writer, and I increasingly became convinced that his history would be a far more authentic one if it were told using his own words. I therefore determined to quote freely from his letters and journals, allowing them to tell the story whenever possible, leaving the unschooled grammar and spelling of his younger years largely as I found them, and making adjustments in his writing only as necessary to facilitate the flow of the narrative. Not only has this allowed for a more genuine account of the events, it also illustrates his growth and development as nothing else could, and paints a much more vivid picture of Royce S. Bringhurst as a complex and conflicted individual, who as a scientist rose to become the world's foremost authority in his field, but who as a man grappled with fundamental issues of doubt, faith, flaws of character and fears of mortality.

Some time after the death of our mother, Pearl D. Bringhurst, in 2013, I took early retirement from my work as a physician in order to dedicate more time to the completion of this history. This has been a nearly full-time occupation for most of the last two years, something of which I suspect my father would never have approved. As I have done so, I have become increasingly convinced of the value of Royce Bringhurst's life story, not just to family members and descendants whose curiosity might be satisfied thereby, but also to his former colleagues and his successors in the fields of plant breeding and genetics, and more generally as an intriguing chronicle of the times in which he lived and worked. This biography should not be construed as a scholarly work in the conventional sense, as I have not had formal training in academic scholarship and have made no attempt at exhaustive research; indeed, this history makes relatively little use of outside sources, relying chiefly upon the vast collection of materials which Royce left in possession of the family. At the same time, my position as a family member and my many years of close contact, both as a child and as an adult, have given me the unique capacity to synthesize these materials into a more authentic portrait than would have been possible from research alone. I consider those segments in Royce's own words to be of greatest value, as they bear the stamp of his intellect, his wit, and his great sense of humanity.

Materials Used

The sources cited in this history come almost entirely from the Royce S. Bringhurst collection, which was donated in March of 2016 to the Special Collections and Archives section of the Merrill-Cazier Library at Utah State University. Most information on Royce's ancestry and early years was taken from family records and from recorded interviews of him and his wife Pearl, supplemented by an interview of his brother George by my cousin (and credentialed historian) Newell Bringhurst, which gives some excellent detail about early family life. Information about his schooling and early college years is taken from a personal diary which Royce kept from 1936 until 1939, supplemented by a few letters. Royce's mission for the Church of Jesus Christ of Latter-day Saints was well documented by journals which he kept faithfully on a daily basis, and by frequent letters home to Pearl and to his parents, all of which are found in Royce's collection. The period of his early marriage is, by contrast, little documented, and had to be pieced together from a few letters from Pearl to her mother, and from later recollections of that happy interlude from letters exchanged between Royce and Pearl during the Second World War. The war itself, like the mission, was vividly documented through journals, letters, and a few other documents.

After Royce's war service, he kept no journals for many years, and wrote few letters except of a professional nature, and only a few of those were available to me. Information from this time came chiefly from letters from Pearl to her mother, which she wrote faithfully from the time she and Royce left the Salt Lake City area shortly after the war. These are occasionally supplemented by letters from Royce to Pearl during extended absences, or by later recorded recollections of family members. Pearl's letters to her mother ceased some time before her mother's death in 1975, but in 1976 Royce began again to keep a daily journal which eventually extended to eleven volumes, ending only after the conclusion of his mission to Chile in 1994. These journals provide the bulk of information for that period. By 1994 Royce had developed a serious tremor and was losing the ability to write, and information on his final years depends mainly on a series of personal journals kept by Pearl until well beyond the time of his death. I have also relied to a limited extent on my personal recollections, as well as recorded interviews of Royce, Pearl, and other family members and colleagues to fill in details of these later periods.

Information on Royce's professional work has been gleaned from a variety of sources. A few specifics were found in the small yellow field journals which were used to make notations during his professional work in Davis, and some came from a very limited collection of professional letters dating from the early 1960s—especially interesting in this regard was Royce's long correspondence with Henry A. Wallace, former Secretary of Agriculture and former Vice President of the United States. However, the most important source of information on Royce's professional work came from the scientific papers he authored or co-authored, both published and unpublished, beginning with his doctoral thesis at the University of Wisconsin. Since these papers were mostly written to be read by trained geneticists and agronomists and are therefore often complex and highly technical, I have endeavored to convey the essence of them in layman's terms so they can be appreciated by individuals without advanced training in the field.

I realize that one disadvantage of using Royce's own writings to tell the story is the exaggerated emphasis it places on those periods of his life when he happened to be keeping a journal, or when circumstances required him to write letters. Royce's mission, for example, encompasses less than thirty months of his life, yet occupies a full chapter, and his military experience, a period only slightly longer, was so long in the telling that it had to be divided into two separate chapters. Yet to me these chapters are among the most important, for not only do they offer a glimpse into one of the momentous periods of world history, they also provide an illustration of how those events helped to shape Royce as an individual. In a similar way, while Royce's later international travels may seem to receive undue emphasis because they were more amply documented in his letters home, they also serve as valuable windows into Royce's character and thought, and some of them marked legitimate transition points in his life as well.

Final Remarks

I acknowledge that in the process of my writing I have failed to do justice to many of Royce's friends and associates, and where some are mentioned, it has often been without a true appreciation of their relative contributions to their field, or even of their relative influence in my father's life. Royce Bringhurst had a broad range of personal and professional associations, and such large numbers of individuals appear as single threads in his written records that it would have made the history impossibly complex to make meaningful mention of the vast majority. Of these multitudinous "threads" I have had to select a few to weave in and out of his story in order to bring some interest and cohesiveness to the whole. Some of these were included (or excluded) somewhat haphazardly, based either on my personal familiarity with them or on a chance emphasis given by Royce in his journals, neither of which may truly reflect the role they played in his story. In like fashion, few members of the extended family appear except as they happen to intersect the story line of Royce's life as I have told it; for example, of his grandchildren, only a few of the oldest are mentioned at all. This again should not be construed as a reflection of their relative importance, but only of the fact that this story is about Royce's life, and in that story all others, including myself, play a secondary role, serving only as pivot points in its telling.

In composing this history, I have constantly faced the challenge of condensing the large volume of written material Royce left into a clear and coherent narrative, and avoid lapsing into a mere chronicling of events. In the process I have had to make hard choices about what to include, always with the awareness that a fair portrayal of Royce S. Bringhurst's life is my principal objective. Family members may be disappointed that the history does not include many cherished family events and recollections, or touches on them only lightly, while those interested chiefly in Royce's professional life and scientific work might be dismayed at the lengthy treatment given to his personal and religious thought, or at the limited mention made of some aspects of his work, which had to be deduced from his publications and often sketchy writings. I have tried to compensate for these limitations by making liberal use of endnotes to include additional information which would have distracted from the flow of the narrative. Photographs have been placed at the conclusion of each chapter individually, in an effort to keep them within their historical context. In the end it became necessary to remove many personal vignettes and recollections entirely to reduce the book to a publishable length, though for the benefit of family members and other interested parties, I have retained an electronic version which includes many of these.

I recognize that the way Royce's story is recounted in this book is partly a reflection of my own history and perceptions, and that it would have been told quite differently from other points of view. My personal experience as a college student and as a missionary, for example, strongly influenced the writing of the chapters on Royce's own college and mission experience, and my religious convictions are obviously reflected in the account of Royce's beliefs, which in fairness are similar to and helped to shape my own. One should not suppose, however, that I have simply superimposed my own values and perceptions upon my father, or attempted to paint an idealized picture of his life. That might indeed have been the case had he not left such an abundant store of written records, for Royce was, despite appearances, a very private man, even to his family, and only his journals appear to have served as the repository for his innermost feelings. When I began to open those journals for the first time, I felt no small trepidation as to where they might lead me, and it was only after reading them that I became aware of the depth of my father's personal struggles and doubts, and the factors which induced him to spend the latter years of his life in a quite earnest quest for personal refinement and reformation, which began, coincidentally, at about the same age at which I have undertaken the writing of this history. I have tried therefore whenever possible to let Royce's words speak for themselves, though to the extent I have introduced any bias of my own, it is likely in a direction sympathetic to his character.

Knowing that it was Royce's original intent to write this story himself, I am also aware that, like me, he was concerned not just with how the story would reflect upon him, but also with how truly authentic it would be. By Royce's later adulthood, the writing of personal history had become almost a tenet of the Latter-day Saint faith, and after one church meeting in which the topic was discussed, Royce had written, "It is a good idea, but when to do it—no time now, I say, but wait until you 'have time' and most of it will fade into oblivion with the fading of memory, and that which gets written often degenerates into telling it the way it would have been in the best of all worlds if we had it to do over again, rather than as it was. They tend to encourage this by telling us to 'accentuate the positive' which on the other side reads out, 'obliterate the negative.' With that sort of telling the human part of us is lost. 'Let him who is without sin speak,' or rather, 'inflate your virtues, mute your vices,' and everyone will think you were great, except those who really knew you and God himself. Our desire to be something much greater than what we are makes 'fiction' of much of what we wish to be known as our 'history.' I am as guilty as the next about this, and yet within my own heart, I am most painfully aware of my fragile feet of clay. The public mask that I have created for myself is but a shadow of the real, weak-willed individual that I am. My only strength is my dogged dedication to routines and patterns, and I have managed to make that into less than a virtue all too frequently. I continue to be that same unfulfilled soul who 'wishes' he were stronger and better." (Davis Journal, volume 2, January 22, 1977, content condensed and edited.)

In this last statement, I believe my father expressed the wish of every decent human being, and the degree of frankness and depth of feeling conveyed in his writings leads me to conclude that he did in a real sense write his own history, and that history is a very authentic one indeed. That he had many imperfections, even serious ones, we of his family are well aware (having spent our childhoods with him in close quarters and having seen those aspects of him that would have remained hidden to the outside observer). But in the end despite his flaws, I believe we all were able to see him for what he was—a man who, though very human and scarred by early life experiences, was also possessed of great human kindness and a thoughtful spirit, who by the time of his death was deeply and deservedly honored by colleagues and friends, and revered and respected by his children and descendants. I hope in some measure this biography may do justice, not only to his life and memory, but to him as a person.

John Royce Bringhurst
February, 2016

Acknowledgments

It is an impossible task to acknowledge all who have contributed in some way to the completion of this book. I am grateful to many family members, both living and dead, who have helped in various ways. Of those in the immediate family, my sister Florence Nielsen read through the preliminary manuscript and made many suggestions, while my other sisters, Jean Andersen (now deceased), Marla Vaughn, Ann Huffaker, and Margaret Dobbins provided valuable recollections and helped to put some of the events into their proper context. My Mother, Pearl Davidson Bringhurst, contributed her substantial store of letters and journals as well as many of her own remembrances before she died.

No one provided greater assistance than my cousin Newell Bringhurst (Royce's nephew), a noted historian and accomplished scholar. Newell treated Royce's history as part of his own, and read the manuscript in its entirety, making countless suggestions which were of inestimable value—I suspect the book would be better than it is had I followed them all. Newell also provided valuable information which helped to fill in the gaps in my father's past, and gave me far more encouragement than my abilities deserved.

I am grateful to Royce's former colleagues such as Douglas Gubler and Robert Webster for their insights, and for the encouragement and kindness of former students such as James Hancock and Daud Khan. Hamid Ahmadi spent long hours helping me to understand some of the details about the later part of Royce's life, and I thank him for his many kindnesses to my father when he was in his declining years. Douglas Shaw, at a turbulent time in his own life, graciously provided information about his own relationship with Royce in the transitional years when he assumed responsibility for the university strawberry program, and filled me in on the subsequent progress of the program. For all these and the numerous professional and church associates of Royce who have offered varying degrees of support, I express appreciation.

Finally, I thank my wife Betty for things too numerous to mention—for her forbearance as I delved through the large collection of my father's papers which were scattered about her normally tidy home for extended periods, for her patience as the writing progressed, for her encouragement and companionship as the book finally took shape, and for the sense of perspective which she gave the entire endeavor. As the book was nearing completion, she took many hours of her time to read through the entire manuscript, making valuable suggestions and correcting questionable grammar. Above all, I thank her for her continuous and selfless service to both my parents, including my father, the subject of this book, during the waning days of their lives. Any tribute I have rendered to my parents through my writing after their deaths, she far exceeded through her kindness and care for them while they were still living.

Contents

Origins and Childhood

Chapter 1

This is the story of a life and of a time. The life was that of a Utah farm boy who would leave home and farm, face adventures and dangers, and rise to become an outstanding scientist and the world's foremost expert on the strawberry. The time was an age of astonishing technological progress, spanning forward from the days when the automobile had only recently become commonplace, and when daring pioneer aviators had just begun to expand the limits of manned flight. The life would be a long and full one as lives go, replete with perils, joys, journeys, struggles, and close connections with the great and the lowly. The time would witness colossal wars, the clash of conflicting political systems, and sweeping social changes, accompanied by life-changing technological advances. That the time helped to shape the life is without question. That the life helped to shape the time is also true, though on a smaller scale, and more broadly in the sense that it was of many such lives, collectively acting and interacting, that the time itself was constituted. The time begins in the waning days of 1918, just after the conclusion of World War I, then referred to simply as the "Great War," or more hopefully, "the war to end all wars." The life was that of Royce S. Bringhurst.

Ancestry and Family Origins

Royce S. Bringhurst was born to a family descended from pioneers who entered the Salt Lake Valley in 1847 with one of the first immigrant companies, though the family history extends back to much

1

earlier beginnings. The Bringhurst family had its origins in a village called Bringhurst in Great Britain, some distance northeast of London. Those in America trace their ancestry to an Englishman named John Bringhurst, whose father, of the same name, became a printer in London in the 1600s. A member of the Society of Friends, commonly known as "Quakers," he was arrested for his publications and suffered religious persecution in England, and consequently he removed for a time to the more religiously tolerant community of Amsterdam, Holland. After his death, his widow and family emigrated to America, seeking religious freedom in the newly-established Puritan colony of Pennsylvania,[1] where some of his descendants, including Royce's direct ancestors, established themselves in the carriage-making trade.[2] It was there, generations later, that Royce Bringhurst's great-grandfather, Samuel Bringhurst, together with his brother William, came in contact with "Mormon" missionaries and were converted to the Church of Jesus Christ of Latter-day Saints, thus alienating themselves from the community of Quakers, who with their family issued a "Declaration of Disownment."[3]

These two brothers relocated to Nauvoo, Illinois in 1845, shortly after the martyrdom of the Church's founder Joseph Smith. Nauvoo, a city on the Mississippi River at the westernmost border of Illinois, had become the headquarters of the Church and the gathering place for the Saints, but by the time the Bringhursts arrived there, friction between the "old citizens" and the followers of the strange new religion had grown intense, and it had become clear that the Church members would not be allowed to remain. As Samuel Bringhurst was a carriage maker and mechanic, having mastered the family trade back in Pennsylvania, he became an important member of the community as rapid preparations were made to flee the storm of religious persecution and move west by wagon into the great American wilderness.

Great-Grandparents
Samuel Bringhurst with his wife, Eleanor Beitler[4] Bringhurst, fled Nauvoo with other Church members during the winter in the opening months of 1846. Crossing the Iowa territory with the Saints that year, they stayed several months in Winter Quarters, Nebraska[5] before coming west in the spring of 1847, in one of the first wagon companies led by John Taylor, who later became president of the Church. After the usual privations of such a journey, they arrived in the Salt Lake Valley on September 29, 1847, and shared the sparse winter conditions of those first settlers. Samuel was 34 years old at the time of his arrival. His wife, Eleanor Beitler Bringhurst, became a close friend to Eliza R. Snow, the famous Latter-day Saint poet and community leader.[6]

2

As Salt Lake City developed, Samuel continued his trade as a mechanic, and opened the first wagon and carriage repair and blacksmith shop in the Utah territory.[7] After living in Salt Lake City for a time, Samuel and Eleanor Bringhurst eventually settled with their family in the area of Taylorsville, somewhat to the south of Salt Lake and west of the Jordan River, which cuts northward through the valley. This was considered a less desirable part of the valley. Royce recounted in the later years of his life: "They ended up in the west side, in a place that's come to be called Bennion, or the Bennion Ward. We called that the dry farms out there." To this, Royce added a bit of family folklore to explain this choice of location: "He had some kind of falling out with the authorities. Part of it had to do—according to my father, now, this is what my father said. It may or may not be true; my father had his biases too—he said that they were putting pressure on him to marry another woman, to practice polygamy, and he said, speaking of Eleanor Beitler, his wife; 'That woman has suffered enough, I shall not impose that upon her.' This is what my father said that he said, and this put him at odds with the Church authorities, and they moved out, because they were there in the very early period of colonization, the first year."[8] Whether or not the story was strictly accurate, it is a fact that while plural marriage was actively practiced and defended by the Latter-day Saints of the time, none of Royce's direct progenitors of the Bringhurst line ever entered into it.[9]

Samuel and Eleanor Bringhurst had three living children on their arrival in the Salt Lake Valley in 1847, and five more were born after their arrival. One of these was Royce's grandfather, John Beitler Bringhurst, born June 13, 1854.[10] As the community around them grew and flourished, Samuel Bringhurst worked as a farmer, but continued his trade as a carriage maker as well. At one time either he or his son John constructed a miniature wagon, complete with spoked wheels, for the children of the family. Royce (a great grandson) remembered having played with the toy wagon until it fell apart many years later. Samuel died at age 75 in April of 1888, and Eleanor survived him only three months, dying at age 73 in July of the same year. Both were venerated as pioneers—in Royce's own words, "The detailed pioneering was done by these folks. They were on the front lines, and they were there in the first few years of colonization."[11]

Royce felt a strong sense of heritage as a descendant of the first Utah pioneers. During his first extended journey outside of Utah, while serving as a missionary in California, he wrote a youthful tribute to his home state: "It's the best place to live in the whole wide world. The people there are stronger and finer than the people of other places because the mountains have made them that way. One is always impressed to see those mountains or pictures of them, but I can't

3

ever see them without having my bosom swell because I know that my ancestors helped to conquer them, and I can't see the pictures of the beautiful Utah valleys without thinking at the same time of those courageous pioneers who indeed made the desert blossom as a rose through their hard work. I have often thought of the wisdom of the Lord in having them settle in that uninviting place rather than coming thru to California, as did Samuel Brannan[12] and others. We would have lost our identity as a people in a relatively short time as many of our people do when they come here even now."[13]

Grandparents

Royce's grandfather, John Beitler Bringhurst, attended school in the winter, and in the summer worked with his father in the blacksmith and wagon shop.[14] In 1876, at the age of twenty-one he married Emma Frances Tripp, a girl of seventeen and daughter of pioneer converts to the Church who like her husband had been born and raised in the Salt Lake Valley. Emma's father, Enoch B. Tripp, originally from Maine, had traveled to Nauvoo, Illinois, as the Saints were making preparations to leave. He initially intended to take advantage of low land prices from the forced sale of property there, but while in Nauvoo he was converted to the new faith and joined the Church. Enoch Tripp was asked by Church leaders to remain behind for a time, but he eventually made his way to Salt lake in the 1850s, where he participated in the institution of "plural marriage." Emma Frances Tripp's mother was Jessie Eddings, daughter of George Eddings, from a family of converts from England, and was the fourth wife of Enoch B. Tripp.[15]

In 1880 John Beitler Bringhurst and his wife Emma moved to Bennion where, on the old Bringhurst farm, John built the adobe home in which Royce was later born. In politics he was active in the Democratic party, an affiliation which continued in the family and which was carried on by Royce.[16] He was active in public life, serving as Justice of the Peace, constable, road supervisor, and school trustee.[17] Royce said of him, "My grandfather I remember well. He was a tall spare man, bent over just a little bit. He was about 6 feet 2. And he made my father look like a midget, because my father was very small. That comes largely from the Tripp family. They were small people."[18] John Beitler Bringhurst was a farmer on the original land settled by the family in Bennion. He and Emma had eleven children, and the fourth of these was Royce's father, John Tripp Bringhurst, who was born on August 11, 1883. John Tripp Bringhurst, known to friends as "Jack", was the first boy in the family, and though well-respected, he was known for his reserve.

The two brothers that followed John Tripp Bringhurst in the fami-

4

ly both became well known, but for quite opposite reasons: William Albert Bringhurst, Jack's closest brother, just two years his junior, became the notorious "Bill Bringhurst" who, after serving as a missionary for the Church, fell into bad company and spent time in the Utah Penitentiary for a rash of robberies, and eventually was executed in California's San Quentin Prison for his part in the shooting of two policemen after a failed robbery in Los Angeles, California. Though his activities were never mentioned in family gatherings, they brought a cloud over the family's name and reputation during Royce's formative years.[19] By contrast, the next brother, Samuel Enoch Bringhurst, six years younger than John Tripp, became prominent in Church affairs and served as president of the Swiss-Austrian mission beginning in the 1940s. Under assignment of Church president David O. McKay, he selected the site for the Swiss Temple, the first LDS Temple to be constructed outside the United States, and later served as the first temple president there.[20] A younger brother, Arthur B. Bringhurst, achieved prominence in his own right. Wounded in the First World War, he later served many years as a highly respected Justice of the Peace in Salt Lake City, though he was known to Royce simply as "Uncle Art".

Royce considered John Beitler Bringhurst's generation to be his strongest link with his pioneer past, and when Royce was himself in his advancing years he would remark, "I can remember my grandfather's generation well, and they remembered the pioneers who were still around. I have the best recollection of my Uncle Sam and Aunt Sarah, then Uncle Louie.[21] The rest are more vague. I do remember my Grandfather Bringhurst, and how big he was compared with his wife (my Grandmother Tripp). In her last years I remember that he picked her up like a child. Her last years were sad because of my father's brother William. I still find that part of our family history difficult to think about and accept, although now it is history. I suppose every family has its bad members."[22]

Parents

On January 7, 1886, some two years after the birth of John Tripp Bringhurst, Royce's mother, Florence Elizabeth Smith, was born to George Fred Smith and Elizabeth Ann Newman Smith, both also descendants of early Latter-day Saint pioneers. Florence's mother died when she was still young, and it was her stepmother, Mary Elizabeth Taubman Smith, whom Royce would remember as his grandmother. The Smiths farmed in an area called Cottonwood in the southern part of the Salt Lake valley, where in addition to growing apples, sugar beets, and livestock they became successful at growing and selling strawberries. Through their farm wound the old wagon

5

road over which the original blocks for the Salt Lake Temple had been transported by ox cart. On the north side of this long winding path, known as Vine Street, they had constructed a comfortable home, purportedly built on a foundation made of granite blocks from the temple quarry itself.[23] It was in front of that home on July 6, 1906, that John Tripp Bringhurst and Florence Smith were introduced to each other, and after a pleasant courtship of nearly a year, they were married in the Salt Lake Temple on June 12, 1907,[24] in a double wedding with Jack's younger brother William, who had recently returned from a mission for the Church, and for whom these were still happier times.

Initially John and Florence lived in Bennion, probably on the old Bringhurst property, where their oldest daughter Naida was born in June of 1908. Subsequently they moved for a time away from the home of Jack's parents to West Jordan where a second daughter Dean was born in 1910. Less than six months after Dean's birth, John Tripp Bringhurst was called to serve a two-year mission for the Church. When asked if he would serve, he reportedly replied, "Sure, but you had better send me someplace I can't walk back from, or I'm apt to come home to my wife and kids." When the call came, it was to New Zealand, and he departed for the mission field on July 10, 1910, returning home in October of 1912. In New Zealand he worked chiefly with the native Maori people, though he was never required to learn the Maori language and did his work mainly in English. He was ordained a Seventy before departing for his mission, and on his return he served as one of the presidents of that quorum in his stake.[25] John Tripp Bringhurst's mission to the South Seas, performed at considerable sacrifice to his young wife and two daughters, was an important event for the family, and almost surely influenced Royce's later decision to serve a mission of his own.

After Jack's return from his mission, the young family moved back to the family farm in Bennion, where the remaining children were born. Royce's sister Rhea was born in 1913; his older brother John Smith (known to all as "Smith") in 1916; Royce in 1918; and his younger brother George, with whom he remained close throughout his life, in 1920. The youngest child in the family, June, was not born until nine years later, in 1929, and hence was thought of ever after as the baby of the family.

By the time of Royce's birth, Jack was in the process of purchasing the family farm from his father, but the proceeds from the farm were insufficient to make the mortgage payments, so he had to supplement them with other types of employment. During Royce's boyhood, his father's chief occupation outside the farm was in the Murray smelter. As Royce's brother George later recalled, "He would be up early in the morning, maybe 4:30, fix his own breakfast and pack his own lunch, then he would go over to the smelter and he would work eight hours over there. Then he would come home and have dinner, and then would actually lay down on the floor because it was more comfortable for him, because what he was doing was really 'pick and shovel work,' so to speak. The flue dust that they had to haul by hand in wheelbarrows was principally arsenic, and he would get arsenic sores. They would tie the cuff of their coveralls and the sleeves of their coveralls and then they would wear gloves, but there was always arsenic dust where you would sweat around the wrists, particularly where it was tied, and I remember the arsenic sores that he would get. He used to say, 'I know they could never poison me with arsenic. I've eaten too much of it already in my job.' Dad worked at the old Murray smelter until they closed it down."[26] In later years John Tripp Bringhurst worked for the Salt Lake City Water Department.

Though not outgoing, John Tripp Bringhurst was amiable and had a wide circle of friends, among whom he was well liked and respected. His life and outlook must have been profoundly affected by the criminal activities of his younger brother William, with whom he had evidently been quite close, and whose widely publicized trial and execution in 1924 cast a pall of shame over the entire family. Jack was no stranger to disappointments, and the Great Depression, which began in 1929, greatly limited his opportunities at a time when traditional farming was proving insufficient to provide an adequate living for his growing family. Although there was always an abundance of food from the farm, John Tripp was obliged to work long hours each week to make ends meet, and relied on his sons to do much of the farm labor. Always a diligent provider, he would lend various types of support to Royce in the years following his departure from the farm to strike out on his own.

Royce's own description of his father, written as part of a presentation some years after his father's death, was poignant and telling: "I am not very large but my father was much smaller. I know that he viewed me as an extension of himself, a potential idealization of some of his dreams of accomplishment. He had but an 8th grade education, but had often told me of his desire to have finished secondary education and attend the state agricultural college. He was a good farmer because he loved the land and the plant and animal life associated with it. He was a good judge of horses, our sole source of energy for tilling the land. He bred good cattle and sheep on a very modest scale. He appreciated and always saw that we had a good hunting dog. He had a fine tenor voice, much more lyric than mine. He was painfully aware of his deficiencies and lack of preparation in many areas, but he loved his fellow man."[27]

7

Royce had great respect for his father, but he revered his mother, who would become a constant source of encouragement, particularly as he grew to manhood and left home. He would later recall, "My earliest memories of her was when she worked all day in the beet fields, blocking beets out with a hoe, and we followed behind, thinning each cluster to one. She still would quit early enough to prepare meals for us, morning, noon and night, plus all the other household work without modern conveniences. Our home was orderly and clean always, and the food was good and ample. It was pleasant then, and is still pleasant to me as a recollection. She was kind and gentle and forgiving to all."[28]

Birth and Childhood

Royce S. Bringhurst was born two days after Christmas, on December 27, 1918, in the old Bringhurst home on Redwood Road. The origin of Royce's unusual first name is now lost to memory, though it was always understood that the middle initial stood for "Smith," his mother's maiden name. At the time of Royce's birth the world had been rocked by the momentous events of the time. The horrific "Great War" (now known as World War I) had just come to a close in an armistice the preceding month, and American Soldiers were at last returning from the conflict in Europe, but in the winter following the war's end the deadly Spanish Influenza had made its appearance in a pandemic that spread the world over, from Europe to the Americas. Hence, although by that time hospital births were commonplace, both Royce and his future wife were born at home, since pregnant women were particularly vulnerable to this devastating illness, and were advised to remain home for their confinement rather than come to the hospital, where they were likely to be exposed to the virus and succumb.[29]

The house where Royce was born was the same old adobe farm house build by his grandfather, John Beitler Bringhurst, on the original Bringhurst property. The address of the house was Box 124, on R. D. 5 (road delivery), in Murray Utah, a community just southwest of Salt Lake City. Royce recounted, "I grew up in Bennion Ward, out in Redwood Road. We had a paved street—my sisters could remember when it was paved. I could not, but we used to roller-skate on the road. It was called Redwood Road because they used redwood stakes to mark it out."[30] Royce's brother George described the house itself: "The original house was adobe bricks, and then they had it finished with a coating of stucco to help preserve the brick. The upstairs was actually part of the attic, then there was about half a basement. There was no heat upstairs. It was hot in the summer and cold in the winter. We used to pick up a hot iron when we went to bed in the winter,

8

and wrap it in several layers of newspaper to put our feet on."[31]

Store-bought clothing was a luxury seldom afforded to the Bringhurst children. Royce recalled: "My mother made all of the overalls I wore for all the time I was a boy, and up until the time I went to High School I still wore home-made overalls, and home-made shirts. My mother could do anything, as far as housekeeping and sewing, and that sort of thing, and she did a good job. I was never really embarrassed wearing home-made clothes. I wore them until I went to High School, then I started to wear the striped blue overalls. I liked those striped ones."[32] Family activities centered around life on the farm. Besides raising small herds of livestock, the family grew a variety of crops, including wheat, alfalfa, and berries, though their main cash crops were sugar beets, which they planted yearly, and apples from the family orchard, which they sold from a stand to passers-by on Redwood Road.

By the time Royce was old enough to help, his father was already working full-time at a day job, and the work on the family farm fell mainly upon the three boys. Royce's brother George later described the daily routine: "My father had a 40-acre farm which had belonged to his father. He would leave for work probably around 5:00 a.m., and then later my mom would see that myself and my brothers were up and moving and getting up and getting the farm animals taken care of. The horses needed their stables cleaned and the cows needed to be milked and the other animals, chickens, pigs, and sheep had to be fed. We didn't have a lot of them, but we had a number, and they kept us pretty busy. The chores were always done before we sat down to breakfast and got ready for school if it was the school year."[33]

The family diet was hearty farm fare, which Royce's mother regarded as the healthiest diet. Years later when Royce had left home, she wrote: "You haven't been used to fancy food but I believe you have been raised on the best food for health it was possible to have: plenty of good milk, butter and cream, bread, fruit and vegetables. There aren't many families as large as ours that has had less sickness than we have and I think the food we eat has had a great deal to do with it. Health is wealth, you know."[34]

The farm work was especially labor intensive, and George, who unlike Royce never planned on making farming his life's work, described it in detail many years later: "On the farm we raised sugar beets, which at that time were practically all hand work right from the very beginning when they were planted. After they were planted, the beets had to be thinned, and my mother Florence would go along and do what they called blocking the beets. This meant cutting out the beet plants so that they were spaced about eight inches apart, so that

9

the beets had room to grow. We would crawl along the rows and where these little clumps of beets were, we would pick all but one plant, so the beet would have a chance. Then during the summer time, there was irrigating to do. Our water came from an open canal, and when we would have our watering turn it would be about 40 hours every 7-8 days, and that one day and night straight through. There was always weeding to do, irrigating almost every week, weeding every two weeks, and then the beets matured in the fall of the year and it was all hand work to get the beets out of the ground. We had a horse-drawn implement that would loosen the beets and pull them out of the soil, and then we would follow up after the beets were loosened up so that we could pick them up. We had what we called a beet knife, that was a knife about 14 inches long with a hook on one end that we would hook into the beet and pick it up, then cut off the top and throw it in the pile. Then the beets had to be loaded into a specially built wagon rack, then hauled down a mile or so away from the home along a railroad track that ran through the middle of our farm. We also raised the alfalfa, and my dad would also raise grain and usually some barley, some oats, and some wheat and that again was cut with a binder and this involved a lot of hand work, a lot of team and wagon work, loading the crops onto the hay rack or the grain wagon to haul it into the stockyard where the grain would then be put through a threshing machine."[35]

At a time when mechanized farming was making its way into common practice, the Bringhursts with their small family farm continued to rely mainly on draft animals, though this was not always a point of agreement. As Royce later recalled, "In those days, we were one of the last people in the area that were farming with animals, using horses as our power source. By that time most everyone had tractors. We did not. My mother had a commentary on the horses on the farm. She said, 'Those horses stand around that barn all winter long and eat hay, and don't do a thing for it.' She would have been very much in favor of getting a tractor, and let the tractor stand around not eating anything when it wasn't in use."[36] Royce never begrudged the work, however, and later reflected in a letter to his parents, "I'm glad that I grew up as I did working. It gave me a strong back and a strong arms and an appreciation for a lot of things I should otherwise have known nothing about."[37]

Royce recalled the animals with great fondness, particularly the horses. "You can't live with animals without getting some emotional involvement toward that animal. And we had a big horse. He weighed close to two thousand pounds, which you recognize is a ton. It was a big old Percheron horse, and he was a huge, powerful animal. My father was very kind to us. He used to let us take his horses, and we always took old Dick because he was so obedient, you just had to turn him down, point him in the right direction, tell him to go, and he'd go there if he'd ever been there before, or had any idea of what we were going to do. I used to load up old Dick, and put about five young people on the back of that horse. I've actually counted that many sitting on the old beast. We'd get them all straddling the same horse. We used to go up and go swimming. In the summer we just rode him bareback. In the winter we always dragged a one-horse sled. We used to start at our place, and go down Redwood Road, and then off on the side streets, and go down to Turpin's Hill, where we coasted on sleds. Turpin's hill was very steep, and ended on the Jordan River, down by the way where my ancestors lived at one time." It was on one such excursion that the beloved horse was struck by a careless driver and killed, a loss deeply mourned by Royce and his family.[38]

Growing up on a family farm meant that Royce and his brothers and sisters experienced the Great Depression somewhat differently from those who lived in the city. Though money was scarce, food was plentiful on the farm, and there was no lack of work to be done. The family sometimes came into contact with some of the many drifters rendered homeless by the depression. Royce in his later life described one such interaction with wistful nostalgia:

"He came on a cold fall day just after beet harvest. I don't remember his last name, and perhaps that is appropriate to the setting and the times. We knew him as 'Alfred,' one of the thousands upon thousands of unemployed, demoralized, despondent and restless men seeking security for the winter; a roof overhead, and warmth and food enough for survival. My mother was home alone; my oldest sister was at work, we were at school, and Dad was at the mill getting wheat ground into flour, a common practice in Utah then. He asked only for a meal at the time, and proudly stated that he would not want it unless he could earn it. He clung to that shred of human dignity with fierce determination—the work ethic was deeply fixed in the men of that generation.

"There was always work to be done on a diversified subsistence farm such as ours, and Mother told him to rest a bit while she prepared him some food, and that she would instruct him in what he could do after he had eaten. When we returned from school, he had done most of our week's assigned special chores. Dad was home, and it was agreed that he would spend the night with us because it was cold and he had no place to go. He spent that night and over 100 more with us, making himself useful as a general handyman, hired hand (without salary, for we had no money to give), teller of exciting adventure stories, and singer of interesting folk songs. And, he wore shoes that did not match, that had no mates.

"He lived in our brooder house,[39] sleeping by the furnace where it was

11

warm and comfortable. He attended church with us and became as it were an adopted member of an already large family of seven children. In our restricted vision, we thought this would go on forever; such are children's happy illusions of permanence in all things. Spring came, and when the push of spring planting was over, he left one warm spring day, as abruptly as he came. He told my mother 'goodbye' while we were at school, and Dad was at the neighbors'. He just bundled his few belongings together and left. When Dad came home he was sorry to find him gone, and said sadly, 'I did so want to buy him a new pair of shoes'. [40]

One of Royce's great pastimes from childhood was swimming, which, even in the Great Depression was a diversion available to all. "There were two canals that came from the Jordan, which came from Utah Lake, down in the south end beyond the Salt Lake Valley. One of them was the little canal, and one was the big canal. It didn't make any sense—they were both about the same size. The little canal was closest to our home, and I learned to swim very young, when my brothers and our neighbors threw me in the canal. I learned to swim in a hurry."[41] As Royce grew older, most swimming took place in the high school swimming pool, where the boys and girls were completely segregated, as the boys swam without suits.

Other important pastimes were camping and hiking, and fishing for the native trout which abounded in the mountains around Salt Lake. These were among Royce's happiest recollections, and he remembered with special fondness fishing expeditions to Blacks Fork of the Green River in Utah: "We could hardly cast a line in without getting a fish, and I mean we got fish, all we could eat, all we could handle. All you had to do was get up and hit the water in the morning, and then you could fish and catch, catch, catch until about 11:00 o'clock, and the fish took time off for lunch, I guess."[42] In addition to fishing, Royce and his family enjoyed hunting, usually of pheasant or rabbit, which they did on the farm and elsewhere. The family kept hunting dogs for this purpose.

As a youth Royce joined the Boy Scouts, which was affiliated with the youth programs of the Church of Jesus Christ of Latter-day Saints. He was never an avid scouter, though of course he enjoyed the camping, hiking, and other activities. One particular event stood out in his mind, and he retold it again and again. "One example of stupidity was when I hurt my foot, swimming in the big canal. I stepped on a can, and the can cut me very deeply between my first and second toe. I had on rubber boots and by the time I had walked home from swimming, I had a boot that was washing with blood, and the worst thing was, the next day we were going on *the* important hike of the year, and that was to go up to a place called Lake Blanche[43], about a five mile walk. I had my backpack ready and fishing pole and everything. I did not tell my folks for two reasons: first, I was presumed to be irrigating at the time, and second, I feared if I told my parents, I wouldn't be going on any Scout trip, which of course I should not have been permitted to do under the circumstances. And sure enough, I developed a roaring infection. The scoutmaster happened to be a good practical man with medicine, and he made some salve out of pine gum, and put it on my foot, and that's how I saved the foot, because I really had a serious blood poisoning situation by the time I got out of there. Four miserable days later I returned home with the badly infected foot. My folks didn't reprimand me, since they realized I had paid for my foolishness, and they proceeded to help with the healing. I went without shoes for the rest of the summer."[44]

An important part of life for Royce and his family was participation in activities related to their membership in the Church of Jesus Christ of Latter-day Saints. In accordance with the prescribed practice of the Church, Royce was baptized at the age of eight and confirmed a Church member by his father. His baptism took place in the baptismal font beneath the Salt Lake Tabernacle, an event which he described in his later years. "I went with the stake up to the tabernacle to be baptized in the font that they have there under the tabernacle. All I remember about the baptism is that it felt like heaven. It was nice and warm, and comfortable there, even though we were just in and out, it was a wonderful feeling."[45] As a young man Royce was ordained to the usual offices in the Aaronic Priesthood, including deacon at 12, teacher at 14, and priest at 16, which allowed him to officiate, to different degrees, in Sacrament services. Of these ordinations, only that of Teacher in the Aaronic Priesthood was performed by his father, who may have been less than constant in his religious activity. In those days the Church held Sunday School services in the mornings in every ward, with Sacrament services (generally referred to in the Bringhurst family as simply "church") in the evenings. Royce generally attended Sunday School but was not as faithful in his attendance at Sacrament services. The family, and Royce in particular, were less scrupulous about Sabbath observance than Royce would be in his later life. He often went hunting on Sundays, and both he and the family planned fishing and camping expeditions that involved staying over the Sabbath. Farm work was for the most part set aside on Sundays, though some tasks, such as care and feeding of the livestock, and the scheduled taking of water for irrigation, were an unavoidable daily necessity.

The Church, in addition to providing the spiritual grounding for the family, was also the center of much of the social life of those times in Utah. Wards and stakes of the Church sponsored dramatic productions and roadshows, dances and dance festivals, speech contests, musical productions, sporting competitions, and a variety of other events which added variety to the lives of these near descendants of the Latter-day Saint pioneers. Cultural activities were often shared from ward to ward, and occasionally they were competitive. Royce was involved in many of these activities, both as a spectator and as a participant. One night each week there was a meeting of the Young Women's and Young Men's Mutual Improvement Association, two related organizations which were designed to keep young people involved in Church activities and progressing in their personal lives. Royce usually attended these evening activities, which were referred to as "Mutual."

High School

Like almost all his generation, Royce grew up in the public schools. From first through ninth grade he attended the Plymouth School, a combined primary school and junior high school then located at the corner of Redwood Road and 4800 South[46]. He subsequently attended Granite High School[47], which was located across the valley at 5th East and 33rd South in Salt Lake City. Family tradition held that Royce faced many disadvantages attending high school. In his earlier years his dress and interests marked him as a farm boy in a school heavily attended by the children of city dwellers. He was also plagued in school by negative family associations. His older brother Smith had a reputation as a mischief-maker in school, and according to family folklore, Royce was often blamed by association for the antics of his brother. One prank specifically mentioned was the placing of a lit cigar in the ventilation system of the school library, with effects which can only be imagined. One telling of the story had Royce performing the deed, but others affirm that it was Smith. Royce must have engaged in some misdeeds of his own, however, as in his senior year he received a summons to appear before a student-run court,[48] which he did not take very seriously; his senior Ag Club picture shows Royce with his hands jokingly about the throat of the student in front of him.

Whatever the reason, Royce appears to have been held in low esteem by some members of the faculty and administration, but he had serious academic ambitions of which they may have been unaware. At one point he described being sent out of a study group, presumably for some offense, and in a brief entry in his diary he recounted both the punishment and his more serious aspirations: "Old Wilcox sent me out of study today...I hope I get through college and realize my ambition in forestry."[49] His desire to go into forestry was no doubt stimulated by his love of the outdoors and his passion for camping

14

and hiking, but it had probably originated with his admiration for an uncle who had become a forest ranger, and whose example had first prompted Royce to consider a college education.[50] His actual accomplishments, of course, would eventually take him far beyond this, but no member of his family had ever attended college, so his was a high aim despite his disadvantages.

Royce's secret ambitions were not always accompanied by strong self-confidence. He harbored many anxieties and misgivings related to his background and family circumstances, and by nature felt an apprehension around crowds which he would never completely overcome, though paradoxically, one of his most obvious gifts was one which would place him repeatedly in front of large audiences. This was his ability as a singer. Many years later in a moment of reflection, Royce would describe himself in those early years as an insecure schoolboy whose one outstanding talent was an excellent singing voice which alone brought him recognition and self-esteem. He distinguished himself musically at an early age. "I'll never forget the first time I sang at the district music festival when I was just starting in junior high school. I couldn't understand why they chose me, and I guess I still can't. I always liked music more than most things. I remember when I was in my last year at junior high, they were planning an operetta and I thought it was sissy stuff, but finally tried out and ended up singing one of the major parts. I wasn't good but it was fun. Later I got so I enjoyed choral works."[51]

Royce especially recalled singing with one very gifted young musician named Meade Steadman, who later sang at his missionary farewell. "He was really talented. His whole family had an orchestra, and they used to come out to the different wards to play. I used to sing all the time with Meade. One time when we were seniors, they had us sing in music contests through the whole Granite school district, and they went through all the schools, going through the rounds judging on them. I was sitting there listening, and I heard the judges talk about what a nice voice, and so on. They went on and on, and I thought, 'well they're talking about Meade you know,' and then they said my name."[52] This was a point of great satisfaction for Royce, for though his considerable talent was never seriously exploited, he had a rich, robust tenor voice, deeper in tone and stronger in volume than that of his younger brother George, who with a lighter, more lyric voice later sang for many years in the Mormon Tabernacle Choir. In high school Royce joined the men's choir, sang solos frequently at church services and other events, and participated in an opera his sophomore year.

In spite of his insecurities Royce appears to have been well-liked by his peers, and developed a number of strong friendships among a

15

student body which was destined to be thrown into the forefront of world events in the ensuing years. During his sophomore and junior years he joined the "Bachelor's Club", whose purpose was to teach boys to cook and interest them in commercial kitchens, and he was later reputed by roommates and companions to be a very satisfactory cook. In his junior year he also joined the Orans Club, dedicated to public speaking, under the encouragement of an influential speech teacher. He became a member of the high school football team as a lineman in his senior year, and while at practice sustained an injury to his right knee which bothered him to some degree in the succeeding years, though it never seemed to limit his activity very much.

His most important association, however, seems to have been with the Ag Club, designed for boys interested in agriculture as a vocation, in which he participated every year. Throughout high school he belonged to the Judging Team, led by a faculty member named LeRoy Hillam, which participated in crop and livestock judging competitions, actually a competitive interscholastic event in those times. The Granite High School crop team had a high reputation, dominating this event against other schools in the region to a degree that would have been enviable for the athletic teams. Royce distinguished himself, at one time taking third at crops judging in a competition involving teams from 48 high schools from Utah and Idaho.[53] Among his friends and acquaintances he counted Jim Faust, a fellow football player with whom he also sang in a men's quartet—Jim later would be better known as James E. Faust, a Latter-day Saint apostle and later a counselor in the First Presidency of the Church. Others of his schoolmates would similarly distinguish themselves in some way, while still others would never return from the battlefields of World War II, through which nearly all of the young men would pass.

The first day of his senior year of high school was destined to be one of the most significant days of Royce's life. Near the end of the day he had just taken his seat for an afternoon Sociology class when he saw a pretty young woman named Pearl Davidson enter the classroom. As she scanned the room for an empty seat their eyes met, and though they scarcely knew one another, on an impulse Royce gestured for her to take an unoccupied seat behind him, and the two struck up a conversation. Pearl was a daughter of Scottish immigrants who had arrived in Salt Lake City in the early 1900s as converts to the Church. She was the second youngest of a large family of eight children who, having lost their father to illness years before, were being raised single-handedly by a diminutive but determined mother with a thick Scottish brogue. Pearl and Royce, who also shared a Seminary class,[54] felt an immediate affinity for one another and quickly became close friends, talking to each other before and after class, and meeting at Pearl's locker during their free moments. One stormy day in October just before Halloween while attending a school assembly together, Royce got up his courage and asked Pearl on a date, an event which both would later remember as the formal beginning of their relationship. From that time forward the two would be high school sweethearts.

Royce and Pearl were in many respects an unlikely couple. Pearl was born and raised in the city with its culture and refinements, while Royce's status as a farm boy was evident from his rougher dress and appearance, his coarser mannerisms, and his colloquial grammar and spelling. While Pearl's family were devout church-goers, and she and her brothers and sisters would all eventually wed in the temple, Royce's family tended to be more spotty in their church participation, and some of his siblings as adults would attend church seldom, if at all. Yet there were things the two had in common which set them apart from many of their classmates. Both were lovers of poetry, and though Royce did not share Pearl's talent for writing verse (and his early attempts at it were dismal), he often memorized favorite poems, and had already compiled an extensive collection of poetry and song lyrics before the two had met. Similarly, they both had a strong love for good music, and while Pearl had nothing to compare to Royce's magnificent singing voice, she had always stood out in her family for her musical sensitivities, and had often gone with her mother to attend concerts of the Mormon Tabernacle Choir in the imposing Salt Lake City Tabernacle. Royce and Pearl also shared a love of nature, and both had a deep appreciation for the austere beauty of the Utah wilderness that surrounded them.

But it was probably their youthful insecurities that played the strongest role in bringing the two together. For Royce, self-conscious of his obvious homespun background and weighed down by negative family associations, Pearl provided an unaccustomed source of confidence and respect, as well as a constant motivation for refinement. Pearl for her part had weathered with her large family the indignities of poverty during the years of the Great Depression, and felt appreciated and accepted by this thoughtful and hard-working farm boy. Perhaps more importantly, Royce seemed to fill for her a void she had felt since the death of her father when she was a young girl, for that event, though a hardship for the entire family, for her had proven a personal catastrophe which left deep and enduring emotional scars. Royce and Pearl would enjoy a close and loving relationship, vividly documented in a long string of letters during the ensuing years.

On New Year's Day in 1937, Royce also began keeping the first of many journals, making succinct entries in a five-year diary which he

17

had probably received as a Christmas or birthday gift. In it he kept a detailed record of his daily activities as he completed his last year of high school, both the momentous and the mundane, which gave a varied portrait of his school life. Just eight days into the new year, for example, he told of his only drinking episode, during a trip to Ogden for a successful stock and crop judging competition, accompanied by his best friend Steve Mackay, who was frequently mentioned as a companion. "Got 3rd high man in crops. Steve first (unofficial). Got drunk like fools. Had a fairly good trip. Stayed at hotel Wilcox. It was lousy. Landlady got plenty mad." Written above the entry were the words "This was about the darndest fool trick I ever done. I mean it." The next day he wrote, "Bad hangover. Gee I'm sorry we drank. Pearl cried when I told her tonite."[55] A few days later, when he realized word of the episode had spread through the school, he added, "After drinking Friday night I feel that I have fell right back down from where I built up to and I have to start all over again. I know it wasn't right."[56] Pearl remembered the episode, and recalled telling Royce that if it ever happened again their romance would be over. It never did, although that summer, when Royce related having been out with a group of three friends, he remarked, "They drank beer. It's hard being the only one not to."[57]

It is probable that Royce's relationship with Pearl Davidson had a refining influence on his character and behavior during this time. Evidently at Pearl's encouragement Royce began to attend Sacrament meetings regularly, which appears not to have been a habit of his, though he was more faithful attending Sunday School.[58] He continued to hunt rabbits and do farm work on the Sabbath, the sort of activities which would certainly be out of the norm later as a grown man, but he also described spiritual feelings as he participated in church meetings: "Went to church tonite. I havn't missed it once since Pearl ask me to go. I enjoy going. It gives me a restfull peacefull feeling."[59]

It appears from Royce's journal that around New Year's the two high school sweethearts made an agreement to hold their romantic relationship within certain limits which, for example, seems to have precluded kissing.[60] However, a rumor spread at school that they were involved in an immoral relationship, and on a Tuesday in late February they were called into the office of Lorenzo H. Hatch, the high school principal, during which this was discussed and some counsel given, perhaps advising them to break off the relationship. Royce recorded in his journal, "Mr. Hatch called Pearl and I in today and talked to us. Somebodies sure spread some awfull rumors. He was decent about it."[61] The couple discussed his counsel the next day, and a subsequent diary entry indicates they stayed home rather than spending the evening with each other—Royce wrote, "I guess Pearl and I have done about what Mr. Hatch wanted us to do. Stayed home at nite (for once). It seems kind of odd."

A week later Royce went to his bishop, Samuel Smith, about the matter. He wrote, "Went to mutual at night. Asked Bishop for advice about Pearl and I. He told me to keep going with her. He said I was wise doing what I am. It's odd, Bishop Smith and his wife done about the same thing."[62] The relationship continued, and later journal entries would speak with increasing disdain about Mr. Hatch's role in the affair. According to family folklore, at one point Hatch told Royce he would never amount to anything, and by one account Royce became so indignant that he refused to participate in the high school commencement exercises. One family recollection had him returning to the principal later with his PhD degree in hand to show that his unkind prediction had come to naught.[63] Whether or not these events ever happened, it appears certain that from then on there was a strained relationship between the high school principal and the farm boy with high ambitions. Royce's senior yearbook bears Hatch's signature beneath his picture, but another classmate, presumably in a show of mutual defiance, subsequently signed his own name over the principal's face.

Royce's bishop, on the other hand, had encouraged his boyhood ambitions to seek a higher education, and Royce would later pay tribute to the influence he and other individuals had exerted on his own life and choices. "The two best educated members of the Bennion Ward had the most to do with turning my life in the direction it eventually went: namely, toward getting the education that eventually qualified me for a UC professorship. They were brothers, grandsons of Hyrum Smith, sons of President Joseph F. Smith, and half brothers of Joseph Fielding Smith.[64] Both had PhD's (rare in that day). They were, first, Samuel S. Smith, bishop of my teen years, professor of mathematics and astronomy at the University of Utah, and second, Calvin S. Smith, then superintendent of the Granite school district. Both were paragons of virtue, intellectually brilliant and humble. Other than my parents, two other men and one woman also influenced my life during crucial periods of the days of my youth: first, Dr. F. V. Owen, a well-known USDA sugar beet breeder, whose curly top resistant US1 saved the sugar beet industry in the west at that time; second, LeRoy Hillam, my high school Ag. teacher, who took me to the USAC campus in Logan as a crops judge, and third, Emma Ray McKay (daughter of David O. McKay[65]), who taught me 'to speak out' if I had something to say and wanted it to be heard and understood."[66]

During the last portion of his senior year Royce became involved

19

in numerous activities, some sponsored by the Church. In January, for example, he got a part in a play which in the ensuing months was presented in various wards around the area. At about the same time he learned that dance instruction would be provided for several weeks after Mutual, and wrote, "I'm glad because then I won't embarass Pearl any more."[67] He continued these lessons for a number of weeks, reporting on his progress ("My dancing has sure improved. I can turn corners now"[68]), and subsequently took Pearl to the Gold and Green Ball, the Senior Prom, and various other dances, writing favorably on the experience. He also tried out for an opera to be put on by his ward, and earned a singing role. In addition to church activities, Royce avidly followed the schedule of the Granite High School basketball team and attended most of its games during a successful season, ending with a 3rd place playoff finish in the state tournament. He occasionally attended the movie theater, usually with Pearl, to see such cinematic titles as *The Plainsmen*, *Garden of Allah*, *The Trail of the Lonesome Pine*, and *The Maid of Salem*. He also went out frequently with friends, some of whom would later attend college with him, but generally he spent evenings and free times with Pearl. Like a typical high school senior, he was often up late and many times expressed being fatigued and suffering from loss of sleep.

As winter gave way to spring farm duties became more pressing, and Royce began missing school frequently to attend to them—shoveling manure, trimming fruit trees, leveling ground, harrowing and planting beets, spreading phosphate, clearing ditches, and eventually taking water by turns which generally ran into night and early morning hours. At times he went days at a time on a few hours' sleep. It appears that at that time Royce was permitted a portion of the farm to raise his own crop, perhaps as a means of paying for his upcoming college enrollment. He was under commission for a certain acreage of sugar beets which he farmed separately from those grown by his father. He also planted his own crop of tomatoes. In April he received a partial payment for the previous year's beat crop, which amounted to $8.05; a few days later, he reported with satisfaction: "Well I done one good thing today I paid my first tithing. I'm glad I did...1/10 of all I earn goes to tithes now.[69]

Royce grew nostalgic as his high school career neared its close, frequently counting the days and remarking on how much he would miss school. He was hardly a brilliant student academically, and his periodic report cards reflected both the lackluster academic standards he held for himself at that time, and the toll that outside activities were taking on his scholastic efforts. On January 20 he had written, "Got our report cards today. I got good marks for once, all in the B's but one that was a C." By March 3rd things had deteriorated: "Got report today. it was awfull 2B- 2 C- & 1 D-." Finally on April 14, as the farm work had grown heavier, he had written: "Got our reports today. All I got was a bunch of Incompletes."[70] On May 3rd he tried out to sing for graduation exercises; on May 4 he reported not being selected, though this may not have had to do with his singing skill, as three days later he wrote: "Received 2 shocks today: 1st my name wasn't on the graduation list. 2nd Pearl is going to Oregon as soon as school lets out. I think my name wasn't their due to a mistake." A few days later he learned that it was no mistake: "Went to school today. I guess I can't graduate college prep. I lack enough basic units. I wish they had told me before."[71] Whether this referred to a special advanced placement status or to graduation in general is unclear; however, in the ensuing days it appears Royce made up sufficient lost ground to qualify for graduation, and this he did in spite of frequent nights up watering, probably with the help of his father, who cultivated his beets for him as the last week of school approached.[72]

The last week before high school graduation was a flurry of activity. On Saturday, May 15 Royce, accompanied by Pearl and his friend Steve Mackay, attended a Seminary "pilgrimage" to the Temple grounds, followed by a fried chicken picnic up Emigration Canyon, where Royce reported, "all I done was chop wood & lift Dutch ovens. Boy am I tired." The next day, Sunday the 16th, Royce attended Seminary graduation where he sang in a quartet.[73] He then spoke at church, and his mother met Pearl's mother, Jane Davidson, who attended the services with Pearl. The next day, Monday, May 17, he rose at 4:00 a.m. to water the farm, then attended school to take final exams, and got a haircut before returning to his watering. The following day he wrote, "I've got exactly 7 hrs. sleep in the last two nites, my back aches between the shoulders in fact all over and tired." Wednesday was the last official day of school, and Royce took Pearl to the honor assembly where each received a letter—presumably Royce for football and Pearl for tennis—and Pearl gave her letter to Royce. Royce purchased a light grey summer suit for graduation exercises. On Thursday yearbooks were distributed and carried around for signing[74]. That day Royce recorded, "Went to practice graduation march today & a capella choir. Checked out of school today. I have an 82+ for this year av. Tomorrow is about the most important day I have ever had."

Friday was graduation day, and contrary to family tradition, it appears Royce attended and participated.[75] Of this day, he recorded: "One of the most important things in my life. Graduated from High school. Gee Mother & Dad gave me a swell watch. Its a Walthrum [Waltham]. I took Pearl to the exercize's. She sure one swell girl. I don't believe I could ever get a better." He added below, "I'll remember Mr. Payne[76] for what he told us. He said we were the ideal couple at school this year. I hope we're still that way 1 year & 50 years from now."[77]

Illustrations for Chapter 1

Great-grandparents: Samuel Bringhurst and Eleanor Beitler Bringhurst.

Family of John Beitler Bringhurst, Royce's grandfather. John Beitler Bringhurst is seated front center with his wife, Emma Francis Tripp. Royce's father, John Tripp Bringhurst, is second from left on the back row; William is forth from left on the back. Samuel Enoch is at left in front, and Arthur ("Uncle Art") is probably the younger boy at his side.

23

Royce's parents, Florence Smith Bringhurst and John Tripp Bringhurst, in photos taken during Jack's New Zealand mission, and the family about the time of his mission call, with children Naida (standing) and Dean.

Family of John Tripp and Florence Smith Bringhurst in a post-war photo (1946). Top, left to right: June, Royce, Rhea, George, Smith. Bottom: Dean, John T., Florence S., Naida.

1968 photo of the Bringhurst home on Redwood Road. Built in the pioneer era by John Beitler Bringhurst, this was Royce's birthplace.

View from the Bringhurst farm, in a 1976 photo. A freeway interchange was subsequently built here.

Early pictures of Royce: At left, with his older brother "Smith" in 1919, at right, with George (left) and Smith (right), around 1923.

25

Early High School pictures of Royce, left to right: As a sophomore, In striped bib overalls as a junior in his Seminary class group photo, acquiring cooking skills in the "Bachelor's Club", and his junior yearbook photo.[78]

Senior year yearbook photos of Royce: Formal and informal Senior photos, as a football player, and as a member of the "Ag Club".[79]

High school crop and livestock judging team, Royce at bottom row, center. Second from left at top is best friend Steve Mackay. Bottom row at left is Sam Oliver, friend and college roommate; between Sam and Royce is Arthur Wallace, who would become a lifelong professional associate.[80]

Pearl Davidson
Granite Jr. High; Year-book '36, '37; Paper Staff '36; Music '37; Home Ec. '37; Sec. of Seminary '37.

Yearbook photos of Pearl Davidson: At left, her senior pictures and activity summary; at right, her photo as Seminary class officer.[81]

Royce and Pearl's 1937 graduating seminary class. Royce is in the second to back row, second from left. Pearl Davidson is leftmost two rows down. In Royce's row, two students to his left, is James E. Faust, later apostle and member of the First Presidency; behind him is Royce's lifelong professional associate Art Wallace, who like Royce would later serve a mission. This is the first known photograph in which Royce and Pearl appear together.

27

College

Chapter 2

Summer of '37

As he completed his senior year at Granite High School, Royce applied for and was granted admission to the Utah State Agricultural College[1] at Logan, Utah. His journal record of the last summer before starting into college in 1937 gives a glimpse of a Utah farm boy coming of age in an era of change for both him and his community, with each holding to elements of the past as they looked forward to the future. Chief in Royce's mind at the time was Pearl Davidson's planned departure for an extended family trip to Oregon to visit the family of her oldest brother, scheduled just three weeks after high school let out. While she remained, Royce visited Pearl frequently, and participated in events important to her family. One of these was the departure of LaMont Toronto,[2] boyfriend of Pearl's sister Helen (described by Royce as "a grand fellow") on a mission to New Zealand. (Helen and "Mont" would eventually wed.) The Davidson family gathered for the occasion, and as was customary for the time, a series of events were held, including a family party and a formal weekday "missionary farewell," which was advertised in the local newspaper, and to which family and Ward members were invited.

Mont left on June 5, and during the days that followed Royce and Pearl saw each other as often as they could, attending social events and getting to know each others' families. As Pearl's departure approached they spent much of their time just walking and talking

28

alone. Finally, on Friday June 11 Royce recorded in his journal, "Well, bid Pearl goodby tonite. She leaves Sunday Morning. We went out to Saltair[3], to the airport & talked & sang. Gosh I won't see her again for about 3 months."[4] Pearl left in the early morning of Sunday the 13th of June. On the day of her departure for Oregon Royce wrote her a letter, and he continued to exchange letters with her throughout the summer, writing frequently of her in his journal during this first separation since their romance began. He continued attending Sunday School and church regularly during the early part of summer, seeing to farm duties during the week, and occasionally going out with his friends to various attractions in the evenings. On Friday, June 25, he visited Pearl's remaining family and went to a ward show with them, writing that day, "They sure treat me swell. I like her family."[5]

In the days after school let out, Royce returned to farming full-time, details of which were recorded in short, vivid lines of his journal. The first order of business was hauling straw and hay for the livestock, then thinning the large planting of sugar beets occupied the family for a couple weeks ("More beet thinning and boy am I weary evry muscle"[6]). Other tasks included taking irrigation water in turns from the public supply ditch, often at odd hours of the night, and directing it into the crops ("Irrigated today. Gee I hate to irrigate. I beleive I'd almost rather thin beets"), setting out "hootch pots" in the apple orchard to capture moths ("...sent my first report on moths in – I caught 110 in 4 pots. The fight's on"[7]), and sundry other farm duties ("Cleaned the chicken coup, hauled the trimmings out of the orchard & planed some furrows in the orchard"[8]). This was a season of the year when the farm demanded constant attention, and a few days before Pearl left, Royce wrote, "Wattered all day & nite. Dad cultivated the beets today. The darn wind is sure blowing. Farm life would be O.K. if it wasn't for Irrigating, beet thinning etc."[9]

The work was physically demanding, and after living the "soft" life of a high school student, Royce quickly returned to being lean and muscular. Standing 5 feet 8 inches tall with a sturdy frame, he wrote not many days after graduation, "I only weigh 144 lbs now, 14 lbs. loss since school closed."[10] Royce's abilities as a farmer and his reputation for hard work were such that he had been appointed a 4-H club leader, under the direction of a county agent. He had done some recruiting during the school year, and on the same Saturday on which he recorded his new weight he also organized his 4-H club and held the first of his meetings, which he continued to hold intermittently through the summer. Later that summer he would enter crops and livestock into the Salt Lake County Fair, and participate in the judging.[11] On a Sunday in June, Royce traveled to Logan with former classmates who planned to attend college with him, and they found a

29

house to rent there at $25 per month, just half a block from campus. He remarked in his journal that it was the first time that year he had missed Sunday School, though as the year wore on it certainly would not be the last.[12]

On July 6 Royce took a break from farming and began a job with the railroad, which he referred to as "working section", probably to supplement his college fund. His first three days were spent cleaning up after flooding had covered a portion of the railroad, and his account was descriptive: "Saw the first real flood I have ever seen. Did it jam the railroad. Passenger trains were 3 hrs late – worked overtime. The water on R.R. tracks waist deep." The following day he added, "Worked on section again and I mean work – mucking out from between rails. It's sure a dirty slimy muddy mess. Overtime again. Every nuckle & bone aches." And the third day, "Worked like a slave on the section again today, she flooded again last nite. I'm sore and stiff. Only worked 7 hrs today all same as 8, just as tired. That darn Railroad can wash away for all me."[13] After the floods subsided the railroad work became less demanding, but Royce found it monotonous and longed to get back to farming. When his predetermined time—about a week and a half—was over, he was happy to give up the railroad job. "Aw, thru with the section. I quit today now back to farming. No more boiling in the hot sun all day on the section."[14] He was paid $24.31 for eight days of labor, enough to cover college fees for one quarter.[15]

Royce's journal in the ensuing weeks gives glimpses of the farm life which had been his lot since childhood, and in which he had become accomplished as he approached full manhood. As July wore into August he took on watering the farm, weeding corn and sugar beets, repairing a binder, acquiring a colt and then some heifers, and overseeing the breeding of various livestock. Many days were spent mowing and raking hay then hauling and stacking it. He tended to the orchard by watering and weeding, cleared ditches, and saw to watering the farm both day and night when his turns came during the hot weather. Finally there was the grain harvest, first of wheat, then of barley and oats, followed by threshing of the grain. Some days he took jobs with neighbors doing similar work to earn additional funds. Pioneer Day[16] came and went with little celebration, as he was occupied watering beets all day, though that night was spent out late with friends swimming at hot springs, one of the many amusements which occupied the evenings during even the busiest days of farming. These included swimming, shooting, shows, and shopping for clothes, and occasionally motoring around town in someone's vehicle, an activity Royce referred to as "going howling". Sundays, besides Sunday School and church, were often spent with friends doing a variety of recreational activities, such as hunting, fishing, shows, and trips to the canyon, which would be far out of the ordinary for Royce as an adult. Accustomed to vigorous farm labor, he began to find the Sundays tedious, indicated by comments in his journal like "ho-hum" and "the same old humdrum", which seemed confined to Sundays when he wasn't out with his friends.

During this time visitors occasionally came out to view the activities on the farm. One of these, destined to play an influential role in Royce's later life, was Dr. F. V. Owen, a geneticist from the United States Department of Agriculture who yearly rented a plot from Royce's father for experimental plantings of sugar beets. Dr. Owen provided advice and encouragement to Royce—and informed him that his growing beets were infested with nematodes. When he learned of Royce's interest in studying forestry at the Agricultural College, Dr. Owen also recommended a colleague for him to look up in the forestry department once he got there.[17] By this time Dr. Owen had become a leading figure in plant breeding through his work on disease-resistant sugar beets, which had helped to rescue a sagging U.S. sugar industry. Royce had come to admire "Doc Owen," and accustomed to the labor-intensive life on a family farm, he viewed with great interest Owen's work as a respected plant breeder, which besides making important contributions to the world of agriculture, also allowed him to travel freely through the extensive beet-growing areas of the western U.S. Though Royce did not know it yet, an important seed had been planted for his own life's work.[18]

After the grain harvest, Royce and his friend Steve Mackay planned a fishing expedition—this was to be fly fishing in the alpine rivers and lakes of northern Utah, an area much beloved by Royce. In his journal he wrote of tying a fly ("a willow bug") in anticipation of this trip, and he purchased clothing and fishing gear for the occasion. On Saturday, August 14, Royce set off with the Mackey family at 4:00 a.m. for Blacks Fork of the Green River. Though the road in was described as a "terror, rough, rocky, muddy, everything", they were camped by 10:15 a.m. and Royce caught 20 fish that day—all trout except one of another species they referred to as a "herring". After a couple of days the water became too high and muddy for fishing so the two set off on a hiking expedition, climbing nearby Mt. Tokewanna and exploring lakes and passes. Their final full day of fishing was the most successful, and if there was a limit on fish they surely exceeded it. Royce caught 49 fish ("50 with the one that got away"), including two large ones ("some fighters up there, game to the core"). They arrived home happy and exhausted on Wednesday, August 18, at 4:00 in the afternoon. Royce wrote, "All I have to show for it is blistered feet, some trout, some pictures and a beautifull

31

memory that will last forever."[19] He still recalled this trip with relish some sixty years later.

As successful as it was, the fishing trip seemed to have broken Royce's pattern of church attendance, as he never attended again before leaving for college. This was in part due to the scheduled necessity of irrigating, which in the late summer heat kept the beets and other crops growing. During these days of late summer, the farm work continued heavy and unabated, each day's work recorded in the journal, generally with a mention of how tired Royce felt or how little sleep he had gotten. Evenings he frequently spent with Steve and his other friends Sam Oliver and "Swede" Stenstrom, driving around or going to shows. On one of these occasions the friends got into an altercation with another group. "Saw *Souls at Sea* tonite. Steve, Sam & I got in a fight after, bloody nose. Some fight. They were lots bigger than us. Did we get battered."[20] Police were not involved, and the fight did not seem to have the same serious repercussions it might have had in a later era. The first of September marked the beginning of the Salt Lake County Fair, and Royce made his mark, winning a livestock contest. He also exchanged letters with the Agricultural College and had his high school credits transferred there.

Pearl Davidson returned from her extended stay in Oregon on September 17, 1937, just days before Royce was to depart for college. He wrote that day in his journal, "At last Pearl got home today. 3 months. I went up to her place at nite. We went to canyon. Hauled hay today. Gee it sure was good to see her home I was nearly thinking she wouldn't make it." The following day, after taking her to a show at the Paramount theater, he remarked "She's still the same Pearl. A little changed I guess. I have some." One of Royce's changes was his lapse in church attendance, and in spite of Pearl's return, he made no mention of church that Sunday. His last three days at home he worked on the farm, irrigating and plowing. He went into town with his future college companions to purchase some clothing and household goods, and sent a last telegram to his new landlord in Logan in anticipation of their arrival there. Finally on September 22, he wrote, "My last day on the Farm for quite a while – plowed all morning & part of afternoon it rained. Good by and hello. Pearl and I said goodby again, just one of the many times we have to do that. She's swell."[21] The next day Royce would begin a new life as a college student.

Starting College

On Thursday, September 23, 1937, Royce Bringhurst left home for the first time to attend the Utah State Agricultural College. His journal reflected his excitement at the prospect: "Logan at last. left today after much preperation. got here at 4:00 p.m. House is swell, guys are swell, everything's swell. I'd hate not to be here. Our house keeping career has begun. I hope we succeed." The following day he paid his college fees. "Registered 47 dollars. The first step now done, 4 years of hard work ahead but they're worth it."[22]

Royce's roommates were the same group of old high school friends he had spent so much time with over the summer, and included his best friend Steve Mackay; Sam Oliver, who like Steve had been on the judging team with Royce; Dave Miller, another classmate from Granite High School; and Emil Stenstrom, referred to by everyone as "Swede."[23] Left alone by his roommates a couple of days and feeling lonesome in his new location, Royce occupied himself that Saturday stacking firewood and assisting the landlord with chores, and on Sunday morning he attended Sunday School in the Logan 5th Ward, his first time to attend a church service in over a month. His roommates arrived that afternoon, and the following day he explored the area, walking twice to campus and twice into Logan proper, then attended a dance with his roommate Swede. He retired at midnight with tired feet.[24]

Finally on Tuesday, September 28 Royce began his college career. "1st day of School. From 8 to 5 – Plenty of asignments. I hope I can make it—worked till 11:30 doing algebra and history – I'll be glad when I get onto it – Bot. lab O.K."[25] The following day, the financial realities of college life began to catch up with him. "College books are expensive. We found another way to save money, used cheese instead of butter. Got monkey suit today."[26] The "monkey suit" Royce spoke of was a military uniform, a necessary commodity since he had enrolled in the college ROTC program,[27] and he wore the uniform on the days that class met. That Friday he took his first test and made his first rent payment. When the weekend arrived, he spent the morning clearing weeds for the landlord in exchange for a portion of the rent, and attended his first college football game. Sunday morning he slept in and "fooled around" with roommates and a visitor, studied, but failed to attend church. In a letter to Pearl not long after school started he commented, "I don't believe I am going to have time to go to the ward much—there is just too many things to do. I'll try to go to Sunday School."[28] By the second week he was also starting to feel the need for greater discipline. "Study is beginning to become a problem. We've got to do more. Too much bull sessions."[29]

At the end of the second week of college Royce struck out for home with his roommate Swede as soon as classes ended on Friday, hitching rides (or "bumming," in the vernacular of the time) both ways. This was an unreliable form of travel, of course, and it took them from noon until 8:30 that night to reach home, tired and footsore. Staying at home on Saturday, Royce spent the day watering, though he took the time to purchase some good shoes and arch supports, as his feet had been bothering him at college. He also went out to see Pearl, though it was a limited visit. "Went up to pearls at nite—hello & good by. I've got to study"[30]. The return trip was a little smoother, taking only half the time of the trip home, but it was no easy journey without reliable transportation, and his occasional weekend trips home were hurried affairs, generally with a bit of farm work and a brief visit to Pearl thrown in. On one such visit at the beginning of November, Royce commented, "Went up to Pearls at nite. I've been going with her now a year haloween. Not a quarrel."[31]

The week after his first visit home, Royce, after some effort, succeeded in landing a paying job, picking apples and working in an apple shed in his spare time for 25¢ an hour. Hard work was no problem for Royce, and he put in his hours with some overtime, but the work quickly began to take its toll, as he found himself more fatigued and with less time to study. He failed one of his algebra tests a week after starting work, and a week later he slept in and nearly missed his classes. His studying became more of a problem during this time, and evidently the need to study created friction between Royce and his roommates, which by November 11 had reached a crisis point. One weekend after "bumming" home and back, Royce commented, "I dont know what to do. I can't study. They don't want to and don't want anybody else to. This is hell. I've got to get to work."[32]

The next day Royce paid a visit to a faculty member who also served as an undergraduate counselor, and they worked out a new study plan for him, which involved going up to the library to study each night to remove himself from the distractions of his roommates. It appears he also either completed or relinquished his job with the apples, as it was never mentioned again. He was able to work out a new system with his roommates for house chores. The weekend "bumming" visits back home also largely ceased, except for the long Thanksgiving weekend, when he spent most of the visit plowing the farm in the absence of his father (who had been hospitalized for a foot ailment) and studying in the evenings between visits to Pearl. Despite his struggles with college, he observed in his diary, "Thanksgiving day. I guess I have plenty to be thankfull for."[33]

By late November, Royce's schedule had firmed into a routine, with classes in the morning, chores or study in the afternoon, and the library at night. Sundays he usually attended either Sunday School or church, though his attendance was not consistent, and he sometimes spent these days studying. Tuesdays and Thursdays he wore his military uniform and attended ROTC drills, except for occasional cancellations because of weather. He was pleased to pass his English placement test ("I don't have to take bonehead English. Passed placement."[34]); he did well in Zoology, struggled, ironically, in Botany, endured his Military classes, and made little mention of his Farm Crop classes, in which he excelled. His nemesis seems always to have been Algebra, in which he resigned himself to receiving a "C" grade, for which he seemed grateful. In December he reflected, "The quarter nears an end and what have I accompplished. Something I guess. It seems like I just got here."[35] The first quarter of his college career ended on December 17, and Royce came home to Murray that afternoon.

Royce's first Christmas vacation home was a pleasant break from the rigorous routine of college life. On his first day home he went up to the hills west of the farm and brought back a Christmas tree, and during the two-week vacation he worked on the farm, plowing until frozen conditions prevented it, trimming trees, and spreading phosphate on the soil. He collected a cash prize for his beets from the previous summer, which had won second place in a local contest. Some evenings were spent with Pearl, and the two drove around town to see the Christmas lights and decorations. Shortly before Christmas, snow fell, turning the city white. Royce spent the night of Christmas Eve at Pearl's house helping to decorate, and had dinner there on Christmas day. Two days later Royce turned nineteen, and took Pearl to a movie to celebrate. New Year's eve was again spent at Pearl's house where he saw the New Year in with her family, and he spent all of New Year's Day with her as well. They attended a dance that night, not returning home until 2:00 a.m., upsetting Pearl's mother. His journal entry for January 1, 1938 began, "Another day another year another month, new hope and new ambition."[36] The next day he returned to Logan exhausted, after little sleep and nursing a bad cold.

The Monday after Royce's return to school he received his grades for the first quarter, and learned that he had earned an "A" in Zoology and in Farm Crops and "B's" in his other three classes (Algebra, Botany, and Military Science), which was probably a relief to him; in his journal he pronounced these marks as "good enough." That same Monday most of his roommates registered, but he stayed home with his friend Swede and "just fooled around," which proved to be a hard lesson for him. Attempting to register the following day, he found the classes he sought were already full, and was only able to register on the third day with great difficulty, and then with a schedule that required him to spend all of Mondays and Wednesdays in class without so much as a break to eat or study.[37] As the quarter proceeded, the consequences of his late registration began to manifest themselves. On the first Wednesday of classes Royce wrote, "School from 8 to 5, not even an hour off for noon. Boy was I a hungry gosling."[38] Between late night study and weekend trips to Murray, Royce often found it difficult to get breakfast before going to classes, and on these days he ended up fasting until dinnertime while attending to his vigorous course load. His most challenging class was bacteriology, and he frequently spent extra hours studying for this course.

During this quarter Royce and his roommates worked out a system whereby one of them did the cooking and cleaning each day of the week—Royce's day was Thursday, and this seemed to work well for him, as he had little class work that day, and it served as something of a break from his difficult schedule early in the week. He generally spent Thursday mornings cooking or baking bread, and cleaning the house. The friction he had experienced with his roommates about studying during the first quarter largely disappeared as he continued his pattern of going on campus to study in the afternoons and evenings. Occasionally he would spend the evening home and talk to one or another of his roommates, an activity they referred to as "bull sessions", but this no longer interfered with his study as much as it had in the previous quarter. His activities were not limited to class work. In February wrote Pearl, "There is a 4-H Club Leaders Short Course going on up here all this week and I have to go to their meetings every time I get time. I guess I'll be a club leader for them again this summer."[39]

Royce continued to "bum" weekend trips back home to Murray as often as circumstances would allow. His usual pattern on these weekend trips home was to hitchhike from Logan after classes on Friday and return during the day on Sunday. He would visit Pearl Friday and Saturday evenings, and do farm work all day Saturday. He and Pearl, living far apart from each other most of the time, made the most of these brief visits, going on dates together and often staying up late; Royce also sometimes found it difficult to return to the rigors of farm work after his less physically demanding life as a student. On one weekend he wrote, "Saturday worked all day halling manure. I'm so weak, I guess I can't take it." A couple weeks later: "Hauled manure this day. boy I either got to start working or stop eating. Getting fat."[40]

In addition to their occasional weekend meetings, Royce and Pearl continued to correspond regularly as they would for years, and the sending and receiving of letters (or the lack of them) were recorded and commented on in Royce's journal.[41] By this time the two were considering marriage and discussing how it would fit into their respective plans, and some of their letters seemed to have particular significance to them as they worked to define the bounds of their relationship. After one trip home, during which Royce had brought Pearl home very late from a dance and the two evidently had felt

36

tempted to overstep boundaries they had set for themselves, Pearl's mother became upset, prompting a heartfelt exchange of letters. Royce's reply, written on March 3, reflected his desire to maintain high ideals despite the enticements the two naturally felt as college-aged youth in love. "I want to be clean and respectable and worthy to be the father of your children some day, and I want to appear as such to you. I don't blame your mother for being angry; she had a perfect right to be. I hope she won't hold it against me. I have brought you in late and I'll see that it doesn't happen any more. Sometimes I feel like I am messing up both your life and mine but I pray with all my heart that I shan't. This has come straight from my heart Pearl, I always want you to see things as I see them and to know what I think and feel. Pearl I'm not regretting anything I have done. It's just that we might improve upon it, that it might be a help to us in the future. I wish I could tell your mother how I feel. I want to do right . If I ever have anything said about me, I want it to be that I was decent. I'm trying to be."[42] Just a few days earlier he had reflected in his diary, "Their are just about two things I desire most right now—to make a success of my college work, and to make a decent go of it with Pearl. I think I can do both."[43]

The final days of the winter quarter brought new changes for Royce, as three of his four roommates decided to drop out of college, leaving only Royce and Dave Miller of the original five. The house they had rented was too expensive for just two roommates, so they sought lodging with an older student named Sherm Gold, whom Royce had met in one of his classes.[44] In a letter to Pearl on a chilly Saturday morning Royce intimated what might have been one reason for his roommates' decision to quit college. "Morning again, I'm the only one up. The rest of the tribe can sleep all day. I always have the pleasure of building the fire on weekends, if I waited for someone else to do it, we would never get up. It's 8:15 now and I want to be at the library by 9: that's when it opens."[45] That same afternoon after a day's study in the library, Royce walked into town and spoke with the landlord, about the situation with his roommates and their plan to leave the house.[46] It had fallen to Royce to deal with the landlord, and during the transition he took responsibility to see that the students' obligations were all met.

Royce caught a severe cold at the beginning of Finals Week, the worst of all times for a college student. On Monday, after his full day of classes and an evening of study, he recorded feeling "very tired." On Tuesday, which was to have been his study day for exams later in the week, he wrote in a barely legible scrawl, "Went to school, boy am I sick come home & went to bed – Of all the times to get sick – End of quarter. Will I flunk test." The next day he took a test for his most difficult class, Bacteriology, and recorded, "Went to school; boy did I have a time getting up the hill. Had a test in Bacteriology, 100 questions. I hadn't studdied a bit. Come home noon went to bed, postponed botany." The next day: "Thursday. Still sick, just a little better went to school at 10 to Algebra then took the Botany Exam. Came home and went to bed, ate a little tonite." And the next: "Friday. Still not feeling tip top. havn't studdied, took. Ag. Econ, Bac. & Algebra Exams." Shortly after finishing these last examinations Royce thumbed his way back to Murray. "Left for home about two, got their at 5:30. Felt pretty bad. Just doctored up and went to bed." The next day, feeling much better, he took Pearl to an early show, and then they talked long into the night.[47]

The break between the winter and spring quarters was no more than a weekend, and on Sunday, March 20, Royce returned to Logan with his roommates, writing to Pearl the next day, "We left at seven o-clock Sunday morning, Sam, Steve & I. We got here about 9:00. They loaded up their things and went right back. I hated to see them move out. I have been here alone since."[48] The third roommate Swede came for his belongings a couple days later, leaving Royce alone in the empty house. From then on he would only see his boyhood friends back in Murray during vacation or on his brief weekend jaunts home. For him, life had moved on, and of his original group of Granite High friends only Dave Miller remained.

When it came time to vacate the living quarters, Royce took the responsibility of cleaning out the house and getting it in order. "It took me quite a few hours to clean the house up so it was fit to live in. It was certainly one dirty mess—I felt like I had done a day's work before I was through." In the process, Royce seemed to have gained the respect of the landlord, and he left the house feeling that he had been treated fairly. Royce was still getting over his cold from the previous week, and his mother, who had frequently provided foodstuffs and moral support, came through with a folk remedy which helped him over the illness. Royce would maintain a close relationship with his mother over the ensuing years, and she would prove one of his greatest sources of encouragement and support.[49]

The final quarter of Royce's first year seems to have gone much more smoothly after the move to the new house. His most challenging course proved to be trigonometry, which, as the term progressed, he found to be increasingly complicated, and to which he dedicated extra attention and study—he eventually earned a "C." Able to study both on campus and at home, he did consistently well on most of his tests, and enjoyed attending school as winter gave way to pleasant Spring weather. Considered by his roommates to be a good cook (having taken cooking classes in high school), Royce frequently baked bread, and occasionally remarked with satisfaction on his cooking skills. One Saturday later that year he wrote, "I baked a batch of bread today that is the real 'McCoy.' I'd stack it against any batch."[50] He found time to take up tennis, perhaps to have something in common with Pearl, who had played tennis for Granite High School—he enjoyed the sport, but admitted he was not very good at it[51]. He also participated in the Ag Club, which among other activities was preparing for an annual horse show.

Throughout this time, Royce participated regularly in Military Science classes and drills with the ROTC, which would continue through his first two years at Utah State. Royce never mentioned why he took these classes, though there were many participants in the program in the days leading up to the Second World War. He often went to shows and sporting events or into town, usually with roommates or friends. Among the number of new friends and college associates mentioned in his journal during this time, at least one of these, Harry Miller, an upper classman whom Royce admired, would play a significant role in his later life.[52] Another close friendship which developed during this time was with Norman Johnson, with whom Royce shared his trigonometry class. The two would later exchange letters, and Norm would figure prominently in Royce's later decisions regarding military service.[53]

Royce's trips home to Murray were only occasional now, though he enjoyed the breaks from school. On one such trip made shortly after moving into his new house in April, he wrote, "Got up early, rode the leveler all day. It sure seems good to put in a day at farming."[54] The following month, in May, he made another trip in order to attend a day-long Saturday meeting as 4-H leader. Returning on Sunday, May 8, Royce was driven back to Logan by his father, accompanied by Pearl, who took photographs. It was the first time his father had been to visit him at the university, and Royce recalled the occasion years later: "I had a nice light suit, Dad wore his suit, and I couldn't help but see the wistfulness in his eye, because he always wanted to go to school. He never had the chance."[55]

Although he was having success in his studies, this year in college was a low point, if not in Royce's spiritual life, at least in his church observance. After the first weeks of January he did not attend Sunday School or church again until the final weeks of May. Sundays were generally spent doing housework, cooking, and studying, or during weekend trips home, traveling back to Logan. Sometimes he attended shows on Sunday. He occasionally remarked on being too tired for church or on not getting up on time to attend, but usually no mention was made at all—certainly he was not in the same frame of mind as when he would later determine to serve a mission. Finally in mid May,

Pearl came to Logan for her first visit there, and that Sunday the two attended a meeting at the Institute together. This happened to coincide with a visit by Heber J. Grant, then president of the Church, who spoke at the meeting. What kind of impact this may have had in Royce's life is not recorded, but he did attend church on his own a couple weeks later for the first time in months.[56]

As the first year of his college career approached its end, Royce expressed more confidence as he undertook the various examinations. Largely gone from his journal were the previous fears of failure. In late May he participated in the yearly horse show with the Ag Club and a military inspection on the same day, and shortly thereafter turned in his military uniform. The last of the exams took place at the beginning of June, and Royce recorded, "Well my last nite in the old shack – I'm sorry in a way – but I must move on. It's been fun here." The next day he added, "The quarter ends, the year ends, and the beginning of my college career is over – I'm going to miss it but it will be swell to be home"[57]. That day he thumbed his way home to Murray, returning the next day with a truck to pick up his belongings and vacate the house. Royce had successfully completed a year of college, the first member of his family ever to do so.

Summer and Second Year

On his return to Murray, Royce jumped immediately back into the routine of farm life which had been his lot since early childhood. His first full day home was occupied with the despised chore of irrigation, and he wrote ruefully, "Sunday 'I think.' Started my vacation out right—took the water at nine—wattered all day."[58] At the time of Royce's arrival home, family life in the Bringhurst home had been disrupted by a temporary interruption in the house's water supply, and the following day Royce recorded, "today 1st day we have had drinking water for 4 days. It is terrible. Then it went off again."[59] Despite vigorous physical labor on the farm each day, it was still one more day before the water was finally restored sufficiently that Royce was able to take a bath.

The Summer of 1938 was occupied much like the previous one, with the usual rigorous routine of farm work by day, punctuated by occasional recreational activities on nights and weekends when Royce was not watering. He again held 4-H meetings, and he helped to judge the crops exhibits at the Salt Lake County Fair. He made another fishing expedition with Steve Mackay's family to Blacks Fork, catching trout by the dozens. Toward the end of the summer he went around the community with a faculty member, soliciting funds for a new building at the Agricultural College.[60] The one major difference from the previous year was that Pearl was in town, and much of

Royce's free time was spent with her, going on dates to shows and other events, or simply visiting and talking. When Royce went with his family on a camping and fishing trip up to "the lakes", Pearl followed a few days later and they hiked together while the others fished. On one weekend in September they visited in each other's homes, each sharing personal treasures with the other.

Shortly after his arrival home for the Summer, Royce began attending church again on a regular basis. On his first free Sunday home he acknowledged, "Went to Sunday School, first time for quite a while"[61]. He attended more regularly throughout the summer, often attending both Sunday School and church, and the church services seem to have become more meaningful to him. He began to reconsider his Sunday activities as well, chronicling what he regarded as the consequences of his sabbath-breaking: "Sunday – Hauled hay all morning. tipped over a load of hay & dropped hay. That's what we get for hauling sundays."[62] A few weeks later: "I knew I should have went to church today. Went out to get pine nuts never got any – broke an axle cost 5.60."[63] On a Sunday in late August, Royce received the Melchizedek Priesthood and was ordained to the office of Elder (the same day he learned that his older brother Smith had been secretly married two days before).[64] Thereafter he began attending Priesthood meetings in addition to Sunday School and church, and his church activity would become much more consistent as he returned to college in the Fall.

On September 22, a Friday, Royce went into town in the afternoon to make purchases for school and took Pearl out in the evening to see a movie, which proved to be the first full-length animated film, Walt Disney's *Snow White and the Seven Dwarves*, which had just been released to theaters that year. The next day was his last full day of work on the farm. He plowed in the morning, hauled hay all afternoon, and in the evening went to take leave of Pearl. Though Royce was fatigued from the day's work, the two made a cake together and "just sat arround". The next Day, Sunday the 24th, he got up early to prepare to leave for school, pulling out at 10:30 in the morning. He arrived at 3:30 p.m. in Logan, not failing to attend church that night.[65] The day after his arrival, Royce registered for his classes, which took all day, and he regretted not having come on Saturday to do so. He signed up for basic science classes: Chemistry, Geology and Physics, in addition to a music class and his Military studies.

On October 6, 1938, two weeks after Royce's arrival in Logan, attention shifted suddenly to the world stage, as Royce and his fellow students were summoned to a college-wide assembly to inform them of evolving events in Europe. The Munich Agreement had just been announced, ceding parts of Czechoslovakia to an increasingly aggressive Nazi Germany, and the college faculty had considered this a matter of sufficient importance to the student body to explain the situation to them in detail. This was part of the historical backdrop of the times, and though Royce made only passing mention of it in his journal, it would prove significant for him and his peers, as it was part of the great sequence of events which would eventually plunge his entire generation into a second great World War. At present, though, this seemed little more than a distant cloud on the horizon.

Royce turned to his studies in much the same way as he had the previous year. The class that gave him his greatest academic worries was Chemistry, and after a couple of weeks he wrote, "Resolved: hereby to study hard and make a go of Chemistry. I have just got to."[66] He spent extra hours studying, sometimes until past midnight, and eventually began to gain greater confidence in the class. Studying was generally done at the library after classes, though with his more studious roommates Royce sometimes studied at home. He continued to participate in the Ag Club and the 4-H Club, and on October 23 participated in a stock judging contest at which he placed third.[67] Some weekends he attended home football games, in which he had an added interest this year since his roommate Sherm Gold played tackle for the varsity squad. On other weekends as circumstances allowed he made the lightning weekend trip back to Murray to work a little on the farm and see Pearl, though these trips were modest affairs, Royce hitchhiking both ways and spending as little as possible to conserve funds. He acknowledged this in a letter in October. "I wish to thank you again Pearl for being so darn nice about going out. It takes just about all the money I can get to keep going up here; about the best I can do is a show when I come down. It doesn't matter to me where we go. I just look forward to being with you."[68]

By this time Royce's studies had worked into a routine: "School again till Noon. then a lab all afternoon. Life is pretty busy every day but still that kind of old sameness. Studdied Chem at nite."[69] Royce may have felt some wistfulness about the sacrifices he was making to attend college. Once, after receiving a letter from his friend Steve Mackay who had quit college during the previous year, he wrote, "Got a letter from Steve today. He had a job and a car now. Sometimes I wonder."[70] Royce did not say whether he was wondering about Steve's future or about his own present, putting off any immediate prospects of such luxuries as a regular job or a car while he struggled through college. For the Thanksgiving holiday Royce returned home to Murray, and on Thanksgiving Day he brought Pearl to dinner at his grandfather Smith's house, a Bringhurst family tradition. This was a large gathering, with no fewer than 54 family members and friends attending, and for the first time Pearl was introduced to much of Royce's extended family.[71]

During the fall quarter, Royce began attending church regularly at the Institute, and on the last Sunday in October he recorded in his diary the first intimations of what would later be an important undertaking of his life. "Went up to the institute in the morning to church. I think I'll take a missionary training course up their. I hope I may go on a Mission."[72] Although a mission was by no means expected, nor was it particularly the norm for young Latter-day Saint men of that era, Royce had watched a number of his acquaintances depart on missions, and was of course aware that his own father had served one before he was born. His journal entries during this time reflected a greater spiritual sensitivity than before. On October 23 he wrote, "Sun – Went to church up at the Institute in the morning – I certainly enjoyed." On November 13: "Sunday – Went up to the Institute to church in the morning – Its getting to be a habit (a good one)." On December 4: "Sunday – Went up to the Institute to Sunday School. It was certainly fine. Went up to the Institute at nite to a program then stayed after." On December 11: "Sunday. Went up to the Institute to the Christmas program. They certainly had a large crowd their. It makes a fellow feel good."[73]

Royce looked forward to the end of the quarter and Christmas vacation. On December 16 he took his last final exam in Physics, and upon returning to the house, was informed that he and his roommates would have to move to a different place for the upcoming quarter. They quickly located a new residence just down the street at 541 East 4th North, and within two hours they had moved their belongings to the new house, which would be Royce's home for the remainder of his stay at Utah State. Returning home to Murray for vacation the same day, Royce went back to working on the farm, going out with Pearl in the evening and catching up with his old friends. On his birthday, December 27, Royce observed, "20 years old. I've seen the last of my teens."[74]

Returning to Logan after Christmas vacation, Royce launched into the one of the busiest seasons of his college career. On January 4 he registered for the winter quarter of 1939, taking on a large class load. On the first day of classes he tried out for the chorus of the yearly college opera production. The same day, perhaps influenced by his roommate Sherm who had been a collegiate wrestler the year before, Royce signed up for a wrestling class, but injured his left knee in the class just a week later. He also signed up for a Book of Mormon class at the Institute, giving him a heavy load of 18 credits, with little free time. In spite of the heavy work load, Royce seemed to thrive in his classes. One of these was an English class, Sophomore Composition, and besides spending long hours at the library writing "themes,"

Royce began to chide himself for careless spelling in his diary and in letters to Pearl, and to pay more attention to his grammar.[75] His rural Utah colloquialisms slowly began to give way, and he would eventually become an excellent and thoughtful writer with impeccable grammar. In the midst of an arduous schedule he repeatedly remarked in his journal on his laziness and resolved to improve his study habits, and occasionally gave an uncomplimentary self-appraisal. In February, for example, he commented, "I'm sort of a dope. 'The world you live in just isn't any better than the person you live with.'"[76]

The beginning of this quarter seemed also to be a time of spiritual growth for Royce, and he continued to nurture his ambition to become a missionary. On his first full Sunday in Logan he wrote, "Went up to the Institute to Sunday School – I hope I get a call to go on a mission."[77] At this time, perhaps in conjunction with his class at the institute, he began to read in the Book of Mormon, and one day recorded in his journal, "I realy had a prayer answered last nite Alma Chap. 38: B. of M."[78] The reference was to a chapter in Latter-day Saint scripture containing an exhortation by a father to a faithful son, and, although Royce did not remark further on how his prayers had been answered, clearly his convictions in his Latter-day Saint faith were coming to maturity. He learned with some regret that his opera rehearsals, held on Sunday mornings, would conflict with his now cherished Institute Sunday School and prevent his attendance nearly all that quarter. That same week he met twice with a mentor at the Institute, a man named Sessions, perhaps to discuss this situation as well as his preparation for missionary service.

The student opera for that year was Sigmund Romberg's light operetta *The Student Prince*, which was scheduled to be performed near the end of the quarter in the Capitol Theater in Logan, and Royce had been selected as part of the 100-voice chorus.[79] This was a major annual community event, and as the performance approached, the usual Sunday morning practices expanded into long evening rehearsals which went late into the night and left him with little time to spare. He studied in the library in the afternoons, but as the dress rehearsal drew near he had to dedicate all his free time to the opera, which finally was performed on March 9 and 10, and was well received. Royce was happy to be done with the vigorous rehearsal schedule, but the opera left a good taste in his mouth, and he expressed regret that it was over. He would often return to participation in musical theater in later years, usually in a supporting or chorus role, or as Royce would say light-heartedly, "spear carrier, third row." He never tired of this, and was especially delighted when he could involve members of his family in these productions.

Royce made a few trips home to Murray during the winter quarter.

44

In January he came back to take Pearl to the Gold and Green Ball as he had the previous year. A month later, when his oldest sister Naida came to visit, he again returned home, bringing Pearl to a family dinner on February 12, when his family was reunited for the first time in several years. Two weeks later his Uncle Lou died, and he made yet another trip through heavy snows to attend his funeral on Sunday, February 26. Royce completed the winter quarter successfully in March, and despite the rigorous schedule and the many opera rehearsals he noted with satisfaction that he had earned his highest marks since starting college, including an "A" in the Chemistry course which he had at first found so daunting.

During spring quarter, Royce enrolled in a class in English literature, in which he excelled, and a one-credit typing class, in which he did not, despite spending many hours practicing at a typing lab.[80] His most challenging class was Plant Taxonomy which, in addition to introducing him to the vast field of plant classification, as the quarter progressed also involved long field trips and endless walking through the canyons and mountainous areas surrounding Logan. Although Royce did not excel in the class (he earned a "C"), the notebook of pressed plants he collected during the field trips became a keepsake which he would retain throughout his life. Royce also enrolled in a chorus, far less demanding than the opera chorus of the previous quarter, and in connection with this participated in a production of Mendelssohn's oratorio *Elijah*, which was presented at the end of the quarter. He again assisted with the Ag Club's annual horse show, and took a field trip with the 4-H club ("Swell crowd, swell time").

During a trip home for Easter weekend Royce had his vision tested, and was prescribed eyeglasses for the first time.[81] A couple of weeks later during another visit home, he had his first run-in with the law, and was pulled over in his car for speeding while on his way to pick up Pearl and her brother Daniel to take them to a banquet. The following day Royce took the speeding ticket to his Uncle Art (Arthur B. Bringhurst), who happened to be a highly-regarded justice of the peace. Attending to Royce's case on a weekend, Justice Bringhurst administered the law to his nephew, though reducing the fine from the original $10.00 to $5.00 because there was no history of previous arrests, and Royce paid the five dollar fine. (Uncle Art would later return to Royce an equal amount as a gift during his mission.)[82]

As the quarter progressed, Royce strengthened his resolve to serve a mission, and spiritual concerns occupied an increasingly important place in his journal entries. On a Saturday before Fast Sunday in May, to fulfill an assignment for Institute, Royce stayed up at night writing out his testimony,[83] which he entitled "Why I believe Joseph Smith is a Prophet of God", and which he planned to give in testimony meeting the following day. In it, he summated his growing faith in the "gospel of the Restoration", giving as his motives his examination of the Book of Mormon, the teachings of the Word of Wisdom, the principle of continuing revelation, and his own pioneer heritage; however, his greatest motive was his own experience with spiritual feelings. "The most important reason for my belief to me is simply, I have a conviction deep down within myself that I know he was a prophet. It gives me much joy just to think about it. This Spring, I stood down there in the tabernacle on Easter Sunday and sang with the group, 'We Thank Thee Oh God for a Prophet.' It just done something to me. My heart was filled with the spirit of things. I was glad, I wanted to shout out, to let people know I believed." In conclusion he wrote, "Yes, I believe he is a prophet, and in a humble way I try to live as he taught, although often I feel that I fail miserably in this respect. I have never been more proud to do something than when I had the opportunity to pass the sacrament at the Institute. It's just a deacon's duty, but I felt ever so honored with having the privilege. I haven't been on a mission but each night I pray to God that I might be granted the privilege of going."[84] The next day he recorded, "Went to Sunday School, passed the sacrament. To testimony meeting[85] at 5:30. I bore my testimony, I pray that it will strengthen."[86]

The prospect of serving a mission increasingly occupied Royce's thoughts. The following Wednesday his entire journal entry was a reflection. "Another day in the swift moving parade of hours & minutes. School draws to a near close. What then? I am hoping and praying that I may go on a mission." The next day he added a quote that summed up his feelings at the time and would remain a governing theme for the remainder of his life: "He that is not spiritually developed is poorly educated."[87] Two Sundays later he wrote, "Sun. a far cry from graduation. Now I look to a mission with all expectancy next summer. I would rather go than anything I can think of. Sunday School in the morning. I have learned to love the Institute." The next day he again met his mentor from the Institute, and made some specific plans to prepare for missionary service. "I had a long talk with Bro. Sessions, figured out my next 5 years. Two years for a mission figured in. It'll be a long old time till graduation now, but I'm glad. I want it that way." Royce had no way of knowing how many years the currents of history would add to that delay, but his mind was set on a mission, and the next day he wrote his home bishop to advise him of the decision and his plans.[88]

The school year drew to its close. The week before finals Royce took his last plant collecting excursions with his taxonomy class. "Wed. Went swamping today on a tax collection trip – A lot of plants, wet and tired." "Sat. Taxonomy collecting trip. Quite a hike, about walked our leggs off – walked up Spring Hollow – Clear to lake at head of providence Canyon. Went swimming in it – very cold."[89] He took exams as always, but on Friday, June 2, the last official day of classes, having no classes to attend, he broke away from his studies to travel to Ogden where he signed up with Southern Pacific Railway to do "section work" on the railroad over Summer. This was evidently part of the plan he had worked out with Brother Sessions in order to finance his mission. The next day, Saturday the 2nd, he took a last grueling three-hour examination in Taxonomy, packed, and left Logan and college for what he expected to be hardly more than a two year absence. He later summed up in his journal the money he had expended on the first two years of college, and the total came to just over $400.00, $219.60 for the first year, and $192.20 for the second, a tidy sum for the waning years of the Great Depression.[90]

Summer and Fall—Preparing for the Mission

Arriving home on Saturday, Royce attended Sunday School the next morning, then spent Sunday afternoon with Pearl, retiring early. The next day he began his work on the railroad, "working section". Unaccustomed to physical labor while at college, he at first found the transition challenging, and the journal entries from his first week give the flavor of his work. "Mon. Started on the section out at Garfield. Tamped ties. I am stiff all over & the hands are blistered; be glad when I get over it." "Tue. No. 2 day on the section. not as bad as yesterday. I'm still stiff tho and the hands are awfully sore. I don't mind tho. It's the dollar that counts; to bed at nite." "Wed. Worked on the section again. not such a hard day. Road up to Pearl's with mother & Dad. Never went anywhere. Came home early." "Thurs. first week is almost past. I don't feel so bad. In fact I feel pretty good all considered. Took mother up to Dean's then went to bed." "Fri. Work again & plenty of it. About done me up; the tie gang came down. Quite a bit of labor. Went to bed at nite, too tired to do other wise." "Sat. Work work work, I thot this day's would never end. I am really tired. I went up to Pearl's at nite. We didn't do much of anything."[91] Although he quickly grew accustomed to the hard labor, the weather conditions at times impacted the work, as his journal on three successive days of the following week indicated: "Wed. Another scorcher so hot it was almost unbearable." "Thurs. Not hot as usual but boy oh boy how that wind blew my eyes are so sore from sand blast." "Fri. Turned off cold & stormy today worked as usual."[92] The railroad work was also tedious, without the accustomed variety of farm life. "Tues – Work again. Started out there today again. The same old monotony." "Thurs – another day in the endless chain of time worked again. Never went anywhere at nite other than to bed."

47

[93] The money was good, though, for the times—for two week's labor Royce was paid close to $40.00, and the money went to canceling debts and saving for his upcoming mission.[94] Even while working on the railroad, Royce spent his free moments attending to chores on the farm. Given the day off for Independence day, he wrote, "The fourth. Irrigated, my usual holiday pass-time."

Royce worked on the railroad continuously six days a week until mid-July, when he was laid off and returned to farming full-time. He launched immediately into weeding and irrigating the beets on the family farm, tending the apple orchard, and shocking the grain. On one Sunday in July he invited Pearl out to dinner and afterwards taught her to ride a horse.[95] As July wore into August he cut, raked, and piled hay, threshed the grain, and hauled both grain and hay. During these months he continued as director of his 4-H club, holding meetings periodically throughout the summer. Evenings were frequently spent with Pearl, often on joint outings with friends such as his "best friend" Steve Mackay or Ralph Willes, a friend from Granite High who had attended Utah State with Royce, and, like him, was preparing for a mission. Sundays he attended Sunday School, often with Pearl. By this time Pearl had also gotten a job at a department store which involved some shifts on Sunday, and Royce would often pick her up from work. In late August he took his yearly fishing expedition to Blacks Fork with Steve Mackay and his family, although in contrast to previous years, the trip was planned to avoid being gone on Sunday.

During this time Royce continued to actively prepare for a mission. On July 6, while still working on the railroad during one of the hottest days of Summer, he passed an important milestone, receiving his patriarchal blessing from Thomas W. Dimond, his former bishop, for whom he had previously done farm work to earn money for college.[96] Finally a month later, on August 7, a Monday, he was called in to an interview, and at last received the longed-for request to prepare for missionary service. He wrote, "Today was very important to me. I was asked to go on a mission. Dad consented; if I pass O.K. now, I go."[97] From that time forward the prospect of a mission was more a reality than a hope, and Royce began preparing in earnest.

In late August of 1939, as Royce busied himself helping to set up the fairgrounds for the annual county fair and preparing for a mission, he paused to take note of momentous world events which daily began to unfold in his diary entries. On August 31, 1939, Thursday, he wrote, "War looks inevitable." The following days he added: "Friday – another day another month, and another War. Germans & Poles began to fight today, no official war yet.... Saturday – War between Germans and Pole's declared as such.... Sunday – France &

48

England declared war on Germany today."[98] So began the Second World War, and events were set in motion that would eventually change the future for Royce and for all his generation, though as yet the war was on a distant continent and affected their day-to-day lives but little.

Royce had been called back to the railroad for a few days in August, and on September 6 he started back to work full-time, which he continued through the month. On September 23 he remarked with some nostalgia, "Sat – If I were going to school I would register today." Remaining at home as the school year got underway, Royce was able to provide some help to his younger sister June, who was struggling with her own school assignments.[99] Knowing that his time in Murray was drawing short, he continued to spend evenings and weekends with Pearl and to see as much of his old friends as he could—no longer mere schoolmates, their lives were quickly moving on as was Royce's, and these occasions became precious to them. Royce continued work on the railroad until September 30, when again he was laid off. "Sat – My last day on the section but I didn't know it till I got off work. Never will I be a section laborer again."[100] He returned to the farm in time to help with the apple harvest.

During the Summer and fall, the mission was never very far from Royce's mind, particularly as he saw friends and acquaintances both leaving and returning. Pearl's future brother-in-law Mont Toronto returned from his mission in New Zealand in July, and her sister Jeanette from the Southern States in October. Royce's friend Art Wallace had just departed on a mission, and Ralph Willes and Norm Johnson would soon follow suit, with Sam Oliver preparing to leave early the following year.[101] On September 22, in the evening after work, Royce had his pre-mission physical exam, and in early October he picked out a suit and had dental work done. On Sunday, October 8, he attended General Conference at the Salt Lake Tabernacle with Pearl. "Pearl & I went to Conference from morning till nite – three sessions. It was fine – a lot of things to remember – It was tiring but I enjoyed it."[102] The next day he went back up to Salt Lake City for a final interview with Joseph J. Merrill, a member of the Quorum of the Twelve Apostles, and was told he had been approved for missionary service.[103] Three days later on Thursday, October 12, the long-awaited letter finally came, and Royce recorded in his journal, "Received my call for a mission – to serve in the Spanish American Mission. It was a thrill."[104] The call letter specified that Royce was to enter the Salt Lake Mission Home on November 14, 1939.

After receiving his call Royce continued his farm labor as vigorously as ever. He completed the apple harvest, then set to work topping beets, first for a neighbor and then on his own farm, a long and

49

tedious chore which he was grateful to finish. After the beet harvest, Royce plowed the fields and hauled manure, taking brief time off to go pheasant hunting in early November. It was during this time of preparation that an episode occurred with one of the animals that Royce would recall with sadness for the rest of his life. One of the horses had been injured and developed a serious infection, and as the rest of the family was departing for a several-day outing, Royce was left with instructions to shoot the horse if it failed to recover. On a Friday in October he wrote, "One of the hardest and saddest jobs I ever had to do in my life. I had to shoot the little saddle horse, Belle."[105]

As the date of Royce's mission approached, a series of "showers" were held for him, this being a custom in those times for departing missionaries to provide them with some of the necessities for their work. He recorded, "Pearl, the girl friend, gave me a shower on Saturday October 28. It was fine. Nice things and nice people. On November the third, our missionary class gave me a shower. It was certainly fine of them. On November the fourth, another shower was held for me at my home, given by Aunt Kate. It touches me deeply to know the fine spirit that exists in the family. The list of things I received at these showers are as follows – 25 pr. socks, 14 shirts, 9 ties, 4 pair garments, 2 pair pajamas, 10 handkerchiefs, a diary and this record book, sewing box, lotion & razor blades, clothes brush, & stamps."[106]

Finally the night of his missionary farewell arrived—this, according to the custom of the time, had been scheduled and announced in the local newspaper, and involved a formal program with a dance afterward. This was an opportunity for well-wishers to see off the missionary and contribute funds. Royce wrote, "My Farewell was held on Friday Nov. the tenth. I only hope and pray every missionary gets the same grand send-off I am getting. 66 dollars were taken in at the door plus $11.50 that has been given me outside. The beautiful music and the kind words that were said were very impressive to me, the memories of which will be recorded pleasantly and permanently in my mind." The music included solo numbers by his friend Meade Steadman and his sister Rhea. His bishop, Samuel S. Smith spoke, as did a relative, Bishop Henry Bringhurst. Finally, Royce himself gave his farewell address. "With the help of the Lord I was able to bare my testimony, and to say the things I wished to say in the way I wanted to say them."[107] After singing the moving and sentimental favorite hymn, "God Be With You Till We Meet Again", the closing prayer was given by Bishop Dimond.

The next day after some last work on the farm, Royce went into town to have a final typhoid vaccination and to purchase a briefcase and an overcoat for his mission. The last day, Sunday, he attended a stake conference, and in the evening went to Pearl's house, where he was to stay the night prior to entering the mission home. That night he made the final entry in his five year diary, allowing his words to spill over into the empty space below intended for entries in subsequent years. "Sun. Went to Conference in the morning, loafed all afternoon, up to Pearls to spend the nite. I now close this book and write no more. I enter the Mission home in the morning and will continue my life record in a different book."[108] The "different book" was his missionary journal, where his first entry ended with the words, "I go in the home Monday and I hope and pray that the spirit of the Lord will go there with me and be with me thru my mission and after."[109]

Illustrations for Chapter 2

Left: Royce with a day's catch of trout at Blacks Fork; Right: climbing the south face of Mt. Tokewanna.

Top: "Old Main", Utah State Agricultural College, winter of 1937-38. Bottom from left: Royce's roommates Steve Mackay, Emil "Swede" Stenstrom, Sam Oliver, and Dave Miller. Of these, only Dave Miller and Royce completed the first year. Sam Oliver later served a mission.[110]

College photos of Royce. Left: Royce in military uniform, required for participation in ROTC, late 1938; middle: with friend Norman Johnson, winter 1938; Right: Royce's sophomore yearbook photo, October 1938.[111]

Left: Royce with his father in front of "Old Main" building in May, 1938; Right: Royce and Pearl at Grandfather Smith's home in Cottonwood, on Thanksgiving day, 1938.[112]

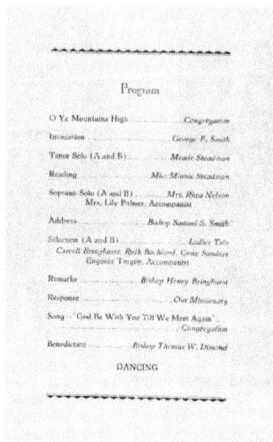

Left: missionary farewell cover, middle: program; right: newspaper clipping.

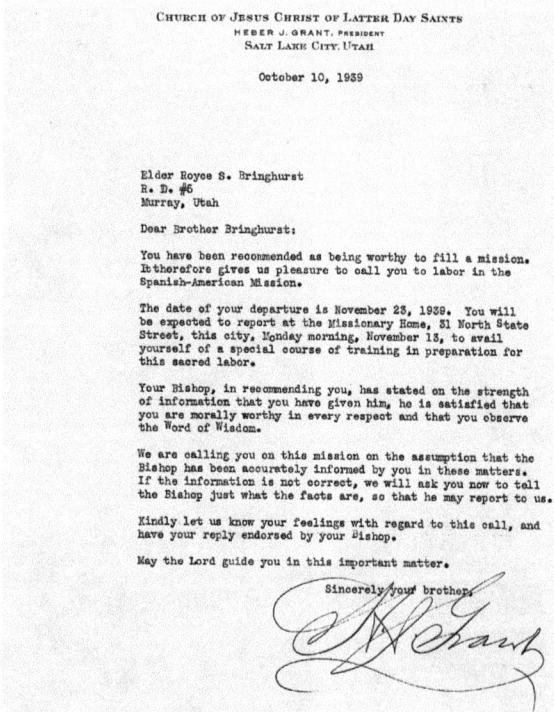

Royce S. Bringhurst's mission call, signed by President Heber J. Grant.

The Mission

Chapter 3

The Mission Home

While earlier missionaries for the Church, like Royce's father, were often married men and left from their homes directly to go into the mission field, by the time Royce received his call, missionaries were single (and expected to remain so), and a "Mission Home" had been established near Church headquarters in Salt Lake City, where they received brief training by Church leaders before departing for their missions.[1] Although sent on a Spanish-speaking mission, Royce would receive no formal training in the language like later missionaries, but would instead be expected to learn Spanish from his companions and native speakers during the ordinary course of missionary service. Because of this expectation, Royce, like other foreign language missionaries, was called for a period of thirty months rather than the two years required of those who spoke only English, putting a six-month delay on his expected return to schooling afterward.

Upon entering the mission home on Monday, November 13, 1939, Royce's initial impression of his fellow missionaries was one of disappointment, perhaps understandably so, since for the most part they, like him, were immature youths sent from various communities and walks of life, most having departed home for the first time in their lives, but his opinion of them quickly improved as their training proceeded. On the day of their arrival, the group was addressed by the president of the Church, Heber J. Grant, and in succeeding days they were taught by various prominent Church leaders, who under-

55

took the task of preparing and motivating the newly-called missionaries. On the fourth day of his stay in the Mission Home, Royce entered the Salt Lake Temple to receive his temple endowment, an entirely new and impressive experience.[2] A couple of days later he and the other missionaries re-entered the temple for instruction and a tour of the interior. "The ceremonies and ordinances and the temple itself were further shown to us today. We were priveledged to stand on many a sacred spot, rooms that had been made hallowed by the presence of the great men that have communed with God in those various places."[3]

At the conclusion of the training on Wednesday, November 22, 1939 Royce was set apart as a missionary, and at the same time was ordained a Seventy by Elder Antoine R. Ivins, one of the presidents of the Seventy.[4] Although the next day, Thursday, November 23, was listed on Royce's mission call as his departure date, since that was Thanksgiving day he was permitted to spend the holiday at home nearby with his family before leaving for the field. The next day, Friday November 24, he was dropped off by Pearl and her sisters at the bus station to begin the journey to his mission headquarters in El Paso, Texas, by way of Los Angeles. The trip to Los Angeles took all night, and by lying across the seats Royce was able to get some sleep. During the ride Royce had his first religious discussion as a missionary, with a "hard-shell Baptist," which he regarded as good experience. "We didn't argue with him, just talked." Arriving in Los Angeles, Royce and his traveling companion were met at the station by the four Spanish-speaking missionaries, and Royce borrowed some stationery to write a quick letter to Pearl.[5] The two new elders ate that day with the local missionaries, then both were taken to the station for the overnight trip to El Paso.

Beginning Missionary Work—Fabens, Texas

Royce arrived in El Paso on the morning of Sunday, November 26 (which he marked in his journal as "day one" of his mission), and immediately came face to face with the Spanish language for the first time. "They took us to church before we had even washed or shaved. I don't know a single thing that was said there. Everything was done in Spanish. We even sang in Spanish."[6] The next day Royce obtained his Spanish language books, and after meeting the Mission President and his wife, he expressed his gratitude and his determination with the language. "I wouldn't trade places with anybody, not anybody. It certainly is going to be a tough old grind with Spanish, but I'll stay with it. I'll be speaking it with the best of them when I am released."[7]

The Spanish-American mission was the Church's attempt to initiate work among the Spanish speaking populations along the border between the U.S. and Mexico at a time when the Mexican government would not permit foreign missionaries residence nor allow open proselytizing inside their country. The missionaries lived and worked in border towns where Spanish was spoken, teaching Spanish-speakers on the U.S. side, and at times crossing the border into Mexico by day to hold meetings, and returning into the U.S. at night. Royce's reaction to crossing the border for the first time was hardly favorable. "I got a look into Old Mexico today, down in Juarez. It's a funny old town. The first thing that impressed me was the awful smell." The next day, however, he attended a meeting at the home of a Calderón family in a community called La Caseta, and a little more contemplatively described the surroundings: "A little one room hut made of mud bricks about 10' by 18', a roof of willow, straw and dirt, walls and inside covered with whitewash, no floor just clay, a very minimum of furniture, hardly any dishes. But within those barren walls there was a spirit I have seen in very few other places, a true Christian spirit. Very little they had but they would give it to you. Humble, sincere, and yet there seemed to be something noble and aristocratic about them." Though he immediately took to the Mexican members, the unaccustomed food was something of a shock. The same day he wrote, "Had my first taste of Mexican food prepared by Mexicans. Tacos & Frijoles. The tacos were so hot I could hardly get it down. The water we had to drink tasted like slough water if I may say so."[8]

Royce's first missionary companion[9] was Wilmer Porter May, from Kane, Wyoming, and the two were assigned to begin missionary work in the town of Fabens, Texas, southeast of El Paso along the Mexican border. Entering Fabens on Saturday, December 2, 1939, they were unable to find a place to live, but on Monday the fourth they finally found a temporary residence. "We moved in with the cockroaches at a tourist camp in Fabens. It's rather a smelly place, 'not the city just the room;' but I guess we will get used to it and be able to stand it for a month. We visited a lot of the good Saints thru here and in Mexico today. They really are fine people despite their poverty. I'm learning the greetings in Spanish now anyway."[10] The two missionaries divided chores between them, and Royce was designated cook while Elder May became the dishwasher.

In these early days of his mission Royce expressed optimism and a determination to work hard, but he was continuously frustrated by his inability to speak and understand Spanish, though gradually he made small degrees of progress. Long hours were spent tracting, and when an occasional person took time to listen to the missionaries and their message it was recorded as a major event in Royce's daily journal. When they attended Sunday School with church members in the

57

nearby town of Tornillo, Texas, Royce was able to bless the sacrament in Spanish for the first time, but during the class time he tended children "who could not understand a word I said" while his companion spoke to the adults. Nearly all the teaching was done by Elder May, with Royce standing by trying to catch an occasional word of the conversation, and wishing he could participate in the preaching. When after a month in the field the two encountered an English-speaking couple willing to listen to their message, Royce's pent-up desire to preach at last found an outlet, and he wrote, "I really had a chance to tell the gospel from start to finish to someone for the first time since I entered the mission. I talked for an hour and a half, and really had the help of the Lord. Words just flowed out. I certainly enjoyed it. Elder May never got a word in sideways."[11]

The bright spot for Royce seemed to be their trips to La Caseta to hold Sunday services with the tiny branch there. On these occasions the missionaries were fed by the host family, and Royce quickly grew accustomed to the local fare. On December 10 he wrote, "These Mexican women can really cook. The meal was excellent—no knifes forks or spoons, just tortillas to eat with." The following Sunday he continued, "Second Sunday School down at Calderon's in Mexico. Had another feed down there and still I can take anything they give us except that very foul water, even the dirt in the beans and it's really there. But the spirit down there I just haven't seen any where else." The following Sunday was Christmas Eve, and in La Caseta Royce was introduced to the Mexican tradition of Tamales as the standard holiday fare. "The Calderon's killed the old black hog that used to be staked out under a tamarac in their back yard. We had tamales made from him for dinner and they were good. I think I can eat anything now—even drink their water. Our sunday school was fine there."[12] A number of investigators had begun attended the meetings in La Caseta along with the small group of members. Some who stuck out in Royce's mind were a man named Bustamante, with whom Royce held long conversations during their visits, and a large family named Sepúlveda, who were later to be baptized.

Royce's first Christmas away from home was spent at the mission home with the Mission President, Orlando C. Williams, and his wife. On Christmas day it snowed, and the missionaries enjoyed a fine traditional dinner of turkey with the trimmings, attending a show together in the evening ("The Lane Sisters, and by the way they are Mormons"). Two days later was Royce's birthday, and he took note of having officially arrived at adulthood. "Today I am a man. I have looked forward for years to the day when I turned 21 and now that it is here I do not feel a bit different." On New Years Day, on a more somber note, Royce observed, "New Years Day, a new year in a war-torn world. It seems just like almost everything broke out in 1939. War in Europe, another black year on the pages of history. I only hope & pray this year 1940 may pass without this great country of ours being drawn in."[13] That prayer at least would be granted, for the time being.

Early in January of 1940 the two missionaries transferred their belongings from their temporary accommodations into more permanent quarters in the home of a local school teacher.[14] Rent for their room was $6.00 per month for Royce, and presumably the same for his companion; food was somewhat more. As shipments of cash from home tended to be somewhat irregular, the missionaries often found themselves out of money and having to make do on very little for days at a time until new funds were sent from home. This pattern of periodic famine seemed to become second nature as time went on, and "broke but happy" was the attitude taken by Royce as he recorded the events. In March he received a welcome package from his Uncle Art, the Justice of the Peace who had fined him the year before for speeding. "The money shortage is broken again, at least temporarily. Uncle Art sent me $5 dollars with a cake and some real farm butter mother sent. I guess he is more or less paying me back the five dollars he fined me once for speeding. He will never know how much his kindness was really appreciated. It just seems that we get things when we need them the most."[15]

There was the full range of missionary experiences as the days wore on. There were endless hours of tracting, or traveling from door to door handing out literature and seeking opportunities to explain their message. They had not been in the area two months before Royce observed that they had tracted nearly the whole town of Fabens at least once, and some of it the second time, usually with unfavorable results. On a day in February, Royce wrote, "We had about a half dozen doors slammed in our faces & were told several times in no uncertain terms 'to go'." The next day he added, "We tracted again today for four hours, a total of 18 this week. It is strange the way they try to dodge us. If they happen to see us coming they shut up tight and just aren't home."[16] One day in Tornillo Royce and his companion administered to the mother of a member who was near death, learning later that she had died peacefully later in the day.[17]

Little by little Royce began to gain some ability with Spanish. The missionaries continued to hold Sunday meetings in the morning in Tornillo on the American side and in the afternoon in La Caseta on the Mexican side, and Royce increasingly participated in these meetings, carefully typing out his talks in Spanish when asked to speak.[18] One Sunday in February when mission leaders came to visit he wrote, "We had two fine Sunday Schools at Tornillo & Calderóns. I took charge of the one at Tornillo, sang two duets at Tornillo & two at Calderóns, gave a little talk at Calderóns, blessed the sacrament at both places & tended the kids at Tornillo, all in Spanish. I really enjoyed it today. There was real Spirit everywhere we went. The people down there in Mexico are just starved for religion. Gee, the good we could do if we could."[19]

The weather often had an impact on their work. On a cold day in January a chill wind made tracting miserable, and the following Sunday a rare snow fell, "Not just a light fall, but a real old fashioned blizzard right here on the border."[20] In March it was not snow but rain that held them back. "The rain which came at last after all the wind cheated us out of our services at Mexico. We could hardly push our bikes against the wind and got soaked as it was. I certainly missed not having the meeting at Calderóns. I had a little speech in Spanish all learned & everything."[21] A week later they made the trip, in spite of the weather. "Boy, what a wind we had to buck today, about the ninety mile an hour variety. It was an improvement over last week, however. Last week we got wet & tired: this week we just got tired. The fine services we are privileged to attend repay us a hundred times for our bit of bodily discomfort."[22] Later that month on two different days he described a genuine Texas dust storm. "Dust, dust and more dust, how the wind blew today and how the dust flew so you could hardly see a thing. Over in La Papelote it drifts and sifts like snow. I guess it could be worse, and to think I used to complain of the wind up home."[23] Eventually the wind died down and the dust was replaced by heat. "Beautiful weather, tho it is awfully hot. It hasn't blown for three days. That's something to talk about. Without sand in my eyes and everywhere else all day I hardly know how to feel."[24]

In Fabens Royce discussed religion with people of various religious backgrounds, a marked change from the fairly homogeneous population where he had grown up. The majority were raised in the Catholic tradition which predominated among the Spanish-speakers, but there were others, ranging from an agnostic Russian with an Eastern Orthodox heritage to members and ministers from a variety of Protestant denominations, who were more inclined to discuss religion with the missionaries. In March of 1940 the two decided to attend a Baptist revival, and Royce, who had never witnessed either the doctrines or the dramatic preaching style of these Protestant meetings, recorded his reaction. "We went to a Baptist Revival meeting tonite. My, it was repulsive. The Rev. Leake of El Paso. His sermon was very boisterous. The house was not a house of order. How could the Spirit of God be there? How thankful I am for the Church, for the priesthood that I hold and for all that it means."[25]

The missionaries attended a whole series of these revivals, and though they felt the meetings had more noise than substance, they were major community events, and they considered them worthwhile, as they did lead to a number of subsequent gospel discussions with people they had seen attending them. Royce saw no redeeming value, however, in a Pentecostal meeting the two were invited to attend after visiting one of its adherents in town. "We went to their meeting at nite and was it a mess. One woman went into a trance and gibbered there for some time like she was possessed with the very devil & I believe she was. They shouted like they were crazy all nite long. They ask Elder May to say a few words then spent the rest of the nite skipping all over the Bible blaspheming the word of God. I was never so glad to get out of a place in my life, although I did not understand everything that went on but I picked up enough."[26]

By contrast, on Sunday, March 3, 1940, a few days after the Pentecostal meeting, the missionaries attended a major missionary conference in El Paso, a quieter event which Royce considered a spiritual highlight. "One of the most impressive days I have ever spent in my life. Conference all morning. We had missionary report & testimony meeting in the afternoon. The Spirit of our Heavenly Father was present to that meeting in a great abundance. It seemed like everyone that got up to talk was just filled with the Holy Ghost. I just could not find words to describe my own feelings and I think everyone else felt the same way. We were there for 3 hours and I just wish there had been three more."[27] The following morning the missionaries set out early as a group for the 150 mile journey to Carlsbad Caverns, where they took a tour of the cave which Royce would recall and recount for the rest of his life. He especially remembered a point in the tour in a large chamber dominated by a geographical feature which had been named the "Rock of Ages", where after entering the lights were turned off, leaving the visitors in total darkness. Awed by the caverns, Royce wrote, "I never dreamed anything could be as marvelous as they are. When they turned the lights out it was so black you could almost feel it. Then the rangers sang 'Rock of Ages' and they started to turn the lights on a half mile away. It just thrilled me all over." That evening the remaining session of the missionary conference, presided over by the Mission President, was held at Carlsbad, and Royce was assigned to speak in Spanish. "My tongue was certainly loosed by a power greater outside of my own. I was able to say everything I wanted to say."[28] Although a spiritual feast for the missionaries, the trip to Carlsbad was grueling, and driving back after the conference they did not arrive in El Paso until 3:30 a.m. the next morning, and had to rest for two days to recuperate.

After the impressive conference the missionaries went back to the slow grind of missionary work in Fabens and the surrounding areas.

The most promising work continued to be across the border in La Caseta, where they were constrained by Mexican law to abstain from open proselytizing. After one such trip Royce lamented, "I only wish with all my heart we could do open work there because I know how much could be accomplished."[29] That day Royce struck up a conversation with an older investigator who requested him to bring him a Book of Mormon the following week, and the next week he described the trip. "We took a Book of Mormon down to the old fellow who bought my songbook, Hno. Bustamante. There will be some baptisms down there before long if nothing interferes. I gave a speech today without notes. It was quite easy for me to say what I wanted to. How glad I am now that I was privileged to come to this mission. I carried the Book of Mormon across under my belt in back of my coat. I guess that makes me a smuggler."[30]

In late May, 1940, the missionaries received word of an upcoming mission-wide conference, during which Royce was to receive a new companion. This was a significant change for Royce, for he and Elder May would part after exactly six months as companions, which was a long companionship by any measure for a Latter-day Saint missionary. The conference was to be a special event involving missionaries from the entire mission, as Royce recalled many years later: "This was a grand trip. The president of the mission decided he wanted to get together one time while he was in the mission, in one place with all of the elders and sisters from the mission."[31] The place chosen was Mesa, Arizona, the site of the only Latter-day Saint temple in the region, and the conference was much anticipated by all the missionaries involved.

On the evening of their departure for the conference, Royce began to feel "an awful pain in my chest" which he referred to as "pleurisy". That night he was unable to sleep. "I never even closed my eyes. Every time I took a breath lying down or a deep one standing up pain about like fire shot thru my ribs."[32] The missionaries left nonetheless the following morning for Mesa, traveling as far as a member's ranch at Virden, New Mexico, where in spite of his illness, Royce, being young and doubtless a little homesick for the farm, went horseback riding for much of the morning before the group continued on their way to Arizona. Arriving in Mesa the next day, Royce attended Sunday services with the rest of the missionaries, though his pain by evening was bad enough that he spent the night up in a chair. Feeling a little better the next morning, he participated in activities with the rest of the missionaries and slept a little that night.

On Tuesday May 28 the missionaries were scheduled to enter the Mesa, Arizona Temple, and Royce seemed to enjoy a reprieve from the pain of his illness for that occasion. He later wrote glowingly of the day in a letter to Pearl. "The whole Mesa visit was climaxed perfectly with the trip thru the Mesa Temple on Tuesday morning and the report and testimony meeting in the afternoon. The temple is beautiful. We watched several couples get married while we were there. I would never be married anywhere else. I certainly look forward to the day when I can pass thru there for that purpose."[33] In his letter home Royce made no mention of sickness.

The following day the conference concluded, and as the missionaries started on the two-day journey back to the mission home in El Paso, Royce was at last struck by the full force of his illness, and endured a miserable car ride with a high fever. By the time they arrived at the mission home in El Paso, he was completely prostrated by a life-threatening sickness, almost certainly lobar pneumonia. He was administered to that day by his new companion, together with one of the office missionaries who allowed Royce to stay several days in his own bed while he recovered from his illness. Royce spent nearly a week in the mission home, much of it in bed, before returning with his new companion to Fabens, and once back in his area it was nearly another full week before he gained enough strength for the missionaries to resume their usual rigorous work.[34]

Royce's new companion was Reid F. Ellsworth, a native of Pocatello, Idaho, who though newer to the mission than Elder May, was also considerably older and more mature. Though they would be companions for scarcely more than a month, Royce considered the time they spent together to be most valuable, and the two became lifelong friends. Among other things, Elder Ellsworth spoke excellent Spanish, And Royce later said of him, "He put me off at least on reasonable pronunciation. He said, 'I'm going to make it so you can talk so at least they can understand you.'"[35] A couple days after their arrival at Fabens, in June of 1940, Elder Bringhurst began to take Elder Ellsworth around town to introduce him to the area, and to continue the work he had begun with Elder May. That Sunday they attended the meetings in Tornillo and across the border in La Caseta, where there were now two serious investigators, Leandro Bustamante and the woman named Sepúlveda. "Things were a little better in Tornillo. In Mexico better than ever. *Hno.* Bustamante bore his testimony and also paid some tithing, it was a thrill to me because I remember the first time Bro. Bustamante attended one of our meetings. *Hna.* Sepúlveda bought her Book of Mormon today too. Both of them desire baptizm. It is really a thrill to see how they grasp the truth when they have the opportunity."[36]

The following week the elders began work in earnest, and accustomed to a more haphazard approach to missionary work, Royce immediately began to gain valuable experience with his new senior companion. "Had my first taste of tracting with Elder Ellsworth today & I liked it. He is a lot more efficient than Elder May. We visited a lot of places and had two invitations in. We should be able to do a lot here. Elder Ellsworth plans his work."[37] In the succeeding weeks, the two missionaries began to systematically revisit the various areas which had been tracted with Elder May. They concentrated some effort in a very Catholic community called La Cuadrilla, finding considerable interest in their message, though their teaching at that time may have been a little too effective—a couple of weeks later, when the interest of the people in the area had cooled considerably, they learned that a Catholic priest had been visiting their homes, warning them not to speak to the missionaries nor accept literature from them. They visited areas that had been previously skipped over or neglected, sought out members on the Church records who had not been heard from, and followed up on previous contacts. In Tornillo, where they held weekly Sunday Schools with a small and struggling congregation, they began to visit members and give blessings, helping to strengthen the group there. They also made weekly visits across the border in La Caseta, where they continued to have serious investigators, though their hopes of holding baptisms there were set back somewhat when they learned that the man who led the congregation was smoking, making them reluctant to proceed when the leader was violating one of the basic Church standards.[38]

Apart from religious concerns, the missionaries had always to reckon with the goodwill of the Mexican authorities, given the unofficial status of their work in that country. On June 30, after one of their many Sunday trips into La Caseta, Royce wrote with an element of concern, "We found out that the Mexican authorities know who we are and what we do. I hope they don't do anything about it and cut us out of our visits there."[39] The following Sunday while attempting to cross the border for their Sunday meetings they were turned away and refused admission into Mexico because of a national election which was held that day; a couple of days later, however, the election being over, they were allowed across to travel to La Caseta. It is possible that the elections in Mexico had resulted in heightened security on both sides of the border, because their problems were not only with Mexican officials. The following week while on a brief visit to Juarez, Royce and another missionary were detained on their return to the border by American authorities, who intended not to let them in unless they could prove their citizenship. They were saved when a relative of the other missionary happened by and vouched for them.

A week later, when the missionaries drove into Mexico by car to seek out some Church members there, they had their most serious encounter yet with the authorities, which Royce described in a letter

64

home. "We cross the border at 'La Caseta' and are supposed to go no farther than thirteen miles in, which is just a little past San Ignacio where we usually go. Well, we went clear down to Porvenir which is thirty miles from where we cross the border for the purpose of hunting up some members. They got us there. The first official wanted to show his authority because we were Americans out of bounds without a permit. He wanted to rush us right up to Juarez and I don't know what else, but finally one came that had more say than the first one. He kind of took a liking to us. He told us he was our friend and he was sorry we had made a mistake and to be more careful after. Then he let us go. We went too before he changed his mind, and really burnt up those 30 miles of rough road back into the good old land of the free."[40]

That summer brought changes to the mission. In June Royce learned that a new mission president had been called to replace President Williams, and at a conference in July, he and Elder Ellsworth were informed that they were to be moved to the city of El Paso, while two "lady missionaries" (or "LMs" in the missionary vernacular of the time) were to take their place in Fabens. The two set about finalizing their work in the various areas and seeking a place for the female missionaries to live. In late July, 1940, three elders came by car to Fabens bearing the two lady missionaries who were to work there. The following day the whole group attended the meetings in Tornillo and La Caseta, which Royce recorded carefully in his journal. "We all spoke in them just what came to us at the time. I also sang a solo in one, my first time at this. Our first was a Sunday School in Tornillo. It was quite well attended and we had a good spirit I believe. In Mexico was the best of all tho. I have certainly learned to love the people there. There is just a spirit there you can't find anywhere else. We will have some baptisms there in time."[41]

The next day the missionaries made a last trip to visit members in Mexico, then they packed and made the trip to El Paso. As he left Fabens, reflecting on the people they had taught particularly in Mexico, Royce wrote, "I shall never forget them, their humble existance and most of all their own humble spirits. They taught me a lesson in humility that I can never forget. In reality, they know they have the opportunity to have everything that really matters. I did learn to love them and think of them as my own flesh and blood, because I know that in the eyes of God there is no race or color, when they are joined in the light of the gospel of Jesus Christ."[42]

No sooner did the missionaries arrive in El Paso than they learned that their transfer there had been merely an interim step to their more permanent assignments. Royce recorded, "My first day in El Paso and allmost my last. I received the shock that must come to all missionaries sooner or later. That I am to be made senior[43] in Tucson Arizona with Elder Ballif Evans. If I ever felt more unworthy or unready for anything I don't know when or where it was. I had just unpacked too." The next day he repacked, and was able to spend some time with his now much-admired mission president, who was soon to complete his own term in the mission. "At nite after we had some ice cream and went up to the mission home we sang ourselves hoarse and then got the president talking on the things we all liked to hear most, the conversion of the Lamanites, etc. He told us how he felt sure mass conversion would come among the Indian tribes and how it would come about."[44] The missionaries did not return to their quarters until after midnight, and the next morning Royce boarded a train to Tucson.

Senior Companion in Tucson

On Thursday, July 25, 1940 after nearly a whole day on the train, Royce arrived at Tucson, Arizona, and was met at the station by his new companion, Elder Evans, who showed him around the area before they retired to bed. The second day, between visits to families in the area Royce retrieved his luggage in the bus station and unpacked, and that night he wrote, "I think I am going to like it here but my is it hot. I just sit and sweat streams. Hot at nite too." Despite the heat, Royce's initial impression of Tucson was a positive one. "Tucson is certainly a nice place. It is almost like a Mormon town. Nice wide streets blocked out fairly well and clean. So much nicer than El Paso. El Paso has little narrow streets running every which way." The missionaries also had the advantage of easy transportation. "We ride the bus line for nothing. Some American members own it. We save about 13¢ every time we get on it."[45]

Royce felt considerable apprehension about his new role as senior companion, which he expressed in a letter a couple of days after his arrival. "I can't think for the life of me why. I am just eight months old in the mission today. There are missionaries older than me that are still 'juniors'. I am still very limited in my Spanish. In short I don't think I am ready to take on the responsibility but if the president has enough faith in me I think I can do it. There are about thirty saints here and lots of investigators. My work has been for the larger part tracting so far. I havn't had any experience in this type of work."[46] That Sunday they attended church in a small Spanish-speaking branch, and Royce, asked to give a talk, again acknowledged his limitations. "Gave it on prayer. It wasn't very good. Other than my talk it was a fine meeting. *Hno.* Dominguez, the branch president, gave a fine talk. I did find out what is the most crying need here right now. It is to get some of the lazier members out & get some of the investi-

66

gators stirred up. I only hope in a humble way I may be able to accomplish it. I need the help of the Lord more & more. My spanish is so limited, I just have to have it."[47]

Royce's new companion, Elder Ballif H. Evans, from Oakland, California, had previously worked in El Paso, and Royce's prior acquaintance with him had not been very positive. His initial assessment was hopeful. "He'll be a fine companion. He has certainly changed since he came in the mission." A couple of days later, however, he expressed more concern as he came to terms with his responsibility as senior companion. "At night to please Elder Evans we went over to visit with some American member girls. Elder Evans has changed a lot but not enough. I can't feel that we should have been there. I don't think the president would have approved of it. I'll have to see just what I can do about things here." Royce's approach to what might have been a problem companionship was the same he that would often take in later associations—he simply went to work and expected the best. Though their companionship was brief, no further problems were mentioned, and Royce's final comment on Elder Evans would be both positive and kind. "I enjoyed my work with Elder Evans. He's a good Hermano, just had a harder time getting settled than most of them."[48]

The two missionaries quickly set to work, visiting and encouraging members of the branch and looking up investigators whom the missionaries had already been teaching, and after the first few days Royce began to feel he was getting his feet on the ground. The visits continued throughout the week, and that Saturday the missionaries were asked to go to a nearby community called Binghamton to administer to the daughter of a member, although this time the roles were reversed from Royce's previous experience in Fabens—his companion anointed the little girl, and Royce performed the blessing. As an untried senior companion Royce adopted the oft-repeated motto, "A busy elder is a happy elder". In spite of their efforts, however, the Sunday services the next day were still a disappointment, and Royce recorded, "Something has got to be done. Bro. Dominguez is getting very discouraged. The members do not come out to church. Tonight we had five out of 30."[49]

That Sunday proved to be the last with Elder Evans, as a change of assignment came, pairing Elder Bringhurst with Elder Quintin V. Christensen from Shelley, Idaho. Elder Christensen at the time was much spoken of in the mission for having been healed by a priesthood blessing when physicians had despaired of his life.[50] The new companions had scarcely met that Monday when they went to a teaching appointment, and the remainder of the week the two made many visits to both members and investigators, and tried to establish regular cottage meetings with them, though this often stretched Royce's still-limited abilities in the language. "We arranged for a meeting for every Wednesday nite at the home of an *Hno.* Contreras. He certainly tried to tie me up on a few things today, and he did in my Spanish. However, he did say that he wanted us to start from the very beginning of our beliefs clear thru. He says the only church that interests him is the 'Mormon' church."[51] The missionaries also worked with several of the non-attending members, including a family interested in Genealogy[52], a woman whose husband had lost faith in the Church, and another whose husband was not a member. On Thursday they held a weekly Primary[53] meeting at the home of one family, and they arranged to begin holding similar Primary meetings at the home of a Duarte family, which began successfully on Tuesday of the following week. The first Sunday Elder Bringhurst and Christensen were together, their efforts were finally rewarded, as the members they had visited began to attend.

With full responsibility for the teaching, Royce's Spanish improved rapidly, and that week he wrote a letter to Pearl describing his progress and feelings as a fledgling senior companion. "We were very much encouraged last Sunday because most of our members were out to services. I had to speak again for my third time in a row here on Sunday night. We had a good week last week. We arranged for three more meetings in addition to the two we were already holding during the week. This really puts me to work because I am the one that has to do the talking at these meetings. Last Monday night I held the first cottage meeting that I have ever held without someone besides me to do the principal part of the talking. I shall always cherish the memory of the moments when I have felt my pulse quicken and a burning within me and my tongue has been loosed for I know that this help has come to me thru the spirit of the Holy Ghost. Sometimes, I am so discouraged about the way I speak Spanish and yet when the time does come when it is really necessary that I speak I am able to do it and I know that I could not do it without the help of the Lord."[54]

The following Tuesday, after their successful Primary meeting at the Duarte's, Royce and his companion decided to attend a movie in Spanish in a local theater, perhaps in an effort to help Elder Christensen with the language. This proved to be a memorable occasion. "At night we went to a Spanish picture show but never saw it all because the rains came for about a half hour. 2-1/2 inches and that's a lot of rain in Arizona. This stopped up the power house. The city was in darkness." The next day the consequences of the rainstorm began to become clear. "What a storm last night. It about washed the city away and the viaducts were full like lakes. We also saw a real river with running water in Arizona. We have light but most of the city does not and won't have till tomorrow nite. It soaked up everything at the power plant. They said it was by far the worst flood in years." The water continued to take its toll, but the missionary work went on. "We went down to Binghamptom first. It flooded very bad there too. They are without water and lights both, because they have electric pumps. We visited them all and encouraged them to come to our conference. We had a very fine Primary at 4:00."[55]

That weekend a District Conference of the mission was scheduled for Tucson, to which the missionaries took pains to invite the local members. On Saturday the missionaries began to gather, and it was that night that Royce first met his new Mission President, David F. Haymore, who was present with his two daughters. The missionary meeting of the conference proved to be an eventful one for Royce. "At 4: P.M. we had a missionary meeting for about two hours. President Haymore gave us a fine talk and some fine advice. I received a transfer to Nogales to take Elder King's place as senior to Elder Bergeson. This puts even a greater responsibility on me. I only hope I am worthy and able to do this work."[56] The day after the conference, Royce packed and wrote up a list of information on the members and investigators for Elder King, with whom he was exchanging areas under the direction of the new Mission President. The next day he held a weekly Book of Mormon study meeting with one of the investigators, then at 10:00 a.m. he boarded the train for Nogales and his new assignment, one which was to last many months and become a memorable episode of his life.

Assignment to Nogales, Arizona

Royce arrived in Nogales on Tuesday, August 20, 1940, and was met by his companion, Elder A. Berthel Bergeson. The two immediately crossed the border into Mexico to make a visit, and later that evening, at the advice of the new Mission President, they cancelled a planned meeting there that was probably in violation of Mexican law, which required all religious meetings to be led my Mexican citizens. After his hot month in Tucson Royce wrote, "It is good to be back crossing the line again. I know I shall like it here. It certainly is cool & nice. The city is built on a lot of hills."[57] The next day they crossed into Mexico again for haircuts and to take shoes for repair—evidently these domestic necessities, as well as laundry services, were less expensive across the border. Their quarters were conveniently located just four blocks by foot from the border with Mexico.

Royce quickly became acquainted with the area, and as senior companion made adjustments to the work there. He found a Primary class illegally in operation in Mexico on Fridays which had to be discontinued, and that Saturday sized up another Primary being held by

69

the missionaries on the U.S. side, which he felt was doing little good. His first Sunday he attended services both in the U.S. and in Mexico, and as at Fabens he was required to participate. "They got me right off. I had to lead the music and give a 2-1/2 minute talk. They have a fine service here. We have a mexican class that I have to teach also. In the afternoon we went over to Mexico and held a service with the Marquez family. Bro. Mahlow who is a citizen of Mexico presided. This family lives up the prettiest little canyon off from Mexican Nogales you ever saw. At nite we went to church. I had to do the speaking and in English. Frankly, I didn't think much of the job I done either. Well it's all in the life of the missionary. However, I'll take the Mexican meetings any day."[58]

As in Tucson, Royce threw himself into the work, and gradually began to organize a regular teaching schedule on set days of the week. He also began to consider holding a street meeting, a very time-honored and public form of missionary work which Pearl had first suggested to him in a letter, and which he had dismissed as impractical in Tucson.[59] However, by the time he arrived in Nogales the idea had grown upon him to the point that he visited the town mayor with his companion to request permission to hold an open-air meeting, which was granted for the following Saturday night. Royce began carefully writing out a discourse in Spanish for this meeting, and since both he and Elder Bergeson were singers they also decided to prepare a vocal duet, and to this end they met twice that week with a member who played the piano. When the day of the street meeting finally arrived, they spent most of the day doing their domestic chores, as was usual for Saturdays, but they also held their weekly Primary meeting, tracted, and made some teaching visits before arriving that evening at the place appointed for their much-anticipated street meeting. This was a significant and daunting prospect for the elders, as an open-air meeting harked back to the beginnings of the Church, when courageous early missionaries used such meetings to gain some of the first converts to the new religion.

The missionaries began the meeting by singing the song they had prepared to attract the attention of onlookers, after which Royce delivered the speech which he had written out in his missionary notebook. This was a well-reasoned discourse addressed to fellow Christians, explaining the need, in view of the many conflicting Christian denominations, for a restoration of the gospel, and declaring that such a restoration had in fact occurred. In his journal Royce wrote of this event, "Saturday nite the 31st of August will be a day I will always remember. We held our first street meeting. It was certainly a knee knocking experience to start with but it was wonderful. We enjoyed the help of the Lord to a very great extent. We were able to give them something and they seemed quite interested. We even sang the Holy City right there and done it all right—the rest was in Spanish."[60] By this time Royce's Spanish had improved substantially, and his discourse and subsequent conversations were evidently understood and well-received; he regarded this as one of the major events of his mission.

Meanwhile, the missionaries continued to contact and teach people of various religious persuasions, and Royce recorded his reaction to these discussions. For example, one Wednesday in September he wrote, "It was quite a day. We were first invited into the home of some Seventh-Day Adventists. They really wanted to argue about that old 7th day but nothing accomplished. Later we went to a Catholic lady's home and had a good discusion with her. She finally asked us to wait for her son to come at noon. He is a Padre named Alatura. I guess she really thought that he would confound us but he didn't. He never ask one question we couldn't answer and we ask him several he couldn't, namely the one about baptism for the dead. He was very friendly and obliging & we had the opportunity to eat with them. It was very interesting the way he blessed the food & they greeted each other. He was really sincere but it ended up by me preaching more Mormonism to him than he preached Catholicism to us. I am enjoying this work more every day I believe. It seems like the more responcibility I have and the more work I have to do, the happier I am in it."[61]

With Royce as senior companion, the missionaries began to organize their work into something of a weekly schedule, tracting in the mornings, and holding weekly "study meetings" (the equivalent of the modern "missionary discussions") in various homes in the afternoons and evenings. They did not confine themselves to investigators, but taught in the homes of branch members as well, and Royce began to gain a conviction that strengthening Church members was an important aspect of missionary work. One Friday in September he wrote, "We are really proving the statement that the busy Elder is the happy elder because we are certainly busy and extremely happy in our work from 5: A.M. till whatever hour we manage to get to bed. The members are really helping us out now. I only hope we can keep them animated. They can do twice as much missionary work among their friends than we can. I think these little weekly meetings we have started with each of them really help. Too many times we take the members for granted and let them drift along rather than keeping them growing, then they become disinterested and start sliding back."[62]

Not long after Royce's arrival in Nogales, he and Elder Bergeson were approached about taking over leadership of a Boy Scout troop in the area, and they eventually consented to run Troop 26, enrolling about twenty scouts. Royce described the venture to Pearl. "It is a group of Catholic Mexican boys under the direction of the Elks Club. I have been made Scoutmaster and *Hno.* Bergeson Assistant. I only hope we can put in the time needed to handle it."[63] Their involvement with the Boy Scout troop started well, with weekly meetings on Wednesday nights, but after a few weeks interest of the boys began to wane and attendance to drop, and Royce concluded that they needed to plan a hike to "stir them up." This had its desired effect, and both interest and attendance increased as the date of the hike approached.

The departure date was set for Friday, October 11, and arrangements were made to camp at a ranch some distance away. On the morning of the chosen day, Royce purchased a pair of overalls, picked up a pair of shoes, and made final plans. He wrote, "We & the Scouts pulled out about quarter to six and went out to the Burt Yoes ranch which is about fourty miles from here. We had supper & then played games & sang etc. untill it was time to go to bed. That nite, some of them tried to get up at 2: A.M. It was a nice how do you do – we finally did get to bed though." The following day he continued, "We rose early and kept on the go all day long. We went through a cave on the ranch there, a pretty big one too. We also played a game of football and done in general the things that the boys like to do. My am I worn out. Some of those kids are just about like handling unruly sheep although all in all they are a pretty fine bunch and I certainly did enjoy the trip with them." The next day was a Sunday, and Royce commented on the after effects. "Got up late this morning (7:30) and I could still have stayed in bed longer. I am stiff in every joint & muscle from that football game yesterday."[64]

At about the same time, Royce received a letter from his mother with news which shook the world of his boyhood, and he passed the news on to Pearl: "My folks are building a house over by my grandparents place on that ground they own in Cottonwood. They will be moved before I get home. I didn't know what to think when they told me. Smith is going to take over the farm. I'll be a lost sheep when I get home if I can't go back to the farm."[65] Along with this change of residence, there was news almost continually of friends and family members being married, including Royce's best friend, Steve Mackay.

Royce and Pearl in their letters discussed their own plans for marriage, which Royce increasingly felt should not wait until he was done with his college studies, but should occur soon after his mission ended. Their plans continued to evolve when, in the fall of 1940, a law was enacted requiring all young men ages 21 to 35 to register for the draft. The week after the camping trip, on October 16, 1940, Royce recorded in his journal , "Quite a memorable day, the first peace time draft in the history of the country. We registered first thing in the morning, then found it was useless to go tracting because they were all down town to watch the parade and hear the speeches."[66] Royce initially hoped his standing as a college student would exempt him from the year-long mandatory training which would result, but he quickly learned that was unlikely. "The registrar said I would be in the army in six months if I wasn't a missionary. That defers me until June of 1942. Then after I go into he army for a year, I'll be about 25 years old."[67] This hopeful calculation was of course based on the expectation of a continued peacetime draft and military service, and Royce's intention was obviously to get back to his studies and on with life. The reality of war would interrupt all plans all too soon, and destroy any immediate hope of getting back to "life as usual."

Before long Elder Bergeson received a transfer to become senior companion in Tucson, and Royce was given another junior companion from Salt Lake City named Milton Vaughn Bitner. By this time Royce had begun to feel his inadequacies as a Boy Scout leader, in spite of the successful camping trip, but the timely arrival of Elder Bitner helped to revive the venture. After their first meeting with the Scouts together Royce wrote of his new companion, "He will be fine here especially with our Scout troop. He is an Eagle Scout and has had experience in scouting that I have not had. We had a pretty good meeting with them at nite."[68] The missionaries' work with the Boy Scout troop continued throughout Royce's stay in Nogales. Although the boys could be a challenge (at one point he referred to them as "destroying angels"), overall his assessment of the venture was positive. "The Scouts aren't so bad, I guess. Some of them certainly get a person down, but all in all it's fun and doesn't cost anything & might do some good."[69]

On a weekend in November, 1940, the missionary routine was interrupted by another conference in Tucson, which proved a memorable event for Royce because it was presided over by George Albert Smith, then a member of the Quorum of the Twelve Apostles.[70] The conference was held on Sunday, and both the Mission President and Elder Smith addressed the missionaries, followed by a three-hour "report and testimony" meeting at which the missionaries spoke in turn. During this conference Royce had the opportunity to spend some time with the apostle, as he later recounted, "He was a thin fellow, a wonderful man to be around, a wonderful man. I traveled with him, and I translated for him."[71] On Monday November 11, the missionaries had a final meeting with the Apostle. "He certainly told us a lot of wonderful things and gave to us some fine advice. Anyone listening to him just couldn't help knowing that he speaks the truth. He just has the certain convincing way about him that makes you know that he is right and knows he's right."[72] That afternoon Royce accompanied the mission president and Elder Smith to the rail station, and had one last talk with them while the train was being prepared for departure.

Thanksgiving day marked the first anniversary of Royce's missionary service, and though the missionaries were amply fed by the members, when he and his companion made one of their teaching forays into Mexico Royce's enjoyment of the holiday was dampened by the living conditions he saw there. "They just haven't anything. Not enough to eat or keep them warm and a terrible house. It makes me feel terribly sorry every time I visit them. The members are certainly fine to us down here. I often wonder if we deserve all that they do for us."[73] The contrast between living conditions in the two countries continued to trouble Royce, and for Christmas he resolved to prepare a shipment of blankets and food to carry to some of the poverty-stricken Mexican members.

Shortly after Thanksgiving Elder Bitner too was transferred, and Royce spent several days alone, attending to appointments and Church matters while awaiting a "green" missionary, fresh from the Mission Home. It was December before he met his new companion, Elder Dean Rowley of Helper, Utah. Elder Rowley had entered the mission field with no more knowledge of Spanish than had Royce at the beginning of his mission, but he had a good attitude and a willingness to work, and the two got along well. After an exchange of letters with their families at home, they learned that their fathers had actually served with each other some thirty years before in the New Zealand Mission.[74] Royce began instructing Elder Rowley in the language, and before they had been companions for a month he had already given his first talk in Spanish. The two got along famously, and quickly became close friends.

In December, Royce and Elder Rowley were approached by missionaries from the English-speaking mission in the area who were soon to be stationed in the city. This removed from Royce's shoulders the responsibility of the English-speaking members, freeing him and his companion to increase their efforts among the Spanish-speakers. For Christmas they held a "dance" for the Boy Scout troop, and Christmas day found them at another missionary conference in Tucson, where Royce met up with Elder Ellsworth, his last senior companion from Fabens, and though the two had been companions for scarcely a month, they stayed up until 4:00 a.m. talking over "old times" in Fabens and exchanging new experiences.

New Year's Eve was spent in the home of the Nogales Branch members eating the traditional holiday tamales, and as the first day of 1941 dawned, Royce became reflective, considering that this was the

74

year he was to have graduated from college. Despite the war that continued to dominate news from abroad, Royce felt optimistic about his personal prospects. "I look now at the year coming up, a year that I know will be filled with more than the last year. I know I will be a better instrument, not because the task will be easier but because my power to do has increased and will continue to increase if I am prayerful and humble. Yes, I look even forward beyond all that to June of 1942 when I will be released and go home to try to take up where I left off, in that life that seems almost like a dream to me now. It seems like I have been a missionary always. It just seems that there couldn't be any other life and yet all the time I know that there is, and only too soon I'll find myself right back into it. That too will be another great experience—more worlds to conquer, more things to do, and all the time I have the comforting assurance that my power to do will increase if I work, study, and pray."[75] Royce's optimism was genuine, though he could not have known at the time how the coming years would unfold, and how many years of "education" still awaited him before he would again set foot in a college classroom.

It was in January of 1941, with four missionaries sharing the work load in Nogales, that Royce faced what was perhaps his most difficult personal challenge yet. Crossing into Mexico shortly after New Year's day, Royce and his companion passed by a bar and were surprised to see the branch president from Nogales, apparently drinking and intoxicated. The four missionaries met and conferred, and tried to broach the matter with the branch president without success, but both Royce and the English-speaking elders thought it was too important a matter to let pass. Finally one day later in the month, the English-speaking missionaries appeared at their apartment, requesting their help in bringing the branch president back from a drinking spree across the border. This they accomplished only with great difficulty, after they had searched several bars and at last confronted the recalcitrant leader. The following day Royce wrote, "We had to do something this morning that I sincerely hope I never have to do again. We went over to see what we could do to straighten out the mess and to help Bro. _____. He broke down and told us everything, that is he and his wife—they were both crying. I never saw anything so utterly pitiful. He smokes all the time, drinks coffee and goes to get drunk on the other side very often. We administered to him to try to give him the strength to take hold of himself. I don't think I shall ever forget this day. I hope & pray it done some good."[76] The branch president unfortunately continued to struggle with his drinking problem, and the missionaries, having reported it to their leaders, in the end simply left the matter in the hands of the district president. Many years later Royce said of him, "Aside from that, this man was wonderful. A

75

wonderful orator. He could talk, he knew the gospel, and no one could question. That's one reason I've always been kind of merciful toward people that have problems within the Church, because I didn't think it was my prerogative to try to change him overnight."[77]

In late February, Royce received a letter from the Mission President advising him of a transfer to San Diego, California, and he and his companion made the rounds to bid farewell to the now beloved members and investigators in the Nogales area. On the last day of February they boarded the train for Tucson to attend another missionary conference and to effect the transfer. The conference proved to be a historic occasion, as the Southern Arizona District of the Church was reorganized for the first time into a "Stake of Zion" by John A. Widtsoe, an Apostle. The next day Royce caught a train to Phoenix, and there he took the opportunity to visit President Williams, his first Mission President, who resided there after his release. He remained there two days, and when his planned train trip to California was delayed, he took a side trip to Mesa to visit the temple grounds where he had been so sick the previous spring. The train finally departed after yet another delay, and Royce at length arrived in Pomona, where he tarried an additional day with a missionary friend and his family before at last being driven down to Los Angeles, where a full week after leaving Nogales, he at last caught the 5:30 train to San Diego, having a gospel conversation with another passenger en route.

Introduction to California: San Diego

Royce arrived in San Diego at 9:30 p.m. on Friday the 7th of March, 1941, at a time when that city was playing a major role in the nation's preparations for war. Royce wrote, "I had to find my way to the address and it wasn't too easy. I never knew that so many sailors & marines existed." The next day he added, "I saw the ocean for the first time this morning. You can see it from the kitchen window. When we went down to get my bags we passed by the docks and I saw several destroyers and cruisers, and two submarines."[78] Royce's assigned companion in San Diego was Elder Harold J. Nielson, and the missionary residence was located at 1893 Irving Avenue, on the top floor of the building used as chapel for the thriving Spanish Branch there. Since they resided at the chapel, the missionaries also served as church custodians, and on his first full day in San Diego Royce spent the day helping to clean up the church yard, a task which left blisters on his hands. The next day, Sunday, he attended church services there for the first time, and was asked to speak.

The branch in San Diego was far more developed than those in which Royce had worked in Texas and Arizona, and most of the or-

76

ganizations and activities usual to the central stakes of the Church were in operation there, at least on a modest scale. On Mondays the missionaries held an afternoon Primary meeting for children at a private home, and on Royce's first Tuesday he attended the other church auxiliary meetings, including Relief Society[79] and Mutual. That first week the missionaries were called upon repeatedly to give anointings and blessings to the sick, twice to a woman hospitalized with pneumonia, and once to a sister who had suffered a miscarriage (both recovered). Royce quickly became aware of another characteristic of the San Diego branch. "There are certainly a lot of returned missionaries down here with Mexican wives," he observed, and a couple of days later added, "There have been several Mexican girls here get returned Elders for husbands, and they seem to think the Elders are pretty fair game. It just means we have to be that much more careful."[80]

As in his previous areas, Royce and his companion set to work with dedication. He noted that it was easier to contact people in their homes than it had been in Nogales, and the missionaries had much to do both in San Diego and in outlying communities such as Julian and National City, where they began holding regular meetings with struggling members and investigators on Thursdays. Royce's abilities in Spanish continued to progress, and his companion informed him that he had even been heard speaking the language in his sleep. Whenever he expressed himself well, he generally attributed his abilities to divine intervention. "Spanish is easier to speak every day I live. It seems like it just flows out of me today. I know it is thru the Spirit of the Lord. Without it I am nothing."[81]

Royce and his companion continued to have frequent contact with people of a variety of religious backgrounds. Having been raised in an area dominated by a single religion, Royce's missionary experiences gave him valuable insights on other faiths, and helped to mature his attitude toward them. Told by a devout Catholic woman to go see a priest, Royce in response arranged a visit, during which the priests impressed him more with their education than with their doctrine. He continued to hold little regard for the loud, chaotic worship style of the Pentecostals, but was struck by the more reserved attitude of a Japanese woman who was Buddhist. "It is the first one I have talked to that was not a Christian, although in practice I rather believe she is a lot more Christian than a lot of people I know."[82] By far the majority of those he taught were Catholics, that being the traditional religion of the Spanish-speaking community, and Royce saw the most promise in the younger generation, who were less tied down to long-established traditions. "It appears in a way that many of the Catholic youth are revolting from the old narrowness of Catholicism and really seeking truth rather than clinging to superstitious traditions. It has a double effect. Some of them swing right arround where they don't believe anything and fear neither God nor the Devil. But most of them that revolt are after truth, and I know we have what they want."[83]

Royce was confident of the value of the message he was teaching as a Latter-day Saint missionary. Speaking of one visit he made while tracting, he wrote, "We were invited in and although they don't know much about us, they have inquiring minds and are searching for truth. That is all we ask. Truth is always discernible to the honest in heart and all Mormonism is truth."[84] By this time Royce had also developed an ability to win over strangers who were disinterested or skeptical, and to deliver his message in such a way that they took genuine interest and parted on friendly terms. "We make it a point not to attack their church but we do preach them the first principles and the Book of Mormon, working out from the things we have in common with them. If the mission didn't do that we would hardly have a Mexican member."[85] Like many missionaries, he often felt it was only his love of the message he bore that gave him the ability to face opposition. "I do appreciate the Gospel, if I didn't I could never face strange doors every day, and in the face of every attitude there is, including scorn (which is about the worst), testify that I know we carry the message of the everlasting gospel. All people don't think, talk and act as we do. We have something they don't have, something that makes the greenest, most unpolished young missionary a worthy instrument in the hands of the Lord."[86]

Along with their usual task as Church emissaries, the missionaries often had to contend with problems among the members, including a rift between some of the branch members and the family of one of the leaders—a matter which contrasted sharply with the usual inspirational experience of missionary work. The conflicts between members often were burdensome for the missionaries, and they were a frustration for Royce, who was conciliatory by nature. After one meeting he wrote, "We held Mutual at nite, and I told them off as diplomatically as I could about gossiping and having petty fights among themselves over nothing. As far as I know, I didn't insult anyone."[87] The missionaries generally found themselves in the center of any crisis involving Church members. When the teen-aged daughters of one member family ran away from home, the missionaries made trips to the police station and later to Tijuana on behalf of the family, inquiring after the girls. A day later Royce recorded their successful recovery, together with his opinion on their adventure. "The girls came home with their folks. The police found them in L.A. I guess they got a scare thrown into them. They need spanking."[88]

78

The missionaries continued to be responsible for maintenance of the chapel, which was also their place of residence. They purchased shovels to help control the weeds that emerged in the spring, and at the beginning of May they took a full week out of their usual missionary schedule to concentrate on the chapel grounds, a task which had to be done without modern power equipment. They dug out the weeds and spaded the yard, then installed a fence formed of metal pipes set in cement, and landscaped the yard with plants donated by a member. Royce's farming experience and agronomy training doubtless helped in the effort, and after one of their days of heavy labor he noted with satisfaction, "It certainly looks 100% better. In fact it's rapidly becoming the best looking lot on the block instead of the worst as it has been."[89]

On May 28, 1941, shortly after completing his eighteenth month in the mission, Royce and his companion while tracting made one of the most important contacts of their mission—a couple named Guzmán. Royce described the husband as "a Baptist fellow who really seemed to be interested, especially in the Book of Mormon. He had never heard of it before. He expressed the desire to attend our services. I was quite favorably impressed by him. He seemed to be really concerned."[90] That Sunday the missionaries were asked to take charge of the services, and the English-speaking missionaries were called upon to assist them. The services were especially good, and Royce recorded them in detail, and at the conclusion wrote, "I bore my testimony and I could really feel the Spirit of the Lord in me as I spoke. The Guzmáns that we met last week tracting came. They seemed impressed by our services. They told us to be sure and visit them, and ask when we held Sunday School also. They are the best we have ran into down here tracting. I only hope and pray that we can guide them aright. Today is one day that I shall remember."[91]

In the days that followed, it became still clearer to Royce that the Guzmáns were an exceptional couple. After one visit he wrote, "About 6: A.M. we went over to Guzmáns where we held a little meeting with them. They are really prepared to receive the gospel, more so than any family I have met in the mission since the Sepúlveda family down Fabens way. They keep the Word of Wisdom to the letter. They don't even use coffee. The *Hna.* told us she had heard so many bad things about us she thought we must have something good. They seem to like everything we teach them. I with all my heart hope and pray that I may see them baptized in the Church. It's the dream of every missionary to see someone baptized or baptize someone they have been the first to contact."[92] That dream would eventually be realized with the Guzmáns.

At the same time, with the encouragement of the mission presi-

dent, Royce and his companion began to take new initiatives in their
area. They began holding weekly Priesthood meetings, which continued the remainder of the time Royce was in San Diego. When some
of the younger men expressed a desire to participate in tracting, the
missionaries permitted them to come along, in part to help them prepare for missions of their own. They also determined to begin holding Primary meetings for investigators in National City. Royce described the first meeting to Pearl: "We asked if any of them could
give us a little prayer, forgetting for a moment they were all Catholic.
Up went the hands, so I called on one. She repeated 'Hail Mary', and
another had to have the chance too. He repeated one of their confes-
sional prayers, pledging allegiance to the Catholic Church over all
else. That kind of took the wind out of our sails for a minute, so to
speak, but we recovered quickly and taught them a simple little prayer
to open with, and all was well."[93] The Primary meetings continued
every Thursday in National City, and the following week Royce wrote
the Mission President requesting that "lady missionaries" be sent into
the area to assist with the Primary groups which were underway.

In July, Royce received word that his parents had completed their
new home in Cottonwood, and had moved out of the old farm house
in Murray where he had grown up. This event meant the end of his
boyhood home as he remembered it, though at the same time it
opened up the possibility of his taking over the family farm, which as
yet remained in the family's possession. The resulting flood of emotion was poured out in a letter to Pearl. "You know how I feel about
the old farm. I don't ever want to see it leave the family. Of course,
there are the 'lean years' but there always is happiness that comes
from turning rich brown soil in furrows and planting things in it, then
watching them grow. It puts you near to God, you have to trust in
him and he is near things that grow. It makes your back and your
mind strong. That is my ambition, to make my living that way and to
be true to our God and his Church. I do have a most fervent desire to
live the gospel all my life to the best of my ability, and to bring my
children up in the atmosphere of a real Latter-day Saint home, and
give them the opportunities I have had."[94]

As summer wore on, Royce and his companion continued to have
the sundry experiences typical in a Latter-day Saint mission. Taking a
lead from local members, they began to boil their drinking water after
bouts of sickness. Royce had a formal missionary portrait taken by a
photographer. He captured a flea and memorialized it under cello-
phane tape in his journal. He assisted the branch president in translating a Relief Society lesson into Spanish for the mission. On July 4th
he and his companion attended an Independence Day celebration
involving both the San Diego and Los Angeles branches at Mission

Beach, where he took many pictures.[95] The missionaries also continued to help members out with personal problems. They visited and counseled a member in National City who had taken to drinking. They took a young woman to the hospital for a mastoid operation, and visited her often while she recovered; later when she began to waver in her commitment to the Church and its teachings, Royce had long talks with her to encourage her to remain faithful. The missionaries also took their promising new investigator, Brother Guzmán, to the hospital for what was to be a routine hernia operation, but his course became gravely complicated, first by a heart problem and later by pneumonia which very nearly took his life before he recovered. They gave him a blessing which he later credited with saving his life.

One night in August, a missionary was transferred to San Diego to become part of a missionary threesome with Royce, and he brought word that Royce would shortly be transferred to Los Angeles to serve as district president. For the remainder of his time in San Diego Royce worked with these two companions, and this period proved to be something of a letdown for him. The branch continued to be torn by infighting, he had to labor with several members who had become weak in the faith, and his most promising investigator clung to life in the hospital. Working as a threesome of missionaries also proved a challenge, and Royce commented, "I can see the wisdom of the Lord in sending the missionaries out two and two now. Though we can get quite a bit done with three of us, it works lots better when there are two only."[96] One afternoon found Royce and his companions visiting a member who had left the Church to attend another, and who Royce felt had never been converted in the first place. He wrote, "While I talked to her for about an hour I had an elder asleep at either elbow. We were all sitting on the same sofa."[97]

The following Sunday there was a heated discussion in Priesthood meeting which put a spiritual damper on the remainder of the Sabbath activities, and the next day Royce received a letter from the Mission President, officially notifying him of the transfer to Los Angeles to become District President of the California District as of the following week, in place of an elder who was completing his mission. Royce wrote, "I hardly know what to think, all I can do is my best. It seems like I have kind of let down a bit lately on my missionary work."[98] The next day was spent packing his things and traveling about to bid farewell to the San Diego members and investigators. He paid one last visit to the Guzmán family, brother Guzmán having finally come home from the hospital. "He is looking lots better. We sang a song to him and talked a while. He said he would never be able to pay us for what we have done for him, but he said that the Lord would pay us some time."[99] The following day Royce finished pack-

81

ing, and was driven by a member to the train station where he caught the next train for Los Angeles.

District President in Los Angeles

Still feeling a little off-balance as a missionary, Royce arrived at his new area of assignment on Thursday, August 27, 1941, and on departing the train took the wrong streetcar, riding 15 miles to the end of the line before realizing his mistake. It was suppertime before he finally arrived at the district president residence, which was located at 3401-1/2 East Floral Drive in Los Angeles. The day after his arrival, Royce made preliminary visits in the area with Elder Karl Buchmiller, the outgoing District President whom Royce greatly admired. The change in leadership was to take place that weekend at a District Conference of the "California District" centered in Los Angeles, and for the first time Royce sat on the stand during the conference meetings. On the last Sunday session of the conference Elder Buchmiller was at last officially released as District President and Elder Bringhurst was appointed to take his place, giving him charge of the Spanish-speaking Church throughout Southern California. A day after the conference ended, a surprise party was held for Elder Buchmiller, and the next day Royce put him on the train, his mission having concluded.[100] Taking stock of his new area of labor, and knowing it to be the last place he would serve, Royce felt determined and optimistic. "I do hope and pray that the Lord will give me the strength and courage to really do a good work here before I leave. Conditions are really ideal here. There is a good branch, good Elders and unlimited tracting fields. It just couldn't be much better."[101]

With Elder Buchmiller's departure, Royce was temporarily left with two companions, but shortly one left for other assignments, and Elder Albert Packer remained as Royce's junior companion. Together they started in on the usual door-to-door contacting, and Royce continued to demonstrate the ability, developed in his previous areas, to speak to strangers who were initially hostile and to establish a good enough rapport with them to deliver a meaningful message and part friends. In addition to tracting, the missionaries also visited the various families who had been studying with the missionaries in that area. This included Royce's landlords, an older couple named Muñoz, with whom Royce was very impressed. Another was a Sister Villalobos, with whose family the missionaries were holding regular meetings. Not long after Royce's arrival Sister Villalobos' husband was killed in an accident, and the missionaries attended to her needs, helping her to arrange for burial, clothing the body for her, and singing at the funeral.[102]

As in other areas the missionaries were required to assist with the

Sunday services. Royce found himself teaching a class of youngsters who had a reputation for being unruly, and who would challenge his patience during the coming weeks. He also found that the members were rehearsing to perform a musical production he called an "opera," in which the missionaries were expected to participate. Royce felt their time would be better occupied elsewhere, but since the project was already underway, he felt obliged to support it. "I don't know the songs, I don't know anything about it, and really I'm scared to touch one of those girls."[103] At the same time, Royce's love of music began to shine through in his leadership of the missionaries, some of whom were capable singers. They began to organize missionary ensembles, including a men's quartet with which Royce practiced for the remainder of his mission. Royce also had developed a love for Mexican popular music, which he had begun collecting and sending to Pearl.

As District President, in addition to work in his own area, Royce's duties now involved working with and encouraging the various missionary companionships under his supervision, which included fourteen elders and four lady missionaries. On one occasion while making visits in the area of Elder Johnson and Elder Rowley (his former companion from Nogales), he encountered a minister whom he would come to regard as a friend. "We went to the house of a Mexican Baptist minister and he had us there for three hours. We didn't argue but we sure did do a lot of comparing. I was able to tell the story of the restoration and about the Church organization; also we loaned him a Book of Mormon. He was really a nice fellow, though he is a bit 'errante' in his interpretation of quite a bit of scripture."[104]

Royce also was responsible for missionary transfers and assignments, as well as reports of missionary activities to the mission. In one such transfer, an Elder Reynold Watkins was assigned as senior companion to his younger brother, George Watkins, and the two would work together as a companionship for months, the missionaries referring to the younger affectionately as "*Hermanito* Watkins." Royce also conducted periodic missionary meetings to discuss the work and build unity among the elders and lady missionaries. His first such meeting was held on a Sunday in September shortly after the missionary transfers, and offered him a point of reflection. "It was the first one for a lot of them. I well recall my first one. When I look back, I have come a long way from there. Of course I would change a few things if I could go back again. The Lord has been good to me. I have really enjoyed that time."[105]

Early in October Royce made his first visit as District President to his old area of San Diego. Arriving when the missionaries were out, he and his companion ate with the Branch President, who hosted them during their visit, and made some visits to members well known to Royce. He was saddened to find that the missionaries seemed to have lost the confidence of some of the investigators he had taught. On Sunday he attended church services and the turnout was sparse, and dissention among the members seemed still to plague the area. After the services Royce and his companion went to the hospital to see Brother Guzmán, who was again ill, and they had a joyful reunion. The Guzmáns had continued to progress in their investigation of the Church, and had even defended the missionaries openly to their previous minister. In the evening the missionaries attended a fireside where, as he had when he was stationed in San Diego, Royce worried that the members tended to carry their doctrine "into pretty deep water".[106] His overall assessment of the members was still positive. "You know when you see how far good 'Mormon' people outshine all the rest it does make you very proud to be one. Mormonism just does something to people that makes them fine and more genuine. The average Latter-day Saint uses his Church as a road map rather than a 'straight-jacket.' He knows where he's going, and I guess that's what does it. Sometimes when I see such fine people, though they arn't members of the Church, it almost makes me wish it were possible by a twist of the hand or something, to make them see as I do."[107]

Returning to Los Angeles, Royce and his companion continued their work, though it was often interrupted by other concerns. At one point they spent a couple of days assisting in painting the chapel; on other occasions Royce had to make special trips to attend to the temporal needs of the members, a task which fell to him as District President. One rainy day Royce and his companion were embarrassed to learn that a pair of "lady missionaries" had stopped into their messy apartment to get out of the storm, and cleaned it up for them. On a Saturday evening they decided to attend a service in an impressive Protestant church established in Los Angeles by Aimee Semple McPherson, a colorful and sometimes controversial religious figure who had pioneered the use of modern technology to further her evangelizing. The missionaries were cordially received after the service and were given a tour of the facility, but Royce's impression of the visit was not favorable. "I just thought that if that's what we have to be to merit heaven and they are the kind of people that will be there, I would just as soon be with the people that I have learned to love and live with, and go somewhere else."[108]

On October 21, 1941, Royce and his companion took a train to San Bernardino to work with the missionaries there. It happened that Pearl's brother Daniel was living there with his wife Mary and their first child, and as it was raining too hard to do missionary work, Royce spent the evening with them talking about their family, and he and his companion spent the night there as well. The next day, attending to his District President responsibilities, Royce had a singular experience which he recorded in his journal. "When we were just ready to eat a girl came over whom the Elders had met one day in their tracting. She wanted us to hold the funeral of her mother at three o'clock. We very quickly consented and hurried over there by three without even having the opportunity to prepare very much. We certainly did enjoy the Spirit of the Lord in these services. All the people there were Catholics as far as I know. I preached the sermon because I could speak more Spanish, then we sang about four of our best funeral hymns. At the graveside Elder Teerlink and I sang 'Nearer My God to Thee,' in Spanish as we did all the rest, and there were not just few eyes that were wet with tears as we finished. I then dedicated the grave. I don't think I ever needed the help of the Lord any more. I had never known a single one of those present at the funeral and among those present very few if any excluding the family understood or knew who we are before the services."[109]

Back in Los Angeles, the Elders continued their missionary efforts. Working with a pair of missionaries, Royce paid another visit to his friend the Baptist minister. "When we were about to leave there was really a spirit of 'I want to know' and open-mindedness present. He invited us back, wanted to keep the Book of Mormon for further examination and wanted more liturature. I know if we never did influence his life a great deal, we will some, because he can't help but feel the same spirit I felt."[110] The minister remained Royce's friend throughout the remainder of his mission, and many decades later Royce still recalled the experience with great warmth. "We never baptized him or anything like that, but on the other hand, I'll bet he never thought unkindly about Mormons. He was a pleasant person. Well, I got to thinking about it. He's not so unlike a Mormon missionary, after all. He's out in his own little world, trying to fight for a niche, and trying to believe in certain principles, and follow those principles, and who's going to put him down for that?"[111]

November 11, 1941 was Armistice Day, the yearly celebration of the signing of the peace accord which had ended the "Great War" in 1918. Attempting to make a visit, Royce and his companion found their way blocked by a parade, which they watched for a time. "It was quite a contrast. First the young soldiers of the present and next war, in the uniform of Uncle Sam now, and then hundreds of veterans of the World War, quite a few truckloads of disabled ones plus some of the Spanish American and some of the Indian wars, etc."[112] War was then raging in Europe, but at the time neither spectators nor participants could have known that within a month the United States itself would once again be at war.

That week the missionaries learned of companionship changes which were to take place in connection with an upcoming district conference. On November 13 Royce picked up the new arrivals at the train station. "They were just like new missionaries usually are, bewildered & confused, wondering what it's all about." That night a major earthquake struck the area, which Royce recorded the following day in his journal, though with a little chagrin: "Another rather strong earth quake last nite, in fact the strongest in years. It done over a million dollars' damage. I slept peacefully thru it all. Oh well, maybe the next one will come in the daytime so I can feel it."[113]

The district conference, held the weekend of November 15 and 16, 1941, was presided over by an apostle, Elder Richard R. Lyman, whom Royce picked up at the train station on Saturday, along with his mission president, President Haymore, though it would fall to Royce as district president to conduct the meetings. That day the missionaries held their "testimony and report" meeting, at which the missionaries in turn each stated individual problems they had faced, and what they had done to solve them. The next day Royce conducted the general session of the conference, which for him was an impressive experience. "This is a day I shall indeed remember. It is the first time I have ever introduced an Apostle in a meeting. We began the first session of conference at 10:00 A.M. We had the missionaries speak first and then Bro. Haymore, after which Elder Lyman talked for the remainder of the time. The main points of his talk were that we should all devote money, time and talent to the work of the Lord. He testified as an apostle of the Lord that the only thing that could bring peace to the war-torn world is the gospel of Jesus Christ." Two other sessions were held, and in the last, an evening session, Royce found a message that resonated with him. "Bro. Lyman spoke on courtship and marriage. It was really fine, especially for the young people of the branch. He reaffirmed the value of chastity in the lives of young people, and said lack of it has been the root cause of the downfall of every nation that has ever been great. This was indeed a fine conference in every respect. In a way it was a relief to me to have them over tho I enjoyed them immensely, because I had to take charge of all of them. I am indeed grateful to my Heavenly Father because everything went over as fine as it did."[114]

That Thursday, November 20, was Thanksgiving Day, and Royce led the missionaries in individually recounting their blessings in a few words. That same night, as District President, he was presented with a leadership dilemma: one of the missionaries, Elder Donald Jones, came to request permission of Royce to return home briefly, having received a telegram informing him that his aunt, who had raised him as a mother, was seriously ill and his presence was requested. As this was an exceptional request, Royce attempted to telegraph the Mission President for guidance, but receiving no answer, he made the decision himself the following day to allow Elder Jones to return home temporarily, sending his own companion to work with Elder Jones' companion, while he remained with Dale G. King, who had recently been released as a missionary. Shortly after, Royce received a telegram from the mission secretary telling him belatedly not to allow Elder Jones to leave, and this left him feeling he had committed a serious error as a leader, though he had tried at the time to use his best judgment. He quickly mailed a letter of explanation in reply, feeling certain he had a severe rebuke in store, but the rebuke never materialized; instead the Mission President sent Royce a kind letter to relieve his mind. When the incident had come to its conclusion, Royce took stock of his own developing capabilities as a leader and reflected, "I felt then and I feel yet that I done right, and I remember Bro. Lyman telling us to make decisions whether they be right or wrong. Things like this make the life of District President just a little harder. The trouble with me is I nearly always let people have their way because I would rather do anything than hurt their feelings."[115]

At length Elder Jones returned, and Royce got his own companion back; the same day a pair of new missionaries arrived from Salt Lake, and Royce assigned them each companions. As life returned to normal in his district he expressed satisfaction. "Everything is just fine here now, all happy and working."[116] Sunday services were both more orderly and better attended than before.

The members and missionaries had planned and prepared a branch program for Saturday, December 6, and the missionaries stayed up that Friday night "making an airplane of considerable size" for the program, an activity which occupied the entire night. The next day, without sleep ("no rest for the wicked"), the missionaries went to work putting things together at the chapel, and the program and dinner went well, though the "quartette" in which Royce sang was hampered by colds. Tamales were served, and Royce commented that they were good, but having helped to prepare them, he appreciated the hard work that went into making them. He returned home exhausted, his grammar regressing as he gave account of the day. "I thot I'd die before nite come and I got to bed. We worked hard and I havn't even lied down since I got up Friday morning."[117]

The next morning, December 7, 1941, began as a typical Fast Sunday. The missionaries, rested from their night up, attended their morning meetings and were just setting out for the chapel in the afternoon when the news arrived that would reshape all their lives. Royce recalled the story more than half a century later: "There was a staircase out of our home, and I had just come out of the building

87

heading for the chapel, and just then the Muñozes who rented our house to us, came out in a hurry, all excited, and in Spanish he told me we were attacked by the Japanese, *los Japoneses*, and we were now at war."[118] He recorded the event in his journal. "A day that long will be remembered. Out of the clear blue sky in the afternoon came reports of a Japanese attack on Hawaii and other U.S. and British possessions. That which we have waited for, dreaded, and thought as inevitable has at last come to peaceful America, War. It seems almost unbelievable that this could be true, but we must face realities. It's something that has been thrust upon us. It sure doesn't help a fellow's future any, but at least we know what to plan now." The next day Royce observed, "It seems like all the dissention and division are gone from America. She is at last united, because we must pursue a course that we did not choose but rather was forced upon us. Elder King went down to enlist first thing this morning. It is what I would do also were I not a missionary. It's what I'm going to do as soon as I am released. They will never draft me if I can get to the recruiting station first. I have no fear in what the future may bring; come what may I know that if I live the Gospel of Jesus Christ I will be taken care of, if not in this life, in the life to come. I think America goes into this war with her head up and grim determination and confidence in her heart that it be the will of the Lord that they win. The war of nerves is over."[119]

The sudden reality of war required Royce to rethink his own carefully laid plans for the future. In his first letter to Pearl after the attack, Royce expressed the reality as he saw it. "My heart aches to think into the next months. I have always felt that sometime you would be mine, that sometime we would go together to the temple of the Lord to be sealed one to the other for all eternity. For the first time I am wondering about it all. I love you, want you and need you, yet I know that when I leave the mission I shall enlist. I can't help but think that that is also what you would want knowing you as I do. I couldn't be a stay-at-home and ask someone else to do that which I could do and am capable of doing just because I didn't want to go to war. Had I been out of the mission I would have enlisted yesterday. I told you at one time that if I were to go to war I would like to have you sealed to me before I went. I still feel the same way, but I will leave that entirely up to you. I certainly haven't anything to promise you on this earth under these circumstances except sorrow. If I die I desire but one thing and that is to go from this life with my heart pure. I do hope I have made this all clear darling, I'm going to leave the rest up to you. It's the only decent thing I can do."[120]

He expressed similar sentiments to his parents a week later. "I too felt before war started that I would do most anything to keep out of the army, but now it is a different thing. Of course it means the shattering of about every dream I ever built and the changing of most of the plans I have ever made. As I see it there is just one thing I can be sure of and that is if I live as I have been taught according to the Gospel of Jesus Christ, no matter what happens I won't have to worry about it. There are a few things that are more precious than life itself, and I believe what we have at stake right now is one of them. I'm proud to be an American, proud to have health such that I can give service to the armed forces of this country and I would be proud to lay my life with the thousands of others upon the altar of freedom. I have never been a coward. If I ever felt fear I always made myself do that which I feared to teach myself I was not afraid. It isn't like we declared war first and then tried to whip men into line. There won't be any of that this time." Referring to his older and younger brother, both married, he added, "I am glad that Smith and George both have families. It has always seemed to me that the tasks that were a bit out of the way came to me."[121]

The coming of war cast a shade over the missionary work, and its effects on the community at large were immediate and profound. The surprise attack on Pearl Harbor had demonstrated the ability of the Japanese to bring air power to bear on distant targets, and there was genuine concern of attack in coastal cities such as Los Angeles, where the familiar noise of aircraft was enough to make the inhabitants jittery. Military barracks sprang up in city parks, and anti-aircraft batteries appeared throughout the city. Radio stations were silenced except for a brief news transmission every half hour, to avoid the possibility of enemy pilots using the transmission as a beacon to target coastal cities.[122] Within a few days strict blackout regulations had gone into effect, and the first complete blackout, on December 10, came during a church meeting. "We were sitting in Mutual when the signal came, five minutes on the fire whistle. Pretty soon the whole town was blacked out except for stray lights that went out one by one. No one would ever have known a city was here, it was so dark. Mutual was dismissed and we had to walk home with a member. It was so dark I would have been lost if someone hadn't been there who knew the way. We heard some planes but I guess they were ours."[123] The blackouts would occasionally curtail the missionaries' evening activities from that time forward, and church schedules were changed to avoid evening meetings. Perhaps a more significant change was that all eligible young men either enlisted or were called up to military service, and this left Royce feeling self-conscious as he and his companion resumed usual missionary service after the disruption caused by Pearl Harbor. "It was just a little difficult out there today for some reason. I felt like apologizing for being at their places. I believe they do wonder why we are doing what we are doing there instead of in the army."[124]

During this eventful time the missionaries made another trip to San Diego. This visit was much more of a homecoming to Royce, who was pleased to see the progress that had been made with the arrival of additional Elders and the presence of the Lady Missionaries he had requested. Church attendance on Sunday was the highest he had ever seen, and an unusual opportunity presented itself which set the tone for the day. "I ate over at Fackrell's after, then we done something I have wanted to do in a long time. I baptized Gloria Fé Fackrell. Today is her eighth birthday.[125] It was really a thrill to me. I done it in the Pacific Ocean. The water was cold but I didn't notice it. When we got back, we went over to see *Hno. y Hna.* Guzmán. They are still the same wonderful contacts. I certainly was happy to see them and to see they are progressing in their study of the gospel. We held church at 4: P.M. to be in accordance with the defence program. After services Brother Guzmán came up and asked me if it wouldn't be possible for me to come down and baptize him. It really thrilled me and sort of makes my mission complete in a way that it could not have been otherwise, because my voice was the first Mormon Elder's voice he had ever heard, and now he is ready for baptism." That evening Royce attended a fireside where a good spirit prevailed, and a sailor was there who had attended services when Royce was still in San Diego. "He is leaving for the war, gone tomorrow with 3,000 men on board the former dollar liner 'Pres. Grant,' now a transport. I certainly did feel for him. In a way he seemed to know he wouldn't be back."[126]

Back in Los Angeles, the missionary work gradually returned to the usual routine of tracting in the morning and visits in the afternoon and evening. On Christmas Eve, the day generally celebrated by the Mexican people, the missionaries prepared bags of candy and helped with a branch Christmas party. Royce and another elder took a side trip to South Concord street to "play Santa" secretly for a family with a gift of food and candy, and later at night they went caroling with the other missionaries. Christmas of 1941 was a subdued time, however, as Royce's journal reflected, "Christmas and our nation at war. It's rather a jest to celebrate the birthday of He who came 2,000 years ago to bring peace and good will when the whole world is locked in armed conflict and men's minds are busy conceiving better means to kill." The Elders found missionary work to be fruitless on Christmas Day, and in the end Royce and his companion hitched a ride to San Diego with a visiting friend of one of the Lady Missionaries named Christian Fink, who was stationed there with the Naval Air Corps, "not a Mormon but he is certainly a wonderful fellow. I got to talk some Mormonism to him as we lie abed." During his long talk that night with the Navy aviator, Royce firmed up in his mind his own resolve to join the Army Air Corps when his mission was completed.[127] He confirmed that to Pearl in a letter the next day. "Lately, I have been thinking about what to do when my mission ends which is only too near now. I want to join the Air Corp as a flying cadet. I have had sufficient schooling to qualify and I am sure I am physically O.K. That would mean we would have to wait about 6 or 7 months before getting married. It would be dangerous, but all war is."[128]

After Christmas, Royce received a Christmas letter from his family who had gathered for the holiday and had each written a few lines. His response reflected the realities of his life as he approached a new year in a world now completely changed: "I used to dream of when I should be a man, what I would do etc. I know always I have wanted to be a farmer. I guess I learned to love the earth and things that grow because I grew with it. While I was in college I never planned to do anything but work with soil but things have changed a bit now. If it were possible, I know I would return and work on the farm but I am almost certain that cannot be. In times like this it really isn't yours to choose except possibly to see where you can best fit into the gigantic war machine that must be built. Smith better stand by the farm. He can make pretty good money there during the war and right after and I know he won't be called from there to go into the army. If I go into the air corp, unless they change their rulings I would have to remain unmarried during the training period, about seven months. If they change it and Pearl wants to I will get married as soon as I get home. I sure don't have anything to offer her, in reality not even myself."[129]

On New Year's Day, Royce and his companion returned from dinner to find a missionary in bed in their apartment, Elder Clayson, whom Royce had known in Texas. He had been sent to begin work in the area, and assigning him with his own junior companion Elder Corbett, Royce found the two an apartment the next day, then went to pick up his new companion at the train station, but no companion came, so once more he was left alone. The New Year had brought frigid temperatures, and the first church services of 1942 were sparsely attended, but that afternoon a baptismal service was held at which Royce baptized Gloria Villalobos, a daughter of the Villalobos family whose father had recently died. At evening services Royce sang in a "men's quartette" with other missionaries. This men's quartet, which met fairly frequently to rehearse, had by that time become quite accomplished, and they were often in demand to sing at church services and at other activities. The following Tuesday, January 6, was Pearl's birthday, and Royce remarked that he had never yet spent one with her, and it looked unlikely that he ever would. Still companionless, Royce spent his mornings studying, writing letters, filling out reports, looking for a secondhand suit to replace his old one, and finally reading the *Improvement Era*[130] cover to cover. Afternoons he kept himself busy making visits with the various companionships under his direction, each day expecting the new companion who never arrived. After nearly two weeks he determined he had neglected his own area long enough, and began to make visits on his own, often with multiple investigators present. That month he wrote Pearl about a new discovery which would impact his later professional life. "Say, I have a new favorite fruit, the avocado. I couldn't stand them when I first came out but now I like them better than anything."[131] Finally, on January 27, after nearly a month alone, Royce welcomed a new junior companion.

Not long after the arrival of his new companion Royce for the first time faced a new challenge, a wayward missionary. "I am afraid I have at last got some difficulty in my district. I hardly know what to do about Elder _____. Aparently, he is taking regular trips over to Van Nuys and other things that he has no business doing and not telling me about it. I don't think he's very much of a man. I hope his companions don't have to put up with him for too long."[132] Though the problem may have corrected itself and was never referred to again in Royce's journal, it seemed to have put a damper on his spirits which lasted several days, and his usual positive entries were punctuated with negative comments. Three days later, frustrated with contacts indifferent to their message, he wrote, "Went out tracting again this morning but really it wasn't very good...I did have several good talks though most of them I talked to were as though I talked to the door sill. I got about the same responce. It wasn't as bad perhaps as it could be, though sometimes maybe it would be better if a door were slammed in our faces rather than all the indifference we meet from day to day."

The next day, still in ill humor, he wrote, "We were eating or just finishing when Elder Clayson came in bearing great tidings of joy. Elder Richardson has the mumps. Every time I get things going good something happens to throw us out again and this time is no exception."[133] Royce took the sick missionary into his own home, sending his companion out with Elder Clayson, and spent the next several days caring for him, nursing and feeding him between missionary visits. Frustrated at the frequent setbacks and interruptions to his work, Royce lamented, "Since I have been here in L.A. I have never been able to systematise everything as I was able to do everywhere else I worked. I just get things going smoothly, tracting every day and doing missionary work every minute as we have been doing this month,

then something comes up like this and I have no companion to work
with or else I have to do something else. I don't know whether I like
being District President or not. I would lots rather be doing just regular missionary work all the time, but someone has to do this job too,
so I guess it's mine to make the best of. I always have to make decisions and I always do, right or wrong."[134]

Not long after, on the night of February 23, 1942, there was a vivid reminder to the community of the reality that they were at war—
not serious, but enough to set the whole city on edge. Royce wrote
the next day, "There was shelling last nite at Santa Barbara. This is the
first of this war." The following morning he wrote again, "Faintly
heard the anti-aircraft fire last nite and then I thought I was dreaming.
Got up at six and turned the light on as usual, and they come and told
me to turn it out again because there was a blackout on. I don't know
whether there was planes over or not. First papers said yes but later
no."[135] The first of these attacks, on Santa Barbara, proved to be genuine, an actual attack from an enemy submarine, while the second
was merely a false alarm. Although both these incidents proved in-
consequential, in the wake of Pearl Harbor the threat of attack
seemed very real, and there were frequent reminders of the war, including news from abroad and the departure of young men to military
service. Once while tracting, the missionaries encountered a family
which had lost a son in the Pearl Harbor attack; another had a son
captured by the Japanese on Guam. Royce also exchanged letters with
his old college friend Norm Johnson, who, having served a mission,
was already in training as an air cadet, and began supplying Royce
with welcome information about the Air Corps.[136]

During his work Royce frequently encountered Japanese immigrants, who at the time were looked upon by some with suspicion
(and who would shortly face interment in relocation camps for the
duration of the war), but who had nonetheless become an important
part of the social fabric in California. In January, after receiving a
haircut from a Japanese woman, Royce quipped with the rather poor
taste characteristic of the times, "That's the first time I ever had a Jap
that close to my throat with a razor. She's a good barber, though."[137]
He spoke with much greater respect when he came in contact with
Japanese who had embraced the Gospel. Staying overnight in the
home of Japanese members in Pasadena, Royce remarked, "They are
really fine people and just as loyal Americans as I am."[138] A few days
later, after singing with his "quartette" in a meeting of the Belvedere
Ward, he wrote, "Two Japanese members talked and I don't believe I
have ever heard finer talks given by anyone. One told how once he
was not even a Christian and that he first heard of Christianity at the
age of 19 and now the Church and the knowledge he has are his most

priceless possessions."[139]

As he approached the last months of his mission Royce became bolder in speaking with people of various religious persuasions. Confident of the principle of the restoration of divine authority taught among the Latter-day Saints, after one discussion he wrote, "I don't know how convincing it is to others, but to Elder Bringhurst it's all convincing."[140] Royce believed that any sincere seeker of the truth would be led to believe the Latter-day Saint doctrines if given the chance. After a discussion with a neighbor whom he and his companion had been teaching, he wrote, "She has a copy of the Book of Mormon and apparently a lot of her own thoughts and ideas coincide perfectly with our teachings. It's strange how people who really think do come to that invariably, that they figure out humbly for themselves the why, the how, and the what without sectarian influence and arrive at the knowledge, though meager, of the gospel of the Savior."[141]

Most of the missionaries' work, of course, was still among Catholics, and Royce developed an increased respect for the Catholic clergy, whom he had come to regard less and less as adversaries. Attending an evening meeting by a *"Padre Misionero,"* Royce wrote, "To be frank and honest I must say that he gave a wonderful talk, especially at the end where he talked about liquor and chastity, etc. I hope it sank into some of those rascals there. That Padre certainly talked beautiful Spanish. It was as hearing English."[142] Some time later the missionaries had the opportunity to talk to the Father personally while out tracting. "We got arround to where the Padre lives and decided to tract him too. We knocked on the door and were invited in. We then explained to him that we were working among his people etc. and told him very briefly what we believe and what we were doing among his people, and gave him some of the literature we pass out. I gave him '*La Restauración*' by Pratt. He couldn't talk to us very long because he is suffering from a heart disease and there was a notice there from the doctor asking people to please make their '*entrevistas*' short. He sure talked beautiful Spanish and seemed to know quite a bit about the Mormon people. He might say something favorable about us in church because of that visit."[143] Royce was thoughtful about his efforts with the people he taught. "They are very simple people. I have wanted to just reverse things and see through their eyes, hear through their ears and understand through their minds just to see and understand a bit more their way of thinking and reasoning. I think if we could do this it would greatly augment our ability to teach them. The work is going forward among them though, and the success we have is always determined by the amount of effort we put into it."[144]

On the weekend of March 7-8, 1942, another district conference was held, presided over by President Haymore, and Royce could not help but notice how tired the president looked when he picked him up at the station. The following day the president consulted with Royce on the placement of the missionaries, and then the missionary meeting was held in Spanish. Royce described the president's instructions. "He complemented us on our work and told us that he was happy with it but then told us of baptisms in San Diego, San Joaquín and other districts, and told us he felt sure that some would be possible here in this district. He suggested that we be more forceful and positive in teaching the restoration and baptism etc. and that we diligently strive for baptisms here. In Los Angeles there hasn't been a convert baptized here in five years despite the many missionaries." The Sunday meetings of the conference centered around the theme of the home, and to Royce's satisfaction the services went smoothly and he found them very inspiring. As part of the evening services he and his quartet sang a song entitled "Unfold, Ye Portals," which Royce had translated into Spanish while awaiting his companion back in January. At the conclusion of the meeting three investigators, all adults, applied for baptism, and Royce's landlord subsequently made the same request. Royce spent as much time as he could receiving counsel from the Mission President before his departure the following day.[145]

After the conference Royce and his missionaries went back to work with good spirits, and his journal entries became long, detailed, and joyful. Royce worked hard, and tracting was often fruitful, leading to conversations, return visits, and new investigators. The missionaries frequently loaned the Book of Mormon to interested contacts (a practice their Mission President had encouraged), which served not only to increase interest in their message, but also gave them a reason to return and follow up with investigators. After one busy morning of tracting, Royce wrote, "Thus passed the morning quickly and happily because we were about our Father's business. I hope some of these seeds we have sown bring forth fruit and I believe that they may." After errands and visits in the afternoon and evening of the same day, he concluded, "I'd like to know just how far we walk sometimes in a day. We went only in the house about twenty minutes between nine o'clock in the morning and ten o'clock at nite today; at that I didn't get done one-tenth of what I wanted to do. I guess life is always that way."[146] To Pearl he wrote, "We have been busy from before sunup till after sundown day after day. The more we do the more we realize there is to do."[147] The missionaries' efforts had even paid off in the once-rowdy Sunday School services, and the Sunday after the district conference Royce commented, "Sunday school was well attended, hall filled and things went over very well. It seems like of late there

95

has been more of a spirit of reverence present there than ever before. I have felt it a lot stronger and I am sure the Lord is pleased with it."[148]

The joy of missionary service was colored by the ever-present reality of war. In March, Royce received news of the first of his close friends killed in the conflict. "One of my best friends from out home went down on the cruiser Houston at Java, Wilson Frost. He was a good clean Mormon fellow. He always used to look me up first whenever he got leave to go home. It made me very sad to hear of him."[149] Royce continued to plan his own course for wartime service. He had learned that married men were now being accepted for air cadet training, and as his mission drew to a close, he outlined his thoughts to Pearl. "I do want to get married before I go into the service. I will most likely be called in June. I do plan on the Air Corp if I can pass the tests OK, but I am not going to do it untill they are ready to draft me, so I can stay home as long as possible. It doesn't look too bright, but we have to remember there are millions of other people making the same sacrifice that it may be necessary for us to make."[150] Pearl was evidently apprehensive about his joining the Air Corps, and Royce sought her approval, sending her copies of enlistment information, including a waiver she would be required to sign were they married when he enlisted. Though a realist regarding the ramifications of wartime service, Royce was well aware of the imperatives of his time and of the perils that faced his country, and felt a sense of idealism regarding the war effort as he determined how to fit his own life into the whole. "Whatever we do let's do it with heads up, eyes clear, and faith in Him who is almighty and with the thought toward doing a part toward building a world in which our children and all generations to come might have peace and the right to live as we would like to live. Times like these require that we sacrifice, sweat, and even bleed. May we as two keep the faith and do our part."[151]

That same month, Royce received a letter which brought another change of plans. "My folks want me to come home a month early. I didn't know what to think because I really do dread leaving the mission field. At the same time it's going to hurt whenever I go, and I do owe that to Dad and Mother because I probably won't be with them very long and there is lots to do in May." Royce informed the Mission President of this request and left it in his hands; on April 2 he received a reply, notifying him that his mission would officially end on April 27, a full month earlier than scheduled. Royce had just received his last missionary companion, Elder Richard A. Allphin, who had been in the mission for six months. That Sunday was Easter Sunday as well as General Conference, and, after listening to the conference broadcast, the missionaries held a baptismal service for the Ruiz family, the three adult converts who had presented themselves for baptism during the district conference. Royce spoke at the baptism, and described the experience. "I never have felt more inspired on my mission than then. The words just flowed out and they were beautiful words and clear. There is quite a contrast between what one can do when he is inspired and when he is not. A missionary in and of himself is nothing, but inspired of the Lord is a flaming torch and a trumpet."[152]

The following Thursday Royce, having received permission from the Mission President, traveled with his companion for a last trip to San Diego to attend the baptism of the Guzmán family. While there he participated in missionary visits, and on Friday with his companion Royce made the rounds of many of the members and investigators he knew, and found the Guzmán family still prepared and desiring to be baptized. The necessary arrangements were made, and that Sunday Royce recorded the event in his journal. "It was the answer to a prayer and a dream for me. I have always somehow known that they would one day join the Church. He said that I was the first Mormon that he ever heard. I had the priviledge of baptizing them. His name was Juan José Guzmán and he was confirmed by Elder Rex Terry. Her name is Rosa Enrique Guzmán and she was confirmed by Elder Henry Gleve. I well remember the first time I ever met them and how they have grown and progressed to where they have known that the gospel is true."[153] In a letter to Pearl informing her both of the baptism and of his early release, Royce expressed the feelings of a missionary nearing the end of his service. "It's kind of a mixed feeling it brings, on one side of great joy and the other side the 'sweet sorrow' of parting from these people that I love. They have enriched my life more than I could ever repay them, and I have realized and saw time and time again that our work among them is not in vain."[154]

During the last weeks of his mission Royce took in hand another project which he considered necessary before his departure, the landscaping of the Los Angeles chapel. This was similar to the work he had done on the chapel in San Diego, and he requested both the necessary permission and funds to do so. Plans were drawn up, and with the help of a member he obtained large numbers of plants at below wholesale prices. During his last week he and other missionaries gathered at the home of the Espinoza family who had a recording device to engrave phonograph records, and the missionaries made a record for Royce as a keepsake, which included a singing duet by him and Elder Donald Jones. His last Thursday was spent digging holes for the plants, which arrived the next day and were planted by the missionaries. Unaccustomed to the sudden heavy exertion, Royce wrote, "We didn't realize how much there was to do till we got started. We finished up quite late but we finished. I sure was worn out. Work can kill a man when he's not used to it." He was well pleased with the resulting appearance of the chapel. On Friday night when the work was done a party was held by the branch in Royce's honor; he would have enjoyed it more had he not been so tired, but it was well-attended.

That night Royce was awakened from sleep by an urgent plea to visit a hospitalized member to whom he and his companion had administered the previous Sunday, and thus in his last days as a missionary he faced yet one more experience entirely new to him. "At midnite I heard of the critical condition of Hno. Manrique so I went up there after sending two Elders up there earlier to administer to him. It was asked in the *unción* that the will of the Lord be done. I found him in a dying condition when I got there. He never regained consciousness. He died at 2:20 a.m. It's the first time I have ever seen anyone die. In a way it was a beautiful thing, one moment he was alive breathing etc., the next moment he was free from pain and suffering. Sister Manrique is a brave woman. She wept or rather sobbed a bit as did the rest, but then felt relieved and grateful to the Lord that he had aleviated the suffering. I got to bed at 4: AM."[155]

Royce arose two hours later at 6:00 a.m. to attend to his duties, which included picking up the Mission President who came by train from San Diego in order to hold a conference which had been timed especially for Royce's release as district president, since his last weekend in the mission had come. That afternoon they held their usual missionary meeting, and Royce wrote, "It was a fine meeting, one of the best I have ever been present at, except for one thing—I got my release effective Monday. It's just a piece of paper, but means that when I wake up Monday morning I will no longer be a missionary in the mission. It's the first time that I have choked up and have hardly been able to speak for a long time." The next day, Sunday, was the last day of Royce's mission, and he conducted his last district conference. The missionaries participated, and he was pleased with the part played by those he had led. "Never have I heard more effective and clear speaking from missionaries, and especially from the younger ones as I did today." In the evening session of the conference Royce gave his parting speech. "I believe I never felt more inspired and at the same time more humble than I did as I spoke. It was easy for me and I was able to say everything I wanted; I had feared I wouldn't. The president said he had never assisted in a conference where he had heard more inspiring talks than the ones the *Hnos.* gave. It was a perfect climax to a mission that has been perfect in every way. We took President Haymore down to the station and I bid goodbye to a man that I truly love and respect as if he were my father."[156]

When he awakened on the morning of April 27, 1942, Royce Bringhurst had officially been released as a missionary. Unable to sleep, he arose at 4:00 a.m. to complete his final leadership reports, and then made ready to leave, though he stayed through that evening to be able to sing at the funeral of Brother Manrique. The next morning he bade the other missionaries farewell, boarded a northbound train, and left Los Angeles, his mission completed. In two weeks a full thirty months would have transpired since he had entered the Salt Lake City mission home.

It is impossible to overstate the influence Royce's mission had on his later life. Not long before its conclusion, Royce had written, "My mission is and will always be a precious thing to me. It's taught me lessons I could not have learned otherwise, and I know it has increased my sense of values a great deal, and the more that increases the more your ability to appreciate and even enjoy life increases."[157] It was largely on his mission that Royce developed the capacities which would make him successful as an individual, as a scientist, and as a professional. Though his missionary service was largely motivated by religious conviction, through it he acquired a personal discipline and a capacity for concentrated study which would serve him in other fields as well. His mission made him an excellent and thoughtful teacher, and a confident and compelling public speaker who commanded respect from his listeners. He had learned to soothe feelings and to reconcile differences. He had developed a richer understanding of religions and cultures, and an ability to win the friendship of people from all walks of life. The fluency he acquired in the Spanish language would lead to opportunity after opportunity as his professional and personal life expanded, and his exposure to the Southwest, particularly California, would later help to direct the course of his life.[158]

Above all, Royce's mission laid a foundation of basic faith and patterns of personal worship which would see him through the troubled times ahead, and which would influence all his future endeavors. Though his faith would at times be severely tested by the cultural trends of his time and challenged by the rapid expansion of scientific knowledge of which his life's work was a part, he returned to it again and again, and remained unflinchingly committed to the Church and all it stood for. He truly believed that the religion of the Latter-day Saints encompassed and embraced all truth, and he came to feel that neither he nor his religion had anything to fear from scientific inquiry or scholarly investigation and scrutiny. Indeed, it may well be said of him that he followed the uniquely Latter-day Saint injunction to "seek learning, even by study and also by faith".[159] That faith would remain a governing principle throughout his life, not just accepted in precept but thoughtfully incorporated into his daily actions.

Illustrations for Chapter 3

Missionaries at Salt Lake Mission Home on November 14, 1939. Royce is
third from left in the center row.

Left: Royce as a new missionary in Fabens, Texas. Right, with first
companion, W. Porter May.

Sunday School in Tornillo, Texas, December 1939. Royce is at left with companion, Elder Porter May.

Left: Missionaries at mission conference in Mesa Arizona on May 28, 1940 with President Orlando C. Williams, front center. Royce Bringhurst, at far right, was developing pneumonia when this photograph was taken. Right: Royce with Elder Reid F. Ellsworth, companion in Fabens from June-July 1940, and later lifelong friend.

The Calderón home at La Caseta, Chihuahua, Mexico, where services were held. This photo was taken on Sunday, July 21, 1940, during Royce's last visit there prior to his transfer from Fabens. Royce is at far right; Elder Ellsworth is opposite at far left.

Tucson flood waters, August 14, 1940.

Left: as senior companion in Tucson with Elder Christensen; Right: packed for transfer to Nogales.

International border at Nogales, U.S. on left, Mexico on right.

Elder George Albert Smith, center with beard, at Tucson train station. At left holding bags is President Haymore; Elder Bringhurst is the first of the three missionaries to the right of Elder Smith.

Left: with Elder Rowley; Right: "tracting" in Nogales.

103

Members of the Spanish Branch in San Diego, Mother's Day 1941.

Left: with companion Harold J. Nielson; Right: Royce (at left) working with members on chapel grounds in San Diego.

View of San Diego from upper floor of chapel.

District of Los Angeles, Royce (2nd from left) as president.

At desk in September 1941, at birthday celebration in December.

Constructing model airplane for activity on the eve of Pearl Harbor attack.

Left: planting trees at the Spanish Branch chapel; Right: mounting train in Los Angeles at conclusion of mission.

Marriage

Chapter 4

When he completed his mission in the Spring of 1942, Royce entered a far different world from the one he had left in 1939. The attack on Pearl Harbor had swept away any remaining sense of normalcy, and the reality of war affected every decision and overwhelmed every personal consideration. Military service was a certainty, and in those early days when nearly all news from abroad dealt with crushing defeats by brutal and seemingly invincible enemies, prospects seemed bleak. Gone were the hopeful plans of completing college and getting on with life. Royce had already deferred such plans three years by accepting a mission call, and the mandatory peacetime draft had added an additional year, but by the time he came home, war stretched ahead as far as the mind could contemplate, and survival itself was in serious doubt. College by now was also out of the question. The mission had served as a temporary haven, but now most of Royce's friends and acquaintances were either in military service or bound there.

The Journey Home

When Royce left Los Angeles on the morning of Tuesday, April 28, 1942, he headed north, bound for Oakland, California, where he arrived at 5:45 p.m. There he was met by Ernie Buchan, husband of Royce's oldest sister Naida, who took him to their home on Rockdale Drive, where Royce saw Naida for the first time in several years. The next morning Royce slept late, a luxury he had not enjoyed in many

Chapter 4

days, and that afternoon Naida took him to see the sights of the San
Francisco Bay area, serving him shrimp cocktail at Fisherman's
Wharf. They met Ernie after work for dinner at a Chinese restaurant
(also a novelty for Royce), and then ascended the Oakland hills where
in the evening Royce got his first full view of the entire bay area, including Alcatraz and Treasure Island, the Bay Bridge, and the Gold-
en Gate bridge. The next day they had planned a trip to the beach,
but rain quickly cancelled their plans, and Naida and Ernie instead
bought Royce a good pair of shoes to replace those he had bought on
his mission.[1] Stopping by the San Francisco train station, Royce determined that it would be possible for him to depart for Salt Lake
City that same night, so bidding farewell to Naida and Ernie, he
crossed the bay on a ferry and made his way to the train station,
catching the next eastbound "Streamline" train.

The train trip to Utah took a full night and a day, and as the train
pulled in to Salt Lake City on May 1, 1942 around 9:00 p.m., Royce
was greeted by a family he had not seen in two and a half years. Much
had changed in that time—his little brother George was now married
with a child, nieces and nephews were no longer infants, and his baby
sister June was hardly to be recognized. Perhaps most significantly,
the house the family took him home to was not the old farm house in
which he had been raised, but a newly-built home in Cottonwood.
The new house, located at 1780 East Vine Street, was in an area familiar to Royce, since it was directly across the street from the home
of his Grandfather Smith, who was nearing the end of his life, and
who it was said remained alive only until he could see his grandson
return from his mission. Arriving at the house late that evening,
Royce took a good bath before retiring to bed.

The following day, on May 2, Royce went to Pearl's house, finding
her unaware of his arrival. Their reunion could only be imagined;
Royce wrote: "I know I couldn't express exactly how I felt. We spent
the day together visiting, etc. Went out to the old place."[2] The following day Royce attended Sunday School in his family's new ward. He
was asked to speak, and he also took some time in the ward's missionary class to recount his mission experiences. Afterward he went
over to Pearl's home and brought her and her mother to dinner with
his parents. it was during this visit that they determined to get married, and set the date.[3]

The decision to marry under such circumstances was not to be
taken lightly, and Royce and Pearl had discussed it repeatedly in their
letters while Royce was on his mission, during which time the prospects for the future seemed to be constantly changing. In truth, few
couples in history have likely undergone such a long courtship during
such a portentous time and under such uncertain circumstances. In

his final letter to Pearl in April of 1942 shortly before his return home, Royce had summarized their situation: "When we parted in 1939 I thought as I am sure you did of what a long time it would be till we should meet again. I don't know about the waiting part. Somehow I always knew that you would wait and that things would not change that way. Of course we will be changed, but we should call it perhaps progressed. Life is made up of changes and improvements. I sincerely wish that we at least in a way might know what awaits us in the months to come but we are a part of this great movement of preparation that the country is now involved in, and have to serve where we are called and where we fit into it."[4]

It would perhaps be difficult for succeeding generations to appreciate all the factors that Royce and Pearl had to weigh as they considered marriage under such uncertain circumstances. Surely some of the older generation questioned their choice. Royce and Pearl many years later jointly described the thought process:

Royce: "This decision to get married at that time wasn't a casual decision, because we knew I was going into the service. I hadn't seen her for 2-1/2 years, and here I am talking to her. 'Well, do you still want to get married?' or words to that effect, 'Do you still want to get married to me?'"

Pearl: "And I said 'Yes!'."

Royce: "And she said 'Yes' before she consulted with anybody."

Pearl: "And I talked to my mother, and she said, 'What if he doesn't come back?'"

Royce: "And this was exactly what we'd talked about, because we thought, well, we'd rather have a family from us than somewhere else, if something happened to me, because I was fully aware of the risk that was involved, and it turned out I spent 2-1/2 years in the military service, including the air force service in Europe on a B-25, and I was fully aware that this might be the end of me. And she was aware...."

Pearl: "I told my mother I wanted to get married, because if I didn't, and he was killed, I would always feel sorry, and we both knew that marriage in the temple was for eternity, and that's what we were looking for."

Royce: "And I didn't think it was so stupid...the only thing I really erred on, was how difficult it was going to be for her, to provide for what she needed."[5]

The Wedding

The date they set for the marriage was May 14, 1942, just under two weeks from the date of Royce's return. This was hardly a precipitous marriage—Royce and Pearl had known each other (and regarded themselves as "sweethearts") for over six years, even if their time together was limited chiefly to their senior year of high school and the couple of summers before Royce's mission—the rest of their courtship had taken place by mail. The immediacy of the marriage date was dictated by their circumstances. Work on the farm was pressing, and war service loomed imminent. If they were going to marry, it made no sense to delay. Having made the decision, they returned the same day to the Davidson home, and then to the home of Pearl's brother Willard, who lay ill. Assisted by Helen's husband Mont, Royce administered to him. The following day, Royce and Pearl went into town to get their premarital physical examinations, and later that day Royce went to the office of Elder Richard R. Lyman, the apostle whom he had met and introduced at the Los Angeles district conference just a few months before. At that time temple sealings were often performed by general authorities of the Church, and since Royce had made his acquaintance, he asked the apostle if he might perform the ceremony for him and Pearl, and Elder Lyman agreed. He spent about an hour with each of them, giving counsel to the couple which was gratefully received.

That Wednesday Royce purchased a diamond ring which he gave to Pearl that evening without ceremony, remarking in his journal that they were "officially engaged now."[6] Because this was to be a temple marriage, Pearl would first need to receive the endowment ordinance which Royce had undergone before his mission, and later the two would re-enter the temple for the wedding itself, which was to be a sealing ordinance, meaning that according to Latter-day Saint doctrine, they would be united in marriage not just for the duration of their lives, but for the eternities. To enter the temple each of them had to obtain a temple recommend which was required to certify worthiness,[7] and as they did so Royce and Pearl discovered that both had been baptized the same day in the same place—in the basement of the Salt Lake Tabernacle on January 29, 1927.

On the evening of Sunday, May 10 Royce spoke in Sacrament meeting at the old Bennion Ward which his family had attended before their move, giving a formal report of his missionary service to his ward. The next day he worked all day on the farm, then on Tuesday, May 12, he and Pearl went to the Salt Lake Temple accompanied by both their mothers and by Pearl's sister Helen, and Pearl received her endowment. That afternoon they obtained their marriage license, and on Thursday, May 14, 1942 Royce and Pearl returned to the Salt Lake Temple for the marriage itself.

The wedding was not to be an extravagant affair—Latter-day Saint sealings in general are attended only by a very small group of family and close friends, partly because of the requirement that every person attending qualify for a temple recommend, and partly because the sacred nature of the building does not allow for the traditional pageantry customary in wedding ceremonies elsewhere. Royce and Pearl's wedding was austere even by Latter-day Saint standards—Pearl was a war bride hastily marrying her sweetheart after a long absence. The short notice and limited means allowed for no elaborate preparations, and Pearl went to the temple wearing a nice skirt, blouse, and hat, while Royce wore his worn missionary suit. In the temple both were required to change into white clothing for the ceremony, and they dressed in the rather plain white outfits provided by the temple. Pearl later recalled that the dress she wore "looked like a nurse's dress". When they saw another couple more suitably dressed, the bride in a fine wedding gown, Royce felt stung at not having provided finer clothing for his own wife, and for a moment was reduced to tears[8].

At length they were introduced into one of the rooms of the temple especially set apart for sealing ceremonies. Two witnesses were selected from among the few present—one was Louis DeYoung, a friend from Pearl's ward whose family had taken Pearl in as a young girl for a time at her father's death, and the other was Royce's friend Ralph Willes, who like Royce had served as a missionary and thus was qualified to be present. At the appropriate time Royce and Pearl knelt at the altar, and there they were legally married and "sealed for time and all eternity" by Elder Richard R. Lyman of the Quorum of the Twelve Apostles. The ceremony itself was brief, lasting only a few minutes, but the apostle was kind and encouraging to the young couple, and took some time to give them advice for the years ahead. The day was a bittersweet one for the Bringhurst family, particularly for Royce's mother, as his ailing grandfather, George Fred Smith, died the same day.

Changing back into the clothes they had worn to the temple, Royce and Pearl made their way to the steps in front of the temple where they took a few pictures to memorialize the event, then they drove over to the home of Pearl's brother Willard for a wedding breakfast. That evening, in accordance with Latter-day Saint custom, a wedding reception was held at the Davidson home—this was the opportunity to greet friends and well-wishers, and to receive much-needed wedding gifts. Here too, some of the more conventional marriage traditions could be carried out. Pearl had two bridesmaids, Elizabeth Dale and LaPriel Russon (her best friend Claire Van Dam was away on a mission to the Eastern States Mission), while Steve Mackay served as Royce's best man. Willard Davidson and Mont Toronto served as ushers, and a Japanese friend of Pearl named June Niki helped with the reception.[9] Since Pearl had no wedding dress, she

111

borrowed a white gown from her sister Laura which, with a white jacket, an imitation pearl brooch, and a small veil she had fashioned from some Mexican lace which Royce had sent her from his mission, served very nicely for the reception. The next day, wearing the same borrowed dress, she and Royce went to a photographer to have a formal wedding picture taken.

The Newlyweds

The move of Royce's parents to the Cottonwood area had left vacant Royce's boyhood home, the farm house on Redwood Road, and there the newly married couple took up residence after the wedding. Pearl assumed housekeeping duties, while Royce took over the primary responsibility for the family farm while his father continued to labor in the copper smelter.[10] Finances were tight for the newly married couple, Royce having no financial reserve after his mission, so for the first few weeks of their marriage Pearl continued to work at the telephone company and they lived off her income. This arrangement was an obvious necessity, though it made Royce uneasy as he regarded breadwinning to be the husband's responsibility.[11] In accordance with his plans, rather than wait for a draft notice Royce enlisted in the Army Air Corps, and was sworn into military service on June 6, 1942, just three weeks after the wedding. However, the delay caused by his mission worked now to his advantage, as by the time Royce enlisted, the training pipelines were already filled with new recruits and draftees, and Royce was informed that there would be a delay of several months before he was called into active duty. This would enable him to farm that season for his parents, which would also provide some financial support to the newlyweds.

In the end Royce and Pearl were able to spend just over six months together before Royce had to report for military service, and despite the uncertainties that hung over them, this became one of the happiest and most fondly remembered times of their lives. Royce was fresh off a mission, he and Pearl both held high ideals and were determined to live by them, and their temple marriage brought them the promise of an eternal union if they lived worthy of it. Pearl quickly adapted to farm life, and Royce, in turn, became accustomed to Pearl's style of cooking, though it may have been a challenging transition, since by the time they were married Royce had been on his own for nearly five years and had a good deal more experience than Pearl as a cook. After some months together Pearl wrote her mother: "Royce gets startled every once and a while at some of the foods I fix up, or the way I do something. He knows a few cooking tricks himself, and we learn from one another. He finally got me to cook with a little garlic and it's really good. I wish I were as good a cook as you

112

are and as good at a lot of other things too, but I'm buckling down and learning."[12]

Not long after their marriage, Royce received a surprise wedding gift from the missionaries of the Los Angeles district which he had formerly led. Royce later recalled: "My missionary friends, headed by the Watkins team (brothers that I had in the mission at the same time), all got together and chipped in, and one day I got a message from the railroad company in Murray, Utah. They said, 'We have a goat for Mr. Royce S. Bringhurst. Do you know how we can get it to him?' So I ended up borrowing a vehicle and getting over to get this goat. She was an educational experience for us, because she was so bright. She could figure out most anything. She got along particularly well with the horses, although we had to be careful that they didn't step on her, and in turn, they both got along with the rooster. This rooster had come from Pearl's sister's husband, LaMont Toronto. He had a rooster and a duck, and the duck kept pushing the rooster into the pond, so the rooster came out to our place. The rooster used to come and stand on the kitchen window sill, and the goat jumped up with the rooster, and they used to stand and look in on us, and wonder what we were doing."[13]

Royce and Pearl had not been married long before Pearl was found to be expecting a child. Speaking years later of their decision to marry with war looming, Royce recounted: "The argument against getting married is, the first thing you know she'll be pregnant, and they were absolutely correct."[14] In late June Pearl began to experience morning sickness, which became severe enough that she had difficulty keeping any food down, which concerned Royce as he continued to work on the farm. Pearl gave up her job at the telephone company, and near the end of July she broke the news to her mother in a letter: "Royce asks do you mind being Grandma again? I hope you don't mind if we have a family. We both wanted one Mother, & we felt that you'd feel all right about it. They released me a week early from work. Royce called and told them he didn't think I'd be able to come in this week so they made my release a week early. It's lots better that way."[15] Pearl began to see Dr. Sanders, a physician at the Salt Lake Clinic. In August she wrote: "I hadn't been feeling so good for the past six weeks but I think I'm coming out of it now because I'm feeling fine again. Royce is just about as happy about it as a fellow could be. He's certainly been kind and good to me too. Gee Mother, we've just been so happy these past months, it's going to make it hard to see him go now but I guess we should feel thankful to have had this long together."[16]

Throughout the summer Royce worked on the farm, sowing grain and sugar beets, and attending to the apple orchard. He let out a por-

113

tion of the farm to a cannery for a planting of tomatoes, caring for the crop as it ripened, but in late July the farm was hit by a fierce summer hail storm which destroyed the entire crop. This was described vividly by Pearl: "Yesterday we had the worst hail storm I have ever seen. Royce said it was the worst he has seen here in the valley. The hailstones were about twice the size of a good sized marble. It wrecked the tomato crop, ripped the plants all to pieces, and ripped open the green tomatoes. Royce said it ruined about half the grain, and the wheat looks like a herd of horses was turned loose in it. It bruised a lot of the apples on the tree. They'll be all right though. It knocked down the ripe ones. You'd have thought fall was here the way it ripped the leaves off the trees." She added, "The farmers out here get so they take things like that pretty well, though. Royce says it doesn't make him feel half as badly as he would if he could have helped it."[17] They made the best of the situation, and Pearl salvaged and canned a few tomatoes from the ruined crop.

In addition to his work on the family farm, Royce helped the neighbors with their own crops, as he had in previous years. Pearl, new to farm life, described to her mother the relationship between farmer families and their neighbors: "Royce is over helping the Dimonds with their threshing because they let him take the hay rake all last week. It's just sort of, 'you help me and I'll help you' out here. It's nice to have it that way, though. The stake out here has a dairy farm as a church project, and Royce has spent several nights over there." She added, "Our little goat follows Royce around worse than ever now. I don't know what the poor thing will do without him. He can't run the mowing machine or anything without the old goat following him up and down."[18]

Amid the rigors of farm work Royce still took some time away for other activities. In August he planned another fishing excursion to Black's Fork with his boyhood friend Steve Mackay. Royce had long wanted to include Pearl on one of these excursions, and the two friends had planned to take their wives, but Steve's wife had injured her foot with a needle which rendered her unable to go. Rather than tag along as the only woman, Pearl stayed with her sister Helen while the husbands made their trip. Royce and Pearl feasted on trout for several days afterward, and Royce helped with the cooking since Pearl was still unwell with morning sickness.[19] One Saturday in August they attended a Bringhurst family reunion, and Pearl remarked: "I guess it was just like any family—some were snobs, some were no-goods, and some were very very nice."[20] Later in Autumn, hunting season came around, and Pearl wrote: "Pheasant season opened here yesterday and we've really had hunters all over the place. People start shooting up the place before daylight. Royce is out hunting now, with his brother

George. Royce got two. If he can get his limit today which is three, he said he'd take some up to Helen and Mont. I don't know what makes him think we can eat two ourselves."[21]

When it came time to haul hay, Royce enlisted the help of his nephew Jay Norberg, the 11-year-old son of his sister Dean, and this led to a memorable experience, recounted by Royce many years later: "We always made alfalfa hay for the horses and the cows to eat during the winter months. The railroad cut our farm in two pieces, and I was driving above the tracks. Jay was up on top of the load, tramping down the hay as I threw it up to him with a fork, and suddenly the ladder on the front end of the wagon gave way, and when Jay fell between the horses, on the tongue of the wagon, the horses bolted and took off across the field. I wasn't anywhere near where I could grab hold of the reins and haul them back, and get them tamed down. They took off at an angle across this field headed for a ditch, and I figured, 'well, they'll turn at the ditch.' So I cut across the field and made it there in time to stop the horses. Jay had made a few sounds, but I thought I had dragged him to death, and I called out to him, 'Jay, are you all right?' He said, 'I'm fine, can you get me out of here?' And, sure enough I was able to halt those horses, I was able to get them calmed down, then Jay and I just walked them out of their harnesses, undid their belly bands, and all the paraphernalia that they wore, and we took them down. Jay had slivers, splinters of wood, all over the front end of his body where he was dragging, and he had bruises and cuts, but no great injury. By the time I got them stopped, the hay rack that I had loaded up with the hay didn't have a spear of hay left on it. It all bounced off, and I never did gather up that hay."[22]

The summer after their marriage was a happy time that Royce and Pearl remembered with fondness. For transportation they relied on an old family truck that was something of an eyesore, and a challenge to keep running as well. Remembering back after a couple of months in the military, Royce later wrote: "Does the old truck still run? Remember, or rather I should say how could you ever forget all the rides I took you for in it. We kind of had to swallow our pride and make the best of it, but there was only once when it didn't bring us back and we had to walk. I guess that old summer we spent out on the farm together will always be just as unforgettable to you as it is to me. That was really where we found out that we did belong together."[23]

During these precious few months of married life, the war was never very far from their minds, the entire community having been affected by it. After the American surrender in the Philippines, Pearl wrote: "I notice by the paper that most of the boys of the 5th Air Base Group that Art Walker was with have been reported missing.

They were the last reinforcements sent to the Philippines. We haven't heard anything from him or his mother...I don't believe there are any fellows left out here in Royce's Ward. The last three went last week. There are no young fellows out here at all now."[24] As the months passed, Royce expected to be called at any time into active military duty, and this became an anxious waiting game as the harvest season approached. Pearl later recounted: "Royce had an apple orchard, and he was going to have to get those apples off, and get them sold, and he was worried about it, and every day we would run to the mailbox, and then we'd breathe a sigh of relief until the next morning, then we'd be upset until the mail came. Finally Royce called Fort Douglas. He said, I have an apple orchard. Do I hire someone to take these apples off, or can you tell me when I can expect to be called up? And they told him to go ahead and take care of his apples, he wouldn't be called up until November."[25]

The summer of 1942 turned out to be a good year for apples on the Bringhurst farm. In August an offer had been made to purchase the entire crop, but in the end it appears most of the apples were hauled to markets or sold at a roadside stand next to the farm. There the family goat, which they had named Nancy, got into the act by climbing atop the cars of prospective buyers and staring down at the startled occupants through the windshield. Royce later remembered this detail of the apple harvest with relish: "It was rather awkward but still a funny situation when we would come out of the house and find that 'no good' perched on top of someone's car and then giving us that insulted look when we chased her off and she mounted to the top of the granary where she must have thought she was just on top of the world."[26]

With the wartime manpower shortage, Royce and Pearl were both kept busy with the apple harvest. It had been a poor crop elsewhere, and the apples were in high demand, and brought a good price. At the beginning of November, Pearl wrote: "We've been as busy as bees in the apples. This week should finish the last few. Royce picked the last of them yesterday and brought them down from the orchard. There was about one thousand bushel of apples on the orchard this year. We've had people clear from Kaysville, Layton, and Bountiful come here for apples because they couldn't find any in the orchards up that way. Last year they barely got $100.00 from the orchard. This year we've cleared $850.00 not counting some that Royce's folks sold, which will average close to $50.00. He gave half of it to the folks. We have $387.00 in the bank from the apples. Royce put the account in my name too so that I can draw it out after he goes."[27] This was not an insubstantial amount for the times, and gave Pearl some security for the coming months. Pearl also put up some apple jelly and apple

116

butter for use during the winter. After the apple harvest Royce was able to harvest the year's beet crop[28] before departing for the military. Royce later wrote in a letter to Pearl, "You'll never know how grateful I was for that apple crop and how thankful I was that we had such success in disposing of every bushel. It made it possible for me to leave home happy even though it meant leaving you."[29]

Despite this joyful summer together the future was uncertain for the newlywed couple. As the time of Royce's departure drew near, Pearl reflected in a letter to her mother: "I've wondered more than once if I was right in getting married now. I might be taking care of you. I tried to think things through and I prayed hard about it. Somehow I felt that things would work out for what was best. I thought that after I was married that I might go on working and later, after Royce had left, still take care of you. Royce and I both wanted a family, so we prayed about that too, that whatever was best for us would work out some way. You had always taught us to live clean natural lives, and that bringing children into the world was a part of marriage. Now our little family is on its way, Royce and I have been so happy about it. It's true the future doesn't look too bright, but everything will work out for whatever is best."[30] Pearl made plans to move back in with her mother after Royce left, using Royce's military income to help support the household. Their income would be supplemented with vegetables brought by Pearl's sister Jeanette, who now lived with her husband Burnell Turnbow on a farm in Tabiona, Utah, in the Uinta mountains. During Royce's absence Pearl would be invited to eat often at the home of Royce's parents, and she would become closely acquainted with Royce's mother.

As for the old Bringhurst farm on Redwood Road that had been their first home together, it was destined to be sold and at last pass out of the immediate family.[31] For years it had provided only a marginal income for the family, which had required supplementation with other forms of work. Royce's father now lived in a new home; his brother George was unwilling, and his brother Smith unable, to manage the farm, and with Royce going into military service, its fate was sealed. Royce later reflected, "I don't think I shall ever be able to go by the old place without feeling a pang, knowing that the old days cannot be recalled and that the place of my childhood and also our first life together can never be ours again. I guess I wouldn't buy it if I could. I don't believe I would want to go back there again now. I'd rather remember all that as it was and go about building our home somewhere else. The orchard was closest to me I think. There wasn't a tree that didn't have my mark on it. I pruned them almost from the first. I dyked all through it and planted alfalfa there to retain the soil. I kept the grass and weeds down. It meant a lot to me and that old or-

117

chard was good to you and me while we had it. It was the only thing that payed off for us that summer. We had to work during those few weeks but I was never happier about anything I've done in my life."[32]

One November day Royce was at work in the beet field topping the last of the beets, and had just completed the final load when a letter arrived from the War Department notifying him that he was on active duty as of November 14, and would be leaving shortly. He reported for duty at Fort Douglas, located on the hillside bench overlooking the Salt Lake Valley from the east, and there he received orders assigning him to airman's training in Santa Ana, California. On the morning of November 20, 1942, almost three years to the day since he departed for his mission, the family once again gathered to see him off, this time at the train station. Pearl recalled the scene: "We waited hours for the train to go. And it was never going to go, and so finally, after I think two or three hours, Mother and Dad Bringhurst and my mother said, 'Well, we'd better go.' So we went back to the car, and I told Royce good-bye, kissed him good-bye, and here I was, you know, five months pregnant. And he turned, and went, and I looked...I watched him as he walked back, and he didn't turn and look. I think he was afraid to turn and look back."[33] Royce and his fellow soldiers waited another couple of hours before the train finally departed, bound for California. Once more he left the Salt Lake Valley behind him, this time for good. He would never again have a permanent home there.[34]

Illustrations for Chapter 4

Newspaper clippings announcing the Bringhurst-Davidson wedding.

Left: Royce and Pearl on the steps of the Salt Lake Temple following their wedding on May 14, 1942. Right: with their respective mothers. Royce's mother, at right, lost her own father the same day.

Royce and Pearl in their official marriage photo, taken after their wedding reception. The dress was borrowed for the photograph.

Left: Pearl with goat which came as wedding gift; right: Outbuildings of the old Bringhurst farm, sold after Royce departed for military training.

Air Corps Training

Chapter 5

Santa Ana—Classification and Pre-Flight Training

Royce and his fellow inductees arrived by train in Los Angeles on Friday, November 21, 1942, where during a brief layover Royce was able to see the missionaries and several acquaintances from his mission before continuing on to the Santa Ana Army Air Base, located several miles southeast of Los Angeles proper, where he officially reported for military duty. There he, along with others who had received their orders at the same time, was assigned to Squadron 18, a preliminary squadron whose purpose was to determine whether cadets would be classified to train as navigators, bombardiers, or pilots. A fair number of Royce's squadron came from the Salt Lake City area, and a few days after his arrival he wrote, "Sixty of the 240 men in our squadron are Mormon boys. 50% of them are good boys in every respect, the rest have some habit or something but they are all good fellows and don't expect others to do what they do."[1] Among the active Latter-day Saints who had accompanied Royce to Santa Ana were Frank Van Limberg, a married man whose military career would follow the same course as Royce's, and Frank Denstad, who had attended Granite High School a year behind Royce and Pearl, and who, like Royce, was married and expecting a child. Royce especially felt an affinity for Frank Denstad, and the two quickly became fast friends.[2] Back in Salt Lake, Pearl looked up Frank's wife and remained in contact with her as well.

Santa Ana was an air base without landing strip or aircraft, and while its main purpose was to provide ground training for air cadets, they also underwent basic training typical of other army units, including the rudiments of military discipline and the art of marching in formation on the parade ground, which greatly impressed Royce. "It's quite a sight to see 15,000 men on the same field all dressed up and really marching. First all of the squadrons line up in order and then the 'Star-Spangled Banner' is played, during which time all salute and hold it untill the last note is played, then the parade begins. They march in squadrons (240 men) and it really is something. Each squadron in turn goes clear arround the field and then does an 'eyes right' as they pass the reviewer's stand where all of the top men of the post are. Then the squadrons are judged and the winner has a big banner to carry in the parade the following week."[3] The new cadets were placed on a two-week quarantine, during which time they received their initial vaccinations and underwent a series of interviews and mental and physical tests, as well as medical examinations. About the time the squadron was to be released from quarantine, two of their number were found to have the measles, so they were quarantined an additional two weeks.

On December 16, 1942, Royce's squadron underwent "classification," designating what training each soldier would receive. Royce wrote, "In our squadron of 240 (now 238) men, about twenty or possibly twenty-five were definate washouts; about 15 or so remained unclassified for the present at least; one was made navigator (his eyes wouldn't pass the pilot's test but he is exceptionally good at math); about 25 were classified as bombardiers, and the rest (between 180 and 170) are pilots." To Royce's great satisfaction, he was listed for pilot training, though because of their quarantine those of his squadron were not scheduled to begin active course training until the following month.

The possibility of "washing out," or being eliminated from flight training, was a constant concern, as Royce explained in a letter: "They told us they want everyone to learn to fly but if we are ever eliminated from 'air crew' training it will be for one reason only: to save our lives. They told us that we are not being prepared to die, but to live for our country, become efficient aircrew men, help win the war and to return once again well and strong to a country we have helped keep free."[4] The most common reason for "washing out" at this early stage was failure to pass the physical examinations, but there were other reasons as well. "There is just one punishment arround here for any insubordination and general misbehavior. The offending cadet is grounded or 'washed out' as they say here. He is broken to the rank of private and sent to 'Minter Field.' They call this the field of lost souls because all that fail are sent there. If I ever get sent there it will be because I just cannot fly, and I'll never know about that untill I try."[5]

After nearly a month at Santa Ana the squadron had a Sunday free on December 20, and the Latter-day Saint soldiers, still in quarantine, determined to hold a sacrament meeting. "We got hold of an unused building here at the post and held our services in it. We had to sit on the floor and it was cold, but I have never been to a meeting that I enjoyed more, not even in the mission field. It was something we needed, and we really got a lot out of it. We are going to continue having them whenever possible."[6] For Christmas the servicemen were allowed a three-day furlough, but one unlucky soldier had to be chosen to stay behind and watch the barracks. When an attempt by the Utah soldiers to charter an aircraft home failed, Royce volunteered to watch duty, and spent a quiet Christmas day in the barracks alone. The loneliness was broken by a kind gesture from one of the officers. "Our commanding officer had two of us go to Santa Ana to dinner today with a family there that wanted two cadets. They really fed us well and treated us as if we belonged there, but it still wasn't like home, though it was nice to get off the post for the first time."[7] The following Sunday, December 27, was Royce's birthday, and Royce called Pearl on the phone, and for the first time since he had reported for duty was able to talk with her directly.[8]

The next day, on December 28, the squadron finally received the welcome news: "We are now officially out of quarantine. That means that we have the run of the post and that we get weekend passes every week. It's just like getting out of prison. We can visit any of the squadrons out of the quarantine, go to shows on the post (they only cost 15¢), and everything. Also we will be able to go into Santa Ana to church every Sunday. All we have to do now is sign our names in a book and go where we wish in our free time."[9] The next weekend, the first in the new year of 1943, Royce took advantage of his newfound freedom and spent his first leave visiting his old missionary haunts in Los Angeles. "I went back and saw a lot of the members and the Muñoz family that we rented our house from there. They seemed to be glad to see me, though I guess it was quite a contrast to see me in uniform after knowing me so long as a missionary. I went down to the chapel to see how the plants and trees we put in were coming. They really look fine now, and it appears that they are taking good care of them all. A little light green hedge that another elder and I put in with our bare hands looks especially pretty."[10]

The new year brought other changes as well. Out of quarantine at last, on January 5 Royce and the other cadets who had been selected for pilot training moved into new barracks and became a part of Squadron 57, enrolled in pre-flight school. Royce described the new living situation: "Over here we have upper classmen, men that have been here three weeks now. We have to sleep in the upper bunks, and the upper classmen in the lowers. We are required to do the cleaning. There is one beauty to it. We are only lower classmen for about three weeks, then we take the lower bunks and some new fellows have to do the work. This goes on until you are ready for primary. We have had to use our footlockers for desks before; here we have tables and chairs."[11]

Pre-flight training presented an arduous course of study for the cadets. "They try to cram into you, in from about six to nine weeks just as much math as it takes them two years to get through you in school. Then they have physics and meteorology and code and naval identification and aircraft identification which includes being able to identify all of the planes of any of the warring nations, friend and foe. I wonder if I have brains enough to get all of this thru my head."[12] These classes were of necessity very fast-paced, and the course material was condensed so it could be learned in as short a time as possible. "The story goes down here that a cadet dropped his pencil in math class, and while he stooped to retrieve it he lost two quarters' work of college algebra."[13] Standards were exacting, and testing scores had to exceed 90% on average for the candidates to qualify for flight training. The stakes were raised when the Air Corps, now with plenty of recruits, made a change in policy that gave cadets only one chance at flight officer training. "I found out today that I have just got to make the grade as a pilot if I want to be a member of an air crew, because I will not have the chance to come back and try for bombardier or navigator if I should wash out. I'll do the very best I can, and if I can or cannot fly they will let me know."[14]

The Sunday after their arrival in the new squadron, Royce was able to attend church in Santa Ana with ten of his group who were Latter-day Saints, and was surprised to see that a number of commissioned officers were also present—church services put them all on an unaccustomed equal footing. The soldiers were all invited to sit on the stand, and one of the officers, a captain, drove them back to the base after the services. Later that same week Royce received a surprise visit from his former missionary companion and close friend Reid Ellsworth, who himself had "washed out" of pilot training and was then completing preflight officer's training as a navigator, and was soon to leave the base. The two spent as long as they could talking about old times and new hopes, and Reid left just in time to make it back to his own barracks before Taps.[15]

As classes got underway, Royce studied hard, and found that he excelled at most of the subjects. The first round of classes included mathematics, including repetition of some concepts Royce had learned in college as well as others that were new to him; "code," which involved learning Morse Code and receiving and recording messages at given rates; naval identification, which required cadets to identify the various classes of ships in the United States and Japanese navy on sight; and maps and charts, which included elementary principles of navigation. When they were done with maps and charts and naval identification, a physics course was added, together with military hygiene and aircraft identification, which included memorizing the length, wingspan, and appearance of airplanes from the various combatant nations. Study for these courses was intense, and occupied most of the cadets' free time. Royce found that he performed well in the class work, scoring grades higher than those needed to qualify. He did well in the naval and aircraft identification classes, and excelled in code beyond his expectations, though on one occasion he described being marked off. "I was taking code O.K. untill a couple of planes or rather a couple of dozen began dog fighting. There were some P-38s and Grummans going at it. We just sat in there and watched. He gave us a 'B' for our behavior in class because of that."[16]

The classes were interspersed with marching drills and other physical activities, and occasionally a work detail. After one of these, Royce, accustomed to heavy farm labor, wrote, "We spent all morning today on a pick and shovel detail at the wing headquarters. I've got a few calouses and blisters now. These work details in the army have only one rival for having a lot of men do such a little amount of work, and that's the W.P.A. However I shoveled all the dirt I wanted to this morning, and when they said quit and go eat I wasn't a bit reluctant."[17] There was rudimentary combat training, including firing machine guns, submachine guns and rifles at a beach range, skeet shooting, and practice in the newly introduced art of judo, in preparation for hand-to-hand fighting.

The squadrons also rotated through kitchen duty, which replaced an entire day of course work and drills with the menial tasks associated with the mess hall. This was hard work, though in itself it was sometimes an adventure. "We had to serve six meals in the course of the day and there wasn't any water all morning. Someone ran into a fire plug and broke the watermain. We really had to keep on the ball to get food on the tables and dishes to put it in. The mess sargent was pleased with our work because he reccomended us all for merits and told us he liked the cheerful way we did an unpleasant job that the army would make us do whether we liked it or not."[18] The "merits" thus given helped to cancel out demerits, or "giggs," which the men received for minor infractions such as having their beds improperly made, or their living spaces untidy. Enough "giggs" and the soldiers had to perform feared weekly walks at attention, an hour for each "gigg" over eight during the week. Inspections were rigorous and Royce received his share of demerits initially, but he quickly became wise to the military's attention to detail and, while some soldiers took the walk almost weekly, he never did.

Calisthenics and physical exertion were a routine part of each day. Royce did not shy from such work, and the physical exertion began to pay off. "Today in the 3/4 mile run we made I came in first by a good twenty paces in our group. I was feeling fresh as could be then. I guess we are getting a bit hard."[19] The hardening of both bodies and discipline were of course part of the principle objectives of basic training. By the time he completed his term in Santa Ana, Royce would write, "We had our final exam in calisthenics today. You are supposed to do better than you did when you began the course here at pre-flight; of course, we were already in pretty good shape then. Today, I chinned myself sixteen times, did fifty-two legg lifts, thirty-five pushups and jumped eight feet six inches. I improved on everything and knocked a whole second off the time of each one of my runs. All in all, it was pretty good. I guess we are better physical specimens than when we got in this outfit."[20]

At the beginning of February, the bunk mates from the upper class completed their training, and Royce wrote, "Today our upperclassmen bowed out as officers and gave their insignia etc. to their successors. Well, as soon as this had transpired and 'fall out' was given, our upper classmen got the cadet captain, first sergeant, flight Lt's, and Sergeants and put them in the showers 'O.D.' [on duty] uniforms and all. There was really quite a riot for a little while and some of the 'long-face order' boys didn't like it very much but they couldn't do a single thing about it so they took their wettings. We of 'I' class just stood arround at points of vantage and enjoyed the show."[21] That day was pay day, and Royce made note of another common form of military entertainment. "You should see the money fly arround here in card games. As soon as they got payed, some of them started games and they are still going on, every spare moment they get. I guess some of them don't care. They haven't any wives or folks to worry about so they figure on gambling away at least part of it, or if they are lucky, and they kind of take turns at that, winning the other boys' money."[22]

Royce and his fellow upperclassmen moved their beds to the lower bunks and made room for the new crop of underclassmen, who arrived a few days later. Most of these were volunteers from the regular army, some of whom had already seen combat in the South Pacific. Although Royce considered them a "good group of fellows," he also noted that those who had been foot soldiers a long while tended to have picked up coarser habits. "It seems like the longer they have been in the regular army the oftener and viler they can swear and the more gutter low their minds get."[23]

During the time Royce was in Santa Ana, he and Pearl developed a pattern of letter writing which would continue throughout the war. Pearl wrote almost daily, making Royce the envy of his bunkmates. In turn, he wrote her every two or three days as his training schedule would permit, though as training progressed, his writing became almost daily as well. Royce sent most of his pay home for Pearl's maintenance, instructing her each time how much tithing to pay on the salary. Shortly after the start of course work he wrote her the happy news that his friend Frank Denstad was the father of a baby girl. "They announced it in front of the whole squadron and had him step out and take a bow. He went red as could be, but he liked it I know. He is certainly happy to know that everything went OK. I'm sure that it will be that way with my little wife too when the time comes for little 'Hoimon' to make his debut."[24] "Hoimon" (Herman) was the name jokingly used by Royce to refer to their own unborn child, and in his letters to Pearl he offered her frequent advice and numerous exhortations to take care of herself and follow the doctor's instructions exactly. In a light-hearted way Royce hoped for a son, and he teased Frank a little for fathering a girl, though he reassured Pearl, perhaps less than convincingly, "Certainly it will be all right if it's a girl. Don't ever think that I'd be dissapointed; honest, honey, I wouldn't. We will take what we get and like it."[25]

As Pearl's due date in March drew nearer, the two discussed possible names for the baby in their letters, settling on Jean if it was a girl and David Royce if a boy, and Royce suggested that Pearl consider moving in with his parents after the baby was born so as not to overburden her own mother, since her older sister Jeanette had a baby due near the same time. As a prospective father, Royce gave voice to the mingled anticipation and disappointment felt by many then in uniform. "As for being thrilled about 'Hoimon,' you guess. How does any man feel to know he is about to be a father? I'm sure that he or she whichever it is will be just what we want, and that some day we will have reason to be proud of our 'war' baby. I'd be there if I could, but there's a war still going on and I'm in uniform for the duration. I'm kind of proud to be able to do that too. It's a priviledge to stand for something and fight for something worth fighting for."[26]

The remaining time in Santa Ana passed quickly, and Royce's group of upper classmen of Squadron 57 completed their course work and made ready to move out for flight training. On March 5, 1943, Royce learned of his new assignment. "We leave Wednesday the eleventh and are supposed to go to Santa Maria. It is about 150 miles north of L.A. We all go together. That is where 'H' class of this squadron went." He added with a note of concern, "About 1/4 of them have washed out all ready. I hope we have better luck than they."[27] On Saturday, March 8, Royce was issued his "dog tags" which included his name, his serial number (19073779), his blood type (AB) and his wife's name and address; this among other things served as a durable form of identification in case he was killed in action.

That Sunday morning, free once again from quarantine and other duties, Royce and Frank Denstad traveled one last time to the chapel in Santa Ana, and the bishop and a captain from the post drove them to a special "young people's convention" in Long Beach. "All in all it was just something that I really needed. Several young people who have been converted to the church talked and bore strong testimonies. I just had to get up and express myself too. I really went away from there feeling happy. After that the Bishop took us out to his place and gave us a good lunch, then took us back to the base, getting us there just in time to keep us from being A.W.O.L."[28] That evening Royce and Frank went to the telephone exchange to call their wives and notify them of their new assignment. Arriving at 6:00 p.m., Royce was unable to get his call through until 9:40 p.m., and he and Frank barely made it back to the barracks for bed check.

The next day, after a marching drill, the squadron attended an informal talk by cadets who had just completed primary pilot training. "They really boosted it, and told us the things we really wanted to know about it, namely, if you really want to fly and do your darndest and are physically able they just can't stop you."[29] The next day they packed their bags and made ready to leave the barracks and Squadron 57. On Wednesday, March 10, 1942, Royce wrote, "Well, today is the big day. A medical exam this morning and more chasing arround. We left the air base (without any tears or regrets) about 3:30 and left Santa Ana about 5:30 P.M."[30] The group did not arrive in Santa Maria until the next day, and Royce, exhausted, wrote of the trip, "I do hope that I never have a more miserable train ride. We didn't get here untill 7:30 this morning. The cars were of about 1918 vintage or earlier and overcrowded. The distance is only about 185 miles so we must have averaged about 13 miles per hour. Spent a good bit of time just parked on railroad sidings. I slept some, but not much."[31]

Santa Maria—Primary Pilot School

Hancock Field was one of many flying schools in which pilot candidates were trained. It housed approximately 500 cadets, as opposed to more than 25,000 in Santa Ana, and the future pilots enjoyed special treatment. After the austere and crowded conditions at pre-flight school, the air base at Santa Maria was a welcome change, and Royce described the new surroundings. "I'm going to college again. You see, all the primary schools are privately owned and operated in conjunction with the pilot training program of the army. This is the 'Hancock College of Aeronautics,' the oldest primary school, and believe me it's all right. The weather is perfect with clear blue skies with just enough clouds to break the monotony. There is an incessant drone of airplane engines because we are only a few hundred yards from the take off field and the cadets are taking off and landing constantly. We take our first flight the day after tomorrow, Saturday. This will be the proverbial 'dollar ride.' The food is excellent. Honestly, I've never had the pleasure of eating food that was much better. There are girls to wait on the tables and we don't have a thing to do with the food except the eating of the same. The grounds are simply beautiful with shrubs and flowers which are now blooming, plus nice green grass everywhere. After so long at Santa Ana this seems like a bit of Heaven."[32] The air base was such a pleasant change, and the anticipation of flight training so inviting, that the obligatory two week quarantine did not seem like much of a burden. Royce and Frank Denstad were assigned to the same squadron, and were housed in opposite wings of the same building.

The next day, Friday, March 12, 1943, marked a key event in Royce's life. "This noon I was sitting at mess enjoying an excellent meal when they called us to attention and gave a few announcements. They called out my name, and said that a telegram awaited me in the orderly room. I tried not to be too obviously in a hurry, but I got right up and went there as quickly as I could. There were five words that really took a heap of a load off my mind: 'Pearl and baby are fine.' I don't think I have ever recieved a more wonderful message than that." The telegram was from Pearl's mother, Jane Davidson, and informed him that baby Jean had been born at 5:30 that morning. Royce was elated, and word quickly spread among his fellow cadets. Royce took no time at all reconciling himself to having a daughter rather than a son. "I knew all the time that the baby would come all right, and I knew it would be a girl. I'm not dissappointed; I couldn't be happier about it. I had to take back everything I said to Frank when his was born. Things are just perfect. Everything was timed just right."[33] Royce considered the timing of the birth especially auspicious, since Pearl's impending confinement had never been far from his thoughts, and he had worried that his concern over Pearl would distract him from the immediate task of learning to fly, which was by then an overwhelming ambition. "It does take a load off my mind. Things worked out just right for us, I believe. I won't have both that and flying to worry about at the same time."[34] The next day he added, "I've taken quite a ribbing from the fellows about the baby. It seems that everyone knew it within a half hour flat."[35]

129

On the day Jean was born, Royce was assigned a flight instructor. At that time the army not only used private flying schools for military pilot training, it also employed civilian flight instructors to provide the training, and Royce was one of five cadets assigned to a Mr. Robert A. Danneberg. That day they received their first instructions. "We went over the planes very carefully today and received our flying equipment, goggles, helmet, leather jacket, and gloves. It's all good equipment. They certainly don't waste any time getting us started. They told us that they are only payed to graduate us and not to wash us out, so not to worry about that. Mr. Danneberg told us to let him do all the worrying because that's what they are payed to do."[36]

The next day, on Saturday, March 13, Royce took his much-anticipated "dollar ride," and got his first taste of flying. "It was wonderful. I didn't get sick, and I didn't get scared. It seemed quite natural to be upstairs looking down, and everything looked so pretty. All the fields are green and growing and everything takes on the appearance of a huge checkerboard with some irregularities. He had me handle the stick just enough to make me realize how much I have to learn. The planes ride much easier than a car altho the air was just a bit rough. I'll really have to think and dream flying all the time in order to make it. If you havn't got it they don't make any bones about sending you away, although they do give you every chance in the world to make a go of it. When you aren't flying you have to be out on the flight line to help guide planes in. One man has to get on each wing and sort of guide it in. When you taxi the plane along you can't see straight ahead so you have to follow a lazy S pattern down the runway so you won't run into anything or anybody."[37] The aircraft used by the cadets were Stearman trainers, easily-handled biplanes with a forward cockpit for the instructor and a rear one in which the trainee learned the rudiments of flying through a steering mechanism interlinked between the two cockpits. As novice pilots in training, Royce and his fellow cadets were given the uncomplimentary title of "gophers," a term used for the underclassman pilot trainees.

On Sunday there was no flying scheduled and, although confined by quarantine to the base, Royce was feeling on top of the world—he had just become a father for the first time, had begun training as an Air Corps pilot, and was now living the privileged life of an air cadet in an idyllic setting. He spent the day happily, writing letters to his family and to Pearl, and that evening had an unexpected treat. "Dinner was good today, and after it was over I was doing a bit of studying when they called me over to the phone. It was Pearl. Oh, it was wonderful to hear her voice and hear her say that she and the baby are all right. I'd give anything in the world just to see her. She cried before we finished talking. I know how she must feel, but the day will come when we can be together all the time. She's been wonderful about everything and we do have lots more to be thankful for than we do to be sorry about."[38]

Royce had to take the call in the presence of other cadets in the Officer of the Day's office, and this did not afford him the privacy to express his feelings to Pearl, particularly upon hearing her cry. After hanging up he sat down and wrote her a second letter, in which he formulated plans for her to travel to Santa Maria after her recovery, though this was based on many contingencies. "If you get well quickly and the baby comes along O.K. and the weather is good and I make out all right here and the Doctor and you think it's wise, I'm going to have you come down here for a weekend or two in about eight weeks. It would be all right here. The town is within walking distance, and I could bring you up on the post on Sunday. I can get off from Saturday evening untill 6: P.M. Sunday nite. Maybe I could pull strings and get a bit more time on Saturday. It would be a good vacation for you. It's wonderful in California at that time. Of course if you come you are going to travel first class both ways and have everything you might need."[39] Royce anticipated an extended stay in Santa Maria and, longing to see his wife and meet his new-born daughter, felt hopeful and certain that this would be possible before long.

The next week brought flight training in earnest. Though the cadets attended ground school to learn the fundamentals of flight, and continued to participate in group calisthenics to stay in shape, they all looked forward to the time when each would get his turn in the air. Royce recorded details of each flying session in his journal and in his letters home. On Monday he wrote, "Flying again in the afternoon. I have never enjoyed myself more in the same amount of time in my life. I am really beginning to get the feel of things. We practiced turns and did a few stalls, power on and power off. I followed thru and did them all after him as best I could." On Tuesday he continued, "He let me, or rather had me practically take it off. I think I'm getting the hang of that O.K. When we got up there we did a few turns and then went clear out to the beach and did stalls, one right after the other. I really enjoyed it but he made me work. I practically landed it too, but we would have ground looped if he hadn't been sharp after the landing."[40] He described to Pearl how the stalls were performed. "You put the plane into a climb and let it pull untill it will pull no further, just like a car going up a hill until the motor won't pull it. Well you hold it up until it starts to stutter and the stick is loose in your hand, then you push the stick forward (that lowers the nose) and she dives; at the same time you open the throttle wide open to swoop out of it. I don't know how to describe it, but it has all the thrill of a good

131

steep ski hill and a roller coaster combined."[41] At the end of each journal entry Royce carefully recorded his cumulative time in the air, as total flying time was part of the requirements for advancement to the next level of training.

On Wednesday the planes were grounded on account of rain, but the cadets flew again on Thursday, and that was when the first signs of trouble began to show in Royce's training. "A beautiful clear morning, just right for flying. I flew fourth and really, I didn't do anything right today. I started out by leaving the goggles up on top of my head. When we had climbed to about 600 ft, he told me to put them on. I was wondering why I was feeling the wind on my face so much. I did some lousy stalls today. I stalled the ship and then let it settle before I dived it to recover. The spins we did were fun. Outside of all this and being jerky on the controls I believe things were all right. Anyway, I did better when we were landing than we have done before."[42] Royce attempted to correct his errors and carry on, but two days later more problems surfaced. "Of course, I had to make some blunders. When it was time to come in we were way over to one side of the field. Well, when the instructor told me to take the ship into the pattern I got into it all right, but I was going the wrong way and I knew it. He just bawled me out and turned her arround at 360 degrees and flew out of traffic, then we turned arround and came into traffic right. When we were landing I made another error and tried to land the ship about twenty feet in the air. Of course the instructor straightened things out then we landed right...I still forget some of the thousand and one things you have to do habitually without taking time to think."[43]

The following Monday, Royce's worst fears were confirmed. He wrote, "Well I'm going to be honest with myself. If I don't start getting things just a bit better I'm going to be one of the has-beens. The instructor told me I am a bit behind. I really messed things up right today. I didn't do hardly anything right. Well, I'll do my best and if that isn't good enough I'll have to see what other part of the war machine I can fit into. It will cut me pretty deep if I can't make a go of this."[44] Leaning on the faith he had acquired as a missionary, Royce had made his desire to succeed a matter of prayer, and in a letter to Pearl the same day he expressed the frustration felt by anyone who has watched a blessing denied them go to someone seemingly less deserving. "Sometimes Honey, I just about can't understand everything. You know I try to live right, Honey. Honest I do try to live the gospel above everything else, and yet I always have or seem to have something to worry about. I see fellows, dozens of them on all sides of us here, who are as immoral and filthy-minded as the day is long, and yet they breeze thru everything without any trouble at all. I do believe and trust in the things I have been taught and tried to teach, and yet at times I just can't help feeling a bit discouraged."[45]

Three days later in another letter Royce analyzed his difficulties with flying. "I was up for a whole hour today and I don't think I did much better than I did yesterday, and neither did my instructor. It is just like he tells me, I fly wonderfully on the ground in that I can describe every maneuver we have had perfectly, etc. but knowing them and doing them are two different things, and I fall kind of short on the doing part. I'm afraid I handle the plane a lot as if it were a farm tool. They kind of need gentle treatment in a certain way." Continuing his letter, he tried to reconcile himself with what now seemed inevitable. "I'd rather not be a flier than just a mediocre one. They are those of whom you read as 'missing in action.' If you are below average in learning, you will be below average all the way thru and the man you meet in combat is average, so anyone barring luck etc. can figure out the result. I'm doing my best, I mean that, and I'm not through untill they tell me I am. I'll keep trying as hard as I can and I'll pray as I have been doing. I know that is what is making me feel all right about facing the facts."[46]

That Saturday Royce went on what proved to be his last flight with Mr. Danneberg, and after the ride action was formally taken, described by Royce in a letter the following day. "We came back from our usual ride yesterday and after landing the ship my instructor told me that he wanted to talk to me. We went over to the side and talked it over. He told me that he was putting me up for a check ride Monday. It's because of insufficient progress. I told him that I respected his judgement in the matter because he has been training students for quite some time and he should know when one isn't up to par. It doesn't mean that I am eliminated yet. If I pass the check tomorrow and the next day, I will be given another instructor and another chance. I hope I pass that check tomorrow, really I do. I do want to fly if it is possible. In event that I don't, I'll just have to make the best of it." At that time Royce had been in flight training for just over two weeks, and had logged six hours and forty-four minutes flying time.

The check rides were to begin Monday, and that Sunday, still confined to base by quarantine and weighed down by his failure with flight training, Royce put in a call to Pearl in the morning. Although the telephone connection was poor, he was relieved to hear her reassure him that everything was all right whether or not he succeeded with pilot training, and as an extra bonus he was able to hear his daughter Jean crying in the background. That afternoon he wrote Pearl a long meandering letter, which reflected the unsettled mood of a Sunday afternoon confined to the base with his entire future hanging in the balance. "I think you can always very easily tell the conditions under which I write letters. Today there is a poker game going on in one corner of the room, and in the other there is a radio playing some extra lousy jitterbug music. It's the kind that I really detest. It is very difficult to write this way. There is one fellow here that averages aprox. $5.00 every day playing poker. He really makes money at it. Of course, someone always has to lose that same money, but they nearly always come back for more. It's illegal to gamble so to do it they use chips, buying so many and then cashing them in after the game is over." He made mention of an event the night before, involving one of the cadets who had a girlfriend pregnant out of wedlock. "Last nite one of the boys went over the fence to get himself married off. He tried to get the commandant of cadets to let him get off just long enough to do that but he wouldn't let him. So he did up his bed and went over the fence, coming back about 5:00 a.m. this morning. He got away with it too. None of the fellows would ever tell on him."[47]

The flight elimination process involved two check rides on successive days: the first with a civilian pilot, and the second with an assigned military one. A candidate who passed either flight check was allowed to return to training with a different instructor. Royce's first check ride was scheduled for Monday, March 29, but before his turn came fog rolled into the airfield, and he was granted a reprieve as all flights were grounded. His civilian check ride was finally performed on Tuesday, March 30, 1943, and he recorded in his journal, "Check ride this morning and I flunked it hands down. I did one thing good and that was the spin. I came out of it swell. I didn't realize until Mr. Durden, the check rider, showed me today how many bad flying habits I had acquired. I know Mr. Danneberg couldn't make me learn them, but I think there would be a different story now, had he specifically pointed them to me as I began picking them up. I can't cry about it now, it's too late I'm afraid. I'll capitalize on what I learned today as much as possible and try to beat the army check tomorrow, then make the best of things."

The following day the army check ride took place. Royce wrote, "This day proves to be one of the so-called turning points in life. I went up with the army check rider and did not beat him. I guess I knew that I couldn't make it before. I just couldn't change my way of flying in one day, which I would have to have done. His main criticism was that I just did not feel my plane, and he was right."[48] This final check ride was a memorable one, and lasted an hour. "We took off and he told me to take the ship up to three thousand feet, to do a ninety degree turn to the right and then one to the left, a couple of power-on stalls and two power-off stalls, then a one turn spin to the right. I did everything as dirrected as best I could. He only interrupted me once, that was after my first power-off stall. He took the stick and did one himself and then said, 'is that what you were trying to do? Try it again.' That shows how lousily I did it. Well, I did everything as directed and then went in to enter the pattern of one of the auxilliary fields. I did that wrong. They changed the pattern and I entered at the wrong side. Well, about thirty or thirty-five minutes had gone by by then, and he said, 'let's go back to Hancock Field, I'll talk to you there.' I cut the throttle and shouted at him, asking him to give me a good ride with the trimmings. He tapped his head, which means 'O.K. I understand.' Now, that Lieutenant can really fly. He really gave me some acrobatics. The first time he did a snap roll, I lost the stick when he pushed it up as far as it would go to the right. Where it went after that, I do not know. I certainly did get a kick out of it. I just sat by and enjoyed it. Quite a bit of the time I found myself hanging by the safety belt looking at the world below me by bending my head way back. Oh, I do love to fly. I can see why I didn't make it, but it doesn't make me like it any less."[49]

That day Royce broke the news to Pearl in a letter. "Well, it's happened. I was eliminated from further pilot training today. I failed on a civilian check ride yesterday and on the army today. I flew as good as I have ever flown but it just isn't right. I guess I just have to say I can't learn to do it the Army way. I knew I could fly, but I can see that I am not up to par. It's kind of a relief in a way, Honey. It relieved me in the way that one is relieved when a very dear one passes away after you have known that it was inevitable. I wish that I might have gone ahead and made it, but aparently it wasn't meant that way. I have prayed about it every day, asking that my Father in Heaven bless and help me to do right and realize my righteous desires and that his will be done in all things. Maybe it is good for us from time to time to be denied something that we desire a great deal. I feel confident that I shall be guided and helped to do that which I really should do. It hurts a great deal to leave those whom I have been with for such a long time, seeing them going on doing that which I desired so much to do. Maybe I wanted to and tried to do it too much. The thing is right now that I again find myself at the crossroads, knowing not as yet which way I shall take. I can take this and still hold my head up anyway. If I couldn't take this and all the rest of the knocks and jolts I yet have to take I'd never amount to anything would I?" Royce called to mind what the cadets had been told about "washing out" back in Santa Ana. "I want you to remember that they eliminated me to save my darn hide. I would never have quit. They either had to eliminate me or let me kill myself, which I could very easily have done, the way I was going. Remember that won't you? I think we would both rather that I go on living."[50]

With his check rides behind him, Royce's fate was sealed, and all that remained were the official procedures. On Thursday, April 1, he appeared before his immediate commanding officer, an encounter which he described with an edge of bitterness. "I went in and saw Captain Bishop this morning. He should have a record made of his little routine speech to 'wash outs.' He started as always with 'We can't always be pilots' etc. etc. I felt worse after leaving there than I did when I went in."[51] The next day was the final step which made his elimination from pilot training official. "I went before the board this morning for the formalities. They commended me on my good academic record, and told me I was recommended for further air crew training of some other type. I said 'yes, sir' and 'no, sir' and 'thank you, sir.' It was all over in about three minutes or so." Realizing that his hopes of bringing Pearl to Santa Maria were now dashed, Royce sought other means to see her. "I asked about a furlough, but they arn't authorized to give furloughs at this post except in case of an emergency of one kind or another. I explained to Captain Bishop what the situation was, and he said that I should ask about it as soon as I get to my new post, and that I stand a very good chance of getting one. Don't count on it too much; I daresen't."[52]

That same afternoon Royce happened to be assigned to duty minding the phone in the operations office at Hancock field, and as he wrote his letter to Pearl, he was able for the first time to quietly observe the activities of the officers as they congregated there. "Most of them aren't more than a year or so older than I am. It's interesting to sit in and hear the officers talk about the cadets and other things in place of hearing the cadets talk about the officers etc. It's about the same line of 'gab' either way. The officers are just boys that have had a bit more experience than the cadets. They think just about down the same alleys and act just about as silly when they are away from other people among themselves." Royce's realized that by now own prospects of becoming an officer had become very slim. "They said that I had an excellent academic record, and as I had requested I was recommended for further 'air crew' training (which I doubt if I will get). If Santa Ana says I can go back for training as a bombardier I want to go. The decision in that matter rests with them; all they can do here, which they have done, is recommend me." His elimination from flight training left Royce looking at the training from the outside in, and with his future completely uncertain, he wrote, "The fellows who started from scratch with me are soloing now. There were a good half dozen of my squadron who soloed today. Well, I can't eat out my heart on that. I didn't make it, so now for the next chapter in my life. I do hope that it proves to be more successful than this one just past has been. I'll get a really good break sometime, of that I'm sure. Perhaps in the long run this experience will prove very valuable to me. I hope so. I have worked hard these months to get up to where I was, and it isn't too heartening to suddenly find yourself with the props knocked out from under you, forced to start the whole climb over again. I guess I was fortunate to get as far as I did."[53]

Fresno—"Field of Lost Souls"

Having washed out of flight training, Royce was moved out of the squadron barracks into new quarters. Designated the "A" Barracks, these were nicer than the regular barracks, and were evidently intended to provide some measure of consolation for cadets who had "washed out" of pilot training. "They really put you up in style. There are two of us to the room. I'm with a fellow named Crawford, from California. I chose him because he does not smoke. We have two three-quarter beds, good springs and nice thick mattresses. There is a desk for each of us, a chest of drawers and a clothes closet apiece. We also have a basin with hot and cold water and a good mirror. The privacy we enjoy here is just something I haven't been used to in the army. I don't know why they ever called a private a 'private.' He certainly does not ever have any privacy, and a cadet is no different."[54]

That Saturday the cadets were finally out of quarantine and their weekend passes became valid, so Royce was at last permitted to leave the base. He explored the town with Frank Denstad and a couple of other soldiers. On Sunday morning he accompanied a group of cadets to the local chapel for Sunday School services and, arriving early, he was invited over to the bishop's home for the morning Tabernacle Choir broadcast[55] prior to the meetings. The cadets swelled the local congregation considerably. In the afternoon Royce again called Pearl and found her feeling quite "blue" at having lost the chance to come and visit him. He wrote her a letter after hanging up, encouraging her and assuring her that they would work out something "just as soon as it is possible." He asked her to hold off having the baby blessed, in hopes of getting enough time off to perform the blessing himself.[56] He was also restless to move on from Santa Maria. "I'll be glad when they move me out of here now. This old life of just passing the time is all right, but I want to be up and doing. It isn't any fun seeing others solo and go on while all I can do is watch from the ground."[57] That evening, having nothing else to do, he returned to the chapel for evening church services, arriving back at the base just in time for Taps.

The next morning Royce's hope of becoming an Air Corps officer at last came to an end for good. "Our moving orders came today. We were told to report to Captain Bishop at 2:00 P.M. and when we got there we were told that we had been relieved of our status as aviation cadets and were to leave the next morning for Fresno, Calif. It kind

137

of hurts not being able to go back to try for Bombardiering or the likes. We spent the afternoon getting our clearance papers fixed up. After that, we were given passes. I went to town only long enough to get me some things I needed, then came back to spend a bit of time telling the boys over in Squadron "B" goodby. That hurts the most."[58] His letter to Pearl that day began, "Well Honey, I guess I am at the bottom of the ladder in the U.S. Army, or at least I will be there by the time you get this letter. After tomorrow I'll be 'Private' instead of 'Air Cadet.' It hurts my pride to know that I have failed at something that I wanted to do so much. I leave in the morning for Fresno from where I shall probably go to a basic camp for enlisted men for the present. I guess you don't feel to good about it either, but let's don't worry about it. I know that there are better men that have borne the name of 'Private' than I, so I'll just have to make the best of it."[59]

It was a cloudy, dreary morning in Santa Maria when Royce and the other eliminated cadets pulled out in a truck for the bus station from which they would depart for Fresno, located in the southern half of California's great Central Valley. They left the Santa Maria station at 10:15 in the morning, but their progress was slowed by many delays and a near-accident; it was not until 10:30 p.m. that they finally pulled into Fresno and, after calling the new base by phone, spent the night at a Hotel. The next day, Wednesday, April 7, 1943, Royce reported at the base and, accustomed to the special treatment afforded the air cadets, was dismayed at what he saw. "We went out to the base at the Fairgrounds. It was really different than what I thought. It is an old jap camp, and the huts are really sad, tar paper shacks. My bags weren't here with us; there was a slip-up. We were assigned the 804th training group, Flight 203 and Barracks 183. The food is really a far cry to what we had at Santa Maria, but it will keep us going."[60]

The makeshift base in Fresno had served temporarily as a Japanese relocation camp, but by the time Royce arrived it had been converted into a base known as the Fresno Army Air Forces Training Center, whose main purpose was to train soldiers for support assignments in the Air Corps, though it also served as a temporary way station for former air cadets who, like Royce, were undergoing reassignment. On Thursday Royce had a routine physical inspection, and the following day, Friday, April 9, 1943, he underwent testing and interviews which would help to determine his ultimate assignment. Though this was a pivotal time in his military career, the pace was less strenuous than what he was accustomed to, which served as something of a rest and gave him time to reflect. That day he wrote to Pearl, "I feel kind of free for a change. Before, there were always worries and pressing matters, and gee, I just havn't any right now. This life here isn't bad for a while, though I will be glad to leave it even though it does mean I am about to prepare myself for the real thing, combat. I'd hate to have to tell my kids that I spent the war just moving arround preparing, though. I guess deep down inside me I do want action, in spite of what it brings with it. Most of the fellows feel just about the same way. I know that I could almost spend the war training for one thing after another, never actually doing the real thing. I also know you wouldn't, and neither would I, want it to be that way in spite of the inherant selfishness that seems to be a part of nearly everyone's make-up."

He described the selection activities he had already been through that day. "I took the general 'G.I.' intelligence test and the mechanical aptitude test this morning. I did well in them. I am qualified for most anything that is open. In the afternoon, we went in for interviews. They told us first of all we would be aerial gunners. They want us for that because they know we are physically and mentally adapted and qualified for flying as pilots. That means we will make good gunners too. Anyway, in addition, we were permitted to choose one of three schools which are open at present. They are armament, aircraft mechanics and radio. I chose radio. I havn't met the board yet, but I think that is what it will be. Finishing both courses would leave me with a sergeant's rating and gunners wings as an aerial gunner and radio man. I'll still be flying when I get all through and drawing good money, as much as I would clear were I to be a second lieutenant on the ground. Of course, as I said before it probably means combat, but that's part of war, and I'll be just as safe as I would have been had I made pilot." Royce also took some consolation at having been joined by other former cadets from his old squadron. "Another group of my friends from Santa Maria arrived this morning. There are dozens of fellows arround here that I know."[61]

Despite the slower pace, the routine of military life continued unabated while Royce awaited his new assignment, with daily calisthenics and obstacle courses, work details, marching drills, and on a couple of occasions, propaganda films. That Saturday and Sunday he was given a weekend pass and, taking a Saturday evening trip into town, he had a dinner of oysters which made him immediately ill. Returning to the base, he was still too sick Sunday morning to go to Sunday School, though by the evening he was well enough to attend church services at the local chapel. Noting the slow pace at the Fresno base and the ease with which passes were given, Royce saw his opportunity to finally see his wife and his newborn child. He wrote a letter to Pearl instructing her to make arrangements to travel to Fresno the following week to meet him there. However, a short time later he was forced to send a telegram indicating a change of plans, for reasons he explained in a letter. "I am sending a telegram today, I hope it does not alarm you. I just want to cancel asking you to come down here. Yesterday they cancelled all three day passes and all furloughs until further notice from the classification board. Then, a couple of the boys got orders to move out, they don't know where to. We are all going out in a matter of days they say. You never know to where until you get there. If we go through Salt Lake when I go, I'll let you know some way so I can at least see you a minute. Don't ever count on anything, though. I just about let myself figure on seeing you next week, but that's out now."[62]

That Friday Royce found his name on the "availability list," meaning that he was to leave soon, and was confined to the post. The next morning, Saturday, April 17, 1943, orders came. Royce had his final physical examination and made ready to pull out, though the destination was unknown, save for the code number "601." Royce called Pearl to advise her not to try to meet him in Salt Lake, as he had no idea where he was heading, or by what route. The soldiers waited until after dark, when at last their train pulled away from the station, and the next day, as daylight illuminated the surroundings, Royce wrote bitterly to Pearl, "Don't believe anyone could feel much worse than I. I'm on my way home and yet I don't believe I'll get to see you at all. Right now I am traveling on the Southern Pacific R.R. just past Carlin, Nevada. We are about four hours from Utah and it's about 7:30 P.M. I don't think we are going to reach Salt Lake at all, and we shall probably be in Ogden bright and early Monday morning, and I can't get a telegram to you or anything. I didn't know what dirrection we were going until I woke up this morning and found us traveling thru the Sierras."[63] In Ogden, Utah, there was a brief stop as the soldiers were transferred from Southern Pacific to the Union Pacific line, and another in Nebraska where they switched again to the Burlington railroad. All told, the journey took four days, stopping daily just long enough for the soldiers to exit and perform their daily calisthenics, until at last they arrived at destination "601," which proved to be Chicago, Illinois.

Army Air Force Technical School—Chicago

Arriving at Chicago on April 21, 1943, Royce was assigned quarters in the massive Stevens Hotel,[64] reputed to be the largest hotel in the world, and one of two Chicago hotels pressed into service as army barracks after the entry of the U.S. into the war, to accommodate the large numbers of men then in technical training. Royce, who was assigned to the 15th floor, had never stayed in a luxury hotel before, and was dazzled by the surroundings. "You should see the set up that we have at this place. Really, it's wonderful. I never expected to fall heir to such a good thing. This hotel was modern and well-equipped throughout. They put four of us in a large room. There is a bath and shower and everything else in a room adjoining. It's a private bath, just for the fellows in the room. There is a nice thick carpet on the floor. The only thing 'G.I.' is the bunks; all else is hotel stuff." During one leg of the trip from California the soldiers had been transported in a train with an old coal-burning locomotive, and they arrived covered in soot with no change of clothing, as their baggage had not yet arrived. Royce still did not hesitate to take advantage of the fine porcelain bath tub. "You should have seen the water after I bathed in it. It was just like after a hard days' work on the farm, only I felt lots dirtier. It took me quite a while to wash the soot out of my hair. It's good to have my body clean again, even if my clothes are not."[65] One difference Royce noticed immediately about Chicago was the absence of nightly blackouts, which out of wartime necessity had become a way of life in the areas on the west coast which he had occupied since his enlistment. "It's good to be in a city where the lights are on. Los Angeles and all the coast is dimmed out. This is bright and beautiful just as it was before we went to war. I guess the song 'When the lights go on again' really is appropriate."[66]

The first couple of days after Royce's arrival were occupied by lectures which he described as "not so good, but such things have to be tolerated I suppose." That Saturday his unit had their first marching drill, and Royce, accustomed to the college-educated air cadets he had lived with in California, was dismayed at the young draftees which now surrounded him. "My morale was really at a low ebb this morning. Drill was terrible. Those young boys in our outfit always act like a bunch of adolescents. I don't know whether it was a mistake to draft them or not. This morning I thought yes."[67] Later that same day, though, the classes in the technical school began, and Royce immediately began to feel more at home. Although reputed to be difficult, the class material was the kind of work to which he was accustomed, and one of the most feared subjects was code, at which he had already excelled in Santa Ana. Courses at the school were taken in shifts, and Royce's group began with evening shift classes, which did not end until 11:00 p.m. As a result, he could not get to bed before midnight, even though he and his bunk mates were expected to be up in the morning for calisthenics, and to keep their plush room spotless for the frequent inspections. Royce found that the course work was relatively easy for him and, as life became busier, his flagging morale quickly rose. His first Sunday was Easter Sunday, and he was able to attend church at the local ward.

With the disappointment of being unable to see Pearl and the baby still fresh in his mind, no sooner had Royce gotten settled in Chicago than he began to make plans for Pearl to travel there to visit him. He set about finding a place where she would stay, and began submitting requests to maximize the leave he would be granted once she arrived with the child. At the same time he was also giving consideration to seeking reinstatement in the Air Cadet Program, which he had learned was now a possibility. There was a natural hesitation in Royce's tone as he wrote about this to Pearl. "If I were to get it, it would mean going back to Santa Ana again, taking the psych. tests over again, and possibly being disappointed again. I don't know whether I would want to risk the old 'heart-break' trail again or not."[68]

Finally, on Thursday, May 13, the looked-for day came. Pearl's train arrived late, and Royce described their meeting. "I was out early this morning to go down to the station to meet Pearl. I had to wait quite a while, but I could never describe the thrill of seeing her coming toward me with the baby in her arms. It is a cute little thing too. All I had time to do was to get the baggage and get in a taxi. I had to get off at the hotel and leave them to go on alone. School just couldn't pass quickly enough but finally did and I went to her as soon as I could. It was so good to be with her again. Somehow it just made everything all right again." There was much to catch up on, and Royce and Pearl stayed up nearly the whole first night talking. The next day coincidentally marked the completion of a year of marriage, and the two slept in, then went out to visit an art gallery, but again spent most of the day talking. Royce wrote, "Our wedding aniversary, and the end of a year of happiness for me. Really it was more than I dared hope that we should spend this day together after being parted so long, but it is so—we are together and so very happy. I guess the baby is everything I ever hoped it would be, you see, I have never had one before and perhaps I do not know just how a baby really is, but I will learn I am sure. It is wonderful to reaquaint myself once again with my wife, and know that she is all right and happy. I hope that when another year moves round we can be together to stay. That is perhaps the dream of all, and it may come true. I know it will if wishing helps."[69]

During Pearl's stay in Chicago, Royce arranged as much leave as he could, spending every evening with her, with extended stays on the weekends. Never having cared for a child of his own, Royce described Jean to his parents. "The little tyke is just as good as she can be. She hardly ever cries, and never does without a good reason. Sometimes she lies for the longest time cooing and trying to talk and moving her little arms and leggs around just having a great time."[70] It took some time for Royce to get used to being a father, and Pearl wrote to her mother, "He was almost afraid to hold her for fear he might hold her wrong. I wish you could have heard her talking to him last night. She talked loud and long It was just as if she understood everything and was trying to tell him everything that had happened up to now. They are certainly happy over one another."[71]

Pearl and the baby stayed in Chicago for just over three weeks, living mainly in private homes. No record was kept of their activities together, which Royce only summarized in his journal at the conclusion of the visit. "It has been a month of joy and happiness because my wife and baby have been with me. I have become reaquainted with my wife and almost feel that I know my little daughter. She has been good, just as good as a little baby can be, and it has really been wonderful to have them here. The days past were rather hectic but they were happy days. Part of my time I was in school, part with them, and the rest getting permission to go be with them." As the day of Pearl's departure at last arrived, Royce wrote, "That awful day of parting. I knew it was coming but now it is here. We didn't think about it all day as we went about getting ready. Maybe I should say that we didn't talk of it, because it did weigh on me so that the thought was with me all the time." Royce and Pearl bought each other gifts, then made their way to the train station. "I got everything checked and ready and saw them on to the train, and stayed with them till we heard 'all aboard.' I then watched them through the window. A part of me went with them."[72]

A week after they had gone, Royce reflected sadly in a letter to Pearl, "Ever since we were married it has seemed like the time we spent together wasn't really ours, but rather borrowed time: time that we had hardy dared hope to be able to have together. I always felt that way about it. We have never once been together when we didn't have the prospect of an immediate parting always before us. Maybe I don't appreciate the things that really count as much as I should. For everything that isn't quite right in our lives there are several things that are right. I suppose that is what really counts. We have a lovely baby and we all have good health and none of us want for anything. Yes, things could be lots worse. I've about got myself talked into that now. I hope you feel likewise. Let's smile and be happy and tell ourselves that everything is all right."[73]

Shortly before Pearl's departure, Royce had learned that his days in the luxury hotel were numbered, as the Chicago technical schools were slated for closure by the army, and the soldiers would have to be shipped to other locations to complete their training. During his last week in Chicago Royce described his progress. "My final marks for this past three weeks were higher than they were in the first three weeks, so I'm doing all right. It hasn't been difficult for me either.

These things come to me quite easily. All we have been doing the last couple of days is trouble shooting. We leave our transmitters and the instructor removes or loosens some part or installs a faulty one, then we have to detect what is wrong and fix it up so it will function right."[74] This type of brain work involving methodical deduction and reasoning came much more readily to Royce than did the immediate decision making and manual skill of flight training. The last day of school in Chicago was Thursday, June 10, and Royce noted with some satisfaction that he was in the upper part of his class, and that he was now able to record code at the speed of 16 words per minute.

Having completed his Chicago coursework, Royce had some time free before traveling to his next assignment, and he took advantage of free tickets offered at the Service Club to see some professional plays in the downtown Chicago area. On Thursday he enjoyed a stage production of the comedy *You Can't Take It With You*, and Friday he attended *The Thirty-nine Steps*, a more dramatic play. That Saturday, perhaps still lonely, missing his wife, and surrounded by thousands of soldiers, many with loose moral standards, he recorded a close brush with temptation. "I don't know what possessed me to do it, but I went to a theater where they showed a show and then had a burlesque show. The show wasn't evil, but when it came time for the burlesque I could think of one thing, and that was Pearl. Finally, before it got underway I up and walked out and oh, I felt good. I'll never lead myself into that kind of temptation again. I'm glad I was strong enough to walk out." The next day, Sunday, he wrote "Made it up in time to get out to the University Ward to Sunday School. I'm glad that I was able. I need the church more than ever before."[75] He visited a museum and the aquarium before returning to his quarters at the hotel to pack up for departure to his next assignment, which he had learned was to be in Sioux Falls, South Dakota.

The next day was the departure date and after rising early, the soldiers spent most of the day waiting for transportation. Royce received a letter from Pearl and another from Frank Denstad, who was still in basic pilot training, and he wrote wistfully in his response to Pearl, "Frank sent me a letter which I received today. He was the second one of the class to solo. I was glad to hear from him, but it hurt a bit to hear how well all the boys are doing there and know of what might have been, but I guess I am over that all right now."[76] Later that day, he wrote, "We finally left in the afternoon. The train was dirty and hot. I got the upper berth for the night. Kind of hated to leave old Chicago, but I guess they will get along all right without us."[77]

Radio School—Sioux Falls, South Dakota

Royce arrived the next day, June 15, 1943, at the Army Air Forces

144

Technical School in Sioux Falls, South Dakota. The arriving soldiers went through what was by now the routine on arrival at a new camp. "We got the usual welcome speech by the chaplain telling us that he knew we would like this place. Of course the medical officer had his say about the prophylaxis stations also. Then we had to listen to the articles of war for about the 'nth' time. I slept thru most of that. We also were fed once. The mess hall is the dirtyest I have ever seen and the food just so-so."[78]

In the days while they awaited formal training, Royce and those who had come from Chicago with him spent much of their time assigned to labor on work details, which were designed to help maintain the military facility at Sioux Falls, but which were despised by the soldiers and avoided whenever possible. On Sunday Royce attended church at the local ward in Sioux Falls, one of eighteen soldiers from the post who were able to attend that week. Among them were several good friends from earlier training or from his mission days, and Royce also discovered a soldier from his own barracks, a Robert Smurthwaite from Baker, Oregon, who was a faithful Church member, and who became a friend during his stay there. The soldiers also met with the missionaries serving locally, and Royce was presented with an opportunity which carried him back to happier times, and which he explained in a letter to Pearl. "The missionaries have asked us to help them do missionary work on our days off. I'm certainly going to do it. We are to work as temporary short-term missionaries. They say that the mere fact that we are in uniform and with them is invaluable to the work they are doing. They are invited in more and find people more receptive and less given to ridiculing them for not being in the army." He added, "We went to a fireside tonite as well as church today, and it did me a world of good. I really had the opportunity to help in a good discussion, and it helps to keep me pepped up. I believe that's been one of the main things wrong with me."[79]

While in Sioux Falls Royce continued to attend church meetings, including firesides and Mutual, whenever his training schedule permitted, and he went out frequently with the missionaries. He considered this joint missionary activity to be highly beneficial, both for missionaries and soldiers. "If those few permanent missionaries weren't here we could do nothing, and they are also doing us a world of good outside that, in bringing us together and seeing that we keep contact with the Church and each other. The value of missionaries in the field during war time will never be questioned by me, I know that. These Elders here are really a good investment and they are doing a whole lot more in my eyes than they could ever do in the army."[80]

On the evening of Sunday, June 27, 1943, after nearly two weeks at Sioux Falls, the transferred soldiers at last began radio classes.

Training was received in shifts, and Royce's group had the night shift, attending classes starting at 6:00 p.m., not getting to bed until 2:00 a.m. the following morning. The technical school at Sioux Falls operated on a different teaching system than that in Chicago, and though it was rumored at first that the new arrivals would have to start over and complete the entire 18-week course, when they began classes, their prior training was taken into account, and the actual time proved to be considerably less. During the first week the soldiers were tested in receiving messages in Morse Code, and Royce passed at fourteen words per minute without difficulty, and within a couple of days he was up to sixteen, the best speed he had ever achieved at Chicago. He was nearly up to twenty by the time formal testing took place the following week. Shortly after that, Royce learned that he had "washed ahead" to another class, shortening his training period by three weeks.

For a while, the study routine prevented Royce from attending Sunday church meetings, but on Sunday, July 18 there was a change in schedule, and Royce and Bob Smurthwaite hurried down to the church where the sacrament meeting was already underway. Royce was asked to give the closing prayer, and as he came to the front he recognized an old friend from Squadron 18 in Santa Ana, Frank Van Limberg, who, like him, had washed out of pilot training, and who had also left a wife in Salt Lake. Together the LDS soldiers remained for a nighttime "fireside" which left Royce inspired and contemplative about his own life, which in a short time had seen many significant transitions. The next day Royce wrote, "I don't feel that I am ever any better than those I associate with here, mainly because of words which I don't believe I have forgotten since I first heard them: 'When much is given much is expected.' I have been given much, therefore my life must be different for me to be even as good as those with whom I associate, judging both them and myself by the sliding scale of 'knowledge received'."[81] These words, as much as anything Royce ever wrote, explain the motivating factor underlying most of his life's accomplishments. The desire to live up to the advantages and opportunities he had been given established a high standard which would increasingly motivate his activities and efforts in the years to come.

In a letter to his parents at this time, Royce described the demanding schedule his training required. "Reveille is at eight in the morning. We then have till ten to shave, etc., clean up the barracks and get breakfast. It takes every minute too. At ten, we have to be at code. We take that till 12:37 p.m., that's just about enough to send you mentally off the beam. (I have passed twenty words/min. and am working now on twenty-five which is indeed fast.) We then have till about 1:25 to eat and get back for mechanics class. (We spend most

146

of that time waiting in a chow line.) We then tune sets and trouble shoot on transmitters and receivers untill six p.m., at which time we have about ten minutes to fall out for calisthenics, which last till eight p.m. After showers etc. we go to eat again which usually takes till after nine. The rest of the day is free time. That's a bit of a laugh I guess. They have nerve enough to call it that. Taps are sounded and lights are out at twelve midnight. That's my day in the army at the present time. We are working with real equipment now too. So far, since we have been in upper division, we have been working only with high frequency sets—they are not like ordinary commercial sets. I guess they would call them 'short wave.' The higher the frequency, the shorter the wave. The things really work."[82]

On August 3, Royce received word that his application for Officer Candidate School had been approved, though by that time he knew enough about Army priorities not to place much hope in returning to flight training, despite the increase in income it could provide for him and Pearl. As he wrote Pearl the next day, "I'm quite happy right where I am in the work I have. If I can finish this and then gunnery school before they call me I would probably take it. I'm saying all this because I know it's a long wait before I ever get called for school, and if it involves staying here after I finish school, I don't think I want it that much. It's not that I just can't wait to get into combat, but I am about fed up on going to army schools and then not finishing them. This did my morale one big lot of good. It kind of helped my ego a bit or something, makes me feel that washing me out didn't flatten me out. Anyway, I might be an officer some day but don't count on it too much."[83]

By this time in the war effort the Army Air Corps had sufficient radio trainees in the pipeline to meet its immediate military needs, and the focus shifted from turning out radiomen in large numbers to increasing their readiness to serve. "Quality over numbers" became the new watchword, and standards began to be raised for radio operators, a change which was to affect Royce's class. The methods for evaluating them became more stringent; for example, code velocity, measured in words per minute, was now tested for a period lasting a full four minutes, a length which evaluated not just speed but stamina with a pencil as well. Those who failed to meet the rigid standards did not necessarily "wash out" as in pilot training; rather, they generally "washed back" to a previous level of training, giving them more time to master the necessary skills. In August, Royce's class was joined by a large proportion of a more advanced class which had "washed back" to swell its ranks. Royce, however, was confident he would succeed, as he was well ahead of his class in code, and had a high aptitude for the mechanical aspects of the training as well.[84]

147

Royce found the course work increasingly interesting as it progressed from the rudiments of radio operation to the actual procedures to be used in military action, which he described in a letter to Pearl. "We work most of the time on tactical procedure. They hook us up in nets and we send things back and forth using call signs and correct procedure. Some of it is quite complicated. There is a procedure to cover any situation or any type of message. There are certain letters which have assigned meanings used as prosigns which we have to learn. There are many types of messages and many different ways to handle the same message under varying conditions. Also, an accurate record must be kept of every transmission that comes over the frequency you are assigned to watch. It's quite a deal. I have had a lot of trouble with nervousness when I have been sending but I am getting the best of it. I'm not the only one that has suffered from 'key fright.' The boring part of it is the standing by the hour listening to dry lectures on circuits and tech. orders. The practical use of it is very interesting."[85]

During their training the soldiers carefully followed news of the war, which after the dark early days had become much more favorable as the tide began to shift both in Europe and the Pacific. An important milestone came on September 9, 1943. "We were called into an assembly the second hour of calisthenics and marched down to one of the post theaters. There, they announced to us that Italy had given in. You should have heard the fellows cheer. It was before anything at all had come over the radio concerning it. It sure hit the spot."[86] Royce's mood, with that of the rest of the country, became more positive as a possible end to the conflict came into view, and he wrote his parents, "For the first time since we have gone to war I believe we have reason to be optimistic, at least some."[87] He also dared to think forward for the first time to what he and Pearl would do when the fighting ended. "One thing I havn't as yet been able to figure out to my satisfaction is 'what to do when the war is all over.' Maybe it's best not to even try to figure out anything definately. Readjustment won't be too easy but I'm sure we will make it somehow. I don't think are going to be the poorest people in the world either. I guess we can't expect a world just the same as the one we knew before the war came, but I'm hoping it will be a better one. If it isn't there will have been lots of blood spilt in vain.[88]

Early in September, Royce's class was switched to the "graveyard shift," and that same day they underwent the first stage of the evaluation to determine whether they would continue and graduate with this class or "wash back" to an earlier stage of training. This involved meeting with Army examiners, referred to by the soldiers as "quiz kids," who performed a check on their ability to tune the various radio sets they had worked with, and answering questions which tested their knowledge of the same sets. Royce did well on the tests and passed without difficulty, but many men in his class were washed back, among them his good friend Robert Smurthwaite. The pressure built as a week later they underwent an evaluation consisting of nerve-racking spot checks while sending and receiving code under the observant eyes of the "quiz kids." For Royce and most others this was the crucial test which determined whether they would graduate on schedule. When the final results were tabulated three days later, fewer than a third of the class qualified to continue as upper classmen, but Royce was among them. Relieved to have beaten the odds after his disappointment in air cadet training, he wrote Pearl, "I don't think you know how much it means to me even though I'll possibly still be P.F.C. Somehow, that doesn't matter too much anymore. The best men in the Army are the enlisted men. If you've never been one you don't really know the Army."[89]

Along with the others of his class who had "beaten the quiz kids," Royce advanced in mid-September from the classroom to the field, where he would receive his final training before graduation. The first week of practical training was done in transmission towers from which the soldiers practiced communicating with each other by both voice and code over the various radios with which they had trained, and by code with blinkers. Royce described his first graveyard shift atop the towers: "There are eleven of us assigned to a single tower and we work in pairs. The first three hours today, we worked with blinkers; that is, sending and receiving code by flashing lights. I carried on a two hour conversation with one fellow a long way away in another tower. He told me he had a baby boy that he had never seen. The rest of the time was devoted to sending and recieving with radio, and coding and decoding messages. We had quite a time of that too. It wasn't boring in the least."[90]

After their week in the transmission towers, the soldiers spent a second week in a "Covert Training Unit" where they lived and trained under simulated field conditions, which Royce outlined in his journal. "We slept on the bare ground with our blankets in tents. It was plenty cold most of the time. We were still on night shift, getting up at 12:00 midnite and going on duty untill ten a.m. at which time we had calisthenics. Following a meal we reported at the flight line at 2:00 p.m. and flew in "Cubs" every day until 3:30 p.m. at which time we returned to camp and went to bed."[91] One night Royce was able to see the northern lights "like giant searchlights blazing across the northern sky."[92] While on duty at night and early morning the men practiced their radio skills during simulated flights which they planned and executed in three B-17 "Flying Fortresses" that remained parked on the ground as they worked and transmitted messages back and forth. These planes were obsolete models that had been retired from active service, but they were of great interest to the men who were training for duty aboard similar bombers, and at least one of them had historical significance as well. "This particular ship has seen a lot of action. It was at Hickam Field [at Pearl Harbor] when the Japs hit it. It was in other actions too. They sure have patched the thing up in a lot of places where shrapnel and bullets have hit it. It was kind of interesting roaming arround it. They are plenty big too."[93]

During his last week of training, Royce engaged in a variety of activities, including manning a control tower ("we got all the planes back onto the ground, here or at other places—it was all make believe of course but we do go thru all the radio work of a real flight"), standing guard over aircraft with a submachine gun ("the gun was heavy after the first hour; till then, it wasn't bad at all"), and surviving a simulated gas attack using tear gas ("I was one of those without a gas mask but I got hold of the thing before the gas reached me").[94] A particular treat for Royce were the daily flights aboard small Piper Cub aircraft to practice radio skills aloft. This gave him his first taste of flying since his brief initiation during pilot training, and he could not conceal his fascination with it. Royce recorded one particularly memorable flight: "We flew at an altitude of 15 feet and less all the time. Time after time our wheels hit the ground. We went between trees instead of over them, "buzzed" everything and everybody. When we landed, there was corn tassels and weeds caught in the landing gear. There was another plane in front of us (follow the leader) and we came a little closer to everything than he did. It was a thrill very comparable to the one I used to get skiing behind a car, going full speed ahead. Dodging those trees was just like dodging rocks & stuff & getting over drifts. Of course it might have been a bit dangerous when we were touching between 80 & 90 M.P.H."[95]

The two weeks of field training came to their conclusion and finally, on Tuesday, September 28, 1943, Royce recorded, "Class 53 or rather I should say what was left (about 1/3 or less) graduated today. I was happy to get that little piece of paper saying I had finished very satisfactorily the work required at Sioux Falls A.A.F.T.T.C."[96] Royce mailed his diploma to Pearl after showing it to his classmates, most of whom would remain behind. The graduation exercises were attended by the missionaries with whom Royce had worked, though Royce was one of only two active Latter-day Saints in his graduating class. A third, a sergeant named Guy F. Anderson from Utah whom Royce characterized as a "Jack Mormon"[97] because he smoked and did not actively attend church, would shortly become a close friend as both continued their combat training together. After graduation, Royce

150

had dinner with his friend Frank Van Limberg. Frank had started pilot training with Royce, and though he now had the good fortune of having the companionship of his wife, who was staying in Sioux Falls, he had also washed back not once but twice in radio training, and Royce felt fortunate by comparison.

The day of his graduation, in a letter to Pearl, Royce described an underhanded complement paid to him by other soldiers in his absence, as related by his friend Bob Smurthwaite. "It seems that one of the boys used some pretty vile language. It was in fact so low that a discussion on that subject began. In the course of the talk one of them said, speaking about me, that they bet I had never said anything obscene or profane in my life and they discussed that for a few minutes. Of course, they don't know me as I know myself but I thought it was about as nice a thing as anyone could say about my present habits of speech and thought. It was because of you more than anything or anyone else that I got hold of myself and decided to eliminate 'swearing' from my speech for good. That is just one of many ways you have left your mark on my life. Of course I never did those things in the presence of my parents, but it was because of you that I quit doing them in anybody's presence."[98]

While still in Sioux Falls, Royce also exchanged letters with a couple of old friends from Santa Ana. One was Frank Denstad, who by then had progressed to advanced pilot training, and sent news of other fellow classmen who had "washed out" subsequent to Royce. Frank had been recommended for single-seat fighter training because of his aptitude and small stature, but opting to train instead as a bomber pilot, he expressed the hope that he and Royce might one day work together on an air crew. Frank had been granted a furlough which allowed him to visit his wife in Salt Lake City, and while there, the two had stopped by to see Pearl and the baby Jean. He spoke longingly of the two families getting together after the war when they could at last be free of military obligations and enjoy each other's company.[99] Some days later Royce received a letter from his old mission companion, Reid Ellsworth, who had just completed training as a navigator and had been shipped to Europe where he was awaiting his first combat mission. Prohibited from sharing any information about his activities and whereabouts, Reid instead provided Royce with detailed news of all the old missionary acquaintances with whom he had remained in contact. Most of them, like Royce, were married and in the service, some with children.[100]

On October 2, Royce learned of transfer orders assigning him to gunnery school in Harlingen, Texas, and after multiple attempts he at last succeeded in phoning Pearl to let her know of the transfer. The next day he boarded a troop train for the trip south, and the soldiers

151

from Sioux Falls had the misfortune of being shipped in an antiquated train car for most of the four-day journey. The day after departure Royce wrote, "I don't know if or when I spent a night more miserably than last nite. We were cold and very uncomfortable; they hadn't turned the heat on in our car. The car is so old the only lighting is gas. There are some antique washing facilities that are both inadequate and unsanitary. We are the only ones on the whole train traveling thusly; the others all have pullmans, the same as our orders called for. We were told that we would change to a pullman at Omaha but we didn't, so we spent the nite in this wreck. Then they told us we would get pullman at Kansas City, Missouri but we had to leave there in the same old wreck and will have to spend this nite in it to. We are the only bunch in eleven cars who haven't a pullman and we have the furthest of all to go. I guess the bad deals won't end until we are completely away from the influence of Sioux Falls."[101] Royce and some of the other men took advantage of a stop in Kansas City to take a swim and grab a shower in a local YMCA before continuing on. Finally in Houston, Texas, the men were transferred to a modern pullman car, and spent the last night of their long journey in relative comfort.

Gunnery School—Harlingen, Texas

The next morning, on Friday, October 8, 1843, Royce and his fellow trainees arrived at the Harlingen Army Air Field, located in Harlingen, Texas, where they were to receive gunnery training in anticipation of being assigned to bomber crews. Harlingen was a town located near Brownsville in the southernmost corner of Texas, not far from the Gulf of Mexico. In a letter to Pearl that day, Royce described his surroundings. "I find that I am very pleased with the place. The barracks are all white with green roofs and there is lawn all over the place. It's really well kept up. There is a nice large swimming pool nearly ready too. It's still very warm here and they say it will be so for some time to come. There is a nice airfield here with lots of planes. It's all so neat, clean and well arranged here that I am glad I was sent here."[102] He added in his journal, "The only bother is the million or so gnats bothering everyone all the time, and the touch of heat."[103]

With his transfer to gunnery school, Royce anticipated a change of rank which would bolster the family finances. "I believe we are pretty sure to get in a class that will graduate with seargent ratings so next time you see me I shall probably be sporting three stripes instead of the one I am supposed to display at present and have never worn. It will make no little difference in my pay too. When I start to fly after that I will get an addition of half my base pay for that too, so we won't be so poor."[104] A more immediate hope was the anticipation of a furlough for which Royce would be eligible by the beginning of December, if all went well in gunnery school. The furlough, a standard benefit for airmen shortly to be shipped overseas, had been long in coming, and it was especially important for Royce and Pearl, since Royce's elimination from cadet training had dashed their hopes of seeing each other with any regularity. Royce had repeatedly ruled out Pearl's coming to stay near him because it would require having someone else care for the baby for extended periods, and Royce was also mindful of the distraction it would entail as he struggled to complete his training. For her part, Pearl, living still with her mother in Salt Lake City, felt cheated whenever Royce made mention of the occasional soldier without children whose wife had come to live near him, and she had difficulty hiding her resentment that Marie Denstad, whose husband Frank had continued in pilot training, had gone with her baby daughter to stay with him. Royce looked to the furlough as a temporary healing of all wounds, and the two corresponded about it and made plans almost constantly the entire time Royce was in Harlingen.

As they awaited the start of formal gunnery school, Royce and his fellow soldiers participated in a variety of training activities. One of these was entering a decompression chamber, designed to simulate high-altitude flying. "We made a simulated ascent to 30,000 feet. I was one of several to leave off the oxygen mask in order to illustrate the effect of anoxia (lack of oxygen) on a flier. You get giddy, loose your ability to think, see and respond to an order or write; at the same time you think you are perfectly normal untill you flop and they have to put your mask on you. I was putting a peg with a square end and a round end from one hole (square one) to the other (round one), thinking I was acting perfectly normal and doing it right and I was barely moving. When I finally dropped the peg and nearly fell on my face reaching for it the instructor told me to turn on the oxygen. I could see the stop cock but to save me I couldn't move my hands to it. I would have blacked out, but he did it for me."[105]

Another activity in which they participated the following day was detailed instruction in First Aid, which reflected the state of medical advances which had come to characterize the treatment of war injuries compared to previous wars. "Morphine has to be used for all wounded men. Sulfanilamides have to be given internally and dusted on wounds, and you must know how to stop bleeding properly and splint a broken limb plus being able to treat burns. The planes are fully equiped for first aid. A soldier in Uncle Sam's army receives all the care that medical science can give him in case of injury. The result is an extremely low death rate among casualties to date in this war."[106]

Finally, on Friday, October 15, a week after their arrival in Harlingen, a list was posted in the barracks containing the names of all those scheduled to start gunnery school the following Monday, and Royce was grateful to see his name was on the list, together with his friend Robert Barry, a fellow soldier who had come with him from Sioux Falls. Their celebration was dampened a little by a breach of military conduct which had occurred in grand style earlier that day. "It seems that the fellows from our barracks went to the wrong theater. They forgot to call us out, so we went over to the other theater and saw the show, 'The Fallen Sparrow.' It would have been O.K. but they wanted six of our fellows for immunization and when they called their names over where they were showing the films on the 'Articles of War,' only two out of fifty were present when they checked. I understand that we are to be punished by walking on the ramp for two or three hours. We will carry signs that say on one side, 'I have not been on the beam,' and on the other, 'Don't laugh, you may be next'." The next day Royce reported, "I have served my punishment for missing that Articles of War film yesterday. You should have seen the procession. It was a bit comical. There were twenty-two of us marching single file six feet apart arround in front of the orderly room. We had to do it for two hours, and hours can be long when you walk and walk and get nowhere. Barry and all the boys were there. We told those who questioned us that we were just working up an appetite for supper. We did too. The rough old bald headed sergeant that watched us had chow with us after we finished. He is quite a nice fellow. His bark is very loud but as usually is the case, is worse than his bite."[107]

That Sunday, Royce set out to attend church in the nearby border city of Brownsville, Texas, where he had learned that one of the missionaries from his district in Los Angeles was serving. "I had no idea where the mexican church or the elders could be, so I resolved to ask the first Mexican I should meet. He told me that his mother-in-law was a Mormon and he was on his way to her home and invited me to tag along. It was time for church so I went along with them and met the elders there. *Hermanito* Watkins[108] and I had a pretty good hugging spree. He is still the same pleasant happy fellow I knew as a fellow worker. We really had a time talking over the respective events of our two lives over the course of the past year and a half. I talked in meeting and lead the singing. I was ever so happy to find that I havn't forgotten my Spanish so very much. I could understand everything very well, and was also able to express myself freely. It was also the first time I have been priviledged to partake of the sacrament in quite some time. After a little while with them and the members, I felt almost as if I were still in the mission."[109]

On Monday, October 18, 1943, Royce began formal training as an aerial gunner. His first day was a grueling indicator of things to come. "They kicked us out at 5:00 a.m. We had to go to calisthenics from 7:00 to 8:00, at which time we started school. We first had two hours of instruction on the caliber .50 machine gun. We really have to know that gun backwards. We have to be able to strip it and reassemble it ready for firing again blindfolded. We also must learn the name of every part, even the most minute. After that two hours, we had an hour of aircraft recognition. We then went out on the skeet range for an hour. You shoot at clay birds with shotguns there. I fired twenty-five shells and hit with nineteen. It was the best score of our bunch. Of course I've handled a shotgun more than any other fellow of the group. Following the shooting (my ears still ring), we went to chow (1 hour). After that, we spent another two hours with our best friend the caliber .50 machine gun after which we had two hours on ballistics (that is the study of the forces that act upon a projectile in flight). They really throw a lot of stuff at us. Later we have to stand retreat every night and have code for two hours every other night. They expect quite a bit out of us."[110]

In spite of the difficult schedule, Royce did well—he continued to maintain a high speed in code, and in segments of the training which involved memorization and mastery of systems, such as gun mechanics and ballistics, he also excelled, at times scoring highest in his class. Some of the training focused on the fundamentals of sighting the guns, which Royce explained. "You don't just point a gun at another plane and fire away. If you did you wouldn't come near him. He's moving, you're moving, there is a gravity drop and the effect of air resistance on the projectile. You have to lead your target and correct for everything so that your bullet and the target will be at the same place at the same time. Your guns have quite a range so that if you are a good shot, you can hold the enemy off to where he cannot harm you. They won't as a rule fly into accurate machine gun fire from a caliber .50."[111]

Training was given in a variety of bomber gun turrets then in use, and owing to his qualifications as a radioman, Royce received instruction in a famous turret found only in heavy bombers such as the B-17 and the B-24. "We are specializing in the 'Sperry Ball Turret.' That is the one that radio operators handle. It's the most complex of the bunch but the best to operate when you learn how. It has on it what is called a self computing sight. You spot your enemy, identify the specific ship, put its wing span on your sight (a little dial). After than all you have to do is keep the ship in your sights and fire when it is close enough. The sight automatically computes and applies the corrections to the guns. I swear it has a brain. It's the belly turret on the

155

ship. It's pretty well armored and you are only in it when you have contact with the enemy so it is all right. Of course we have to learn to fire from other turrets and use the flexible guns but we specialize in the ball turret. You fit in it a bit tight but comfortably. The guns are caliber .50 and you have one firing at eather side."[112]

As the classes went on, Royce continued to do well, for not only did the academic training come easily to him, he also had enough experience with firearms from hunting in his youth that he excelled in shooting as well. During this time he developed a close friendship with Guy Anderson, the inactive Church member he had met during radio training in Sioux Falls, who like Royce was thriving on the difficult course work. As the academic work wound to a close, Royce expressed confidence, despite the increasingly stringent requirements imposed on the trainees. "We had two examinations today. Among other things, we had to detail strip the caliber .50 machine gun down to the last piece, make some adjustments on it, and reassemble it with blindfolds on all the time. Anderson and I finished first with a time of sixteen minutes. We were allowed forty minutes. Some fellows barely made it. I did pretty good in the skeet range again too. It wasn't the best score in the world nor the worst. I still lead our group in shooting anyway." He added on a less positive note, "By the way, they washed out some seventeen members of our class today because of low academic grades. They don't fool around here anyway."[113]

Gun training was accompanied by work with the radio, and the evening classes in code required Royce to exercise the skills he had worked so hard to acquire in Sioux falls. After one of these night classes he wrote, "My hand is a bit tired. I'm pretty well back in the groove on my code again. I can take it just as well as ever. A code key is some weapon to fight a war with, but it sure looks like a caliber .50 and the old code will be my tools for this war. There is quite a difference between the two. The key works me a lot the harder even if it is the smaller of the two."[114]

On Friday, November 5, Royce and those of his class took their final comprehensive exams to complete the academic portion of their training, and Royce, as expected, passed without difficulty. After the exams, the men traveled the same day to nearby Laguna Madre, where the actual shooting range was situated. This was within walking distance of the Gulf of Mexico, and was uncomfortably hot and humid, recent rains having left swampy areas in the surrounding fields. The air was punctuated constantly by the sound of gunfire as the soldiers trained in their various shifts. "It sounds like a war going on here all the time. From morning till nite, you can hear the lighter fast chatter of the caliber .30's and the deep, reverberating echo of the .50's. It's something you can't describe very well. I'll never forget the

156

sound though."[115]

The following day the men got their first experience in practical training with real weapons and live ammunition. "Today we went out on the machine gun range in the morning for a couple hours. Some of the boys fired 200 rounds of .30 caliber machine gun bullets. The others fired 50 rounds each in the caliber .50 machine gun. I fired the caliber .50. It really talks. You have to stuff your ears full of cotton to be able to take it. My ears still ring a bit. That gun is really a powerful tool and you have to sort of get acquainted with it and its way of behaving before you can handle it well. Every fifth shot is a tracer. That is a projectile that glows as it goes so you can see the thing in flight. I know which end of those guns I want to be on, they really are mean weapons. You really have something when you have one on either side kicking them out at the same time."[116] Exposed frequently to the noise of the guns, Royce for the first time in his life began to experience hearing problems. "Today I was helping one of the follows fire a caliber .50 and the cotton fell out of my left ear. The fellow with the gun at our left fired about three bursts and the concussion from his fire nearly beat my brains out. Result is that I hear bells and more bells in that ear right now. It's just like my ears used to get every 4th of July. Cotton in the ears is standard equipment arround here. They supply it at all the ranges."[117] The cotton balls the soldiers were given provided only modest protection, and with continued noise exposure, Royce was destined to have partial deafness the remainder of his life.

In addition to regular firing, the soldiers took turns shooting at a "malfunction range," a training ground where the men attempted to fire guns which were defective in some way, and then employed the skills they had acquired during the academic part of their training to identify and correct the problem so the weapon would fire properly. Occasionally they were also placed on detail to clean the machine guns, which was a time-consuming, filthy and unpleasant task. Much of the gunnery training was with shotguns, with which Royce was very familiar from his early days hunting pheasant. Royce described the use of shotguns in their training. "Every other day we fire shotguns from turrets at targets thrown from high trap towers. You have to use shooting judgement there using a sight instead of holding the gun and then manipulating the turret with the controls in your hands. You fire it electrically. I found I could do quite well with that when the sight and gun was in perfect harmonization. They use shotguns so much because the problems involved are the same as those encountered with machine guns. Both throw a definate pattern and you have to correct for lead and so on."[118]

The men also received training for possible exposure to chemical weapons, and this led to an adventure which Royce described unhappily to Pearl. "As usual, something is interfering with my letter writing. This time I'm wearing a gas mask. About an hour ago, they got us out of the barracks and marched us close to a mile from here where a plane dived on us and sprayed gas all over us. It was a potent mixture of tear gas and phosgene (a gas that makes you violently sick in the stomach). We marched back to camp and found that there was just enough wind to blow the gas back and contaminate the camp, but not enough to blow the stuff away, so here we are sweating out a miserable evening wearing these lousy masks and hoping that the gas will clear up before bedtime."[119] The men did ultimately get to sleep, but traces of the gas were still present in low areas the next day.

On November 15, Royce and his companions got their first taste of firing from an aircraft in flight. "We went on a total of three missions and fired some six hundred rounds each at sleeve targets. We don't know yet how well we did, I have fears of the worst. I don't think I did so well frankly. Neither do any of the other boys. It isn't easy to stand up there and brace against the slip stream firing at the same time at a target that looks about like a little piece of spaghetti in the distance. I sure do hope I scored well at it but you never can tell. We had our 'pay dirt' ride third trip. It's the one for the record. It's hard to reload up there too when the ship is doing turns etc. and you have a 125 m.p.h. wind trying to beat you down at the same time. All in all, it was quite thrilling, but it didn't last very long."[120]

The following day brought ground training with a variety of gun turrets. Royce exceeded the required standards in each of the guns, and continued to gain in confidence. "I'm quite sure I shall graduate all right unless something drastic happens between now and then. I'm getting so I really love the feel of two jarring machine guns at my side and the smell of burning powder."[121] The next day he wrote, "I am now all but positive of graduation. My average percentage on firing all ground machine guns, turret and hand held is 33% which is good. That's one hit out of every three shots. We have the results on our 'to date' air firing also. This is all the flexible firing we will do in the air except for some practice shooting at targets on the sea tomorrow and that doesn't count. I got over ten per cent on my air firing which is better than good enough. The only thing I have left to do now is fire from turrets at sleeve targets. I think I'll make out all right with that too. Somehow I have kind of known I would get thru this all right. It seems to be something I can do almost instinctively." Though Royce enjoyed the shooting, it also had its down side. "I'd like to get a cal. .50 gun with endless belt in it and fire just to see the bullets go at the target and then have some one to police up the spent cases and belt links. I believe I have neglected to tell you that we spend about half our time picking up those two repulsive articles.

Every time you fire 200 rounds you have 200 spent cases and 200 belt links to pick up, and how quickly 200 rounds go through a machine gun."[122]

By November 19 the men moved back to Harlingen to enter their next phase of training, which increasingly simulated combat conditions. After some instruction, the men went aloft in a plane equipped with actual gun turrets and manned the various gun positions. "I went on two firing missions and one camera mission. First, we went on a gunnery mission. There were five gunners in each ship and we each fired 300 rounds – 100 went through a waist gun and the other 200 went through the two turret guns. On that mission I had 50 hits which was about 18 per cent and good. Next, we went up for a half hour and sighted at a single engine ship which attacked us from different angles coming in fast. As it came in range we kept the sights on it and pulled the trigger, and a movie camera took pictures of the ship coming in. Later, they show them to us so we can see how well we did firing at it. After that, we went up for another gunnery mission. I hit 67 out of the 300 rounds which I fired that time, which was better. It gave us a pretty good idea of what will be expected of us in combat all in all. We were in the air a total of 3 hours and really had a good time at it."[123]

On Sunday Royce attended church one last time at the Spanish branch in Brownsville[124] before embarking on his last week of training, by now confident of graduation and of earning both an anticipated furlough and a promotion. The following day he flew his last three scored missions, achieving firing scores high enough to assure his graduation, and Tuesday, having fulfilled all his requirements, he was given something of a bonus ride. "I was called with a group to go up and fire from a B-24 because I had all of my gunnery missions complete. A B-24 is a big 4 engine ship. The ship that towed the target was a B-26, a fast twin engine bomber. It attacked us from all different angels and we in the four turrets and at the two flexible waist guns fired at it when it came in range at our respective positions. I fired in all 500 rounds of .50 caliber ammunition. The ammunition I fired would be worth just about 150 dollars. You imagine how much it cost for the 1800 rounds we fired collectively. We had to be very careful that we didn't hit the tow ship because it moved so fast. We also had to avoid putting holes thru parts of our own ship."[125]

That Thursday was Thanksgiving Day, and Royce, though consigned to a military camp far from his wife and loved ones, and soon to be deployed in a war zone where he or any of the men he trained with could easily forfeit their lives, still was able to find the positive. "I guess I can truthfully say that we have more to be thankful for this year than ever before in our lives. We have a good portion of the

159

things that are really of value in this world. We and ours enjoy good health and I believe we are happy, at least just as happy as we could be under the prevailing conditions. We have our baby and everything has been well with her. I guess we couldn't ask for very much more and be right about it. We are two of the most fortunate people that live in the world."[126] Even misfortune in the chow line did not deter his spirits. "We had turkey today but I didn't get much. They gave me a neck. You take what you get here. I enjoyed everything though. They had some extra good apple pie and ice cream along with a good lot of the trimmings that go with Thanksgiving Dinner. They even had table cloths on the tables which made it right homey."[127]

That weekend Royce and the other graduates were kept busy changing barracks to the shipping section in preparation for departure, and at last, on Saturday the 27th of November, the day came for graduation from gunnery school. Royce wrote, "I have my wings and my rating now. The graduation ceremony wasn't very elaborate, but the captain and master sergeant (just returned from combat) gave some very nice talks. We just marched accross the stage and were each presented with a pair of wings. All in all I feel good about it. It's pretty nice to have those three stripes on your sleeves."[128] The joy of graduation was dampened a little when Royce learned that his good friend Robert Barry had washed out of air firing, not having achieved enough "hits" on firing trials. Royce would not continue to train with him, and would stay in touch only by mail.[129]

With graduation behind him, Royce's attention turned to his imminent furlough. Royce looked on the furlough as something of a reward for success in his military training, and indeed such furloughs were routinely given as soldiers completed training and before they shipped to their final assignment to areas of operation. But the furloughs had another important purpose, one which was often better left unspoken, for all were aware that such furloughs marked the dividing line between military training and war, and that for many of these men and their families, it would be the last time they would see each other alive. Royce and Pearl, like all the others, knew this was so—they just did not know whether such would be the case for them.

Royce arrived home early in the month of December, and though he made no written account of the furlough, he referred to it frequently in subsequent letters. Royce found that Pearl had taught their daughter Jean to recognize her father by means of photographs. Recalling the visit, he later wrote, "One of the biggest thrills that came to me when I was home was the way she held her arms and came to me the first time I saw her. It made me feel that I hadn't been left out of her life, and she knew me even though I had been away so very long."[130] Royce stayed at home with Pearl about ten days, and then departed eastward from Salt Lake City by train, evidently on Wednesday, December 15th, bound for Columbia, South Carolina. In his first full letter home he described the parting. "It seems like such a long time since I left you standing at the station at Salt Lake. I guess that somehow that was just about the way I wanted to leave you this time, not having the time to think too much about things while we waited for them to say 'all aboard.' I felt terribly lonely for you then and still do, but things did work out for us pretty well in the ways that really count. I couldn't begin to tell you how much I enjoyed my furlough and how deeply in love I am with my wife and our darling little baby. You have given me so much and made me so happy."[131] As the days of the furlough receded into the past he reflected, "It's been so long since my life has really been my own that it seems like almost a vague beautiful dream of yesterday. I hope I shall always appreciate that as I should when I have it."[132]

The trip to South Carolina took longer than expected, owing to train delays. Leaving Salt Lake on Wednesday, Royce did not reach Houston, Texas, until Saturday, and he and the other men sent a telegram to the commanding officers at Columbia explaining the situation. While in Houston, the men had a long enough delay to shower, have dinner, and take in a show. The train ride east was another grueling one, "about as miserable as it could be. We only had seats part of the time and they weren't much good either."[133] Scheduled to arrive Sunday the 19th of December, the men finally pulled into Columbia, exhausted and a day late, the evening of Monday the 20th. Royce sent a quick telegram to Pearl, and then at last retired to a good night's sleep.

Bomber Crew Training—Columbia, South Carolina

As the day dawned the next morning Royce had the opportunity to observe his surroundings, and he found them agreeable, describing them to the family back home. "I wish you could see the beautiful country I am in right now. The camp is about five miles out of town in some small hills which are all over this country, that resemble our foot-hills. There are pines all arround our barracks and as far as you can see. They are the long needle kind. There are also big leaf oaks that have their leaves turned to gold but haven't yet lost them. The dirt is nice clean fine sand, almost white as snow. The water is cold and as soft as it can be. All in all, it's a pretty nice camp and I do believe that I shall enjoy my stay here."[134]

Royce had also learned a little about his assignment in Columbia. "This is a rather interesting outfit I'm getting into. The planes used are B-25's. They are two engine, medium bombers. They are built by North American Aircraft and are sometimes called 'Mitchells.' Be-

161

tween the two of us I have always hoped to get on one of them but have always reconciled myself to being put on one of the great big ships because that's where they told us we would land. These ships don't make far inland raids or fly in the huge concentrated attacks. You can kind of forget about the belly turret business, because as the wind blows now I rather definately will not be in one. If I make this I will be radio operator and fire flexible waist guns of which there are two, one on either side."135 The B-25 Mitchell was a sturdy aircraft which had already proven itself in battle on multiple fronts, and had a reputation for remaining airworthy even when damaged. Royce noted that his best scores in gunnery school had been from the flexible guns like the waist guns he would use on the B-25. He later added, "The outfit is a new type one. The planes they have have a 75 m.m. (3") cannon mounted in the nose and are for low altitude bomb and strafe attacks. That's the largest gun ever mounted on a plane."136

After Royce's arrival at Columbia it took some days for him to be assigned to training, and on Christmas Eve, the fourth day after his arrival, he once again found himself alone and far from home for the holidays. His recent arrival meant that he was still in transition, and not yet having been assigned to a squadron or gotten settled in permanent quarters, it was difficult to arrange much of a celebration. He acknowledged, "This was a pretty punk time to get assigned to a new place. It makes Christmas week a lot different than it would otherwise be."137 On Christmas Day Royce had planned to travel to town to locate the chapel for Sunday, but he was already nursing a cold, and the day dawned dismally cold with a chill rain that lasted all day, putting a damper on his plans. Instead, he spent a frigid Christmas in the barracks with other "boys" in the same situation. "They had a special meal fixed up but we just about froze eating it. It was dreary in the barracks and we had to practically sit on the stove to keep warm."138 Even so Royce was able to find something good to write home about. "There is a pretty side to all this. The view from our window is beautiful. You see, it is freezing at the same time the rain is falling with a result that the long needles of the pines which are all over here are sheathed in gleaming cases of ice. The trees have taken on the appearance of extra large Christmas trees. Maybe I can't have my 'white Christmas' so I'm getting the next best thing. The barracks were not forgotten either, there are icicles hanging from all of them and we have a rather cheerful fire inside, though there are wide cracks that continuously let in enough of the outside air that we don't forget that it is plenty cold."139

Royce and the other men had little to do as they awaited squadron assignments, and they entertained themselves as best they could, attending shows and fulfilling minor assignments. The tedium of waiting was broken by an unfortunate event involving one of the soldiers who had just spent Christmas away from home. "One of the boys from our gang who moved over to a squadron told an officer today that he refused to fly. He was immediately broken down to a private and grounded. I guess he told him off pretty heavily. It makes them pretty well down in the mouth when a soldier gets this far along in the game, just about to where he is useful after all the training he has had, and refuses to do that which he has been trained to do. This fellow probably had his reasons, but they were probably the same as those any of us could fix up right off. I think we mean as much to our respective loved ones as he does to his."[140]

Finally on Thursday, December 29, the long-awaited assignment came. "I have been assigned to a squadron at last. First thing this morning, all of nine a.m. they came over and informed us that we would be moving to a squadron. Andy and I got in the same outfit. He has the bunk above me here. We are in training for combat duty over there. How long we will be here working before we go over I do not know. I do know we will be here six weeks before anything happens. McNally and Bergun, two of the boys from our gang, have both been flying today. They like the ships plenty."[141] John R. McNally was a soldier from Cincinnati whom Royce had known since cadet training, and with whom he had grown close while in Harlingen. "Andy" was Guy Anderson, the inactive Church member from Utah that Royce had first met in Sioux Falls; together they had been assigned to the 377th squadron of the Third Air Force. These three would remain inseparable friends as they continued their training in Columbia.

On New Years Eve, Royce underwent some of the preliminaries to flight crew training. "Today I checked out to the squadron. This is to say that I passed my checks in code, sending and receiving, and blinkers. I go over onto the flight line every day now to work and fly a bit. I haven't been up flying here yet. We have to go to a school here before we are assigned to combat crews."[142] Royce considered calling Pearl for the New Year, but still feeling sick, he decided to forego the four-hour wait at the phone, and sleep through the New Year. He was awakened, however, at midnight by the shouts of celebration among the soldiers, and again an hour and a half later when Andy (Gus Anderson) came home "with a little too much under his belt, in fact quite a bit too much."[143] In his letter to Pearl that night Royce touched on a now-familiar theme. "I hardly know what other wish to leave with you as a New Years wish, except that this whole mad mess might terminate and that before another year has drawn to a close we might find ourselves together to stay."[144] With some reflection on the turning tide of world events which swirled around them, Royce acknowledged, "Battered and bloody 1943 goes out a lot more hopefully than she came in."[145]

For the New Year, Royce purchased a new journal, and on January 1 he wrote, "Today as it is probably fitting, I flew for the first time in a B-25." He described this flight in a letter home: "I was up for four hours. We were on a practice bombing and strafing mission. They fired that 75 m.m. cannon about eleven times. All it did was make the ship shudder a bit. I got in a position where I could watch the shells hit. They are 3 inch shells and tracers. They really hit with a bang. We also dropped about ten bombs too. I just went along for radio experience; we don't start school untill next Sunday. It's pretty nice riding in these ships. I have a pretty comfortable seat. We flew low all the time today and for the first time I saw the Atlantic when we flew over it."[146]

The men at Columbia received training in discrete phases as part of a Replacement Training Unit (referred to by the men as "R.T.U. School"), designed to prepare crews for overseas combat service, replacing crews which had either completed a tour of duty or been lost in action. Royce was not initially assigned to an R.T.U. school, but he continued to be assigned to training flights as weather permitted, and now found himself doing radio operation in earnest, contacting the base with position reports and requesting weather and direction-finding information for his crew. On January 5, he flew in a long practice mission in formation, and the next day, on Pearl's birthday, he flew in an aircraft practicing "skip-bombing," which he described in a letter. "We would come in at the target right low and then release the bombs in front of the target. The object is to have them hit flat ahead of the target and then have them bounce or skip into the target. It works too. Out of the bombs we dropped two of them knocked out 6' by 6' targets. The last one was a dead center bulls eye. It tore the canvas out of the center and never even broke the frame arround it. That kind of bombing is used on ships and hard to get at places. They try to hit ships about at the waterline and a bit of that really plays havoc with them. I did pretty well on the radio too so things are coming all right."[147]

Royce considered the delay in R.T.U. assignment as working to his advantage, since his continued training flights gave him the practical radio experience he lacked, despite his long training. Having been placed on permanent flying status, as of January 7 he was eligible for "flight pay," an additional monthly bonus amounting to half of the base pay, given to aviators and their crews owing to the increased hazard involved in flight duty. This, coupled with the increase in pay from his recent promotion to sergeant, meant that he and Pearl had a more comfortable income, and they wrote each other about purchasing some basic furniture and accumulating some savings in hopes of some day owning a house. At the same time, Royce was concerned that ten months previously, when the baby was born, he had applied for a dependent stipend to which he was entitled, but Pearl had yet to receive a single check. Both Royce and Pearl looked into this benefit and wrote letters, but as the months wore on the money was still not forthcoming.

Not long after his arrival in Columbia, Royce had run into an old classmate from Granite High School, and on Sunday, January 9 the two set out together through a snow storm to attend church in Columbia. There Royce learned for the first time disturbing news about his mission companion Reid Ellsworth, whom he had last seen in cadet school at Santa Ana. He relayed the news to Pearl in a letter the same day. "Honey, I'm almost afraid to tell you this because I know it isn't going to make you feel good to hear it. It certainly hit me. Reid Ellsworth is 'missing in action.' He may still be alive but he was shot down in combat. He has been missing for sometime. That's why my last letter wasn't answered. I was wondering because he was so faithfull about writing. There was a young fellow from church who is from Pocatello, Idaho and just returned from a furlough. He lives only a few blocks from the Ellsworth home there and he found out about it when he was home. It was a shock to me. He was one prince of a fellow. Not only that, he wanted so much to get married before he went over after he got his wings, and didn't even so much as get a day home. That cuts kind of close to home. Reid will be all right I know alive or dead because he had lived in such a way that it would be that way, but it is such a shame that a man's life should be cut short in the very flower of it."[148]

In the same letter, Royce responded to a concern expressed by Pearl about the moral ramifications if he were to kill anyone during the course of his military service. This thought came to the forefront as the time for his departure overseas drew near, and he explained how he came to terms with the same issues. "Don't worry about killing, Pearl. I shall never kill an individual. Circumstances destroy life as man is forced under them to strike at a principle. God does not hold as murderer the man caught in the net of these circumstances."[149] As the theme continued to come up in their letters, Royce wrote reassuringly, perhaps as much to himself as to Pearl, "You can know that I shall most probably never have to kill a man myself. Air war isn't fought that way. I don't even have to get in a turret. The radio is my primary concern all the time, though I do have to man waist guns for defensive purposes only."[150] News of Reid's loss had continued to weigh on Royce's mind, and for the first time in his letters he opened the possibility that even if he did return, he might be a changed man. "I do want to come back to you so very much. I'll do my best to realize it and I do pray about it always. I still hope to hear that Reid is still alive. He had so much to live for too. War, in a way, is a game, but an awful cruel game that strikes the players from the board never to be replaced. I think I'll be back to you, Honey. Just take me for what I am then. There may be a few changes we might not like at first but we will get used to them too."[151]

On Sunday, January 16, Royce officially began his Replacement Training Unit schooling, beginning detailed checks of each component of the radio equipment he had already been trained with. The radio phase of R.T.U., which ran daily for two weeks, consisted largely of a review of concepts and skills already mastered; there was no flight duty involved, and for Royce, accustomed to the rigor and pressure of radio and gunnery school, it was an easy life. In addition to the classes, he received numerous vaccines and made trips to the skeet range to practice firing. In free moments he learned to bowl with Andy and McNally. Saturday was their day off, but they had to arise early anyway, as improved weather meant weekly parades on the parade ground, an activity which Royce enjoyed in spite of himself. "The parade was pretty good though it was cold. In fact, though I wouldn't admit it to the boys, it felt good to march to a good military band again. (It isn't stylish to like those things. You are to complain about them long and loud, mostly loud.)"[152]

Royce also continued attending church meetings and Mutual in nearby Columbia, which he increasingly looked forward to. "Being with people who think as you do and act as you prefer to have people conduct themselves is really good for you. As far as I have been able to gather, there is no other organization in the world that consistently influences people to the extent that they are like our good Latter-day Saints. It was Jesus who said, 'By their fruits ye shall know them.' This has never yet failed me. I have always been able to feel that they whom I have met in Latter-day Saint meetings were a visual manifestation of the truth of that statement. Good Mormons are the fruit of a superior tree. I only hope that I shall always be able to recognize that. I want to be a better Latter Day Saint than I am."[153]

On January 30, Royce entered the gunnery phase of R.T.U. training. The first two weeks of this training were conducted at the armament buildings on the post, and the remainder took place in aircraft flown from Myrtle Beach, located some distance away on the coast. The gunnery phase consisted mainly of training in the gun positions unique to the B-25, which Royce explained in a letter to Pearl. "We are spending our second day on the Bendix gun turret which is used on these planes. We will most probably never have to operate one because in the ship, it is on one side of the bomb bay and we are on the other but we have to know how just in case. It's the most effective defence from attacting air craft that the ship has and in an emergency, anyone on the ship has to be prepared to operate it. The guns the radio man usually has to handle are flexible waist guns located right in the radio room. One of them shoots out either side. The ship also has a rear turret which we have to be able to handle in an emergency."[154]

On February 13, a frigid winter day, Royce and the members of his group were transported to Myrtle Beach to begin the next phase of training, which involved air-to-air and air-to-ground firing from actual B-25 ships. Winter storms had played havoc with the flying schedule, and even when the weather cleared briefly, Royce's group had to wait for other trainees who had been delayed in finishing their own training. Finally, after waiting through most of the week, on Friday, February 18 the weather cleared enough to begin flying the gunnery missions, which Royce described. "We would fly along the edge of a road and fire at a big cloth target with a fighter painted on it. There were four of these, each one a bit closer. With us flying a straight course it simulated the attack of a fighter plain on the ship. You have to aim quite a way in back of the target to allow for the movement of the plane through the air. I didn't do so bad at it either. One mission I pretty well riddled one target. I caught it squarely in the center of my bullet pattern. I guess that's about the way you would have to catch a fighter plane to knock it down. You have the advantage on them because they have to turn their nose on you in order to damage you and you can turn your guns on them without that difficulty."[155] To his parents, Royce described the sensation of these low-altitude maneuvers. "When a ship does a sharp bank it pulls you toward the floor so hard that you have a difficult time trying to stand up, and it's almost impossible to pick or hold anything up. I've never been on anything on the ground that gives you a similar feeling at all. You see, we can walk all around on this ship while it is in flight, which is quite different from the planes I have been in before."[156]

The next break in the weather came a few days later. "We had a beautiful day today which meant flying. I was up two times. One mission was a tail turret mission, firing from air to ground. It's a gun, or rather two guns, I will not have to know how to operate under normal conditions, but then conditions aren't always normal. It is kind of a different set up than other gun positions and I'm not too good with it but I can handle it all right if it ever should be necessary for me to do so. The other mission was air to air firing using waist guns. We fired at sleeve targets towed by another B-25. When we fired, they were so far away that they looked like a little piece of chalk. We don't find out if we hit it. I think we scared it quite a bit anyway. I saw our tracers cutting pretty close to it."[157]

By the time Royce flew his last assigned gunnery mission, weather had prolonged his stay in Myrtle Beach so long that all his spare clothes were dirty and he had had to borrow money to stay afloat. On Wednesday the 23rd he wrote, "I got my last scheduled flight in this morning. It was a turret mission firing from air to ground, two hundred rounds. It was sure rough up there so we had quite a time keeping the guns on the target as we fired. Up and down drafts just kept the ship tossing like a boat in rough sea. One time, I had a burst going right into the target and before I finished it I was firing yards and yards over the top of it, sending my shot out to sea. It doesn't much matter here whether or not you hit the target. They are concerned mostly over whether you know how to load and fire the guns correctly. I'm kind of glad that I do not have to use a turret normally. I much prefer the waist guns on this ship, mostly because I can handle them better."[158]

The next morning Royce was assigned to the dreaded K.P. duty in Myrtle Beach, but he was pulled away at 2:00 p.m. to return to Columbia Air Base. On arrival, he and Andy learned that they had already been assigned to air crews. Royce had looked forward to selecting his own crew, and had been approached a number of times by officers requesting him to fly with them, but in the end the selection was made for him, probably due to the delays caused by the inclement weather. Royce's pilot was William H. Brassfield, a tall, handsome airman originally from Ann Arbor, Michigan, whom he described as "the kind I hoped for, big and clean living." The navigator/cannon operator was Frank G. Harper from Pittsburgh, Pennsylvania; the engineer/turret gunner was Robert Wendell of Maline, Illinois, and the tail gunner was James Robinson of Newark, New Jersey, whom Royce would come to call "Robbie." Royce at the Radio position completed the standard crew of five men.[159] These men would become close friends in the coming months.

The morning after his arrival back at Columbia, Royce flew for the first time with his new crew, and they flew again the same night, which proved to be an adventure, described by Royce in a letter the next day. "The weather closed in tight over Columbia so that landings were impossible." Looking for an acceptable landing site, the pilot settled on Greensboro, North Carolina. "It was raining here too but we made it in all right. We landed about 1:30. We bedded the ship down out at the field and then they brought us in to town and we got hotel rooms here. It was to have been our day off, and here we are stalled up here with nothing but flying clothes." Because soldiers were prohibited from being seen in public in flying suits, Royce and his roommate, the engineer Wendell, were confined to the hotel, unchanged and unshaven. While the engineer slept, Royce wrote to

Pearl and expressed his relief at the conduct of his flight crew. "I consider myself pretty lucky getting on this crew. It's just the kind of a set up I was hoping for. I was so afraid that I would get a drinking pilot. I guess that was the one thing I watched for mostly. This fellow can fly too. Last night, he brought us in with instruments and made a beautiful landing even though it was raining and dark."[160] The men had to spend another night in Greensborough before weather cleared enough to fly to Columbia, where Royce arrived in time to attend church.

Royce and his crew were training for service in the B-25H, a cannon-bearing model designed for the low-altitude combat necessary in the jungles of the South Pacific, where attacking from higher altitudes was mostly ineffective. In the ensuing weeks, they would fly numerous training missions together, gaining experience in the various conditions they might encounter in combat. They flew a night mission, which allowed the radio operators to communicate by code over blinkers, then a gunnery mission back to Myrtle Beach, where all the men, including the pilot and navigator, fired from all different gun positions on a ship based there. They flew a long-distance mission to the Bahamas and back, circling the islands before heading back to Columbia, a flight of over eight hours. They were trained and tested in loading bombs into the aircraft. When weather precluded flying any distance from Columbia, they practiced formation flying, and did landings and take-offs on a nearby auxiliary field. They flew another low-level "skip-bombing" mission over the sea in rough weather. During a "gas mission" they practiced laying a smoke screen and discharging tear gas.

As the training missions came now in rapid succession, Royce began to observe the effect on his hearing from his constant work at the headset with the volume of his radio high enough to hear over the famously loud engines of the B-25. "It's very tiring to work on that radio about four hours at a stretch in the air. My ears are ringing all the time from being pounded with code and my head spins the rest of the time. On the plane you have to strain about every fiber to discern what is being said, and then you have to keep requesting that they say it over again. The voice is distorted and background noise is so loud it usually hurts your ears. The engines affect both transmission and reception. My ears get pounded till I am embarrassed because I have to keep saying 'what' for a while after we get back on the ground. I guess we do earn our money after a fashion. I'm always worn out after every flight anyway."[161] The heavy noise exposure would continue to be a problem for Royce, and would eventually deprive him of most of his hearing.

Royce learned that, owing to a combination of lack of flying at

R.T.U. school and the weather delays at Myrtle Beach, he had forfeit-
ed his "flying pay" bonus for the previous month, but on March 6 he
was promoted to the rank of Staff Sergeant with an increase of base
pay which, in the long run, more than compensated, and he arranged
to have a larger allotment sent monthly to Pearl. He also continued to
address the many concerns Pearl expressed in her letters as the time
neared for him to go overseas. In response to her worry over his exposure to the habits of soldiers with different moral standards he
wrote, "If anything changes it will be through injury and purely physical. I know I shall not take up habits I have never had. I'm a man
now Darling, not a boy, at least about things like that. I have no desire to drink or smoke or anything else. As for being true to you, you
just have to trust me for that. I have never yet been otherwise and I
promise you I shall not be now, no matter what." He also responded
to her concerns about the elemental risks of combat service. "If I
have to kill, try to understand that it was not my desire to do so. If I
am hurt, I'm going to come back to you no matter what it is, trusting
that you will stand by my side and help me if it is necessary until I am
able to do things myself once again. You must know the many things
that have run through my mind when my thoughts carry me to the
days to come. Some of these things arn't pleasant I know."[162]

Royce also expressed to Pearl his thoughts about the theological
aspects of the war which dominated their lives and clouded their future as it continued to stretch out endlessly before them. "You ask if I
thought the war would end if two people prayed hard enough. I'm
afraid Darling that there are much deeper issues involved than the
lives of a few individuals. Too many times, I think we forget that it is
man not God who is working out his free agency here upon the earth.
It was planned that way. If he wished it to be so, I'm certain that God
could cause that this whole mess should suddenly end but he hasn't
worked that way in the past nor does it seem likely that he will work
that way in the present. Men, through their disobedience and persistent disregard of natural laws have time after time got themselves into
a seemingly hopeless muddle involving all of mankind. God has always left man to work his way out of the muddle just as he let him
work his way in. It doesn't mean that he stands by heedless of the
cries of the righteous and sufferers going up to him. Always He has
stood as He yet stands with hands extended offering every assistance
that man is willing to accept and apply, but he does not and will not
do all this for them. Individually we are responsible for our own way
of life, but we cannot help being an inseperable part of all mankind. It
is perhaps as a huge body of which one part cannot suffer without
the other parts being also affected and caused to suffer. We are our
brother's keeper whether we like it or not."[163]

170

At the same time, Royce felt the need to explain to Pearl why he was doing little in the line of missionary work among his fellow soldiers. "If you could be in my boots for a few days or maybe even one I believe you would understand all right. Contrary to what a lot of well thinking people think concerning soldiers there aren't very many of them who are concerned enough about religion to want to do anything about it. This goes for the majority of those I know who have been in combat. Maybe I'm not doing right. I don't know, but I never have had the desire to be singled out by them as a crank on religion, picking up a name like the 'Deacon' or something similar. If they want to ask me why I live as I try to and why I do what I try to do, that is entirely different, and at times I have been able to explain. It just doesn't seem fitting that I should start a conversation about the gospel with fellows who but the moment before were going over in detail a weekend of debauchery that they had in some nearby town. I hear so much of it that sometimes it almost makes me afraid that I am a part of it, and really I desire none of that." He hastened to add, "I don't feel that I am better than those with whom I associate. I have been brought up differently and I doubt very much if I live proportionatly better for the way I have been taught and for the concept of life that has been laid before me at nearly all times. As a people, I doubt very much whether we live proportionately better for the much greater light that we have."[164]

Royce and his crew continued their training, and on March 13, the day after Jean's first birthday, they dropped live bombs for the first time. "We had a regular medium altitude bombing mission today. We dropped demolition bombs and anti-personnel fragmentation bombs. I watched them all the way down thru the camera hatch. They really burst when they hit too and make quite large craters. We came pretty close to the target. It's the first time we've dropped live bombs. The ones we have been using are filled with sand with about 5 lbs. of black powder in the rear which explodes so you can see where your bombs hit. The ones we dropped today only weighed 100 lbs but they really are quite terrible. I surely wouldn't like to be on the receiving end of them anyway. It isn't pleasant to think of them dropping on people."[165]

By this time Royce had become quite proficient at handling the radio on the bombers. After a formation flying mission near Columbia, he remarked, "The radio work was simple but I do believe the time drags more when I haven't very much to do than when I am really busy. I handed in an excellent log but it was simple to make contact. I'm getting so I can do most everything with very little effort now. We were in a new ship, an 'H,' the same one we had yesterday. It has two big curved windows out of which I can look." However, the same day Royce got news which had important implications for his overseas assignment. "We now have six men on our crew. They gave us a copilot today. I don't know about this for sure but we think we are going to get a new 'J' model B-25 to take over. They do not have the cannon but carry a lot more armor plate for protection, including some for the radio man. They have about the same number of machine guns and have 2,000 horsepower engines. These ships operate at a higher altitude than the 'H's which is all right. You're nearly always high enough to bail out if the ship is disabled anyway. Another thing, it probably means the European theater."[166] The copilot was reassigned a couple days later, so the future assignment still remained in doubt. Based on rumors in circulation Royce expected that it would be in the South Pacific, but he had Pearl send addresses of her relatives in Scotland just in case he were assigned to Europe and happened to spend some time in Great Britain.

As he waited for news of his own combat assignment, Royce received word from Pearl that his friend Frank Denstad had already shipped overseas to the European theater as a copilot aboard a B-24. Worried about the dangers inherent in service aboard the heavy bombers, Royce wrote, "I hope Frank makes it all right. He will probably have a harder deal than I will. They have to fly for hours at extremely high altitudes using oxygen all the time."[167] Royce had kept in touch with Frank through his training, and had kept up on his progress by talking with other pilots who had trained with him. Ever since Santa Ana, he had held Frank in high esteem. "Frank always seemed the most considerate and genuine of the bunch. I think that is why I always thought such a lot of him and went everywhere with him."[168]

Royce and his crew continued to complete training missions almost daily and, as the training neared its end and the time for final assignment approached, they were run through their last preparations for departure. This included a full combat physical (an uncomfortable experience as the men had to wait standing partially clad in cold weather) as well as the issuing of a new set of stainless steel dog tags containing personal information. They were required to stencil their name on all equipment that would go overseas, and Royce sent home any items that were surplus or worn out, including unnecessary identification and cards from his wallet. His crew attended a camouflage training in which they were taught to disguise themselves using netting, face paint, and any materials at hand.

Finally on Saturday, March 25, Royce's crew was placed on alert, which meant that all passes were cancelled as they prepared to ship to Savannah, Georgia, where they would receive their aircraft and make final preparations before flying overseas. The same day they flew their last mission, dropping clusters of incendiary bombs on a target from a shiny new B-25H, the ship they anticipated flying in combat in the South Pacific. They went through final briefings, and on Wednesday March 29, the crew was treated to a nice dinner, courtesy of the pilot and navigator. "We went to a small night club near here for a steak dinner and the evening. The pilot brought his wife, the navigator his girl and the tail gunner his. The Engineer and I brought each other. It really was good. My pilot has a very nice wife. She is from Oxnard, California. He met her when he was taking primary training there. (I think she is expecting.) She and I were the only westerners in the group so we got on very well. The rest were all nice too. I really have a swell crew, Honey and I'm not fooling. They are nice to be with anywhere. The steak was very good too though I was one hungry soldier by the time they finally brought it in."[169]

The next day they made ready to ship out, and on Friday, the 31 of March, the crew departed Columbia for Savannah, Georgia, where they were to receive their combat aircraft and at last learn their destination. Royce was aware that this was the last step before overseas duty, and that literally any day his crew might be called upon to leave the U.S. and start the trip abroad, though he felt more than ready. As the day approached to leave Columbia, he had reflected to Pearl, "I guess I'm not really sorry that the time has come when I am about to do my part to help win this war. I hope I do know what I am fighting for. If for nothing else it's so that I'll be able to return to you and baby. I guess that would about sum up what every soldier is fighting for. They don't really hate anything except maybe the injustice of it all and perhaps a bit for those responsible."[170]

Outfitting—Savannah, Georgia

The final base assignment before shipping overseas was Hunter Field, located in Savannah, Georgia, which was to serve as no more than an outfitting station from which the crew would be sent to their overseas assignment. Here the men were given comfortable quarters and good food, with a schedule which afforded them plenty of sleep and ample free time. The day after arriving being Saturday, Royce went with a couple of other soldiers to check out the city, but they found it unattractive, particularly with the weekend crowds, and they returned to the service club for dinner and some rounds of bowling. Andy and his crew hit the "hot spots" in town, returning drunk some hours later. The next day Royce received a letter from Elder Corbett, one of his mission companions, with some unexpected and welcome news about his friend and mission companion, Elder Ellsworth: "Reid Ellsworth's folks received a telegram from the government dated March 9th saying that Reid is a prisoner of war in Germany. He was shot down on the 10th of November. Apparently it is some time before word actually gets through. I can imagine how thankful his mother must feel about it."[171] Royce felt similarly relieved.

On Monday, April 3, 1944, Royce's crew was assigned their aircraft and, in accord with the rumors that had circulated back at Columbia, rather than the cannon-bearing B-25H on which they had trained this proved to be a new B-25J, the plexiglass-nosed model used in the European theater. They also learned that they would fly to their theater of operations rather than being shipped by boat, which was a relief to Royce, though it dashed any hopes of being able to see Pearl one last time before departing. The next day, April 4, the crew took their new ship up for a test flight and formally accepted it, even though it had some minor hydraulic problems which would require repairs, and would continue to cause occasional difficulties all the way to the front.

In the meantime the men underwent an overhaul of their clothing and equipment in anticipation of departure overseas. "They took some of our things and gave us others, limited the weight and number of bags. Our worries are just beginning. It's not easy to figure out how to put twice as many things in something that half would fit in comfortably."[172] The men also purchased supplies of items they heard would be unavailable once they arrived at their combat station, from razor blades to cartons of candy. "It is rather exciting and interesting getting ready to go over. We all carry guns with us all the time because they have been issued to us and you have to have them with you. It's pretty serious if you lose one. Of course we carry no ammunition. They are big .45 caliber automatic pistols, and we pack them in shoulder holsters. They are plenty heavy too. Mine is new and I don't think it has ever fired a shot."[173] As part of their preparation, the men had their papers checked, and the matter of Royce's dependent pay came up. It had now been over a year since Jean's birth, and Pearl had still not received any of the dependent allowance to which they were entitled, in spite of repeated inquiries from both her and Royce. Royce was at last referred to an officer sympathetic to his problem, and a telegram was issued to the appropriate authorities.

On Friday, April 7, with departure imminent, Royce made one last phone call to bid Pearl farewell, then wrote her a letter containing the home addresses for his crew members and enclosing $15 to cover the cost of the call. The very next day the plans changed. "Our engineer [Wendell] is in the hospital. This morning when it was time to get up he was near burning with fever. He wouldn't go to the hospital so I doctored him up with about four aspirin tablets and covered him with all our blankets and did he ever sweat. He was soaking wet from head to foot. I went over to the PX to get some fruit juice to feed him and also to check with the pilot on things. He went back with me and we got a truck and took him to the hospital. He has to stay until his temperature goes down to normal and stays for twenty-four hours. We will be about the last of our bunch to leave."[174]

The next day was Easter Sunday, and as their departure had been delayed, Royce got a release from his pilot to go into Savannah and attend Sunday School at the local branch of the Church. This proved to be a fortuitous trip, as he met numerous old acquaintances there. Though the branch was small, the members were hospitable, and Royce was invited for dinner at no less than three different homes. He accepted one of the invitations. "It was really a typical Mormon family. We had chicken and all the trimmings. There were lots of children there, including one just about the size that Baby Jean must now be. They hid eggs and we watched the children find them."[175]

As the other crews departed, the weather grew hot and humid, and Royce found the waiting uncomfortable. "I'm getting very restless now and wishing we could pull out. Andy has gone and so have all the rest. Wendell is still in the hospital. I took his mail over today, and although they would not let me see him because no visitors are allowed, they told me he is all right and that he should leave any time."[176] Even at this late date Royce had only the most general idea of where they were going, based in part on the type of aircraft they had been assigned, and in part on the location of their immediate destination. "As far as I know, I'm not going to go to the Pacific theater of operations, so you can come as close as I can to determining just about where we are going. I really do not know exactly. I can be Italy, India or China after North Africa."[177] Royce asked Pearl to keep confidential the fact that they would be passing through North Africa, since until he was assigned an A.P.O. address his letters were not subject to Army censorship, and he feared giving too much information.

Pearl, in turn, had written letters which betrayed a sense of depression about his approaching departure. This brought a response from Royce, who encouraged her with his own pragmatic approach to the uncertainties the war had brought into their lives. "I hope you aren't letting things get you down. I'm afraid that won't do very much to alter the facts. I'm here about ready to go over. You're there with the baby. It's rather a cruel thought and yet even it has its bright side if we but remember how good life together has been, work hard today, and look to the time when we shall have that same life together again. Don't let it make you bitter. Life hasn't singled us out to be especially cruel to. There are many less fortunate than we. I see them and hear about them every day and I am sure you do also. I know that thinking this way very possibly does not make the burden any less, but I feel sure it makes that burden more bearable because there are so many of us in the same boat. Keep your chin up and your hands and mind occupied in keeping your every faculty from drifting toward despair and hopelessness. I don't think God has forgotten us by any means. He has blest us very much. Look at Baby Jean and you know that. I don't know whether I have ever told you how conscious I have felt of the thinness of the line that separates us from the next life. I feel it every time we leave the ground, and yet I always feel secure at the same time, and know that we will feel the wheels touch the ground in safety as we terminate our mission. It's been that way Darling and I'm sure that it will continue thus. My prayers have always been to that end."[178]

Wendell, the flight engineer, was released from the hospital on Wednesday, April 12 and the crew spent part of that day stowing their personal gear into special racks which had been installed in the bomb bay for that purpose. Finally on Thursday, April 13, they took off from Hunter field and flew south to Homestead Field just south of Miami, Florida, which was to be their point of embarkation. Here the ship and crew were given their final checks before shipping out, and Royce took the opportunity to write one last letter to Pearl before leaving the United States and flying off to war. Unable to mention where he was or where they were bound, he told her what little he could, that the weather was hot and that his next letter would be from far away. He followed with his final farewell to the wife to whom he had now written for almost seven years, and with whom he had lived for scarcely six months. "Now I'm going to tell you what I am able to say. I want you to know that I shall be praying always for you and also for me. I'll be thinking of you at all times. You will be near me and I will feel you there and know that your dreams, your thoughts and also your prayers have gone out with me. I will gain strength from them. I shall be comforted when I feel alone and shall feel a special protection when we are winging our way above. My prayer shall be always that my return to you will be certain and as soon as possible. I dreamed of you last night, one of those beautiful almost real dreams that you hate to leave. I walked with you, talked with you and held you close to me. I love you Darling with a love that I hope will reach out to you accross the miles and tell you I am well."[179]

Finally on the morning of Saturday, April 15, 1944, the new B-25 stood in the pre-dawn darkness at the Homestead Army Airfield awaiting their departure. "There was no fanfare, no one to bid us goodby and wish us luck. We were one crew of many and everything was strictly business. We were awakened very early in the morning and went to the mess hall for our last meal. We checked out our bedding and things and obtained our final clearance. By that time we knew our destination and course. We were driven out to the ship where we pulled the pre flight inspection and checked our equipment and packing for the 'nth' and last time. It was barely light as we taxied out to the runway. The engines were run up for the final check and a call was made to the tower to clear us for takeoff. We started down the runway into the dawn, felt the wheels leave the earth, circled a couple of times to get the prescribed altitude and then headed out to sea. In a few minutes land was fading and had faded out of view. We were on our way to war."[180]

Illustrations for Chapter 5

Left: Royce as air cadet in Santa Ana; right: Frank Denstad, Royce's closest friend during flight training, who later became a pilot in a B-24 squadron.

Royce's squadron at Santa Ana shortly before completing ground training. Frank Denstad is on bottom row at far left, with Royce directly behind him.

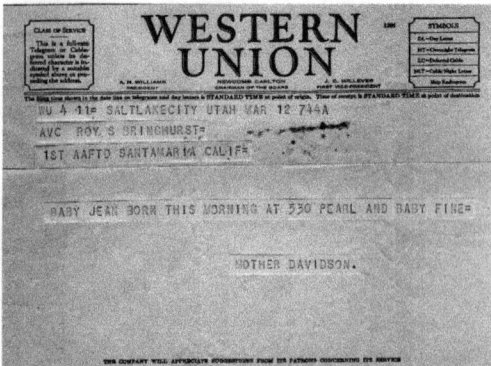

Left: Telegram received in Santa Maria; right: Royce with Jean in Chicago.

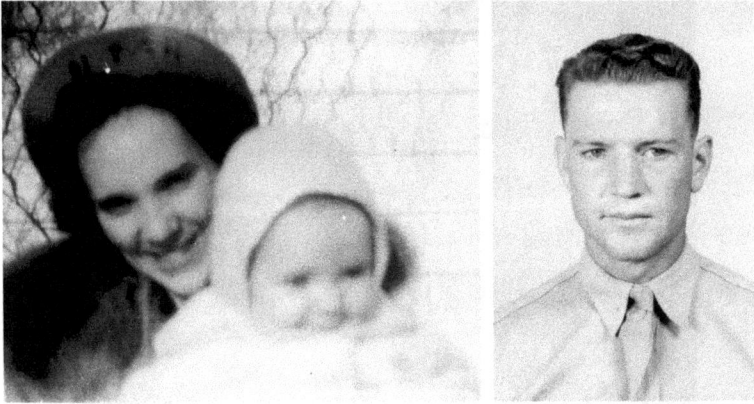

Left: Pearl with Jean in Chicago; right: Royce's official photo at Sioux Falls.

With Pearl during furlough, December 1943.

During crew training at Columbia: Royce (left) with friends
Guy Anderson (Andy) and John McNally.

Crew at Columbia: (standing) Pilot Brassfield, Co-pilot,
Bombardier Harper, ground crewman; (squatting) gunnery
instructor, gunners Wendell, Bringhurst, and Robinson.
The aircraft is a B-25H, the bomber used in training.

Combat Airman

Chapter 6

The Journey to the Front

Between the time Royce enlisted in 1942 and the time he finally entered combat in the Spring of 1944, the war situation had changed dramatically. Where before the Axis armies of Germany, Japan and Italy had seemed numberless and invincible, the tide had slowly shifted in favor of the Allies. The Japanese had suffered defeats at sea, and a long and relentless island campaign had begun to dislodge them from strongholds in the South Pacific. The Germans, after occupying nearly all of continental Europe, had been halted in their attempt to conquer Russia, and were now suffering heavy losses in the east, while the cities in their homeland were being devastated by Allied bombings. German and Italian armies had been defeated and expelled in both North Africa and Sicily, and the Italian Fascist government had capitulated. Italy itself had been invaded by Allied troops, and a long grueling campaign had begun to drive the Germans and the Italians still loyal to them out of the peninsula. Although the worldwide military situation was now much more hopeful for the Americans and their allies, the final outcome was by no means assured. The war continued to rage, and though there was now reason for optimism, no one knew exactly when or how the conflict would end, and it was certain that much of the worst fighting still lay ahead.

When Royce and his combat air crew left the United States to begin their long journey to the front on April 15, 1944, their final destination was still shrouded in secrecy, though they had orders to make their way to North Africa, and they assumed from the type of aircraft they had been issued that they would enter the European theater. Their progress was marked by succinct entries in Royce's journal. "April 15: We completed the first legg of the journey landing at Borinquen, Puerto Rico.[1] I lost a trailing wire (250 ft.) when we landed. It seems that I forgot to take it in. I was very mortified. This is a honey of a base. All the buildings are of concrete and very cool. It's got the best P.X. I've seen anywhere. Slept mighty well.

"April 16: Left Borinquen for Atkinson Field, British Guiana today. We made it to a little past Trinidad when we picked up instructions to return to Waller Field, Trinidad because of bad weather on the route to Atkinson. We flew in a storm for a time in a sky with a very low ceiling but made it all right to Waller Field. This is malaria country. The camp is whacked right out of the jungle. I wouldn't want to be 'permanent party' here. Slept very well for a strange bed. It cooled off at night though it is really humid.

"April 17: We flew from Waller to Atkinson Field, British Guiana today. Part of it was over jungle, the rest over water along the coast line. This camp sets right in the jungle too. Not too bad a place though I would not like to stay.

"April 18: The most dangerous part of our trip today. We flew from Atkinson Field to Belem, Brazil by the Amazon River. It was mostly over dense jungle but some of it was over Savannah land, a name they have for swampy meadow land. We let right down when we crossed the Amazon and meadowland because of a low ceiling. It was most interesting. I never realized that the Amazon was so large. We crossed the Equator too without ceremony. It felt no different on the other side. It was nearly 1300 miles.

"April 19: Flew from Belem to Natal, Brazil, the jumping off place for the other hemisphere. We passed over some jungle, a lot of mountains and found Natal a pretty hot spot right on the point that sticks farthest out into the Atlantic. We will probably stay here a few days. We have a pretty bad hydraulic leak in one of the engine nacelles. I've had more than a little trouble with the radio. I had to have my liaison receiver changed here."[2]

Royce and his crew spent a few days in Natal while repairs were made, then the journey continued. "April 22: Left South America on the first leg of our transatlantic flight. It was all over water to the rugged lava islands, the Ascensions. I didn't think to find mountains there. I watched the sun set into the sea tonight and thought of what a small point we were on. I wish Pearl might have seen it.

"April 23: Flew from the Ascencions to Roberts Field in West Africa (Liberia). It was quite an uneventful trip. We ran into traces of bad weather but saw nothing very striking. We landed at a rather pleasant place and saw our first natives. They do about all the work at the camp. It's still malaria area.

"April 24: Went from Roberts to Dakar, the first French territory we have touched. The trip wasn't much; water and coastal area which might have been most any coast line except for native villages. They grounded Lt. Brassfield here for four days because of some sulfa drug he took several days past. We didn't have much of a time. The tents wern't so good and natives roamed all over the camp."[3] The delay in Dakar while waiting for their pilot allowed the men some unexpected free time and gave Royce a chance to catch up on his letter writing. The soldiers swam at the beach, and for the first time Royce saw a domestic camel in its native habitat. "This old boy had a big load piled on the back of his camel and then he was sitting on top of everything, and the old camel was plodding faithfully along with that swaying funny walk that they have. Its nose was held way up in the air, giving the impression that he was above his surroundings."[4]

When after four days the pilot was released to fly, the journey continued. "April 28: Went from Dakar to Marrakech today. Most of it was over desert but we did fly through some rugged mountains toward the end. Saw our first Italian prisoners. They do all the work at the camp. On the other hand the ones the French have are kept locked up in barbed wire enclosures like so many sheep.

"April 29: Went from Marrakech to Algiers. By far the most pleasant of the whole journey. We hit the coast line at Oran and followed it all the way along to Algiers. We were supposed to have flown to Telergma but bad weather was ahead and we were instructed to land. Not a bad camp, and the food is good. We went into town several times. You must watch your pocket book and be most careful where you go. Our ship was grounded several days in order to replace an actuating cylinder that has caused a hydraulic leak all the way in the left nacelle. I could write about what most of the boys did but I would as soon not do so."[5]

Finally after many delays, Royce and his crew made the one-hour flight to their destination, more than a week behind schedule. "May 3: Flew from Algiers to Telergma, the end of the trail at least for the present. Here we are supposed to receive our combat training. They took our ships to the modification center. We have four days school here and normally would be here about two weeks."[6] Having at last made their arrival at the forward training base, the men were once again issued new supplies. "Our tent looks as if we were going to have a rummage sale featuring G.I. equipment. Seems that I get more and more as time goes on. They certainly have seen that we get everything we need plus some things I wonder if we will ever need."[7] The food was excellent with bread similar to the homemade style Royce had loved back in Utah, with good quality butter. The only thing he missed from home was fresh milk.

Training passed quickly, and with their final combat instruction complete, the time approached for Royce's crew to move on to their combat station. By this time Royce had learned that he was part of the 12th Air Force, and that he would likely be based on the island of Corsica, off the west coast of Italy. On May 11, with departure soon at hand, Royce and his friend McNally decided to make the 30-mile trip to the nearby ancient town of Constantine, perched atop a plateau flanked by precipitous gorges. Royce was impressed by the sight. "I wish you could see the huge bridges they have there spanning a large gorge that must be better than 1,000 feet deep. We saw about everything there was to see, behaved ourselves and returned in safety."[8]

Friday, May 12, was the scheduled day to ship out, but once again mechanical problems came up, delaying the departure of Royce's crew for two days. Any disappointment they may have felt was immediately overcome by the arrival of the mail, which had finally caught up with the overseas training station. Royce related to Pearl, "There were twelve in all from you, four from the folks, one from Naida and one from Uncle Sam."[9] Among other things, Royce learned that his telegram from Hunter Field had had its desired effect, and Pearl would shortly receive $200 in back payments of Royce's allotment for a dependent child. This brought him no small relief, and he eagerly read about events at home.

On Sunday the day for departure finally arrived. It was Mother's Day and, by coincidence, also Royce and Pearl's anniversary. He wrote in his journal, "May 14: Our wedding anniversary and I didn't even get a letter written to Pearl. I couldn't help it though. We left Telergma about 11:00 a.m. for Corsica where we are assigned. We flew over to Sardenia, then down the west coast of it to Corsica. We landed first at Guisonaccia [Ghisonaccia] on the east end of the Island. It is the headquarters of the 57th Bomb Wing to which we are assigned."[10] From headquarters, the crew was sent on to Alesan airfield, the location of the 340th Bomb Group, where they would become part of the 489th Bomb Squadron and would enter combat service. The men could hardly have known as they waited in North Africa that the two-day delay at Telergma had probably saved their ship, and possibly their lives as well.

Combat Duty in Corsica

As the plane made its approach to land at the Alesan airfield which would become their home for many months, a scene of devastation met their eyes. "The Germans raided this field last night, and really did the damage. It's the first time the squadron's had a raid here and they were caught napping. Over a hundred of the boys were wounded, many very seriously. Eleven were killed outright and one more died this morning. This is just from our squadron. They first hit the line where all the ships were with 500 lb. bombs. Only five ships of our squadron were able to take the air on a raid today. Several were blown to bits by direct hits. Others were riddled by bomb fragments and strafing. Most of the casualties occured there because all the mechanics slept there. Many lost their tents and everything they owned. They then bombed and strafed the bivouac area, hitting the 489th much the hardest. Very few had slit trenches, none were covered and many were caught in bed. It really made a mess out of things. The raid started at about 4:30 a.m. and lasted for an hour." Royce added, "Needless to say, after putting up our tent, the first job was to dig a deluxe fox hole with cover right in front."[11]

The 489th squadron, of which Royce was now a member, was a distinguished but somewhat star-crossed unit. A "medium bombardment" squadron flying the rugged twin-engined B-25 bombers, it had flown missions against German military targets first in the North African campaign in Tunisia, then in the Sicilian campaign, and for many months had been flying over Italy during the long drive to dislodge the Germans from the peninsula. Less than two months before, the squadron had been operating out of an aerodrome in Pompeii, Italy, when the 1944 eruption of Mount Vesuvius forced evacuation of the camp, and falling cinder and rock had resulted in the destruction of nearly all the unit's aircraft, though no lives were lost. It was scarcely a month after the squadron had been moved offshore to the island of Corsica that the German air force launched their surprise attack in the early hours of May 13, 1944, once again destroying or severely damaging most aircraft, and inflicting heavy casualties as well. When Royce and his crew arrived the following day, cleanup from the raid had scarcely gotten underway, and during the days and weeks that followed, the entire camp was set on edge whenever an occasional observation plane caused the air raid alarm to sound. The raid had left the squadron crippled, with few airworthy ships. Although its few undamaged planes had been sent as a token force on a mission the day after the raid, the lack of aircraft meant that the new arrivals had time to establish living quarters and accustom themselves to their new surroundings before being assigned to fly combat missions.

Royce had few acquaintances on the base aside from his own crew. Though his friends "Andy" and McNally were both already on Corsica, they had been assigned to a different bomb group and airfield on the island, and he had yet to locate them. Royce and his air

185

crew had arrived at the height of Spring, and in spite of the marks of war, he was struck by the beauty of his surroundings. "The sea is of a blue that seems a little more pleasing to my eye than any sea water I have ever seen. It seems so calm and peaceful when you look at it. The terrain and mountains around are most beautiful. Everything is so green and fresh. If one didn't know I think it might be hard to realize that there is a war going on."[12]

During the days before Royce's crew was assigned to combat missions, he observed with interest the activities of the squadron. The 340th Bombardment Group, of which Royce was a member, was composed of four squadrons which usually flew together. These squadrons were the 486th, 487th, 488th, and the 489th, Royce's squadron. Planes from each squadron were identified by a number on the tail indicating the squadron (a "9" in the case of the 489th), followed by a letter which denoted the specific plane. Most of the missions flown were against railway installations and bridges, with the intent of disrupting the flow of personnel and supplies to enemy positions. Notwithstanding the German air raid just before their arrival, by this time the allies had achieved air superiority in the region, and the few attacking enemy fighters were concentrated against the larger strategic bombers which flew against industrial targets far into the enemy's heartland.

The smaller two-engined aircraft of Royce's bomb group were flying against targets much nearer at hand, and they were often protected by escort fighters flown from nearby airfields. The bombers flew in "boxes," tight double-vee formations of six aircraft each, three in front and three behind and slightly lower. These formations were designed to concentrate the pattern of bombs on a target, and the defensive fire from their gunners against any attacking aircraft. Royce wrote, "Raids are launched daily from the field and sometimes twice. Targets are the supply lines feeding the Germans south of Rome. Very few fighters are encountered and for some weeks none at all. Flak is usually heavy however, but few of our ships are ever damaged. We will not fly with the crews we came over with, at least for the present."[13] The practice in the squadron was to split up newly arrived men among the more seasoned crews for the first few missions, to avoid having completely green crews face combat situations.[14]

At the time of the German air raid, the squadron had recently suffered the loss of experienced pilots who had completed their tour of duty and rotated back to the States, and with less experienced flight leaders at the helm, they had had a frustrating series of failed missions. This had just begun to turn around by the time Royce flew his first bombing mission on May 24, ten days after his arrival. He was assigned to a plane designated 9-X, flown by a pilot named Thomas,

186

and of his original training crew, only the bombardier Harper was aboard his ship. In a pattern which continued throughout his combat service, he wrote a detailed account of this successful mission in his journal: "My first combat mission. Maybe, I should say 'milk run' because we met no opposition; no flak & no fighters. We knocked out both the primary and secondary target. We dropped ours on a viaduct at Ponggibonsi Italy. It's inland, and just a little south of Leghorn. The target was knocked out by a previous box from 487 squadron. Our lead box (6 ships) hit the secondary, an ammunition dump, and it really blew up. We could see flames hundreds of feet in the air from it. The concussion must have been terrific."[15]

The next day, though Royce was not assigned to fly, the hazards of combat flying were brought back into focus. "We lost a bomber and crew today from the squadron. They were brought down by flak over the target. I knew the tail gunner. His name was Scott. We hope they parachuted out safely. Several chutes were seen to open. It's a tough deal."[16] The following day Royce flew on his second mission, another "milk run" in which a railroad bridge was demolished. The third mission, three days after the first, proved to be a little more eventful. "Mission #3 this afternoon, definately not a 'milk run.' We ran into flak so thick you could nearly get out and walk on it. Out the window I could see it bursting all over. They had our altitude very good. It was pretty accurate. We were under fire for a full five minutes, which is tough. I'm not sure about the target. I sure pulled my steel helmet down and made sure that my flak vest was covering me. The target was a road and railroad Bridge in Northern Italy, Castiglione, inland from Pisa."[17]

The greatest danger the airmen faced was the anti-aircraft fire, and Royce described in a letter to Pearl both the flak and the measures taken to protect against it. "We have flak suits which we wear. Call it armor if you like. The design of it varies somewhat to suit the position you occupy on the ship. The kind I wear is a three piece deal, two parts in front and one in back. It comes down past the crotch in front and to the belt in the rear. It weighs quite a bit but we are glad to have them to wear. It's good for morale. We also have special steel helmets. They are designed in such a way that we can wear head sets inside. They look about like a regular G.I. (M1) helmet with special hinged earflaps. We aren't a bit reluctant about wearing them either even though they also add weight. I have ridden thru flak too a number of times. It's kind of pretty, in a gruesome sort of way. Some of it is black and some white. It bursts out like mushrooms and looks harmless enough, but...."[18]

Some of the missions involved planes from multiple squadrons and groups. Such was the case with mission number 7: "We bombed a road bridge near Ternia Italy to make a road block. We were caught in one barrage of flak on our way to the target. It was very well concentrated and very accurate. We very abruptly changed altitude and no ships were hit. I was scared for a minute (and more). That stuff was really close. We really knocked out the bridge. That valley has been bombed effectively from one end to the other. We saw B-26's and P-47's working over there too. Fighters were said to be sighted. It was hard to tell, there were so many aircraft. None attact anyway. Our escort (Spitfires), shot up the flak batteries that opened on us." Royce was conscious of the destructive power their bombs carried, and was aware of the danger they brought both to the people and places below. Perhaps to console himself he wrote, "I don't think we hit many people with our bombs if any. We bomb strictly military objectives. Bombs are brought back if they cannot be released at the target or an alternate."[19]

After each mission there was a debriefing followed by a light snack, which always went over well with Royce. "We go first to interrogation and then we go over and have donuts and coffee if so desired. I always did like my donuts. The red cross supplies and serves these." In general Royce felt good about the combat missions at this early stage of his service. "I rather enjoy going on them. It is a good feeling too to know that you are actually contributing to the war effort in an active way. We don't have things too tough here. I suppose I can say that."[20]

Between missions, Royce and his tent-mates, among whom was James Robinson ("Robbie"), the tail gunner from his training crew, set to work raising the standard of living in their military tent. "Gradually, we are getting some of the comforts of home around our little tent. Today, Robbie and I installed electricity in our little homestead. It is good to have good light again. If we only had a radio, it would really be A-1."[21] The wiring for the project had been stripped from disabled bombers which had been damaged beyond repair. Some days later, Royce reported on the ongoing construction efforts. "Eikhoff and I went down to the motor pool and sweated out a jeep so we could go and get some boxes to make shelves and lockers out of. This afternoon, we got busy with picks and shovels and worked over our tent floor again. We've got it nearly straightened out now and when we started there was quite a steep slope to it. After the pick and shovel detail we proceeded to use the boxes we obtained in the morning. I built myself five nice shelves and a foot locker. I also obtained a piece of hardwood about 2-1/2 feet square which has a smooth finish to use for a writing table. I have it fixed as such and am using it now. The edge of my cot is what I use for a seat."[22]

Royce took special delight in describing the animals with which they shared their living quarters, some of which were welcome and others less so. "We've had quite a time with the ants here. They have been rather obstinate about giving up this particular bit of earth and letting us take over. We have used various methods of persuasion or discouragment, whichever way you think about it, to convince them that they arn't happy here. The gasoline-fire method seems to pay off the best. At least the colony that had set up their happy home on our tent floor and the other group that had their hill out on the side of our fox hole have decided we arn't good neighbors and have moved out. As for our other neighbors, the lizards and few turtles, we don't mind them and evidently they don't mind us because we see plenty of them. One evening, I was writing a letter and felt something nudge my foot. I looked down and saw a turtle who had been interupted when he found my foot in his pathway. There is a pretty brown dog that has adopted our tent for sleeping quarters. She doesn't have very much to do with us in the daytime because she has social affairs to look after with the other squadron dogs down by the mess hall, but every night along about 8:30 she very faithfully makes her entrance and settles down for the night. Some of the dogs even have several missions to their credit and most of them, though it is forbidden officially, made the long trip from the States by air, living on "K" rations. They all have quite a fussy outlook on food; nothing but the best. Such is the dog's life at an overseas air base. There is "Rookie," "Snaps" and about any name you can think of."[23]

Since missions were sporadic and often did not require the full squadron, there was plenty of free time to occupy. "It's not too hard to adjust your living over here as long as you do something during your leisure time. We have quite a bit of that too. Our duties consist of flying missions and cleaning the guns we fire when we get back. The rest of the time is ours. I nearly always have a book which I am reading. Time passes quite agreeably."[24] The task of cleaning the guns came to be routine. "It's not much of a job. The assembly and disassembly of a gun doesn't amount to very much. We did that so many times at gunnery school I can do it with a blindfold. They are important to us so we take good care of them. Water isn't the universal solvent over here. It's 100 octane gasoline. We clean the guns and most everything else inside and out with it. It really does a good job too. It's good for cleaning clothes also."[25]

To relax, the men spent their time writing letters, reading the books that circulated in the camp, and swimming at a nearby secluded beach. Royce considered swimming to be a basic human skill, and in their setting, it was a potentially lifesaving one as well, since stricken aircraft occasionally had to ditch at sea. When Royce learned that his crewmate "Robbie" Robinson did not know how to swim, he took him under his wing and undertook to teach him. In letters home, Royce described the swimming area. "We have an ideal set up with the sea and a pond of clear fresh water side by side. It's sort of our private beach. We go swimming without suits. It's nice to get out in the water a ways and let the gentle waves rock you as you float. The fresh water is an ideal place to learn to swim in. The bottom is all sand and slopes down gently from a few inches to well over your head."[26] After several weeks, Royce was able to report, "Robbie has finally learned to swim. He took his first real strokes today. It was mostly a matter of overcoming his fear of going under. He is still so tense in the water that it tires him out in a very short time. I'm glad he is getting it now. We have kind of sweated him out."[27]

Some evenings the men had the option of going to see a motion picture, which was shown out of doors, with the soldiers seated on a hillside. "Usually, the shows are pretty good here. They set up the projector at the foot of a hill and the fellows come from nearly all over here to attend. We usually carry something along to sit on."[28] The movies shown were popular features brought over from the U.S., and Royce remarked, "I don't know what the fellows would do without the show to go to. It changes every other night. You can go there and take up the evening which is the time when your thoughts just naturally shift more toward the things you want to go home to."[29] He described the taste in cinema among the soldiers in the squadron. "We see lots of shows, good and bad. The ones I dislike most are those about Hollywood people touring overseas army camps, or I guess any pictures about the army for that matter. They always seem phony and stupid. The only shows about the army anyone enjoys are the ones which make a burlesque of the army and make it all show up in a ridiculous light."[30]

When Sundays came around, Royce found himself completely alone as far as religious observance was concerned. He did not know any other Latter-day Saints on the island, and since there were no branches of the Church in Corsica, he inquired about other worship services on the base. Not long after his arrival, he learned about Sunday evening meetings which were being held by the squadron Adjutant, a Baptist named Captain Anderson. These meetings chiefly involved the singing of hymns, which Royce especially loved; he also had considerable respect for the captain, a former preacher who had started the meetings of his own initiative because he felt there was a need. Royce became something of a booster, in spite of their doctrinal differences. "I attended the hymn-sing services that Capt. Anderson conducted tonight. It was good but will be better if they can get more to attend. I suppose I should say 'we.' I made the suggestion that we have a discussion in which all participate. So far, it has been like a most typical protestant service, in that they sing a lot of songs and then the captain preaches. I can't agree with some of the doctrine but so far I have kept silent." Regarding their religious differences, Royce reflected, "Sometimes they seem so I don't know hardly what to say, maybe illogical would be the word. You can't get men who deal strictly in realities to believe in things that are explained to them in such a way that it sounds like a wonderful fairy tale. I know that's hardly the way to say it but I do so for want of better words. What I believe in is as real and as tangible as the ground I walk on."[31]

June 6, 1944, began as a typical day with another bombing mission, which Royce reported on in his journal. "Mission #8. This was really a milk run, short and no flak or fighters. We made a road block in the little town of Vetralla on the present Bomb line." It was while monitoring radio traffic from his seat in the aircraft during this mission that Royce learned the truly important news of the day, first in a propaganda broadcast from Germany, then in a report from England. "Today was an historic day. Our forces successfully established beachheads on the coast of France (Normandy). Le Havre was the point of attack." To Pearl he added, "It really made me feel good when I picked the report on the radio this morning that the continent was being invaded. I guess that no one hopes more that things will go well than we do. The whole world has waited for this day." The day of the allied invasion also marked an anniversary of sorts for Royce. "It was just two years ago that I was sworn into the army. Quite a bit of water has flowed under the bridge since then. It's been good experience, but it's much nicer to look back at it and be glad you havn't done any worse, than to look forward to sweating it out. I hope I am a better man for having been through it."[32]

The next day, Royce learned that he had been assigned back to his old crew from Columbia, or at least what was left of it. "I am again on Lt. Brassfield's crew. He and Harper are the only others of our old crew on it. Wendell, being a mechanic, works on the line. Robinson is on another crew. We have a ship too. It isn't the one we brought over, but just about the same except for color. This one is a silver unpainted one. We havn't named it yet but must do so soon. I have ideas about names, but the whole bunch of us will decide including the ground crew. In a way, they are more important than we are because they keep the ship in the air. It will probably be something a bit silly, most of them are." The ship, designated "9-R" would eventually be christened "Snot Nose," and rather than the racy nose art for which the squadron was famous, would bear the picture of a baby boy in a cowboy hat, accompanied by a puppy.[33]

On Thursday, June 8, "9-R" flew on its first mission with the new crew, to a place called Bucine. "Mission #9. We bombed a viaduct at

191

Bucine, Italy. I flew with Lt. Brassfield in our ship with our crew – copilot Walsh, Bomb. Harper, Turret Gunner, Brown, and tail gunner, Carter. Ship 9-R. We flew #3 in the second box. Our box was the only one that hit the target. The other two boxes both hit short. We knocked out the viaduct. I took pictures. We met no fighters or flak."[34] Royce's next mission, flown five days later on June 13, in "9-R" with Lt. Brassfield at the controls, had a less successful outcome which was sobering to Royce. "We bombed at a bridge at Perugia. At least we were supposed to have bombed it. We were third box and didn't drop ours. We took the alternate, a bridge below Leghorn on the coast. We dropped 24 1,000 lb. bombs, they hit quite a bit to one side on some peasants house and the road. I hope no one was in the house. The main target at Perugia was knocked out. We ran into flak four times, once going to the target, at the target and two times coming back. Several bursts were accurate but most of it was wild and not concentrated. Two ships were hit. No one was hurt. All were scared."[35]

A couple of days later, Royce flew on a different type of mission. "Mission #12 – this one was what is called a Nickling mission. In other words, dropping propaganda leaflets on the enemy. We flew at 13,000 feet. We dropped the leaflets just like bombs, eight big bundles to the ship (four ships). I threw smaller bundles out the ditching escape hatch. We dropped them over Piambino, where we saw a little flak but not enough to worry about, and Leghorn Harbor which usually puts up a heavy barrage but didn't today, and over a town about 40 miles from Leghorn, Cienetta."[36] These propaganda leaflets, or "nickeling sheets," were designed as a small folded newspaper-like bulletin. Scattered by the bombers over large areas occupied by enemy troops, the leaflets were designed to weaken the morale of the retreating German army with news of Allied victories on all fronts.

On the same day as the leaflet mission, Royce finally received news of Guy Anderson, the inactive "Mormon" from Fairview, Utah, whom Royce almost always referred to as "Andy," and with whom he had been close friends since leaving radio school at Sioux Falls. "I heard that Andy got hit in both legs by flak his second mission and was sent back to the states. It must have been bad. I was more than sorry to hear it."[37] Having received only the barest information, Royce wrote to Pearl asking that she write to Andy's mother to inquire about him. It was only with time that he would learn the severity of Andy's leg wounds, which would leave him hospitalized for many months and incapacitated for life. His wounding after so short a time was a sobering blow, as Andy, McNally, and Royce had been close friends throughout most of their training and had looked forward to serving together.

Royce's next mission was his thirteenth, the survival of which, in the ceremonial superstition of bomb crew lore, was a landmark referred to as "getting over the hump." In Royce's case, the time leading up to the "hump" involved an anxious waiting game, as stormy weather forced cancellation of missions for several days. Finally, on June 21, Royce wrote, "Mission #13 and over the so-called hump. We have sweated this out four days because of bad weather. In the meantime the French have taken Elba, the only enemy territory we could see from camp.[38] We bombed a place called Savaiona up between Florence and Bologna. We really caught the flak going into the target. We knocked out both the primary and alternate targets (RR viaducts). We hit the primary. Several ships were holed. No one was hurt."

Despite the milestone, Royce by now had seen enough of combat flying to realize that his tour of duty was going to be a long haul, particularly as he learned that the maximum tour of duty for combat crews had been increased. "The going is getting rougher. I hope I make it. They say we have to fly 70 missions now, or a year of combat. It was 50."[39] The next day's mission did little to dispel Royce's concerns, though he was not assigned to fly. "They really caught it today. I didn't go. They were over by Florence. In one Box of the 310th squadron flak knocked down five out of six ships. The other limped as far as our field on one engine and bellied in. The pilot and copilot as well as the others on the ship were quite badly wounded. One of the 487th's ships came in on one engine with wounded men. I saw one other ship with a prop feathered. All of our ships had holes in them. 9-P had the hydraulic system shot out and hit a tree at the side after landing. It was really a rough day."[40] The following day, spirits lifted a little when three of the airmen from the squadron, who had been shot down over enemy territory on May 25, returned after making their escape to the Allied lines.[41]

Despite the frequent danger and the sense of gloom that hung over the squadron whenever planes and men were lost, Royce wrote encouragingly to Pearl, and outlined the mental approach he took to the realities of war. "There really is no reason to worry and fret about it. I'm sure I don't. It wouldn't help in the least and so far just accepting things as they come along without question or anything else has worked out fine on this end. Our ships go out and they come back. One day I go along another day I don't. We live quite a happy life. We manage to keep nearly the whole thing on the humorous side and that helps plenty. When we take off I don't feel any different than I used to feel going on practice missions back in the States even though I am fully conscious of where and why we are going and the fact that the enemy isn't laying out the 'welcome mat' for us."[42]

As the weeks passed, Royce settled into a routine which was governed by the rhythms of camp life. After two months in Alesan he wrote, "I really believe that I notice the passing of time less over here than I did over in the States. We sort of measure time by the coming and going of ration days and they seem to be quite frequent. We get P.X. rations once a week and that usually comes to two or three candy bars and maybe some lifesavers and a package of gum. We get enough to eat but I always seem to have a craving for something like that. The result is that they go awfully fast when we do get them."[43] In June Royce received news of a promotion to the rank of technical sergeant, which brought a welcome increase in pay, and he at last paid a visit to his friend McNally, who was still awaiting his promotion, and had teased Royce during training about having received the first promotion there.

As June wore into July, the missions became both more frequent and more intense, and Royce was scheduled for four hazardous missions on successive days. On June 30, he wrote, "Mission #14 today, and definately no milk run. We ran into flak four different times. We were supposed to hit a tunnel at Prato north of Florence. We caught our first flak as we crossed the bomb line.[44] As we made our way to the target we were supposed to hit we were under almost continuous fire. When we were on the run it was so hot they couldn't sight. Everywhere I looked it was a solid mass of black puffs. The whole sky was covered. I don't see how we got thru. I could see the bursts right off the wing. We could feel the concussion and smell the smoke. Several ships were hit but no one was wounded. 9-H had the hydraulic system shot out. We dropped our bombs on a bridge near Leghorn. I was really scared today. It's not much fun when you have to stand up there & watch that stuff follow you without being able to do a thing."[45]

The next day he recorded, "Mission #15. We bombed the tunnel we were supposed to get yesterday at Prato. We tried to outfox the flak by a different approach but we caught it as soon as we crossed the bomb line. They followed us all the way to the target. We got two holes in the ship. One you could put your fist through was in the nacelle of the right engine. It came close to knocking it out – 1/2 inch one side and it would have hit the feathering motor, a half inch the other side and it would have hit the hydraulic lines. As it was, it hit a brace. It really jolted the ship. The other hole in our ship was in the bombardier's compartment. We didn't hit squarely on the target and were under fire again on the way to the coast. Just south of Leghorn by the coast, as we were going home we saw a B-25 of another outfit knocked down in flames by the target they were attacking. I saw it hit. It's not a pleasant thought. I was plenty scared today."[46]

194

July 3 marked the fourth of these missions for which, with all the participants, Royce would receive one of six air medals. "Mission #17. We had our introduction to the Po river valley today and really caught it. We hit oil storage tanks at Ferrara, on the banks of the Po River with incendiaries & demo's. The only flak was right on the target but it knocked down 3 of our ships and put holes in every ship that went over. We flew #5 and got one hole. The ship alongside (9-M) got 18 and one in front got so many they couldn't count them and a wounded radio man, Hunt. He got a bad wound in the upper leg. That ship nearly crashed on landing. One of the 486th's ships did but no one was hurt. It was really terrible. On the bomb run it burst right under us so loud and with such force we thought every one was the last. We got the target. One enemy fighter made one attack but failed to do any damage. Three ships came in with single engines. 9-H caught a burst in the bomb bay while it still had a load of incendiaries. Fire started but they managed to salvo the bombs. The bay doors wouldn't close." Even in the face of mortal danger, as a lifelong farmer Royce couldn't help admiring the fertility and natural beauty of his surroundings. "By the way, the Po valley is one of the most beautiful and productive looking valleys I've ever seen."47

Because Royce's mail was censored to assure the secrecy of his squadron's operations, he was unable to write about these missions in his letters home except in the vaguest terms. His letters still found ways of capturing the essence of what it was like to fly on a bombing mission over enemy territory. "I've been on another mission and have the same thing to say as before; 'back safe and sound.' I don't want you to think I don't sweat them out because I sure do. I can't help feeling that certain something inside when I see flak bursts. About that time, I find I'm never cold in the least. You'd have to see it to understand I think. I caught on right quick. I don't think of myself as really green any more. All of us know pretty well what the score is."48 On another occasion he wrote, "Sometimes, I think I have pretty well reviewed my life while we have been cruising along on the way to some target. Of course we are always watching and keeping our mind on our job, hoping the flak won't be too bad and that we get to the target all right. Observations must be reported and a lookout must be kept for fighters. I do notice the cloud formations. We see all kinds, over, under and on both sides of us. I like the big fleecy white ones. We steer to avoid them and they tower alongside like huge mountains of whipped cream. We kind of sweat them out too because if we find them obscuring the target, we cannot bomb. The little villages and valleys below look so pretty and peaceful that at times it is hard to realize than an enemy lurks below who might open up on you any time with 88 mm guns. Of course, about that time we always manage to present a most elusive target and we pull the 'getting out of there' act very quickly. It gives you a feeling of strength to see all your ships in formation roaring through the sky toward something which you know must be knocked out to make the path and job of our boys on the ground a bit easier."[49]

Independence Day in 1944 came and went with no special observance aside from a brief entry in Royce's journal, "July 4th. No mission, just another day." There was mail, however, and Royce was pleased to get a much looked-for letter from his friend Frank Denstad, with whom he still felt a close bond. "I was so happy to hear from him, it's been a long time since we last met. He's still in a B-24 and says that he is wishing yet that it could have been B-25. I can understand that well enough. They have to fly seven to ten hours at a time with three or four of those hours on oxygen. That's no fun. It isn't only that either. It gets very cold up there. It's fairly comfortable at the altitude we operate at with a leather summer flying jacket and a pair of light gloves. I wish I could get over to see him but there is a war going on and maybe our visiting will need to wait."[50]

By this time, Royce had received the additional assignment of mission photographer during the missions he flew, which he described in his letters home. "Did I tell you that I get a camera every mission I go on now? It gives me a little something more to do while we are flying. I take pictures over the target through a glass camera hatch right under one of my seats. I also check the bomb bays to make sure our bombs all get away all right and that the doors are properly closed."[51] He later explained, "I take target photos as the bombs burst. It's by the photographs that our bombing accuracy is judged. We do well, but we are always trying to do better."[52] As photographer, Royce seldom flew with his old crew or ship, but was instead shifted around according to the needs of the squadron. As he later explained, "I have rarely flown with the same crew twice in a row. It's the same situation with airplanes. I have flown missions in just about every one the squadron has. People always hear about the way the heavy bombers fly. The same crew goes in the same ship nearly all the time. That isn't so here. It just works smoother for us the way we operate."[53]

On July 12, the squadron was assigned a difficult and dangerous target, a series of heavily defended railway bridges over the Po River near the city of Ferrara. Royce was in the second of two missions flown that day, and on this mission a ship was lost, not to enemy fire, but due to an engine problem which forced the pilot to ditch the plane at sea. The crewmen were rescued the next day, but two gunners, including the radio operator whom Royce knew, had already parachuted out over the sea and were lost.[54] The following day the squadron returned for another attack. "Mission #22. Our primary target today was the much feared city of Ferrara. Only one box bombed. The flak came up on us on our bomb run so accurate and so concentrated that had we continued to the bomb release line I'm sure they would have knocked out half of us. Our box brought all its bombs back. One of the 487's ships limped back way late on one engine. We thought it went down."[55]

The mission flown the following day, Friday, July 14, 1944, gave Royce his first exposure to attack by enemy fighters. "Mission #23. This, I guess was 'the' mission so far. We were hit by F.W. 190's and Me. 109's.[56] They really pressed the attack. We had no escort and they knew it. Our target was a Po river road and two railroad bridges. We really knocked those bridges out plus the alternate, a fuel dump nearby. Every box of bombs was right in there in spite of the fighter attacks. On our way into the Po Valley from the west, we made contact with enemy fighters almost as soon as we entered the Po Valley. They attacked several times on the way to the target. We ran into flak at the I.P.[57] Fighters again attacked after we left the target. I fired about 100 or so rounds out of each gun. I think I may have got some hits. The gunners of our box succeeded in beating off all the attacks on our box. There were three probable Jerry's downed. One was confirmed. Our other box wasn't so lucky. Three ships were shot up, one badly. Three gunners were wounded, one very seriously. A tail gunner named Fetherstone was hit in the face with two shots. He lost an eye and his face was all torn open. The other two were top turret gunners. Porter, a radio operator, was the luckiest. A shot come thru the plexiglass beside him, grazed his head and went through the lense of his sun glasses. He was unharmed. He bent over & picked up the round where it had fallen after failing to penetrate the ship on the other side. I don't know just how many fighters there were. I saw 15 or 20. I don't believe I thought of being frightened while the attack was on. It was rather exciting to shoot at something living. It's no soft war."[58]

Finally, on July 15, Royce recorded, "The Group went after the Ferrara railroad bridge twice today. They missed in the morning. On the evening our squadron knocked it out. It's laying in the Po now. It cost the Group quite a bit. Two ships washed out landing this morning. There was one man wounded. This afternoon, there were 5 wounded on the 486 ships. 9-J caught over 50 holes; none of them were wounded. I didn't go today. That is the 7th mission we have run on that bridge. I went on 3 of them. It's the most heavily defended target we have ever had."[59] These dangerous missions proved the value of the North American B-25J flown by the squadron, for, in spite of the fierce opposition by antiaircraft guns and fighters, the only deaths in the squadron had been of the two airmen who parachuted into the sea after mechanical problems. The "Mitchells" continued to live up to their reputation of sustaining heavy battle damage and still bringing their crews home.

The tension from combat missions had begun to exert its effects on the men, and tempers flared more easily than before. With Royce, this was especially true when he failed to receive mail from home, which in his overseas station came only irregularly. At times he lashed out at Pearl. "Today, I received a "V" mail[60] from you. I hope you won't write too many of them. They are better than no letter but not much. It's just about like getting a post card back in the states when you expect a letter. Please don't send them unless you can't send anything else. I guess I was kind of disappointed today because it's been several days since I received a letter from you and I expected to get a nice long one today. It's all right. You've heard the same old story from me before. Letters mean so much more to me over here. It's the only real link we have with the life we would like to be living." The close living quarters which were imposed upon the men did not help matters. "We in the tent very often find our tempers getting rather hot and arguing over silly things that don't matter at all. It's nothing serious but it does happen. Usually things are forgotten almost as soon as they flared up. It's just a form of diversion. I think I told you once before that it surely isn't good for man to be too long without women. I suppose that has been a problem in warfare ever since man first raised and organized armies. Of course some of them always seek out and find the questionable substitute. Most of the fellows don't see things that way so they try to find some other diversion. We make out pretty well most of the time but the old letters from home really fill the spot. If we fly in the afternoon, the first thing we ask is always if any mail came for us that day."[61]

Some of the tension that Royce felt was due to the fact that his own standards of personal conduct were so different from those of the other men, that he felt he had no one with whom he could share those things of greatest importance to him. "I know I have much less to complain about than most soldiers but sometimes you just can't help feeling alone over here. I have no one here whom I am very close to. I have a lot of friends but it's been so long since I had someone to be around like Frank. I always hate the vulgar conversation that goes on on all sides. I feel embarrassed because I just refuse to have part in it. When I am included in remarks of that nature, I try as tactfully and firmly as I can to express my distaste for it. That is the position I find myself in quite often. I don't want the boys thinking of me as a crank reformist or a 'never-never' boy, so most of the time I keep my mouth shut and ignore conversation in which I desire no part. Sometimes, it's a very difficult situation. They are my friends and

198

I like them and hate to insult, enter into arguments or provoke ridicule upon me. I think that most of the fellows deep down are 'darn' good. They have just drifted with the general tide in the army as it were. When I came in, I had certain standards and distinct ideas about moral values which I knew to be good, and I made up my mind that I should not be separated from those things no matter what. How well I have succeeded in all this most likely remains to be seen. I hope I have done well."[62]

Royce shared his tent with men who formed something of a typical cross-section of American life experience, and he described his tent-mates in a letter home to Pearl. "Eikhoff is from Chicago. He's big and blond and just out of high school when he was drafted a little over a year ago. You already know Robbie, or at least of him. There is also Philip Suskind who comes from Pennsylvania. He's been in the army about six years. In all seriousness, I think that is where he found a home. I don't have much to do with him for reasons I won't explain in a letter. The other in the tent is Daniel Mallicoat. He comes from Detroit. He's about the same age as Eikhoff only he comes from different circumstances. Eikhoff must come from a real home. His father is a rather accomplished musician and his boy is very much inclined artistically. Dan came from a broken home and I think he has rustled for himself ever since he was old enough. He is all right to live with. Robbie and I have the best deal in the tent because we arrived here first and we put it up."[63] During breaks between missions the five men set aside their differences and worked on improvements to make their crowded military-style pyramid tent more hospitable.

Still unaware of any Latter-day Saint services on Corsica, Royce began to attend regular Sunday morning services held by the base chaplain, a modest red-haired man named Cooper whom Royce came to admire. "Our group chaplain is a very fine man. He's not much older than many of the soldiers so he does see eye to eye with us. I think he's been with the Group all the time overseas. He does talk very good sense in his sermons each time, and I like his taste in the hymns we sing. Many of them are hymns we've sang all our lives, and some I have learned attending chaplain's service. I never fail to get something out of his talks. They are always worthwhile, and there is good attendance at the meetings. It's a good sign too. A large percent of the fellows do realize that there is a power greater than that of man. Some think of it in different terms but the thought is the same. I am quite sure that nearly everyone has felt quite close to it at some time or other on missions. I know I have."[64]

In mid-July, Royce learned that Robert Wendell, the engineer from his original training crew whose illness had delayed their overseas departure, was once again in the hospital. Because of Wendell's special training in aircraft maintenance, he was required to work a number of months in the maintenance line before being rotated into flight duty, and that had resulted in a hernia. "He says he did it lifting on a propeller. When I saw him this morning on my way to breakfast, he was kind of worried and asked about it, so I told him that it would be very similar to apendectomy and he would be on light duty for a long time. He didn't mind that idea. If he ever flies over here now, it will be a long time from now. He's been a good hard worker on the line. I don't believe I ever passed his ship when he wasn't busy. He was that way all the time when we were flying over too. I am sorry to see him laid up."[65] Royce liked and respected his old crewmate, and both he and his pilot Brassfield visited him while he was in the hospital.

More sobering was news from Pearl about fellow airmen from Utah who had been reported missing in action. The first of these was Norm Johnson, Royce's Utah State classmate and returned missionary under whose encouragement Royce had volunteered in the air cadets. "I was sorry to hear about Norm. He has a very good chance of being all right. They are all listed as missing 'till something definate is known. He'll probably turn up. I do hope so."[66] Scarcely two days later, he learned about another friend, Frank Van Limberg, who had been with Royce at Sioux Falls, and who with his wife had invited Royce for dinner when he graduated from radio training. "I certainly am sorry he is missing. The last I heard from him, he was still at Sioux Falls trying to bat his way through Radio School. He is a darn nice fellow and has a fine little wife. I hope you see her. I'll say the same about him as I did about Norm. I believe he has better than a 50-50 chance of still being alive."[67] Perhaps encouraged by the news from his missionary companion Reid Ellsworth, and by the returned crewmen from his own squadron, all of whom had been listed as "missing in action," Royce chose to be hopeful about his friends, though both would later be confirmed to have joined the growing list of men killed in action. The fate of some of his old companions was not in doubt. A couple of weeks later news would come of the death a fellow missionary from the Spanish American Mission. "I lived with C. Fred Schumann. He was in charge of us when I lived at Tucson, and was there when he received his release to go home. He was a very fine fellow. I'm so sorry to hear about him. I knew that he was in the army but I didn't know that he was killed in the Solomons. Tell his wife that I knew him well. I don't know what else I could tell you about that. It's pretty difficult to say anything that would make it easier for her."[68]

For his own part, Royce was well aware of the danger they all faced, and in his letters home tried to give some semblance of meaning and purpose to the work he had been called upon to perform. He had gradually come to develop a personal brand of pacifism, based not so much in a belief that war was of itself immoral, as the purely pragmatic fear that it would be without purpose, as the previous "war to end all wars" had seemingly been, and that the sacrifice he and his friends and fellow soldiers were making would bring no lasting benefit. "Do I sound unhappy? I don't mean to because I'm not really. It's just that – well, I think you understand what I mean. The job over here has got to be done I know and I'm really not sorry that I am able to do my share. It's just that I've never yet seen any sensible reason why wars should be. Nothing has ever been completely settled by one so far. I hope this one will be different. If I thought that my children in another twenty-five years were going to have to go through this I don't know what I would do. It would almost make you decide not to have any."[69]

At the same time, Royce's military service and his brief experience abroad had broadened his perspective about what it meant to be an American, and if he was cynical about the war itself, he was at least as idealistic about his nation that was fighting it. "It always seems to me that in America we have every good thing you find elsewhere, plus many more that do not seem to be found in other countries. I am proud of my country. In the few hundred years it has been a country, it has proven the most provoking problem in the history of mankind, that of human relations, is not without solution. We have proven and are proving every day that men of all creeds, races and colors can live together peaceably for the common good of all. The very nations which are at present at each other's throats and have been for centuries are each strongly represented in any cross section of the American people. They live together, work together, marry one another and are all Americans. Of course, we have our heels and radicals, but America is a free country. I hope every fellow is conscious of this who is now serving his country."[70]

Toward the end of July, Royce reached another milestone by completing his twenty-fifth mission. "We successfully attacked a hard one today, the Ostiglia rail bridge, the only remaining bridge in partial operation on the Po River. They have the heaviest flak we know of there. There were about thirty-three 88 mm guns, some 40's and 105's. They can shoot too. 310th was last over it. They had 18 ships holed and nine wounded men. We really worked them over today. Twenty P-47s led the attack by dive bombing and strafing the flak batteries then 3 of our ships went over dropping frag bombs and throwing out chaff (little metal strips that mess up radar sighting). I was in the #2 chaff ship. I surely beat my right hand up throwing that stuff out. We went in there doing about 270 air speed and 300 ground speed which makes it more than difficult to do. I was so busy with that, that I hardly noticed the flak which was intense and accurate. We caught two holes in the tail. The bombers followed dropping 1000 pounders. Two of the boxes walked right across the center of the bridge. Pictures showed good bombing. We had about ten ships shot up some (out of 14). Three made emergency landings. 9P had more than 100 holes in it. No one was injured in our squadron. The 486th had a wounded man on board. We were plenty lucky. I don't know just how the other squadrons made out. No ships went down."[71] To Pearl he made the mission sound routine. "I've been on another mission which brings the count up to twenty-five. Everything is O.K. I'm pretty well accustomed to them now. They are quite a large part of my life these days. We think of them in nearly the same way we used to think of our training missions back in the States. They are different of course but we feel better about them too because we can see we are doing something to bring the end closer."[72]

The next day at noon, the 340th Bombardment Group received intelligence of a possible planned enemy invasion of Corsica from bases in either France or Italy. Royce wrote in his journal, "We were put on immediate double alert today because of an expected attack by German paratroops or sea and air forces. We have been ordered to carry our guns and gas masks at all times, and certain of us are organized into groups to go immediately to the airdrome to help defend our ships in case of paratroops." Three days later he wrote, "We are still on general alert. We watch the machine guns in the area all night now. I was sergeant of the guard early shift tonight – 9:00 to 1:00."[73] The men continued these precautions for a number of days, carrying their weapons always and doing guard duty at double strength, and patrolling fighter planes were seen streaking over the island day and night in anticipation of an attack, which in the end never materialized.

At the same time, Royce was looking forward to taking a rest leave to Rome, as members of the squadron had been allowed to do in shifts according to both seniority and crew position. Robinson, the tail gunner, had already returned from his rest leave, though Royce noted that he had failed to see many of the historical and artistic treasures which were of interest to him, and he looked forward to his own furlough, which he expected to come at any time.

Flying over France

At the beginning of August, 1944, there came a sudden change in the routine—rather than the usual attacks on northern Italy to the east, missions were assigned over German-occupied southern France to the northwest, and they came with even greater intensity and frequency. "Mission #27. We attacked a target in France today. It's the first time for the 340th. Our target was the Var River Bridge #4, located near the coast a little west of Nice. We had a steel bridge, double decker which we knocked out. I've never seen thicker flak. Ships attacking other bridges further down bore the brunt of the flak. No ships went down. One of the 486th's ships had a direct hit from an 88 m.m. shell on one wing. You could put your head & shoulders through the hole. It rolled over on its back (the aileron was shot off and the one engine out) and started to go into a spin. The pilot recovered and limped it home. There were five emergency landings. One of our ships was shot up." The next day the squadron returned to the same area. "Mission #28. We succesfully hit a bridge just south of yesterday's target. There was lots of flak again but it did our Group no damage. The 321st caught it today at a bridge near the water. We had P-47's to strafe & dive bomb today at the gun positions plus chaff ships but due to a mistake on the part of the 321st, they hit the wrong place and helped us not at all." Royce added, "I flew with Capt. Dyer to photograph installations around here in the afternoon. I got pretty worn out. To top it off, I was Sergeant of the Guard again and had to stay up all night."[74] Guard duty was still a necessary evil because of continued rumors of a German paratrooper invasion.

After these first missions to Southern France there were two days of stand-downs,[75] but on August 6 the squadron returned to press the attack. "Mission #29 today. We went back to the French Riviera once more and finished up a road & rail bridge over the Var. Our flight went in first after the chaff-frag ships. We hit the R.R. Bridge in the center & on the approach. The chaff worked fine for us. There was lots of flak but none hit us. The other flight got shot up some. The lead bombardier of the 487th box was wounded in the head on the bomb run but went in to release his bombs. We really expected to catch it today because we flew right over the flak batteries. We were plenty lucky. The clouds were all around but not over the target. I saw one whole box of bombs from the other flight hit right on top of a crowded bunch of houses. I guess it's pretty terrible. It would leave nothing but rubble. I hope the people were out. It's pretty hard to aim true all the time when you're trying to dodge flak to save ships and crews. They really had lots of it there."[76]

Royce flew again the following day. "Mission #30. Target for today was two bridges, one accrossed the Rhone River at Lavoulta (France), a steel suspension deal. The other was a masonry bridge accross a Rhone tributary at Livron. We attacked the second. Three spans, 1/2 of the bridge, were pulverized. The other bridge was hit too. This was one hours' ride from the coast into France. We met no fighters or flak. We had P-47 escort. Two boxes of the 486th got off the beam and crossed the coast line out to sea at the wrong place. They were shot up. One pilot got it in the shoulder." Royce was not assigned to the following day's mission, which proved to be a costly one. "Our Group really caught it today. I didn't go. They attacked and hit a bridge over the Rhone at Avignon. The 488th had one ship go down in flames over the target. Five chutes were observed. Another was so crippled it had to ditch 10 miles after leaving France. Two others made emergency landings, and one smashed up. Every ship was hit. Two 310th ships landed with wounded. One copilot had a bad one in the chest. Some of our gunners got jumpy and shot up one ship of the escort today."[77]

On August 9th, tension increased as a new announcement was made. "We are restricted to the island untill further notice. That means no rest camp for now. There's something cooking, maybe the invasion of southern France. We've been expecting it for some time. Corsica is to be the main springboard."[78] On August 12 and 13, Royce flew two additional missions, both against gun emplacements on the French coast, heightening rumors that their missions were to soften up defenses for an impending invasion. Finally, on the evening of Monday, August 14, the rumors were confirmed. "We were called down to Group tonight and told that tomorrow is the big day. "D" day for the invasion of Southern France. We were shown the plans for it. "H" hour is set for 8:00 a.m. We precede the landings, bombing shore installations then follow up in the day where we are needed. There are some 3,000 ships massed for the attack. We've seen much of the preparations and have bombed up and down the coast to confuse. Most of the troops are American. There are a few French. I hope I'm on the mission."[79]

The next day, August 15, 1944, was "D-day", and though the invasion of Southern France garnered less international attention than the one at Normandy two months before, it was still a huge operation, opening up an additional front to hasten the departure of the German occupiers from France, where they had been under assault from the north since the more famous invasion in June. Two missions were flown by the 489th Bomb Squadron on "D-day," and Royce flew in the second and costliest. "Mission #33. We really had a rough one today. There were 72 ships that went after three targets at Avignon. The 489th had a road bridge accross the Rhone which we knocked out with a beautiful pattern. The 488th got theirs too but the 486th missed the big R.R. bridge. Now the sad part; one ship had a wing shot off over the target, went into a crazy spin, caught fire and exploded on the ground. I saw it just as it hit. No one got out of it. It was a 486 ship. Our box was pretty lucky. Our ship got no holes. The flak was so thick we couldn't see half the ships trailing us. They were diving and twisting all over the sky to escape the flak. 13 of our ships were holed, some of them will be inoperative for some time. 9-P

204

failed to return. No one seems to know what happened to it. I knew all the crew.[80] They may have ditched safely. They are listed missing. Just off the coast as we broke out, a 488th ship which was badly hit, caught fire and started down. I saw the last chute leave it then it went into a vertical dive and exploded when it hit the water. That makes 6 ships we've lost over Avignon counting the other day. Only one man was picked up who jumped from the 488th ship. The rest were believed to have drowned. I knew the Bombardier, Wustein. He is missing. Everyone felt quite badly about it. It's pretty tough seeing fellows you know get it." Then, with a sense of the historical import of the day, Royce added, "The big show started this morning. We had an early morning raid before "H" hour (8:00 a.m.). I didn't go, it was a milk run. We saw the show in the afternoon though, flying the whole length of the invasion coast from Nice to Marseille. I never expect to see that many ships again. There were innumerable types and numbers. I couldn't even count all the battleships. We could see some of them firing at the shore. The whole of the area was a mass of smoke and explosions. Bombers were coming and going. I'll never forget it."[81]

Royce was scheduled to fly two days later, but the mission was called off at the last minute because of weather. The following day, August 18, Royce's ship almost met with disaster. "We did take off but nearly cracked up. We were in 9-S and its rudders were shot up over Avignon the other day. When they put the rudder trim-tabs on, they reversed one. We nearly slipped into the ground on takeoff. Lt. Walker was pilot. He and the co-pilot had to push as hard as they could and hold on right rudder while they brought it around and in with four 1,000 lb. bombs. We didn't miss a mission. The whole flight had to turn back half way because of bad weather."[82] Just a few days later, another bomber Royce was assigned to had a similar close call due to mechanical problems. "We were briefed and started on a mission after the big railroad bridge at Avignon. We got as far as the Alta field and our right engine started to throw oil, smoke, and cut out. We turned back and came in. It's a good thing we didn't get farther. The engine was in bad shape."[83] Royce related to Pearl the engine problem and early return, and without mentioning the target location, he quipped about the heavy anti-aircraft fire they had encountered over Avignon. "One of the officers says that the flak gunners there go around in a crouch because of the weight of the medals they have won for marksmanship. I believe it too. Last time I was over there I watched back and could hardly see the ships following us over the target because of the smoke from bursting shells."[84]

Between these two abortive missions, on August 19, Royce flew a mission to its completion, and once more was able to see the unfolding of history beneath him. "Missions #34 & 35. We were up at 4:15, briefed and sent after bridges at Orange north of Avignon. We gave the bombs a round trip, for the whole Rhone valley was clouded in. We returned in the afternoon to the same targets and did the work. Ours was a R.R. bridge accross a tributary of the Rhone River. It was destroyed. The other three boxes went after a big suspension road bridge at Montfaucon. About half of it was down after the bombing. They put quite a bit of flak on us. Quite a few ships were holed, ours wasn't. As we went by Toulon harbor, we could see our battleships shelling it. There was some return fire. We flew all along the battle line which was easily seen because of smoke and explosions. I also saw the gliders in which our airborne troops landed during the invasion. They were in fields above Cannes. We already have airstrips built near there. Things really look good. The bridges we destroyed today were most vital to the Germans. It was about the only reinforcement route open to Jerry. He's in a bad way with Paris now being threatened too."[85]

Royce flew one other mission against fortifications on the French coast, but on Friday, August 25, there was a general stand-down of the entire 340th Bombardment Group, for reasons Royce explained to his family. "Today is an anniversary for this outfit. It was organized two years ago at Columbia, and from there shipped overseas. They are having a celebration today, that is, as much as can be had. There was a presentation of awards, a baseball game is going on now, and there is a show tonight. So far, I've been for about a 2-1/2 hour swim in the sea for my celebration. The water was really good. We stayed there until we were just about worn out."[86]

That same day news came that Paris had been liberated by Allied forces. This fueled a growing wave of optimism that the war in Europe would reach a speedy conclusion, which helped Royce and others to reckon with the uncertainty regarding the number of missions that now constituted a tour of duty. He had expressed this uncertain optimism in a letter home just the day before. "Mother keeps asking me how many missions I must fly over here. In case you're wondering the same thing I'll tell you now that I do not know. It's all rather indefinate. I do rather expect to be over here and still flying when Germany throws in the towel. We don't have anything to say about how long, where or when we fly. That is all determined for the most part by Operations and the Flight Surgeon and they take their orders from farther up so there isn't much we can do outside of what we are told. Our job is combat and we can be kept at it just as long as it seems advisable." Royce added, "For us, it isn't so much the action that wears you down but rather the time between when you've made the mistake of not having it planned and devote it to nothing. Don't get me wrong, I nearly always have something to keep me busy. It's just that you can't help wanting everything at once; the end of the war, home again, and all that goes with it."[87]

During this time Royce continued to maintain contact with his old crew. The tail gunner Robinson who shared his tent had long since become a good swimmer. The engineer, Wendell, had left the hospital after his hernia operation and at last been assigned to flight duty and begun accumulating missions, having done his time on the aircraft maintenance line. The pilot Brassfield and the bombardier Harper had both received their promotions to first lieutenant, which was cause for celebration. Although in the crew rotations of the squadron Royce seldom flew with his original crew, he still admired his old pilot. "I've flown with lots of pilots over here but never have found one I like to fly with more than Brassfield. He really knows how to handle that ship. I don't mean to infer that the other pilots arn't good. They all are. I've been with most of them and always felt confident in their ability to handle the ship no matter what should come up. The pilot is the man on any crew. The rest of us just help out."[88]

On August 27, two days after the anniversary of the 340th, Royce nearly had his chance for a rest leave. "I almost went to Rome for five whole days. I was scheduled and nearly got everything ready, then it had to be cancelled. I ended up flying another combat mission. There was a good fellow to go with scheduled to go with me too. Oh well, better luck next time. The war has to come first."[89] In his journal Royce recorded the details. "I ended up flying mission #37. We were briefed to attack a bridge above Orange on the Rhone but before we got off a report came in saying our forces had taken it. That was early morning. We took off at 12:30 to hit the islands in Marseille Harbor. Jerry still has gun positions there causing much trouble. We plastered the whole island with 1,000 pounders. I observed smoke and explosions after. I don't think there is much left there now."[90]

The frequent missions and associated duties had often kept Royce from attending the chaplain's Sunday services, but his letters frequently focused on spiritual aspects of life. In August, Royce had given his Book of Mormon and a pamphlet to a soldier stationed with a fighter squadron on the island who was seeking "a religion he could believe in," and Royce had a long discussion with him about his faith. In a letter to Pearl the same month, he summed up his own religious feelings, particularly in the setting of his present circumstances. "I've always appreciated the literature put out by the Church because it is so sensible and reasonable. It makes you think, and brings religion out of the realm of fairy tales that seems to be the usual way of expressing it. If all the theological reading I could get was the type put out by most churches, I think I should be a near atheist or at least very soured. Too much is left up to lip praying and calling for help and not enough is left up to the individual. When it comes right down to it, we will have a better world just as soon as we build it, and you can't build it without working. Some of the stories they peddle about soldiers suddenly finding religion when they realize that death is staring them in the eyes don't go over very well with me. We cannot spend our life trifling with immorality, debauchery and the other evils that drag men down and expect to have all that thrown from us in a moment of terror and weakness. The clay has been formed and has hardened, a remolding is necessary. Right is and always shall be right."[91]

A Rest Leave and Restless Weather

At last, on August, 30 Royce was granted his five-day rest leave in Rome, accompanied by another radioman/gunner from the squadron named Robert Saxton. "His idea of what to do in Rome was the same as mine so we went together. His home is in St. Louis, Missouri and he is a Catholic boy." The two made the most of their time, visiting the Roman forum and the ruins of palaces build by the Caesars, the ancient catacombs where Christian saints were said to be buried, and the Colosseum, which impressed Royce with its size and its masterful engineering. With his Catholic companion Royce also made a visit to Vatican City, and described to Pearl an audience with the Pope. "It was very much like the newsreels you see, but it made me think of the old medieval and post medieval royalty they show in the movies. We waited for him in a long hall with a passage down the center. He was carried in on a red carrying chair by his four chair bearers who were dressed in ancient uniforms of bright red. They carried him down the passageway to the end where they sat him down and he mounted a beautiful throne. He was dressed all in white and really looked royal. He spoke a few words in English to the effect that he hoped the war would soon end. It was all over in about five minutes. Most of it was in Italian. There were soldiers of all nations there."[92] Royce was struck by the size of St. Peter's Basilica and the lifelike art depicted in its paintings, mosaics and sculptures.

Besides seeing the sights, the soldiers enjoyed unaccustomed good food and stopped by the Red Cross for ice cream. "It was only as good as canned milk can make it. Of course it does taste mighty good after not having it for such a long time."[93] They were able to associate with Italian citizens and with soldiers from other branches of the service, but they had to walk everywhere they went. "I did enjoy it but Oh! I have a sore pair of feet. I have worn nothing but G.I. shoes since coming to Corsica and like a fool, I had nothing but low quarter shoes to wear over there. I have many blisters. We had only one way to get around and that was on our own two feet. We walked from morning until night."⁹⁴

While in Rome, Royce was also exposed for the first time to some of more sobering aspects of the war, as experienced by those on the ground, including those whose lives had been affected by aerial bombing. "They feel the pinch of the war in Rome yet, and most of them have known the horror of it. I noticed so many of them either dressed in black or wearing a black ribbon which is their way of showing that they are mourning for their dead. War has been pretty cruel to the Italian people. One day in the ice cream line at the Red Cross one of the girls who was serving noticed our wings and asked us in Italian if we flew in bombers. I could understand that. I told her yes and she said it was very bad (the bombing of course) and I asked her why. She showed me on her leg and arm where she had been hit by bomb fragments and said there were other places she couldn't show me. She also said that several of her family were killed. It is a pretty sad thing. I surely hope it ends soon."⁹⁵

After Royce's return from Rome, stormy weather moved in and grounded the squadron for several days, but by September 9 the weather cleared, and the squadron was sent on a series of daily missions against heavily fortified German troop emplacements in the rugged mountainous areas of northern Italy. "Mission #38. We went after defence positions on the Gothic Line above Florence with frags and 500 pounders. The flak was heavy, moderate and very accurate. A number of ships were shot up. 9-Z was forced down in Italy. It made it to one of our fields. 9-V landed with a shot out engine. 9-G had a runaway prop due to flak. 9-R, Brassfield's ship, had about 50 holes. We got one in the tail flying right alongside 9-R."⁹⁶ Miraculously, none of the planes went down, and no men were wounded or killed. This was only one of a number of missions in which other planes in Royce's box or squadron were damaged while his remained relatively intact, and he observed that when planes did go down, they were most often from other squadrons in the Group. Amid the seemingly random fortunes of war, Royce began to feel either charmed or blessed. The same day he wrote to Pearl, "I do feel sure that I will safely see the end of this war and come home to you. I've been afraid many times but things just don't seem to happen to me. The ship flying next to me will get hit and we won't except for a hole or two that arn't too bad." He continued, "This is a strange war we fight. People at home worry and write about the fact that soldiers think of nothing but going home and seem to think that it means that they don't know why they fight. We do know what we are fighting for and what could better sum it up than to say 'for the right to go home again.' To me, it embraces all else."⁹⁷

The daily missions against the Gothic Line were followed by a September 12 attack to finish off an already damaged bridge across the Po River, after which Royce observed, "I surely find myself sweating them out now. We hit prop wash landing and nearly spun in."[98] The next mission was scheduled for a heavily defended target, a tank distribution center in Bologna. Tension mounted as bad weather began to interfere. Royce described the scenario to Pearl, omitting details of the target. "Two days ago, we were first briefed for the mission. It was to really be a hot one. Special means of attack were planned to make the flak less effective and I guess we all got pretty well keyed up about it, whether we wanted to or not. We were about ready to crawl in the ships and take off when they came around and told us the mission was off because of weather. Everyone was kind of relieved and pretty happy though I think nearly all secretly wanted to have it over with, at least I did. Next morning, we were up at 4:30 and got ready to pull the same mission only to have them cancel it, or rather postpone it until the afternoon for the weather. The same thing happened again in the afternoon. Well, after sweating it out two days, we were up long before the sun this morning and did go. By that time everyone had thought about it enough to feel just a bit afraid of what was coming. Right now, we are all back safe and sound. The mission turned out to be just about what we call a 'milk run.' The flak was scant and very inaccurate and we very effectively bombed the target. I'll remember that mission (#41) because it really came the hard way."[99]

Royce was not assigned for the next couple of days, and after that, rain clouds moved in again and the squadron went back on standdown. The chill of fall was in the air, and heavy rain tested the field tents used by the airmen. Most began to look at reinforcing their tents, and building materials became valuable items on the base. Royce wore flying clothes at night under his blankets, hoping to receive a G.I. issue sleeping bag, which had been promised but not yet delivered. Royce took the discomfort in perspective. "I realize that the boys on the line over in Italy usually sleep on the ground with an overcoat and shelter half to roll up in and that they go for a week at a time without removing their clothes so we don't have it very rough. Our beds are dry, we have clean clothes and are even able to have a warm shower just about anytime. I'm afraid I don't envy the boys in the ground forces. On the other hand, I have spoken to a lot of fellows who say they didn't envy us either. It's a long way down if things go wrong and you've no fox hole to crawl in or cover that will do much good to get behind when the shooting starts. I guess it's every man to his taste. You do what you like if you can, or try to like what you do if you can't."[100]

Finally, on Friday, September 22, the weather cleared and the men flew. "Mission #42. We finally had good enough weather to pull this mission. We raided Nervessa di Battaglia, north of Venice. The target was a large railroad bridge. We really gave it the works. One pattern hit the center and the other two hit on either side of the center. Flack was moderate and accurate. Our box got very little but they really shot up #1 box. 9-J had an engine out and it limped to a field in free Italy. Eikhoff [Royce's tent-mate] was on it. Several other ships were quite badly hit. There was one man wounded, a bombardier (Steward). Our left engine smoked all the way over & back but it didn't cut out."[101] The delay between missions was taking its toll on Royce. To Pearl he wrote, "They aren't going up very fast now, and I seem to notice them a whole lot more than I used to. It isn't that I ever feel that I'm not coming back. I've always felt sure I'd return every time I've gone out, but you can't help being scared about it, and the more you see of it the more it bothers you."[102]

With news of Allied advances on every front and a major offensive underway in nearby Italy, Royce shared in the general expectation that the war in Europe would soon be over. "All of us are hoping the war will end before winter. It won't stop a whole lot of the suffering in store but it will alleviate it some. Germany has nothing to gain by prolonging it, and much to lose."[103] Adding to the general sense of optimism was a string of successes which the 340th Bomb Group had enjoyed with difficult targets. Royce's unit prided itself on its effectiveness in striking its objectives and hampering the enemy's ability to wage war. "This outfit has a record for bombing accuracy that is really high. I don't know of any other as good. A bridge is a mighty small thing from the altitude we fly, and they are just about the hardest type of target to knock out. It's got to the point where we just don't miss. We always feel good when we go over a tough target and really knock it out so that it won't be necessary for us or anyone else for that matter to go back and run through it again. That happens too. It's always good to see your bombs go home and then have the smoke clear away before you're out of the area and see several sections of a big bridge down. You can see that you're helping the war effort."[104]

Writing to his parents about the ever-changing number of missions required to complete a tour of duty, Royce explained, "I think I can figure on returning home after the European war ends. I expect to fly a couple more months anyway. I'll let you know just as soon as I have been grounded from combat flying. There was a time when they used to be through after the fiftieth mission, but that doesn't hold over here as I have told you before. I don't think our missions are any easier than they used to be. I guess they have decided that fellows can take more than they used to think they could. It varies with the men now as always. They have to ground many of the fellows early because they show signs of cracking up." Speaking for himself he added, "I'm getting along all right. I guess I notice things a whole lot more than I used to but that is only natural. I used to go to the window all the time to watch the flak but now I make myself as small as I can on the big piece of armor plate they have in the floor of the radio compartment, that is, I do that just as soon as I finish taking pictures of the target. That helps out too because I don't notice the flak so much while I am busy."[105]

The last days of September brought cooler weather with more rain, which grounded the squadron, bogged down ground offenses, and dampened spirits. Optimism for an early end to the war was quickly passing. The enemy was proving itself to be a determined and disciplined foe, and Royce wrote, "I'm afraid the war is going to last longer than we had hoped. It looks as if the Germans are going to make us pay just as dearly as they can for every gain we make. I wish it might be otherwise. This surely isn't any way to live, and so many are under worse conditions. For a time, I had almost let myself hope that I would be home for my first Christmas in six years, but I'm afraid that will have to wait too. I can't help but call myself down for feeling sad about it because many fellows in the outfit have been overseas over 25 months."[106] A few days later he reflected, "Six months ago today, we received our new airplane back at Hunter Field and took the thing up for a test hop, preparing our things the same day to leave for our port of embarkation. It seems like almost a lifetime has passed since then. I did look forward to going over, and wanted it. I don't know what you call it, but it always seemed to me to be a form of curiosity, and a desire not to be outdone by others as far as the war was concerned. I would hesitate to call it a patriotic feeling, but I did want to come over and have my share of the war. I'm not sorry about it now. Right now, the strongest desire I have is to see the end of it. I would never have felt right about the whole thing if I couldn't have seen combat."[107]

As October approached, the storms relented enough that the squadron was sent to bomb heavily guarded targets in rapid succession. On October 1, two missions were flown, and Royce was on both of them. "Mission #s 46 & 47. We had it pretty rough today both missions. The flak really came up. We destroyed a gas plant at Piacenza on the first mission. Other ships were after other targets there. The whole region was a valley of smoke. We got one hole in the tail out of it. Others hit close. In the afternoon, we hit a bridge at Magenta southwest of Milan. They threw up plenty there, though we got no holes. I was in the chaff flight. We dropped white phosphorus on gun positions. Some of the flak they threw up was red in color.

Two of the 487's chaff ships were hit quite badly but they made it to Leghorn. There were wounded."[108]

After a day's "stand-down" for bad weather, the squadron returned to Magenta on October 3. "Mission #48. This was one of the roughest missions the Group has had in a long time. Our squadron never lost any ships but the one I was riding in and one other were the only ships not holed. 9-B came in on a single engine. We were the third box thru the flak and for some reason never caught it so bad. Our box stayed fairly intact. We doubled back and were out of range as the other flight made their run. They had already had accurate heavy flak twice on the way. As they hit the worst of it, the sky was black with flak. The boxes broke up and ships went every which way dodging and twisting to avoid the flak which followed everywhere they went. One went down in flames and I don't know if any others didn't make it. If that bridge was hit it was nothing but luck. I was with the 486th and they bombed the Camani bridge just upstream with beautiful hits. Several of our men had minor wounds including Harper. It was his last mission, #60. Somebody watches over me."[109] Royce related to Pearl the news about Harper, the bombardier on his original training crew, and the first of the crew to complete his quota of missions. "Harper has been grounded. He has had enough for this tour of duty. He flew much faster than I have and has quite a few more missions. I don't know when he will return to the States. I wasn't going to, but I just as well tell you that the magic number is 60 missions and I have 48. How long it will take to get in twelve more I do not know. It depends on the weather and a number of other things. You can be sure that it will be soon if I have any say about it."[110]

Wing Bombardier

The next day, the squadron was scheduled to return to the same heavily-guarded target at Magenta, but as the crews left the briefing area and mounted the trucks to the waiting ships on the flight line, Royce was about to embark on a new assignment for which he had just been certified. Rather than climbing up the customary rear hatch into the familiar spacious radioman's compartment between the waist guns, Royce mounted the forward hatch, and after takeoff slid himself through the narrow metal chute that led forward to bombardier's position in the plexiglass-covered nose of the airplane. In his journal he wrote, "I was on as bombardier for the first time today. We were briefed and took off to go after the same target as yesterday (Magenta road and R.R. bridge). We were barely on our way when they called us back due to bad weather. I would have been #6 in the last box over the target (Purple Heart position). We nearly cracked up landing when we undershot the field."[111]

Royce explained the new assignment in a letter to his parents. "I am starting next mission to fly at least part of the time as bombardier. I have been attending a navigation class nearly all last week and have been checked out as a wing bombardier. What we do is all that most bombardiers do overseas. Very few of them become lead bombardiers. It isn't a difficult job. You ride up in the nose where you can see everything. I fly that position on my next mission."[112] He later clarified, "I have nothing to do with aiming the bombs. All I do is release when the lead ship does. They call us wing bombardiers. We fly on the wing of a ship that does sightings, open the doors when he does and release the bombs when he releases his. The idea is to put six ship loads of bombs in as nearly the same spot as possible. You can imagine what that does to bridges."[113] He explained to Pearl the reason for the change. "We have quite a number of enlisted men doing it now. They work just as well as officers. I have been checked out and we have been spending about an hour and a half a day lately in a navigation class. It happens that the squadron is short of bombardiers and there are plenty of radiomen. Lt. Brassfield talked to me about it and I decided it would be pretty good to add one more job. I'll probably still fly nearly all the rest as a cameraman as I have done so far, but I would kind of like to release bombs on a couple of missions."[114]

Royce also explained why it was necessary for him to take classes in navigation. "Don't get the thought that I'll be navigating planes around here. It's just something that bombardiers must understand, enlisted or otherwise. If a ship is in trouble and has to leave formation on single engine and make its way back alone, it is the responsability of the bombardier to see that the right course is taken to get home. A navigator is carried in the lead ship only. Flak areas must be avoided and many other things have to be done." Royce had studied basic navigation during his air cadet training, and this practical field training built upon what he had already learned there. "The job of navigating a plane is a pretty complicated affair. There are so many things you have to remember. Computers are used to solve all the problems. In order to be successful, you have to know certain facts and then apply them correctly." He described what a "computer" meant to an airman in 1944. "A computer is a little instrument that will solve practically all navigation problems with reasonable accuracy if you know how to work it, and that's what we are learning. They look and work on the same principle as a circular slide rule. You can do ordinary division and multiplication on them too. It's a right handy little tool."

As a part of his training, Royce also was given at least rudimentary instruction in bomb sighting, though this was a highly specialized task which he was not required to master. "I spent an hour on the bomb trainer today, using the bomb sight. Of course, I'll never use a bomb sight, but it does help you to understand a few things and makes the Squadron training report look good. The sight is mounted on a stand which is on wheels and has a motor to drive it. One person steers it watching a needle and the other does the sighting on a paper with a target drawn on it which is placed on the ground up ahead. After sighting is complete and bombs are away, a marker comes down, marking the paper where bombs would have hit had you been bombing."[115]

The unsettled weather which had caused cancellation of Royce's first bombardier mission dragged on for over a week, leaving the squadron on stand-down. This was the longest period of inactivity since Royce's arrival in Corsica, and it was not entirely welcome. Although it provided the air crews some temporary relief from the dangers of air combat, it also prevented them from doing what they were trained to do, and prolonged the time it would take for them to complete their tours of duty. Like the others, Royce had to find something to do with his time, and he described one typical day during the stand-down. "This morning, I went down to the navigation class again. We were down there about two and one half hours. That pretty well took care of the morning for us. There was a meeting of all radio-operators at 1:00 p.m. and that took some of the afternoon. There were some slight alterations in our procedure on operational flights which we had to be instructed about. After that, I just got back to the tent and started to read when the whistle sounded, and being obedient I went down only to find myself sucked in on a bit of work. It seems that the operations tent fared rather badly during the current stretch of bad weather. Some of the ropes were down and the rest were loose so that the thing sagged in a very sad way indeed. We killed a bit more of the afternoon fixing that up. After that, I payed a social call over at Lt's Harper and Brassfield's tent where I found Wendell. He and I amused ourselves for a time throwing a bayonet at a cork tree. I have learned to throw a knife fairly well since I came overseas. From there, I found my way back to the tent onto my "sack" and proceeded to utilize the other two hours of the afternoon in finishing up the book 'Rogue's Company.' Exciting day, wasn't it? Some of them are even worse than that. There was no mail today. The shower is out of order. It's too cold for swimming and you just have to find something to keep the mind at work. I use books."[116]

With the coming of cooler weather, Royce and his tentmates also found it necessary to prepare their tent for the coming winter, and they took advantage of the break in flying to install a wood framework and adjust the canvas. When the gunner Robinson returned from a rest camp in Malta, he brought a new luxury item which had become more necessary as the days grew shorter. "Robbie brought some good bulbs and sockets for us. We junked the old airplane lights we were using and installed the ones he got. They are British made and just a bit different from ours. We have 40 watt ones. It certainly is going to be good to have a good light like that. A couple of candles did just as well as the little lights we used to have."[117] Eventually, they would add a crude stove heater for the tent, which with winter weather had come to be regarded as a necessity in the camp. "We made the thing out of a 5 gallon oil can and used bomb fuse cans to make the chimney. It isn't much, but it takes the chill off the room and makes it much more livable. It's surprising how many things you can make out of the crudest material when the need arises."[118] Eventually the tent would sport a complete door and frame, and after a particularly fierce storm, Royce would report, "My corner is now very snug. When there was a heavy wind, the air used to howl through."[119]

As day after day passed without flying, Royce began to feel edgy. "I may be foolish but I feel like something is going to happen to me next mission. It's probably because I'm on as bombardier for the first time." At last, on October 11, the weather relented and the squadron took to the air to bomb a railroad bridge at Canneto, Italy. Royce's forebodings about the mission proved to be ill-founded. "Mission #49. A milk run at last. I flew as bombardier for the first time and got my bombs away all right.[120] To Pearl he wrote, "I've been on my first mission as bombardier. I think I'll fly all the rest of the time in that position. I got my bombs off all right today and it's always half the fight to get off to a good start." Regarding his new position in the exposed nose of the aircraft, he added, "I rather like flying up there although I think it is just a bit colder. I took along my winter flying boots so that made it quite comfortable. I wear heavy woolen underwear, two woolen summer flying suits, a leather jacket and the boots. It was just a couple of degrees below 0° Centigrade which is equal to 32° F. A breeze that cold coming in on you is uncomfortable unless you are bundled up for it."[121]

The next day the squadron flew again, in a much larger operation against troop positions near Bologna. This was also a landmark mission for Royce, as not so many months before it would have marked the end of his tour of duty. "Mission #50. We were in on quite a big show this morning. Our flight of 25's and some of the heavies were first over the target area, so we got most of the flak that was thrown up. One heavy was seen to go down. All they did was put holes in some of our ships. Two of our box were hit. P-47's and P-38's were working on the gun positions with white phosphorus. We dropped 500 pounders on troops and storage areas. The sky was full of ships, and after we left, the ground was covered with smoke. I flew bombardier in a radio box. My bombs wern't released because of a broken connection in the circuit. I'm glad it wasn't because of something I did or failed to do."[122] The "radio box" Royce referred to was an innovation made within the 340th Bomb Group, in which all bombs from the ships in the box were to be dropped automatically by means of a radio signal from the lead bombardier. Royce with his experience as radio operator and bombardier was involved in the development and testing of this system, which was designed to reduce the natural imprecision which resulted when six separate bombardiers attempted to release bombs at the same moment.[123]

The weather continued to change rapidly, causing long delays between missions, and hampering the accuracy of bombing with cloud cover. After one of these delays, the squadron was mobilized to complete the destruction of the road and railroad bridges at Magenta which they had attempted some time before. Royce had not been scheduled to fly, but was recruited at the last minute. "I wasn't on the original schedule and when mission time came, one of the radiomen was missing and I was handy so they had me fill in. I was glad to get the mission. They come slowly now."[124] Once again, the weather played a role. "Mission #52. We were after our old hot one, the Magenta road & R.R. Bridge but we didn't get to it. The only bit of land we saw over there was the mountain tops which were level with us and covered with snow (the Alps). They towered out of the sea of white clouds which completely obscured the Po Valley. We dropped no bombs though we did catch one vicious barrage of flak, I don't know just where."[125] This was the last mission Royce would fly in his original crew position as a radio operator.

Finally, the next day, October 19, the weather cleared enough for the Group to make their bomb run over Magenta, and Royce was once more on the schedule as a wing bombardier. "Mission #53. It was really a rough one today. The Group sent out 8 boxes with 1,000 lb. bombs after the Magenta road and R.R. bridge. We were the 5th box over and the box ahead of us dropped chaff so we didn't get it too bad. The box ahead of us got it about the worst (487th). A copilot of one ship was hit in the head and killed. Four others on the same ship were wounded. The pilot brought it back O.K. There were a number of others wounded. No ships went down but there were plenty of emergency landings. There was flak bursting all around us and I could see it all, as well as hear and feel it. There were two holes in the tail. I flew bombardier again and got my bombs off OK. I hope we don't have to go back there again, but I'm afraid there are other targets just as hot."[126]

The weather permitted just one additional day of bombing, on

Friday, October 20. "Mission #54. We had a real milk run today. No opposition. It was a R.R. bridge north and east of Milan, just below Lake Diseo at Palazzolo. We hit it. I was in the radio-release box again. Some of the other outfits got shot up today."[127] That day Royce wrote Pearl about the expected duration of his tour of duty, which by then seemed to be following an unwritten norm of topping out at sixty missions. "I have one more mission in. I may only fly about six more. I don't decide that but if my case goes like the rest, that will be it. Personally, I feel like I could fly another 20 or 30 but I'm not going to stick my neck out. I don't know how much longer it will take me. Flying bombardier has made it so I get nearly every one. I wish I could tell you that I'll be coming home soon. I don't know how long it will be after I finish. It can be months." Determined to get his combat service done with, he added, "I am not going to rest camp again if I can help it. I'd rather fly right on through. The missions don't come so often that it's hard on you."[128]

After the unopposed "milk run" to Palazzolo, the weather closed in again, and the squadron remained grounded the rest of October and into November, as storm after storm swept through. The rain fell heavier than most of the men had ever seen, and played havoc with the roads and the tents, though the one Royce and his companions built remained secure, and like the other men, he took the storm in good humor. "It poured all night long and nearly all morning. We still made out quite well, though it did its best to wash us off the hill. The drain ditch on the uphill side nearly filled with mud so it overflowed and quite a bit flowed under my floor, doing no harm. This afternoon, when we went down to navigation class, we mushed through water flowing across the road in many places about two feet deep. I thought it might carry the jeep off the road a couple of times but they are pretty rugged little cars, and we made our way along with no trouble."[129]

It was in October, during one of the stand-downs because of weather, that Royce at last met another Latter-day Saint on Corsica. "Something really good happened to me today. An LDS chaplain looked me up. I find that there are other Mormon fellows in the group including the line chief of this squadron. His name is Christenson and he comes from Idaho. I have seen him around quite a bit but have never met him. The chaplain came here to get me to attend a meeting they are holding on the island tonight. A fellow named Parkins whom I have not met yet is calling for me to take me down there. He joined the Church over here. I guess I can't but blame myself for not having known about a lot of these fellows before. I certainly havn't made a secret of where I come from and so on but till now I just haven't met up with anyone from the Church. The chaplain seemed just about as glad to see me as I was to see him."[130] That night Royce met Stanley Parkins, who told him of other Latter-day Saint soldiers on the island, including a B-26 pilot who had informed them of Royce. "Parkins is the 1st Sergeant of the 487th Squadron. That gives him a jeep, so I can travel with him to meetings and things. It brightens up my life over here a whole lot."[131]

That Sunday, Royce drove with Stanley Parkins to the church service, where he met up with Keith Russon, a mutual friend of his and Pearl's, who turned out to be the flyer from the B-26 squadron who had sent the chaplain. "I'll be able to go up there to church every Sunday evening now. There were about a dozen of us there. I do not remember all their names or where they all came from. Services are held in a tent by candle light. We had a most interesting discussion on the Priesthood. The chaplain wasn't there but that makes no difference. Almost any one of those present could have handled the meeting all right."[132] Royce quickly struck up a friendship with Parkins, and the next day visited him over at the 487th Squadron. "Sgt. Parkins just recently joined the Church. A Latter-day Saint chaplain over in Italy worked with him and later baptized him. His wife lives in Salt Lake."[133] In addition to Parkins, Royce met a Technical Sergeant in the 486th Squadron named Clyde Miller who was from Utah and an active Church member, and they were joined by an officer from the 488th Squadron, a bombardier named David L. Atkinson from West Virginia.

The year 1944 was a presidential election year, with Franklin D. Roosevelt making his bid for a fourth term, and Royce acknowledged that both information and interest about the candidates and issues was limited on the front. "I don't believe too many of us over here worry much about the election. Possibly we should be more concerned, but getting the war over with and returning home again is so much more important to the fellows over seas."[134] He requested an absentee ballot from Utah, but admitted to Pearl, "We hear next to no politics so we will have to rely pretty much on what judgement we have. I don't know enough about the issues involved to even talk about it."[135] In the end Royce cast his ballot by mail, but determined to keep his voting to himself. "I was going to write and tell you how I planned to do but I won't. You have your own decisions to make though I know how you will vote for President. I don't think I'll be cancelling your vote this time. Just don't vote the one party way. I plan to mix them up some."[136] Pearl was a staunch Republican and if Royce, with a family background in the Democratic Party, sided with her in the presidential race, it was because he held Roosevelt partly responsible for the decisions that had led to the war.

At the same time, Royce felt little sympathy for the Germans against whom they were fighting, having seen first-hand the results of their occupation. When Pearl expressed in a letter the concern that the Allies were dealing harshly with Germany in their prosecution of the war, his feelings were clear. "Don't ever let yourself feel sorry for the Jerries. They have been directly responsible for more misery and human suffering than they could ever experience no matter what we do to them. I am sure that any time Jerry wants to throw in the towel, we will be more than ready to accept and stop the war and they would receive just treatment far better than any they ever gave to the countries they overran. We as a nation don't know what it is to suffer as people have been made to suffer under them, and had they been successful in all, our country would have met the same fate. As a people, we don't know about starvation and all the terrible things that have cursed every country they have occupied. For the first time in modern history, Germany is having her cities layed waste just as she has been doing all along to weaker nations. I don't claim to understand the workings of international politics but I do feel that this war must be pushed to a successful close and that the pathway will once more be open to the building of a new world of peace."[137]

As for his own part in the war, Royce was still grappling with the uncertainty caused by a shifting standard as to the number of missions that constituted a full tour of duty. When Royce had first arrived on Corsica, crews were being rotated home after fifty missions, but as the year wore on. the limit increased to seventy, though this was thought to exceed the endurance of most crewmen. It was left to the squadron commanders and their medical officers to determine when the men had had enough, and for a time, an unwritten limit of sixty missions seemed to have become the norm for the Group, and Royce had begun to calculate his time remaining by that standard. But when he learned that Atkinson, the Latter-day Saint bombardier from the 488th, had already hit sixty and was still flying, he wrote, "I frankly do not know how many missions I will have to fly. If they go on being able to take it and it looks like they will, I'll be over here for the duration. I've always wished for a short war but I do more than ever hope and pray that this thing will end soon. Expect me home after that. I feel all right about it. This is our war and I guess I still want my part of it."[138]

This uncertainty played havoc with the men's nerves, as it left them with no clear indicator of when their time was done. The winter weather aggravated the uncertainty by making the accumulation of missions slower, and by adding repeated false starts and stand-downs, which exaggerated the sense of fearful anticipation that was natural before each mission. There was no way of knowing in advance whether a mission would be hazardous or routine, and each mission was itself an anxious waiting game, with hours of uneventful flight punctuated by a few minutes of danger over the target, which was sometimes extreme. This game of nerves was trying for everyone, but it seemed especially unfair to men like Atkinson who had already made plans to return home. "He was grounded at 60 and then had to go back once more to flying. He has sixty-three in now. There are a lot of boys in the same position he is, and that is pretty rough in a way. Many of them had written home saying they were through with combat missions, and now that all has to be recinded."[139] Royce's old crewmate Harper, who had finished his missions ahead of the rest, now found himself in the same situation.

Gone, too, were the high hopes of the previous summer for a quick end to the war. As the men waited restlessly for the rain to clear sufficiently to resume bombing missions, Royce wrote to Pearl, "We are still weathered in, Darling. I don't expect to be home before spring. I'll still have to fly at least sixteen more missions and I'm not making any progress in that direction at the present time. There for a time, I had myself believing I could be home for Christmas. I won't fool myself any more. I guess I was right some time ago when I said I expected to see the end of this war before finishing up."[140]

Finally, on November 4, the weather cleared and the squadron began another series of daily missions over Italy. The first day proved to be a difficult one. Royce's box was assigned to bomb a bridge at Villa Franca di Asti, while a second was sent to destroy a railroad bridge not far away. "Mission #55. Our box got through without trouble. The other box flew with the 488th and went after the Casale Monferrato R.R. Bridge. They got lots of flak there, and then flew over Allessandria where they really caught it with barrage type flak. 9-E went down in a spin right into the city where it exploded. No chutes were positively seen. The crew was as follows: Pilot – Rossler, Co-pilot – Gittings, Bombardier – Newman, Radio – Harris, Top turret – Corle, Tail gunner – Danny Mallicoat. Newman was on his 61st mission, having been grounded once. They were all good guys. A 488th ship caught it so bad it had to ditch in the sea (2 men drowned). Both the 310th & 321st landed wounded at our field. It was a sobering day for all of us. It's no fun knowing that your friends die. We hit our bridge. I guess this was a bad day for everyone. Gittings has his tent in back of ours."[141] For Royce it went without saying that Danny Mallicoat, the tail gunner on the lost ship, was Royce's tentmate and friend. In fact, Royce had been on missions in the downed aircraft and had flown with most of the men on the ill-fated crew, and knew them all.

The next day, another mission was flown, a "milk run" to destroy two bridges near Padua. The third day, November 6, brought an entirely new target for the squadron—they were sent on the first of a long series of missions against targets in the Brenner Pass, the vital German supply line linking supplies in Austria to the German forces in Northern Italy. The pass, which was heavily-defended by anti-aircraft batteries, would be an important but much-feared target for the men of Royce's squadron in the coming months. The first mission there was a large massed attack, which Royce described. "Mission #57. There was a concentrated attack on the Brenner Pass today. Our job was the main electrical power station at Trento. It was pretty well defended but we got through all right. Our ship (9-C) had to go to the service squadron after the mission. A piece of flak about 2" x 2" cut a couple of spars in the left wing right by the engine nacelle. You would put your fist in the hole it made. I thought I felt us get hit but didn't know. We hit the target. There was a hospital very near it. I'm afraid a bomb or two hit it. There were so many ships of all kinds going and coming it made me think of 'D' day. Some of the fellows saw Jerry fighters attack B-26's. They seem to leave us pretty well alone."[142]

The following day, the squadron returned to Brenner Pass. "Mission #58. We attacked another small bridge on the Brenner line today at Ala. We really hit the thing, though it was very small, 100' x 20'. They threw up some flak at us but we weren't hurt. No fighters bothered us. Up there, we dodge around snow covered mountains, many as high or higher than we fly. Today's target was just east of the north end of Lake Garda, south of yesterday's. We had to pump the wheels and flaps down because of a failure in the hydraulic system. We sweated it out for a while."[143] These winter missions against targets in the north required a change in flight clothing. "I wear an electrically heated suit now. It's a snappy blue affair that looks like very thin long underwear. I just put a summer flying suit on over it. For me, the lowest heat you can turn it to and have it on is just right. I have on my fleece-lined boots over my shoes so everything is quite comfortable and still not bulky. I quit wearing my leather A-2 jacket because I was snagging it going through the crawlway to and from the bombardier's compartment. It was with me on 54 missions. I rather prize it."

This mission also brought a change in Royce's official assigned crew. "I'm on another crew now. Lt. Brassfield flies in a position that requires a trained bombardier. He fills in for the lead ship if it falls out. I can only fly on the wing as bombardier. I'm on Lt. Rouse's crew. I've flown a number of missions with him before. He's a good pilot." Although assignment to a "crew" gave the men a sense of identity, in actual fact, Royce would continue to fly with various pilots and crews as each mission was assigned. "I've flown with every pilot we have in the squadron I think. They are all good."[144] He later added, "The ship I am supposed to be assigned to now is called 'Sweat and Pray.' I've flown in it several times."[145] The nose art on the ship featured an airman kneeling wide-eyed in frantic prayer, and though Royce made no mention of it, it was perhaps indicative of the way he increasingly felt as his combat tour lengthened toward its end.

During another stand-down brought on by stormy weather, Royce responded to a series of letters from Pearl, commenting on their growing list of friends who had been lost or widowed by the war. Pearl had expressed reluctance about sharing bad news with Royce, given the pressures he was under, but he responded, "You just as well tell me all the sad news as it happens. I find out in time. We have quite a bit of that over here. You don't get used to it but you can always take the sad along with the good news." Among other things, Pearl had seen Lola, the wife of Royce's friend Frank Van Limberg, who had long been listed as missing in action. "I understand how you feel about seeing Lola. There isn't much you can say. You can't tell her that he's probably still living, simply because a man is always listed as missing if there is any possible doubt in regard to death. They have to know it or the War Department won't announce it." Even as he wrote, his thoughts turned to Frank Denstad, whom he greatly admired and still regarded as his closest friend since entering military service. Frank was notoriously bad at answering letters, and Royce had repeatedly asked Pearl to find out from his wife Marie how he was doing. He had heard nothing, and now gave voice to his worst fears. "Pearl, you havn't said anything about Frank Denstad for ever so long. Something has happened to him too has it? It isn't any worse for me to know about my friends being killed or injured than to have things happen to people you eat, sleep and fly with."[146] In spite of his fears, he still hoped the best for Frank, and looked forward to a reunion with him.

The presidential election came during the stand-down, and when Royce learned that Roosevelt had been re-elected for an unprecedented fourth term, he wrote, "I wasn't surprised. I figured it out that way just judging by the general drift among the soldiers in the squadron. I believe they make up a pretty good cross section of America and American opinion. I guess we can expect another four years about like the last in administration. As always, my big interest is what's over here. I do hope this ends in the near future. The Reconstruction of the essential things that have been destroyed over here is going to take such a long time. This must not happen again." He then added cynically, "I guess men have been saying that for generations since time began. It's like the habitual drunkard who swears off on every 'morning after,' only to go on another binge just as soon as he

223

settles his head and stomach."[147]

When the weather cleared, planes from the squadron took off at last to bomb the railroad bridge at Casale Monferrato, the same target near which they had lost a plane the previous week. The mission proved to be a frustrating one, and Royce wrote, "Mission #59. This was a messed up mission. We got shot up and didn't do any good. We attacked a R.R. Bridge at Casale Monferrato. A malfunction of the release on one lead ship caused an early release and a miss. 9-R and 9-S came back on single engines. Brown got hit in the cheek today. Fighters hit the 321st Group today and got 3. Three of their ships made emergencys on our field. It was a rough day all around."[148] He described to Pearl what he could about the mission. "I can't tell you very much about it except that it was surely a cold one. My electric suit kept my body warm all right but my hands and feet were nearly petrified. When it was time to release the bombs my hands were so cold I could hardly press the toggle switch. I think the cold was blowing in off the Alps because we surely had a strong wind and they are covered with plenty of snow. The ship I was in had quite a few cracks up front where the breeze whistled through so that helped too. I'm back and everything is all right so I guess that is what counts."[149]

The next day was November 11, and Royce flew again, describing a degree of danger which had now almost become routine. "Mission #60. Armistice Day, but not for us. We went after the R.R. bypass at Cittadella again today with 500 pounders. We hit the thing too. We caught flak again in several places. Down by Carbola on the Po, I could see the flashes of about twelve guns really working on us. They didn't do so badly either considering how far away we were. One piece hit the pilot's windshield on our ship. Rouse thought his day had come. I saw that one burst right in front of us." This time the usual fear of danger was aggravated by the winter chill, and the fact that Royce had once looked forward to this mission as his last. He wrote miserably, "I once thought I'd be through at 60 but now, I find myself looking for at least ten more. We nearly froze today. It was minus 10°C and my heated suit went out on me. I wish the whole mess would end."[150] Weather closed in again, and the waiting game continued to work on the men's nerves. The squadron braved one more mission through cloudy weather, but the target was obscured by overcast, and Royce's plane returned with the bombs still loaded.

The Final Missions

On Thursday, November 16, 1944, the weather cleared, and the squadron was briefed and sent on another mission, to bomb a road bridge near the town of Faenza. This was a relatively brief mission, lasting just over two hours, but it was destined to be a fateful one for Royce, and proved to be an experience he would relive for the rest of his life. "Mission #62. Today was my close one. I don't think I could come nearer to a purple heart and not get it. If I'd been inches either way I would have been hit, and it came through like a rifle shot. Right after the breakaway, I felt the first burst. It lifted us. I had been bending over observing the bombs and I sat back on the armor plate and crouched down. About that time one burst right in front of us at eleven o'clock a little low. It holed our ship from one end to the other. I had ten holes up in the nose. One piece exploded a round of ammunition in the box right beside my face. Another cut all the wires to the control panel box. Plexiglass splinters put little welts all over one side of my face. My helmet stopped most of it. The pilot's windshield had two holes. There was one four inch hole in the left engine nacelle. It hit a bulkhead instead of the wheel. Another piece hit the steel band holding the left gas tank in place. The left engine was hit with two pieces and we lost most of the oil. We landed before it went out. There were about 30 holes. We were in 9-J, the old flak magnet. We bombed right near the bomb line. Captain Bennett saw the bombardier's compartment and wanted to know what happened to the bombardier. That is the closest I've come to death over here. I was just about petrified and still shake."[151] More than fifty years later Royce was able to recall details of the event. "I thought we'd had it. Our communications systems were gone when the pilot called up. I was up in the nose, flying as bombardier, and the pilot sent the waist gunner up into the nose to see if I was still alive, because it blasted out part of the plexiglass."[152]

This close scrape with death was a defining moment in Royce's military service, and it left him severely shaken. He had often been on dangerous missions before, but when he had flown it always seemed that his own ship, his box, or his squadron came through relatively unscathed even when others around him were damaged or destroyed. Now his was the ship that was hit, and it was his own bombardier's compartment in the nose, the position most exposed to the outside world, which had borne the brunt of the damage. It was not that Royce had lost his faith in divine protection. To the contrary, he was very aware that just before the shell exploded he had been leaning forward watching the target, and had he not moved his head at that very moment and assumed the position he had, in a crouch with his back against the armor plate, he almost surely would have been killed or seriously wounded. He would later use this as an example of having heeded what he regarded to be a divine prompting which saved his life, and he would keep the shattered .50 caliber cartridge which had exploded next to his face as a reminder of it for the rest of his life.[153] But the near-miss had left its mark, and altered his view of combat flying and of his ability to continue it indefinitely. To Pearl he made no mention of the event, save to say, "I do hope that it won't be too long now before I can say I won't have to go on any more. Right now, I have an urge to fly and fly and get them over and done. I'm still most interested in seeing this whole thing over with so that all of the fellows can say they are finished up. It does work on your nerves a lot, especially after you've had a rough one." He concluded the letter by saying, "In the Stars and Stripes the other day, Bill Mauldin had G. I. Joe saying that he felt 'like a fugitive from the law of averages' and I'm afraid we feel that way too at times."[154]

Royce had little time to think about how his harrowing experience had affected him. Missions had to be flown as long as the weather was clear, and in the bombardier's position he was required on nearly every flight. The next day brought another mission to Faenza. "Mission #63. We went back after the same bridges we damaged but failed to destroy yesterday. I don't think the flak was quite as bad. At least, our ship came through it all right. I really sweated this one out after yesterday." The following day, Saturday, he flew a mission which was unique for this squadron, and proved to be the longest mission he or most other squadron members had ever flown. "Mission #64. We really had a long one today. We flew all the way to Yugoslavia and halfway into it to bomb a railroad bridge at Novska. Everyone was really tired after that trip. The only flak we saw was thrown up at someone else. I have 15 missions now as bombardier." The mission lasted just under five hours, and though his mission had gone well, it had been a hard day for other squadrons. "The 310th caught it today. One ship bellied in at our field with a dead turret gunner and a bombardier critically wounded."[155] The stricken plane had been hit on a raid over Casale Monferrato, the same target where Royce's friends had been shot down a couple weeks before. It was just another of a long series of reminders of how near death was for the combat airman.

On Sunday the squadron did not fly, and Royce was able to attend church services with Parkins and Atkinson, but the next day he was assigned to another mission, which was cut short because of overcast. Day after day of unstable weather ensued, dashing Royce's hopes of quickly completing his remaining missions, as the old stand-down waiting game continued. Royce looked for ways to pass the time. He had a long talk with his old pilot Brassfield, who was fighting his own war of nerves. He took part in the endless games of volleyball which now took place whenever the weather allowed. He read the books he had received by mail from Pearl, then passed them on to other soldiers. He discovered that the Group had a skeet range, and he quickly went back to what had been one of his favorite pastimes in gunnery school. His skill with a shotgun was quickly recognized, and he was selected on a team to represent the squadron in Group-wide competitions.[156]

In spite of the diversions, the stand-downs kept Royce's nerves on edge. Any sense of living a "charmed" life was gone. He no longer felt he could continue flying indefinitely, particularly as soldiers with whom he had come overseas began to be grounded from combat flying. The increased number of required missions was considered beyond the endurance of most of the crewmen, but it put them in the humiliating position of having to go to their superiors to request being grounded from combat service. Royce explained to his family, "I know you are looking forward to the time when I write and tell you I am through with combat flying. I will say that I have had just about enough. I can take it as long as I must. Lt. Harper is through and he will soon be going home. Lt. Brassfield is also through. At least I don't think he'll fly any more. It's still indefinite in his case. I know that you are wondering right now just why it is that they have finished and I have not. The number you go on varies and we don't determine just what that number will be. You are supposed to fly just as many as you are able to fly. To the satisfaction of those who decide, Harper has done that and I think Lt. Brassfield has about reached the point too. It won't be very much longer for Robbie and me. I don't think I'll ever be home for Christmas but I have hopes that it won't be too long after. I never looked forward to anything in my life as I look forward to seeing everyone once more. I have wondered if I ever would at times."[157]

To Pearl he was more direct. "As far as combat is concerned, I've had just about all I want to take. I have almost gone in to see the major and tell him I've had enough but I just keep telling myself, 'come on, only six more.' I'm on the next schedule now so there is one of them coming up then there will be only five to go. I definitely will not have to do over seventy. I am trusting that I'll get through them all right. I may sound unpatriotic and everything, but that is the way things are."[158]

Royce's feelings about the war had developed an increasingly bitter edge which cast a noticeable shadow on his letters to Pearl. After attending another patriotic movie on the hillside theater he wrote, "I can't even remember the name of it but all the fellows were wondering why they ship trash like that over here. It was a lot of corn about some of the poor boys getting shipped overseas and having their sweethearts hanging onto them untill they left, just like it never happened to any of us. I wonder why in the world the people who make those things don't get wise. If anyone ever dies, he volunteered for a dangerous assignment or mission and a lot of medals are sent immediately. They ought to come over and see how simple and unspectacular it all can be. War isn't anything but a foolish mess. If they would make a point to teach it as such maybe we wouldn't have it so often. There's a lot more I'd like to say."[159]

As Thanksgiving came to Corsica, Royce characteristically looked to his blessings, though in doing so he related to Pearl for the first time how close he had come to death. "Tomorrow is Thanksgiving Day and I have more to be thankful for than ever before in my life. Someone has watched over me and I know your prayers and mine have not been in vain. I have had the bombardier's compartment laced from one end to the other by flak with me in it and I'm here to write about it. I think I shall always think of flak when I hear a door slam. It sounds just like it when it bursts close to you."[160] Writing his parents, he said simply, "I have had time to feel thankfull although it doesn't take a special day for me to feel that way. I have felt it very strongly every time I feel our wheels touch the ground after we have been on a mission."[161] The Saturday after Thanksgiving, Royce flew on a low-level practice mission, the sort which had been routine during his training back at Columbia. However, accustomed now to the higher elevations used for combat missions in Europe, he now found the low-altitude flying nerve rattling. That Sunday Royce had work duties which kept him from church, and the bad weather continued, dampening everybody's spirits. "It has tried to rain nearly all day long with considerable success several times. It has been enough to make it dreary. There wasn't any mail and no candy in the 2-weeks P.X. rations. There was beer and cigarettes and I don't consume either."[162]

Royce's spirits lifted when he learned of an opportunity to attend an upcoming meeting for Latter-day Saint servicemen to be held in Foggia, Italy, in early December. "I may be going over to Italy shortly for several days. There is a conference of the Church being held pretty soon there and we are planning to go if the weather permits. Parkins is lining up a ship from his squadron to get us over there in. There is a chance I'll see Frank if he's still there. I certainly hope so. If I go, I expect to meet a number of fellows I know."[163] Royce looked forward to this conference of the entire theater as an opportunity to get away and see old friends whom he knew had been stationed in Italy, most particularly Frank Denstad, whom he now regarded as his closest friend, even though they had only had contact in occasional letters. As the days ticked by, another reason for attending the conference became clear. "This will go down as a rest leave for me. I am long overdue on that. It will be best for me to get it in, in case I get stopped before I hit 70 missions. The Flight Surgeon might decide that I better go to rest camp and then come back and fly some

228

more. It's happened here, including Lt. Brassfield. He's at rest camp now and is supposed to fly more when he comes back. I don't know if he will and wouldn't blame him if he didn't."[164]

The increase in the required number of bombing missions affected Royce as it did everyone else. As the weather cleared in late November, he anticipated another mission, but once again there was a stand-down. He was by now sleeping poorly, and the same day he wrote to Pearl, "I didn't sleep very well last night. I wasn't cold or anything, I just couldn't seem to get comfortable and relax. I'll try to make up for it tonight."[165] Finally, on December 1, the weather cleared enough to allow the squadron to dispatch a single box of planes to attack a target, and Royce was assigned in his usual position as wing bombardier on the flight. The mission proved to be a disaster that left the entire squadron sick at heart. "We went after a bridge at Villavernia, Italy, up above Genoa. Mission #65. I'm afraid we killed some people today. We missed the bridge and as near as I could tell destroyed the whole town. We covered it from one end to the other with 1,000 pounders. We saw quite a bit of flak, but it wasn't thrown up at us."[166] It was the lead bombardier who had aimed the errant bombs, but Royce knew that he had thrown the toggle that released the bombs from his own ship. That day he wrote to Pearl, "If it makes you feel better, I won't have to fly more than five more missions if I fly that many. I have sixty-five now. They are coming pretty slow now but I should be all finished up and waiting to go home by Christmas time. It's been a tough old grind and I've had about all I want for right now."[167] He did not know it yet, but this would be his last mission of the war.

The next morning Royce left for Foggia with Stanley Parkins and Clyde Miller, the only other LDS men from the 340th Bomb Group who were able to make the trip. The first thing Royce did after he arrived at the conference was to look for his friend Frank Denstad, whose heavy bombardment squadron Royce hoped was still stationed in Italy. Not seeing Frank, Royce ran into the LDS chaplain and inquired of him. "I believe the first thing I said to Chaplain Cooley after I met him was a question about Frank Denstad. He put me off at first, told me where his bomb group was and what a fine fellow Frank was, then changed the subject. Just as soon as I could I worked the conversation back and asked again if Frank was all right. He then asked me if he was a close friend of mine and after I told him he said that when I first spoke of him he decided to get me off alone and tell me about it. He showed me all the information. Frank was copilot on a B-24 that was observed to blow up near Rheims, France after an attack by about 18 F. W. 190's on July 12th. One chute from the aft part of the ship was observed. It hit me pretty hard when I was told.

A better fellow never lived than Frank. I can't very well express how I felt on hearing of this. I've never had a better friend."[168]

The shock of learning of Frank's death was compounded by the fact that Pearl had known for months that he had been shot down, and had written nothing about it, even though Royce had pumped her repeatedly for information about Frank from his wife Marie. Royce felt deceived, hurt and foolish. His next letter to Pearl began, "Why didn't you tell me about Frank? You must have known that I would find out about him the very first opportunity I had." Knowing that Frank was still listed as "missing in action," Royce thought of Marie. "I don't know what Marie has been told but Frank is almost certainly dead even though he is still listed as missing in action. The ship was hit by fighters and observed to explode. There was one chute observed and it came from the rear of the ship. The pilots are always the last out. There is nothing I could write and say to Marie that would help things any. When I come home, we will go to see her. I think she must be taking it about as well as it can be taken. I've never known a more decent, clean fellow than Frank."[169] Pearl's failure to notify Royce that Frank had gone down remained a sore spot between them ever after, and if Royce did not express anger, neither did he excuse her, though she reasoned, perhaps rightly, that the news might affect his ability to perform his duty on the air crew.

The conference for LDS servicemen did have its happy moments, as Royce ran into old friends from Granite High School who had been fighting in the Italian theater, and spent many hours catching up on their respective lives. Most of these were infantrymen who rejoiced in the opportunity of being able to remove their clothes to sleep, a luxury never afforded them on the front lines. Royce wrote, "Their living conditions are of the very poorest up front. It made me realize how fortunate I have been in that respect. I hope more than ever for the end of this war after talking to them and to others." Although the reunion was a happy one, it did little to calm Royce's nerves, and his description of it concluded with the words, "I did everything over there except sleep.[170]

The conference meetings were held on Sunday, December 3. "We had boys from all over the theater and all branches of the service present. There were about a hundred forty of us. We had three sessions in all, plus a party on Saturday night and I want to tell you that it was a tonic and inspiration for me that I have needed for a long time. That's the cleanest bunch of fellows I've associated with since I left the mission. It does something to you that is good."[171] He recalled the event more than fifty years later. "The conference was standard fare...the idea was to cheer up the boys. They were cheery talks, stiff upper lip, prepare to do your duty, be a good soldier, obey orders, keep the Word of Wisdom, you'll come back safe, maybe. They always covered their tracks a little bit. But they did the best they could. There were some wonderful general authorities in those days."[172]

On Monday, December 4, Royce and the other men returned to their base in soupy weather, stopping at Bari and Naples en route. Arriving in Corsica, Royce found letters from home, including one from Guy Anderson, who was still in a hospital in Brigham City, Utah, recovering from his flak wound. "He says that his leg is still dead from the knee down, but he hopes it will get over it. He said that he thought he had the last operation on it."[173] Royce was fond of the hard-drinking and smoking inactive "Mormon," and he proposed to Pearl that they go visit him when he got home. "He'd enjoy it as much as I would, I think. He's a good fellow in spite of everything. He's never pretended to be what he isn't to anyone and that's something I admire."[174]

The next day was another stand-down. Royce kept himself busy, helping to fashion a plastic window for the tent door to let in some light, and he made out a money order for Pearl, though he held on to some of his pay in anticipation of completing his service. The prospect of waiting out the remaining missions weighed continually on Royce, and the trip to Foggia had done little to settle his nerves, especially after the news about Frank. He acknowledged, "I'm afraid that I haven't been sleeping so well lately. I guess it's nothing."[175]

Wednesday, December 6, 1944, was the eve of the third anniversary of the fateful day when the attack on Pearl Harbor had brought the United States into the war, but for the men of the 489th it brought a round of celebration as a large shipment of mail arrived. "We are having Christmas. I wish you might have seen all the packages come in today. There were four truckloads of packages unloaded here today and I think there is more to come tomorrow. I received five today. I think I've already overdone it on the eating. If I don't get sick, I've missed a good chance."[176]

Surrounded by celebration and by the passing around of treats from home, Royce that day arrived at a quiet decision. "I went in and saw Major Kaufman (our C.O.) today about being grounded. He passed me O.K. I'll have to see Major Brussells, the Group medical officer now. I didn't know I was going to do it untill I passed his office and saw him in there. On a hunch, I went in. I felt I should."[177] Two days later he wrote, "I saw Major Brussells today and got myself grounded from combat flying. I had a pretty long talk with him and that finished it."[178] He wrote Pearl with the news. "The Flight Surgeon grounded me. That is all that has been done to date. My orders have to go through Group and Wing, and then I will be sent to a place for transportation back to the states. The time involved varies from weeks to months."[179]

Royce had no regrets about asking to be grounded. He felt he had done his share, having flown longer than most members of the squadron, and most of his crew was either grounded or nearly so. Royce's switch to bombardier had caused him to accumulate missions faster than all the others except for Harper, the bombardier on the original crew. "Lt. Brassfield isn't grounded yet but he will be in a day or so. Robbie is still flying too. I don't believe it will be long before he will be grounded either. That will leave only Wendell of the old crew, and he has a long way to go yet. I guess Harper is about half way home by now. I'd like to be right along with him, but everything is all right now that I know I'm going home soon."[180]

It took some time for Royce to get used to the idea that he was no longer required to fly missions. In order for his orders to be submitted, he was obliged to document his combat service, and this proved to be a daunting chore. "I spent today checking and listing all of my missions and the crews I flew with. I didn't realize how much work I was getting myself in for, and it was a job I had to plow on through after I had started. It's still pretty hard for me to realize that I am all finished. I feel a bit strange when I hear the engines roaring and watch a flight take off on a mission. It is a left out feeling that you just can't help." He added, "Since I've been grounded, I haven't had any trouble sleeping nights. I hadn't realized what was the cause of my sleeplessness. Very often I would be there thinking of tomorrow's mission and not sleeping a bit. It wasn't anything too serious but I couldn't go on. I am glad that I have all that off my mind."[181]

At the same time, Royce had no doubts that the step he had taken was the right one. He concluded that ultimately it was hearing about Frank that drove him to the decision. "I had a long talk with Chaplain Cooley when I was over at Foggia. Perhaps you noticed that I flew no more missions after I returned. I have been bothered for some time but always figured that a few more wouldn't hurt. Chaplain Cooley said that lately all he had been doing was writing letters of condolence to wives and parents. I didn't want him writing one for me. When I got back I saw the C.O. and then the Group Flight Surgeon. We have no definite number to fly, just as many as we can. I felt I had flown about all I could now. At most, I could have flown only a few more and I had reached the point where I was afraid I'd do the wrong thing if anything were to depend on me. You get so you do things in a strange way when you see flak.[182]

A few days after he was grounded, Royce got news from Pearl of the death of another close friend, Meade Steadman, the gifted musician who had sung at Royce's missionary farewell. "I was very sorry to hear of the death of Meade. He couldn't have been overseas very long. It seems like Mother kept mentioning his singing over the radio all the time I was in the States. I don't ever recollect that I knew he was in the Army. What a fine fellow he was. It seems like only yesterday when we were in school together. Back there, we never dreamed we were growing up to this. They taught us about the last war to end war. I think I'd rather not have any boys if I thought they were to reach manhood for this."[183]

Time passed slowly. Royce received a visit from his friend McNally, who like Royce had been grounded and was awaiting shipping orders. A week before Christmas, Royce attended the Sunday morning chaplain's service, and sang the Christmas carols he always loved. He again represented his squadron in a skeet tournament, and nearly took top honors. As Christmas approached, the weather again turned stormy, and life was confined to the now snug serviceman's tent. Royce began a letter to Pearl, "And still the rains came. If this keeps up we will need some boats to get around the place."[184] The weather reminded Royce of Christmas the year before, which he had spent shivering in the barracks back in Columbia. Christmas Eve was Sunday, and Royce went to the LDS services with Dave Atkinson and Clyde Miller in a jeep provided by Stanley Parkins, who was unable to attend. That Monday marked Royce's sixth straight Christmas away from home, and on Christmas Day he wrote, still with an edge of bitterness, "It's pretty hard to think of 'peace on earth and good will toward men' while the earth itself is filled with hate and the most terrible war ever. Perhaps another year will find a better world. I hope so anyway. I know we could hardly help being involved in this war but that certainly doesn't make the whole mess any less sorry. The present creed of 'hate thy neighbor, kill thy neighbor, and destroy thy neighbor's property' is a far cry from what He taught."[185]

On the 26th of December, things took a more cheerful turn as another shipment of packages arrived from home. "We just finished a feast. It's nearly eleven p.m. and time for bed. I didn't get any packages with things to eat but the neighbors did as well as some of the fellows in the tent. First of all, we went down to visit Brassfield where we had toasted cheese sandwiches and fruit cake. After that, he, Robbie and I went over to visit with Wendell where we put away a large can of olives. At nine, we picked up the bread we swiped at the mess hall tonight and went over to the neighbor's tent where we toasted bread and ate cheese, lobster, tuna fish and candy. These were all from Christmas packages. I really abused the old stomach. I hope there will be no after effects. Robbie finished up his missions at last. Wendell is the only one of the crew still flying. He'll be at it for a long time yet."[186]

December 27 was Royce's 26th birthday, and he received enough packages from home to celebrate it in style. By chance, the birthday cake which his mother had made a yearly habit of sending him arrived that very day. Royce wrote to thank her. "I want you to know how happy I was today when I opened one of the packages I have received and read "Happy Birthday" on top of a cake. You have sent a lot of birthday cakes to me and this one perhaps had the least chance of any of reaching me on my birthday but it did just that. It arrived in splendid condition."[187] The same day he received a card from Frank Denstad's widow. "I received a pretty card from Marie. I don't know whether to write her or not. I'll decide tomorrow. It's the first time I've had her address. She said that she felt she knows me because of what your friendship has meant to her. It makes me sick every time I think of what happened to Frank. It must be terrible for Marie."[188]

The day after his birthday Royce at last received his shipping orders, notifying him that he was to depart Corsica in 72 hours. He made his final preparations, and bade farewell to his tentmates and what was left of his former crew. As New Year's Day, 1945 dawned, he boarded a plane and felt the wheels leave the runway at the Alesan airstrip for the last time, bound for Naples, the famous port in southern Italy, where he would await the troop ship which was to take him home to America. It was a year to the day since he made his first flight aboard a B-25 bomber, and this would be his last. His tour of combat duty was over.

Post-Combat Service

The flight to Naples was a bumpy ride through stormy weather. In Naples the men waited a number of days for the arrival of the troop ship in which they were to sail to the United States, which proved to be the luxury liner *America*, which had been requisitioned by the U.S. Navy as a troop carrier and renamed the *West Point*. Among the hundreds of returning American soldiers aboard the ship was Royce's old friend McNally and Benjamin Angland, another radioman/gunner from the squadron with whom he had formed a friendship. This ship was a speedy ocean liner, fast enough that a Navy escort was not required to protect it against enemy submarines, though the ship took evasive action constantly during the crossing.[189] On arriving in the United States at the end of January, Royce placed a call to Pearl, the circumstances of which they jointly described many years later:

> Pearl: Royce wrote home, and told me that he would be on a boat at such and such a time, and be headed for home. I got his letter, and that day a troop ship was sunk in the Mediterranean, and I thought, Oh, dear, Royce was on it. I was crying, and Willard [Pearl's older brother] came in, and asked me what was the matter. I showed him Royce's letter, and I

said, "Have you heard the news?" He said no, and I turned on the radio, and of course they were talking about the war news all the time, then it told about this boat that went down. And Willard didn't say a thing, he just put his arms around me and patted me, and comforted me. Well, the next day was my mother's birthday, and that's the day Royce called me, from Boston, and I told the operator, "Well, where is he?" She says, "I can't tell you." I says "Where is he?" She says, "Don't you want to talk to him?" I told her to put him on.

Royce: I was calling. Everybody just lined up and took their turn, because everybody had a free call. I was calling from Boston, Massachusetts...

Pearl: So I let his mother and dad know.

Royce: ...and that was really something. I remember how good the Corn Flakes tasted. (Laughter.) You wouldn't think that. I still like Corn Flakes.[190]

Having returned from a tour of duty overseas, Royce was entitled to a three-week furlough, and in February he arrived in Salt Lake and he and Pearl were reunited. This time they took time out to visit the families of soldiers Royce had known, including Marie, the widow of Frank Denstad, whom Royce met for the first time. They also drove up to the hospital in Brigham City where they visited Guy Anderson, still disabled from the flak wound he had sustained on one of his first combat missions.

At the conclusion of the furlough, Royce was assigned to a base in Santa Monica, California, where he was to undergo classification for his next assignment. Although technically a military base, Santa Monica amounted to a rest camp for the men, as they were allowed to take their wives and were given ample free time. Leaving Jean, now nearly two years old, in the care of Pearl's mother, Pearl at last accompanied Royce to California, where they spent an additional three weeks together. They toured local attractions with Ben Angland and his wife Della, visited some of the Spanish-speaking Church members whom Royce had known in Los Angeles, and enjoyed a dinner together with Lt. Brassfield, Royce's pilot, and his wife. They also went deep sea fishing off the coast, where after several hours without fish Royce at last hooked a halibut. On their final Sunday they attended the Spanish branch in Los Angeles where Royce had served as a missionary. "They have no missionaries now but it certainly was pleasing to me to see them doing as well or better than they were when last I saw them. They are on their own and really going all right. They made me feel almost as if I'd never left them."[191]

Royce and Pearl's meeting was an undeniably joyful one after so

long a time apart, particularly as they recognized that Royce was unlikely to return to combat, and as they considered the many friends
and acquaintances who would never return. During their time apart
they had shared their deepest feelings in letters, and recalling the
pleasant months they had spent together after their marriage, they
had doubtless anticipated that when they met again they would have a
happy reunion and things would be as they had been before. But
Royce had returned from the war a changed man. He was restless and
edgy, jumped at loud noises, and slept poorly. He had an explosive
temper. His hearing was so damaged by gunfire and aircraft engine
noise that he had trouble discerning ordinary conversation, and he
turned the radio volume up to uncomfortably loud levels. Royce was
conscious of the change, and would acknowledge it to Pearl by letter
shortly after the furlough and rest leave were completed. "I'm sorry I
was disagreeable to you at times. It wasn't because I wanted to be,
Darling. My nerves are going to be quite some time going back to
normal. I never want to hurt you and I know I have at times. I'll work
with myself and try to keep hold of myself better than I have been
doing. I think I'm beginning to sleep a little better, only the smallest
thing will still disturb me without fail."[192]

As their time together neared a close, Royce at last received his
classification, and was assigned to attend a gunnery instruction school
at Laredo, Texas. He had hoped for better, and wrote to his parents,
"I told you that I would very likely be sent to radio school back east. I
didn't have any choice in the matter. You see, the Army still puts you
where you are needed at the moment you are ready to be put. I ended
up here just where I didn't care much about going. We took a comprehensive gunnery examination and I did pretty well so they did this
to me." Initially disappointed to be assigned to gunnery rather than
radio, Royce quickly warmed to the idea, knowing that the Army left
him little alternative. Before leaving for Texas, he put Pearl on a train
to San Bernardino to visit her brother Dan and his wife Mary, with
whom Royce had stayed while he was a missionary in Los Angeles.
Royce then took another train to Laredo, passing on the way the
town of Fabens, Texas where he had begun his mission. That now
seemed a lifetime ago. "It looked just about the same as when last I
saw it. I'd like to have stopped and visited for a while. I think many
of them will still remember me after six years have gone by."[193]

Arriving in Laredo, Royce found the weather "hot and getting hotter," and after a brief evaluation he began instructor training, whose
purpose he briefly explained. "The primary idea is to have us release
men to go overseas by taking the jobs they now hold as instructors."[194] His training involved learning about an entirely new gunnery
system. "I am specializing on a different airplane than I flew on.

None of our class is specializing on B-25's. I chose the A-26. It's a new plane that does either medium or light bombardment work. It's faster and everything than a B-25. The turrets are remote control turrets like the ones on the B-29. That's the principal reason we are here, to learn them."[195] He described the new modern turrets in detail. "There are two of them. One is an upper and the other a lower turret. They fit almost flush on top of the ship and are remotely controlled by one gunner who sits on a seat in back of the bomb bay and looks through a periscope sight which covers both the top and the bottom when the person desires to fire in the lower or upper hemisphere after firing in the other. The guns are trained automatically, and the sight changes through a mirror which reverses itself. It certainly is a honey of a thing. I wouldn't mind operating it at all."[196]

On Sunday, Royce attended the only branch of the Church in Laredo, which was in Spanish. "I certainly enjoyed it. I'm pretty rusty on my Spanish as I discovered last week at Los Angeles, but I get along all right. I understand everything fine. They asked me to talk on the Easter program. I'm happy to have the opportunity. About fifteen soldiers go out there to church, even though they know no Spanish. They do the best they can on the hymns and everything, then hold their own class so it works out very well. The chapel is a pretty humble one, but it does very well. All of them seemed very disappointed when I told them I would only be here for six weeks. They're the same good people you find anywhere in the Church."

In Laredo, Royce was also delighted to renew an old friendship. "You've heard me speak of McNally before. I've been with him ever since Santa Maria California. When I transferred to the school section, I met him here. He's in my class and taking the same thing I am. We seem to hang together no matter what. If Andy had made it back intact, he very likely would be here too."[197] Not long after, Royce and McNally were also joined by "Robbie" Robinson, Royce's tent-mate and tail gunner from his old crew, who had completed his missions while Royce was still on Corsica but had not shipped home until February. The gunnery training school brought men together who had not seen each other since leaving for overseas, and stirred many recollections. Royce, on learning that his sister Dean had moved to Greensborough, recalled, "One of my friends back in radio school, Baker, was from there. I'm not sure about him but I believe he was killed. It's been impossible to keep in contact with all the boys I've been with in the Army. Some of them are prisoners or M.I.A. We expected that though. I've met four or five fellows here whom I havn't seen since radio school. We've been in all parts of the world since that time."[198]

As before, Royce excelled in his classes, and though he realized his work with 0.50 caliber weapons would be of little use after the war, there were other ways in which he thought the training beneficial. "It most certainly isn't a profession that you can follow after the war but I will be instructing, and I am pretty certain that the experience as an instructor will be of value to me. I hope I'll be able to get me a spot in it where I can last the war out. I definately don't think I'll ever have to go back over. We are replacing men who have never been to combat and there are plenty of them still."[199] As news of allied victories in both Europe and the Pacific continued to roll in, the likelihood of returning to war became more and more remote, which was a relief to Royce. "I feel all right about it because I can in honesty say that I did just about all that I could possibly do while I was overseas. Toward the end over there, I was in pretty bad shape and going downhill all the time. I want to see how I take flying again when I start on Monday. I haven't flown since January 1st and I was a pretty jumpy character about it then."[200]

Royce handled the flying part of his training without difficulty, and before his course was even completed he received gratifying news. "Today they published the list of fellows they would like to have stay here. Every graduation, they pick out the top 20 or 30 men and ask them if they'd like to stay here. They put the list out and I was on it. It's entirely voluntary. You do not stay unless you desire to do so. I couldn't see it that way, so when I went over to see the officer in charge, I told him that I didn't want to stay. Housing conditions down here are the worst I've ever seen in the Army and I know I'm almost sure to get sent to a better field than this when I ship out. The cost to get Pearl there would be about the same and I of course travel at government expense. It's hot, dry and dusty down here and getting worse all the time. I'd much rather take a chance on getting shipped to a better place."

During the month of April, the war was winding toward its conclusion in Europe, and news from the front brought back memories for Royce. "All these places that have been falling to the 5th Army in Italy lately are very familiar to me. We've bombed nearly every one of them at one time or another. I guess the boys in my outfit are really in on all of it. I've had several letters from the fellows there and they say it's pretty rough." There was also news from home that was especially cheering to Royce. "Pearl sent me a clipping saying that Reid Ellsworth was among the prisoners released by the Russians. He's already home. I certainly was glad to hear about it. He was shot down about a year and a half ago."[201]

On April 30, after six weeks in gunnery instructor training, Royce finally wrote, "Today was the day I've been waiting for around here. We graduated and also found out where we go from here. As things

238

stand right now, I go to Florence, South Carolina. I am pretty happy about it because I was afraid I might have to go to Florida and I'd just about as soon stay here as go there."[202] The next day brought even more welcome news. "I'm coming home on furlough! We are getting 15 days plus traveling time. It was all a big surprize to us. We are the first class ever to receive it after graduation, and we hadn't even heard any rumors. You should have heard the fellows shout when they told us about it. I couldn't believe my ears at first."[203] A few days later Royce boarded the train for Salt Lake, and it was about the time he arrived home that the news came that the Germans had surrendered.

Final Assignment

Royce spent two weeks in Salt Lake; then he and Pearl set out together on the long cross-country trip to his new assignment in Florence, South Carolina, where they found conditions crowded, and were obliged to board with a family at a high price while they looked for an apartment. The weather was warm and humid, and by that time Pearl was already expecting a second child, and spent time looking for maternity clothes. The end of the war in Europe had completely changed the military situation, and since there were large numbers of trained servicemen already available for the continuing war in the Pacific, instructors like Royce and McNally, who had once more been assigned together at the same service post, were left with little meaningful work to do. "I'm not flying nor do I think I will be. McNally and I were both assigned to the turret department of the ground school here. All we did the few days I was there last week was keep some turrets running and teach a few men coming through how to operate them. This next week, we are going to school again. I guess I'll never get away from that as long as I'm in the Army. The only reason I can see for this school is that they have so many of us out there that all they can find to do is to put us to teaching each other on things we already know. I don't much care. We just go out and put in time as our own bosses and everything. I have to report on the job at 8:00 a.m. and am free to leave at 4:30 p.m. It takes about 15 minutes to make it home to Pearl. I don't believe she finds it lonely with Jean here."[204]

Although Florence was a pleasant town, there were no LDS Church services, leaving Royce and Pearl with little to do on Sunday. "We have so far hesitated about taking Jeanie to one of the Protestant churches in town because she gets so unruly anyplace where she knows that it's difficult for us to handle her." Royce had returned a total stranger to his daughter Jean, and he was annoyed by what was probably typical toddler behavior. "She is pretty badly spoiled but we are trying to take it out of her as much as we can, and we do seem to be having some success. She's eating better and everything. If only people would let her alone and not try to make so much of her all the time, our job with her wouldn't be difficult at all. I'm surely not going to have her thinking she is supposed to be a little side show all the time."[205] Pearl would later comment on how harshly Royce treated Jean when he returned from the war, and the two of them would never have a close relationship.

Although the nation was still at war with Japan, the capitulation of Germany meant that far fewer soldiers were required for military service, and a point system was developed to prioritize discharge of veterans based on time of service and the number of actions in which they had been involved. The system was favorable to Royce and other airmen who, by flying support along a whole front, were credited with participation in many battles, but the administrative wheels still turned slowly. Royce received a letter from Wendell, the turret gunner from his old crew, who, after finally completing his missions, had returned home and been immediately released. "He hit it just right. I've heard from him since then. He likes being a civilian." The uncertainty left Royce and Pearl reluctant to look for better lodgings, even though they were paying high rent. "We rather hesitate about getting too firmly established here because of the possibility of a discharge. They have orders here that no men with over 80 points are to be transferred from the field pending discharge. That may or may not mean something for me. I hope so." Royce added, "I believe we're getting more non-essential every day. They keep bringing more and more men in. For the present, I'm on the skeet range where we spend our time shooting clay pidgeons. I know it's not helping the war effort much if any. I also know that with all the men being processed and moved now, it's impossible to give individual immediate attention to all the boys who are eligible for discharge. They tell us to be patient, and that we try to do."[206]

Just a couple of days later some truly "essential" work did come Royce's way. The large number of new and returning soldiers who had come to Florence to be processed had placed a strain on the normally quiet town, and Royce and most other instructors were assigned to serve as Military Police in the community. "This week, I'm on night patrol here in town. I go on at 7:00 p.m. and am off at 12:00 p.m. The rest of the time is my own. I report at the police station and don't even have to go out to the base. It's just five minutes from here to the police station." Royce had held the usual G.I. attitude towards the M.P.'s and this was a task he would have disdained earlier in the war, but the circumstances were now much changed, and he found the work quite agreeable. "My on duty hours are the best I've ever had, and the work so far has been interesting. It takes a bit of time to get accustomed to an M.P. band on your sleeves and white belt and leggings."[207] McNally was the only one of the instructors not assigned to police work, and they supposed it was because he had a stained record, having himself once spent the night in a guard house during their training at Sioux Falls.

The assignment led to one last stint in a military school, for even while he worked his shifts as a Military Policeman at night, by day Royce attended a ten-day course entitled "Law Enforcement Officer's Training School," for which he received a certificate on July 27th, 1945. By that time he had already accumulated quite a bit of experience in police work; and even before the course began he observed, "I know all the cops in town and in the county now, plus the mayor and the railroad police. Our work is at times rather interesting. We just put one of the beer joints in town [out of business] and now every other cafe owner in town just about bends over backwards to please us. They will beg us to come in and order anything we want for free. We take them up on good watermelon and other things like that. They all know now it pays to cooperate in helping to keep order in town. They would most of them go broke without the military trade. We don't take so many in. Most that do get taken in get it for being drunk and disorderly. We play nurse to lots of drunks. Most of them, we just send back to the base with orders to go to bed. If they put up a fuss, we let them sweat the night out in the guard house. They don't bother anyone there." This assignment to police duty did more than merely keep the gunnery instructors occupied. The use of battle-experienced veteran instructors like Royce for Military Police duties also proved to be a wise move in maintaining security at a challenging time, for reasons Royce explained. "Nearly all the M.P.'s in town now are fellows like myself so we don't have to take very much of anything. We have the rank and overseas service. Nearly all the boys respect that."[208]

As the summer progressed, Royce made daily trips to the base to check on his discharge, and he began to think seriously about postwar plans. "If I get home in time, I believe I will try to make it back to school. I'm pretty sure I can get by all right. Pearl wants me to, and I do think that if I hold off for a while I might not ever go back, because I would get started at something else."[209] As July turned to August, still no word of discharges came. Royce and Pearl had finally located an apartment, and Royce wrote hopefully to his family, "My chances of getting out are just as good as ever. It is only a matter of time until they do release a lot of us. There is nothing we can do to hurry it up. Now that we have our apartment, I do not mind so much. Pearl is so much happier too. We have a yard for Jean to play in."[210]

One day in August Royce and Pearl were seated on the porch when the voice of President Truman came over the radio, making the startling announcement that a powerful new atomic bomb had been dropped on the Japanese city of Hiroshima. Not long after, it was announced that Japan had agreed to surrender, and that hostilities had ceased. The final signing of the peace agreement would take place just a few days later, and at last the war was over. As a wave of jubilation swept the nation, on August 19 word finally came through that Royce was authorized to be discharged from the Armed Forces, though the discharge itself would entail some delay, as Royce explained to his family on August 22: "You see, they have 500 of us on their hands right now and the Army wheels sometimes grind a bit slow. I'm going up to Fort Bragg, North Carolina (one day), then I'll come back here and we'll go home. Otherwise, I'd have to go down to Fort Benning, Georgia and wait for a troop train and orders to take me to Fort Douglas for a discharge, and Pearl would have to travel alone. This way, I'll come back from Fort Bragg free as a bird. We are waiting for Fort Bragg to wire back and say they can handle us. I'll likely be out of the army shortly after you receive this."[211] It was not until August 30, 1945, that Royce was officially discharged from the United States Army at Fort Bragg, and finally he and Pearl with their daughter Jean boarded a crowded train westward toward Salt Lake City, where they arrived in Early September, ready at last to begin life on their own terms.

Epilogue

Who can gauge the impact of a great war in the life of an individual? The Second World War was a defining and transforming time for Royce Bringhurst, as indeed it was for the entire world. From the backdrop of life as a whole, the war was a world apart, and the relentless march of events had swept Royce along it its path as it had virtually all his generation. He had entered military service voluntarily, motivated not by patriotism alone so much as by a sense of responsibility, nurtured through his early years as a growing farm boy, to make his own contribution, and in a sense, to represent his family and community as a soldier, just as he had as a missionary. There is no question that he desired his place in combat, a sentiment he expressed both before going overseas and as he neared the end of his tour of duty—this was not just an urge to experience warfare firsthand, but also a desire to be an active participant in what everyone increasingly realized was the key event of their time.

Royce entered the war with faith and confidence in Divine protection which he had developed in the mission field, but also with a clear awareness of the ever-present possibility that he might not survive.

Like many others, he passed through extreme danger, felt gripping fear, and witnessed the violent deaths of others and very nearly his own. He lost close acquaintances whose vibrancy and youthful vigor had been cut short by the war, and for the rest of his life his mind would carry the images of friends like Frank Denstad, Norman Johnson, Frank Van Limberg, Meade Steadman, Wilson Frost, and a host of others who had, like him, hazarded their lives, but unlike him would never return. The chance nature of their sacrifice had hardened and embittered him as he watched the inexorable tide of events claim such truly good men while sparing less deserving ones, and he would ever after look upon war as a monumental, arbitrary and unjust evil. Royce felt satisfied he had performed his duty in the war, but he never regarded himself as a hero, though as the years passed he became increasingly aware that his was a heroic generation who, in subjecting themselves to the seemingly random fortunes of war, had not only triumphed in battle but had fought the defining struggle for human freedom of their age. He would never boast of his war service, save to acknowledge that he had flown sixty-five missions on a bomber crew, which was saying enough.

The outcome of the war would verify Royce's greatest fears, but also his highest ideals. Having seen the utter failure of the "Great War," concluded just a month before his birth, to free the world from military conflict, his constant underlying worry had been that this new World War would merely set the world up for further injustices and lay the groundwork for future wars, as the last one had done. In the aftermath of the war, the world would indeed watch as Joseph Stalin went on to become one of the most brutal tyrants of the 20th Century, and as all of Eastern Europe and much of Asia fell under the withering influence of communism, a brand of utopian totalitarianism every bit as pernicious as fascism had ever been. During Royce's lifetime he would see the ensuing struggle over communism lead to two major wars and a host of minor ones. But he would also watch as the very principles he believed in, those of "love thy neighbor" and even "love thy enemy," embodied in the nation's post-war policies, would render Germany, Italy and Japan not just lifelong friends and allies of the United States, but bastions of freedom in their own right.

Royce's war experience had indelibly changed and refined him to such a degree that it is difficult to imagine his later successes without it. The war had been a study in extremes. During training, Royce had experienced the keen humiliation of personal failure as he was eliminated from pilot training, but also the maturing process of accepting a secondary role as an enlisted man and performing it well, ultimately gaining the respect and friendship of officers and enlisted men alike. The war greatly expanded the physical and mental discipline he had gained as a missionary. The long succession of technical schools, though they did not teach him concepts which would be of substantive benefit after the war, proved nonetheless a valuable training ground, as coupled with the accelerated urgency of wartime need, they provided him with the drive, stamina and discipline which would enable him to excel in his future academic and professional pursuits. Even the aggravating inefficiencies which naturally resulted from subjecting individual lives to the rapidly changing needs of a massive military organization and to the vagaries of war had taught him patience and resilience which would later be of great value to him.

Perhaps equally significant were the refinements to Royce's character, and the marked changes in his relationship to the world around him. As a missionary he had already been exposed to people and ways of life beyond the insular Utah environment in which he had grown up, but during the war he had been made to live, eat, sleep, train, and fight with men of all backgrounds and walks of life, and had experienced, to a far greater degree, the world beyond his own. While he still regarded Latter-day Saints as a whole to be the finest and cleanest living of people he had encountered, he had also come to recognize that there were virtues in men of all faiths and backgrounds, and that in the naturally coarse world of a military camp, even the most vulgar soldier had a certain measure of basic human goodness.

Although the war had left marks of refinement, it also left its scars. Royce's obligatory confinement to crowded military camp life made him wary of crowds and jealous of his privacy, and he would often experience anxiety in crowded public places. His case of "war nerves" would continue to trouble him for the rest of his life, and his older daughters especially had to learn to walk softly around him, knowing that if they made a sudden noise or let a door slam he might jump up with a loud and scathing rebuke. They would remember him sleeping poorly, and waking up at night shouting out in his sleep.[212] Royce would retain a sharp and, at times, explosive temper which was much feared by his children. On more than one occasion Pearl would comfort a sobbing child who had been harshly treated by saying she wished they could have known their father before the war. With the passage of years, both his temper and his nerves mellowed, but an event or sound could bring them back to him in an instant. For example, while driving one day along a two-lane highway in Chile more than twenty years after the war ended, Royce suddenly and violently swerved the car to the side, then explained to his startled wife and family that a passing truck had made a noise that "sounded just like a shell."[213]

Royce would never recover normal hearing, and was nearly deaf to noises behind him, as well as to higher frequencies. Although he continued to sing, he could not hear the upper tones of a standard piano, and perceived only a clicking noise when those keys were played. His hearing loss was severe enough to qualify him for a 10% disability rating, for which he received a pension from the Veteran's Administration for over ten years until, as a university faculty member, he requested that the pension payments be terminated, since the hearing loss no longer affected his earning ability.[214] Beyond his nerves and hearing, his military experience had left him with a cynical edge which, in contrast to the idealism of his early years, was often reflected to varying degrees in his attitudes, particularly toward governments, the military, and authority figures in general.

Like many of his generation, Royce almost never spoke to his children of his war experiences, though they would remained a palpable and ever-present shadow over his life. On the rare occasions when a friend from his years in the military came to visit, rather than drawing the guest into the family circle Royce would disappear with him into the yard to engage in long and quiet conversation away from the family. It was only in his waning years that he became willing to speak openly about his experiences, especially after his war journals, which had been lost for years, reappeared. By then his memory of the details had faded, though the feelings they evoked were still vivid, and there were times when he would cut short his narration when emotions came too close to the surface. Royce would one day refer to World War II as "the great adventure," and though he could not bring himself to speak much about the war he had fought, in numerous small ways he succeeded in leaving with his children a sense of the weighty impact it had had both on him and on the world at large, along with the imperative that it never be forgotten.[215]

Illustrations for Chapter 6

Wreckage resulting from German air raid of May 13, 1944, one day prior to Royce's arrival at Alesan, Corsica.[216]

Planes of the 489th over the bivouac area, Alesan, Corsica.

246

489th Squadron camp in Alesan.

Royce and tentmate with one of the camp dogs, Corsica.

Royce in sleeping quarters and at makeshift writing desk. The men made
continual improvements on their tent during the stay on Corsica.

Planes of the 489th over the Alesan airfield.

Aircraft from the 489th Squadron departing for a mission. In the background are six planes in "box" formation.

A B-25J from Royce's squadron. In the background at right is a P-51 fighter, part of a fighter escort which sometimes accompanied the bombers to guard from attack by enemy planes. The insignia "9Q" distinguished the bomber as belonging to the 489th Squadron.

A ship of the 489th performing evasive maneuvers after a bombing run. The bombed bridge in the background is near Milan, Italy.

"D-day" invasion of Southern France, August 15, 1944, showing warships and beachhead. Royce's bomb group lost three aircraft on this mission.

Royce's crew for the "D-Day" mission. From left standing: Royce, pilot Brassfield, bombardier Harper, Co-pilot Walsh; seated: tail gunner Robinson, turret gunner Beasy. Four of the six men were on Royce's original crew.

Mission over Var River in France, early August, 1944. The four black spots in the foreground are bursts of "flak," generally fired in a barrage of four shells. Royce kept this photograph with the shell casing shown below.

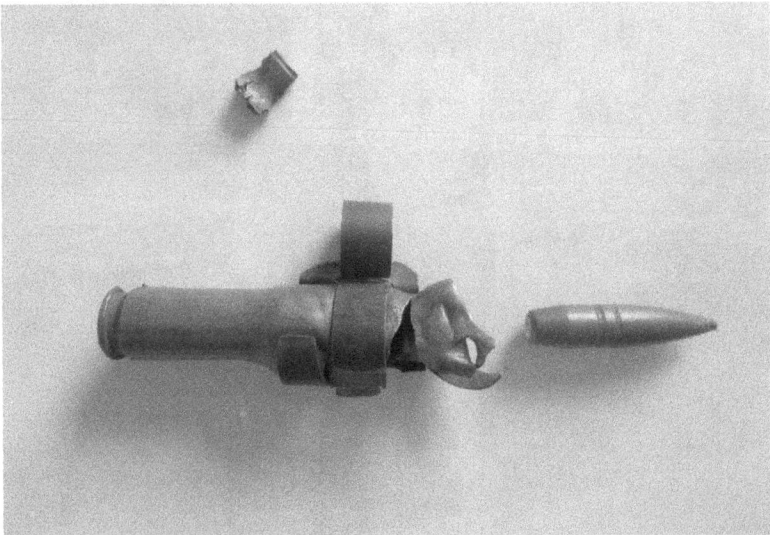

Fragment of German flak shrapnel, and exploded .50-caliber cartridge and round from Royce's plane, recovered from bombardier's compartment after the Faenza mission on November 16, 1944. Royce kept these artifacts the remainder of his life.

Royce checking out the anti-aircraft armament on board the troop carrier *West Point* (formerly the luxury liner *America*) as he prepared to ship home.

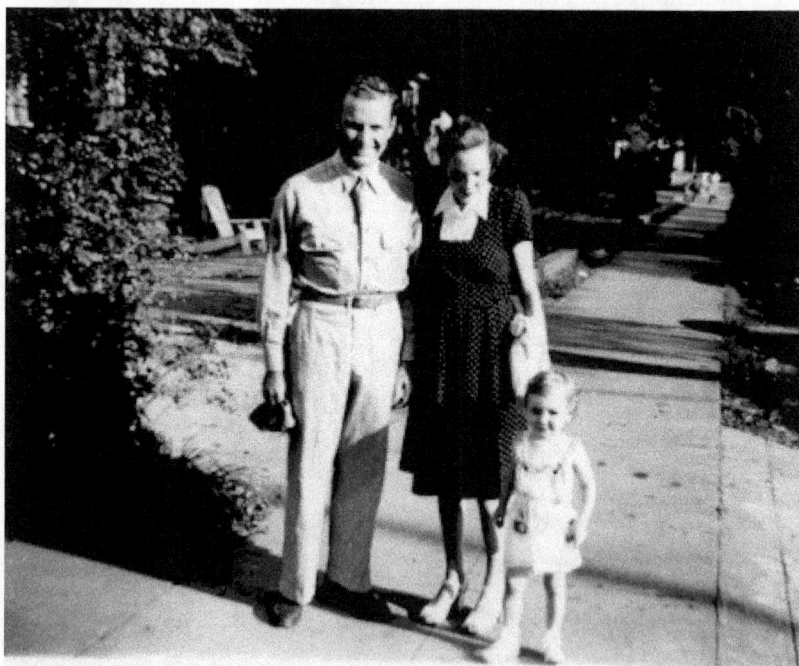

Royce and Pearl with 2-year-old Jean in Florence, South Carolina, Royce's final military assignment. By this time Pearl was expecting a second child.

College and Graduate School

Chapter 7

Post-war Plans

While still overseas flying combat missions in 1944, during the optimistic weeks after the invasion of France when it looked like peace was at hand, Royce had begun to give some serious thought to what he might do after the war. His entry into military service had forced the sale of the old family property on Redwood Road, so his early ambitions of returning to work the family farm were dashed, aside from the few months he and Pearl had spent together there after their marriage. Royce still longed to make a living in agriculture, and when word came through military circles of new opportunities being considered by the government to allow military veterans to homestead and farm land in Alaska under very favorable terms, he wrote to Pearl to sound out her feelings on the subject. Pearl did not respond very positively to the idea of farming in Alaska, but shortly afterward Royce learned of recently enacted legislation to benefit veterans who had contributed to the war effort. For Royce, the most appealing part of this legislation was a law which would come to be known as the "G.I. Bill,"[1] which Royce described to Pearl in a letter home: "It seems that the government is recognizing the fact that many men are losing the training that they would normally get in our colleges and that it's bound to effect the country in an adverse way. Because of this, they are making it possible for a man who has served to receive his schooling free at the college he chooses. If you have served two years, and I have, you will be entitled to two years schooling. I would very much like to go back, finish up my two years and get my degree. I would have a pretty tough time at first getting back to it but I think I could do it quickly enough. How would you like to go to college with me? I think I'd like to go back to Logan. You would love to live there."[2] This time Pearl's response was nothing but positive, and from that time forward, finishing a bachelor's degree at Utah State Agricultural College became Royce's post-war goal.

When Royce and Pearl arrived in Salt Lake City in early September of 1945, Japan's surrender had just become official, bringing a final end to the Second World War. This became a time of enormous transition for the country as a whole, as nearly the entire young adult male population had been engaged in military service, and most industrial, agricultural and social enterprises for the previous three and a half years had been focused on the successful prosecution of the war. Now that the war was over, Royce joined a huge wave of discharged soldiers returning home to rejoin civilian life. The overwhelming task of demobilization had proceeded in an orderly, if sometimes arbitrary fashion typical of the military, and if Royce was not among the first of his peers to be discharged, neither was he among the last. The transition from war to peacetime would occupy Royce's and Pearl's lives that entire year, as it did for everyone else.

Although Royce had planned to return immediately to Utah State Agricultural College, his discharge came too late in the year to make that a realistic option. Arriving within a month of the start of fall quarter, he and Pearl had to attend to the endless details necessary to re-enter civilian life, and having been away from home for nearly three years, there was much to catch up on. Pearl was by then in the later stages of her second pregnancy, and that may have been a deciding factor, since the couple had little time to look for adequate housing for a family in Logan under those circumstances, much less arrange for good obstetric care. Royce took a temporary job as a lineman's helper for the Vance Electric Company in Salt Lake at $208 per month, which he continued through the month of December.[3] Royce and the family stayed with Pearl's mother in the Davidson home at 3112 South 7th East in Salt Lake City, which had been Pearl's place of residence while Royce was away in the military.[4]

On Monday, November 26, 1945, Pearl went into labor and gave birth to another baby girl. Having been absent for Jean's birth, Royce had vowed to be present when the next baby was born, and this time he was able to be with Pearl at the hospital in Salt Lake City. They named the new baby Florence, which, besides being a tribute to Royce's mother, also called to mind the town in South Carolina where Royce and Pearl had finally been permitted to live together in the military after over two long years of separation. The following month, for the first time in seven years, Royce was at last able to spend a Christmas at home, and to be with his family for his own birthday, which fell two days after Christmas. It was on Pearl's birthday, on January 6, 1946, that Royce performed the baby blessing for Florence, in the Sacrament meeting of the Springview Ward, Grant Stake, where Pearl's mother resided.[5] By then he and Pearl were already making final preparations for their move to Logan to begin school.

Completing College

In early January of 1946, Royce arrived in Logan with his young family to register for classes for the winter quarter at the Utah State Agricultural College. More than six years had elapsed since the relatively carefree days when he had completed his second year there, though for Royce, who had since been through a mission, a marriage and a war, it must have seemed a lifetime. The college was itself a changed place. The war had drained the campus of nearly the entire male student body, and for the past few years it had been essentially a women's institution. At the same time, portions of the campus had been requisitioned for the training of male military recruits, including members of the air corps, civilian pilots, army specialists and navy radiomen; in fact, many uniformed soldiers had spent time there, and some of these had actually been trained by college faculty.[6] Although some former soldiers had, like Royce, already found the opportunity to enroll and go on with their schooling, others were still completing their military service, and by far the majority of students in Royce's class were women, in sharp contrast to a few years before. Some men would never return—the college yearbook for that year listed nearly 140 students or alumni who had lost their lives in the war, among them Royce's friend Norm Johnson. Those who returned were an older and more mature group, many of them already married, and a large proportion of them had seen their share of combat during the war.

Royce and Pearl moved into an apartment located some distance from campus, at 162 West, 3rd North in Logan. The weather was cold, and they had difficulty keeping the apartment warm enough for the children until a neighbor showed Royce how to operate the stoker on the coal-powered heating system, and in short order they had turned their little apartment into a comfortable home. With the large influx of returning servicemen and their families, housing was at a premium in Logan, and when they occupied their apartment, the cost of rent had already been raised multiple times and continued to go up.[7] Royce and Pearl were not the only student family affected by these price changes, and complaints had been filed with a government

255

agency for relief. Learning that a faculty member he was acquainted with was on the local committee, Royce himself lodged a complaint in February, and eventually hearings were held between the city council and the O.P.A., the agency charged with overseeing pricing.[8] Although some families were dismissed from their apartments for filing such a complaint, Royce and Pearl were allowed to remain, but at the same time they began to look around for other living accommodations.

Royce registered for a full load of course work for winter quarter and, in accordance with his plans, took a variety of upper division agricultural classes. These included fruit production, which involved newer pruning techniques and orchard care practices, feeds and feeding, which included actually providing a balanced diet to assigned livestock (in Royce's case, sheep), classes in cereal and root crops, and an elementary architecture class on the side, which together amounted to seventeen units. This was comparable to the course loads Royce had taken before his mission, but the years of maturity and the study habits and discipline acquired in his long series of military schools had yielded their benefits. In spite of his long time away from college, Royce studied hard and earned "A's" in every subject.

Life at home had also taken a turn for the better. Royce and Pearl had talked over their childrearing practices when they realized that Jean seemed to be nervous and eating poorly, and they decided that Royce would back off on scolding Jean for errant behavior. Jean, in turn, responded with better conduct and an improved appetite, and Royce began playing with her and reading books to her. The new baby Florence, for her part, was a carefree and smiling child, and Jean, ever observant, began imitating her mother in every particular, doing everything her mother did with a doll she carried, to which she also gave the name Florence. Owing to the cold weather, Royce and Pearl took turns going to church meetings and staying home with one or both of the children until the weather grew warmer. For Pearl, who had spent her entire life in the Salt Lake valley, the winter in Logan seemed cold and windy.[9]

Spring came at last, and for the spring quarter Royce enrolled in courses in irrigation, forage crops, and dairy husbandry, and also took a required advanced English composition class. To round out his schedule, he indulged his love of music by enrolling in a men's advanced chorus class, and was also invited to sing tenor with a local men's quartet.[10] By this time Royce had developed a rich, robust solo tenor voice, and despite his scant musical training and lack of opportunity while serving in the military, he still loved to sing. His chief handicap was the impaired hearing resulting from the noise exposure while flying in the bombers during the war, which made it difficult for him to hear the musical accompaniment to which he was singing. This would continue to trouble him to some degree throughout his life, but he adapted to it as best he could, and he sang whenever the opportunity arose.

Royce continued to pursue his studies with vigor and discipline, and remained at the top of his class, even as he became actively involved in other activities. He was elected president of the Ag Club, which that year had revived the traditional springtime horse show and livestock judging in which Royce had taken a part years before, but which had been abandoned during the war years. There were also choir concerts in which Royce sang, and for which he attended rehearsals. During the spring quarter family funds were tight, so Royce took a job with a local professor to bolster the family finances,[11] and Pearl did some babysitting as well. With the permission of their landlord, they bottled apples from the apple tree and raspberries from vines on the property, and Royce planted a garden in the yard. He also considered the purchase of a secondhand car, but they were unable to find any of sufficient quality for a price they could afford, and in the end Pearl talked him out of it, though this left the family without reliable transportation.

After completing the spring term, Royce and Pearl remained in Logan rather than returning to the Salt Lake area as Royce had done in earlier years while attending college. There was no longer a family farm demanding Royce's attention, and continuing school through the summer term would place him back on schedule to graduate the following summer, despite his late return from military service. Additionally, Royce calculated that if his record of straight "A" grades continued for another quarter, he would qualify for a scholarship, which would take additional pressure off the family finances. That summer quarter Royce enrolled in another full schedule of courses, including classes in soils, weed control, vegetable production, poultry, and elementary veterinary science. Aware that Pearl's brother-in-law, Burnell Turnbow, was still farming acreage on a family farm in Tabiona, Royce also collected agricultural extension bulletins for Pearl to send to him, reasoning that he might as well also benefit from Royce's college experience. Up to this time Royce had not yet given serious consideration to graduate school; rather, his goal was still to complete a bachelor's degree and immediately find employment in an agricultural field.[12] By the end of the summer, he had completed his junior year course work with perfect grades, and it appears that he did receive the desired scholarship, for that fall he and Pearl moved from their apartment to a rental home located at 60 North 4th East in Logan. This was an older building, but it put Royce closer to campus, and provided a little more room for the growing family.

By the fall quarter of 1946, the Utah State Agricultural College was at last coming into its own as a major postwar university. Many returning student veterans had by now worked their way through the military system and were returning to continue their education, and many others had taken advantage of the G.I. Bill benefits to enroll in college for the first time. Not only was there rapid expansion of the college due to record enrollment, but the veteran students like Royce also came with the increased maturity and discipline they had acquired during military service, which raised the academic standard for everyone. It was a time of booming optimism for the students and for the country at large. The oppressive days of the Great Depression were now a fading memory, and in the academic setting there was growing confidence in the power of science and technology to transform the world. Indeed, there was an awareness that the Allied victory in the war had largely been due to the accomplishment of American scientists and industrialists, coupled with the courage, confidence, and unprecedented technical training of the soldiers themselves, who were now flocking to universities in large numbers. Royce Bringhurst was one of these, and having excelled in his postwar studies, he entered his senior year of college with supreme confidence.

During the first quarter after his return to USAC, Royce had come under the influence of Del Tingey, one of the pioneers in wheat hybridization,[13] who happened to have taught his cereals class. It was with Dr. Tingey that Royce had taken the part-time job in March of 1946, assisting in a weed control project at $35 per month. Dr. Tingey, sensing Royce's aptitude and interest in plant breeding, began encouraging him to consider graduate studies in the field, and recommended appropriate programs to which he might apply. Royce's relationship with Dr. Tingey perhaps brought to mind the early admiration he had held for F. V. Owen, the famed beet geneticist and plant breeder who had done field work on his father's property.[14] With Dr. Tingey's assistance and encouragement, Royce submitted applications to three different graduate programs in plant genetics. Of the three, the one that most appealed to him was at the University of Wisconsin at Madison.

Meanwhile, as Royce continued into the winter quarter, life became busier. Besides work, school and church, he found time to be involved in a variety of other activities. He had become a member of Alpha Zeta, an honorary fraternity for students of agriculture and forestry, which held periodic meetings. He continued to serve as an officer in the Ag Club, and busied himself planning the annual horse show, of which he would serve as chairman for 1947. Additionally, he continued to sing in the advanced chorus, and had passed the tryouts to become one of the singers in the university opera production of

Bizet's *Carmen*, which was presented that spring, and included performances in both Idaho and Ogden. There were duties at home as well, not only with the two growing girls, but with Pearl, who was by then expecting a third child. The demands on his time piled up, and amid these many activities a "B" grade slipped through, the only flaw in an unbroken series of "A's" since returning from the military. In spite of this, and in spite of Royce's less-than-perfect lower division work before his mission, he would finish the year among the top graduates in the School of Agriculture for 1947.

As spring approached, Royce waited anxiously for word from the three graduate programs. He received a letter in March from the University of Minnesota offering him a research fellowship for the following academic year, and for a time he was in a quandary, as the offer required a response within a given time, and Royce still had not heard from the University of Wisconsin, which was his program of choice.[15] Pearl and Royce discussed the situation, and Pearl advised that he give the matter some prayer, and delay his response to the other programs. At last in early April the long-awaited letter came from Wisconsin informing him that he had been accepted into the genetics program there, and with some relief Royce sent to the other universities declining their offers. As graduation approached, Royce was informed that he had been elected to Phi Kappa Phi, a national honor society limited to student of high integrity who are ranked scholastically in the top of their class—he was one of six students in the School of Agriculture so honored, and one of eleven of the entire graduating class of nearly four hundred to received a special "A" award granted by the university for scholarship.

At length the spring quarter wound to its close, and Royce completed his final undergraduate course work. Graduation ceremonies were held on Saturday, June 7, 1947, in the Utah State Field house. Dressed in cap and gown, Royce stepped forward as one of a large graduating class to receive his degree as a Bachelor of Science. The degree had been a full ten years in coming from the time he had first entered the campus as a hopeful freshman. From the very beginning, some had doubted his ability to succeed in college work, and Royce himself had had second thoughts, often assuming during the long interruption for mission, marriage, and military service that he would never have the opportunity to return and complete what he had started. Yet not only had the ambitious farm boy from Murray become the first member of his family to earn a college degree, he had done so with distinction, graduating third in a class of sixty-two from the School of Agriculture. By the time he received his diploma his prospects had greatly expanded, and more years of schooling still lay ahead.

Graduate School in Wisconsin

Shortly after his graduation, early in the summer of 1947, Royce traveled to Madison, Wisconsin, to begin graduate studies. His work there was to be chiefly under the tutelage of Dr. William K. Smith, an assistant professor in the Department of Agriculture and Genetics, for whom Royce was hired as a half-time research assistant at $95 per month, which would be supplemented by his continued "G.I. Bill" payments.[16] Dr. Smith had been working since the late 1930s with sweet clover, an important forage crop for the Wisconsin dairy industry, but one which was capable of producing serious illness in cattle because of its high coumarin content. He was engaged in challenging genetics research in an effort to develop coumarin-free strains of sweet clover, and it was this research in which Royce would be taking part as a graduate student, under the sponsorship of the Wisconsin Alumni Research Foundation.[17] Dr. W. K. Smith was a native of Scotland and spoke with a thick Scottish brogue which was difficult for many of his American students to understand, but not for Royce. When his colleagues asked him how he could understand his research professor's speech so readily, he replied that he had heard the same accent for years from Pearl's mother, who was herself a Scottish immigrant.

Royce was initially advised to come to Madison alone, as housing for married students was in short supply due to the large influx of veteran students. Once again the timing was difficult for his family, as Pearl was already in the advanced stages of pregnancy and would soon be giving birth to her third child. Royce began his summer courses while arrangements were still being made for Pearl and the children, who were staying in Salt Lake with Pearl's sister Helen. The housing he finally obtained for the family was in Badger Village,[18] a group of barracks-style apartments which had first been constructed for munition workers during the war, and had been rapidly expanded in order to accommodate the huge influx of married student veterans attending college under the G.I. Bill. Badger Village was a self-contained community located near Baraboo, Wisconsin, in a rural area some 35 miles from the campus in Madison. Open only to married student veterans, the village included a community center, a student-led governing body, a community garden, and an elementary school. A system of antiquated buses, mostly retired school buses, transported students to the campus on a regular schedule throughout the day.[19]

When Pearl and the girls arrived from Utah later in the summer to move into the village, work was still going on to make the apartment habitable. Pearl wrote to her mother, "The maintenance men tore out our walls the day before I came in and straightened them. They tore up the floors the day I came in and replaced them because they were warped. Someone went in and put down our linoleum when they were through."[20] The apartment had an ice box rather than a modern refrigerator, and ice was delivered on a regular basis to keep food cooled. Milk was delivered daily in glass bottles, and in the winter the milk would freeze, suspending the cap over the bottle on a frozen column of milk. Heating was from a coal burning pot-bellied stove, which residents were required to keep burning throughout the winter months to keep the water pipes from freezing. These village apartments were hastily built and very modest for the standards of the day, but they were far more economical than more traditional housing, which was also more difficult to come by. To student couples who had been reared during the Great Depression, and to men who had been subjected to military conditions during the war, they were more than adequate, and there existed a strong sense of community there.

By August, Royce was nearing completion of his summer term classes as Pearl's due date approached. On the morning of Friday, August 8, Royce had just left on the bus for campus when Pearl began to have labor pains. She sent for a neighbor who had agreed to take her to the hospital if she went into labor while Royce was away from the house. Driven to the hospital in Baraboo, Pearl remained there, while back in Badger attempts were made to contact Royce at the university. All alone in the hospital, as the birth grew close Pearl was grateful for the presence of a Catholic nun who coached her through the labor, and at 3:00 p.m. she gave birth to another baby girl with dark brown hair and long, slender limbs. Attempts to contact Royce through the day had been unsuccessful, and he did not learn that Pearl had even gone to the hospital until about the time the birth was taking place.[21] Royce later wrote, "I was in at Madison attending summer school when the call came that she had gone to the hospital at Baraboo and had a baby girl. I 'rushed' back on the first bus, arriving after a tedious 25 mph, 30-mile drive to Badger Village. I got in the old Buick and went to the Ringling St. Mary's hospital. Trying to make a joke, I asked Pearl what the idea was coming in 2 weeks early and having another girl."[22] Pearl was in no mood for joking, and seizing her pitcher of water, she threw the entire contents at Royce.[23] The two named the baby Marla, a name Royce had encountered and admired during his mission in Los Angeles.

Royce gave Marla her baby blessing on Sunday, September 7, 1947, in the meeting house in Madison. Although Royce and Pearl were officially part of the Madison Branch of the Church, they were allowed to meet weekly as a Sunday School group together with a few other LDS families in Badger Village, avoiding the need to travel all

261

the way to Madison to participate. Royce and Pearl were the heart of this group, which met either in an activity room in the village, or in one of the member homes, depending on the number who attended. The adults took turns teaching the lessons, and on those weeks when they were the only ones there, Pearl simply read Bible stories to the children. A few months after her arrival, Pearl helped organize a small Relief Society for the four Latter-day Saint women who wanted to participate,[24] and these women set about making dresses for the Church's welfare system.

Meanwhile, Royce continued with his school work, which included both academic course work and experimental work with sweet clover and other plants, under the direction of Dr. W. K. Smith.[25] Royce's graduate program was designed in two stages, leading first to a Master's degree and then to a PhD, a practice common at that time. Royce explained the rationale years later. "You could go and you could get a Master's Degree en route to getting a PhD degree, and that's what happened to me, and with a lot of others. It was a good way to go, because you didn't end up with nothing if things went badly, and not every graduate student succeeded."[26]

The experimental work involved both coumarin-containing and coumarin-free species of clover, and the complicated process of attempting to hybridize the two. Royce had to master the difficult techniques involved in pollinating, transplanting, and grafting these plants. It was meticulous and demanding work, and sometimes unexpected challenges would arise. For example, Royce later recalled caring for a greenhouse in which the plants were segregated, with coumarin-containing plants on one side and coumarin-free plants on the other. At the end of a busy week the greenhouse was locked up for the weekend, but unbeknownst to Royce and the others, a rabbit had made its entry undetected, and took up residence in the greenhouse for the weekend. When Royce arrived on Monday, the rabbit had feasted on sweet clover, annihilating the plants on the coumarin-free side of the greenhouse, but leaving those on the coumarin-containing side untouched. One plant, however, on the coumarin-free side had been left alone, and when it was examined, it proved to be a coumarin-containing plant. Mistakenly categorized as coumarin-free by the researchers, it had been correctly identified by the rabbit.[27]

Winter approached, and the school work continued. Winters in Wisconsin were even harsher than the ones in Logan had been, with blizzards and winds which caused the snow to pile up in large drifts, and with temperatures that remained below zero for months at a time. Despite the severe cold, the little apartment in Badger Village remained warm and cozy and the family lived comfortably. For Christmas, Pearl and Royce purchased gifts for each other and Pearl sewed dolls for the girls. They dined on a chicken for Christmas, and on a small ham for New Year's, and on New Year's Eve they gathered with several other couples at a friend's apartment, where they had what Pearl described as "probably the only dry party in the village." Marla had proven to be a pleasant baby and easy to care for, while Florence was lively and good-natured and had begun to talk constantly, and Jean, still bursting with curiosity about everything, was old enough to be a help with household chores.

As Spring approached the family had another setback. Jean developed a rash and was diagnosed with scarlet fever, and the whole family was placed under quarantine except for Royce. For the period of the quarantine he was not permitted to sleep in the apartment for fear of contagion, and was only allowed to visit the family once a day. When the quarantine was finally lifted, the family was required to wash down the walls and repaint the apartment, a task which largely fell to Pearl, since Royce by then was involved in a busy round of transplanting at the university. The paints issued by the village were dull white in color and showed the children's handprints badly, so Pearl experimented with the addition of pigments, and painted the rooms in marbled colors which made the apartment the admiration of their friends. For his part, Royce between school responsibilities was able to find time to plant a garden in the community plots to supplement the family's food supply, and this eventually provided a bountiful harvest, including bushels of tomatoes for canning, and a supply of potatoes which extended well into Autumn. The family also gathered puffball mushrooms in a field near the village, which Pearl sliced and fried in butter.[28].

Royce completed the requirements for a Master's in Agronomy and Genetics at the conclusion of his first year in Wisconsin, and in June he attended his first graduation there. The university at Madison was much larger than the Agricultural College in Logan, and Royce was impressed with the scale and good order of the graduation exercises. In a letter to his parents he wrote, "I received my degree without much trouble. It was a rainy, rainy day but all went well. I never dreamed a thing so big could take place with such an utter lack of confusion. With over 2900 graduates, that is quite a bit to say. They had no rehearsals or anything. They just had an efficient usher system to pull it off smoothly. The governor, president of the university, and General Omar Bradley (Army chief of staff) all spoke. There were no unessential frills at all. We took the kids in and left them with Dr. Smith's wife while Pearl attended the services. I think she was glad to get away from Badger for a day."[29] A few days later, on Tuesday, June 22, Royce presented a paper to the American Society of Agronomy, held that year in Madison, probably the result of his research for the

263

Master of Science degree.

After graduation, Royce continued on with the summer term, which in addition to advanced classes in genetics included a course in German, to be followed by French later in the fall. These languages were required to enable him as a doctoral student to be able to review literature written in other languages by scientists in the field. His course work was rigorous, and he was required to score at least a "B" grade in all his classes in order for them to count toward his doctorate. The course work was to be completed midway through the third year, to allow him to use the remaining time to complete his research and write his doctoral thesis. To complete both his study and his experimental work, he took to arriving at the university by 6:30 a.m. and returning home at 11:00 at night. Not only did he have to study for his classes, he also had to attend to the plants which formed the basis for his research. It had been a dry year for farming in Wisconsin, and by the end of the summer Royce was under necessity of carrying water by hand in order to keep his and Dr. Smith's plants alive.

During the two week break between the summer and fall terms Royce continued going in daily to work with his plants, but he took one day off to take Pearl and the family on an outing to the Wisconsin Dells, a series of rock formations on the Wisconsin River. They enjoyed the pleasant summer weather for which Wisconsin was famous, but though they found the Dells beautiful, both Royce and Pearl concluded they were no match for the beauty of the canyons back home in Utah. However, when later in the fall Royce took Pearl to the University of Wisconsin stadium to watch a football game, they decided that sporting events in Utah were no match for what Wisconsin had to offer. The University of Wisconsin sponsored one of the most impressive football programs in the nation, and accustomed to the far more modest teams and facilities at Utah State, they found the game a thrilling experience.[30]

Meanwhile, the girls continued to grow. In the fall of 1948, Jean was old enough to begin kindergarten, and Florence was sent to a nursery school, where the teachers considered her a precocious child. Marla, meanwhile, had learned to walk early, and trooped about the house imitating her older sisters. "Little Marla thinks she is as big as the other two and she acts it. She gets along especially well with Florence and imitates everything Florence does. Florence imitates everything Jeannie says and does, so we just about have three of a kind."[31] Pearl, who of necessity had become an accomplished seamstress, took to making the girls matching clothing with contrasting trim, which added to the charm. Royce by now had become more accustomed to the constraints of parenthood, and had become a favorite with the girls, as Pearl described to her mother: "Royce still tosses his kiddies around. Little Marla who can barely say a few words, goes up to him and holds up her hands and says, 'Swing.' They all think they have a pretty nice Daddy. I am very glad that Royce is good at helping me discipline them. He may seem severe at times but I guess that's a man's way. He is really gentle to them, but firm when he disciplines."[32] Marla especially was sensitive to scolding. "She has delicate feelings. If her Daddy speaks cross at her she sobs as if her heart would break."[33]

For Thanksgiving, the family was invited to dinner at the home of W. K. Smith, Royce's professor, and as Christmas came around, they improvised once again, and enjoyed a very pleasant Christmas. For the girls Royce had made a doll house out of apple boxes which Pearl had carefully painted, and each of the girls received a doll as well. He also made them a set of shelves out of orange crates to go with their toy dishes. Pearl gave Royce a warm shirt to wear on his long bus rides to campus. Dr. Smith, who felt a special kinship with Pearl because of her Scottish heritage, gave the family a box of Scottish shortbread with a bunch of heather attached, and Pearl sent a sprig of the heather to her mother in Salt Lake. Royce's birthday, two days later, was marked by amusing family incidents described by Pearl: "I made a cake and had the two layers out of the oven cooling on the table. I guess little Marla got a little hungry and had both layers down on the floor eating them. Royce caught her at it and we all had to laugh. She tried to hide by ducking her head. Florence was chewing some gum today that she got out of her stocking. Some way she got it in Marla's hair and proceeded to pull with all her might and main, trying to get her precious gum back out of her hair. Of course Marla was screaming at the top of her voice. Well, it had to be cut out with scissors. Boy! There's never a dull moment with these little girls. What one doesn't think of, the other two do."[34]

Meanwhile, Royce's work at the university continued on schedule. In October, he passed his German course, and on Pearl's birthday in early January, he successfully completed his course in French as well. Most of his graduate class work was already behind him, and as other graduate students he knew were completing their degrees and preparing to move on, Royce began to give some thought to where he would want to locate following his own graduation, and to discuss it with Pearl.[35] Neither of them wanted to go east, and Royce had by this time grown weary of the midwestern winters. Increasingly his eyes were drawn westward to the areas he had served during his mission, and he would eventually send applications there.

In the meantime, Royce continued to attend classes through the winter months, and he took to carrying an old army blanket with him to keep him warm on the long bus ride in to campus. The winter was milder than usual, but in Wisconsin that still meant serious cold, and after a January rainstorm the rain froze on the ground, leaving a treacherous layer of ice on the sidewalk, though Pearl took advantage of the weather to freeze ice cream outside on the porch. As the winter dragged on, sickness came once again to the family. One Sunday morning as Spring approached, Marla awoke from a nap after Sunday School with gasping respirations and a high fever which alarmed her parents. A doctor was summoned, and diagnosing her with bronchial pneumonia, he administered the first of three sequential injections of penicillin, a new and seemingly miraculous treatment for bacterial infections which had been developed during the war. Marla made a rapid recovery, and before long was running about the house as before.[36]

In the spring, between course work Royce became part of a village committee, jointly sponsored by a Catholic organization and a Protestant fellowship, to prepare and perform an Easter cantata entitled "The Crucifixion." He began to attend regular rehearsals after school, and as his outstanding tenor voice came to be recognized, he was assigned several solo parts. Three performances were given over the period of a couple weeks, the first in Sauk City, then in Baraboo, and finally at Badger Village. Royce meanwhile was studying for preliminary examinations, which took place in May, and he continued working on his sweet clover experiments in the greenhouse.

As the summer of 1949 came on, the busy schedule continued. Royce again planted a garden in the village plots, and from it Pearl was able to bottle dozens of quarts of tomatoes, besides beets and pickles. Royce later obtained bushels of peaches, apples and pears, and Pearl bottled these as well. With the help of Royce's parents, the family had purchased a car, a well-worn 1939 Buick with running boards and a round receptacle for a spare tire. On Independence day they packed the girls into the car in their pajamas to go and see the fireworks, but as they watched, one of the university students let off an errant Roman candle which headed straight for the car. Entering the left rear window, it struck Jean, catching her pajamas on fire and scorching the car interior. Royce quickly rolled Jean in the grass to put out the flames, and she was relatively unhurt, though the pajamas were a total loss, and the burned car trim remained as a permanent reminder of the incident.[37]

Late in the summer Royce was able to take off enough time for a vacation to Utah. Despite having a family car, the trip was an exhausting one, as both the Bringhurst and the Davidson families were by now scattered over Utah and the nearby states. A highlight was a visit to see Pearl's sister Jeanette and her husband Burnell, who ran a small ranch in Tabiona, high in the mountains, where the girls were able to mount horses. For Royce, who had grown up around horses from his earliest years, this was a real treat as well. While still in Salt Lake City, Royce paid a visit to Dr. F. V. Owen, the beet geneticist whom he had admired when he had been tending the beets on his father's farm. Royce's father had continued to allow Dr. Owen to grow beets on his new property in Cottonwood, and the veteran plant breeder spent time with Royce, giving encouragement and discussing his genetic work with sweet clover.[38]

As the end of 1949 approached, the pace of Royce's life increased. Besides his academic courses and greenhouse work, in October he traveled with Dr. W. K. Smith first to Green Bay and then to Milwaukee to attend Agronomy meetings. Between classes he continued his study of sweet clover, with cycle after cycle of breeding, seed collection, planting, and grafting. That Christmas was snowy and cold, but the family was happy to spend it together, though Florence and Marla were frightened by Santa Claus, who arrived at a holiday gathering. In January rain fell once again, and the sidewalks became icy and treacherous. For some it was an opportunity for ice skating, but the conditions were dangerous, and one woman from the village slipped and fell, sustaining a skull fracture. For his part, Royce had just completed the last of his classes, for which he took the final exams in January. With his course work done, he was left to dedicate his entire time to completing his research, compiling his findings, and writing his doctoral thesis. This was an impossible task in the close quarters at home with the children, and he continued to leave for school at 6:15 in the morning, not returning until 11:00 p.m. at night, including Saturdays.

As the deadline for his thesis drew nearer the pressure mounted, and Royce grew nervous and tense. Submitting the thesis to his professors for correction, he undertook a revision and a rewriting, which expanded the thesis to some 75 pages. Pearl helped as she could, proofreading the manuscript for spelling errors, which had always been a weak spot for Royce, but she could not understand the highly technical details of the writing. Royce became so anxious that he worried constantly and had a difficult time sleeping and thinking clearly. Arriving home at 11:00 p.m., he would stay up and attempt to write through the night. Finally, in April, Dr. Smith sat down for a talk with his student. He had worked with graduate students before, and he reassured Royce that things were going well and on schedule, and gave him new encouragement. At his suggestion, Royce began to take a walk with Pearl each night before going to bed, and he stopped trying to stay up late at night to write.

As Royce got into the final stretch of his graduate work in May, sickness struck once again. A severe strain of chicken pox swept through Badger Village, and Florence was taken ill, followed by Marla. Once again the children were placed under quarantine as Royce continued the last stages of his thesis. At the same time a severe windstorm hit, which caused considerable damage in the barracks-like dwellings. Pearl wrote, "The roofs of six units blew off and the walls of two others blew in. Our children were quite frightened. I was afraid several times that our walls would blow in but it managed to hold."[39] At the elementary school attended by Jean and Florence, all but the kindergarten room was unroofed by the wind, requiring some adjustments to find room for the various grades.

Royce was putting the final touches on his thesis when he learned that he had been accepted for a faculty position at the University of California at Los Angeles. Realizing they would soon be moving, Royce and Pearl made an appointment with a photographer to have formal pictures taken of the girls. As luck would have it, before the picture could be taken Jean lost her two front teeth, and the day before the appointment Florence discovered a pair of scissors and gave herself and Marla a haircut. Pearl did the best she could to smooth over the effects by curling their hair, but it was a wet rainy day, and the two younger girls were photographed with an odd hair style that Royce and Pearl would chuckle about the remainder of their lives. It was probably that same week that Royce at last submitted the final draft of his thesis to the typist.[40]

Genetic Research and Doctoral Thesis

Royce's research was linked to Dr. W. K. Smith's work with sweet clover. Since the early 1940s, Dr. Smith had been working on the problem of high coumarin levels found in sweet clover, whose scientific name was *Melilotus alba*. Coumarin itself, although a little bitter, was not toxic to grazing animals; however when the clover was cut and stored, the coumarin could be converted by mold into dicoumarol, which was a potent anticoagulant and could cause a potentially fatal "bleeding disease" in cattle.[41] An obvious solution was to create a hybrid from the closely related clover species *Melilotus dentata*, which came from Asia and Eastern Europe, and was relatively coumarin-free. Dr. Smith had run into a problem, however, as the hybrids which resulted from the cross of these two species turned out to be severely deficient in chlorophyll, a condition which was lethal to the hybrid plants. Pale in color, they were unable to photosynthesize to sustain growth, and they died within a few days of germination.

Some years before, Dr. Smith had developed an ingenious solution to this dilemma in order to allow the defective hybrid plants to produce seed and continue the breeding process. His method was to plant the hybrid seeds, then cut the young defective seedling plants just below ground level and graft them into healthy sweet clover plants. In this way, the healthy plants could provide enough energy for the defective plants to produce flowers and seed, which would allow further crosses to be made. The ultimate goal was to transfer the genes responsible for low coumarin levels in *Melilotus dentata* into plants of *M. alba*, while still allowing the offspring to have normal chlorophyll levels and reproduce freely. (This was, in effect, the objective of all plant breeding, to carefully make crosses and select among the offspring in order to obtain the desirable combination of traits in a single plant or breeding line.)

At Dr. Smith's suggestion, Royce undertook the task of examining the genes which were involved in the lethal loss of chlorophyll in the crosses between the two species. To accomplish this, he would use a method called backcrossing, which meant taking the defective hybrid between the two species *Melilotus alba* and *Melilotus dentata*, and crossing it with the healthy parent species *M. alba*, in an effort to isolate the effects of some of these genes. By first crossing the defective plants with the normal ones, then breeding the resulting crosses with themselves and with each other, Royce hoped to systematically observe the effects of the chance combinations of normal and defective genes in the offspring, in such a way as to be able to describe the action of the individual genes.[42]

This was no easy task. It was completely unknown whether the genes involved were many or few, or how they interacted with each other. The field of genetics itself was still a relatively new one, and the earliest experimental work to identify the nature and structure of DNA was still ongoing. DNA sequencing and analysis, as we know it today, was still decades in the future. Instead, the genes involved had to be identified by observing the physical characteristics (known as phenotypes), which they caused in the plant offspring. This would give the clues necessary to deduce the specific gene combinations themselves (known as genotypes). It was known that each cell in a species contained two sets of chromosomes which carried the genes, one inherited from each parent, and that the combination of genes in the offspring would determine its physical characteristics, or phenotype.

In the case of the defective sweet clover, the phenotypes were indicated only by different shades of green color, depending on the number and type of chlorophyll deficiency genes that were in operation. As Royce made the various crosses, many involving defective plants grafted onto healthy ones, he knew he would need some kind of reliable scale to measure the various shades of green which resulted in the offspring. This would allow him to describe the phenotypes of each plant in a reproducible way, bringing order to what would

269

otherwise be a chaos of different shades of green. In this way he might be able to see the patterns of inheritance in chlorophyll deficiency, and report them in a reproducible way.

For his color scale, Royce at first used a published horticultural color chart showing different shades of green, and he divided the shades into six classes, Class 1 being pale yellow, and Class 6 being "leaf green," a normal deep green color. In this way Royce could describe a plant by a class number according to the color the plant most closely resembled, and if it was a little darker or lighter than one of the colors on the scale, a plus or minus sign could be employed. For example, a plant with completely normal color would be described as "Class 6," while a paler plant might be "Class 3+" or "Class 4-." However, as crosses continued to be carried out in the greenhouse in both winter and summer, it became clear that growing conditions, as well as heredity, could have some effect on the color of the offspring, such that a given plant might appear paler at one season of the year than at another, which rendered the color chart less reliable.

This difficulty was rather ingeniously overcome by developing true-breeding seed lines of plants in each of the six representative shades ranging from pale yellow to normal green. Thereafter, whenever a new series of crosses was planted, a row of seeds from each of the six indicator plant lines was planted in the same flat. This gave Royce a living plant scale of six plants, ranging from pale yellow to dark green, which he could use as a measuring stick to judge the color of the experimental plants. Since the indicator plants were subject to the same environmental conditions as the experimental ones, the comparisons of color would be equally valid in any season of the year.

Royce began his work by collecting several commercial and experimental varieties of sweet clover, *Melilotus alba*, and attempting to hybridize them with four varieties of *Melilotus dentata* which had been obtained from the Soviet Union, China, and Czechoslovakia. Most of these crossed readily as long as pollen from *M. dentata* was used to pollinate *M. alba*; attempts to perform the cross in the opposite direction was very difficult and resulted in too few seeds to study. The resulting hybrid seeds grew when planted, but, as expected, the resulting plants were pale in color, (classified on Royce's color scale as Class 1 and Class 2), and survived only a few days if they were not grafted. When Royce grafted the seedlings into healthy plants, he was able to get them to mature, but the resulting flowers had pollen which was very infertile, and when he examined the cells under the microscope they were found to have chromosomal abnormalities. He made many attempts to fertilize these plants with their own pollen, but found it nearly impossible to obtain viable seeds from the self-pollinated flowers.

270

The next step in the experiment was to backcross some of the defective hybrids of *M. alba* x *M. dentata* with one of the normal parents, either *M. alba* or *M. dentata*. This backcross of a defective plant with a normal one would help to clarify what genes were involved in the color defect. Because of the very low fertility of the defective hybrid plants, this cross was difficult to make successfully, as almost none of the pollinated flowers set seed. Dr. W. K. Smith had encountered the same problem years before, and with great effort he had been able to obtain only six seeds from a backcross with *M. alba*. When he had planted the seeds, four of the resulting plants had the lethal defect, but the other two had enough chlorophyll to be raised to maturity, and both had set seeds after being fertilized with their own pollen. The seeds from these two original plants were available to Royce when he began his research, and he gave these two backcrossed hybrid plants (which had been grown before he ever met Dr. Smith) the names of A1 and A2 ("A" being for "*alba*," the species name of the parent plant). The plant he called A1 had been a normal green color, while A2 had been somewhat pale, but still able to grow to maturity. Royce knew that each of these crosses contained some of the genes responsible for chlorophyll deficiency, and that these genes would be randomly distributed among their self-pollinated offspring, according to the rules of probability that were familiar to every geneticist. It was the self-pollinated seeds from plants A1 and A2, previously collected by Dr. Smith, that Royce used for the next part of his research.

Royce first planted the seeds from plants A1 and A2, and described the phenotype of the resulting plants using his newly-developed color scale. He then began carefully crossing and recrossing the offspring plants in successive generations. As he did so, he noted the color of the resulting seedlings, then grew them to maturity so he could collect seed for the following generation while observing the results, in order to deduce the characteristics of the separate genes. This was an unusually challenging task, for while the more normal plants high on the green scale could reach maturity on their own, the more pale offspring had to be meticulously grafted onto a normal plant in order to reach maturity and yield seed.

As Royce carried out these crosses and listed his observations, from among the seemingly random shades of green that resulted, clear patterns began to emerge, with specific colors of plants in specific ratios. For example, among the descendants of plant A1, just over half the plants were Class 6, with completely normal green color. A small proportion of the plants were Class 3, and unable to survive without grafting. Between these extremes were populations of different sizes which were Class 6-, Class 5, Class 4, and Class 4-, all in certain ratios to each other. By carefully studying the relative proportions of these different offspring, Royce was able to deduce that the differences in shade were consistent with the presence of two separate genes affecting chlorophyll production which he named Ch_1 and Ch_2, each with their defective counterpart, which he called ch_1 and ch_2, according to the conventions set by early geneticists. His color scale enabled him both to predict and to verify the presence of the two genes, as well as their relative strength in causing chlorophyll deficiency. Once he had established the presence of the two sets of genes, he found that he could breed the offspring plants of various shades with each other, and could predict exactly the proportion of their offspring which would be of a given color.

With the descendants of hybrid A2, the slightly pale parent plant which Dr. Smith had bred, Royce found an entirely different and much simpler pattern. A quarter of the plants were normal Class 6 plants, and a quarter were lethally pale Class 3 plants. The remaining one-half were a uniform intermediate, measured as Class 5. This he immediately recognized as the result of a single pair of genes, which Royce classified as Ch_3. Plant A1 had simply inherited a normal gene (Ch_3) from its normal *Melilotus alba* parent, and an abnormal gene (ch_3) from its chlorophyll-deficient hybrid parent. Its descendants had then inherited the two genes in predictably random fashion. The 1/4 that inherited two normal genes (Ch_3Ch_3) were normal green Class 6 plants, the 1/4 that inherited two abnormal genes (ch_3ch_3) were Class 3 lethals, and the 1/2 that had inherited a normal gene and an abnormal gene (Ch_3ch_3) were an intermediate Class 5, like their parent. Once again, when Royce crossed these plants with each other, he found that he was able to predict the offspring, though with one specific exception, as the Ch_3 gene seemed also to interfere with pollination under certain conditions.

Royce performed crosses between the plants derived from A1 and those derived from A2, so he could observe the effects of the Ch_1, Ch_2, and Ch_3 genes in their various combinations. Using his color scale, he carefully catalogued the color combinations which resulted from each cross, explaining them in terms of the predicted genotypes, and the results fit well with the predicted behavior of the genes.

In addition to offspring from the original plants A1 and A2 previously bred by Dr. Smith, Royce finally succeeded in making his own backcrosses, which he named A3, A4, and so on. Because of the time required to breed these hybrids, he was not able to observe as many generations of offspring as he did with A1 and A2, but he was able to ascertain that at least four other genes were present among these hybrids which affected their ability to express chlorophyll. He succeeded in producing backcrosses between the lethal hybrid and the *Meli-*

lotus dentata species, and found a somewhat different pattern of inheritance involved in the chlorophyll defect in that species. He also made a series of other difficult crosses between various of the hybrids, and described the gene patterns which he observed. Before completing his work, he developed a system of making the backcrosses with much greater ease, by making use of a scale he had developed to ensure the presence of a hybrid—this greatly simplified the process which had been so laborious at the beginning of his study.[43] Throughout this process Royce made careful study of the microscopic appearance of the pollen produced by the hybrids, noting those instances in which the hybridization appeared to result in chromosomal abnormalities which affected fertility.

Through his research with sweet clover, Royce had made considerable progress in describing the genetic patterns that caused the crosses between the two species to be lacking in chlorophyll and unable to survive. He had shown that this was the effect of a range of genes of varying potency. Three of these he had actually isolated and had characterized their actions, and he had been able to ascertain the presence of at least four others. He had shown that the genes for chlorophyll deficiency behaved differently in offspring of *Melilotus alba* than that of *Melilotus dentata*. He had identified fertility problems in the offspring, and had described the microscopic findings associated with them. He had analyzed his findings and compiled them clearly, and his thesis, when it was completed and retyped, measured just under a hundred pages.

In a broader sense, Royce's graduate work had prepared him to be not merely a plant breeder, but one of a new and growing class of true plant geneticists. In his course work he had learned the rudiments of genetics as far as they had been developed, but in his research he had also learned to approach a complex genetic problem systematically, to perform multiple experiments concurrently, and to record his data consistently. He had learned to recognize clear patterns of inheritance from a seemingly random distribution of findings. He had learned the importance of objectively measuring traits, and when he encountered limitations to his measurements, he had learned to adapt new and ingenious means of measuring. Through dozens of experiments he had learned the basic skills of the agronomist, including pollinating, transplanting, grafting, and caring for plants. He had also become skilled in examining his materials in the laboratory, and under the microscope. All these would be second nature to him by the time he was doing his own genetic work and training his own generation of graduate students. Throughout his life, he would keep himself abreast of new advances in genetics, and incorporate them into his own work.

All the time and meticulous labor spent coaxing seeds and hybrids from chlorophyll-deficient plants would eventually pay off. For Dr. W. K. Smith, it would eventually lead to the release of a healthy green variety of sweet clover with extremely low coumarin levels, due to the successful transfer of genes from *M. dentata* to *M. alba*. For Royce, it would lead to a Ph.D. degree, which would pave the way for a long a satisfying career as a plant geneticist and breeder. Dr. Smith and his wife would remain cherished friends to Royce and Pearl as long as they lived.[44]

Royce's final oral examinations were scheduled for May 31, 1950. As fortune would have it, a viral illness was circulating through the village at that time, and Royce became sick just before the exam and spent May 30 in bed with a fever. He was still sick on the 31st, but there was no accommodation made for illness, so with everything at stake he made his way to the university to face his orals. Entering a room with a panel of five professors, Royce was probed with a long series of technical questions related to his field. For two hours the questioning went on, and at last the professors withdrew to deliberate. After only five minutes they returned to congratulate Royce on the successful completion of his Ph.D. degree.[45]

The remainder of the time in Badger Village was spent making preparations for departure. Royce made arrangements to have his doctoral thesis retyped and bound. Meanwhile, feeling a large debt of gratitude, he continued to go in to the university every day to help Dr. Smith with his ongoing research work. Pearl sold the canning jars which had served them so well, as they would no longer be of any use to them in Wisconsin. Spring had come, and in the midst of their preparations the family attended separate picnics with the Genetics Department, the Agronomy Department, and the Madison Branch of the Church. Royce took a day off to give the apartment a final cleaning. The girls were full of anticipation—Jean marked the departure day on her calendar, Florence took to counting the days, and Marla merely acted excited, reflecting the mood of her older sisters.

Finally, on Friday, June 16, 1950, Royce made his way to the University Field House, accompanied by Pearl and his mother, his father having remained behind in Salt Lake to take the place of a co-worker who had died suddenly. Near the end of a long ceremony, Royce stepped forward in cap and gown to receive the hood and diploma granting him the degree of Doctor of Philosophy, one of 129 awarded that day. It was nearly thirteen years since Royce had first entered college, longer by a year than it had taken him from start of grammar school to graduation from high school. That Saturday, Royce and Pearl shipped off their larger belongings, and after resting Sunday, they spent a day packing the family car. On Tuesday, June 21, they

274

left Wisconsin for the last time, bound for Salt Lake City for a brief vacation before moving on to California. Royce spent only a short time in the Salt Lake area, since he was expected at UCLA by the first of July, and he planned to leave Pearl and the family behind while he went ahead to make living arrangements. During the few days he was in Utah, he drove to Granite High School and, leaving his family in the car, walked into the office to show his diploma to Lorenzo Hatch, the principal who had once expressed doubt of his ability to succeed in college and in life.[46] He had succeeded, and from that time forward, bearing the title of those who had risen to the top in their field of study, he would be known to the world as "Doctor Bringhurst."

Illustrations for Chapter 7

Royce as chairman of USAC horse show, 1947.

With USAC graduates, 1947. Royce is kneeling, second from right.

F. V. Owen, beet geneticist who inspired Royce to pursue genetics as a career. The beet plants in the background are bagged to control pollination.[47] This photo was taken in Utah in 1958.

Graduate students and faculty, University of Wisconsin. Royce is third row from the front at center, in white shirt. W. K. Smith is on the back row, leftmost of those seated in front of the doors.

Pearl and children in the snow at Badger Village.

Family portrait just prior to leaving Wisconsin. From left: Florence, Royce, Jean, Pearl, Marla.

Royce on day of graduation with his advisor, W. K. Smith.

University of California at Los Angeles

Chapter 8

Applying for a Faculty Position

While Royce was working on his graduate research during his last year at the University of Wisconsin, he had begun to make inquiries among his professional contacts at Utah State and at Wisconsin, in hopes of finding an open faculty position following his graduation. While about halfway through his graduate training in January of 1949, he was surprised to receive his first offer from an agronomist at the University of Hawaii agricultural experiment station in Honolulu, offering him a research position in forage crops and pasture research. He had been recommended for the position by Dr. D. W. Thorne, an agronomist at Utah State Agricultural College who was familiar with Royce's work and abilities. Though the Hawaii offer was intriguing, Royce sent a reply that he intended to complete his doctorate work, which would not be done until the summer of 1950, and he could only consider the position if it was still available at that time.[1]

As his graduate work drew closer to completion, in October of that same year (1949) Royce sent an inquiry to Dr. Thorne himself regarding a faculty position opening up at Utah State in the farm crops section of the Agronomy department.[2] This would involve teaching as well as research in field crops such as potatoes, sugar beets, wheat, beans, and forage crops, which were all familiar to Royce and had some appeal for him. But at the same time he contact-

280

ed an old friend from Granite High School, Art Wallace, who like Royce had served on a mission and in the military, had gone to graduate school, and now held a position on the faculty of the University of California at Los Angeles.[3] Royce asked about any upcoming faculty positions, and Art, knowing of his work in genetics, informed him that a position would be opening up the following summer for a junior plant breeder in a program involving subtropical fruits, chiefly avocados. Royce must have recalled the love of avocados he had developed during his mission, and he still had fond recollections of the time he had spent in Los Angeles, both during and after the mission. He expressed interest in the UCLA position, and before long a letter came from Dr. S. H. Cameron, a professor of plant physiology, who described it to him.

There ensued a long series of letters between Royce and Dr. Cameron. One problem was the fact that Royce's graduate work was with forage crops rather than fruit, and while this did not disqualify him, Dr. Cameron wanted to be assured that he would be willing to transfer his area of interest to fruit crops. Royce responded that this would not be a difficult transition for him, since as a youth he had taken responsibility for the small commercial apple orchard on his father's farm for many years, and had enjoyed the work very much. In fact, the thought of shifting his research to fruit crops increasingly appealed to Royce. He was told that working with avocados was to be the central focus, as the avocado industry had a serious need for new varieties to extend the fruiting season as well as the range over which avocados could be grown. There would also be an opportunity to do breeding work with other tropical and subtropical fruits, such as the cherimoya, white sapote, and feijoa, as well as the guava and loquat, depending on the interest of the breeder.

Because of the long intervals involved in breeding tree fruits, Dr. Cameron suggested that the person hired would require some secondary interest such as cytogenetics, since almost none had been undertaken with these fruits. Cytogenetics, or the study of the microscopic appearance of cells during cell replication, had been an important part of Royce's work at Wisconsin, and this fit perfectly with the needs of the UCLA position. As these letters passed back and forth, Art Wallace began to take a hand as advocate and intermediary. Dr. Cameron clearly wanted to hire Royce, and used Art to unofficially prompt him in as to the type of information he still needed to provide and what responses would be most favorable to his hiring, and eventually even what level of salary to accept.[4]

Meanwhile, in February Royce received a letter from Dr. Thorne at Utah State offering him the agronomy position which he had applied for there. By that time his negotiations with UCLA were well advanced, and by the first of March Royce was confident enough that he wrote to Dr. Thorne declining the position at Utah State, though doing so left him feeling somewhat torn, as he still felt a close affinity to Logan, where he had begun his college career.[5] It was not until May, just over a month before his anticipated graduation, that Royce was officially informed that the University of California Regents had approved his appointment as a Lecturer and Assistant Specialist in the Agricultural Experiment Station at a salary of $4800 per year. Once he could demonstrate a PhD degree, his title would be changed to Instructor and Junior Plant Breeder, which reflected higher qualifications. Royce wasted no time in replying, and on May 11, 1950, he wrote a final letter to Dr. Cameron accepting the position.[6]

Royce received one additional job offer before he left Wisconsin. Back in January of 1950 he had sent a letter to F. V. Owen, the sugar beet geneticist he had known since his youth, and with whom he had visited the previous summer. In his letter, he requested information about an automatic seeding machine used by Dr. Owen, which he thought might be useful for the work Dr. W. K. Smith was doing in Wisconsin.[7] Dr. Owen sent a reply, and Royce thought little more about it, but in late May he received a letter from the U. S. Department of Agriculture, offering him an opportunity to apply as a plant breeder at the Natural Rubber Research Station in Salinas, California, based on a recommendation from Dr. Owen. Royce replied that he had already taken the job with the Subtropical Horticulture Department at Los Angeles, but he was grateful for the expression of confidence from the veteran plant breeder whom he still admired.[8]

So it was that by the time Royce began his own career as a plant breeder at UCLA he had already enjoyed the close acquaintance, mentoring, and encouragement of three of the pioneers in the field of plant genetics. When still a boy he had worked in sugar beet fields with F. V. Owen on his father's farm in Bennion, and had first dared to imagine a similar career for himself. As a college student he had worked alongside D. C. Tingey at Utah State, where he was encouraged to pursue graduate training in genetics. Finally, as a graduate student in Wisconsin he joined W. K. Smith in his complex genetic research into the problems of hybrid sweet clover. He had earned the respect of all, and was at last poised to begin making his own contributions to the field.

Beginning Life in Los Angeles

The end of June, 1950, found Royce alone in the old family Buick, making the long drive from Salt Lake City to Los Angeles, California. Arriving there, he met Dr. Cameron at the campus of the University of California at Los Angeles, located in an area called Westwood, at some distance from Los Angeles proper, and just west of the upscale neighborhoods of Beverly Hills. Dr. Cameron introduced him to the campus facilities, which were limited in office and laboratory space, but were well stocked greenhouses and teaching laboratories, and had good equipment to perform cytologic study. In an area rapidly being overrun by urban development, acreage for experimental plantings was in relatively short supply,[9] though the campus in Westwood Village boasted a ten-acre orchard of citrus and subtropical fruit varieties, which included an impressive array of specimens of many named varieties of avocado and of breeding stock, as well as avocado relatives which had been collected from various countries for use in a breeding program.[10]

The University of California, like Utah State Agricultural College, had been established under the federal land grant program, and the university was under a similar obligation to conduct research to train and benefit local industries. Royce's position was in part to fill that obligation for the avocado industry, which had grown to prominence in California in the early part of the 20th century. The industry was represented by an organization called the California Avocado Society, which had been formed in 1915 to represent the needs of avocado growers in the Southern California area. Some of Royce's work would be published in their annual yearbook, which served as a professional journal for experts in the avocado field.

Royce had been hired to assume the continuation and expansion of a subtropical fruit breeding program begun some ten years earlier by Dr. Walter E. Lammerts, a plant breeder who had resigned from his university position a few years before. The program had been discontinued at that time, except that some of Dr. Lammerts' breeding stock had been maintained so it would be available for his successor. The title of Royce's position was that of Junior Plant Breeder,[11] but in his publications he would bear the title of Assistant Geneticist, and would work in cooperation with Dr. J. W. Lesley, a senior plant geneticist nearing retirement, who had been assigned to work initially with Royce while he became established. At Dr. Lesley's retirement Royce planned to assume full responsibility for the avocado breeding program.

While he awaited the arrival of Pearl and the children, Royce secured a ground story apartment for the family in a 14-unit apartment complex which had been made available at 10755-1/2 Strathmore Drive, within walking distance of the campus, and which they were allowed to rent for up to a year until they could make more permanent arrangements. After an eventful train journey delayed by a derailment,[12] Pearl and the girls arrived on July 20, and began unpacking and settling into their temporary home. Not long after their arrival,

283

they were invited to Sunday dinner in the home of Dr. Cameron, who to Pearl's delight was an immigrant of Scottish ancestry, just as W. K. Smith had been.[13] The family began to take stock of their new California surroundings. Having heard of "sunny California," they expected hot summer weather in July, but their home was only four miles from the ocean, and to their surprise the weather was cool and spring-like, such as they might have enjoyed in May in Salt Lake. Royce brought home as much citrus fruit as the family could eat, and excellent fresh vegetables were available at low prices unheard of in Wisconsin or Utah. Drive-in movie theaters abounded, and children were allowed free admission, so Royce and Pearl began taking the girls to see movies, which would probably have been impossible in a standard theater.

Although the weather was agreeable and the surroundings pleasant, there was an edge of worry in the air. War had broken out in Korea just as the family was making preparations for their move to California, and the United States had entered the conflict shortly thereafter. With the possibility of war with Communist China looming, there was once again talk of the danger of bombing attacks along the Pacific coast. One Sunday a jet bomber flew low over the apartment building, making a noise "like a terrible explosion" which left the whole family shaking, and doubtless brought back unpleasant memories for Royce.[14] His feelings about war had not changed, and to see a new war begin so soon after the last had ended brought back a tinge of bitterness. "Some people are jittery about war down here now. I hope there is a sensible solution. War certainly isn't. I don't think we'll open war on China because that is what Russia desires most. It's unfortunate that things get so muddled but a messy war certainly won't solve anything. The last one did not and it was bad enough."[15]

The family remained as active as ever in the local congregation of Latter-day Saints. Accustomed to attending church in a tiny branch as they had in Badger Village, they now found themselves members of the large and fully functioning Beverly Hills Ward, located in one of the most upscale neighborhoods in the Los Angeles area. Although they found the ward members friendly, Pearl, who had grown up in a poor family during the Great Depression, initially felt a little suspicious of what she described as "practically the wealthiest ward in the Church."[16] Her prejudices were broken down a little when she saw the well-to-do Church members working hard alongside her one late night canning peaches at the local Church cannery.[17] The same week Royce had a similar experience as he labored at the Church welfare center hauling cement with other men from the ward.

When the Bringhursts arrived, a new ward chapel had just been announced, to be built on the same property as the planned Los An-

284

geles Temple, and once ground was broken Royce began going in on Saturdays to the chapel grounds to labor with pick and shovel on the foundation.[18] Meanwhile, Pearl was assigned to teach a class in the Primary, and Royce taught a group of young priests at weekly Priesthood meetings, and he also sang in a church chorus. They took the girls to Sunday School in the morning as other families did, but in a ward that was not oriented to young families such as theirs, they found it made something of a stir when they took them to evening Sacrament services as well, though they persisted, and after a while other families began following suit.

The girls rapidly adjusted to their new environment. Seven-year-old Jean, who was beginning second grade, had to be bussed to her school, and was a precocious reader. Florence, who was starting kindergarten, was a pleasant and good-tempered child, and filled the house with her drawings, while Marla, the youngest, who would remain at home, had developed a loud voice with which to make her presence known. The mildness of the Southern California weather took them all a little by surprise, and when for Thanksgiving Day they visited Pearl's brother Dan Davidson and his wife Mary, still living in San Bernardino, it was so warm they did not require coats or even sweaters, and Pearl sent Jean's snow suit to one of her sisters who still lived where there was snow. During Christmas season the traditional decorations looked a little out of place, sparkling as they did in the bright sunlight with a backdrop of green palm trees. California brought other changes as well. It was earthquake country, and at school the girls were becoming accustomed to earthquake drills in preparation for the occasional earthquakes in the region. They had not been there long when they felt their first earthquake, a minor tremor which was strong enough to arouse excitement in the girls.[19]

Royce and Pearl had not been in Los Angeles long before they began to look into purchasing a new and permanent home. Royce now had a reliable income, and as a war veteran, he qualified for a loan with no down payment and only $65 per month in mortgage payments, which was sufficient to purchase a fine home for their family at less than they were paying per month for their small apartment.[20] With the postwar growth in the population, vacant houses were difficult to come by, but Royce and Pearl were able to find a neighborhood in Reseda, located in the San Fernando Valley to the north, where new housing was being built, and they arranged to purchase a home that was under construction on a small cul de sac which was to be called Le May Street. During the winter months they took frequent drives out to the house site to check on progress. On one of these visits they were shocked to find that the back yard had been replaced by a huge trench, the contractor having needed the dirt for

285

road construction. In short order the trench was replaced with fine soil, and the house construction proceeded rapidly—indeed, perhaps a bit too rapidly, for the builders in their haste failed to place necessary braces under the flooring, and a couple of years later would be obliged to replace the kitchen floor when Pearl fell through it.

Because the move to Reseda would require traveling longer distances for basic services, driving became a necessity, and Pearl, who had never learned to drive a car, began to take driving lessons from a community adult education program so she could drive the old family Buick. (This was harrowing in the heavily-trafficked streets of Los Angeles, and Pearl reasoned if she could learn to drive there, she could drive anywhere.[21]) At length the house was completed, and on Saturday, April 28 the family made their move, assisted by two of Royce's colleagues, Art Wallace, Royce and Pearl's old friend from Granite High School, and Art Schroeder, another professor in Royce's department.[22] It had been a mild winter with almost no rain, but it rained on the day of the move, which proceeded anyway without incident. The house was a grey stucco home located at 18153 Le May Street, on the turnabout at the end of a short street in Reseda. It had long been Royce's dream to live in a place they could call their own, and although he and Pearl had few furnishings for their new house, they quickly made it a home. A couple of weeks after the move, the entire department from the university came to the house for a housewarming party.

The Avocado Program at UCLA

In joining the faculty of the University of California at Los Angeles, Royce was assuming command of an avocado breeding program which was still in its infancy. Avocados had first been introduced in California in 1871, and in the early 1900s a fledgling industry had developed to exploit the avocado as a fruit crop. The avocado, widely used in Latin America and the Caribbean but relatively unknown in the United States, had slowly begun to gain popularity north of the border. The California Avocado Association (later known as the California Avocado Society) was organized in 1915 to help the growers to benefit from accumulated experience in avocado horticulture, to identify the best varieties, and to promote the fruit in a unified fashion.[23]

Although the basic principles of genetics were fairly constant from one crop to another, the breeding of tree fruit such as the avocado involved very different considerations from that of forage or cereal crops with which Royce had worked in the past. Field crops were propagated from seed, and the plant breeder had to develop true-breeding inbred seed lines with the desired characteristics in order to establish a new variety, which would then be put into use by growing the crop and harvesting the seed. In the case of hybrid crops, this would involve an additional step of establishing two parallel inbred seed lines with certain parent characteristics, and pollinating one with the other, producing seeds which consistently combined the desired traits of both parent lines.

Fruit breeding was different, in that, unlike field crops which are planted and grown from seed, commercial fruit crops are propagated clonally. This means that once a variety is selected, it is reproduced by taking a portion of the plant, usually a bud or a limb cutting, and grafting it onto a compatible root stock to produce a full fruiting tree of the desired variety. In this way, every tree of a given variety originates from the same seed, and each one is in theory genetically identical with every other, and produces identical fruit. For example, every "golden delicious" apple originated from the same seedling tree, which was first bred and selected, then propagated by grafted clones to create vast orchards of trees which produce that fruit type.

The same principles of plant breeding and hybridization applied to both kinds of crops. By means of controlled pollination, the breeder would cross two different known plants with certain desirable characteristics, and then plant out the seeds which resulted. Each seed would have a random assortment of traits from each parent, some desirable and some less desirable, and from among the offspring the breeder would select plants in which the most desirable traits were combined, and would discard the inferior ones. The selected plants would then be propagated and tested, and if they proved to be of value they could either be given a name and released for use, or used as breeding stock for the next generation of crosses to continue the process. The selected traits included not just fruit quality, but other factors such as productivity, disease resistance, ability to ship, and adaptability to different agricultural regions. In fruits such as the avocado, years were required for the seedling crosses to come into fruiting, and still more time would be required to evaluate the fruit and other characteristics, so this was no small undertaking.

When Royce took over the breeding program, virtually none of the commercially grown avocado varieties in California had resulted from breeding work, but they were instead chance seedlings of unknown parentage which had happened to do well. Some of these had been propagated from existing trees in Mexico or Central America, which had been discovered during exploring expeditions, then introduced to California growers by astute pioneers such as Wilson Popenoe.[24] An example of these was the variety "Fuerte," which then dominated the industry in California. Others were random seedlings of these earlier imported varieties, which had been discovered and selected in fields by California growers or hobbyists. Among these was a variety known as "Hass" which though recently introduced, was already becoming an important variety.[25]

The UCLA avocado breeding program, the first of its kind, had officially begun in 1939 with the hiring of Dr. W. E. Lammerts, a geneticist who, with a subtropical fruits expert named Dr. R. W. Hodgson, began collecting appropriate breeding stock that year to begin his experiments. At first sight it appeared that avocado breeding would be a simple matter, since good varieties already existed, and careful cross-breeding of superior cultivars would be expected to result in rapid improvements. However, Dr. Lammerts quickly discovered that it was very difficult to make the crosses. Cross-pollinating the flowers by hand, as was commonly done with other fruits, was both difficult and inefficient, with a mere handful of fruit resulting from thousands of laborious hand pollinations. He also experimented with screened cages around the trees to keep bees from introducing unwanted pollen, and tried enclosing two varieties of avocado in a single cage and introducing a hive of bees to complete the task of cross-pollination.[26] This resulted in a mix of cross-pollinated and self-pollinated seed, but was still better than what had been done before. The lack of an adequate pollination technique was aggravated when a large number of Dr. Lammerts' resultant seedlings succumbed to a soil fungus.

About six years into the program, Dr. Lammerts had announced his plans to resign to pursue other interests. In five years of work, he was able to report just over 150 seedlings and seeds with known parentage from controlled hand pollinations, and the first of these were just beginning to bear fruit. Many of the specific crosses were so few in number as to be nearly impossible to evaluate. In reporting his resignation to the California Avocado Society, Dr. Lammerts expressed his frustration with the plant. "I regret to say that due to the cantankerous nature of the avocados, I haven't nearly as large a number of seedlings of these crosses as I wanted because I have never worked with a plant that was quite as cantankerous and difficult to handle from all points of view." [27] With Dr. Lammerts' departure, the breeding program had ground to a halt, except that his hand-pollinated progeny plants had been maintained for future use.[28]

In one of his letters to Royce during the hiring process, Dr. Cameron had summarized the task that awaited him. "Our fruit breeding program has in the past involved work on avocado, cherimoya, white sapote and feijoa. Of these subtropical fruits, by far the most important from a commercial standpoint is the avocado. At present the industry is largely dependent upon one variety (Fuerte), which is excellent in all respects except that it is erratic in bearing behavior. The major problem is to find or produce a Fuerte or Fuerte-like fruit that is more dependable in bearing behavior than the one we now have. A second, and very important, problem is to produce a variety (preferably Fuerte-like) which will mature its fruit either earlier or later to permit year-round marketing of fruits which can be sold as Fuerte. There is also definite need for new, high quality varieties that may succeed in areas in which our present varieties are not successful." He added, "Breeding work with the avocado has proved to be very difficult. As yet we have not discovered a satisfactory pollination technique. Growing and fruiting of the seedlings after the seed is obtained is not difficult."[29]

Royce was no stranger to difficulties in pollination, having had similar problems with the sweet clover hybrids he had worked with in Wisconsin, and with his entire professional life before him, he felt confident taking on the challenge. Starting work in Los Angeles, he quickly acquainted himself with the various varieties of avocado used in the industry and grown in the experimental orchard, and reviewed what work had been done with them. He studied the physiology and the fruiting characteristics of the different varieties, making a record of the size and shape of the fruit and seed by slicing fruits in half and making an avocado imprint on a dark sheet of paper.[30]

As the trees came into flower Royce began, like Dr. Lammerts, to make crosses both by hand pollination and by introducing bee hives into enclosures containing selected plants. Royce was quite sensitive to bee stings, and quickly learned to move the hives at night to minimize his chance of being stung.[31] He kept careful records of the hand pollinations, noting the time of day and conditions under which each were made, so he could determine the optimal time to do this laborious work.[32] As seeds developed from the cross-pollinations, he planted these out, and as they grew he used a cataloguing system developed previously by Dr. Lammerts to keep track of each seedling, using a number which started with the year the cross was made.[33] Royce would find such a system essential in later years.

Royce had several objectives in his breeding work. When he began, the industry was built around the variety "Fuerte," which was a beautiful fruit with a pleasant pear shape, a shiny green skin, a good size for marketing, and excellent flavor. The industry had used this variety in its advertising and the public had come to regard it as the standard. But Fuerte was a difficult avocado to grow. Grafted trees took a long time to come into production, and once they did the bearing was inconsistent, with heavy crops one year and very light ones the next, and the variety had a limited season, so that much of the year there was nothing to sell. There was also the matter of cold hardiness. Although Fuerte was more cold-tolerant than some other varieties, a bad frost could still wipe out a crop or an entire orchard, and hence the range in which they could be grown in California was

289

very limited.

Among the many other varieties available, Royce was especially interested in two. A newer variety called "Hass" had begun to be widely grown; although it was considered to have the "wrong" appearance for marketing, being less pear-shaped and nearly black in color with a thick, pebbled skin, Hass grew very well in a wide range of areas and was much more consistently productive than Fuerte. One objective of the breeding would be to attempt to combine the appearance, size, and shape of Fuerte with the more consistent bearing habits of the Hass.

Another variety of great interest to Royce was one called "Mexicola." This was an entirely different type[34] of avocado, with a shiny black skin that was thin enough to be edible. Too small and strong-flavored to be marketable, it nonetheless had excellent possibilities, since it bore early, heavily and consistently, and it had a unique nutty flavor of which Royce was especially fond. Perhaps more important, it was a very cold-hardy tree, and could be grown in areas of the state where no other avocado would survive. Royce was very interested in crossing it with Fuerte, Hass, and other varieties, in order to develop marketable varieties which could be grown in areas hitherto closed to avocado cultivation.

As he worked on the hybrid crosses which he hoped would eventually lead to new and better varieties, Royce also undertook a careful study of avocado pollination, which had been such a problem in previous breeding efforts, and which also accounted for difficulty in getting avocados to fruit consistently. It had long been observed that avocado flowers had very unique properties, in that while they had both male and female parts, these parts did not operate in unison, making it difficult for an individual tree to self-pollinate. Each flower would make two openings: in the first, known as stage I, it would open as a female flower, with a prominent stigma but no shedding of pollen, then after reclosing it would open the next day in stage II as a male flower, with good shedding of pollen but with female parts no longer active or receptive to pollen. The term for this situation was "dichogamous," and the result was that avocado flowers often could not self-pollinate.

It had also long been noted that in a given tree stage I and stage II openings seemed to occur separately, and there appeared to be two distinct patterns, known as Class A and Class B. Class A varieties had been observed to have stage I flower opening in the morning and stage II flower opening in the afternoon, while in Class B varieties the situation was reversed, with Stage II flowers present in the morning and Stage I in the afternoon.[35] This seemed to imply that ideally Class A avocados should be planted near Class B avocados to assure good pollination of both, though the tendency in the industry at the time of Royce's arrival was to attach little practical importance to pollination in avocado fruit production. It was also known that some avocados were self-fertile, so that there must be some overlap between Stage I and Stage II openings, and some researchers had observed that the timing of the stages seemed to be modified by changes in weather.

In concert with Dr. J. W. Lesley, the senior geneticist working with him, Royce undertook a series of experiments to better define these characteristics. Royce and Dr. Lesley began making simultaneous observations of the flowering behavior of various avocado varieties at the experimental orchards in both Los Angeles and Riverside, and compared them to the relative temperatures measured at the two sites. They found that the "Class A" and "Class B" designations greatly oversimplified the flowering habits of the trees. They noted that cooler temperatures tended to inhibit the presence of stage I flowers, and on some cool days stage I opening did not occur at all with some varieties. Warmer temperatures by contrast favored stage I opening of the flowers. There were times and conditions when Class A trees behaved like Class B and vice versa, and it was frequent to see them behaving in a fashion intermediate between the two. They noted that the most important variety, Fuerte, did not open its stage I flowers very completely or for very long, which perhaps explained why that variety had inconsistent fruiting. They concluded that if crosses were made, it would be best to use Fuerte as the pollen donor rather than the recipient, since they would find few female flowers open. Finally, they found that large insects such as bees were more important to pollination than had been previously realized, as trees enclosed with a swarm of bees tended to set fruit, whereas enclosed trees from which bees were excluded did not set fruit at all, unless they were hand pollinated.[36]

At the same time, Royce undertook his own experiments comparing the trees' behavior outdoors with their behavior under more controlled greenhouse conditions. He selected potted avocado trees of two varieties (Hass and Anaheim), and divided them into four groups. One group was kept outside continuously, while a second was kept in the greenhouse continuously with controlled temperatures. A third group was kept outside at night and brought into the greenhouse in daytime, and a fourth was reversed, being kept inside at night and brought out during the day. He was able to observe the flowering behavior of all four groups, and compare them with that of established trees of the same varieties. The difference was marked and immediate, with the trees in the greenhouse exhibiting flowering behavior opposite that of the trees kept outside. Once again outside temperatures played a role, and Royce was able to observe that the most important time for the temperature effect on the trees was at night, since trees kept outside only at night behaved like the outdoor trees in their blossoming, whereas trees kept in the greenhouse at night behaved like the greenhouse trees.[37] These observations had important implications both for Royce's breeding efforts and for the growers in the industry, who were dependent on successful pollination to produce avocados. Royce published both these studies in the Avocado Society's yearbook in 1951.

Home in Reseda

While Royce continued his research and breeding work, he and Pearl quickly settled into their new home in Reseda. This was a time of rapid growth in the greater Los Angeles area. The pleasing climate combined with ample employment opportunities had drawn young families into the area in large numbers, and the post-war "baby boom" swelled the number of young and growing children. When the Bringhursts first moved to Reseda, their home was about a mile from the elementary school, and Jean and Florence quickly learned to walk the entire distance, being accompanied by their mother only across the busiest street. During the following two years they would be transferred to first one then to another new elementary school, each more convenient than the last, while the city would continue to grow up around them, transforming small semi-rural lanes into busy suburban streets. Their own home was located on a quiet cul de sac, surrounded by good neighbors.[38] Royce wasted no time in putting in a garden, which was followed by a lawn and fruit trees; these did not include an avocado, though Royce was determined to breed one hardy enough to survive the frosty weather common during Reseda winters.[39]

The day after the move to Reseda, Royce spoke and sang one last time in the Sacrament Meeting at the old Beverly Hills Ward, but the family quickly grew to love their new ward in Reseda, which was full of young families like their own, which were starting their married lives in the suburbs of Los Angeles. The population growth had greatly impacted the new ward, which in a couple of years had grown from 200 to about 1500 members, and when they attended church on Fast Sunday they found there were so many confirmations and baby blessings that there was no time for testimonies.[40] Royce and Pearl quickly became actively involved in church assignments. It was not long before Royce was teaching a large Sunday School class while simultaneously taking assignments from his Seventies Quorum[41] which required him to travel to other wards within the stake; for her part, Pearl was assigned to conduct music for Relief Society, and to direct a women's choir. Ground was about to be broken on the new

Los Angeles Temple, and during the coming months they would make contributions to its completion, while at the same time paying tithes and offerings in their own ward.

That summer Pearl was able to take the children to Salt Lake for a visit to family members there.[42] Royce, being busy with his research obligations, was unable to accompany them, but before the start of school he took a couple of weeks of vacation from university duties so the family could travel about the area together and see many of the sights around Los Angeles. In September, Royce left by himself for a short stay in Salt Lake City on his way to meetings of the American Society for Horticultural Science in Minneapolis, where he was to present a talk on his research with sweet clover in Wisconsin. His absence was hard for Pearl, who was well into a fourth pregnancy and feeling ill, but the two oldest children pitched in, and Jean took over the mothering role when her own mother was unwell. Not long after Royce's return in September Pearl developed bleeding and cramping and a doctor was summoned, who informed her that it was probable she was undergoing a miscarriage. A couple of days later, the bleeding and pain became alarmingly severe, and Royce stayed home and waited on Pearl constantly, attending to the children while he tried to locate a member of the bishopric to assist him in giving her a blessing. After many hours the bishop was finally reached, and soon one of his counselors arrived, and after kneeling in prayer with the family, he assisted Royce in administering an anointing and blessing to Pearl. Her pain immediately subsided, the bleeding slowed, and after sleepless nights Pearl and Royce were at last able to get a night's sleep. After the blessing Pearl quickly began to regain her strength, but the pregnancy was lost.[43]

The family spent Thanksgiving that year at home by themselves, and the weather being cold and dreary, they built a fire in the fireplace and dined on stuffed turkey with vegetables from the garden and some of Pearl's excellent pies, a setting which reminded the children of Wisconsin, and Royce and Pearl of their old home in the Salt Lake Valley. That evening the girls began dancing to music on the radio, and after a while Royce got up and joined in. Dancing playfully, he gave a high kick, landed awkwardly on one foot, then with a howl of pain came crashing to the floor, where he lay pallid and shaking. Pearl warmed him and gave him aspirin, and the next day took him to the doctor, who pronounced the injury a foot sprain, and taped up Royce's foot. With this rudimentary treatment Royce was able to go about his business with a limp, but it was many months before his foot felt normal again.[44] For Christmas, Royce brought home a poinsettia plant which he had coaxed to full bloom in the university greenhouse, and the family painted a Christmas scene on a front window of their new home. Royce's parents had come to visit from Utah for the holidays, and on a cold New Years Day Royce and Pearl took them to Pasadena to watch the famous Rose Parade.

California at this time was in the midst of a drought of seven years' duration, but just a few days after Royce's parents returned to Utah heavy rains began to fall, and before long such a storm swept in as had not been seen in Southern California for many decades. Water quickly rose in the Los Angeles River which ran a few blocks from the Bringhurst home, and the large cement drainage canals with but a trickle of water that Pearl had smiled at when they first moved to Reseda were transformed into raging torrents of churning water which soon overflowed their concrete embankments and rushed in shallow torrents down city streets. The children made their way bravely to school, but returning home Florence had to cross a street wading through water deeper than her boots, and arrived home looking "like a little drowned mouse."[45] The children bravely continued making their daily march to school until one morning a policeman turned them back home to inform their parents that school had been cancelled until the flood subsided.

As the waters rose whole neighborhoods were evacuated, Red Cross shelters were set up not far from the house, and the ward made preparations to use their chapel as a refugee shelter if the need arose. The Bringhurst home remained high and dry, though the flood waters eventually turned LeMay street into a waterway. It had become impossible for Royce to travel to his work in Los Angeles, and perhaps in an effort to do something useful, Royce joined a neighbor in a rowboat, rowing about the neighborhood to observe the flood. Things went well enough in their own little dead-end street, but when they reached the intersection they were quickly swept downstream toward the swollen Los Angeles river, and were only saved by the fast thinking of a truck driver who, observing their predicament, stopped his truck crosswise to the street in front of the boat, allowing them to stop their downstream course and wade sheepishly back home against the current with the rowboat in tow.[46] As the flood continued, the soil became unstable, and landslides blocked major routes. Local grocery stores were mobbed by residents stocking up on supplies and food.

Eventually the waters receded and life returned to normal, but the flood appeared to have brought one enduring benefit to the Bringhurst family. Royce was finally persuaded that the old black Buick was too risky a vehicle to continue to rely on for his daily commute through the hilly terrain, particularly when Pearl refused to drive it after a stall. He began a search for a new vehicle, and in the spring he put down a bid on a retired university vehicle which was in good condition with low mileage. Royce calculated what the winning bid was likely to be and bid one dollar more, which gained him the vehicle, a white 1948 Pontiac with a rounded back.[47] This car was a major step up in reliability, and would serve the family for years to come. No one was sorry when Royce sold the old Buick to a neighbor boy for use as a jalopy.

During the summer of 1952, the rapid growth of the Church in the San Fernando Valley led to the realignment of ward boundaries, and Royce and Pearl found themselves new members of the Encino Ward, where Royce was immediately put to work as the ward welfare director, with Pearl as the ward welfare secretary. This gave them both the additional responsibility of overseeing work on the Church-owned welfare farm, where Royce had already headed a project raising onions. Royce felt obliged to participate in the welfare assignments, and he and Pearl made plans to encourage larger numbers of members to participate, to take some of the burden off the more faithful members. There was a substantial welfare assessment for the family to help fund the program, and they found that in addition to their usual tithing, they were simultaneously contributing to building funds for a ward chapel, a stake center, and the Los Angeles Temple, all of which were then under construction. Royce would also contribute his labor to the ward building. In addition to their welfare callings, Pearl was assigned as music director in the Primary, and Royce was teaching classes, and would continue to do evening work as a missionary. Although church activities kept the family busy, Royce felt committed to serving in the Church wherever assigned. He did his assignments faithfully, and paid tithes and other contributions willingly, only complaining when there seemed to be an excess of unnecessary meetings, or "meetings about meetings," as Royce viewed them.[48]

On Sunday, July 20, while at Sacrament meeting, Royce was called up unexpectedly from the congregation to speak to the ward about the need for families to accumulate a year's supply of food storage for use in case of emergency or disaster. This had been the counsel of Church leaders since the Great Depression, and as ward welfare director, one of Royce's tasks was to promote the concept among the members. This he did as best he could, but it caused him and Pearl to consider their own situation, for although they had planted a home garden and canned surplus food each year, something which was considered a necessity in both their families growing up, they had not yet accumulated a year's supply of food. The recent flood has served as a reminder to everyone of how rapidly a family could be cut off from outside sources of help, and within hours they would have yet another vivid reminder.

It was in the predawn darkness the very next morning, on Monday, July 21, 1952, that Royce and Pearl awoke to flashes from electrical fixtures outside, and became aware of a strong rocking motion within the house, followed by the shouting of neighbors as they poured into the street, and a loud crashing noise from within the house, which they later learned was caused by a falling ironing board. Realizing that they were experiencing a major earthquake, Royce sprang to the window to see if the neighbors were hurt, then outside to check the gas and water valves, while Pearl woke the children and had them stand in a doorway until the shaking stopped, as they had been instructed. The shaking, which continued for about a minute, reminded Royce of the movement of the troop ship during his trans-Atlantic crossing during the war. When it ended, Royce and Pearl checked the house and found nothing amiss, aside from an outage of the electricity and a rocking chair in the girls' bedroom which had been moved across the floor. Quieting the excited girls, they sent them to bed and returned to bed themselves, but they had scarcely settled in when the first of the strong aftershocks came rolling through. This was enough for their neighbors, who set up camp in their yard, although Royce and Pearl, reasoning that the house had come through the original quake without harm, stayed in bed and told the girls to remain in theirs, hoping to get back to sleep. By the second aftershock the girls had had enough, and Royce and Pearl, realizing their night's sleep was over, allowed the girls to climb into their own bed until daylight.[49] This earthquake, which was referred to in news reports as the Tehachapi earthquake, had measured 7.7 on the Richter Scale, and since it occurred near both population centers and scientific research centers, it was widely reported and studied.

Genetics Work, and a Turning Point

Amid these events, in the spring and summer of 1952, Royce continued his work at the university. As the avocados came back into flower he again set about making the painstaking series of crosses which he hoped would rapidly lead to varietal improvements, though he realized that at best this would be a work of many years. Meanwhile, he took advantage of the impressive collection of wild avocado relatives which had been planted at the UCLA tropical orchard, and made a series of crosses of avocado varieties with them, realizing that the offspring of these interspecific (cross-species) hybrids might prove useful to the breeding program by contributing genes for disease resistance or for tolerance for different kinds of soils. Even if these crosses did not improve fruit quality, they might introduce traits to allow commercial avocados to adapt to a wider range of growing conditions, and they might be useful in eventually providing superior

296

root stock on which to graft commercial varieties. As he made these crosses, Royce examined under the microscope the cells of these various avocado relatives while in meiosis[50], and was able to confirm for the first time that all of them possessed twelve chromosome pairs, as did the domestic avocado.[51]

In addition to conventional breeding, Royce tried still another technique in hopes of improving avocado quality. Other geneticists had already experimented with applying a substance called colchicine[52] to plant tissue in order to disrupt normal cell division in such a way that the chromosomes intended for the two daughter cells would be retained in a single cell. The resulting condition was known as polyploidy. Normal cells, with two full sets of chromosomes, were known as diploid, whereas those with four sets, double the normal complement of chromosomes, were known as tetraploid. In other fruit crops, inducing polyploidy had resulted in larger fruit size, and Royce had noted that the Mexicola variety of avocado, though an excellent bearer with a uniquely pungent flavor, was not commercially viable partly because of its small fruit size, and he looked to polyploidy as a possible solution.

Beginning in 1951, Royce began treating the shoots of Mexicola seedlings with colchicine in variable concentrations, with or without splitting of the growing tip of the plant with a razor blade to increase penetration of the colchicine. By observing the results, he was able to determine what concentration of colchicine was necessary to induce polyploidy in the avocados, and was able to establish tetraploid avocado plants as persistent clones. Once successful with the Mexicola seedlings, he tried the same procedure on some Fuerte shoots, and was successful in establishing both tetraploid and mixed diploid and tetraploid clones of that variety. He got some of these clones to flower, and although his attempts at cross-pollination did not result in any fruit, he did not consider that of great importance, since they had flowered during the winter, and attempts at pollinating even regular varieties at that time had been unsuccessful. He did note that the pollen grains of the tetraploids, which were much larger than diploid pollen grains, germinated readily when placed on the stigma of a receptive flower, suggesting that they would likely bear fruit without difficulty. Under the microscope, he was able to confirm on leaf smear preparations that the cells contained approximately 48 chromosomes, or four sets of twelve, consistent with a tetraploid condition. He eventually published the results of these experiments.[53]

In addition to his work on avocados, Royce began to do cytological study on the various subtropical plants in the orchard. He observed, for example, that the white sapote (*Casimiroa edulis*, a distant citrus relative) appeared to have about 20 chromosome pairs.[54] He

297

also continued on a very limited basis the work with sweet clover which he had begun in Wisconsin. He had described his thesis work with chlorophyll-deficient sweet clover to a colleague at UCLA named Dr. Appleman, who as a biochemist had a special interest in lethal albino plants and their relationship to varying levels of an enzyme called catalase. Royce wrote Dr. Smith in Wisconsin to request packets of seed from various sweet clover varieties, so that he could do some hybridization work in cooperation with Dr. Appleman.[55] For the same purpose he wrote to Dr. Rollo W. Woodward, a professor with whom he had worked at Utah State, for seeds of lethal albino barley which he had collected.[56]

Additionally, Royce continued to work on the problem of avocado pollination even as he continued his laborious hand pollinations of selected varieties. In 1952 in the California Avocado Society's annual yearbook he published a summary of his observations on the relationship of pollination with fruiting.[57] At the same time, he continued to travel throughout the avocado growing regions, establishing the necessary relationship with growers to further his research, and meeting with them to share the results of his investigations.

It was at such a meeting in La Habra, held under the auspices of the Avocado Institute on October 18, 1952, that an event happened which was to prove a turning point in Royce's life and his career. Royce was the second speaker of the morning meeting, and gave a talk entitled "The Sex Life of an Avocado," explaining the observations he had outlined in his paper that year. Scheduled to speak after lunch that day was Dr. Alfred M. Boyce, who was then serving as director of the Citrus Experiment Station located at Riverside; his topic was, "What the Citrus Experiment Station is Doing for the Avocado Industry." Royce later recorded his recollection of what happened that day: "Al Boyce sat by me at lunch, then proceeded to patronize me to a sickening degree and I wondered why. As he spoke in the afternoon, I understood, for he spoke of hiring Peter Peterson at Riverside to replace Dr. Lesley in breeding avocados. He tried to explain how the two programs did not overlap, etc. – hardiness etc. – all complete drivel and nonsense. This shocked me and made me feel double crossed since I had been informed that I would have sole responsibility for avocado breeding when Dr. Lesley retired."[58]

This decision to hire another breeder for avocados, probably a political decision made at Riverside, was an unexpected and bewildering blow for Royce . Ever since coming to Los Angeles he had envisioned a satisfying and lifelong career in avocado breeding, but now in effect another man was being assigned to take over the very task he had been hired to do. As he drove home from the meeting that evening unhappy and sick at heart, he pondered the situation. By the time he pulled into the house at Reseda, he had arrived at a decision to leave UCLA and seek a job elsewhere. In accord with this new resolution, he began to make inquiries about open positions for a PhD in genetics. Before long he was made aware of a job opening for sugar beet breeding with the U. S. Department of Agriculture, which doubtless brought to mind the work of F. V. Owen, the much-admired "Doc Owen" with whom he had worked in his father's beet fields before college, and with whom he had consulted during his graduate years. Royce had grown up with sugar beets, and he began to consider the offer very seriously.

As Royce was exploring his options, life continued to press forward. Pearl's mother came to visit during Christmas of 1953, and by that time Pearl was once again expecting a child, and was feeling unwell. A large cyst which had required an operation during the war had reappeared, and Pearl, though pregnant, had to undergo a surgery to correct the problem. Meanwhile, both Royce and Pearl were busy with new assignments in Church. Pearl was now a leader in the ward Relief Society, and Royce, who had received an assignment as a stake missionary in the San Fernando Stake, was out doing visits nearly every night, and was often assigned as a speaker in other wards. The early spring was a busy time at work as well, for although Royce had not changed his resolve to leave the university, he continued his work on avocado crosses, and travelled to avocado growing areas to give talks to growers.

In the spring of 1953 Royce learned from Dr. Chandler,[59] a retired assistant dean at UCLA, of a new opening for a geneticist which had recently been announced at the northern branch of the School of Agriculture, which though administered through the main campus at Berkeley, was located in a little town called Davis, in the great Central Valley of California not far from Sacramento. This position, which would involve strawberry breeding, immediately caught Royce's attention. In his work at Los Angeles he had begun to enjoy his new identity as a fruit breeder, and because the position was within the University of California system, this would involve a transfer rather than a re-hire. He immediately made inquiries into the position, and was invited for an interview.

In mid-May, Royce took a flight from Los Angeles to Oakland at university expense to look into the Davis opportunity. Staying the night with his sister Naida, he took a train in the morning to Berkeley to interview for the position, and from Berkeley he traveled to Davis to look over the area. Originally called "University Farm," the Davis campus had been purchased in the early 20th century as a site for UC Berkeley students to study agricultural sciences, and had already expanded into a much larger campus housing the northern branch of

299

the UC School of Agricultural, the southern having recently been converted into UCLA. Davis was a small but pleasant rural town, and it was of a far different character from the massively urbanized and heavily-trafficked Los Angeles area. A move here would be a return to country life, well away from the bustle of the large city, and Royce would travel to work on a bicycle rather than a car. Royce returned home with a job offer, and he was allowed until the end of the May to communicate his decision.

Back in Reseda, Royce and Pearl had long discussions about the pros and cons of moving to Davis. On the one hand, the family had come to love their home in Reseda, were happy with their church associations, and looked forward to the completion of the Los Angeles Temple, which for the first time since leaving Utah would at last bring a temple to within easy traveling distance. Royce's department at the university pressured him to stay, as there was no provision made for his replacement, and his departure would be a blow to the department. Dr. Cameron, who had originally arranged for Royce's hiring, strongly urged him to remain at Los Angeles, pointing out that he had already distinguished himself professionally, and was well liked and respected by his university colleagues and looked up to by the avocado growers. Also, on a personal level, Dr. Cameron and his wife had sincerely come to like Royce and his family.[60]

On the other hand, not only had Royce's professional plans been dealt a serious blow, but there was also uncertainty about the future of any agricultural work at UCLA, since there were already plans underway to move the Agricultural College from the UCLA campus, which had been hard pressed by urbanization, to the Riverside campus farther inland. It did not make sense to sink their roots even further in the Reseda area only to have to move a few years later, and the rapid urbanization and the resulting traffic jams were increasingly a burden. There was still another factor, perhaps unmentioned but certainly present. Since his military service Royce had always felt anxious around crowds, and the question of moving to Davis was in large measure a choice between remaining amid the increasingly crowded environs of Los Angeles, or moving to the much more quiet and semirural environment up north.[61] After long deliberation Royce finally made up his mind, and on Friday, May 29, 1953, he sent a wire to Davis informing them that he had accepted the position.[62]

Although university positions customarily started the first of July, Royce had negotiated delaying his arrival at Davis until September, to avoid traveling too close to Pearl's due date in July. With mixed feelings, the family made ready to put their new home up for sale, and with the rapid influx of families into the area, within two weeks they already had a buyer and were in escrow. As Pearl's due date approached, Reseda was hit with swelteringly hot weather which was so severe there were fatalities from the heat, and reports of people collapsing even on the sea shore. After one of Pearl's prenatal visits, the family tried to escape the heat by taking a morning trip to the beach, but though it was not yet noon, the sand was already too hot to walk on. Nine months pregnant in the summer heat, Pearl was envious of a neighbor woman due at the same time who had already given birth, but her envy disappeared when the child died at three days of age.

Finally on the afternoon of July 26, Pearl went into labor, and was driven the 45 minutes to the Queen of Angels hospital in Los Angeles, where at 6:00 p.m. she gave birth to another baby girl. For the birth she received for the first time a spinal anesthetic, and for Pearl, who had mothered three children without anesthesia, the experience of pain-free birth was nothing short of miraculous. She described the baby as "a little like the others and a little like herself."[63] The three older girls, though not excited about moving, were entranced with their new sister, whom Royce and Pearl named Ann.

The remaining time in Reseda was spent making preparations to leave. Royce was kept busy finishing up his work at the university and making trips to Davis to make living arrangements for his family, and his mother came from Utah to help Pearl take care of the baby. The relentless urban growth was continuing around them at a rapid pace even as they made ready to move, and their packing was accompanied by the rumble of heavy machinery only a couple of blocks away, where a walnut orchard was giving way to the construction of a new high school. On a memorable day not so long before, Royce had driven the family to a vantage point overlooking the San Fernando Valley, alive with orange groves as far as the eye could see, and Royce had said, "Take a good look. This will soon be gone." By the time the family left Reseda for Davis, many of the pleasant citrus groves had already given way to busy suburban neighborhoods,[64] and in the ensuing years the valley would indeed be covered with cities, but the older girls would always have fond recollections of their time in Reseda, where they had passed through many adventures and survived two major natural disasters. On September 9, 1953, the family locked up their home on LeMay Street[65] for the last time, and bundled into the Pontiac for the two-day journey northward to Davis.

The Aftermath

In his work with avocado breeding, Royce had begun to lay what he felt was a solid foundation for a breeding program which he was certain could result in superior new avocado varieties with improved traits and expanded range, and which would prove of great benefit to the industry. His departure from UCLA marked a turning point, not

301

only in his own life, but for the avocado breeding program as well. Dr. Peter Peterson,[66] who upon Royce's departure would head the program out of the Riverside campus, would only remain with the program a couple of years before he too resigned. In 1960, Dr. B. O. Bergh would take over the avocado breeding program. Faced with the same difficulties in pollination that had dogged his predecessors, Dr. Bergh would decide to abandon the well established breeding method of hand-pollinating selected cultivars, and turn instead to planting blocks of selected trees and allowing them to open-pollinate, not recognizing perhaps that that was essentially the same system which had been in informal use by growers long before the breeding program ever existed, and which had resulted only in a long succession of disappointing and inferior varieties.

Royce, in a treatise on fruit breeding strategy some thirty years later, would use the avocado as an example of a commercial fruit in which no effective breeding work had ever been accomplished.[67] Although a large number of named selections had been released, the only commercially viable ones were the same old chance seedlings of uncertain parentage which had been there all along. Fuerte had been imported as cuttings from a dooryard tree in Mexico, and Hass had been discovered accidentally, having been planted as a rootstock in a private grove where it outgrew its graft and bore unexpectedly good fruit. In a time where rapid advances had been made in fruit breeding elsewhere, decades of misguided effort had led to no new varieties of commercial importance, and for practical purposes, the list of usable cultivars was little changed from the time Royce had first arrived at UCLA back in 1950. The much sought after Fuerte-like fruit with broader range and better fruiting characteristics never materialized, and eventually the industry repositioned itself around the less handsome but far more reliable Hass variety. The public, left with no true alternative, gradually adapted to the appearance and its flavor of Hass, which though once regarded as inferior to Fuerte was still excellent. To the present day, this accidental mongrel seedling remains the most important single commercial avocado cultivar in the entire world.

Two of Royce's avocado crosses would later be released as named varieties by others. Creelman, a cross of Fuerte x Hass, introduced in 1961, resulted from one of Royce's pollinations ten years earlier. Nichols, introduced in 1962, was a Fuerte x Mexicola seedling from a cross made by Royce in 1953.[68] Neither had any commercial importance, and it is unlikely that Royce would have released either had he remained with the program, though they might have served as breeding stock for a subsequent generation of crosses.

Rapid urbanization, coupled with the needs of a growing university, eventually dealt death to the UCLA College of Agriculture. Dr.

Cameron, who had hired Royce in 1950, would oversee the dismantling of the college in 1960, and most of its programs would be transferred to the Riverside campus, which had itself become a full-fledged university with the rapid postwar expansion of the U.C. system. The only remains of the old tropical fruit orchard were a few trees near the Health Sciences building, which Royce's daughter Marla would recognize when she worked there many years later. So it was that the last remnants of the UCLA avocado breeding program slipped quietly out of existence. But what had been a loss for the avocado would in time become a gain for the strawberry.

Illustrations for Chapter 8

Royce shortly after arrival at UCLA, October 1950.

Avocados in the experimental orchard at UCLA.

Pearl and the girls, at completed family home in Reseda.

Bringhurst family just prior to move to Davis. From left: Royce's mother
Florence Bringhurst, Florence, Pearl with Ann, Marla, Jean.

305

University of California at Davis

Chapter 9

Arrival in Davis

It was the afternoon of a warm Wednesday in September of 1953 when Royce and Pearl finally left Reseda and headed north with their family in the white Pontiac. Marla was irritable, having had a high fever the day before, but newborn Ann slept most of the way in a hammock-like device the family had acquired, which was suspended from the rear of the front seat. Arriving that evening as far as Bakersfield, at the southern end of California's great Central Valley, they spent the night with some old friends from Los Angeles before continuing on the next morning northward toward Davis. The temperature was unseasonably hot as they made their way up the valley, measuring 106 degrees by that afternoon, which, in a day before automobile air conditioners, made for a long and weary drive. It was 5:00 p.m. when they finally pulled into Davis and checked into a motel for the night, and Pearl was grateful to find that Ann's car hammock also functioned nicely as a bed in the motel.

The next morning, on Friday, September 11, the family was invited to the home of Dr. Warren P. Tufts,[1] the chairman of the Department of Pomology, where they enjoyed a breakfast prepared by Mrs. Tufts, to be followed by lunch at the Browns, another family from the department. Between meals they checked into an apartment, located on East 9th Street, which would be their home for the next several months until they could make more permanent arrangements. Later that day the movers arrived with their furnishings, and they began the task of turning the empty apartment into a home.[2]

The family, accustomed to the bustle of Los Angeles, quickly became attached to their new surroundings, which Pearl described as "a sleepy little country town," with a few businesses in the town center, surrounded by quiet country roads lined with olive trees. The family arrived just in time for the girls to begin school, and that Sunday they attended church for the first time, at a ward located in Woodland, about eleven miles to the north. The Woodland Ward was a modest-sized ward compared to the large, rapidly growing wards they had just left in the Los Angeles area, or to the long-established ones in Utah where Royce and Pearl had grown up, and it was meeting temporarily in a rented Odd Fellows Hall while a new chapel was under construction. As he had in Los Angeles, Royce spent some of his free time working on the chapel, which would go into use that December and be dedicated the following March. The month they moved in, Royce and Pearl were asked to speak in Sacrament meeting, and it was not long before they both held positions in the ward Sunday School.

The Bringhursts had not been in Davis very long before it became clear that Royce's job at the university there was much more demanding than the one he had left behind at UCLA. Although his office was close enough to home that he rode to work on a bicycle, the strawberry program itself included projects which spanned almost the entire length of the state, and Royce was gone frequently, and for extended periods of time. As soon as they arrived in Davis, he and Pearl began looking for a home to purchase, but affordable houses were not to be found, and they decided to rent while they saved money for a down payment on a new home. Even a rental home was slow in coming, but by early 1954 they were able to make arrangements to rent the house of one of the senior professors who was scheduled to be out of town for seven months. This proved to be none other than Dr. Ledyard Stebbins, a renowned botanist and plant geneticist who had already distinguished himself as one of the world's foremost pioneers in evolutionary botany, and who would later strongly influence Royce's work.[3]

The family moved into the Stebbins' spacious modern home on the first of February, and were pleased to find a piano in the house. Pearl took the opportunity to begin teaching the rudiments of music to the girls, and soon the whole family, including Royce, were taking their turn at the piano. Eventually the family purchased a piano of their own, an old Lester spinet which, though it had its flaws, was good enough for the children to practice on. The girls began formal piano lessons, followed by Pearl herself, and before long Jean and Florence began taking violin lessons at school as well. Their first

307

efforts on the violin were predictably sour, and in a good-humored way reminded Royce of "a squeaky gate," but they rapidly improved, and music increasingly became an important feature of the Bringhurst home as the years passed.

During their first summer in Davis, in July of 1954 the family took the first of many annual trips "back home" to Utah to visit the family there. While in Utah, Royce made a discovery which would prove to be a landmark in his professional career. Between visits to relatives, the family took a side trip up the Big Cottonwood Canyon to Brighton, a beautiful mountain resort beloved by Pearl's family, where they had often gone camping.[4] While hiking with Pearl and the girls in Brighton, Royce paid special attention to the small colonies of wild strawberries which abounded in the Wasatch Mountains, some of which proved to be of a type which Royce identified as *Fragaria ovalis*,[5] a species genetically compatible with the commercial strawberries which had now become Royce's livelihood. As a newly hired strawberry geneticist, he determined to take a sample of these plants back to California, but lacking a permit to transport live plants, he instead took some of the tiny ripe berries and smashed them onto a paper towel in order to collect the seeds. Back in California, he germinated the seeds and kept some of the progeny as part of his growing collection of wild plant material which he would systematically interbreed with commercial strawberries. Though he did not yet know it, these few seeds he had gathered from the mountains of Utah during his first year as a strawberry geneticist carried a gene which would later lead to one of his greatest contributions as a plant breeder.[6]

By the time Dr. Stebbins returned to Davis in September, Royce and Pearl had located a more permanent residence where they could remain until they were able to build a home of their own. This was a pleasant rental house belonging to a man named Jim Stiles. Located at 530 E Street, it was an established older home with a shrubbery, a lawn, and orange and fig trees. The family moved in at the beginning of September, and nearly a year after arriving in Davis, they finally felt they had a home of their own.

The University of California Strawberry Program

Royce had arrived to begin his work with strawberries in September of 1953, and as he had done previously with the avocado, he began to acquaint himself with the history and peculiarities of the strawberry as a cultivated crop. Strawberries, all of which belonged to the genus *Fragaria*, had been known in Europe since ancient times, though the European varieties were only distantly related to the commercial varieties of today. The most well known was the woodland strawberry, *Fragaria vesca*, which bore a small, soft, fragrant fruit which was usually picked in the wild, and was seldom grown in gardens. Important from Royce's standpoint as a breeder was the fact that *F. vesca* was a diploid strawberry, meaning that it had only two sets of chromosomes, while the commercial strawberry had eight sets, making it an octoploid. Although early references to *Fragaria vesca* came from European sources, as a species it was widely distributed throughout the Northern Hemisphere, in America as well as Europe. Royce would encounter it constantly in the wild, and it would play an important role in his work, both as a botanist and as a plant breeder.

After the discovery of America, a completely new type of strawberry was introduced into Europe. This was *Fragaria virginiana*, closely related to the species Royce had collected in Brighton. This strawberry was an octoploid (containing eight sets of chromosomes), grew widely across what is now the United States and Canada, and had been imported to Europe from the earliest colonies on the Eastern Seaboard. Much later, in the 1700s, a French military agent named Frèzier brought back to Europe yet another type of strawberry which he had observed in Concepción, Chile, where it produced fruit of unusual size, though it did not fruit very well in Europe.[7] This new plant was the coastal or beach strawberry, *Fragaria chiloensis* (another octoploid), which grew wild all along the sandy Pacific coastal areas of North and South America. Eventually, these two octoploid species were crossed in European gardens, producing first hybrids, then eventually self-fruitful varieties, which were planted and replanted, crossed and recrossed. The result was the modern garden strawberry, which amounted to no less than a new domesticated species (often referred to as *Fragaria* x *ananassa*), and which includes virtually all commercial strawberries now grown.[8] Some of the cultivated strawberry varieties eventually made their way back across the Atlantic to the Americas and were planted in gardens and farms throughout the United States, becoming the basis for new generations of varieties developed in America. In effect, the garden strawberry, sired in America and born in Europe, had returned to its homeland where it would come of age.

The history of strawberry breeding in California extended back to the 19th and early 20th centuries, most significantly with the work of Albert Felix Etter, a child of German-speaking immigrants to California who had dedicated much of his life to the breeding of improved varieties of apples and strawberries in Humboldt County, where his family owned a ranch. Unlike many other breeders, Etter felt that there was value in making crosses between the usual garden varieties of strawberry and the wild colonies of *Fragaria chiloensis* he found growing widely along the California coastline, in order to make his selections better adapted to the local conditions in which they grew. (Royce would later regard this as a decision of greatest importance, and as the basis for much of the success of his own work.) Etter's varieties were used by other plant breeders both in the United States and abroad, and he later may have donated his strawberry materials to the University of California. Some of his cultivars provided breeding stock for the many successful varieties which were to follow.[9]

Meanwhile, although California would one day dominate the world in strawberry production, strawberry farming in the state had gotten off to a rocky start in the 19th and early 20th century. The varieties of strawberry then available had almost all been developed in the eastern United States, and most were poorly adapted to the drier conditions which prevailed in the west, and they also tended to be very susceptible to the diseases and insect pests which abounded in California. Although rapid systems of transportation were being developed, strawberries were still far too perishable to ship to distant markets, and farmers had to be content with local sales. Another limitation was the fact that nearly all strawberry varieties produced their fruit during a narrow window of time in the early summer. This meant that fresh strawberries were available for only a few weeks of the year, and any that could not be sold fresh had to be channeled into the less profitable frozen or processed fruit market. The challenges of strawberry farming were so daunting, and the requirements for success so exacting, that few Californians ventured to do it. Of those who did, the most prominent were a number of tenacious Japanese-Americans farmers who, with great care and perseverance, were able to establish a successful and highly specialized niche as strawberry growers in California coastal areas until World War II brought a sudden halt to their efforts.

When Royce arrived at Davis in 1953, the University of California strawberry breeding program had already been in existence for many years, though it had undergone periods of upheaval. UC Davis, like UCLA, was part of the Land Grant university system, and was under the same mandate to provide scientific investigation for the benefit of local industries. It was under this mandate that the strawberry breeding program was initiated in the late 1920s as part of a small fruits program, under the direction of William T. Horne and A. G. Plakidas. After a short time the program was taken over by Dr. Harold E. Thomas, a professor of Plant Pathology at the University of California at Berkeley, and Earl V. Goldsmith[10], an agronomist who worked under Thomas. In their case, it was the agronomist Goldsmith rather than the professor Thomas who led the way in strawberry breeding, and their combined breeding efforts ultimately

310

culminated in the development of five new "University of California" varieties of strawberries, the most important of which, named after California's two famous volcanoes, were called "Shasta" and "Lassen." These new varieties at last gave growers cultivars which were well adapted to the California climate and conditions.

These seminal UC varieties had been released in 1945, but with the displacement of the Japanese-American farmers to relocation camps during the Second World War, there was scarcely an industry left to make use of the new varieties, and support for strawberry breeding work had naturally taken a low priority during the war. The same year the new varieties were released, Thomas and Goldsmith both resigned from the university to join a nonprofit organization called the Strawberry Institute, where they would continue their breeding efforts in association with the Driscoll Associates, a private strawberry producing firm. Using plant material from their breeding work at the university and now with ample private support, these two would continue to develop strawberry varieties, and in a limited way would compete with their successors in the university program during the ensuing years.[11]

After the departure of Thomas and Goldsmith, the University strawberry program, underfunded and in a state of disarray, was placed under the direction of Dr. Richard E. Baker, a UC Davis pomologist, and it was at this time that the program was shifted from Berkeley to the Davis campus. Baker undertook to reorganize the California program, and working with the plant material left by his predecessors he was able to release of a couple of new varieties in the late 1940s, but neither of them succeeded commercially.[12] As growers returned to farming after the displacements caused by the war, the strawberry industry began to slowly revive, making use of the 1945 releases, Shasta and Lassen. However, there were still many problems facing the struggling industry. Yearly variations in harvest could produce huge market fluctuations, rendering even a bountiful crop unprofitable. In some years the weather conditions caused the crops to be strongly affected by plant diseases, particularly *verticillium* fungus which caused a severe wilt in strawberries. There were also basic horticultural problems, such as the high salinity caused by furrow irrigation, which was toxic to the plants. All these factors became serious challenges for the farmers, who were fighting for their very survival. The strawberry growers in turn brought pressure to bear on both the University and the state legislature for applied research to help their struggling industry. New funding for strawberry research was finally approved in 1952, but the road had been a turbulent one, and in March of 1953 Dr. Baker announced his resignation, opening the way for Royce's hiring.[13]

311

Shortly after arriving in Davis in 1953, Royce set off with Pomology Department chairman, Warren P. Tufts, to acquaint himself with the various research projects and facilities which were part of the extensive university strawberry program. Nearest at hand was the Wolfskill Experimental Orchard (WEO) located in Winters, a few miles west of Davis, where virus-free stocks of plants were kept for testing throughout the state, and where Royce would make many of his initial experimental plantings.[14] In the far north, near Redding, were nurseries where strawberry plants would be propagated, grown, dug up, and stored for later planting. South of the San Francisco Bay was a university field station near San Jose, which represented growing conditions for that area, and which Royce would use for experiments in plant diseases. Further south, near the Monterey Bay, were extensive commercial plantings around Watsonville and Salinas, where Thomas and Goldsmith had concentrated their efforts, though the university had no official presence there. Inland in the San Joaquin valley near Fresno was another important strawberry growing region, and in the far south, near San Diego, was the USDA Experimental Station at Torrey Pines, where research was being conducted under conditions which prevailed in Southern California. Royce got along well with Dr. Tufts as they toured these distant sites, in spite of the department chairman's penchant for cigar smoking, which Royce had always found disagreeable. Dr. Tufts gave Royce complete liberty in setting up his program, making neither stipulations nor suggestions as to how he should go about it. As chairman of the Pomology Department, Dr. Tufts had borne the brunt of the strawberry growers' collective discontent with the University, and his only instruction to the promising new associate professor was, "Just keep those people off my back."[15]

It may have been while visiting the Torrey Pines USDA station that Royce first met Victor Voth. Victor was perhaps the most positive thing to emerge from the years of turmoil preceding Royce's arrival. Hired as a lab technician in 1946 to assist Dr. Baker in his work, he had quickly begun to master the intricacies of strawberry horticulture, and with the new state funding secured by the industry, he had recently been appointed assistant research specialist over strawberries at the Torrey Pines station. Victor was an affable man of medium height who liked to keep his dark hair cropped in the short, flat-topped style popular at the time, but his most distinguishing features were a pronounced stutter whenever he spoke, and a perpetual twinkle in his eye. Victor had a passion for every aspect of strawberry cultivation, and when Royce arrived he was already working on solutions to the problems that plagued the strawberry industry.

These two men, a plant geneticist and a horticultural scientist, found themselves paired together at a crucial time, when the industry was reeling from a series of setbacks which threatened to bring an end to commercial strawberry production in California. As they started working together, Royce took the lead in the breeding and the genetics aspect of their work, while Victor concentrated on horticultural practices which would allow the new selections to be successful, but both became full participants in every aspects of the work, Victor doing his share of variety selection and Royce making ample contributions to what would become revolutionary growing practices. Indeed, the two would work so well together that it would be difficult for colleagues or growers to imagine the success of one without the complementary work of the other. The two quickly became close associates and devoted friends, and would remain so during most of the next forty years.[16]

Speaking of his own role as geneticist Royce later remarked, "They needed some breeding work in the worst way, and that was the expectation that they had of me in hiring me, to do strawberry breeding."[17] Of the two early U.C. varieties used by the industry, Shasta bore a fruit which was of adequate quality and suitable for freezing, but it did not do well in the southern growing areas of the state, where warm winters and growing conditions made the growers dependent on the better-adapted but otherwise inferior Lassen variety. Reflecting on their breeding objectives at that time, Royce wryly observed, "The first thing was fruit quality. The strawberry that was then grown was the Lassen variety for the most part, along with Shasta, and 'Lassen' had terrible flavor. It was just cold and wet, that was the best that could be said for it."[18] In consultation with Victor, Royce began the long and complicated process of expanding and developing the extensive breeding program required to produce better strawberry cultivars for the state.

There were several things that needed to be accomplished before this objective could be realized. The first, to identify what breeding stock was available to use, was relatively straightforward at this early time. Royce had available to him both the named varieties and the breeding material left behind by Thomas and Goldsmith and later by Dr. Baker, and Victor was already familiar with these. Royce expanded the breeding stock to include parent material from good cultivars from outside of California, but he quickly came to realize that the success of the UC program depended on varieties adapted to the state, and he would depend heavily on the better material left by his predecessors. As time went by, he also worked to breed the commercial varieties with wild strawberries which had desirable traits and were well adapted to the conditions in California.

313

The second task was to identify the actual objectives of their breeding efforts, meaning what traits the selected plants needed to have to succeed commercially. Good flavor was an obvious objective, as was attractive fruit, but there were many other traits which had to be considered and bred for. Productivity was the main factor as far as the commercial growers were concerned, since even a strawberry with an excellent flavor would not succeed if too few fruit were produced to cover the expense of growing them. Large fruit size was also important, not just to impress consumers, but to make picking the fruit easier and therefore less expensive. Fruit firmness and durability were essential if the fruit was to be shipped to distant markets. And there were a number of other traits, such as disease resistance, adaptability to adverse growing conditions, and earliness or lateness of fruiting, any one of which could mean the difference between commercial success or failure of a given variety. The identification and selection of appropriate traits was the most difficult task, since what would make a variety successful was often hard to predict, and could quickly change depending on advances in agricultural methods on the one hand, and new environmental challenges on the other. Hence, the goal of creating the perfect strawberry was always an elusive one, and would shift from year to year depending on changing consumer preferences and on a host of factors outside the breeder's control.

The process of breeding itself was a well established one. Once parent plants had been decided on based on favorable traits, cross-pollinations were carefully performed, taking care to isolate the female and male reproductive parts of the flowers to avoid self-pollination, as well as uncontrolled pollinations from unwanted parents. When the fruit from the crosses matured, the ripe berries were processed in a blender to collect the seeds, which were carefully preserved and prepared for planting out.[19] From among the offspring seedlings that resulted, the ones with the best characteristics would be selected, either for use as a variety or to serve as parents for the next generation of crosses. In this way, by careful selection generation after generation, more and more favorable traits could be combined in a single plant. This process, referred to by Royce as "recurrent mass selection," was the basis of modern plant breeding science. Once a variety was selected, it would then be propagated on a large scale for formal planting.

The final task was to subject the new varieties to rigorous field testing to learn what planting and farming methods would make them most productive, and to determine what role if any they might play in the overall industry. This was an essential step, and it was an area in which Victor's horticultural experience would play an important role.

As they worked together, Royce and Victor were constantly aware that their work was of little benefit to the industry they had been hired to serve unless the varieties they released had been rigorously tested under actual growing conditions, so that their release could be accompanied by specific information on how to grow them productively. This would require astute observation and the careful application of scientific principles, as well as a good relationship with the growers themselves, whose livelihood often depended upon the success or failure of whatever they planted. Victor especially would work constantly with growers large and small, and in time some of the growers would come to play an important role in the testing of new varieties, and even in helping determine which ones should be released for commercial use.

Just as it had taken Thomas and Goldsmith years to develop the first university varieties, so it would take Royce and Victor years of breeding and selection before they would begin to deliver new and improved varieties of strawberry, but it was a process they would refine as the years went by. Whether Royce realized it or not when he accepted the transfer to UC Davis, strawberries were in a sense the ideal crop for the plant breeder. With avocados, the best he could have hoped for was a few hundred hand-pollinated crosses per year, but with strawberries he could make and plant out thousands. Where avocados would have required many years to come into fruiting and even more to evaluate the quality and quantity of the resulting fruit, strawberries could begin to be evaluated within a couple of years of breeding, and as information was gathered, successive generations of crosses could be made very quickly. The work of a lifetime for a tree fruit would be a matter of a few years for the strawberry. Royce Bringhurst and Victor Voth could not possibly have known when they met what a transformational impact their work would eventually have, not just in California but throughout the world, but they both felt full of energy, ideas, and supreme confidence in the power of science to solve the most perplexing problems, and it was in that spirit that they began their work together.

Life in Davis

It was September of 1954 when the Bringhurst family moved into the little rental house on E Street. The older children began attending school, and Pearl busied herself with sewing while Royce worked at the university. Royce had been assigned to teach a class in small fruit horticulture, and had to prepare two lectures a week, for which he began to assemble a collection of slides and other materials to assist him in his teaching. University salaries at that time were austere, and Royce planted a garden, as he had in Reseda, to supplement the

315

family table. During pheasant season he obtained a hunting license, and began to go out weekends as he once did in Utah, bringing home pheasants for the family to eat in addition to the produce he brought back from the university or grew in the garden. The family continued to work hard in their church assignments. Pearl was called into the Relief Society presidency and began to make frequent visits, while Royce was released from his Sunday School calling and was called on a "Stake Mission," in which he was assigned a companion from Woodland. Although he loved missionary work, this was a difficult calling for Royce, as he had to travel frequently to attend to his strawberry experiments, and his companion, a police officer, had to work evenings which made it difficult for them to get together, but within a few months they had baptized their first convert, whom Royce confirmed.[20]

By late 1955, Pearl was again pregnant, with a child due in January, at a time when three other women in the ward were also expecting babies. As her date approached, a couple of weeks before Christmas the weather grew stormy, and heavy wind and rains descended on the valley and continued for days. Water in the rivers began to rise, and on the Friday afternoon before Christmas, the university received a call from the Highway Patrol advising them to send any employees from Sacramento home immediately, and by 4:00 p.m. the highway to Sacramento was closed by flooding. For the Bringhurst family the storm was little more than a nuisance, making the inside air damp enough that laundry took days to dry on indoor clothes lines, but there was too much excitement in anticipation of Christmas to worry much about the rain.

On the morning of Christmas Eve the family turned on the radio expecting to hear traditional Christmas music, and were instead surprised by an urgent series of emergency broadcasts. During the previous night the swollen waters of the Feather River had burst through a levee fifty miles north of Davis, devastating the town of Yuba City and the surrounding countryside. Floodwaters had swept through quiet neighborhoods, destroying and damaging everything in their path. Many people were stranded and had to be rescued by helicopters from a nearby Air Force base, while others less fortunate were killed outright, including a counselor in the Yuba City Ward bishopric with his entire family. It was not long before everyone in Davis was aware of the disaster. Evacuees poured in, and the university buildings were pressed into service as emergency relocation centers, housing dazed survivors. Royce took the family on a short ride to see normally placid Putah Creek just south of the university campus. Now a raging river, it had completely submerged the road bridge, whose location was now marked only by tangled masses of uprooted trees swept downstream by the flooded creek, which had swelled to a torrent which dwarfed even the flood-swollen Los Angeles River back in Reseda.

Christmas day was Sunday, and the family drove to church in Woodland between flooded fields which had the appearance of geometric lakes with the road passing between them. Arriving at the church building, they discovered that it, too, had been pressed into service as a evacuation center, and the recreation hall was filled with cots and blankets. Many of the evacuees joined in the morning services, though by the evening Christmas program most had departed, heading back to Yuba City to see what had become of their homes. As the flood waters receded, Royce joined other men from the Woodland Ward who traveled to Yuba City to help residents shovel mud from their flood-ravaged homes. It was the first time Royce had seen the Church welfare program in action, and he was pleased with what he saw. The family later learned that it was only the newly-completed Folsom Dam which had spared Sacramento from major flooding, and they were also told that had the rains continued, state authorities had considered evacuating Davis and blasting the levees which protected it in order to divert floodwaters away from Sacramento. Gratefully, the waters receded before such drastic measures became necessary, but Davis remained isolated from Sacramento by floodwaters for several days, making this an anxious time for Pearl, who was eight months pregnant and planned to give birth in a Sacramento hospital.[21]

On a rainy Sunday morning just four weeks after the great Yuba City flood, Pearl went into labor and was driven to Mercy General Hospital in Sacramento, where in the early afternoon she gave birth to her fifth child, a baby boy with a dimple in his right cheek and a "right husky yell." Royce, who returned from the hospital in time to attend church in Woodland, had been scheduled to sing a solo in church that day, and as he rose to sing, he announced to the ward with an obvious flush of pleasure that he was at last the father of a healthy son. They named the boy John Royce, after his grandfather and father.[22] Not only was this Royce and Pearl's first male child, he was the first child born under anything close to ideal circumstances. Jean had been born while Royce was in military training during the war, Florence as he was preparing to begin school in Logan, Marla as he started graduate school in Wisconsin, and Ann as the family was making their move to Davis. It was not until the fifth child was born that Royce had a stable job and the family was settled, and Royce's mother once again traveled from Utah to help care for the baby.

As the baby grew, life went on in the home on E Street. Jean, already a teenager, was now tall enough to be able to fit into some of her mother's clothes. With a house of their own, the family began to keep pets, starting with a pet dog, as they had in Los Angeles. One day Royce brought home a baby crow which had fallen from its nest, and the girls reared it to maturity; on another occasion, he brought home a wild baby rabbit, which became a tame pet after Pearl nursed it back to health from a sickness. The family continued to live frugally, harvesting food from their garden, and canning large quantities of fruit that Royce brought home from the university. Royce and Pearl, having grown up in the Great Depression, had both adopted the practice of growing and preserving their own food, and canning became a family institution, though it was one the girls sometimes looked forward to with dread, especially when Royce brought home particularly large quantities of fruit.

Life was busy at church as well. Pearl now had a calling in the stake which required her to travel frequently, sometimes as far away as Placerville, nearly sixty miles distant. Royce, while still making occasional visits as a stake missionary, was also increasingly involved in preparations to locate a branch of the Church in Davis, for although the Woodland chapel had only recently been completed, with the rapid growth of the university in Davis the ward was quickly becoming too large for the building. The Bringhurst family began hosting groups of Davis students who met at their home, and the bishop, authorized to look for land for a chapel in Davis, asked Royce to assist him in driving around town looking for an appropriate lot, and for a building which might serve as a meetinghouse in the meantime. At the same time Royce and Pearl continued to search for a permanent place to live, and eventually they cast their eye on a new series of houses which were planned for construction just west of Anderson Road, then at the western edge of town.

During this time the family weathered a series of near-disasters in their rental house on E Street. One afternoon while Royce was out of town on university business, the girls were excited to see fire trucks on their own streets as they walked home from school, but their excitement turned to alarm as they realized that the fire trucks were parked in front of their own home. A fire Pearl had lit in the fireplace had ignited the ivy vines which had climbed the chimney, setting the roof ablaze, though gratefully the fire was extinguished with little damage. On another occasion, a still-burning match one Sunday morning found its way into the kitchen garbage where wax had been discarded after canning jelly the day before. The wax ignited, and the whole kitchen would have been aflame had not Royce, who uncharacteristically had stayed home from church because he felt unwell, smelled smoke and extinguished the blaze.[23] (Royce had never missed church as long as the girls could remember, and the family regarded his sickness that day as an act of Providence.)

An even closer call occurred in November when baby John, then nine months old and prone to put things in his mouth, suddenly began gasping and turning blue. Pearl, realizing the child was choking, in desperation stuck a finger down his throat and tried to dislodge the obstructing object, and succeeded in repositioning it so the baby was able to breathe. With the baby in arms, she raced out the house and ran the several blocks to the doctor's office, calling to a neighbor as she passed to take care of Ann and watch for the other girls as they came home from school. At the office of the family doctor, an x-ray showed the object to be what appeared to be a metal campaign button. The doctor quickly gathered his instruments and prepared to extract the object, but realizing that a misstep could result in instant asphyxiation, he withdrew his instruments and hastily wrote Pearl the name of a specialist in Sacramento, telling her to drive the baby there immediately. Pearl interrupted Royce in a meeting at the college campus where he was scheduled to speak, and quickly rescheduling his talk for the end of the meeting, Royce accompanied Pearl to the car, and the two of them raced the baby at a high speed along the two-lane wooden causeway bridge which linked Davis with Sacramento, fourteen miles away. Entering Sacramento, Royce continued at a high rate of speed, running stoplights in a vain attempt to pick up a police escort, before finally arriving at an office across from the Mercy Hospital, where they were met by a physician who specialized in conditions of the respiratory tract. The doctor took the emergency in stride, and quickly and expertly removed the object, which proved to a small round metal insignia button with a bendable tab which Marla had brought home from school.[24]

By this time Royce and Pearl had begun construction on the new family home, which was to be situated on an inside corner of a new U-shaped street called Mulberry Lane, located a block west of Anderson Road, at the western edge of town. Perhaps remembering the hasty workmanship of the Reseda house, Royce made frequent visits to the construction site after hours, driving in additional nails to shore up the framing, and pouring DDT insecticide around the foundations to ward off termites.[25] By January of 1957 the new house was nearing completion, and the family planned their move for Saturday, January 26.

As in Reseda, the weather took a bad turn just in time for the move. On Friday, January 25, Davis was hit by a rare snow storm, and a north wind brought bitter cold on Saturday as the family made multiple trips to carry their belongings from the old house to the new. A few small items had been left behind at the old residence, and on Sunday after church services Royce returned with Jean and Florence to the E Street house, taking with him a pile of old papers and newspapers to dispose of in the fireplace there. The papers made a merry blaze on the hearth which warmed the house nicely, but as Royce and the girls pulled away to drive home, Florence looked back and saw smoke pouring from the roof, which then burst into flame. The high winds had sucked the burning paper up the chimney and onto the roof, setting it on fire, and the stiff wind was now fanning the flames. Turning quickly back, Royce jumped into action and had the fire out by the time the fire department arrived, but it was the third house fire in just over two years, prompting the owner, Jim Stiles, to remark with a laugh that Royce didn't need to feel he had to burn the house down just because he was moving away.[26]

The new house on Mulberry Lane was a pleasant and roomy home with a unique and award-winning design. Built in a rough "L" shape with an attached garage, it had three bedrooms, as well as a study which could be partitioned off from the living room by an accordion screen, to serve as bedroom space to accommodate the growing family. The house had solid hardwood floors, and a large free-standing closet just inside the front door formed a sort of entry hallway. The house's location on the inside corner of the street gave it a small front yard, but an ample back yard enclosed by a fence, which at a point farthest from the house formed a sharp angle. Royce partitioned off this fenced corner as a sand box for the younger children to play in, and near it he planted a redwood seedling scarcely taller than his son John, then a toddler, though it would quickly grow to become the tallest tree in the neighborhood. The spacious back yard was ideal for gardening, and as Royce began filling in the yard with a lawn, trees, and a garden the following spring, he carefully worked around the California Poppies which he found growing wild there.

Finances were still tight for the Bringhurst family. Royce's parents had provided a loan for the down payment to purchase the house, and had helped out further by canceling their debt on the old Buick. Though the house was roomy, it lacked furnishings, and Royce was unwilling to purchase furniture on an installment plan, so the family had to make do with what they had. Pearl acquired some pieces of secondhand furniture, which she quickly learned to sand and refinish to a high standard. Royce used the carpentry skills he had learned as a Utah farm boy to construct a nice kitchen table with an end which hinged downward to make more space when not in use, and a long matching bench on one side for the children to sit on, with storage space under a hinged seat. He also constructed a workbench in the garage, and a set of cupboards to accommodate the canned fruit which was an integral part of the family diet. Pearl sewed curtains for

320

the bare windows, and the house was gradually converted into a comfortable home. When the family first moved in, Royce did not allow fires in the fireplace, remembering their series of near disasters in the E Street house, but gradually he relented, and the hearth often had a cheery blaze which brought relief from the cold on dark winter evenings.[27]

In February, 1957, just two weeks after the Bringhursts moved into their new home, the Woodland Ward of the Church was divided, and a new Davis Ward created. Royce, who had distinguished himself as a fine teacher, was immediately called as Sunday School superintendent of the new ward, and Pearl as Sunday School chorister and Primary teacher. The new Davis Ward, with an attendance of 105 people the first week it met, was large enough that the house on B Street which Royce had helped to locate for it was scarcely adequate, and the congregation was moved to the more spacious Davis Odd Fellows Hall, the same solution which had been resorted to in Woodland while the chapel was being built there.

As soon as the ward was formed, the members turned their energies to raising money to build a chapel. As in Reseda, a large proportion of the building cost was to be borne by the local congregation in the form of a ward building fund assessment, which had to be collected before construction could begin, and Royce was assigned as co-chairman of a ward finance committee charged with collecting these funds. The committee made assessments for contributions from individual families in accordance with their financial circumstances, and they also arranged a long series of fund-raisers ranging from rummage sales, to paid dinners, to sales of roast chicken in the central park, barbecued in steel drums and sold to the public. Members also performed services in exchange for contributions into the fund, and Royce and Pearl hired ward members skilled in cement work to lay an attractive patio in their back yard, the payment for which went into the building fund. By 1960 enough had been collected to authorize construction on the new chapel, and an empty lot on the corner of West 8th Street and Elmwood was purchased and dedicated to that purpose. Once again the members went to work, this time in the construction itself. Royce began to spend evenings and weekends at the chapel site working as a laborer with other men in the ward, in addition to scraping together $300 as his family's assessment for the building fund.

Besides his university and church obligations, during this time Royce also found time to involve himself in music on a larger scale than before. As he sang for church and community events, his remarkable tenor voice, now developed to its full strength, began to be better known, and in 1959, he was invited to join the Sacramento

Symphony Chorus, led by Alex Gould, a tenor who, after making a name for himself as a popular singer, had settled in Sacramento and taken on directorship of this high-level community choir.[28] The Symphony Chorus exposed Royce as never before to challenging and powerful choral music by master composers, greatly deepening his already strong love of classical music. For his part, Alex Gould was quick to recognize the quality and operatic strength of Royce's voice, and Royce began to take vocal lessons from him, and often practiced singing at full voice while working in the back yard, something his children learned to take in stride. His voice was powerful enough that Alex felt it required a full orchestral accompaniment, and he arranged for Royce to sing some solos with the Symphony Chorus.[29] Pearl also eventually joined the Symphony Chorus and began taking voice lessons of her own, and as Florence became old enough and developed sufficient musical skill, she too would join. Music became an important focus for the entire family, and they participated, either jointly or individually, in many organized musical events in the Church and community. Royce was frequently in demand to sing for special occasions, often at short notice, and he nearly always obliged these requests.

With a growing family, life became increasingly busy. The older girls were now attending the high school, which was held in an aging red brick building in the heart of town, and were involved in other activities as well, such as musical groups and 4-H projects. Ann had begun school at West Davis Elementary School, two blocks from the house, while John was started in a nursery preschool adjacent to the university campus. Saturdays were occupied by yard work and household chores, often including the hated task of cleaning the garage, which always seemed to be in a state of complete disarray. The one time the family was able to get together free of other concerns was Sunday, when they attended church together, and in the evening held "family nights" when they would talk together while eating snacks or sipping hot chocolate, at times gathering around the piano to sing.

Every summer they continued to make their annual pilgrimage to Utah to visit relatives, generally alternating between the home of Royce's parents on Vine Street, and the home of Helen and Mont Toronto, Pearl's sister and brother-in-law, who now housed Pearl's mother in a basement apartment. On one of these trips, in 1957, the aging family Pontiac, whose cooling system had developed a tendency to overheat on an uphill grade, finally boiled over completely as they were climbing to the summit over the Sierras with Royce at the wheel, stalling the family on the two-lane highway with traffic backed up behind them. This was the last straw for Royce, and shortly after their return to Davis he began a search for another vehicle. This time he was determined to purchase a new one, and he finally found one at a bargain price: a '57 Chevy station wagon, complete with the iconic tail fins, matching winged nose cones on the hood, and twin rubber-tipped cylindrical projections from the front bumper. Probably getting a discount for the color scheme (pink with white roof and trim), Royce had no way of knowing that he had bought a vehicle which would one day be regarded as a vintage classic of the "hot rod" era. The pink Chevrolet station wagon proved to be a wise purchase, and would serve the family well for many years.

The family continued to have their share of minor mishaps, just as they had in the E Street house. One windy day in late Spring, the front screen door slammed shut on Marla's finger, severing the tip, so that the doctor was obliged to remove a piece of the bone, leaving her with a mildly deformed fingertip.[30] Not long after, the screen door claimed another victim when three-year-old John, returning home from a walk alone down the street, caught his foot in it while his protective sister Ann was pulling him back to the safety of the house, breaking a bone in his leg. The family doctor once again was called upon and came to the rescue, placing John's leg into a long-leg walking cast which barely slowed the energetic boy down.[31] A year later, Florence broke her leg in a diving board accident, an injury which proved to be much more serious, resulting in a long hospitalization and a dangerous bout of infection. Her leg had to be pinned and casted for nearly half a year, so that even when she returned from the hospital Pearl was obliged to drive her to and from school for many months. On that summer's trip to Utah, unable to fit in the car seat with the other children, Florence occupied the rear compartment of the station wagon, with her casted leg stretched out diagonally.[32]

Royce's professional duties had by now increased considerably. The strawberry program, already large and complex, was undergoing expansion, and new strawberry varieties were being tested and prepared for release, while at the same time Royce was about to embark on a major new study of wild strawberries. He continued to teach two undergraduate classes at the university, and was supervising graduate students who assisted him in his research. He began service on a scholarship committee, charged with setting the standards for scholarships, selecting the candidates, and holding social gatherings for the recipients,[33] while at the same time he served as an individual faculty advisor to undergraduate students. A new building was under construction for the Pomology Department, and preparations had to be made to move all the academic offices and laboratory operations into the new facility.[34] Meanwhile the Bringhurst house continued to

323

serve as a home away from home for Latter-day Saint students at UCD, who were frequently invited for dinner.

By 1960 Pearl was pregnant with her sixth and final child[35]. It was a busy time as the birth approached—most of the family members were involved in musical productions, and at the university the new department building, Wickson Hall, had just been completed. Royce, who had already been struggling with a fungus disease in that year's crop of strawberry seedlings, was still transitioning to a new office and lab facilities as the birth approached. Pearl's growth during her pregnancy was more rapid than expected, and for a while the doctor suspected twins, but on April 2, 1960, she gave birth to a baby girl weighing 9 pounds 2 ounces.[36] The birth was a difficult one, and the doctor was obliged to assist with instruments, but the baby was healthy and robust, with silky dark hair. They named her Margaret.

Margaret's birth marked something of a transition point for the family. Pearl, now in her forties, was not destined to bear any more children, so with five daughters and one son, the family had reached its full size. Jean had already surpassed Pearl in height, and would graduate from high school the following year. Rather than remaining in California, she was determined to leave home to attend college, and perhaps remembering her childhood in Logan, she applied and was accepted to the same college Royce had attended, which had since been expanded and renamed Utah State University. Royce made the long drive with her in September of 1961, and after he dropped her off in Logan, Jean would never live at home again, aside from yearly visits to Davis for summer employment. Florence and Marla were by that time in high school, and it would not be long before they, too, would be applying to college and beginning the process of moving away from home. The younger three children, Ann, John, and Margaret, would be raised almost as a second family, and unlike the older girls, none of them would remember any home but Davis.

A few months after Margaret was born, another event marked the passing of generations. On February 1, 1962, Royce received a call informing him that his father, John Tripp Bringhurst, had died unexpectedly at the age of 78. Royce hurriedly made arrangement to make the long drive to Salt Lake, accompanied by Pearl and the two youngest children. He had dreamed about his father's death not long before, and now learned that his father, still in reasonably good health, had simply left the house to talk to a neighbor, and while returning home, had collapsed near a tall pine tree he had planted some twenty years before, dying instantly. Snow had fallen as he lay there, leaving an imprint of his body on the ground which was still visible when Royce arrived there. His father had designated Royce and his younger brother George to take charge of family affairs at the time of his death, and they found them in perfect order.

The funeral was held on February 5, and was a time of deep reflection for Royce. He later wrote, "The last impression I have of my father is of his hands as he lay at rest in his coffin. His hands, gnarled and worn from a lifetime of hard work, were folded across his chest. I regretted that I could not tell him what was in my heart at that moment, how I appreciated the many personal sacrifices he had made for me."[37] Royce, together with his mother and the rest of the family, was surprised at how many people attended the funeral, and the large number who regarded his father as a kind personal friend. As Royce returned with Pearl and the children to California, a winter storm struck the Sierras, and Royce was obliged to bind ill-fitting borrowed snow chains to his tires with wire to make the perilous transit over the summit through the driving snow.[38] John Tripp Bringhurst's death marked an important personal transition point for Royce, though back in Davis his life would carry on as busily as before.

Early Work in the Strawberry Program

In approaching his work as a strawberry breeder, Royce began, as he had with the avocado, addressing the difficulties involved in making the parent crosses. In the case of strawberries, the main problem was not pollination as it had been with the avocado, for it was a relatively simple matter to make controlled pollinations and harvest the fruit that resulted. With strawberries, the main difficulty was in getting the seeds to germinate once they had been collected. From the published literature, Royce was aware that strawberries germinated better if pre-treated with potent chemicals such as sulfuric acid, but he also knew that like many plants of the same family, strawberry seeds required a period of "stratification," or exposure to cold temperatures, in order to germinate adequately. Since there were no published guidelines as to how to go about this, starting on his arrival in September, 1953, Royce began a series of experiments to determine how long a period of stratification was required to germinate seeds from the various varieties. He found that about four months of continuous chilling was required to produce consistently acceptable rates of germination, and that to avoid disease in the seedlings, he had to use a lower temperature than had been described by other researchers. Royce and Victor published these results in 1956,[39] and they served as guidelines for Royce as he continued his breeding work.

It would take years of meticulous breeding and selection before Royce and Victor would be ready to introduce new varieties of their own for the use of growers, and in the meantime they turned their

325

attention to other problems that plagued the industry. One of the first of these was high salinity, caused when the perfectly flat fields traditionally used for strawberry growing were irrigated in conventional furrows. The irrigation water in the furrows tended to leach salt and minerals from the soil to the surface, which, as the water evaporated, were deposited in high concentrations toxic to the plants. This had become a serious problem for strawberry growers, and as a remedy for the situation Victor and Royce began to recommend sprinkler irrigation for young strawberry plants until they began to set fruit. Unlike conventional furrow irrigation, which leached soil minerals to the surface near the roots, sprinkler irrigation had the opposite effect, leaching the excess minerals deeper into the soil where they did no harm to the growing plants. Although based on good experimental data, sprinkler irrigation flew in the face of conventional wisdom: furrow irrigation was much simpler and cheaper, and in a crop highly susceptible to fungal diseases, it also kept the foliage dry. It was a challenge to persuade growers to adopt the practice, but when they did, the problem of high salinity disappeared virtually overnight, and there appeared the unexpected benefit that strawberry crops could now be planted and irrigated on rolling hillsides, rather than on the carefully flattened fields traditionally required for furrow irrigation.[40] The immediate success of sprinkler irrigation gave growers new respect for the University scientists and opened the way for the introduction of other advances.

Another similar innovation was the use of polyethylene mulch.[41] Royce and Victor were aware of experiments which had been performed with other crops using plastic sheeting as a ground cover, and in the 1950s they began experimenting with its use for strawberries, trying out various colors of polyethylene and observing the results. The polyethylene sheets were obtained in long rolls which were applied by tractor the full length of a crop row, with the edges turned under the soil and with perforations to allow the plants to grow through the plastic. Victor and Royce found that under the right conditions, clear polyethylene sheeting greatly improved fruiting qualities of the plants by warming the soil around the strawberry plants during the cool winter months, allowing much more vigorous growth as the plants were preparing to bear fruit. The result was a significantly earlier start to the spring crop, which was a primary goal in the Southern California growing areas, which were the focus of their work. As with sprinkler irrigation, there were other benefits to the practice as well, for the plastic did an excellent job retaining moisture in the soil, cutting down on the need for irrigation, and further limiting the accumulation of salts from evaporation.[42] Most importantly, it greatly improved fruit quality, since rather than developing directly on the soil where they were prone to rot, the strawberries ripened neatly on top of clean plastic. Once again, an expensive innovation led to greatly improved fruit yield and quality which more than offset the increased cost. The polyethylene mulch system was quickly adopted by Southern California farmers, and with greater experimentation and a little persuasion from Victor and Royce, it was not many years before the entire California strawberry industry had turned to the use of plastic row covers. Strawberry fields in California became easy to spot by the long parallel lines of clear plastic sheeting which marked the rows.[43]

With similar careful experimentation and the application of sound scientific principles, Royce and Victor began recommending other changes which were just as revolutionary and which challenged long-established practices. Because strawberries are a perennial plant, the customary practice had always been to make an initial planting, then allow the plants to remain in place for two or more years, continuing to harvest the fruit while allowing runners to develop to provide a new generation of younger plants. Royce and Victor, observing the decrease in yield which invariably occurred after the first year through the depletion of soil nutrients and the accumulation of plant diseases, began recommending yet another costly innovation, which was to plant out an entirely new set of plants on an annual basis, not allowing them to remain in the ground a second year. This necessitated the increased cost of more frequent planting, and required nurseries capable of providing large stocks of disease-free plants on an annual basis, but once again the benefits far outweighed the increased cost, and growers began more and more to turn to annual planting of strawberries.

Yet another area of investigation for Victor as well as Royce was the timing of strawberry planting. All commercial strawberry fields were sown with plants of a given variety, purchased from nurseries which had grown the plants during the previous months, then harvested them and stripped off the foliage, leaving only a crown with roots which could be stored, shipped, and planted out by growers. The traditional system for strawberries, which had originated in areas of the country with much more severe winters, was a spring planting system, in which these purchased nursery plants were laid out during the spring months, with the expectation of fruiting the following year. The mildness of the California climate meant that strawberries could be planted at other times of year, and in the decade prior to Royce's arrival, growers in mild coastal climates had shifted to a winter planting system, in which plants were set out from November to January, allowing fruit to be harvested the following spring. Royce and Victor began experimenting with different planting times for

327

these winter plantings, and found that the timing of planting greatly affected the overall yield of the strawberries, as well as the time that the fruiting occurred. The timing of harvest was especially important for Southern California, whose mild winters enabled the ripening of strawberries much earlier in the year than other areas, at a time when even a smaller initial crop would be profitable.

Very early in these experiments, Royce and Victor recognized the importance of winter chilling in the subsequent fruiting behavior of the plants. In making their early comparisons, they noted that nursery plants which had been grown at higher elevation nurseries often grew with greater vigor and had higher yields than those from low elevation nurseries where they had received less winter chilling. Royce and Victor conducted experiments to determine how early in the winter nursery plants could be dug up for winter planting and still have enough vigor to produce fruit the following season. At the same time, they began to experiment with an entirely new summer planting system, in which nursery plants dug during the winter were placed in cold storage until the following summer, when they could be planted out. This innovative system, which allowed a harvest both in the fall and the following spring, greatly increased yields in Southern California, and also opened the possibility of growing superior varieties in the south, where previously only the inferior Lassen had been successful because it had a lower chilling requirement.[44] The success of these different planting systems, and the timing of the harvest which resulted, was strongly dependent on the variety of strawberry which was planted, so as new varieties were developed, prior to releasing them Royce and Victor would subject them to rigorous testing in different regions using the different growing methods, to see how they responded, and where in the industry they might fit in, if at all.

Of all the challenges addressed by Royce and Victor during the 1950s, none was more daunting than the problem of plant diseases. There were a multitude of these which abounded in the mild California climate, some fungal, some bacterial, and some viral. The bacterial and fungal diseases, which generally lived and multiplied in the soil, could be kept at bay in part by controlling the growing conditions and by planting strawberries in uninfected soil. Viral diseases, on the other hand, infected the cells of the plants themselves, and were passed on to daughter plants as the plants were propagated. Spread by insect pests, the viruses usually did not kill the plants outright, but often caused them to lose vigor and become unproductive.

The main defense against viral diseases was to ensure that nurseries which supplied the growers used plants which were certified to be as free as possible of virus, and the state agricultural department, in consultation with experts like Royce and Victor, maintained an official certification system to perform this vital function. Another important measure was Royce and Victor's recommended practice of annual planting, which kept viruses from accumulating in any given field, and the control of insects such as aphids, which carried the viruses, was also important. Once virus was present in nursery or breeding stock, it could sometimes be inactivated by heating the plants, and Royce began experiments with heat inactivation as early as 1954.[45] A later complementary method for eliminating virus from nursery stock was by use of meristem culture, which involved growing new plants from virus-free cells at the growing tips of infected plants. Royce used the tips of strawberry runners as early as 1959 to establish virus-free clones,[46] and this would become an important technique in the years to come.

To monitor for the presence of virus in the field, Royce and Victor made use of the plants of the alpine strawberry, a variety of the diploid *Fragaria vesca*, which were highly sensitive to virus diseases, as indicator plants in some of their plantings. If a virus was present in a given field, the indicator plant would get sick or die. It was a more difficult matter to tell if an individual plant carried a virus (a process known as "indexing"), for this required grafting a portion of a stolon, or runner, from the affected plant onto a stem of the alpine *Vesca* plant, and this could be done only if both plants were in vigorous health, which was often not the case when a virus was present. Royce had developed good skill at grafting during his graduate work at Wisconsin, and working with Victor, he developed a much better technique which involved grafting a leaflet from the sick plant in place of one removed from the indicator plant. This enabled him to tell in as early as a couple of weeks if the plant was infected. Among other things, this method allowed those working in the fields to simply send leaf samples for testing when they suspected a planting was infected with virus. The technique could also be used to check mother plants for the presence of virus prior to propagation, and to certify nursery stock as free from important viruses. This new virus indexing system was quicker, more adaptable, and more effective than systems which had been in use prior to that time. It proved to be a significant advance, and quickly became the standard not just in California but around the world. Royce and Victor published a description of the technique in 1956, which received recognition that year as the most important piece of research to come out of the UC Davis Pomology Department.[47]

Of the various soil-borne diseases, none was more serious than *Verticillium* wilt, a fungal disease which in the right environmental setting could devastate a field of strawberries. This fungus could persist for years in any soil where a susceptible crop, such as strawberries or tomatoes, had been previously grown, and this required growers to to shift the location of their fields for each new strawberry planting. Much of the early breeding work was aimed at developing plants resistant to the fungus, but this always seemed an elusive task. In the end the problem was solved by new methods of soil fumigation, pioneered by a plant pathologist from Berkeley named Steve Wilhelm.

Fumigation was another complicated and expensive technique with huge potential benefits. A field was completely covered with a plastic tarp, and the soil was infused with a combination of a natural but highly toxic gas called methyl bromide, and a synthetic one called chloropicrin, a potent form of tear gas. After a day or so when the gases had completely saturated the soil, the tarp was removed and the field prepared for planting. This method essentially sterilized the soil, not only eliminating *Verticillium* fungus, but also controlling weeds, nematodes, and other pests. Even more importantly, it allowed the same soil to be planted with strawberries year after year, without the usual accumulation of diseases and pests which had troubled the industry in the past. This made it no longer necessary to locate virgin land for strawberry plantings, which was increasingly difficult as coastal agricultural areas underwent urban development. Fumigation was highly successful, and it was not long before it was adopted throughout the state.[48]

Hence, a long series of revolutionary technological innovations, developed or adapted, investigated, and then promoted by Royce and Victor during the 1950s and early 1960s, began setting the stage for the tremendous resurgence of the California strawberry industry even before the first of their strawberry varieties was released. The introduction of sprinkler irrigation, of polyethylene mulch, of annual planting, of new seasonal systems of planting, of careful attention to chilling, of soil fumigation, and of new methods of disease detection and control, combined with basic research into plant nutrition and growing techniques, all laid the groundwork for the success of a long stream of new varieties which would shortly emerge from their breeding program.

Perhaps most importantly, Royce and Victor had gained the respect and cooperation of the growers themselves. Both men had grown up around agriculture, and both felt a sense of accountability to the growers who provided the funding for their work, and whose livelihood was at stake whenever they chose to break with longstanding traditions and practices to follow the scientists' new and often radical recommendations. As they continued their work, Royce and Victor communicated frequently with the California Strawberry Advisory Board, an organization established by the state legislature to coordinate the activities of the university, the government and the industry.[49] This sometimes contentious group provided the main financial support for their work and kept them abreast of the problems and challenges of those who made their living growing strawberries, and in return Royce and Victor submitted periodic reports of their activities to the board as a means of accounting for the expenditure.

While confident in the work they were doing, Royce and Victor also both possessed a willingness to listen, an eagerness to help, and self-confidence and graciousness in the face of sometimes harsh criticism, that generally won over their critics and made friends of potential adversaries. The astonishing success of their work, coupled with the humble practicality with which they went about it, made them champions of the growers, and the resulting strong partnership between the growers and the University program would gradually come to serve as a model throughout the country and the world.

Early Breeding Work

When he arrived in Davis in late 1953, Royce quickly took stock of the breeding material which had been left by his predecessors, and consulted with Victor Voth on what the objectives of their breeding work should be and what parent material should be used. The needs of the industry in the Central Coast area around Watsonville were largely being met by Royce and Victor's predecessors Thomas and Goldsmith. Building on the success of the 1945 varieties and making use of breeding stock they had developed in the university program, they had already made much progress on developing everbearing strawberries to complement the short-day Shasta variety which was still dominant in that region. The University program lacked any research station in the Central Coast region, and rather than attempting to compete directly with the work of Dr. Thomas and Earl Goldsmith at Driscoll Associates which was headquartered there, Royce and Victor turned their attention to the needs of the growers in Southern California, which was increasingly becoming an important area of strawberry production.

Compared with the Central Coast, the Southern California area had a milder climate, which presented both opportunities and challenges. On the one hand, the mild winters allowed growers in Southern California to put strawberries on the market earlier than their competitors further north, when they could command a higher price. On the other hand, the lack of winter chilling meant that they could not make use of superior varieties such as Shasta, and instead

331

had to rely on the other major 1945 variety, Lassen, which, though extremely productive and well adapted to the milder climate, produced fruit of a much poorer quality. Royce and Victor realized that there was a serious need for superior fruiting varieties for Southern California, and Royce immediately recognized the key importance of Lassen as a parent, despite the inferior quality of its fruit. Starting upon his arrival in 1953, he began to breed crosses in an attempt to combine Lassen's good bearing capacity and its adaptation to the Southern California climate with the better fruit quality found in other varieties.

As Royce and Victor worked to select improved cultivars, they also evaluated some of the selections left over by Thomas and Goldsmith, and in 1958 they decided to release a new variety, designated 35.93-11, which had not performed well in the Central Coast, but which bore high quality fruit in Southern California. They called the new variety "Solana," and though it had drawbacks in earliness and productivity, it at last gave Southern California growers a berry adapted to their climate which was of high enough quality to compete with Shasta farther north.[50] Meanwhile, Royce and Victor continued field evaluation of some of the first crosses they had made using Lassen as a parent when Royce first arrived in 1953. Their breeding program involved a number of steps over several years. In this case, the seed from the original crosses made in 1953 were chilled, then sown on special growing medium, and after germinating, were planted out and eventually established in experimental plots in Winters, Lancaster,[51] and Torrey Pines. By 1955 the plants had been evaluated sufficiently to determine which ones had the most promising fruiting characteristics, and these were selected out and propagated as clones for further testing. Plants from each clone were planted out in plots at the different university experimental stations to test them under conditions that existed there, and they were carefully observed for their fruiting and other characteristics, which were then recorded and tabulated. Those clones that performed best in these preliminary tests were then distributed to willing commercial growers, who planted them out in experimental blocks in their fields.

By the end of the 1950s a number of promising clones had made their way through this long process to the final stages of field testing. In 1960 some of these were provided to the nurseries for propagation, and in 1961 Royce and Victor officially released three new varieties, which they named "Fresno," "Torrey," and "Wiltguard."[52] Fresno and Torrey both had originated from the first series of Lassen crosses Royce had made back in 1953, and had been carefully tested in the mild winter areas of Southern California. Fresno, named after a city in the Central Valley where it had

332

performed well,[53] was just as early and as well-adapted to Southern California's climate as the old Lassen variety, but it had larger and more attractive fruit which not only tasted better, but was also firmer and therefore shipped better. Torrey, named after the Torrey Pines field station where it had been selected, fruited even earlier than Lassen in the coastal areas of Southern California, and showed similar improvements in fruit quality. Even though neither variety was as productive as Lassen, the improved fruit quality, coupled with the new growing methods pioneered by Royce and Victor, made their release a major advance in the Southern California strawberry fields. It was not long before Fresno had replaced Lassen as the dominant variety in Southern California, except in a few areas near the coast where Torrey performed better. Lassen, with its large crops of insipid fruit, at last began to fade into history as a commercial crop.

Wiltguard, released at the same time as Fresno and Torrey, was part of an ongoing effort in the University program to breed for resistance to *Verticillium* wilt, which had caused such problems with commercial strawberries during the 1950s. The original cross which led to Wiltguard had been performed by Dr. Baker in 1952, a year before Royce's arrival, but in evaluating the new seedlings, it was Royce and Victor who made the selection and performed the testing on the variety, a process which bore the distinctive stamp of Royce's methodical mind. In their earlier efforts to breed for resistance to the *Verticillium* fungus, Dr. Baker and Victor Voth had simply planted resistant and sensitive clones in soil known to harbor the disease, an inefficient and cumbersome process. Consulting with Stephen Wilhelm, the Berkeley plant pathologist who had developed the fumigation method used by the growers, Royce devised a much more efficient method to determine resistance to the fungus. Using six different strains of *Verticillium* provided by Steve Wilhelm, Royce experimented with directly inoculating each plant with this potent mix of strains prior to planting, to make certain that each had approximately equal exposure to the fungus, then planting them in fields which had been fumigated to eliminate any previously-existing fungus in the soil which might confound the results. He also developed a well-defined scoring system for the plants as they grew, and scored the plants not just once, but repeatedly during the growing season. This new, more efficient testing method was used to test Wiltguard and other *Verticillium*-resistant selections, and Royce published his results in 1961, the same year the new varieties were released.[54]

While Fresno and Torrey succeeded commercially, Wiltguard did not. Although it had some good qualities, Wiltguard had an unfortunate tendency to produce fruit which split during ripening, making it too high a risk for commercial growers, who were now able to control the *Verticillium* threat with soil fumigation. Royce continued to work with *Verticillium* resistant plants and to study the factors which gave them their resistance, and found that it was relatively easy to breed for disease resistance, but he also discovered that high levels of resistance were correlated with poorer yield, which explained why nearly all the clones which made it through the selection process were sensitive to *Verticillium*, even when both parents had been resistant.[55] As a result of these experiences, Royce gradually came to the conclusion that while disease resistance was important, it should never be the primary emphasis of a breeding program. His experience with *Verticillium* resistance had clearly shown that if disease resistance were used as a primary screen for new varieties, it was probable that the desirable fruiting qualities which are essential for success might never be observed or selected in the offspring. In the end, Royce advocated breeding disease-resistant parents whenever possible (and he did continue to maintain and use wilt-resistant breeding stock), but to breed specifically for fruit quality rather than disease resistance, while making use of good cultural methods to control plant diseases.

As he made and tested these early crosses, Royce also began applying the developing tools of the geneticist to his work on strawberries. As he had with avocados in Los Angeles, he experimented with colchicine to artificially increase the chromosome number of strawberries, to see if this technique might hold promise in future breeding. He also began experimenting with crosses between commercial strawberries and the various species of wild strawberry, including clones of wild coast strawberries which clung to the California coastline. He experimented with crosses between the octoploid commercial berry and the diploid *Fragaria vesca* species, which resulted in sterile but vigorous plants,[56] and he also performed crosses with other octoploid species including *F. ovalis* plants such as the ones he had collected in Brighton, Utah,[57] and even made crosses with the closely-related genus *Potentilla*, in an effort to explore different genetic possibilities.[58] He also tried crosses with wild and domesticated strawberry selections sent to him by other researchers. One cross, with a South American strain of *F. chiloensis* cultivated in Ambato, Ecuador, produced fruits of impressive size, nearly twice as large as the best commercial strawberries, and although the plant and its fruit had many deficiencies, Royce and Victor's published report of it prompted a flood of requests for plants from breeders all over the United States and from many other countries, who were trying to incorporate large fruit size into their own breeding programs.[59]

Royce and Victor's breeding program had by now grown to an

impressive scale. The crosses they carried out yearly resulted in as many as fifteen thousand seedlings, of which at least a couple of hundred would be selected for further work. A major challenge they faced was compiling and processing the massive amounts of data required to make appropriate selections of clones which were likely to lead to improved varieties, for the sheer volume of information threatened to overwhelm the capacities of the researchers. Royce realized that work of this magnitude required something more than even experienced human judgment, and he worked to devise systems to objectively measure and score the various performance traits that went into an excellent strawberry cultivar, including size, color, firmness, vigor, and even flavor (in the form of acid and sugar content). He developed a series of data sheets[60] to tabulate this information in the field, and then had the data punched onto IBM cards[61] for analysis on a computer. Although computer systems were in early stages of development and were primitive by later standards, they greatly simplified the formerly laborious process of analyzing data, and Royce became a strong and early advocate for their use in plant breeding.[62]

As Royce and Victor reached the point of actually testing and releasing new University varieties, they increasingly felt the need for permanent experimental stations in the Central Coast area of California, the traditional stronghold of the strawberry industry. The lack of a good experimental station limited their ability to test their strawberry selections or to investigate what cultural methods would be most successful in this important region. They eventually succeeded in acquiring a portion of a sugar beet research station in the Salinas area, and in leasing land from a Seventh-Day Adventist school near Watsonville. The Watsonville land, being owned by a religious group, came with the stipulation that no smoking take place on the property, and that no manual work be performed on Saturdays, and both conditions were more than acceptable to Royce.[63] Royce assumed command of these two new field stations in addition to the Wolfskill Experimental Orchard in Winters, near Davis, while Victor continued to run the South Coast Field Station, which had been moved in 1956 from Torrey Pines to Santa Ana, the same area where Royce had undergone his pre-flight training during the war.

With this new expanded capacity, Royce and Victor continued field testing on their first generation of crosses which had led to the successful release of Fresno and Torrey, and in 1964 they released still another strawberry variety which they named "Tioga," after one of the famous mountain passes in the high Sierras of California. Tioga was actually a sister plant of Fresno and Torrey, though unlike them it had originally been grown and tested at the station in Winters rather than in Santa Ana. Of all the University varieties, Tioga was by far the most productive, outperforming even its parent Lassen, and its fruit was larger and firmer than nearly all the other varieties then in use. Although originally developed as a replacement for Lassen in Southern California, Tioga performed so well in the Central Coast region that it was recommended for use there as well.[64]

With the release of the Tioga variety, Royce had at last come into his own as a plant breeder. The first true breeding triumph of the Bringhurst-Voth team, Tioga quickly became the dominant strawberry variety in the entire state, replacing even the time-honored Shasta variety, a change which greatly increased the already high yields in California. Its value was recognized in other regions as well, and before long it became the most cultivated strawberry variety not just in the United States but throughout the world, bringing increased recognition and respect to the California breeding program.

Wild Strawberry Investigations

As Royce worked with his strawberries, his interest turned to the various colonies of wild strawberry which inhabited the different regions of the state. There were three distinct strawberry species found growing wild in California. Two of these were octoploid (having eight sets of chromosomes), including a high-altitude strawberry then referred to as *Fragaria ovalis*, the type Royce had collected in Brighton, Utah, and which was also found scattered in the Sierra Nevada mountains, and the coastal strawberry, *Fragaria chiloensis*, which was found only along the Pacific coastline. The third species, the diploid *Fragaria vesca*, (with only two sets of chromosomes) was the so-called woodland strawberry, which was found widely in the Northern Hemisphere of both the Old and the New World. Between his pressing duties, Royce collected and studied these plants whenever he could, occasionally sharing specimens or seed with other experts in the field. One of these was Günter Staudt, a German taxonomist who was involved in an ambitious project to describe and categorize every species of strawberry worldwide. Royce had hoped Staudt would visit California to observe the strawberries in their unique habitat there, but when circumstances did not permit the visit, he sent plant samples of the various species, and the two corresponded about the appropriate classification of the species. Staudt eventually came to the conclusion that the mountain strawberry generally referred to as *Fragaria ovalis* really belonged to the original *Fragaria virginiana* species found in the eastern United States, though because it had unique properties he designated it as a subspecies which he called *glauca*, meaning bluish grey or green. Royce agreed with this designation, and from that time forward he referred to the plants he had collected as *Fragaria virginiana-glauca*.[65]

Of special interest to Royce was the wild coastal strawberry *Fragaria chiloensis*, the octoploid which grew in tenacious colonies all along the California shoreline. He became increasingly convinced that it was the inbreeding of these wild plants by his predecessors which had given the University strawberry varieties their unique adaptation to the arid climate, high salinity, and warm winter conditions in California, as well as resistance to the viruses and soil diseases which abounded there. (He would later subject many of these wild strawberries to his standardized *Verticillium* challenge, and found clusters of clones which were highly resistant to the fungus.[66]) Passing by the coast on his trips up and down the state to supervise the strawberry program, Royce stopped as often as occasion would permit to observe and collect wild plants. He became fascinated by the many ways in which they had adapted themselves to their often harsh surroundings, and felt a strong desire to study this in greater depth, though in the midst of his pressing responsibilities with both the strawberry program and the university, he lacked the time or resources to do so.

In January of 1960, Royce received a visit from a man from the flower industry named Neff, who came inquiring into the strawberry breeding program on behalf of Henry A. Wallace, the former U.S. Secretary of Agriculture whose name was doubtless familiar to Royce. Royce gave a description of his breeding program, and at Mr. Neff's request, sent a letter to Wallace giving a brief outline of the program and its objectives, and including a few reprints of articles which had been published about it. He then returned to his work, likely thinking little more about this episode. Royce frequently received visits from interested individuals like Neff, and while he was always cordial and exchanged information whenever he could, he probably did not assign the visit much importance.

A year later, Royce received an unexpected hand-typed letter from Henry A. Wallace himself, requesting some plants of the wild coast strawberry, *F. chiloensis*, as well as some of the winter-bearing clones used in California. In the same letter, Wallace made an unusual offer. "Incidentally I might be in a position to furnish you or the Station $100 or so if you would care to make a few esoteric or exotic crosses which might not have any value at any time in the near future but might be of theoretical interest."[67] In his reply, Royce promised to send the requested plant material, and described in detail some of the esoteric crosses he had already made in the laboratory between cultivated varieties and wild strawberries of various species. In response to Wallace's offer of financial help, Royce wrote, "Regarding your suggestion of possibly furnishing a hundred dollars or so to help us with this work, I would be receptive since we operate on a rather tight budget and appreciate help of this sort. Naturally I cannot solicit, but when checks are sent to my department specifying that the money is to be used for specific strawberry work, we are pleased to accept it."[68] So began an unlikely collaboration which would lead to some of Royce's most fruitful work in the scientific study of the wild strawberry.

Henry Agard Wallace was an interesting figure by any measure. Son of a prominent college-educated farmer and writer, Wallace showed an early interest in plant breeding, and as a pioneer in the breeding of hybrid corn, he established the highly successful Pioneer Seed Company, which brought him both recognition and wealth. He was also an accomplished statistician and economist, and recognized the importance of applying statistical analysis to agricultural problems. In 1933, in the midst of the Great Depression, he was selected by President Franklin D. Roosevelt as Secretary of Agriculture, perhaps partly on the strength of the reputation of his father, who had held the same post in a different administration a decade before. An outspoken writer and a dedicated and thoughtful idealist, Wallace strongly influenced Roosevelt's policies, and as Secretary of Agriculture he instituted controversial programs which compensated growers for non-production in an effort to balance supply and demand, to save farmers from financial ruin. He was a strong believer in the ideals embodied in Roosevelt's "New Deal" programs, and when the president ran for an unprecedented third term in 1939, he chose Wallace as his vice-presidential running mate. Wallace was a strong and influential Vice President, but his opinions and actions, and even his unconventional religious beliefs, fostered great public controversy. He would have succeeded Franklin Delano Roosevelt as president had Roosevelt died just a few months earlier, or had not Democratic leaders, independent of the president, insisted on the nomination of a little-known senator from Missouri named Harry S. Truman as a replacement candidate for Vice President in 1944. Clashing with Truman after Roosevelt's death, Wallace was dismissed from the new president's cabinet, and after a failed bid for the presidency as a third-party candidate in 1948, he retired from public life and returned to his former successes as an agriculturalist.[69]

By 1960, Henry A. Wallace was in his seventies, and renewing an interest from his youth, he had begun to dabble in strawberry growing and breeding. He was very interested in incorporating the flavor of wild strawberry varieties into standard garden strawberries in spite of the genetic difficulties involved. He liked, for example, the famous flavor and fragrance of the alpine strawberry *Fragaria vesca*, and he had a particular fondness for the flavor of the musk strawberry, *Fragaria moschata*, which Franklin D. Roosevelt had introduced him to during a trip to Europe, it being a favorite of the president's. Although he knew the alpine strawberry was diploid (two chromosome sets) and the musk strawberry was a hexaploid species (having six sets), he hoped to find a way to introduce their unique flavors into the standard octoploid berry, and this was doubtless one of his motivations in contacting Royce.[70] Another motivation was to solicit information for what he hoped to be a definitive book on the strawberry, which he had persuaded Dr. George M. Darrow, a scientist with the USDA whom he considered to the world's foremost authority on the fruit, to write.[71] Over the ensuing years Wallace and Royce corresponded, and he would ply Royce repeatedly for information on the history and status of strawberry breeding in California, and for plant materials for use in his own greenhouse. He also repeatedly sent checks of $500 to $1000 to assist Royce in his investigations, and Royce turned these funds over to the University to be applied to his program.

The funds and encouragement he received from Henry A. Wallace gave Royce both the means and the motivation to begin a more systematic study of the wild strawberries of California. In the fall of 1961, with further funding from Wallace, Royce began a series of trips up and down the Pacific coast, locating and documenting every major accessible colony, taking photographs and making collections of the plants. Of special interest in this investigation was the adaptability of the wild strawberries to sometimes extreme local conditions, traits which could prove important in later breeding efforts. Using a yellow field journal, Royce made careful notations on the location of colonies based on local landmarks, recording where he had photographed or collected specific plants.

By December 1, 1961, Royce was able to report to Wallace, "We have already started the first collections under the accelerated program. Part of the purpose of the trip was to locate wild colonies. In this I was very successful indeed. I encountered well over 30 colonies of *chiloensis* near the sea between the city of Santa Cruz and the Golden Gate Bridge." Royce described some of the conditions he saw during these excursions: "In one case I found extensive colonization of a hillside with *F. Vesca* adjacent to the *chiloensis*. This was very interesting because there had been extreme drought in the area and the diploid *vesca* plants were almost all dead whereas the octoploid *chiloensis* appeared to be in excellent condition, indicating the relative ability of these two species to tolerate drought conditions in that area. The explorations that I made that day were very interesting, and in a way, exciting. At one point I walked for a great distance over sand dunes approaching the sea, and in the distance I

339

could hear the barking of seals. Well over in an area where almost all of the native plants were near death from drought I encountered a very thrifty colony of vigorous *F. chiloensis*. In a number of instances I found that the runner plants had actually crept almost down to where the high tide sweeps over the sand."[72]

Royce continued these excursions through the winter and spring of 1962, traveling along Highway 1, a narrow strip of asphalt winding along the beautiful and austere Pacific coastline, which for many years would become his field for study.[73] He drove wherever his car would reach, and when the car could go no further he would hike the shoreline or the rugged nearby terrain, cataloguing the wild strawberries as he went. When he encountered promising areas on private land where entry was prohibited, he contacted owners to gain access. He also contacted the State Parks department to request permission to make plant collections in areas owned by the park system, where disturbing of plant life was ordinarily prohibited. Finding likely habitats for strawberries became instinctive for him, and he became skilled at spotting them at a distance. He made note of the location and appearance of familiar colonies, and over intervals of many years would watch for them and visit them as though they were old friends.

While on one of these excursions, on May 16, 1962, Royce pulled his University vehicle into Point Sur, a wind-swept beach area extending out to a point occupied by a small hill with a lighthouse, about 25 miles south of Monterey Bay. Spotting some wild strawberries by a fence near the road, Royce exited the vehicle and began examining a colony of octoploid *Fragaria chiloensis* which was growing adjacent to colonies of diploid *Fragaria vesca* between culverts under the road. As he jotted down notes and took photographs, he noticed a third colony which appeared to have characteristics intermediate between the *chiloensis* and *vesca* colonies, and which he immediately recognized as a hybrid between the two species. Surprised, he used calipers to take careful measurements of plant parts from each of the three colonies, and found that the measurements confirmed his suspicions.[74]

This was unprecedented. Such hybrids had been created in the lab before under carefully controlled experimental conditions, but they had never been observed or described to occur naturally in the wild. Yet like the hybrids which Royce had bred in the lab, this colony, whose flowers demonstrated it to be a female clone, appeared to grow vigorously and to runner freely, competing successfully with both its parents. Exploring the area, he discovered apparently identical colonies of the hybrid on both sides of the road, indicating that it had probably been developed and established there since before the road was constructed.[75] Beside himself with excitement, Royce collected plants from each colony to take back to the lab for analysis. When the chromosomes were counted, the *F. vesca* colony had the expected number of fourteen chromosomes (two sets of seven) and the *F. chiloensis* colony had fifty-six (eight sets of seven). The hybrid colony had thirty-five (five sets of seven), exactly intermediate between the two, making it virtually certain that it was a natural cross between the two species of strawberry.

The discovery of a natural pentaploid strawberry hybrid in the wild was a significant and unexpected finding, and Royce immediately deduced that if he looked hard enough he was likely to find other such hybrids. As soon as he could, he set out for an area he had explored the previous October where he knew that colonies of *Fragaria chiloensis* and *Fragaria vesca* were growing together, at a site in the San Bruno Mountains about a hundred miles further north. Exploring the area with a guest sent by Henry A. Wallace, Royce was just preparing to get back in the car to leave when he spotted a male colony of strawberries with the same intermediate characteristics he had seen in the Point Sur hybrid, and he immediately collected plant specimens. When these were analyzed in the lab, the second clone proved to be a pentaploid hybrid just like the first. Both these clones appeared to be sterile, as the female plant from Point Sur bore no fruit during fruiting season, and the male plant from the San Bruno Mountains appeared to have little or no viable pollen. This was consistent with genetic principles, for an even number of chromosome sets were required for a plant to consistently express fertility, and these hybrids, having five sets, were expected to be genetically unstable and so generally infertile.[76] Royce wrote Wallace excitedly with news of the discovery, and mailed him a specimen of the Point Sur hybrid.[77]

The existence of two such hybrids within a hundred miles of each other had several important implications for strawberry genetics. First, it proved that hybrids between strawberry species of different chromosome numbers occurred naturally, and apparently with some frequency. Second, it showed that such sterile hybrids were able, by vigorous growth and runnering, to compete with the two parent species for space, and to form persistent colonies. The size of the Point Sur hybrid colony, and the fact that it had evidently been quite large even before the road was built in 1928, led Royce to estimate that it was at least forty years old. It was even possible that new traits acquired through hybridization had allowed the pentaploid hybrid to occupy niches unsuitable to either of the parents, an example of natural selection.

Finally, the long persistence of the hybrid colonies had important

341

implications for strawberry evolution. It had already been suggested that all the polyploid strawberry species (such as the octoploid *Fragaria chiloensis*) had originated from a single primitive diploid species, and had resulted from a series of natural redoublings of chromosomes, which occasionally occur as a chance event, combined with the sort of interspecies hybridization which Royce had observed. The long existence of these pentaploids raised the possibility that a further redoubling of chromosomes might already have occurred in the wild, producing clones of decaploid strawberries (with ten sets of chromosomes), which would likely be completely fertile and capable of intercrossing and replicating in the wild. By definition, this would mean a new species of strawberries had emerged. In effect, Royce could be witnessing the progress of strawberry evolution before his very eyes.

With the help of a bright young Pakistani graduate student named Daud Khan,[78] Royce wrote up a description of the find for publication, but rather than submitting it to the agronomy and pathology journals in which he had made his previous publications, he sent it to the American Journal of Botany, a more prestigious journal which published articles on the basic science of plants rather than their agricultural applications. In his letter advising Henry A. Wallace about the publication, Royce wrote, "I have acknowledged the support of the Wallace Genetic Foundation in the paper. Frankly, I appreciate the assistance you have given more than you perhaps realize. It has stimulated a realignment of a considerable portion of my program, in such a way that I am confident that we can make fundamental contributions to the knowledge of the genus. We are now attracting good graduate students, and this is very important."[79] In response, Henry A. Wallace wrote Royce requesting that his financial support of Royce's investigations remain anonymous, a request Royce honored for the remainder of Wallace's life. The paper, submitted in November of 1962, the year the discoveries were made, was published in 1963.[80]

Royce continued both his correspondence with Henry A. Wallace and his field study of wild strawberry populations, but it was not until 1965, the year that Wallace died,[81] that he was able to take a six-month sabbatical leave, which gave him time to make his most detailed investigations in the wild.[82] As Royce had predicted, he was able to collect and identify more than twenty different wild clones of pentaploid hybrids, and he also discovered a wild clone of hexaploids[83] (6 sets of chromosomes) and one of enneaploids[84] (nine sets). This was a remarkable find, as these colonies could only have occurred by a combination of hybridization he had already observed, together with gene multiplication through unreduced gametes (the failure of cells to reduce their chromosome count normally by half during the reproductive process).[85] These two processes were the very building blocks of new species formation, and once again demonstrated a step in the ongoing evolution of the strawberry. In a published report of this investigation, prepared with the help of a graduate student from Ceylon (Sri Lanka) named Y. D. A. ("Don") Senanayake, Royce outlined the possible genetic combinations leading to the different chromosome numbers he had observed, and others which could be predicted.[86] Taking their investigation one step further, Royce and Don Senanayake performed a series of experiments to determine the origin of the genes in some of the modern octoploid species of strawberry, and were able to make progress in describing how these species had developed.[87]

By this time Royce had discovered that some of these natural hybrids which he had previously reported as sterile were actually somewhat fertile, and by collecting the fruit and planting out the seeds, he was able to study the genetic make-up of the plants which resulted. He found a wide variety of chromosome counts in the offspring, and among them was able to demonstrate cases of "double-unreduced gametes," meaning plants which in two successive generations had failed to halve their chromosome counts normally during reproduction, a very unusual finding, and a significant one in evolutionary biology. Among the offspring he discovered a few that had reverted to a normal octoploid genome (a process known as introgression), and in some of these he was able to demonstrate that there had been gene flow from the natural diploid (*Fragaria vesca*) into the resulting octoploid. He described many other variations among the offspring, and their implications in strawberry evolution, and much later published the results with the assistance of yet another graduate student.[88]

In these landmark studies, Royce had expanded his expertise in strawberries from the purely applied sciences of plant breeding and agronomy, to the more basic scientific fields of plant genetics and evolutionary botany, in which his work had now made him, for the strawberry at least, a leading world figure. In the first of these papers, published in 1966 after Henry A. Wallace's death, Royce was at last able to include a grateful public acknowledgement of the financial assistance he had received from the former Secretary of Agriculture. And in later years, the German taxonomist Günter Staudt would recognize his discovery of the wild hybrid colonies by designating the natural pentaploid he had discovered as a separate species, which in his honor Staudt would name *Fragaria bringhurstii*.[89]

343

Consultant Work at Home and Abroad

As the success of the California strawberry breeding program became more widely known, Royce began to be sought out by other strawberry breeders and experts in the biology of the species. He valued his association with these colleagues, who were in a position to understand the complexity of his work, and to benefit from the advances he and Victor had made in California. He looked forward to seeing them both at national meetings on strawberries and small fruits, and on their occasional visits to Davis. He exchanged information and promising plant material freely with them, and willingly tested some of their crosses in his fields. He was increasingly called upon as an authority on strawberry breeding and culture, and received inquiries from such diverse areas as South America, Vietnam, and Rhodesia.[90]

In late 1962, Royce was approached by a representative of the Agency for International Development (known as A.I.D.), who enlisted his participation in a short-term mission to help develop a struggling strawberry industry in a region of Mexico located in the state of Guanajuato. Accepting the offer, Royce began to brush up on his Spanish, which had declined since his mission more than twenty years before, making lists of agricultural and scientific terms which he began to commit to memory. Although this would be Royce's first long-term stay in Mexico, it seemed like something of a homecoming for him, since his work as a missionary had been almost exclusively in Spanish, and among Mexican people.[91]

Arriving in Mexico City on January 28, 1963, Royce spent a few days in and around the capital, being oriented to the program. Aside from his brief sojourns in border towns during his mission, this was Royce's first visit to Mexico proper, and his overall impression of the capital city was positive. "I must say that this city has good order, many police, and you have a constant sense of security, except on the crosswalks etc. The pedestrian must proceed with utmost caution because the automobile has the 'right of way' when the issue is in doubt. The pedestrians would fare well in L.A., but the drivers wouldn't last a week."[92] When Sunday arrived, having no church services to attend, Royce spent the day taking in some of the great cultural treasures in the city, but he was also struck by the contrast of prosperity and poverty which had so troubled him on his mission. That evening in his hotel he jotted down some of his Sabbath-day musings. "You feel compassion strongly, and yet it is of a helpless sort because you shrug the shoulders and say, 'what's the use, they are too many and it won't really help.' I thought of the story, 'Inasmuch as ye have done it unto the least of these my brethren, ye have done it unto me.' Really, I can get wrapped up in my little world and

344

rationalize my position most any day of the week."[93]

In the second week Royce traveled to the city of León in the state of Guanajuato, which was located in the area called *El Bajío*, where he would be doing most of his work. Accompanied by an interpreter fluent in English, he toured the region, taking stock of the climate and growing conditions, and of the agricultural practices, which in most areas were still quite primitive. He found that some aspects of the strawberry industry were functioning well, but noted that the growers were struggling to make use of ill-adapted plant varieties from Florida and California which gave poor yields, and the local nurseries, which had not learned to follow the principles carefully worked out by Royce, Victor, and others, compounded the problem by providing inferior plants with very low vigor. As Royce met with the men involved in the various stages of strawberry production, he gradually worked out in his mind the type of system which might be successful in the area, and he prepared a group of lectures to be presented at the university located in Guanajuato, a beautiful and picturesque colonial mining town located a few miles away. Still insecure with his missionary Spanish, Royce began drafting his lectures in English, planning to rely on his excellent interpreter to make them understood.

Royce spent two weeks in Guanajuato, alternately visiting strawberry growing sites to set up experimental plantings, and preparing and delivering his lectures to relatively small groups of students and industry representatives. Shortly before one of these lectures his interpreter was suddenly called away, forcing Royce to stumble through his lecture in Spanish, using the notes he had prepared in English. He immediately went to work translating his remaining lecture notes into Spanish, which he found especially challenging because of the highly technical terminology which he had never had to express in that language, but by the time of his final lecture at the university on March 1, his preparation had paid off, and the talk was very well received.

In the end, the loss of the interpreter may have been the best thing that happened to Royce while in Mexico. After his lectures at the university he had scheduled a series of talks for growers in the nearby city of Irapuato, in the heart of the strawberry growing area, and here Royce had a much larger audience than he had in Guanajuato, with over twenty growers in attendance. Feeling increased confidence in his Spanish, Royce was able to give a presentation of over an hour and answer many questions. His second lecture the following day was even better attended, and Royce wrote, "A very successful evening – 29 growers present. I really hit the nursery thing hard. I know that my Spanish delivery was adequate this time—I felt completely at home

345

with it. It has done a lot of good to translate or rather prepare in Spanish."[94] Royce had become convinced that most of the problems he had seen could be solved with a simple combination of good strawberry varieties, well-timed plantings, and better nursery practices to provide more vigorous plants. His final lecture in Irapuato was on Thursday, March 7, and Royce was astonished to have an audience of nearly fifty growers, who listened with great attention and asked many questions. After the presentation, Royce was treated to dinner by the president of the strawberry growers' union, a man who had been cool to Royce when he first arrived, but had been completely won over by Royce's lectures, and before leaving León, Royce made certain to leave him a copy of a strawberry handbook he had prepared.

Royce spent his remaining time in León writing up the report of his work, though he took time out to visit an experimental station and give a couple of lectures there, which left him exhausted. The success of his work in Guanajuato had begun to be overshadowed by problems within the hosting organization, whose officials had threatened to resign in the face of conflict with their central office; indeed, some of them had only stayed on to see Royce through his project. For his part, Royce had concerns of his own from California, for he learned that members of the Strawberry Advisory Board, which had recently increased funding for the University strawberry program, were disgruntled at his excursion to Mexico, and both Victor Voth and the chairman of the Pomology Department had had to respond to their concerns. He returned to Mexico City where he would remain until the time of his departure while he completed his official report of the project.

Royce felt satisfied that his work in Mexico had been worthwhile, and had left a foundation of sound principles which formed the groundwork for real improvements in strawberry production. If the visit had been beneficial to the Mexican strawberry industry, it was equally so for Royce, who for the first time had been able to combine his professional knowledge and skill with the ability to express it fluently in Spanish. This would repeatedly open up new opportunities for him in the years to come, and would enable him to have a much greater impact in the Spanish-speaking areas which he would frequently visit as a consultant.

During his two-month stay in Mexico, Sundays had been a particularly forlorn time for Royce, since there were no church services to attend and no work to be done. He had been away from church longer than at any time since his military service abroad, and he felt lost and alone, and had felt keenly the absence of his family. He spent the time as best he could, writing letters, visiting cathedrals and other cultural sites, and on some Sundays walking through the countryside and observing the botany of the region. For his last Sunday in Mexico, Royce was determined at last to attend a Latter-day Saint church service, and arriving at the meetinghouse after a long ride, he was pleased to meet a couple there whom he had known as fellow missionaries more than twenty years before.[95]

Though happy to see old friends, Royce found that two months without his accustomed weekly Church attendance and daily patterns of worship had taken their toll, and he had reached a spiritual low point in his life. Where his first Sunday in Mexico had been full of thoughtful contemplations reminiscent of his mission, on this last one he found the Church services shallow and lacking in substance, and his personal dissatisfactions seemed magnified. He contemplated how his work in evolutionary biology left him with little patience for a literalist view of the Old Testament, and he felt that in a similar way the history of the Church had developed a kind of mythology all its own which of itself tended to obscure the truth. "It would be hard for anyone to convert me to the Church now," he wrote. "I may be very wrong in all this and perhaps I wish I were, but the neat little world I dwelt in as a missionary just isn't real any more." His years as a scientist seemed to force him to examine his beliefs with greater scrutiny, and he feared some of them amounted almost to a form of "wishful thinking," which he feared would not stand up to close examination in a modern world. He also felt disillusioned with the tendency he saw in the Church to fill the calendar with programmed activity which failed to address basic problems of good and evil. Concluding his account of the day, he remarked, "May I never again say 'I believe' when I do not, or rather when I should say 'I would like to believe.' May my thoughts be guided into paths of development that will enlighten and not obscure. My I use prudence in what I say to others that I may help rather than impede, and may I not forget that my own mental capacity is very limited."[96]

His experience at church left Royce in ill humor the following day, and the easy pace of life in the Mexican office now seemed an annoyance as he worked feverishly to complete his reports. Everything about the country seemed to bother him now, and he concluded his account of the day by observing, "The novelty is gone. It's time to go home."[97] However, his feelings softened in the days that followed as he consulted with the Mexican officials who had collaborated with his project. It was clear that they were sincere in recognizing the value of the work he had done, and were hopeful of putting his recommendations into effect. On his last day in the office, he was presented with a fine leather briefcase beautifully embossed with an image of the famous Aztec calendar stone, and this became ever after a treasured possession.[98] On Saturday, March 31, Royce

347

boarded a plane bound for the United States, ready to return to his family and his work.

New Changes and New Opportunities

Royce faced a heavy work schedule on his return to Davis in the spring of 1963. He was grateful to be back with Pearl and the children, but work had piled up, the experimental stations awaited his attention, and he still had to mollify the California Strawberry Advisory Board for having taken time off for the trip. At the same time, the Sacramento Symphony Chorus had already been rehearsing for a performance to be given shortly after his arrival, and Royce had determined to continue his voice lessons with Alex Gould. The more relaxed work schedule he had enjoyed in Mexico gave way immediately to a flurry of activity as he struggled to catch up with his many obligations and duties, both at home and in the strawberry program. In the midst of his busy schedule Royce developed a painful infection in a finger of his writing hand which had to be surgically drained,[99] but work could not be put off; the demands were pressing, and Royce continued to take trips each week to the different strawberry growing areas. To compensate for the time he had spent away from home, he took Pearl on one of his trips to Southern California, the first time since the war that they had both been away from their children.[100] That summer Royce was advanced from the rank of Associate Professor to that of Full Professor, having completed fourteen years in the University of California, but by that time he was busy putting in summer plantings at the experimental stations and was completely exhausted.

Despite the spiritual misgivings he had expressed while in Mexico, Royce made no mention of his personal struggles to his family or to others. Though he lived up to his determination to avoid claims of conviction where conviction was lacking, as he returned to the accustomed patterns of worship at church and at home his faith took a definite upswing. The first few weeks after his return included a stake conference, a General Conference, and the long-anticipated dedication of the new and magnificent Oakland Temple in May, which brought with it the possibility of regular participation in temple ordinances, those ceremonies which most embodied the spiritual aspirations of Latter-day Saints. Royce had long since committed himself to church activity and service, and this remained unquestioned, regardless of his personal misgivings. If there were still areas of doubt, in the Church Royce at very least continued to find a clear and positive pattern to live by, and he found enough areas where his convictions were strong to carry him forward. He resumed his usual church activity after his return from Mexico, attending meetings, giving lessons and talks, and leading family prayer at home. He had been assigned the previous summer as one of the seven presidents of the Seventy in the Sacramento Stake of the Church, and he continued to function in that capacity, which included visiting the various wards in the stake. In August of 1963 he was invited to speak at the Spanish Branch in Sacramento, and having renewed his language skills while in Mexico, he gave his first sermon in Spanish since his mission, which brought him no small satisfaction.[101]

At that time work was underway on a new chapel addition to the Davis meetinghouse, which was to be an impressive structure with high vaulted ceilings and a formal podium and pews. In spite of his other obligations, Royce dedicated much of his free time to work on the chapel construction site, which included climbing high atop the roof trusses to nail in the long cedar planks which were to form the ceiling for the new chapel, an activity which would become memorable for all who performed it. The expanded building, comparable in scale to the large Latter-day Saint chapels back in Utah, was dedicated in March of 1964,[102] and Royce would always feel himself a part of the meetinghouse, where he had dedicated so much of his own labor.

Life was busy for other members of the family as well. Jean, who was continuing her studies at Utah State University, had become distant from her parents and wrote little, causing Royce and Pearl to worry enough that both made occasional trips to visit her. They were dismayed to learn that she had been dating a student who was not a Church member, to whom she would eventually become engaged, though the engagement would later be broken off.[103] Jean continued to come home to California during the summers to work, but she felt no inclination to return home on an extended basis, and indeed would never live in California again.[104] Florence, who had graduated from high school the same year as Royce's Mexico trip, enrolled at UC Davis where she did well enough to make the honor roll, though her studies were interrupted once more by a bout of encephalitis, which left her bedbound in 1964, and from which it would take her months to recover.[105] Like Jean she began to wear an engagement ring, given her by a young man named Larry Nielsen, who shortly thereafter departed on a mission to Brazil. Her younger sister Marla meanwhile had become involved in student government at the high school, and would eventually also enroll at UC Davis. She shared in the family love of music, and while still in high school she joined the chorus in a performance of Handel's *Messiah*, in which Royce sang a solo and Florence played in the orchestra.[106] The younger children were quickly growing up as well. Ann, who was nearly finished with grammar school, was a bright and meticulous child who distinguished herself as a sixth grader by winning a city-wide spelling bee. John, less disciplined but still a good student, had begun to take lessons at the piano, at which he would eventually become as accomplished as any member of the family. Margaret, the youngest, a shy child with a great love of animals, was enrolled in the old nursery school near campus, and would shortly be starting kindergarten.

Royce meanwhile continued in the various aspects of his professional work, making and evaluating new strawberry crosses, supervising research, and hosting the frequent visitors, many from other countries, who came to observe his program. His work in Mexico and his mastery of technical Spanish had opened new opportunities for him, and in March of 1964 he made a ten-day visit to Argentina and Chile as part of a three-man team of experts sent by the Ford Foundation to help stimulate development in those countries, a trip he would repeat the following year.[107] For 1965 Royce had planned a six-month sabbatical leave to concentrate on his study of wild strawberries, and he began this sabbatical with another trip to South America, for which he departed at the end of February. This trip took him through remote areas of Colombia, Chile, and Peru, where he investigated, collected, and photographed wild plants.[108] Returning to California, he began a series of excursions up and down the Pacific coast,[109] continuing the investigations he had begun under the encouragement of Henry A. Wallace, who would die that same year. When it came time for a family camping trip that summer, the place chosen was Patrick's Point, a rugged, fog-shrouded coastal area which afforded Royce ample opportunity to continue his search for wild strawberries on the north coast.

1965 proved to be a significant year for the Bringhurst family in other ways as well. In January Jean called to notify her parents that she had broken off her engagement, and that summer she announced her plans to marry Bruce W. Anderson, the son of a couple Royce and Pearl had known from their days at Utah State. The Andersons had lived in Davis for a time, and Bruce had been the first full-time missionary to be called from the new Davis Ward; Jean and Bruce's wedding would likewise be the first temple marriage for the ward. The family launched into a flurry of preparation for the wedding, and with the help of ward members Pearl baked the cake and prepared large quantities of delicacies for a reception in the family's back yard which was to follow the wedding. The couple were married in August in the Oakland Temple, dedicated just the year before, and after the marriage they took leave of the family and returned to Logan where Jean was employed teaching school.[110]

Shortly after the wedding, Royce's sabbatical leave came to an end, and the family took one last camping excursion before his classes began in the fall, this time to Yosemite, the national park famous for its sheer granite cliffs and domes. While there, Royce had a run-in with the notorious park bears, which Pearl described in a letter the next day: "The bears came into our camp last night four different times and ate a watermelon right off our table, dumped dishes and pans around, and in general made a loud noise. Royce threw rocks each time to frighten them away, but the fourth time they weren't to be frightened away. They started howling; pretty soon the wolves and coyotes started howling too, until the whole forest was yelling. I was afraid the bears would attack Royce. He finally lighted a lantern and they left. We left the lantern burning the rest of the night."[111] This camping trip was memorable to Royce for another reason, for he had the opportunity to renew his lifelong love of fishing, and it was there he caught his first golden trout, a beautiful native fish found naturally only in those alpine California waters.

The following summer in 1966, Royce was scheduled to present a paper in a meeting at Beltsville, Maryland, and he and Pearl decided to make a family vacation out of the trip, traveling across the country by car.[112] Departing in early August, they traveled the breadth of the country in a few days to Maryland, where they stayed with the family of Pearl's sister Laura, whose husband, Harold Carlson, was stationed there. Here Royce, with his family, had the opportunity for the first time to visit important historical sites in and around Washington D.C. Their visit to Arlington Cemetery to see the changing of the guard at the Tomb of the Unknown Soldier was especially meaningful for Royce, whose war experience was still a vivid memory despite the passage of more than two decades. They visited the Civil War battlefields at Bull Run and at Gettysburg, and on the return trip they passed through the areas of Illinois and Missouri which had played an important role in the history of the Church, including the old jailhouse in Carthage, Illinois where the founding prophet Joseph Smith had been assassinated. Royce felt a keen sense of identification with these areas, where his own ancestors had borne the brunt of anti-Mormon persecution.

Before departing on this cross-country trip, Royce learned that he had been granted an assignment to spend a year abroad in Santiago, Chile, under an agreement known as the *Convenio* ("covenant") between the University of California and the University of Chile, to begin in September of 1967. This was an exciting prospect for Royce, for not only would he be able to assist the Chilean strawberry industry to get on its feet, but he would also have an extended opportunity to study wild strawberries in the very location where the most important parents of the commercial strawberry had originated. Although the departure date was still a year away, planning and preparation began immediately. Pearl started lessons in Spanish, and she and Royce began working on the house, painting the exterior and interior, and making repairs on the bathrooms. They also set about looking for someone to occupy the home in their absence, and eventually settled on a young Latter-day Saint family that needed somewhere to live the following school year.

Royce meanwhile continued his church participation as strongly as ever, and in the spring of 1966 he was ordained a high priest and called to be the High Priest Group Leader for the Davis Ward.[113] This placed him in a leadership role in the ward, including the planning of lessons and the organization of pairs of "home teachers" who were assigned to visit individual households of Church members on a monthly basis. The local ward had undergone rapid growth as the university had expanded, and a third and final phase of building construction had been approved for the Davis chapel, to include a kitchen and a recreation hall and stage. At the same time, a new meetinghouse was under construction in south Sacramento to replace the aging stake center then in use, and Royce and the family became involved in fundraising activities for both these projects,[114] in addition to their usual volunteer work in the Church's welfare farms and canneries. For its fundraising, the stake took advantage of a labor shortage resulting from the discontinuation of the "bracero" program,[115] and enlisted Church members to pick tomatoes for commercial growers, with their pay to go into the stake building fund. In a few hours Royce managed to pick thirty boxes at 20¢ per box.[116]

As preparations were made for the year abroad, life events continued to multiply for the Bringhursts. In January of 1967 Marla transferred from UC Davis to UC Berkeley to continue her studies, while Florence remained in Davis. In February Pearl was involved in a car collision which left her with only minor injuries, but did major damage to the family car.[117] Later, in April, she underwent surgery for removal of a cyst that had troubled her since the war, to avoid the possibility of it flaring up while the family was in Chile.[118] Meanwhile, Florence's on-and-off fiancé, Larry Nielsen, returned from his mission in Brazil, and after spending some time together once more, the two decided they were meant for each other, and set a wedding date for June.[119] That May marked the twenty-fifth anniversary of Royce and Pearl's own wedding, and they celebrated by going to see the newly released movie, *A Man for All Seasons*, which they both found satisfying, despite a dislike Royce had developed for cinema.[120] The family busied themselves with preparations for another wedding, and as the date approached, Jean and Bruce arrived from Utah. On Thursday, June 22, Royce and Pearl again drove to the Oakland Temple, and Florence and Larry were married and sealed, then returned to Davis for a reception in the back yard.[121] The couple moved into a small married student apartment in Sacramento, as the rest of the family prepared for their move to Chile.

The date of departure approached, and the final arrangements were made, including passports, visas, and vaccinations. It was a time of turmoil for the University, which like many college campuses in that era was racked by anti-war student protests, while at the same time struggling to adapt to the austerity measures imposed by the new California governor, a controversial political outsider named Ronald Reagan. The country as a whole was undergoing its own transformation, shaken by a series of social changes which would continue to sweep over it in the coming decades. The assignment to Chile marked something of a transition for Royce as well, a demarcation point from his early career, when he was seen as a new and rising star in his field, to the productive later years when, as a recognized world expert, his scientific and genetic work would bear their most abundant fruit.

On September 21, 1967, after last-minute preparations which kept them up until midnight, Royce, Pearl, and the three younger children made their way to the Sacramento airport and boarded a propeller-driven aircraft to Los Angeles, where they were met by Royce's old friend Art Wallace, with whose family they were to spend the night. The flight was a novelty for the children who had never flown in a commercial aircraft. Though as guests in a strange home they did not sleep well, they had a pleasant time the following day with the Wallaces, visiting the grounds of the magnificent Los Angeles Temple for which construction had begun while Royce and Pearl had lived in Los Angeles, and that evening Art returned them to the airport, where they climbed aboard a much larger modern jet liner. As night fell, the plane thundered down the runway and rose into the evening sky over the great Pacific Ocean, then banked southward, bound for Chile. A new episode in Royce's life was about to begin.

Illustrations for Chapter 9

Royce with labeled strawberry specimens in a greenhouse at UC Davis.

Colleague Victor Voth in an experimental field. In this case, different nursery planting times and periods of cold storage are being compared. This is an early photograph, as polyethylene mulch was not yet in use.[122]

Plant harvest at a strawberry nursery, 1955. Strawberries of given varieties were propagated at these nurseries, and this equipment was used to dig up young plants for planting into commercial fields.

Polyethylene mulch being applied to a strawberry field. The plastic warmed the plants during cold weather, affecting yield and timing of harvest. It also helped to retain moisture, and protected the mature fruit.

Early experimental strawberry plots, comparing winter and summer planting systems, using various types of polyethylene mulch.

An early fumigation rig, used to sterilize the soil of plant diseases. The gasses used were methyl bromide and chloropicrin (tear gas).

Strawberry farmers at a field day, as Royce was making a presentation, in this case about the effects of different colors of polyethylene in the winter and summer planting systems. Most of the growers shown are Japanese-Americans. These field days were essential to communicate the results of experimental work to the growers.

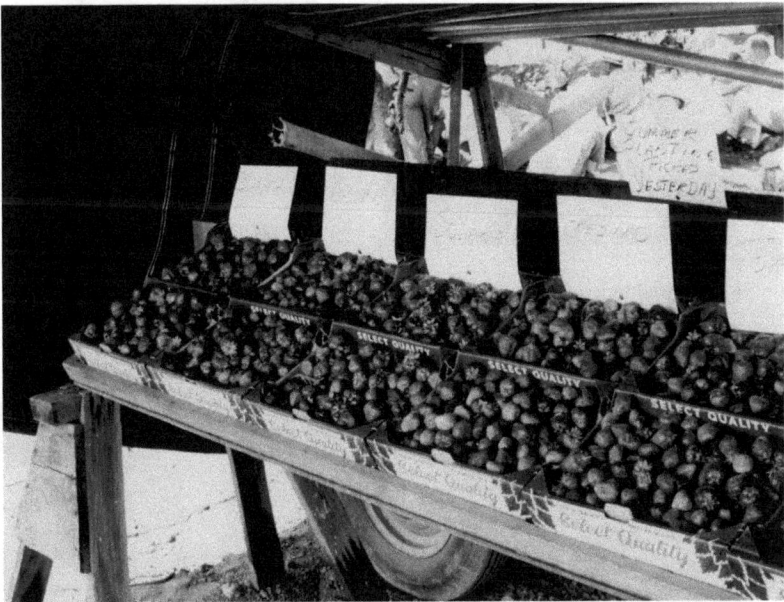

Display of harvested fruit at a field day. In this case, the early varieties Shasta and Lassen are shown with newer releases Solana (1958) and Fresno and Torrey (1961). Fresno largely replaced Lassen in Southern California.

357

Hillside at Point Sur where Royce first discovered natural pentaploid hybrid crosses between *Fragaria chiloensis* and *Fragaria vesca* in 1962.

The original hybrid, and leaves and plants comparing both parent species.

Site at Pacifica where additional natural hybrids were discovered, with notations showing where individual colonies were located.

Bringhurst home on Mulberry Lane, shown under construction in 1957.

Family photo in 1964. Back row: Jean, Florence and Marla; front row: John, Pearl, Royce, Margaret, and Ann.

Davis and Abroad:
An Expanding Career

Chapter 10

A Year in Chile

When the Bringhurst family, consisting of Royce, Pearl, and the three younger children, departed from the Los Angeles airport in a modern jetliner on the evening of September 21, 1967, they were already suffering for lack of sleep, and there was little rest to be had on the overnight flight to South America, which made several stops and took them through turbulent tropical storms. It was the afternoon of Friday, September 22, when the plane finally made its approach at the airport in Santiago, Chile, where the family would spend the next eleven months of their lives. Met at the airport by the wife of another professor in the Convenio program, they were driven to the house assigned to them, arriving late that afternoon. The youngest child, seven-year-old Margaret, climbed on a bed as soon as they arrived, and could not be aroused for dinner, but slept through to the following morning. The entire family might have slept in the next day had not Royce set an alarm.[1]

In the morning the family, finally rested, had the opportunity to take stock of their new surroundings. The house which had been provided for them, and which was to be their home for the coming year, was located at 1586 Luis Pereira Street, a paved and tree-shaded lane in a middle-class neighborhood of Santiago. A beautiful two-story Spanish colonial-style home with a red tile roof and a balcony emerging from one of the upstairs bedrooms, the house was well lit on the ground floor by a series of French doors on one side, which afforded a view of a side lawn and rose bushes which were just coming into bloom, since September was springtime in Chile. There was a beautiful courtyard in the back containing fruit trees and a grape arbor, which provided shade for a long walkway. For a family newly arrived from a suburban-style home in the United States, the house seemed idyllic.[2]

However, in some respects this home away from home took some getting used to for the family. Where their yard in Davis was open to the street, here a tall, spiked metal fence with a locked gate enclosed the portion of the property facing the street, and around the remainder of the perimeter loomed high walls topped with shards of broken glass embedded in cement, to deter intruders. The house, garage, and interior structures such as desks and cupboards were all under lock and key, and there were wooden or iron bars covering all the windows of the house. Another unaccustomed feature was the presence of a maid, Isabel, who lived in a small apartment separate from the house, and a gardener, who came periodically to maintain the outdoor portions of the property. These servants were provided with the home, and were considered an essential component of life in Latin America, though to a family born and bred in America and accustomed to providing for themselves, their presence was a novelty. The gardener was seldom seen, but the maid Isabel quickly came to be a part of the Bringhurst household. A quiet, almost timid woman who spoke no English, she proved to be an excellent cook and an able housekeeper, and despite the language differences she and Pearl worked well together. Isabel was sometimes bemused by the family's cooking and eating habits, and Pearl in turn learned how to adapt to the local conditions by working with Isabel in the kitchen. Isabel appreciated the lack of alcohol and especially of tobacco in the Bringhurst home, and she was grateful to be given every Sunday off, as Royce and Pearl did not feel they could require her to work on the Sabbath.

Santiago was in many ways typical of a Latin American city. A progressive capital with large and modern buildings, it also harbored pockets of extreme poverty, and large public areas were crowded with beggars, a new sight for the children. Automobiles were everywhere, but were much smaller on average than those in the United States, and most inhabitants still did not own a vehicle, but instead traveled about on the city buses, which were plentiful, inexpensive and reliable, though often crowded. The family awoke in the early morning to the cries of street vendors selling newspapers and other goods, and on market days it was not uncommon to hear the unmistakable clip-clopping and grating sound of a horse-drawn cart with metal-rimmed wheels being pulled along the paved street on the way to the market. The twice-weekly local market, or *feria*, was located only a couple of blocks from the house, and presented a kaleidoscope of color, sound and activity. The abundant wares, fruits and produce displayed there were a feast for the eyes, though the stalls of the butchers and fish mongers with their strong aromas and swarms of flies took some getting used to. Excellent fresh-baked bread could be purchased at a *panadería* not far from home, and the children were able to buy penny candy at a small *tienda* built into a house on the corner.

When Sunday arrived, the family set out to attend church meetings with the local congregation. They found that they were part of a large branch of the Church called the Nuñoa Branch, which met some six blocks from their house. The Church was at that time still in its early development in Latin America, and formal wards and stakes had yet to be organized in Chile, though the Nuñoa Branch was as large as a typical ward. It met in a modern chapel similar to those found in the United States, but they later learned it was the only such chapel as yet constructed in the entire country. Although Royce felt immediately at home in the church meetings, Pearl and the children, who spoke no Spanish, were at first much more apprehensive, though the pattern of the meetings and the sound of the hymns were familiar to them, and they were quickly made to feel at home by the other branch members, a few of whom spoke rudimentary English, and all of whom embraced them as brothers and sisters. They would soon come to be an integral part of the Nuñoa Branch.

School had just begun in California, but in Chile most of the school year was already past, and it was necessary to enroll the children in their respective grade levels at mid-year. Public schools in Chile, being reputedly very rudimentary, were considered out of the question, as it was the norm for middle-class families to enroll their children in *colegios*, or private schools, of which there were many in the city. In order to minimize the disruption in their children's education, the Bringhursts followed the lead other families in the Convenio program and enrolled the three children in a 12-year private institution called *Nido de Aguilas* ("Nest of Eagles"), an English-speaking "international" school run by the American embassy. As in other private schools in the country, the children were required to dress in school uniforms, so Royce and Pearl purchased matching white blouses and blue skirts for the two girls, and slacks, blazer and a tie for John, and this became their standard wear on school days.

The children quickly adjusted to their surroundings. 14-year-old Ann, who as an adolescent in junior high school had been opposed to the Chile trip from the start, quickly developed a circle of friends both at school and at church. For 11-year-old John, who had by now shown some promise on the piano, an upright piano with a revolving stool was rented, and a piano teacher was located. He was soon practicing more diligently than he ever had at home, and before long was playing hymns for church meetings. Seven-year-old Margaret, still the most timid, had a more difficult time until she discovered a girl her age living down the street who spoke excellent English, having attended a British school. Before long they were fast friends, and Margaret began teaching English to still another girl living in the same neighborhood. Pearl, in turn, enrolled in beginning Spanish classes sponsored by the embassy.

Chile was a strikingly beautiful and diverse country. Stretched out along most of the western coastline of South America, geographically it resembled a reverse image of the Pacific coast of North America, with a fertile central region, a volcanic area to the south with abundant lakes, islands, and inlets, and an arid desert region to the north. Santiago, the capital city, sat in a valley very similar to the Central Valley of California where Davis was located. However, the Andes mountains near Santiago were much taller and nearer at hand than those in California, and from the house were reminiscent of the mountains which loom over the Salt Lake Valley where Royce and Pearl had been raised. Far from their home and new to this distant country, the family began a weekly pattern of day trips each Saturday, either to areas of amusement or to places more distant from the city, to which they drove in a Ford station wagon provided by Royce's program. One of the first of these trips took them to the hill of San Cristóbal, from which they had a stunning view of the entire city of Santiago. Near the summit Royce discovered California poppies blooming in unaccustomed colors, and he collected seed to plant when he got back to California. Later trips took them all about the countryside, from the steep mountain canyons which lay to the east, to the rugged wind-swept beaches that lined the Pacific coastline to the west, and the varied places and monuments that lay in between.

The Convenio program of which Royce was a part had resulted from an agreement between the University of California and the University of Chile, funded by a grant from the Ford Foundation. Its objective was to strengthen the University of Chile as an academic center in South America and to promote the interchange of students and faculty between the two institutions.[3] Much of the support for agriculture came from UC Davis, and it was to this end that Royce and his family traveled to Chile in 1967. Royce's work in the program was in the field of pomology, and it kept him busy touring the country checking on orchards, nurseries, and plantings. Drawing on his earlier experience at UCLA, he gave a series of lectures on avocado

363

culture and served as an expert for avocado growers.[4] But his principal work as always was with strawberries, and, taking stock of the
situation in Chile, he set to work as he had years before in Mexico,
evaluating and training local experts in the various aspects of the fruit
which had made it so successful in California.

Although the ancestors of the cultivated strawberry had originated
in Chile, the commercial strawberry industry there was still in its infancy when Royce arrived, and made use of chiefly native local cultivars which, though good flavored, were soft and unproductive, with a
limited harvest season. Growing conditions were also very primitive,
with poor and inefficient methods of irrigation and fertilization, and
there was no nursery system to provide disease-free planting stock;
rather, farmers propagated their own plants locally. Working in cooperation with a motivated young agronomist named Vilma Villagrán,
Royce set to work laying the groundwork for a truly modern strawberry industry based on the practices used in California, to which
Chile's climate was nearly identical. He immediately began the introduction of the superior California varieties which were found to
thrive in Chile, and along with the plants came the now well-tested
methods required to grow them. With Royce's help, a system of
nurseries was established to provide high quality disease-free plants
for annual planting, a system of pumped well irrigation was set up to
provide an adequate supply of uncontaminated water, the use of double beds was introduced to improve yield, and the winter and summer
planting systems which had so been so dramatically successful in California were initiated. Work was also done to identify the major strawberry diseases in Chile and to develop methods to combat them.[5] The
system which resulted would later be regarded as one of the key accomplishments of the Convenio program in Chile.[6]

While Royce's main focus in his work was to provide the materials
and methods to make the Chilean strawberry industry a success, he
also continued his studies of wild strawberries which he had begun in
California with the support of Henry A. Wallace. The year in Chile
was a particularly fine opportunity for him, for it enabled him to
spend extended time studying the wild populations of *Fragaria chiloensis* which were native to Chile. Royce was well aware that it was from
among these colonies of wild Chilean strawberries that the progenitors of nearly all the world's commercial strawberries had originated.
As in California, during family outings he was constantly on the lookout for colonies of wild strawberries, which he collected, catalogued,
and photographed.

Of particular interest to Royce was the fact that in Chile there was
but one natural species of wild strawberry, the octoploid *Fragaria
chiloensis*, while in North America there were two octoploid species

plus the diploid *Fragaria vesca*, in additions to the hybrids between them. In California he had found *F. chiloensis* occupying chiefly the coastal areas, while the mountainous regions were dominated by the other octoploid, *Fragaria virginiana*. In Chile, by contrast, since there was only a single species, the wild octoploid *Fragaria chiloensis* populations had adapted themselves to thrive in all these environmental regions, taking over the niche occupied in North America by *F. virginiana*.

Since the ancestors of the modern commercial strawberry had originated in Chile, Royce sought to determine whether the first parent strawberries which the French agent Frèzier had carried to France were similar to the wild populations which still grew in Chile. Making careful measurements of leaf characteristics and flower morphology, and comparing these with early botanical drawings of the original plants taken to Europe, he was able to show that the plants carried by Frèzier had likely originated with wild colonies within a hundred miles of where he obtained them. He also noted that while all North American octoploid strawberries tended to have five-petaled flowers, those from Chile ranged from five to seven, with an average of six, and he noted that virtually all modern commercial strawberry cultivars averaged at least six, confirming their Chilean parentage.[7]

That October, as Royce was still beginning his work in Chile, he and Pearl received a telegram from Logan, Utah, informing them that their daughter Jean had just given birth to a baby girl, their first grandchild. The following month, as Thanksgiving day approached, the family invited all the missionaries in the area to a traditional Thanksgiving feast, and the missionaries, long away from their homes in the United States, were delighted with the meal, and at the conclusion were allowed to take what remained of the pies home with them. The frequent presence of the missionaries in the Bringhurst home harked back to Royce's own experience as a missionary, and if his Mexico trip had marked a spiritual low point in his life, the stay in Chile became something of a highlight, not just for Royce but for those members of the family who were with him.[8]

As Christmas approached, the days grew longer and hotter, and Royce and Pearl considered what they could do to celebrate the holiday in a country where none of their accustomed traditions were even observed. Since both the university and the children's school let out for summer vacation in the middle of December, they decided to plan an extended family trip to the southern part of the country where they could escape the summer heat, and where Royce would be able to continue his study of wild strawberries. Through his connections with the Convenio organization, Royce obtained use of a large pickup truck with a camper shell, and on December 16 the family set out southward from Santiago along the Pan-American Highway. Royce had already passed through much of this country before, but for the family it was the first glimpse of the stunning waterfalls, the majestic smoking volcanoes, and the pristine lakes and rivers of southern Chile.

Royce had arranged for the family to spend Christmas at a private farm, the Fundo Chan Chan, which was nestled at one end of a lake called Panguipulli. The midsummer climate in these southern reaches included rain, mist, and cool mornings reminiscent of Christmas back home in California, and there, without lights, presents or a tree, the family spent a memorable holiday, caroling on Christmas Eve to the baffled occupants in the homes of farm workers. On Christmas Day, Royce and his son John set off with the caretaker of the farm to ascend a nearby volcano where strawberries were reputed to grow, and high up the mountainside, thriving under sheets of snow and ice, they found the tenacious strawberry colonies, from which Royce took specimens to add to his growing collection. Down near the lake he found other strawberry colonies, some with fruit white in color, though sweet and richly flavored.[9] The day after they left Chan Chan Royce celebrated his 49th birthday, and the family stopped by the side of a river long enough for him to catch a fine dinner of fresh trout. Royce's fluency in Spanish made him comfortable taking the small meandering country roads that wound through the lake country on their return journey, and he expressed to Pearl that of all the skills he had acquired, his ability with Spanish was one of those for which he was the most grateful.[10] The family returned to Santiago in time to attend church the following Sunday, and to watch the midsummer fireworks at midnight on New Years Eve.

The new year of 1968 brought long hot summer days to the Santiago area. As the year progressed, Royce and the family received a series of visits from friends, including the Marble family from Davis who were staying in southern Chile, and later Bruce Anderson, Jean's father-in-law who had attended Utah State Agricultural College with Royce. Royce was also visited periodically by colleagues from his department from UC Davis, and he took these professors around the country to inspect the pomology work which had been underway during his stay there. Royce continued to make frequent trips up and down the country much as he had back in Davis, giving valuable guidance in the strawberry growing areas, while continuing his field investigations of the Chilean wild strawberries. As the school year began at the University of Chile, he taught classes and gave lectures, being one of the few visiting professors who had mastery of Spanish.

The family continued to be closely involved with the Nuñoa Branch of the Church. About halfway through their stay in Chile

Royce was called as a counselor in the branch presidency,[11] and he remained a leader in the branch for the rest of his time in Chile. Pearl took an active part in the Relief Society, helping the women in the branch as best she could with her limited Spanish. When John turned twelve in January, Royce ordained him to the office of a deacon in the Aaronic Priesthood, performing the ordinance in Spanish; this enable John to participate in passing the Sacrament during church meetings. John also joined a Chilean Boy Scout troop sponsored by the branch, and having made good progress in his piano studies, he played the piano for church meetings. When Margaret turned eight a few months later, Royce baptized her in the chapel baptismal font, also in Spanish. Despite her shyness, Margaret began to form friendships, and became willing to give talks in Spanish. Ann in turn took part in a church play, and attended a nationwide Church youth conference in the coastal city of Valparaíso.

The summer in Chile was memorable for the family in other ways as well. John, who had already contracted the mumps and missed two weeks of school, became sick again in early 1968 with chills and a high fever. When these failed to resolve as expected, a doctor was summoned, and John was diagnosed with typhoid fever, and was begun on the strong antibiotics needed to treat the disease. There followed a long convalescence, during which all John's laundry and bedclothes had to be boiled to prevent contagion.[12] John's prolonged recovery prevented him from participating in a long-awaited fishing trip to the south which Royce had planned that summer, so Royce made the trip alone, returning with a fine catch of large trout which the maid Isabel prepared as fish steaks.[13]

1968 proved to be an eventful year, not just for the Bringhursts but for the world at large. Though far from their home in the United States, the family followed news of the escalating war in Vietnam, and of a growing anti-war protest movement that was sweeping the nation. They were disturbed by news of the assassination of civil rights leader Martin Luther King, Jr., followed only a short time later by that of Robert Kennedy, who being Catholic, was a popular political figure throughout Latin America. These news reports were widely circulated in Chile and strongly affected public feelings towards North Americans. At the same time Chile, traditionally the most stable democracy in Latin America, was becoming increasingly racked by its own political turmoil, which would shortly lead to the first free election of an openly Marxist chief of state.[14] Strikes and protests, long a fact of life in Chile, became more frequent and severe. A mail strike which started in March of 1968 became so prolonged that the Convenio had to establish a pouch system using returning American families as couriers in order to get mail through to the United States. Students at the University of Chile went on strike and shut down the university, and though the striking students did not bother Royce when he went into his office and worked on his experiments, the situation still kept Pearl on edge.[15]

On one of their many family trips, the Bringhurst family had their own first-hand experience with this political unrest. One day while driving near a college in Santiago with Royce at the wheel, they suddenly found themselves surrounded by a loud group of student protesters who had poured out of a local college and marched directly into their path. Trapped in a stream of shouting students, the car began to be buffeted from the sides, then one of the male students, seizing the window handle on the back of the station wagon, rolled down the rear window, exposing the family to the noise of the protesters. Royce called out sharply in Spanish for the window to be closed, and when no response came, he threw open his door and strode briskly out among the marching students who, seeing the angry expression on his face, gave him a wide berth. Rolling the rear window sharply shut, he returned to the driver's seat and proceeded to drive the car directly forward into the mass of oncoming students who were still blocking the road. They divided and scattered at his approach, and the family was soon out of harm's way, but they were shaken by the ordeal, having had their own taste of what an unruly crowd could do.

As winter approached and the weather grew colder, another unfortunate and near-tragic event occurred one chilly Sunday as the family was leaving church. They had given a ride to a Church member named Irma Ortiz and her two children, and as they returned to the car, the Ortiz family were just climbing in when Pearl, unaware that the transmission was set in reverse, turned on the ignition to start the heater and warm the car. Feeling the car begin to move backward, she reached with her foot for the brake, but she hit the accelerator instead and the car shot backward, slamming the still-open door against a tree by the side of the road as Sister Ortiz and her children tumbled out. The children miraculously were unhurt, but Irma Ortiz sustained a compound fracture and other injuries to her leg, which required an extended treatment and eventually skin grafting to cover the defect.[16] Royce and Pearl visited her frequently during her long recovery, and were grateful to see her walking by the time they left the country. Just a couple of weeks after this event, Pearl was robbed by a pair of thieves as she stepped off a city bus, one jostling her while the other deftly opened her purse and removed her billfold. A school pupil later found the billfold, and it was returned to Pearl with all her documents intact and only the loss of some money, but these events, coupled with other frustrations, left Pearl anxious to leave Chile and return home.[17]

At length Royce's term in Chile drew neared its end, and the family prepared to depart. They attended church at the Nuñoa Branch for the last time, and on a morning in August, 1968, they bade farewell to members of the branch who had come to the airport to see them off. Ann, who had initially objected to the Chile trip, burst into tears when fellow students who had become some of her closest friends appeared at the airport. For his part, Royce left the country satisfied that he had introduced not only his superior varieties of strawberries, but the technical knowledge to grow them productively, and a well-established nursery system to propagate them effectively. But the Chile they left behind was a country in turmoil, stricken by the worst drought in a century and racked by social unrest. In the years to come, the pleasant and beautiful South American republic which the family had come to love would face wrenching change, political upheaval, and ultimately violent overthrow.

As they left Santiago, rather than returning directly home, Royce had arranged for a side trip to give the family a once-in-a-lifetime visit to the remains of the legendary Inca empire in Peru. After spending a few days in the capital Lima, they boarded a plane over the rugged and snow-clad heights of the Andes to the high-altitude city of Cuzco, which had once been capital of the old empire. There they visited impressive ancient buildings in and around Cuzco, then took a one-day train ride to the famous ruins of Machu Picchu, a series of stone buildings and terraces perched atop sheer crags of rock sweeping almost vertically upward from the valley floor. They spent hours exploring the various buildings, stairways, and trapezoidal windows and doorways there, and Royce, armed with a camera, ascended alone the treacherous path up to the topmost pinnacle to take photographs. A few days later, on August 28, 1968, the family at last made their way home to California.

Return to Davis

The world the Bringhurst family returned to in 1968 was in many ways strikingly different from the one they had left barely a year before. The war in Vietnam had grown into a large and seemingly endless conflict which now dominated the news headlines. It was increasingly taking its toll in dead and wounded servicemen, and Royce would shortly learn that the son of Steve Mackey, the best friend of his childhood, had become one of its many casualties. An unprecedented war protest movement had begun to sweep college campuses across the country, in some cases in clear sympathy with the communist foes of the United States. This had already created a spirit of unrest on the Berkeley campus, and was beginning to make itself manifest at UC Davis as well.

369

Profound societal changes were also making rapid headway among the children of the World War II generation as they came into adulthood. There was an upsurge of illicit drug use among young people, and special meetings of the P.T.A. were held in Davis to address the issue. Longstanding moral values were increasingly questioned and discarded by children of middle-class families in ways that would have been unthinkable a few years before. These sweeping societal changes were paralleled by changes in taste and fashion, and the trim conservative dress of post-war America was rapidly giving way to the extravagant excesses of dress and grooming that would mark the decade of the 70s. While Royce remained more conservatively dressed than most, by the mid-1970s he too, like many university faculty, would be sporting sideburns and bright polyester pants and shirts.

The events of the time had impacted all the members of the Bringhurst family as Royce and Pearl set about to pick up where they had left off prior to their year in Chile. Jean and her husband Bruce Anderson were still living in Logan, but Bruce was now on active duty in the Air Force, and was awaiting orders to go overseas to take part in the Vietnam war effort. He would eventually be sent to a base in Turkey as an intelligence officer, and Jean and the baby would be permitted to join him there. Florence and her husband Larry Nielsen were soon to move from Sacramento to Davis, where Larry would pursue graduate studies in History. There they would live in married student housing, and Royce and Pearl would see them regularly and have them over frequently. Marla was still attending college at UC Berkeley, and the following year she experienced at close quarters the People's Park demonstrations, a series of confrontations between activists and police which came to a head in May of 1969. Graduating with a degree in microbiology and immunology, she would later transfer to the UCLA campus for a year's internship to become a medical technologist. The three younger children continued to live at home with Royce and Pearl. Ann was beginning her sophomore year at Davis Senior High School, while John was beginning junior high school, and Margaret was entering the third grade of elementary school, and had developed a passion for horses.

On their return from Chile Royce and Pearl welcomed another family member who had come to live in Davis for a time. This was Newell Bringhurst, eldest son of Royce's younger brother George, who had begun studies at UC Davis. Royce and Pearl, having lived most of their adult lives in California, felt distanced from their relatives who had remained behind in Utah, and Newell, the only member of the extended family who had ever lived near them, became a frequent guest at the dinner table. Thoughtful, pleasant and engaging, he became a favorite of the Bringhurst children, and the entire family attended his marriage to his wife Mary Ann in September of 1969.[18] Newell, like Florence's husband Larry, would eventually earn a degree in history at UC Davis, and although he was not a believing Latter-day Saint, he would go on to become an acclaimed scholar of Mormon history. He would remain close to Royce and Pearl and their children throughout their lives, frequently sending samplings of his writings to Royce.

There had also been changes in the Davis Ward while the Bringhursts had been away. In their absence, a new and final wing had been added to the meetinghouse, which now featured a recreation hall with a basketball court, a fine stage for dramatic productions, and a kitchen. With the expansion of the university and the addition of new colleges, there had been a large influx of new Church members. The ward had grown almost too large for the building, and there was beginning to be talk of dividing the ward and creating a new one just for students at the university. Royce and Pearl were each given callings in the ward on their return, and Ann and John became involved in Mutual activities, which in addition to weekly meetings now included cultural activities such as roadshows, speech contests, and dance and music festivals. John was now a member of a Church-sponsored Boy Scout troop, in whose campouts Royce would be an occasional participant. The three younger children were growing up quickly, and the family would increasingly be kept busy with their activities.

With all the children now attending school, Pearl decided it was time to pursue her own ambition of a college education, and she enrolled that fall in classes at the American River College, a junior college in Sacramento, which she attended part-time for several years before eventually transferring to UC Davis to work toward a bachelor's degree. One day, while visiting the LDS Institute, Pearl met a rather shy young Latter-day Saint student named Pat Galleger, whose conversion to the Church had resulted in an estrangement from her parents and had left her without a place to live. The Bringhurst home was more empty than before with the older daughters out on their own, and after discussing the situation, Royce and Pearl decided to offer her a place to stay. Quiet and unobtrusive, Pat would remain in the home as part of the family for nearly two years, while she completed her college education.[19]

Strawberry Program—Emergence of the "Day-Neutrals"

On his return from Chile, Royce wasted no time in resuming his accustomed work in the statewide strawberry program. As he was preparing to leave the country, he and Victor had already been doing final evaluations on their next generation of new strawberry cultivars,

371

and this had led to the release in 1966 of the variety "Aliso," which was noted for its early harvest, as well as "Salinas," another *Verticillium*-resistant variety which was not destined to succeed commercially. In 1968 an additional variety was released called "Sequoia." Of all these, Sequoia was the most significant, having unusually large fruit size and good flavor, and being well-adapted to Southern California. However, it also had drawbacks, as it lacked firmness, and was somewhat less productive than Tioga, which continued to dominate the industry. Having released these new varieties, the next task was to work with the commercial growers to subject the new varieties to expanded field testing, and determine where in the strawberry industry they would fit in. At the same time, evaluation was already underway for the following generation of strawberry crosses which had grown to maturity and become ready for evaluation. The most promising of these would be ready for release in 1972, and would be given the name "Tufts," after Warren P. Tufts, the Pomology Department chairman who had hired Royce when he moved to Davis from UCLA in 1953. Tufts would eventually prove an excellent selection, and would finally begin to challenge the worldwide dominance of Tioga in the strawberry industry.

As Royce went through the process of evaluating this new generation of plants after his return from Chile, he became aware of an important development which would eventually mark a milestone in his career. This had to do with the wild strawberry plant whose tiny fruit Royce had collected during a family vacation in Brighton, Utah, in the summer of 1954, shortly after his arrival in Davis. The plant was a common meadow strawberry, *Fragaria virginiana glauca*, an octoploid which could be crossed freely with commercial varieties. Royce had germinated the seeds from the fruit, and the following year, in 1955, he had performed a cross of this wild strawberry with the standard commercial variety Shasta. The resulting plants and fruit were not very impressive, but they did have unusual fruiting characteristics, in that they bore flowers and fruit out of season, at times when other strawberry varieties would only produce vegetative growth. Using the best offspring from this original cross, in the subsequent years Royce made a series of backcrosses[20] of the offspring with some of his commercial-quality plants. By the time Royce returned from Chile, some of the offspring of these backcrosses had matured and undergone field testing, and though the plants still did not yet measure up to commercial standards, the improvement in fruit quality and yield had been impressive. But more important by far was the unique fruiting pattern of these crosses, and it was only now, in the early days of 1969, that the potential value of that tiny strawberry from the Wasatch Mountains of his native Utah had at last begun to dawn on

372

Royce.

To this point, all of the strawberries from the California breeding program had been "short-day" varieties, meaning that the shorter daylight hours during the winter months would induce flowering, and subsequent fruiting could then be manipulated and prolonged with the good cultural practices that he and Victor had been developing for years. Royce was also familiar with so-called "everbearing" strawberry varieties then in use, and he knew them to be technically "long-day" plants, meaning that their flowering was simply induced by a different day-length than the short-day types, which under the right conditions would allow them to fruit over a longer period of time. However, they were still responsive to day length, and when the day length changed they too would cease production, even if weather conditions were optimal. By contrast, these new strains of strawberries seemed to have no response to day length at all, but simply began fruiting a few months after they were planted out, and continued to do so for as long as weather conditions were favorable. This tendency appeared to be an autosomal dominant trait, which meant that it was easily transferred from one generation to the next. The potential benefits of this trait were obvious and exciting. That year Royce made a third backcross with the best of the plants carrying the trait, which he began to refer to as "day-neutral," descriptive of the flowering and fruiting characteristics of the plants.[21]

By this time the breeding program had progressed to the point that Royce and Victor felt confident that they would be able to release successive generations of improved strawberry varieties every few years. With his return from Chile, Royce set about refining the complex breeding process. Realizing that a single generation of the breeding cycle, from the planting of seed from the previous generation, to the evaluation and selection of the seedlings, to the production of seed for the following generation, took about three years, he had divided his breeding work into three concurrent primary programs, each of which he could concentrate on for a single year of the breeding cycle. This would enable him to do his work systematically, without the confusion of overlapping generations.

The first of these primary programs he referred to as the pyramid program, which focused on crossing plants from distinct lines in an attempt to combine as many positive traits from different parents into a single plant, then applying the recurrent mass selection process on the offspring. The second was an inbreeding program, in which plants of a similar lineage were crossed with themselves or with a closely related parent, in an attempt to further concentrate its desirable traits. The third program was a species hybridization program, in which plants from related strawberry species, or from the related genus *Potentilla*, were bred with conventional strawberries in an effort to incorporate new and desirable traits into the breeding stock. Examples of this technique had included Albert Etter's introduction of salinity tolerance and disease resistance through interbreeding with coastal wild *Fragaria chiloensis* plants, as well as Royce's introduction of the day-neutral trait from the wild *F. virginiana glauca* from Brighton.

Initial selection from among the seedlings was now performed in two locations: in Santa Ana, at the South Coast Field Station run by Victor Voth, and in the Central Valley, at the Wolfskill Experimental Orchard near Davis, which was under Royce's direction. More advanced testing was now carried out at three separate locations on the Central Coast, two on the south coast, and two in the Central Valley. The main two objectives of this breeding program, yield and fruit quality, had not changed, but the specific components of these objectives had continued to evolve over time. Improvement in yield, for example, now meant not just increasing the already high absolute yields which were being achieved in California, but expanding the production patterns so that strawberries could be made available for a greater portion of the year, or perhaps all year long. This was one reasons the day-neutral plants held such promise.

The definition of fruit quality had also changed. Large fruit size continued to be important, not just for market appeal but for ease and economy of picking, but now a long conical fruit was considered superior to a blocky or wedge-shaped fruit, and a neck below the calyx[22] was considered desirable, both because it made for a more attractive fruit, and because it offered the possibility of mechanized removal of the calyx. Firmness was still important, to allow for increased durability when the berries were shipped to distant markets, and to permit the berries to be picked at full ripeness. To measure firmness, Royce used a standardized instrument called a penetrometer, which measured the pressure necessary for a small metal rod to pierce the skin. Since Tioga was now the standard commercial variety, the firmness of Tioga was taken as the minimum standard for future varieties. Color had also become an important factor; a uniform-colored fruit with bright orange color was now considered ideal; and new varieties with paler flesh or with very dark skin color were no longer acceptable. The most elusive fruit quality, however, continued to be flavor. Some elements of flavor, such as sugar, acid and solid content, could be measured, but the overall quality of flavor was impossible to quantify—it varied according to the preference of the taster, and it could change considerably within a given variety, depending on when in the season a strawberry was picked and what ripeness it was allowed to attain before harvesting.[23]

By 1973, when Royce had completed twenty years as head of the

California strawberry breeding program, he could report having originated and evaluated about 200,000 seedling crosses, from which eight commercial varieties had been released. On average, 150 to 200 new selections had entered the inventory each year as potential varieties or breeding stock, and a similar number had been discarded yearly after adequate evaluation. This process of evaluation and elimination meant that in some years, few of the seedlings had remained, but seedlings from other years had been released for commercial use, or had been retained as breeding stock because of important desirable traits. Among these were a growing number of cultivars that contained the promising new day-neutral trait. Royce was able to say with confidence at the conclusion of his report for that year, "For the twenty years, there is a total of 2382 progenies from 449 parents. The present populations contain most of the traits that are necessary to generate the range of varieties that will give California strawberrymen the capability and option of producing quality fruit somewhere in California throughout the entire year."[24]

Along with his breeding work, Royce also continued his work on the wild strawberry populations, and on basic strawberry genetics. With the help of a graduate student named Tarlock Gill, he completed his chromosome determinations on the offspring of the natural hybrid plants he had identified years earlier along the Pacific coast, and he carefully analyzed the results and prepared these findings for publication.[25] He compiled the information he had gathered on the wild *Fragaria chiloensis* colonies in Chile, and presented the results of this work at an international meeting of strawberry specialists.[26] He continued his work with polyploids, with particular attention to decaploid plants, which could be produced in several ways by a variety of methods, and which Royce felt held particular promise for the transfer of genes from other strawberry species.[27]

Royce also continued his work on strawberry plant diseases. He continued to do breeding and testing for *Verticillium* resistance, and he performed annual experiments in the station in San Jose, using the inoculation method he had developed in cooperation with Steve Wilhelm. He also continued his efforts to overcome the usual virus diseases using meristem culture and heat treatment to produce virus-free clones of the important cultivars. Additionally, he participated in a study on the twospotted spider mite, which is an important insect pest in the strawberry industry. It was found that resistance varied among strawberry cultivars, but in general those that had thicker pubescence (leaf and stem hair) were more susceptible to the mites, and the growth habit of the plants seemed to have some effect on susceptibility. Derivatives of the variety Lassen tended to be relatively resistant to the mites, which was fortunate from Royce's standpoint, since most of the important California varieties had been bred from Lassen.[28]

Agricultural Mission to Egypt

In late 1970, scarcely two years after his return from Chile, Royce accepted a request from the Food and Agricultural Organization (FAO) of the United Nations to spend two months in Egypt, then officially known as the United Arab Republic[29] as an agricultural adviser. This was of special interest to Royce, since Egypt had an arid Mediterranean climate, similar enough to that of California that his experience with strawberries could be of value there. It would also give him the opportunity to see some of his former graduate students who were from Egypt, and who now lived and worked there. Though Royce was excited at the prospect of making the trip, it would prove to be far more eventful than he could possibly have imagined, and it would also mark something of a turning point in his own life.[30]

On the first day of September, 1970, Royce flew from the San Francisco airport to New York, and there he boarded another airliner to Rome, Italy, where he was to report to the FAO office to arrange details of his assignment before proceeding on to Cairo several days later. As he drew near to Italy his plane passed over southern France, and Royce suddenly began to recognize landmarks he had seen long before from the air. His thoughts were carried back to his life as a young bomber crewman more than a quarter century before. "It was clear as we crossed part of France and started over the Mediterranean just off Nice. I saw the Rhone and Var Rivers and passed over the general area where I believe Frank Denstad died so long ago. Later we passed by Corsica and I could see Bastia (the capital) and I believe I made out our old runway area and the beach I knew so well at Alesan (our 'home' then). I saw the island of Elba, known to history because of Napoleon's exile there for a time, but remembered by me and most crewmen that fought there as the pivot point when our formations swung into the path we followed to attack targets in Italy. Needless to say, this brought memories back to me, and I felt a catch in my throat as I thought about the young men that I knew then when I was one of them. My life was changed by that experience in so many ways. My boyhood was terminated then if nothing else, and I was older than some of my comrades, so they had less carefree days than I did. You cannot shed blood in warfare and remain innocent and free of the consequences."[31]

Arriving in Rome, Royce went over his assignment at the offices of the FAO located there, and attended meetings related to his upcoming work in Egypt. In his free hours, he took the opportunity to walk through some of the ruins of ancient imperial Rome which he had seen once before as a young soldier, and which were within walking distance of his hotel. Scheduled to leave for Cairo on Monday, September 7, he completed his preliminary business, then spent some leisurely time in the ancient capital. Unable to locate an LDS chapel to attend church services on Sunday, Royce instead spent a quiet day walking among the pines and writing letters. As he relaxed in Rome and awaited his departure to Egypt, the world was suddenly rocked by startling news from the Middle East. That same day, on Sunday, September 6, 1970, Palestinian gunman in a carefully coordinated attack had successfully hijacked three different commercial jets, two of which were ordered to fly to a remote Palestinian-controlled airfield in Jordan. The third jet, a massive Boeing 747, was thought too large to land there, and was ordered instead to the airport in Cairo, where after a swift evacuation of the passengers it was blown up with explosives before a stunned world. An international crisis quickly developed as Jordan declared martial law and mobilized its military to combat the Palestinians within its borders. Other nations in the Middle East rushed to take sides in the conflict, and the entire region appeared to be on the brink of war. Royce knew nothing of this, but Pearl, hearing the news back in California and knowing of Royce's travel plans, was beside herself with worry.[32]

By the morning of his departure the following day Royce had at last become aware of the rapidly evolving situation, but he proceeded undeterred to the airport as planned, and boarded his scheduled Monday flight to Cairo. His plane was delayed for a number of hours in Athens as the emergency progressed, but at last it took off, and finally late in the evening of September 7, Royce arrived in Egypt, where he was met at the airport and taken around to various offices to make arrangements for his visit. He also met that evening with Khalifa A. Okasha, a former graduate student at Davis who was well known to Royce and Pearl, having been a guest in their home with his family, and who was now working as a professor in a university in Cairo. Arriving at his second-floor office to begin work the next day, Royce was pleased to find the first letter from Pearl waiting for him on the desk. To his surprise, there was also a letter from Dillon Brown, the chairman of the Pomology Department, informing Royce that he had just been elected as a fellow of the American Society for Horticultural Science, a national professional organization of which he had long been a member. (Dr. Brown had also been elected, as had another Davis professor named Julian Crane.) This was a considerable honor, and Royce suspected his old friend Art Wallace at UCLA of being behind his own nomination.[33]

Royce felt well-received in Cairo, both in the FAO office run by the United Nations, and by the Egyptian officials, who were both

377

friendly and hospitable. He was housed in a pension which was clean and pleasant despite having a common bathroom, and he quickly struck up a friendship with a Swede named Terneman who shared the same lodgings. They were also accompanied there by a group from the Soviet Union, whom Royce described as "not nearly so closed-mouthed as most Russians I have met in the past." One of these was a stout man who spoke little English but had a beautiful robust singing voice which Royce appreciated. In addition to the Russians there were Poles, Yugoslavs, a Libyan, and several Egyptians housed there, and gathering for meals before and after working hours, this odd international assortment of personalities slowly came to know one another and to develop a peculiar comradeship.

During the first few days Royce was invited to travel to some of the sites for which the country was famous, including the Giza pyramids, the antiquities museum, and a traditional wedding celebration, which he observed from atop a camel.[34] In spite of the tense international situation which was developing when he had arrived, Royce had a positive initial reaction to Egypt as a country. "My impression is that things are better than I thought they would be. There is no evidence of a military presence here at all. I think that I see about as many soldiers on the streets of Sacramento. In general, politeness is the order of the day. Many strangers greet you on the street, and you get the impression that people care about one another. I believe that is stronger here than it is at home. Poverty is a relative thing, of course, and by our standards, there are many poor. However, happiness is a relative thing too, and by our standards I would judge them very happy."[35]

Meeting the Egyptian authorities, Royce quickly set about his work as an adviser. This would include attending meetings and touring Egyptian agricultural facilities, as well as making presentations on strawberries as an agricultural crop. For this purpose, Royce began to outline a series of lectures, which he felt would cover the basics of strawberry production. The first of these would be on propagation and the establishment of competent nurseries, which had been key both in Mexico and Chile. This was followed by a lecture on the different planting systems (winter vs. summer in California), then one on land preparation, which included the different types of bed system, planting, mulching, irrigation, and fertilization. The next lecture would cover plant diseases, as well as insects, weeds, and other pests. This would be followed by a lecture on harvesting and marketing fruit, including the handling and storage of harvested strawberries. The final two lectures would cover, first, techniques in plant breeding, and subsequently, the origin of strawberry cultivars then in use.[36] This would give Royce the opportunity to review the California breeding program, and to discuss the relative merits of the various varieties.

At that time Egypt, under the progressive leadership of its president, Gamal Abdel Nasser, was in the process of making a series of centrally-planned agricultural expansions in an effort to convert desert regions into useable crop land. Royce's work called for him to travel to evaluate strawberry growing possibilities in the area around the ancient city of Alexandria, on the Mediterranean coast. With Egyptian authorities he explored several proposed sites, but found most of them unsuitable, with heavy soils and high salinity. It was not until his last day in the area that he found a spot which looked suitable for strawberry growing, which proved to be the historically important area of Rosetta (known in Arabic as Al-Rashid), where the famous Rosetta Stone had been discovered. Here all the right conditions seemed to be present, and Royce calculated that strawberries could even be cultivated beneath date palm trees, giving the land double use.

On his return to Cairo on September 24, 1970, Royce once again found himself near the center of rapidly developing world events. The hijacking of the three planes just prior to his arrival in Egypt, and of another two days after, had triggered a crisis in Jordan, where open warfare had broken out between the Jordanian government and Palestinian militants who had been involved in the hijackings, and who now occupied much of the country. There was sympathy among other Arab states for the Palestinian cause, and the conflict threatened to engulf the entire region, including Israel, in war. At the same time the United States, concerned that some of its own citizens were still being held as hostages in Jordan after the hijackings, was threatening military action, and the Soviet Union had responded to the U. S. with its own show of military strength in the region. In the midst of this growing crisis, President Nasser of Egypt, a charismatic and widely respected figure in the Arab world, had called a summit of the Arab League in Cairo, to be attended by Arab leaders from throughout the region.

Royce quickly caught up with news of the unfolding situation, and had a chance to watch and comment upon the events as they developed before his eyes. "Downtown, they have the Nile Hilton sealed off from everyone, with police surrounding it at about 12 foot intervals. The Arab Summit conference on the Jordanian Crisis is in progress. I hope that they can get the fighting in Jordan stopped. They are really losing on that. Prestige means a lot over here, and they know that the world is looking on. The U.S. gets its share of blame from here, of course. We are always conveniently available as a scapegoat, so it is also in our interest that it end quickly without any participation."[37] As near as he was to the action, Royce realized that his view of the world events was limited, as the Egyptian press, including the occasional paper published in English, was controlled by the government, and was looked upon with universal suspicion.[38]

The Arab League summit concluded successfully two days later, on September 27, and Royce wrote optimistically, "The news today is good: peace in Jordan and calm throughout. Even the hijacking business is ready to be a part of history only. They are finally doing something effective about it, and it is about time." Responding to Pearl's concerns about his making the trip in the midst of the crisis, he added, "Again, please feel all right about my being here. I strongly feel that I am where I should be right now. The news today is particularly good since the Jordan war is over, or at least will be. There is no other big crisis on the horizon that I can sense from here. Everyone seems to feel as I do on this, and these are people who lived through the 6-day war and all that."[39]

Within hours of writing these words, Royce, with the rest of the world, was stunned by news of the death of President Nasser of Egypt, who had suffered a heart attack shortly after bidding farewell to the last of the Arab leaders attending the summit. Egypt and the entire Arab world reeled with the news, and the next day Royce wrote, "It is a sad day in Egypt with the death of President Nasser last night. Forty days of mourning have been declared, as is customary in Arab countries. Three days of solemn mourning effective today will be observed throughout the land, and then his funeral will be on the 3rd day (Thursday). The consequences of his death will only come with time. He was the same age as I was." Witnessing the events from the heart of the city, he added, "I can only speak of the genuine sorrow that one detects wherever you go. Some have been or are weeping, others just look unhappy. As a public figure among Arab leaders he was without question number one. He was respected and highly regarded as a powerful voice for the various groups. I don't know what his death will do in the continuing strife and controversy in the area. I am convinced that he wanted to conclude a just peace with Israel, and I don't think that even his enemies have cause to rejoice in his passing."[40]

The death of Nasser brought an abrupt halt to Royce's work in Egypt, as his office would remain closed until after the funeral, which was to take place on Thursday, October 1, three days after Nasser's death. From his pension, located on an island of the Nile River, Royce once again described the scene. "This morning, I tried to walk down south toward where the funeral would start, and met a solid phalanx of police stationed about 3 feet apart. I walked back up to the Nile bridge to the north of us. It was sealed off. Failing there too, I went home and got together with my Swedish friend, and we walked down to the Nile bank as far south as we could go and observed the action across the river. The crowds were tremendous, and we saw the helicopter carrying Nasser's body to the main square. When we got back to the pension where we live, they set up a TV set and we saw most of what happened." Royce, by this time wearied of the customary loud public mourning, continued, "For the people it was one long emotional 'trip,' each trying to out-mourn the other very loudly, or team shouting of slogans or praise for Nasser. I suppose that the emotional outlet is good for people, but it was tiresome to me after a time. I'm glad these three days are now about over and we can return to that which is called normal around here."[41]

Life was slow to return to normal after the funeral, and Royce took advantage of the lull to complete the obligatory written reports of his trip. His time became increasingly occupied as his two-month stay neared its end. His series of lectures had been delayed, and were delivered on an accelerated schedule to attentive Egyptian experts. He again set out with Egyptian authorities, touring possible sites to establish a strawberry nursery, but once again most of these had poorly suited soil and water conditions, though during one of these visits Royce observed some strawberry plantings in the Nile delta area where California varieties were being grown, and he was pleased to see that where Tioga and Fresno were struggling from mineral deficiencies, the new California variety Sequoia was thriving. For the last two weeks of his stay in Egypt Royce was kept almost continuously busy. Besides his regularly scheduled lectures at the U.N. vegetable station, he was also invited to give lectures at the Ain Shams University, where Khalifa Okasha taught. These lectures were all well attended, and Royce found himself fielding many questions about strawberries and their cultivation. Shortly before his scheduled departure Royce was at last shown a site which held some promise to house a strawberry nursery, though in a patronage-steeped society, Royce had begun to wonder whether the Egyptian officials would be able to pull the task off. "I was out by the Giza Pyramids again today to look at a sandy tract of soil that might be used for strawberries. It looked quite good to me, but I doubt whether the effort will be made here to do the job right, though they might surprise me. They tend to want to send someone to do everything instead of helping them do it, just like Chile. How you break this type of vicious cycle I do not know."[42]

This two-month stay in Egypt represented one more period in Royce's life when he was isolated from the usual spiritual influence of his accustomed church observance. As in Mexico, he had no nearby LDS congregation with which to meet, and he filled the void with quiet contemplation. "I only have one more non-sabbath Sunday here. It isn't easy to change routines like this when you switch the week around as I have had to do. Mentally, I am used to it but actually I adjust by merely telling myself that it ends soon. I don't believe that I am any less of a person not attending these few weeks. Sometimes we mistake form for worship and forget that some of these things are only accessories to life and not life itself. I sort of am what I am here, as I would be at home, and no worse I would say. I do have more quiet moments, and a sharpening of my perspective of values. That much is good."[43] For Royce these became times of self-examination, and contemplation on fundamental issues that otherwise escaped his notice, such as the lot of the very poor, which he observed constantly during his stay in Egypt. "How little we think about the truly unfortunate of the world. I think that when we do, it is in the form of patronization for the most part. We think (deeply inside) that some exist to serve our needs, and we are sure that we have a right to our superior position. I sometimes wonder just how superior it is in the long run. Who values useless things the most, those with a lot, or those with very little? I fear that it is the former. Who is more charitable within really? I rather suspect that it is those with the least. Christian virtue indeed comes hard to those who value things of little intrinsic value highly. I keep thinking I'll do my best to overcome it, but not much happens. I view myself, and discover that I am still the same, and worse luck, older and not much wiser."[44]

At last the time of departure arrived, and Royce boarded a plane in Cairo and flew to Istanbul, then took a second airliner to Adana, Turkey, where he had arranged to spend a few days with his daughter Jean and her family near the Air Force base where her husband Bruce was stationed. Jean's oldest son Scott was still a young child, but was nearly as big as his petite older sister Gina, and Royce spent a restful few days getting to know his grandchildren, and at last attended church with Jean's family before flying on to Rome the following Tuesday. In Rome he completed and filed his final report of the Egypt trip with the FAO office, then he flew to Miami, where the annual meetings of the American Society for Horticultural Science were getting under way. There, together with other colleagues from UC Davis, he was honored as a new fellow in the Society, giving a talk on the last day of the meetings. At last, on Thursday, November 5, 1970, he boarded a plane for California.

Chairmanship and Bishopric

Just before departing for Egypt, Royce had been approached by the dean of the College of Agriculture, Alex F. McCalla,[45] and informed that he was being nominated to become the new chairman of the Department of Pomology, replacing Dillon S. Brown, who had served in the position for the previous seven years. Already making final preparations for his trip to Egypt, Royce had asked that the nomination be kept confidential until after his departure, to spare him the necessity of having to respond to the torrent of questions and congratulations which would inevitably follow. Two weeks after Royce left Davis, a letter arrived at the Bringhurst home from newly-appointed U.C. Davis chancellor James H. Meyer, containing the appointment. The letter read in part, "You have been nominated by Dean McCalla, and I take pleasure in appointing you chairman of the Department of Pomology for the academic year July 1, 1970-June 30, 1971. While appointment to chairman is for one year, it is expected that a chairman will normally serve not less than five years, and not more than seven consecutively."[46] Pearl forwarded a copy of Chancellor Meyer's letter to Royce in Egypt, and he responded with a written acceptance from Cairo.

Although Royce was a logical choice, given his experience, seniority, and reputation, it was not a position he had sought—his statewide strawberry program required extensive travel and was very demanding of his time, and he was involved in several areas of research and had a considerable teaching load on campus, besides his family and church obligations. The appointment came at a time when the university itself was grappling with problems ranging from the student protest movement, to emerging minority rights concerns, to an increasing public consciousness of environmental issues which called into question many agricultural practices which had been considered essential by Royce and his colleagues. This was also an era of financial strain on the university, brought about in part by changes in state funding of higher education under California governor Ronald Reagan. All of these would become part of the backdrop during the five years Royce would serve as department chairman, though the immediate and most visible consequence of the new assignment was a move of Royce's office to a location nearest the entrance of Wickson Hall, and access to a secretary to help him handle department business after his return to California.

Simultaneous with the department chairmanship, another major responsibility was also awaiting Royce. The day after his arrival home from Egypt, he received a call from Homer N. Stephenson, president of the Sacramento Stake of the Church, requesting him and Pearl to come for an interview the following day. Royce and Pearl accordingly arrived at the president's office that Friday evening, and President Stephenson informed them that the decision had been made to divide the Davis Ward and form a new ward for students and their families. He then read them a letter from Church headquarters, calling Royce to serve as the inaugural bishop of the newly-formed ward. Royce was asked to select counselors from among the members of the new ward, and at a meeting of the Davis Ward on Sunday, November 15, the University Ward of the Church was officially formed as part of the Sacramento Stake, and Royce was sustained as its first bishop, with Erv Ingraham and Will Wright, two mature married students, as first and second counselor. Two weeks later, on November 28, Royce drove with his entire family to a stake center in north Sacramento, and there he was formally ordained and set apart as bishop by S. Dilworth Young, the senior president of the Seventy.[47]

Thus began one of the busiest periods of Royce's life. As department chairman, he was now responsible for a broad range of academic functions, including directing the planning of curriculum and research objectives for the department. He was also responsible for faculty recruitment and selection, and his years of service would see some faculty turnover, but also a host of visiting scholars from around the world, who would come to bolster the university's standing in their areas of expertise.[48] Royce was responsible for enforcing the university's increasingly stringent academic standards, and in taking action when faculty members failed to maintain them. As chairman, he assumed responsibility for the department's budget, and had to help it live within its means during stringent times. He was also responsible for governance issues, which now included maintaining university standards regarding race and ethnicity, placing unaccustomed pressure on university authorities to hire members of minority groups preferentially, rather than base hiring strictly on merit. This policy, though well-intended, ran counter to Royce's strong sense of fairness, and he continued to favor merit-based hiring whenever he could. On one occasion, for example, when Royce informed the clearly outstanding candidate for a secretarial position that Affirmative Action policies would probably preclude her hiring, he was delighted to learn that she had American Indian ancestry, which fulfilled the "minority" requirement.[49] This was an important issue for the university, and one with which it would repeatedly grapple in the years to come.

At the same time, Royce's new calling as bishop required him to form and staff an entirely new ward, and attend to its various functions. He was assisted in this by capable and committed counselors, but it was he who had to set the spiritual tone for the ward and train its young and relatively inexperienced officers, most of whom were busy students who would have to learn as they went along. He also shouldered the burden of counseling with the young ward members as they struggled with personal issues or made important decisions at the most formative stage of their lives. As the only member of the ward over thirty years of age, Royce became their counselor, mentor, and friend. His religious focus was no longer merely on his own faith, but on bolstering the faith of others who had been placed under his charge, and his success in discharging this responsibility would be attested to by the devotion and faithfulness of many who had come under his influence as bishop. Continuing a practice which had begun long before Royce's call to preside over the new ward, the Bringhursts opened their home as a refuge for Latter-day Saint students, who became frequent guests around the dinner table. Sundays became the busiest day of the week for Royce, as he went to the chapel for meetings at 6:00 a.m., and often remained there in interviews and other obligations until after 9:00 p.m.[50] As he conducted the meetings from week to week, Royce, whose powerful voice could overwhelm the chapel when aided by a microphone, began to lead meetings from one side of the podium with the microphone directed away from him, a habit he retained for most of his life.

In December of 1970, a month after Royce had assumed these new responsibilities, major illness once again befell a member of the family. While Royce was in Egypt, his daughter Marla, who had just completed her internship in medical technology at the UCLA medical center, had been pouring human serum into test tubes when a defective tube broke, piercing her thumb and exposing the cut to fluid known to be infected with hepatitis. When Marla came home to visit the week after Thanksgiving she was already experiencing joint and back pain, and a couple of weeks later she called to inform her parents that she had been hospitalized with a classic case of serum hepatitis,[51] and was seriously ill. She remained under hospital care for several weeks, and on Christmas Day, which that year fell on a Friday, Royce and Pearl got the family up in the early hours of the morning, and after a quick breakfast they piled into the car and made the long drive to Los Angeles, arriving at 2:00 in the afternoon of Christmas. The drive was through rain much of the day, but when they arrived the weather was beautiful and sunny, and they had a pleasant visit at the hospital with Marla, who was already over the worst of her illness and welcomed the visit. She would be discharged within a week, but would be required to convalesce for an additional month.[52]

Back in Davis Royce quickly adapted to his new responsibilities, and before long had settled into a routine similar to the one he had maintained prior to his trip. Although he had many administrative meetings to attend, he continued to supervise experiments in the field stations, and planned his yearly series of strawberry crosses. He also continued to pursue his other interests at home. When it was announced that the Broadway musical *Music Man* was to be performed in Davis, he attended the tryouts and was given a part as a member of the barbershop quartet featured in the play. He organized his work around the rehearsal schedule, sometimes driving long distances to a field station by day, then attending a play practice in the evening. The musical, which was performed in May of 1971, was a notable success for the community.[53]

That summer, a five-day strawberry breeding conference was scheduled in Scotland, which Royce had hoped to attend, and since Pearl's parents were both natives of Scotland, she and Royce decided to travel together. They departed for the United Kingdom on Wednesday, July 21, 1971, and traveled to the college town of Achincruive, on the west coast of Scotland. By now Royce was a world figure in strawberry breeding, and he presented a series of three papers at the conference, one on his observations of wild strawberries in Chile. He also listened with interest to the other speakers at the conference, including his German colleague Günter Staudt, who described a couple of new species of strawberry he had identified. Of special interest to Royce, Staudt had succeeded in characterizing some of the chemical components affecting the aroma and flavor of the European native strawberries, and comparing them with the very different compounds found in the American species from which commercial strawberries were derived. There was, of course, an interest in incorporating the unique flavors of the European species like the diploid *Fragaria vesca*, but Staudt pointed out that the compounds that made those berries so excellent for fresh consumption rendered them essentially inedible when processed, causing him to question whether it really was desirable to try to incorporate them into commercial varieties.[54]

Leaving Achincruive at the conclusion of the meetings, Royce and Pearl traveled through Scotland, making their way northward to the Moray Firth, where they found the tiny town of Hopeman, the fishing village where both of Pearl's parents had been raised. They followed directions to the home of Aunt Maggie, a relative of Pearl's who, though nearly a hundred years of age, received them with great hospitality. Pearl had heard stories from Hopeman all her life, and for her it was a dream come true to stand in her ancestral home and look out over the sea. After visiting Scotland, Royce and Pearl spent a week in England, passing through the London area before driving to Kent to visit the East Malling Research Station, a world-renowned center for the study of fruiting crops, where Royce had the opportunity to speak with the researchers.

That fall brought changes to the Bringhurst household. Ann, following in Marla's footsteps, enrolled in the University of California at Berkeley, where she would eventually major in Entomology.[55] John advanced from junior high to high school that year, as Margaret entered her final year of elementary school. Among the older children only Florence was still living in Davis, and she had taken a part-time job at the English department while her husband Larry continued graduate studies in history at the university. Marla, who had continued her employment at UCLA following her bout of hepatitis, would soon take work at the medical center in San Francisco, where she would spend the next several years. Jean and Bruce came home to visit that year for Christmas, and after a short time in Logan they moved to the Las Vegas area, where they would remain. Meanwhile, Pearl was continuing her studies at American River College, though she found this increasingly a challenge with Royce so busy at work and at church.

The new school year brought changes for Royce as well. In addition to his usual responsibilities as department chairman, with its many meetings and committee assignments, that year he took on an additional responsibility, dreaded by faculty members, of teaching Genetics 100, a basic course in genetics which was required as a prerequisite for many students entering programs in biological and in healthcare fields, and particularly for the infamous pre-medical students. Royce, accustomed to teaching relatively small classes attended by students with a strong interest in Botany or Agronomy, now found himself in front of large classes of students with little interest in the subject, save as a necessary step toward their long-term goals. Although Royce was well-liked as an instructor, the intense competition for grades among the students seemed to take much of the joy out of the course, which he nonetheless continued to teach for years to come.

After his first year teaching the genetics class, Royce participated in a statewide conference about the teaching of basic genetics, which had become an issue throughout the university system, as every campus had observed a similarly low morale among students and faculty.[56] Royce came to feel that the competitive environment which had arisen around the genetics requirement was emblematic of a "trade school" mentality which had crept into university education, which on the one hand tended to cheapen the educational process, and on the other hand, was not likely to result in the best candidates ultimately being qualified for professional schools such as medicine. Having observed the situation first-hand, he was led to suggest unconventional approaches to the problem. "The ruthless competitive drive for a few more grade points is so intense in the large classes required for admission that it is unpleasant for everyone involved. The almost paranoid attitude displayed by many who manage to rise to the top in the grade reports prompt one to fear for the future of that profession. There is no easy solution to the problems generated by this crass situation. One possibility I am tempted to suggest is to

387

require a certain minimum level of achievement for eligibility and then simply select those to fill the available spots by open lottery among those qualifying. I think it would be more fair and I believe that the students thus chosen would be at least as good as those selected by the present process."[57] This dialogue would eventually lead to a system-wide reassessment of the requirements for professional school admission.[58]

In his position as chairman of the Pomology Department, Royce also made changes in emphasis which would have an enduring impact on the department. One of these was related to the University Extension Service, whose task was to convey scientific advances to agriculturalists, and provide training to individuals and groups not formally enrolled in course work. The Extension Service was essential to the success of applied scientists like Royce, and he had long recognized that, despite the key role the many extension scientists played, they often felt alienated from the department they served. Royce began working on a solution to this problem, organizing regular meetings between the department professors and the extension specialists to encourage communication and collaboration. In time, attitudes changed, and the extension specialists began to regard themselves as a more integral part of the department, a change which continued long after Royce's term as chairman came to an end. The Pomology department became something of a model for other agricultural departments in their relationship with the Extension Service, and Royce would later regard this as one of the principal accomplishments of his chairmanship.[59]

During the busy years Royce served as bishop and as department chairman, he continued to remain as active as he could in music, which had become a treasured part of his life. Although he had had to relinquish membership in the Sacramento Symphony Chorus after his trip to Chile, he continued to participate in church and community choirs in Davis whenever the opportunity presented itself. By this time Davis had grown in size, and was becoming something of a cultural center in its own right. Royce continued to be involved in community musical events, and increasingly began to take minor roles in local musical dramas, including American musicals, as well as Gilbert and Sullivan operettas such as *Trial by Jury* and *Iolanthe.*

Royce's taste in music, particularly classical and vocal music, had continued to mature over the years, and he seized upon any opportunity to participate in musical events. This was especially true when his son John began to pursue music as an avocation as he approached adulthood. Royce's demanding career had kept him from spending much time with his only son during his growing years, but John's love of music was something that brought the two together and allowed them a sense of closeness that they had never before experienced. John had not inherited Royce's magnificent singing voice, but he possessed the same love of music, and began to show some real skill. Royce had never had the opportunity to develop his musical abilities beyond the rudiments, and John's musical pursuits (like that of his sisters) became something of an embodiment of Royce's own unfulfilled desires. He found it especially satisfying at singing events when John accompanied him on the piano.

In 1973 Royce attended a community meeting at which he became one of the founding members of the Davis Comic Opera Company, a new community troupe dedicated to the performance of Gilbert and Sullivan operettas and similar works of light opera. Although Royce was never cast into leading roles, even in minor parts he proved to be a natural showman, often winning the hearts of both audience and cast, and his popularity soared when he began appearing at pre-performance receptions laden with crates of strawberries. He continued to sing in a variety of church and community choirs whenever the opportunity presented itself, but his true love was vocal performance, and he listened faithfully to New York Metropolitan Opera broadcasts on the radio on Saturday mornings as he worked in the yard. Whenever he could he watched evening television broadcasts of grand opera, either at home, or from his hotel room during his many trips to oversee the strawberry program. This became one of his great pleasures. "Somehow when I listen to great music done well, I feel that my life is fulfilled, even though the joy is vicarious."[60] Royce knew that his own voice was powerful enough to have sung opera, and these performances sometimes made him wistful. "This always serves to remind me that if I had started early enough with training to get the technical proficiency and gain self confidence I never developed, I might have been a good singer. I have always known that I had the basic equipment to do it."[61]

Return to Chile

As a department chairman, Royce frequently attended meetings regarding the university's ongoing activities in Chile, and in 1973 he once again had the opportunity to travel there, to follow up on the work with strawberries he had begun five years before. Royce planned his trip for November, 1973, and once again the timing coincided with dramatic world events. Chile had been racked by turmoil since the election of Salvador Allende, a prominent Marxist, as president in 1970.[62] Allende's sweeping socialist policies had been met with growing political opposition and accelerating social and economic chaos, and in 1973 Chile, long regarded as the most stable democracy in Latin America, faced a constitutional crisis as political philos-

389

ophies clashed between branches of government. Finally, on September 11, 1973, the combined Chilean military and police forces staged a military takeover, and when Allende refused to yield the government, units of the military attacked *La Moneda*, Chile's presidential palace, with rockets and with ground troops. Within hours the takeover was complete and Allende was dead,[63] and in the days and weeks that followed, Chile was placed under martial law and a strict curfew imposed. The military takeover had at last brought some measure of stability to the country, but measures against those who resisted the new government were severe, and life had become very dangerous for people who opposed the military action or who sympathized with Allende.[64] This was the situation when Royce arrived in Chile, scarcely two months after the military takeover.

Arriving in Santiago on the morning of November 19, 1973, after a long overnight flight which left him exhausted, Royce was issued a car, and while driving to his apartment he began to take note of the changes which had come over the country since the last time he had visited. "There is less traffic than there was five years ago, and the vehicles seem to move slower and perhaps with a little more care (not much). Everything is quiet and orderly right now. We saw few soldiers and police, but they say that even the 'no parking' signs are to be obeyed." Aside from the curfew, there was little sign of the turmoil that had recently rocked the country, except at night when, from the safety of his apartment, Royce heard the occasional sound of gunfire and of automatic weapons in the distance.[65] However, the events had left a strong mark on the economy, which was still reeling after price controls were lifted. "The exchange right now is 800 escudos to the dollar. That is real inflation. I have not had the opportunity to compare prices as yet, but I understand that they have gone up tremendously. When we were here the poorest were working for as little as eight escudos a day."[66]

Royce had been housed in a spacious apartment with an eastward view of the Andes. Obliged to fend for himself for meals, on his first day in the country he was invited for dinner at the home of Ledyard Stebbins and his wife, who were then staying in Chile, and he accepted a standing invitation to eat with them every evening.[67] Royce took real pleasure in being able to spend time with the famed botanist and geneticist, whose abilities as a teacher were legendary. Royce wrote, "In the little spare time that I have while here I am trying to get out with Stebbins several times just botanizing. You can learn an awful lot from him in a single afternoon. I hope to take an overnight trip with him the weekend before I come home."[68]

The day after his arrival, Royce set to work in his familiar task as a consultant, teaching classes in strawberry cultivation, and assisting

Chilean professors in designing experiments to determine the best growing systems for the country. On his previous visit Royce had laid the groundwork to establish a successful strawberry program in Chile, including the establishment of a strong system of nurseries, and this visit was a continuation of that same work. In contrast to his previous experience in Chile, however, Royce now found a new sense of urgency and efficiency in the agricultural program there. "I have been busy almost every minute. It doesn't appear that I shall have any time on my hands without something worthwhile to do. It is much different than it was before. I wait for no one and there is something important and interesting coming up all the time. I shall be home again long before I tire of it all this time. They need help here, and I am in a position to give them some, at least."[69]

Returning one evening from a long day's work at the university, Royce decided to drive by the house where the family had stayed in Santiago some five years earlier. "It's just like it was when we arrived; beautiful flowers and in good repair. I stopped there a moment wondering whether Isabel might be there still, but saw no one. I then drove down Luis Pereira toward the church, and sure enough, Irma Ortiz (the woman injured in the car mishap with Pearl) was out in the front yard tending her flowers. They were beautiful as always. I stopped, went in for a while and talked with her. You will be pleased to learn that her leg healed perfectly. Really, there are hardly any scars that one would notice, and she says that it causes no problems whatsoever." Referring to the chaotic conditions leading up to the recent events, Royce wrote, "She told me about how terrible it was just trying to get food before the '*Golpe*,'[70] and said that she was glad it happened and everything was very well now."[71]

That Saturday Royce had business downtown, and to satisfy his own curiosity about the recent coup, he traveled to the central square in front of the presidential palace at *La Moneda*, to see the place where the military takeover had happened just a few weeks before. He described the scene: "Actually, the walls were relatively intact. The damage was mostly inside, and repairs are proceeding rapidly. The principal damage came from the attack of two aircraft, firing about 18 wing-mounted rockets each, and from ground to ground rockets that were fired from the Mapocho railroad station. A lot of shots were also fired by soldiers, and from various vehicles. The trees and the surrounding buildings will no doubt bear the scars the longest, including the Hotel Carrera, which has a lot of bullet marks on it. The area has been cleaned up there now, and it is hard to realize that a small war was fought there only 2-1/2 months ago."

Writing home about the takeover, Royce wrote, "Here, one does not speak of the *Golpe*, but rather of the "11th" (meaning September

11). Except for a fairly prominent presence of armed men with rifles and machine guns, the city is about the same except that it is cleaner now than when we were here five years ago. All is not completely right here, of course, and the government (via newspapers and radio) reminds everyone continually of the relatively short time that they have been in power. At least there is more order here than one would have thought possible a few years back. The stores are well stocked now, but the prices are really high for the Chileans, although for us they are relatively low because of the very favorable exchange rate. The other thing I noticed is that there is almost feverish activity in construction, repair, and almost everything that you see going on. You will recall the rather leisurely pace of such a short time back. If this continues, you won't recognize much of Santiago in a few years."[72]

In his letters home, Royce avoided mention of his own close scrape with the military authorities in these tense days following the coup. During his visit downtown he had taken a few photos of the repairs to *La Moneda*, and of the damage caused by small arms fire. He had just changed the film in his camera when he was detained by a police officer and taken into a nearby building. Asked to surrender the film in his camera, Royce willingly turned over the unexposed film he had just loaded into it, avoiding mention of the roll of film he had just removed, which remained in his possession. He waited anxiously while his credentials with the university were verified, and was at last released with a stern warning to be most careful, and to avoid taking pictures, as "things are very dangerous right now."[73]

That Sunday Royce attended church at the old Nuñoa Branch meetinghouse, and reported on the progress of the Church to his family. "A stake has now been organized here, so the branch is a ward, and they have a bishop. We didn't know him, but John's old scoutmaster is one counselor. I also saw Pres. Ogaz and the very talkative lady that taught the Sunday School, plus several other people. The building is cleaner than it was when we were here, and is in good repair. The singing has improved, and although the piano is the same, I believed they must have tuned it. It sounds better. They tell me that the Church is growing rapidly but they still lose a lot of the converts. The lesson materials are better than when we were here. That should help."[74]

Royce had planned on making a trip to southern Chile, but in the tense days following the coup, it was first necessary to obtain a safe-conduct pass, to avoid confiscation of the vehicle. The pass was issued on Monday, November 26, and Royce and his colleagues had to be certain to leave by the early afternoon, to avoid arriving at their destination after the 11:00 p.m. curfew. The curfew seemed in such contrast to the cultural practices that Royce and Pearl had observed when they lived there, that it cause Royce to remark, "You can imagine how night-life has changed here. Having them all in their houses with all the action closed down by 11:00 p.m. is just about like having to be off the street in Davis by 8 or 9 p.m. Things really quiet down. I don't mind at all, it does away with all that all-day, all-night stuff that used to plague visitors here. Perhaps that is why workers seem more energetic, they get a full night's sleep since they have no place to go."[75] Royce spent a week in southern Chile, following up on some of the wild strawberry investigations he had performed earlier there. One day while there he accepted an invitation to go trout fishing, and caught a 14-pound German brown trout, possibly the largest he had ever taken.[76]

Returning to Santiago the following Monday, Royce spent his last week teaching another series of classes in strawberry production, beginning with the origin of cultivars and the concept of polyploidy, continuing with nurseries and the use of meristem culture and the conditioning of plants, and concluding with a talk on plant diseases, and on the cultivation of other small fruits. On Friday he traveled to the nearby community of Melipilla where an important strawberry industry was in its developmental stages, and gave a talk there as well. One of the important Chilean figures Royce met with was Vilma Villagrán, the agronomist who had worked with him during his trip to Chile five years before, and with whom he would have future associations, both in Chile and in California. Capable and intelligent, Vilma would become an instrumental figure in establishing the strawberry industry in Chile, mostly with cultivars developed in California, and she would later look upon Royce as a father figure in her own work. On his final weekend in Chile, Royce traveled to a mountain called Campana, northeast of Santiago, famed for having been climbed by Charles Darwin during his famous voyage to the Galapagos. He also kept his resolve to make a botanical excursion with Ledyard Stebbins, examining and collecting plants at a mine near Santiago. Shortly thereafter he boarded a plane for California, having helped to establish the strawberry as a viable commercial crop in its native Chile, and having once again witnessed history in the making.[77]

Strawberry Program—The Advent of Strawberry Patents

Back in California, even as he shouldered the administrative tasks of running the Pomology Department, Royce continued his efforts in the strawberry program. The new variety Tufts, released in 1972, had proven to be a major commercial success, and was quickly gaining a prominent place in the industry. With larger fruit and a more prolonged harvest than Tioga, it had at last begun to challenge the

393

longstanding dominance of that time-honored cultivar in California and elsewhere. Royce and Victor meanwhile had been working on the next generation of strawberries, three of which would be ready for release by 1975. For the best of these, a selection numbered 63.125-39, Royce chose the name "Aiko," which one of his Japanese associates had suggested as "a lovely Japanese name."[78] Two other varieties were given the names "Cruz" and "Toro." Although none of these varieties was destined to dominate the industry as Tioga had for the last ten years, Aiko showed promise as an excellent variety for summer planting, even though it had not done well when tested in the winter planting system. Toro and Cruz, both superior for winter planting, were designed to fill that gap, and particularly Toro, which had the additional advantage of a very early crop, would come to see considerable success.

Royce and Victor continued to experiment with the day-neutral selections, which seemed to show special promise, even though it was still uncertain what role they would play commercially. The success of these varieties depended on many factors, not the least of which was the inclination of the growers themselves to be willing to abandon time-honored varieties and dedicate space in their fields to the newer ones. Royce and Victor kept a careful eye on the new releases, and worked closely with some of the growers who were willing to try out new varieties and techniques. A farmer in the Watsonville area who Royce had especially come to depend on was a Japanese-American farmer named Kuni Shinta, a skillful grower who had proven himself willing to try out many of the new and experimental varieties. One day in the fall of 1976, after working in his day-neutral plots, Royce decided to pay Kuni a visit, and was astounded by what he saw. "Only a week after 4 inches of rainfall, he is harvesting and selling 'Aiko.' I took a crate with me. He tells me that he is about to 9,000 trays [per acre], expecting to hit 10,000. This is certainly a California (and world) record. It is almost beyond belief, between 50 and 60 tons per acre."[79]

By this time a new factor was emerging in the California strawberry breeding program which was shortly to assume major importance. This was the matter of strawberry patents, whose origins dated back to events in the 1960s. After the astounding success of the Tioga variety, which following its release in 1964 had been widely adopted throughout California and in many other areas of the world, Royce had inquired with university officials about obtaining patents for the university varieties, which could generate funding for the strawberry program as well as a cash return for the university. Although laws allowing the patenting of new plant varieties had been in place for many years, none of the university varieties had ever been patented, and the various cultivars, including Tioga, had been distributed to nurseries and growers at cost, without any royalties to the university or to the researchers involved. For Royce, plant patents seemed to offer a good remedy to the financial limitations of the California strawberry program. Although he and Victor received their salaries as paid as employees of the University of California, the funding for their research came almost entirely from an industry "marketing order," a government-brokered arrangement whereby the California strawberry growers collectively contributed the funds necessary to run the program, and in return benefited from the development of new strawberry varieties and the technical assistance to learn to grow them. The level of funding, which had to be negotiated year by year with the industry, depended on the financial standing as well as the good will of the growers, and at times during the history of the program Royce and Victor had had to struggle to make ends meet. Royce, perhaps noting the use of plant patents by the Driscoll operation in Salinas, had brought up the possibility of patents as a more secure and independent means of funding the program.

At first Royce's query seemed to have fallen on deaf ears, but in 1967, shortly before departing with his family for his year-long stay in Chile, he had been called to discuss the matter with the university's patent officer, and it was determined that had the previous varieties been patented, the returns would have indeed been very substantial.[80] Based on these findings, plans were made to begin patenting university strawberry varieties released to the market, a complex process that would involve multiple stakeholders, including the Strawberry Advisory Board, the State Department of Agriculture, and the university. This process would require California growers to pay royalties on the strawberry varieties which the program developed, but the agreement would also give them a competitive advantage by allowing them to license plant propagation, and granting them a substantial discount over out-of-state farmers who used the same varieties. Because strawberry farming was by its very nature a high-cost operation, the amount spent on patent royalties would be a very small percentage of the whole, and would only accrue as new varieties were developed and released.

When Royce was given details of the new patenting program after his return from Chile, he was dismayed to learn that the plan called for a high proportion of the patent moneys to go to him and Victor as "inventors," rather than directly to the program and the university. He objected to this arrangement on the grounds of equity, since most of his university colleagues were involved in work which, though not likely to lead to patented products, in many cases had just as much scientific or even commercial merit, and Royce felt that this could

395

lead to disunion and hard feelings. Pearl, for her part, still scrimping to furnish a home which she and Royce had purchased more than a decade earlier, had no problem with the prospect of patent income, and declared that Royce must be "allergic to money."[81] In the end, the patent agreement was finalized by the university with inventor shares to Royce and Victor intact, as was required by law, and Tufts, released in 1972, became the first patented university variety.[82]

The use of plant patents created an entirely new set of rules for the propagation of the varieties, and a system had to be developed to protect the university's valuable breeding stock on the one hand, and the interests of the growers who collaborated with the needed variety testing on the other. At a meeting of the California Strawberry Advisory Board in August of 1975, it was proposed that those cultivars undergoing detailed advanced testing be given a distinct university number, and their use was restricted by an official UC patent protection form, to make clear to all that they were property of the university and that unauthorized propagation was prohibited. A variety would then be named and patented only when it was reasonably probable that it would succeed commercially.[83] Accordingly, Royce compiled a list of all the most promising cultivars, and gave each of them a new variety number, beginning with the letter "C" for the conventional short-day varieties, or with "CN" for the day-neutrals.[84]

This new patent process seemed to run fairly smoothly until early 1976, when strawberries were discovered shipped from Florida which appeared to be Tufts, now a premier patented variety, for which no authorization had been given and no royalties paid. This patent infringement caused no small consternation among the California strawberrymen, who while scrupulously following the rules themselves, now faced competition from out-of-state growers who had sidestepped them. Royce was summoned to Watsonville to look over two crates of berries which had been shipped there for examination, and felt convinced that while one crate contained completely legal Tioga berries, the other did indeed contain the patented Tufts.[85] He pointed out that to make definitive identification of the errant fruit, he would have to go to Florida and observe the fruiting plants in the field, which he offered to do the following month. Realizing the unrealistic desire of the California growers to maintain a monopoly on the propagation and distribution of the new varieties, Royce struck on a clear solution in his own mind: "I think they ought to make peace with the people concerned, license out-of-California nurserymen, and be done with it. They want to do the impossible."[86]

On February 14, 1976, Royce flew to Florida in company of Victor Voth, and the two quickly confirmed the patent infringement. "We found the 'offending' growers quite open and easy to talk to, telling us quite openly what has gone on, except I suppose concerning the details on origin of the plants 'accidentally' in with Tioga plants, then increased here. It may be true, but there are an awful lot, and they have been increased rapidly. I doubt whether we shall ever find out exactly how they got there. The fact is that they are there, and now the only thing they can do is to give official permission and start collecting royalties in an orderly manner as possible. I doubt if it is worth the cost to unravel everything and 'punish' those who were responsible in the first place."[87]   A series of meetings ensued among the various representatives of the California strawberry industry, and while opinions swayed back and forth, ultimately the decision was made to sue the offending nurserymen in Florida over the breach, and to enforce the patent agreements strictly, though Royce was concerned about the fairness of the methods proposed for calculating the royalties from the patents.[88]

Royce continued to advocate for licensing nurserymen in other states to legally propagate the plants and process the resulting royalties, for he was convinced that failure to do so would simply create a "black market" for the badly-needed plants. The situation became even more complicated when the foreign demand for the strawberry plants was taken into account, and after meeting on campus with a visiting French nurseryman, Royce observed, "He spoke the truth on one point. We either must license foreign nurserymen, or simply consent to having our material stolen and used anyway. I have known this all along."[89] The matter of licensing continued to be argued by the members of the Strawberry Advisory Board, and gradually Royce's opinion began to hold sway. After one patent board subcommittee meeting he reported, "Fairly reasonable good sense now prevails, and with a bit of luck and compromise the case might be settled out of court. It is obvious that only the lawyers stand to gain from prolonging this very much. It could run for years, and who can afford that."[90] It was not until a year after the issue arose, in March of 1977, that the Patents Committee of the board finally decided to recommend licensing nurserymen both out-of-state and abroad, and the decision was taken to a meeting of the entire board on March 31. Of this meeting Royce wrote, "All was approved in principle, authorizing the patent committee to negotiate the rest through the attorney. At long last wisdom has indeed prevailed, and hopefully the rest will be easier to negotiate."[91]

The original lawsuit over the sale of Tufts plants in Florida continued drag forward, however, and Royce was repeatedly called upon to provide supporting information, and later a formal deposition as an expert in the case, a process which he found frustrating and time-consuming.[92] It took another full year before the parties in the case finally agreed to an out-of-court settlement which Royce considered long overdue, though in the end the case proved to be less clear-cut than he had imagined, for the defendants had their own side of the story, and the ultimate settlement of the case would require a complete change in patenting and reporting procedures to keep future patents from being challenged and potentially invalidated.[93] Finally, on August 2, 1978, in a lavish Berkeley attorney's office overlooking the San Francisco Bay, Royce attended a meeting with the defendant and his attorney, and most of the remaining details of the settlement were worked out. Royce commented, "The old ways of doing things are going and almost gone. Our test procedures must change and we must file papers soon on C38, 45, 51, 52 & 55 to avoid possible invalidation (this according to the lawyers who created the problem in the first place and now stand ready to solve it, for a fee of course). I am sick of the whole thing." Although frustrated with the attorneys, Royce felt no antagonism at all for the defendant, who visited him the next day at the Watsonville station with his wife and some fellow strawberrymen from Florida. Royce received them hospitably, and spent the rest of the morning showing them around his experimental plots, then took them over to Kuni Shinta's fields to observe his spectacularly productive Aiko plantings. "They were impressed, to say the least."[94]

The final step in settling the Florida Tufts case was to designate an independent expert on the East Coast to perform the necessary testing to confirm compliance with the law, which Royce arranged within days of the Berkeley meeting.[95] It was not long before the Tufts suit was a thing of the past, but the process had been an exhausting one for everyone involved, and there was clearly a need to simplify the patent process. In December of 1978, Royce and Victor both attended a meeting of the California Strawberry Advisory Board with representatives of the State Department of Agriculture and the university patent administrators. "The subject was termination of the patent agreement with the Advisory Board, and recovery of the same by the University of California. Sink or swim, a new day comes for the strawberry patents and in the future, things should be easier for us, at least I hope so. There are many things to be taken care of, but I am certain that it is best this way."[96] A few days later Royce met with representatives of the university patent office. "Apparently the patent office will handle the strawberry patents directly. Hopefully they can do a reasonable job without infuriating the industry. My fingers are crossed."[97] This change took patent decisions out of the hands of the Strawberry Advisory Board, an often contentious institution which was required to vote on policy issues, and placed it instead into the hands of university administrators, who had no such constraints, and were able to handle the patent issues more professionally and systematically.

Even while the details of the patent system were still being worked through, the first of the royalty income began to roll in. In February of 1977, Royce wrote, "I learned that the patents on strawberries brought in $64,000 last year (a lot of money). I guess that technically about $8,000 should fall to me. It will be tempting not to follow my resolve to use my share for the project. I think that I shall still do it. That would support an additional graduate student."[98] A few weeks later Royce received his first installment of patent income. "This week saw the arrival of the first royalty check for 'Aiko' (1975 year). I got a bit over $250.00, the first I have received. No 'Tufts' money is to be given up because of the Florida law suit. I have decided to simply donate the money to the project for travel, student support and miscellaneous supplies. This way, I shall be putting it where it should go, since I still oppose the whole patent idea in principle, and yet I will be able to maintain control."[99]

During the ensuing years, Royce's share of patent income would continue to grow, and eventually would come to exceed his professor's salary.[100] From this unaccustomed income, he would begin making sizeable contributions, chiefly to the University of California at Davis, but also to Utah State University and the University of Wisconsin, the institutions where he had received his undergraduate and graduate training.[101] The remainder would be tucked away in investment portfolios, while Royce and Pearl, both children of the Great Depression, would continue to live as frugally as before, making few changes aside from updating some of their worn household furnishings. As for the University of California, the strawberry patents would become for decades the entire statewide system's most lucrative series of patents, and would eventually bring many millions of dollars into university coffers.

After the Chairmanship

In 1975, after five years of service, Royce relinquished chairmanship of the Department of Pomology to a colleague named Noel Sommer, marking a major turning point in his university career. The months that followed proved to be a landmark time for the entire Bringhurst family, during which most family members would go through major transitions in their lives. On May 24, 1975, Royce's third daughter, Marla, was married to John Vaughn, an old high school friend whom she had known since childhood. Because John was not a member of the Church, the wedding did not take place at the temple, but was performed instead in the backyard of the Bringhurst home by Ernie Westover, a family friend and bishop of the Davis Ward, with all the family present.

Two months later, word came of the death of Pearl's mother, Jane Davidson, and the whole extended family traveled to Utah for the funeral of this venerable Scottish woman, who as a young widow had successfully reared a family of eight children during the Great Depression. The chapel was filled with her descendants, and at the conclusion of the funeral a recording was played of Grandmother Davidson herself, reciting a poetic version of the 23rd Psalm with her thick Scottish brogue, a recitation which she had given at Royce's own request at a family reunion many years before, and which left many of those attending the funeral in tears. The only member of the extended family unable to attend was Marla, who was on a hiking excursion in Alaska with her new husband and could not be reached. However, Marla was not left without a premonition of the death. In the Alaskan wilderness, she dreamed that Royce had called her to notify her that her grandmother had died, and after her return home the dream was verified.[102]

On October 24 of the same year, word came of another death in the family, that of Royce's older brother John Smith Bringhurst, known to all as "Smith." This death, too, had been accompanied by a premonition, for the night before Royce had dreamed of the death in great detail, except that he had confused Smith for his Father, who had died years before.[103] Smith had lived a tragic life, marred by alcohol addiction and the breakup of his marriage. He had suffered a crippling head injury ten years before when he had been thrown from a horse, and in his last years he had been only a shadow of his former self. Royce traveled to Utah to attend the funeral, and was the one chosen to deliver the eulogy, a challenging task which Royce performed with the gentle thoughtfulness and grace for which his public speaking had come to be known on occasions such as this.[104]

That fall brought other family changes as well, when Royce's fourth daughter, Ann, who had earned her bachelor's degree at UC Berkeley, announced her engagement to Ray Huffaker, a young man recently returned from a mission to Italy, and son of a UC Davis Agronomy professor who attended the Davis Ward. The wedding was set for November 26, 1975, in the Oakland Temple. Shortly before the planned wedding date, Royce's son John received a mission call to serve a mission in Guatemala. When the day of Ann's wedding arrived and the family prepared to depart for the temple, it was found that John, who was to receive his endowment in the temple that same day, was ill with a fever. Royce fetched his consecrated oil and administered to John, who immediately felt relief of his illness.[105] The family took this as a wedding-day blessing, and traveled to the Oakland Temple, where Ann and Ray underwent the same simple sealing ceremony that Royce and Pearl had received many years before. The newlywed couple would remain in Davis while Ray completed law school and subsequently received his doctorate in Agricultural Economics at UC Davis.

John's decision to serve a mission was an important event for Royce, whose own mission had marked one of the pivotal points in his life. Although missionary service had by now become much more of an expectation for young Latter-day Saint men than had been the case in Royce's day, he had been careful not to pressure John about a mission, even when he had turned nineteen and become eligible. John seemed deliberately to delay, and had taken most of the year to prepare, but when he finally submitted his application and received a mission call to Guatemala, Royce was visibly pleased, and doubly so when he learned it was to be a Spanish-speaking mission. On December 31, 1975, Royce and Pearl accompanied John to Sacramento, where he was set apart as a full-time missionary by Homer N. Stephenson, president of the Sacramento Stake. The following day, New Year's Day of 1976, they dropped John off at the Sacramento airport and bid him farewell. That same day Royce renewed a practice from his own mission some thirty-six years before, and began keeping a daily journal which he would continue until well after his retirement. This record left behind telling glimpses of his activities and thoughts.

As the year of 1976 dawned, other changes occurred. Pearl was registering for her final quarter at UC Davis. Florence's husband Larry had just received his doctorate in history, and the two of them were about to depart for a two-year teaching assignment in Brazil.[106] Ann's marriage and John's departure on his mission left only Margaret at home, now halfway through her sophomore year of high school. The last of their six children, Margaret was intelligent and musically talented, with an increasingly impressive soprano singing voice, and a love of the violin, which she had learned in school, following the example of her older sisters. But she was also more distant from her parents than any of the children except perhaps Jean, and more reserved than her older siblings. The remaining years home alone with her parents would prove challenging for her as well as for them.

For Royce also, life had taken a distinct turn. Having relinquished his chairmanship, he was finding the transition back to his previous professional life more difficult than he had imagined, and recorded in his journal: "Only now can I realize how much I lost professionally by accepting the job. I am out of touch with many things that I did so easily before, and seem to be in different thought and habit patterns; not necessarily bad. The question is, how to turn it all around again and get with it. It is sort of like asking to be young again when you

401

are already almost old—it will be most difficult."[107] Though he no longer shouldered the administrative burdens of the department, his work as chairman had given Royce exposure to the extended workings of the university, and he would be called upon repeatedly to participate in committees both on the UC Davis campus and in the statewide university system. He struggled especially to organize his duties, and manage the endless stream of paperwork which continued to cross his desk. After an entire day spent cleaning his office, he observed, "I still have a backlog of 'must be dones' nagging my heels. One should really handle things but once in the business I'm in, dictating a letter if it is called for, or filing in a spot where it can be retrieved upon demand, or throwing it away. All things easily said, but difficult for me to do. For some reason I seem to prefer operating from a crisis position, and that makes no sense at all."[108]

One positive change which brought Royce a sense of accomplishment at this time in his life was in the area of physical fitness. He and Pearl had both been recruited as experimental subjects for a long-term university study of the effect of exercise on health, and as part of the study, they had begun to participate in regular exercise classes on campus. Thus, at the age of nearly sixty, Royce, with Pearl, began a daily practice of "jogging," or running two to three miles in the early morning hours. Royce settled quickly into this rigorous routine, which after a short time he found to be invigorating rather than a chore. "It's a good way to start the day. I feel *alive*. I don't really want to live forever, I just want to be healthy and feel good while I live."[109] Royce would continue this practice for many years, rain or shine, and a couple of years later, after his morning run, he wrote, "I ran again as usual, feeling very good about it. Perhaps someday I will slow down on that sort of thing but not yet. Personally, I'd rather not live as long, as some suggest with exercise, than to get out of shape again. I can indeed 'run and not be weary, walk, and not faint,' as the good books say, and intend to keep it that way."[110]

At the beginning of 1976, Royce was assigned as instructor of a class on citrus and subtropicals, a class he had not taught in years, but which drew on his early experience with avocados and other subtropical fruit at UCLA. His students included all levels from freshman to graduate, but despite the challenges of teaching a course to such a varied group of students, Royce enjoyed the class, and continued teaching it year after year. Since the course was taught in the winter quarter, Royce began a tradition of treating the class to an annual field trip to the Wolfskill Experimental Orchard in Winters, where he allowed students to forage for fruit to take home from among the wide array of citrus and other trees which ripened during the winter there. He also made a point to pick up unusual fruits such as cherimoyas and avocados during his trips to the South Coast Field Station in southern California, and in future years he would come to the first day of class in January with a load of sweet, ripe tangerines which grew on a tree in his back yard. Royce's excellent teaching and the edible samplings of the subject matter proved to be a good combination, and the course became a popular one with students, who Royce came to look on with particular fondness.[111]

Though widely respected as a teacher and scholar, in his journal Royce tended to be ruthlessly self-critical about his teaching and his work. After a strawberry conference in San Jose, he wrote, "Our presentation and question-answer seemed to go quite well although I am never fully satisfied with most things I do, including this. I should concentrate on excellence more since I am basically committed to a quest for perfection in everything. Still, my greatest enemy is time, coupled with practiced procrastination that I find difficult to root out of my system."[112] Royce's frequent trips out of town on strawberry business meant that work tended to pile up in his office in his absence, and there was almost perpetually a riot of papers arranged in sundry stacks on his desk, which presented an endless chore, and resulted in much personal dissatisfaction. "I still waste a lot of time in and about the office. I really should change my work habits and systematically reduce the backlog of things to be done. I am really hurting myself, not to mention those associated with me. I always plead 'no time' and 'overwork,' but the problem is really how I organize and use the time, not the hours I put in."[113]

In March of 1976, Royce's world was shaken when he received word from his brother George that the family in Utah had decided to place his mother in a nursing home. Although he knew that his mother's memory had been declining for years, Royce found this news deeply disturbing. He had always regarded himself as the one to step forward and take on responsibilities on behalf of the family, something he had done many times on the farm, in the mission field, as a soldier, and as a student. This had been a particular point of pride for his mother, who had faithfully written him, kept him aware of family events, and encouraged his activities. Now that she had reached her declining years, Royce, with an established life and career in far-off California, felt helpless to do anything at all to help her. The thought of his mother in a 'rest home' was a particularly bitter one. "The old system was kinder. I think that Pearl would have kept her until the end. I feel badly about it because I haven't really lifted my hand, and now I feel impotent and not in a position to do a thing but accept it. I hope that this is the right thing to do." The next day he added, "Life is complex and I wish sometimes that I could just erase much of what is, along with my memory of it, and return to the simple uncomplicated, naive world that I knew as a boy. Unfortunately, that can never be, so I will face this day and many more, playing the several games that are habit to me now, and wistfully reflecting back on another world that never really was, except in the mind of a boy with a clear conscience and few worries."[114]

The following month Royce drove to Utah and paid a visit to his mother, whose mental decline was obvious. "I am sure that she knew me in a general but not in a specific sense. She was getting ready for dinner, already seated to eat. I found the sight (all those old people waiting to die) depressing, and though they invited me to eat and the food looked all right, I could not eat it, and didn't accept." Lost in thoughts of the past, Royce made his way to his old neighborhood. "I drove out to Bennion Ward to look at the land of my youth. The old house remains, forlorn and lonely beside a freeway that doesn't seem to go anywhere. Nearby is the old Dimond home, and little else that is familiar. Henry's and Uncle Louis's old pioneer home are gone. Truly 'you can't go home again'[115] except in dreams as Mother now seems to do." Returning to pay his mother one last visit, he wrote, "I still doubt whether she knew me specifically, but I suppose that matters but little at this point. Her room is nice and everything seems to be in reasonable order. I don't know whether there were real alternatives available rather than having her live there. I begin to think not, and certainly won't suggest otherwise. Mother talks nonsense a good share of the time and seems to erase as rapidly as things pass through her mind. I am certain that she retains but a very vague memory of my visit as I write this."[116]

Royce never reconciled himself with his mother's decline and dementia, and particularly with her placement in a nursing home, and he would always feel a sense of failed responsibility in connection with it. Even after the passage of years, almost beyond all reason he could not avoid the feeling that his mother might have retained her mental powers with an appropriate degree of nurturing. "I am convinced that aging can only take its toll if one wills it to by refusing to sustain one's mental capacity. Mother wouldn't be in the condition she is in had she continued to do the things she had been doing. I only regret that we don't have her in our home in her old age. Regret of course will not change that."[117] Besides the sense of personal accountability he felt, Royce's professional understanding of heredity made him keenly aware of the implications his mother's condition bore for his own future, and he gave voice to this apprehension. "I hope that I can maintain my mental faculties as I age. I would rather die than lose my mental ability. How sad to see mother as she is versus what she was."[118]

That June, Pearl completed the last of her undergraduate college courses at UC Davis, and on June 19, 1976, Royce went with her to campus to attend her graduation. "It was long and hot and Pearl was almost at the end of the line. She looked good and was happy. I'm not sure that I would have had the spunk and drive to stay with it as she has done. It was worth it though, because her whole view of life has been changed, and she will continue to use what she has learned. Her major was English Literature. I'm really proud of her, and happy that I did have sense enough to stay with her during those seven years she has been a student. There were times that I questioned the value of it all (to myself) but I never said anything. I am completely convinced now that it was the best thing that she could have done with herself, and that her life is enriched permanently, and so is mine by association."[119]

Pearl's college degree, attained when she was 57 years old, was a significant milestone in her life. Born in a family of modest means, and part of a generation of women for whom domestic responsibilities generally took precedence over educational opportunities, she had long had a desire to attend college, and had once made a promise to her mother, who had died only the year before, that she would one day earn a degree. Even when attending college at American River and then at UC Davis, Pearl had taken pains to limit her class work to those hours when the children were away at school, regarding her role as mother as first priority. Despite her devotion as a mother, Pearl had always found it difficult to find her place in the family—Royce's authority in the home had always been unquestioned, whether because of the severity of his personality which resulted from military service, or because of his educational attainments, and Pearl was often treated less seriously by family members. Even Royce, whose affection for Pearl never wavered, realized that he frequently did not give her the respect she deserved, and though he valued their relationship, he always felt a strong reserve about expressing his feelings. "I don't know why it is difficult for me to loosen up sufficiently to say the things that I know I should say. I think about them and then don't. One day maybe my tongue will be free enough to be completely honest with her. I suspect that she thinks that I am frequently displeased with her, and the opposite is really true."[120]

The Church, Science, and Faith
Royce's church calling as bishop of the University Ward in 1970 had also marked a landmark event in his spiritual life. Although he had never wavered in his commitment to the Church, and was always willing and faithful in his discharge of assignments and duties, he had frequently struggled to reconcile the simple religious faith of his youth with the complex and dynamic world of scientific inquiry

405

which was the basis for his life's work, and which permeated the intellectual world around him. This environment had repeatedly challenged his faith, though he had stayed the course, retaining at least outward fidelity to the basic foundational principles of the Church, even when his inner convictions wavered at times. He felt a strong commitment to the moral teachings of Christianity as taught within the Latter-day Saint faith, and he reflected on them frequently, though he often felt that he fell short of living them. His spiritual feelings, which fluctuated between tentative and turbulent, found something of a home among especially the male members of the Davis Ward, many of whom were university professors like himself, and faced similar intellectual and spiritual challenges. Their discussions in church meetings were often perceptive and insightful, and Royce felt safe there expressing his own views, and even challenging from time to time what he thought were simplistic approaches to religious ideas.

When he was called to preside over the University Ward, his point of view of necessity shifted. Rather than concentrating on his own faith and spiritual well-being, his focus was redirected to the spiritual lives of the sizeable group of intelligent, in some cases brilliant young people who made up his congregation. These were a varied group, some single and some married, a few with young children, and they included among their number his own married daughter Florence and her husband Larry. As bishop, Royce became their spiritual leader, and in many cases, their personal counselor. To him they brought their weighty personal decisions, and sometimes their own trials of faith, or their transgression of major Church commandments. Royce was looked to by all as an example and a source of strength, and he rose to the occasion, thoughtfully discharging his duties and providing mature leadership. The new University Ward quickly became a unified and vibrant group, whose members, an ever-shifting group of students who came and went with the passing of academic terms, supported one another as they faced the most challenging and formative years of their lives. Royce met with his counselors to staff the various organizations such as Priesthood, Sunday School, and Relief Society. A small Primary was organized for the young children of the married couples. Royce prepared some of the young men for missionary service, and corresponded with them as they served. He also presided over the Sunday services, which included Priesthood, Sunday School, and Sacrament meetings. From time to time, this required him to exercise considerable tact and wisdom, as when a young couple in the branch grieved publicly after losing a child to meningitis, or when a young man who had been influenced by an apostate group stood up in testimony meeting to declare that he had lost his faith in Church leaders.[121]

Royce's tenure as inaugural bishop of the University Ward left him a respected figure in the university Latter-day Saint community, and when he was released after about three years, he continued to play a role with young adults of the Church. Back in the Davis Ward, he was assigned as a teacher to a Sunday School class for young people making the transition from youth to adulthood (a class attended by his son John, and later his daughter Margaret), and gave thoughtful lessons that stimulated self-examination and fortified faith. At the same time he continued to struggle frequently with his own faith, and observed that although his students did not know it, his carefully prepared lessons were as much for his benefit as for theirs.[122] Royce was also called upon from time to time to make presentations to the students at the local Institute about scientific topics which touched on matters of faith, something his own experience enabled him to do with unusual clarity and wisdom. By far the most important of these topics was evolution.

Evolutionary theory was not new in Royce's lifetime, but it had always been controversial, as many religious people, both inside and outside the Church, considered it to be inimical to religious belief in a Creator. Royce had grown up with a strong foundation of religious faith, and it is probable that he remained largely aloof to the evolution controversy in his early life, until as a graduate student of genetics at Wisconsin, he came to understand how completely evolutionary theory served as the underpinning not just of genetics, but of the entire fields of biology and geology, and how consistently it was supported by scientific evidence. This new understanding likely conflicted with the simple religious convictions of his childhood, and it was certainly hard to bring into harmony with the beliefs and statements of many Church members, including some leaders, who rejected the theory outright, and characterized it as one of the great evils of the time. It was in this setting that Royce had done the soul-searching required to reconcile his genuine religious impulses with his equally genuine scientific convictions. This had for years been an underlying theme in both his scientific and his religious life. As a researcher, Royce was unapologetic about embracing what he saw as a clear scientific truth, and he had little patience for so-called "creationists," whom he felt distorted scientific evidence to further their own ends. But he also had compassion for Church members who, like himself, honestly grappled with the issue.

In his genetics classes at the university, Royce had addressed the religious and moral implications of evolutionary theory in the setting of human genetics, particularly since mankind had progressed to the point of being able to alter his environment rather than solely being shaped by it. As he addressed Latter-day Saint audiences, however, his objective was to enable them simply to understand the scientific principles involved, in order to allow them in an informed way to reconcile their own beliefs with the existing scientific evidence, as he had spent his own lifetime doing. A talk he gave in 1974 exemplified his approach to this:

"I do not wish my remarks to be construed as strong advocacy of my particular position *vis á vis* evolution—how you adjust your own thinking concerning the real world of biological thought is your own personal problem. However, 'Man cannot be saved in ignorance,'[123] and I am convinced that we should be acquainted with some of the hard evidence upon which the case for evolution in general, and human evolution in particular, rests. We should also be aware that almost every day adds something more convincing to the accumulated store of organic evolutionary knowledge. It is not a casual decision, or some vague universal devil-inspired deception, or faulty understanding that has convinced virtually all biological scientists within and without the Church (including your speaker) to accept the basic dogma of evolution—the evidence is there, the facts support it, and it won't go away. Furthermore, I assure you that as with all truth, it can be lived with most comfortably, once you get adjusted to the idea and appreciate the beauty and order of it.

"I have no word of condemnation for those within and without the Church who choose to see otherwise. Within the context of their understanding, they have made a different adjustment. I do not understand them or their rationale, but they are entitled to their views. I would no more raise my hand to support those who would force the issue on evolution as a test of one's standing or qualifications to teach in the Church, than I would raise my hand to support a counter-test from those of an adversary point of view. I might believe that they have made a fundamental error in judgment, but don't we all? We are set apart from animals by freedom and reason, and we should never place ourselves in a position where we force others, or are forced by edict, to compromise either freedom or reason.[124] I am convinced that God has not and will not force the issue, nor will he condemn the honest in heart who, following his conscience to the best of his ability, makes a mistake."

Royce then presented some of the compelling evidence favoring the principle of evolution of species, which he regarded as compatible with doctrines unique to the Church, such as eternal progression. For him, the fact that species evolved in no way implied that there was no purpose in their creation, or that God was not the moving force behind the process. "We can certainly believe with a good conscience that God, the master builder (and 'Creator' means 'organizer') brought it about. The mechanism is very stable, but subject to change, including very rapid modifications. The Creator has seen fit to build in the change apparatus, with the capacity present but masked. How? How long? We can ponder and wonder. The circumstantial evidence clearly points to evolution, in the Darwinian sense, as the way, and millions of years as the time." But Royce included the essential conclusion: "Man, however, is a special "creation" of this process, not just one more animal."[125]

After giving one such talk, Royce remarked, "The young people of the Institute had me present what has become my annual talk on 'Evolution.' The reception is generally good, and some of them definitely benefit from what I say, at least it seems so to me. It is difficult to tread the narrow line between outright refutation of what some teach as the word of the Church (outright 'Evolution is a false doctrine' idea) and a 'here is my impression of evolution and you are welcome to it for what it's worth' attitude. If I knew more, I probably would not say as much, but 'fools walk in where angels fear to tread.'" Although he believed strongly in the scientific evidence for organic evolution, Royce by no means rejected belief in a Creator, and indeed felt that such a belief was supported by the same evidence, as he recorded privately in his journal. "There is no question in my mind whatsoever about the reality of evolution—the time element dealing with pure chance makes it necessary to assume a strong directional force operating on the process. The real problem is reconciliation between evolution and fundamentalist concepts of origin in the Genesis sense that many feel they must retain. I accept that in a very liberal and remote allegorical way only. That leaves you free to ponder more important matters."[126] For Royce, the "more important matters" included the moral teachings of the Church, particularly the way people lived their lives and treated their fellow beings.

Royce especially rejected the notion that there was a religious or spiritual truth which was separable from natural truths which were subject to scientific investigation. "This really bothers me when they keep repeating the old bankrupt idea that there is a difference in knowledge, one that you get from reading the standard works and the books that many of the leadership feels compelled to keep writing, and the other, 'the learning of the world,' which seems to embrace everything else, and is by definition bad."[127] Although he took a more conciliatory posture in his public speech on the topic, in his private thought Royce was far more severe with those in the Church who resisted what he regarded as compelling scientific evidence, whom he characterized as "hard line fundamentalists who, supported by certain general authorities, live in a 19th or 18th century world as far as reality is concerned, unrealistic, unrelenting and dogmatic. They attempt to freeze everything in tune with something that never was and never will be; rather, something that they hoped for. Unfortunately for them, it hasn't worked out, and they are like eloquent ghosts flinging words to the winds, missing the whole point of the Restoration by Joseph Smith. He would have spurned their rigid, tradition and myth-bound nonsense."[128]

Surrounded by the intellectual environment of the university and by a culture which habitually challenged authority, Royce allowed his strong feelings on evolution to extend to other religious topics such as Church history, which he worried had been sanitized to avoid critical discussion of its more controversial aspects. Though still faithful to the Church as an institution, he had begun in his mind to assume a more cynical, and at times almost belligerent attitude toward some of its practices. Royce especially came to dislike the published lesson manuals, which he felt gave a bland treatment of important topics, and as a religious teacher he took pains to prepare his lessons to engage and inspire, making liberal use of scriptures and statements of Church leaders, but also materials from scholars and literary figures. A highly-regarded teacher at church, he imposed a high personal standard on his own teaching, but in tending to apply the same standard to the teaching of others, he was frequently disappointed. He was, however, in earnest in aspiring to a scholarly standard of church teaching, and insisted in his own mind that Joseph Smith, the Church's founding prophet, "envisioned the Church as a community of scholars."[129]

In spite of his excellent and thoughtful lessons, his active participation in church activities, and his willing discharge of his church duties, on a personal level Royce had once again begun to flounder in his spiritual life. It was not that he faced a "crisis of faith"—if indeed he had ever had any such crisis in an immediate sense, he had weathered it long before. But his devotion to the inherently skeptical field of scientific inquiry had not been counterbalanced by a similar devotion to spiritual things, and the imbalance had taken its toll, leaving him little room for a belief in the miraculous. When, for example, a special fast was called by Church leaders to pray for rain for the drought-stricken region, Royce wrote resignedly, "Faith should move mountains and maybe bring the rain, but is there enough, and is it really so? We are schooled in a naturalistic view of all things; it is hard to think 'supernatural' regarding them, at least for me. It rains and fails to rain or whatever upon the 'just and the unjust.'"[130]

Royce's choice to assume a cynical posture toward the Church, though intellectually gratifying, also did not bring out the best in his character. In public he was civil and generally kind to other Church members, but in the recesses of his own mind he often thought of them critically, at times even demeaningly, and freely recorded these

410

thoughts in his personal journal. On one such occasion, after giving an unflattering account of two women's testimonies during a Sunday testimony meeting, one of which he regarded as absurdly humorous and the other as over-pious, he rather flippantly acknowledged, "How hypocritical of me to write this down, but that was my reaction, so I am at least being honest too. Perhaps, I should just die, too, if this were taken back to those I sort of slander. They probably meant well and had no intention of amusing Philistines of my ilk. 'Twill matter little once the curtain is down on our little day anyway, so I guess I don't care. I even feel happy being a confirmed irreverent cynic."[131]

This attitude of cynicism left Royce with a sense of lethargy when it came to spiritual things. He continued to be "active" in the Church, and kept its major commandments, but few things seemed to inspire him, and he allowed himself to indulge in private moral indiscretions which he only hinted at in his journal, but which clearly weighed him down spiritually. One Sunday after church services, he wrote, "A typical Sunday with few unusual things. I attended my meetings with minimal enthusiasm. I'm really not on a very high plane right now, but I recognize it as my fault, no one else should be blamed, and in particular Pearl. She is always there in a very positive way. We have been doing all right on the surface, but Pearl does better than average and I do worse, hence we come out middle-middle."[132] On another Sunday some months later, he recorded, "Priesthood meeting was fairly interesting what with our little games of partial belief and disbelief, held at bay with the apparent 'strong belief' of others which I tend to suspect privately as something assumed or just hoped for but not fully felt. Can that be true faith? I suppose so, for sometimes I feel it too."[133]

Royce clearly felt a longing for the spiritual strength which had accompanied him in earlier years. After hearing talks in church by a particularly sincere couple whose faith was still strong, he wrote, "I envy people with simple faith such as theirs evidently is."[134] Similarly, when his son John wrote home from the mission field, Royce remarked with a tinge of sadness, "We had a fine letter from John today. I would like to have the feeling he has about his work once more."[135] But there were impediments within his own character which left Royce with a sense of unfulfilled yearning. "We all share a piece of the nostalgia of the past, a desire to turn back and become as we were in the age of all prevailing belief and innocence, and yet are too cowardly to make the move except in our dreams. Thus the story never really happens. We remain our same timid, reluctant selves, hoping that somehow someone else will seize the moment and make everything right again. How sad the regrets that we carry through life; lost opportunity unfulfilled dreams; studied, deliberate unwillingness to stand up and be what we could be. Oh to change so much, set aside the facade and live the dream. I fear I shall never have the courage to do it. I am not even certain that I know what the dream is or if there is a dream, and the world moves on without me."[136]

At some level, Royce recognized that the source of the problem lay within himself. "I am lonely and a bit unhappy with myself. Pearl does so much for me and I don't really respond. I guess I should concentrate on others and forget myself. I do wish that the Church meant more to me deep down. I have my problems there. Sometimes I really feel it, other times it just isn't there for me, and I suppose the fault is with me. I certainly can't lay it on anyone else. Sometimes I think I am alone with this problem, other times I am sure that most of my church friends have it too."[137]

This spiritual malaise, coupled with his professional focus on the natural world, at times left Royce almost bereft of hope in an existence beyond this life. After the serious illness of a friend in late 1976, he wrote with a tinge of bitter sadness, "Truly it is said that it is not given to man to tread the earth for long. All shall pass away, and this too. I don't expect much beyond that. Perhaps I did once, but now it doesn't even matter to me any more. I have few friends, and even fewer who really care about what happens to me or mine, other than Pearl and the children."[138] A half year later, after the death of an acquaintance, the same thoughts emerged, "I don't fear death, although oblivion frightens me a little, for I know how soon all that we are fades into the shadows and is lost. Remember, remember, remember..."[139] A short time later, after talking with childhood friends during a summer vacation in Utah, his thoughts assumed the form almost of an existential cry in the night: "I still feel like 'the hunter and the hunted' joined as one in fear of impending darkness that is sure to cover all. 'I wouldn't change it even if I could,' and my fear is only a shallow fear, largely a fear of oblivion. Does anyone out there see me or hear me? Does anything really matter? Will our tomorrows enlighten us about our yesterdays? Will we 'sink and be forgotten like the rest?'"[140] As that year drew to its close, Royce summed up his spiritual misgivings. "The old year passes, oh so quickly, and something within me dies with it. All the things I might have done, all the thoughts I might have brought into reality. Perhaps, it is just as well. I failed in some things and accomplished others, but I think not. I have this sense of doom that haunts me and rides me toward the end when only oblivion prevails and I shall forget and be forgotten. Hence the struggle to avoid the inevitable."[141]

Although free with his feelings in his private journal, the responsibility which Royce seemed to feel to respect and safeguard the faith of others prevented him from expressing such thoughts to anyone, least of all his family. He had always been reserved about his innermost feelings, and seldom if ever shared them, even with Pearl.[142] Though Royce was sociable enough in public settings, his was in some respects a lonely existence, a fact which he freely acknowledged. "I have no real close friends, although I am friendly with most every one. I think this is out of long-term personal preference of a high degree of privacy."[143] This sense of aloneness was accompanied by a dislike for crowds, which made him uncomfortable; it was also aggravated by his poor hearing, which left him feeling isolated and detached in many social situations.[144] Most of the time Royce's professional and personal life kept him busy enough not to dwell on these existential matters, but they were never far from the surface, and repeatedly emerged in his contemplative moments. "I am a dull sort with few outside interests and addicted to work every day, even if it kills me, and some day it may just do that, leaving me to ponder (if the dead do) over the fact that there is little of value that resulted from my brief existence."[145]

At the same time, in spite of his doubts, Royce continued to value and even to treasure his Latter-day Saint heritage. Invariably, when exposed to religious practices of other denominations, he would relate them to the practices of his own religion, which he always viewed in a positive light.[146] He still seemed to regard the Church as the most authentic and legitimate of the Christian religions, and he felt a keen sense of satisfaction and pride in its history. On one Pioneer Day (a holiday which went uncelebrated in California), he wrote, "Our heritage has become much more respected as a vital, important part of the colonization of America than ever before. Our heroes are now national heroes and not 'deluded fools leading ignorant people who have been deceived.' Without the epic journey of the Mormons and even more important, their persistent, well-organized community building centered around self-sufficiency, the west (and even California) might be quite different from what they are today."[147]

This sense of identification was fortified when the Church, under the leadership of President Spencer W. Kimball, began a series of landmark changes, beginning in 1976, when a reorganization of some of the leading quorums of the Church took place. "All 'assistants to the twelve' were released and the 'First Council of the Seventy' was reorganized and a 'First Quorum of the Seventy' established, taking in most but not all of the membership of the former groups. This makes sense to me. The organization was clumsy, indirect in function and unwieldy. Now if they would reduce the number of conferences it would help. I know I have no right to recommend things, but I do have a right and a need to think them. We are a worldwide power, and must face up to that fact. A lot of change will be required down the road if the Church is to remain truly viable."[148]

Even more significant was the 1978 announcement ending a longstanding practice of denying the priesthood to blacks, which had long been troubling to Royce. "To me, and in my opinion the whole Church, this ranks as the most important day in Church history during my lifetime. I have never been happier about a Church announcement. This is one thing that has bothered me tremendously for as long as I can remember. I have never really accepted the doctrine because it made no sense no matter how I viewed it. The policy was never clear and unambiguous when you consider South Sea islanders and other 'black' skinned folks. Genetically, it made no sense either. How very pleased I am that we can now relegate this to the museum of other strange views already discarded. It matters little to me whether it was in fact a 'revelation' as stated publicly, or whether it was a reaction to building pressure put on the Church from all sides and from within, to change a policy born in the era of slavery and unfortunately perpetuated. I tend to favor the latter view, but it makes no difference. The result is still the same. I have always felt that this had to happen in my lifetime, and so it has."[149]

Still later, a consolidated meeting schedule was announced, confining most church meetings to a three hour block on Sundays, rather than the exhausting array of meetings which had been held on weekdays and throughout the day on Sundays. Royce had long wearied of the busy Sunday schedule which seemed to allow little time for anything except shuttling back and forth to church meetings. Hearing of the change, Royce wrote, "I welcome this and hail it as overdue. How many more pleasant surprises will come before Pres. Kimball's work is done?"[150] Royce savored the new-found free time on Sundays. "How good it is not to bounce back and forth to church like an errant yoyo."[151]

But Royce continued to face very personal spiritual struggles. In one year-end journal entry, he summarized his spiritual standing at this turbulent time. "I am not too proud of a number of things. To me my real accomplishments are rather meager, and I must admit to playing the rather cynical way of life that I have come to follow in recent years. How can one be absolutely frank in dealing with people and circumstances and still maintain a few friends? The answer is always within. Most are willing to tolerate and forgive most everything within reason once the first step is taken by one or both parties. I prefer to think that I have exercised more wisdom than unwisdom, and that I have done more good than harm. I always wish to think that some of the things I have done will live for a little moment after I am gone, and that someone somehow will believe that my life has been worth living.

"Whoever might have occasion to read this can pass his own judgment on me as an individual and as a member of the LDS Church. Likely, they will have many of my weaknesses and feel a certain kinship with them. Truly all of mankind are brothers, and we cannot escape common destiny. Our little hour is such that we scarcely make more than the tiniest ripple on the stream of life, and yet collectively that is all that there is here, and human progress and achievement is the sum total of all our struggling and striving on through the generations of time. If this were our only immortality, I believe it would be enough for me, but I like to believe that there is more beyond that. Whether the simple plan we have in the Church is the whole thing, I have my doubts because we are not capable of really comprehending the significance of the whole scheme of things. Death takes up long before our insight develops sufficiently to have more of a complete understanding, and much of our knowing is mere verbalization of desires and hopes. Perhaps this is faith."[152]

Strawberry Program—New Cultivars and Old Colonies

Even during his term as department chairman, Royce had continued his work in the expanding strawberry program, overseeing the testing of the new varieties and selections in the Northern California experimental stations, in collaboration with Victor Voth, who continued to shoulder similar responsibilities in Southern California. After his term as chairman had elapsed, the work on the strawberries continued and expanded. For Royce, this involved a grueling routine of travel, sometimes including several trips a week to experimental plots in San Jose, Watsonville, and Salinas, or to the nurseries around Redding, often leaving early in the morning and returning late at night, or staying overnight in Redding, San Jose, or Watsonville. He frequently worked in the nearby Winters plots on Saturday mornings before doing his yard chores at home. Though the travel was exhausting, Royce loved his work in the fields, which for him embodied every aspect of agriculture that appealed to him—the open spaces, the growing of the crops, and the working of the soil, without the constant tasks and financial pressures of actual farming.

The new strawberry releases continued to succeed beyond all expectation, and now the next generation of seedlings were entering advanced field testing, and were beginning to show even more promise. In the plantings of some of the experimental day-neutral selections, a new milestone was reached in Southern California, which Royce recorded after touring the plots with Victor. "We observed the day-neutrals in the first real commercial tests. They look good to me. There should be some excellent fruit within several weeks. In all cases they are picking already, and this means that they overlap harvest with

415

'Aiko" and some 'Tufts' in the Central Coast, giving year-round harvest of UC varieties in California. We found the same in Orange County, including our plots at the field station."[153] The prospect of a year-round California strawberry harvest, made possible by the day-neutral selections, had been one of the goals of the program. The experimental short-day varieties were no less impressive. "They have never seen better fruit in the Santa Clara Valley. Several of our advanced selections are simply spectacular to see compared with Tioga, and some are outyielding it. Victor and I then drove to the Watsonville plots. Again, the fruit there was superb. We have never had nicer large fruit than on the winter planting."[154]

Royce and Victor had by this time become increasingly efficient in running the complex breeding program.[155] Royce had become skilled at the time-honored method of selecting plants with optimum traits to become the parents of the next generation, and by now each new generation of plant selections had features which made it clearly superior to the previous generation. The next round of releases was scheduled for 1979, and an unprecedented six new California varieties would be named and patented that year.[156] These included three important new short-day varieties, which were called "Douglas," "Pajaro," and "Vista."[157] Of these, Douglas, which had early fruiting and extraordinarily high yields, and Pajaro, which was known for firm, high-quality fruit, would become major commercial successes and begin to replace the older varieties. Also released that year were the first of the new day-neutral varieties, which bore the names "Hecker," and "Brighton;" a third variety, named "Aptos," was later added.[158] The role of the day-neutrals had still not been established in the commercial plantings, and it was as yet uncertain how they would fit in. Noting that they had the advantage of moderate *Verticillium* wilt resistance, Royce introduced them as "likely candidates for home gardeners, as well as of possible interest commercially."[159] With time and experience, these day-neutral varieties with their unique fruiting characteristics would begin to fill in gaps in the California harvest season, and both Hecker and Brighton would come to be minor but commercially significant crops.

The release of the new university varieties had been accompanied by continued advances in growing techniques, for which Victor Voth's skills and instinct played the leading role. The two basic planting systems, winter planting (of plants from high-elevation nurseries, planted out immediately after digging) and summer planting (of plants stored and chilled from the previous winter) were still in use, and Royce's new varieties, bred and tested specifically for these two systems, gave the growers very good results. However Victor, with Royce's help, continued to experiment with new growing techniques, and during the early 1970s a series of new innovations had been introduced which increased the yields even more. One of these was the use of planting slot fertilization, in which fertilizer was applied during soil preparation, in a slot adjacent to or beneath the plants, a practice particularly important with some varieties and planting systems. Another innovation was the use of wide beds, prepared to be 52 or 60 inches wide, rather than the conventional 40 inch beds of the past. Victor and Royce had determined experimentally that the wider beds enabled an increased density of planting (more plants per acre) without affecting the yield per plant, and therefore resulted in a higher yield per acre. The third new innovation was the development of drip irrigation through specialized drip tubes, which resulted in better control of irrigation, and greatly decreased fruit loss. This was yet another example of a costly growing innovation recommended by Victor and Royce which more than paid for itself in overall yield.[160]

Meanwhile, Royce continued the process of pollination and seed collection for the next generation of crosses. This was the crux of his work as a plant breeder, and still required meticulous care to ensure that the pollination was done correctly. Once the pollinated fruit had ripened, Royce had to collect the minute seeds, "stratify" them by chilling, and treat them with a strong acid solution to assure germination. This process always seemed a little miraculous to Royce. "The new planting in the greenhouse is just great. Every time we treat the seeds with concentrated H_2SO_4, I think that we are about to lose the whole lot, but they always come through and that is what counts. This is the next round of day-neutrals and the *Potentilla* hybrids and derivatives I worked on in my office last year. All the good seed is popping right up with almost complete germination. So much for my pessimism. Why do I always doubt myself?"[161] After germination, the seedlings were set out in labeled pots in the greenhouse, and later transplanted in the field at Winters or in Southern California, where they would be grown to see which ones had enough promise to warrant further evaluation. As the process became refined, the results became more and more impressive. One Saturday, after looking at some of these still unselected plants at the Wolfskill Experimental Station, Royce wrote, "Up early and into the field at WEO to take notes on the strawberry seedlings. Without doubt, these are the best we have ever had. They could be harvested as is, and sold competitively with what is seen on the local markets."[162] After taking the notes necessary to select plants for further testing, he added, "I have selected over 200, and some are truly outstanding, the best group of crosses we have ever made."[163]

At the same time, freed from his chairmanship obligations, Royce began to spend more time working with newer laboratory techniques which he thought might show promise in his work with strawberries. One of these was in the emerging field of plant tissue culture. The strawberry program had already used one form of tissue culture by making use of plants derived from meristems to help eliminate viruses from the various plant lines.[164] However, when newer and more effective methods of plant culture and storage were introduced at a professional meeting, Royce immediately recognized the possibilities. "It occurred to me that there are two practicable and practical applications: first, for storage and maintenance of breeding stock, inactive but of possible interest or use later. This is the answer to the perennial problem of wanting to keep many things that practical considerations dictate should be discarded for lack of space to accommodate. And second, in the actual propagation of non-running forms for commercial or home use. The trait (running) is both a curse and a blessing. It costs so much to cut the things off in the fruit plantings, and yet good running is essential to standard nursery propagation. I like this idea, and intend to follow up on it. We may have a few items right now that almost qualify (day-neutrals)."[165] Yet another related possibility which captured Royce's interest was the use of somatic (non-sexual) breeding, a method which combined genes from parent plants by directly fusing their cells, without resorting to the usual method of pollination.[166] This would enable crosses of plants which would at one time have been impossible because of incompatibility of the parents.

The other new technique which Royce began investigating with strawberries was electrophoresis, a method for separating complex mixtures of biochemicals into their various components by applying an electric current across a specially prepared gel. The different components of the mixture would migrate at different rates across the gel, allowing individual components to be separated and identified.[167] Using an extract from strawberry plants, Royce could perform the electrophoresis, and then stain the resulting gel with a material designed to make visible a single strawberry enzyme, such as phosphoglucoisomerase (PGI),[168] or Peroxidase (Px). When the stain was applied, a series of bands would appear on the gel, in a pattern which was always the same for a given plant or variety, but which often differed between varieties. By staining for enough different enzymes, the resulting patterns could serve as a kind of "fingerprint" for an individual variety or clone of plants.

Royce saw two potential applications for this technology. First, it opened the possibility of being able to identify specific strawberry varieties purely by laboratory means, without the need to examine plants and fruit, as Royce had had to do in the Tufts case. When this was first put to the test in 1976, the results were very promising.[169]

418

The second application was in the study of wild plants. Not only did the different band patterns serve to distinguish one wild strawberry population from another, they also held the promise of being able to show the genetic relationship between different populations, and even different species of strawberries. Royce engaged the help of graduate students in this work, particularly S. Arulsekar, a student from India whom Royce began to refer to simply as "Sekar." Here, too, definite patterns began to emerge, showing distinction between different clones which Royce had growing in the experimental orchard at Winters.[170]

Royce also continued his study of wild strawberries in the field, which he had begun during the 1960s. In August of 1973, he had finally had the opportunity of taking his friend and colleague, the German Günter Staudt, on a tour of the wild strawberry colonies which he had identified in California. He visited many sites with the world-famous strawberry taxonomist, who informed him that some of the plants he had considered to be *Fragaria virginiana* were actually natural hybrids of *F. virginiana* and *F. chiloensis*, or else descendants of strawberry varieties which had escaped from cultivation. Staudt referred to these as *Fragaria cuneifolia*, and taught Royce how to recognize them. The two scientists collected specimens of many of the plants, including probable species hybrids from Cape Mendocino and Point Sur, and Staudt had the opportunity of visiting the original hybrid colonies which Royce had previously discovered and described.[171] Royce's duties as Department Chairman had kept him from visiting these sites as frequently as he desired, though in April of 1975 he once again visited the Point Sur location to observe the various colonies of native plants and hybrids which grew there. This time he took extensive leaf and flower samples from the different colonies in order to perform electrophoresis, which he hoped would shed some light on the origin and interrelationship of the different colonies on the site.[172] He returned a couple of months later to collect ripe fruit from the various colonies, and while there, he identified two additional hybrid colonies which had not been noted previously.

Royce had always been impressed by the remarkable adaptability of the wild strawberries to the great variety of natural environments in which they were found, and he had long sought the opportunity to make a more careful study of these plant populations. In 1975, as his term as chairman drew to a close, he began working with a new graduate student named Jim Hancock,[173] and the two began a much more systematic study of the wild strawberry populations. Visiting the different sites where Royce had already identified wild colonies of the two octoploid species, *Fragaria chiloensis* and *Fragaria virginiana*, and of the diploid species *Fragaria vesca*, they made an extensive collection of

plants of each species present at each of the sites, as well as representative soil specimens for analysis from each site.[174] In a series of
experiments, they were able to show how the different colonies had,
through the process of natural selection, adapted themselves to different growing conditions. For example, plants of *F. vesca* from for-
ested areas had better growth in shady conditions than those which
had come from unshaded open areas. Similarly, plants of *F. chiloensis*
which had been collected in salty soils at the coast were able to with-
stand much higher salt levels in the soil than those which had come
from meadow environments. Particularly impressive were some of
the adaptations of the wild *F. chiloensis* colonies along the coast, which
under the harsh salty and arid conditions among the sand dunes tended to develop a long "tap root"[175] leading to a deeply buried root system, along with small thickened leaves to conserve water, and extensive networks of long runners leading to satellite plants which, rather
than putting out roots of their own, depended entirely on the deep
root system of the mother plant. By contrast, plants of the same species that grew in nearby meadow areas with ample water and soil
tended to have extensive dense root systems, larger and thinner
leaves, and to reproduce more sexually than by runners, all adaptations to their very different environment.[176]

The greenhouse collection of strawberry plants which they had
systematically gathered from the wild also gave Royce and Jim Hancock a good opportunity to study the phenomenon of sexual dimorphism, or gender separation, in the strawberry. They noted that all
octoploid strawberries (such as *Fragaria chiloensis*) were generally accepted as having descended from simpler diploid species such as
Fragaria vesca. However, whereas all diploid strawberries were mainly
hermaphroditic (meaning that their flowers contained both male and
female parts), *F. chiloensis* was dioecious (meaning that in most cases,
flowers contained only male or female parts, and plants of both genders had to be present to produce fruit). By studying the sex characteristics of their greenhouse collection, they were able to draw a
number of conclusions about the effect of separation of sexes on the
development and survival of the plant colonies, but they also believed
that the small proportion of hermaphroditic plants which were always
found among the octoploids gave them a survival advantage in extreme conditions. They were also able to show that the octoploids
had a greater variety of traits which helped them to adapt to a wider
variety of environments than the diploids, and by subjecting the various plant species to electrophoresis, they were also able to demonstrate the increased variation chemically. Royce and Jim Hancock
were able to conclude that the octoploid strawberries were able to
thrive in a broader range of conditions than the diploids, partly be-

cause their more complex genetic structure allowed them to differentiate genetically to adapt to different environments, and partly because it made them more stable to changes in growing conditions.[177]

During the process of these investigations, Royce took his graduate students on long drives to visit wild strawberry colonies which had become like old acquaintances. Royce also visited the sites where he had discovered and classified the natural hybrids years earlier. At one of these, he observed, "At Wrights Beach, I found my old 9x plant still thriving after about 14 years."[178] These remote and beautiful sites represented some of the most significant of Royce's work, and he relished having the time once more to visit them. After another such visit, he wrote, "Decided to visit Point Sur today just because I felt that it would be 'good for my soul,' so to speak, and because I haven't been there for some time. It was a perfect day weather wise. Since I had ample time to do as I wished, I climbed to the top of the mountain just above the hybrid colonies. I have never climbed that high before at Pt. Sur. I found *F. chiloensis* and *F. vesca* up on the top, but no hybrids."[179]

Protecting the Germplasm

As Royce continued these studies, particularly of the isolated coastal populations of *Fragaria chiloensis* and its hybrids, he increasingly noticed the damage to the fragile ecosystem which had been caused by human traffic over the years. The coastal strawberries grew in regions of great natural beauty which also attracted tourists and developers, and as the state's population grew, so did the wear on the area from recreational use and development. Of the handful of plant communities which he had identified in the state with natural hybrids, the important site at Pacifica had been completely eradicated by expansion of the city[180], and another at Bodega Bay would shortly succumb to human activity. The hybrid strawberry colonies he had found there would survived only in his own collection of live plants, and most of the remaining wild colonies were in serious jeopardy as well. Royce recognized the value of these colonies not only for their own preservation, but because some of them contained genes which might be essential to future strawberry breeding efforts; in fact, he had already incorporated some of the desirable traits from these wild populations into his advanced breeding lines.

These losses brought out the conservationist in Royce. In response to the loss of the Pacifica colonies, in September of 1977 he authored an article for *California Agriculture*, in which he championed the plight of the embattled coastal strawberry. The article read in part, "A narrow strip of the western coastline has furnished California with a rare, unique source of wild plant germplasm that has helped the

421

strawberry industry to flourish. Ironically, the agricultural wealth this beach strawberry species has helped to generate is contemporaneous with forces driving the plant itself toward extinction in many areas. Subdivisions, commercial development, and people have intruded into the niche the berry had carved for its balanced existence beside the salty Pacific Ocean. ... Much of the area once occupied by *F. chiloensis* has been built over or converted to uses incompatible with the survival of the *Fragaria* species. Recently, for example, in the expansion of the city of Pacifica, most of the important colonies we have studied there for many years were destroyed completely. Our modest collection at Davis of some of the most valuable stock is all that remains. All the existing California strawberry cultivars have California *Fragaria chiloensis* ancestry, and this has contributed much to the high adaptation of these varieties that has resulted in the highest strawberry yields in the world. Unfortunately, the California *F. chiloensis* with the greatest breeding potentiality are those colonies native to the most vulnerable southern habitats, and appropriate steps should be taken to preserve them."[181]

The article, written by Royce but bearing also the names of Jim Hancock and Victor Voth, was widely enough read that it created something of a disturbance in the communities Royce had mentioned. "I have another tempest in a teapot going over the article I wrote on *F. chiloensis* and its possible disappearance in parts of coastal California. The people in San Mateo County didn't like what I said about Pacifica. I say, I'm glad that I stirred them up. Perhaps it was overkill, but that's what it takes to get attention these days."[182]

The article had had its desired effect. Before long, Royce was contacted by a representative of the California State Parks system, and negotiations began that same year for the formation of a nature preserve specifically for the wild strawberry populations. "I met with Andy Manus about the Año Nuevo strawberry preserve idea. If we work it right and convince the right authorities, we should have limited access there, along with the elephant seals."[183] That December the interested parties met at the site of the proposed park. "I drove to Año Nuevo and met most of the folks who were expecting me. We drove into the reserve area and inspected the strawberry colonies. Except for drought injury, the plants were in quite good shape, as extensive as ever. I think that I can pretty well do what I wish. As we circled by the island we encountered two groups of elephant seals (three and four) lolling in the sand. They are very unlikely creatures out of water, huge bags of blubber in oversized sealskin uniforms, but graceful as fish in the water. I took a few photos."[184] These meetings ultimately resulted in the establishment of the Año Nuevo State Reserve, which gave official protection to one of the most endangered strawberry areas. Later, similar measures were taken at the Oso Flaco Dunes, located at the southernmost end of the beach strawberry habitat, and a popular area for off-road vehicles which had begun to damage the colonies.[185]

Partly because of the destruction of these wild strawberry populations, Royce also realized the importance of preserving the many individual plant clones he had collected over the years, some of which now only existed in his collection. This included not only the varieties he had developed and the important parent lines he still used, but also older varieties of historical interest, such as those developed by Albert Etter in the early 20th century. There was also the multitude of plant materials he had collected from the California Coast and during his trips to South America, including the rare but important natural hybrids he had discovered.

At this time there was increasing national and international interest in providing for the preservation of genetic material (known as "germplasm") from valuable and unique plant varieties. In 1974 a National Plant Germplasm Committee had been formed, and had recommended the creation of a group of storage facilities, known as repositories, where plant materials could be systematically preserved. On July 2, 1977, Royce was contacted by Sam Dietz, the western coordinator for the system, asking him to serve along with Dick Converse, a plant pathologist trained at Davis, as western representatives of a small fruits germplasm committee. Royce agreed, and on attending meetings of the committee, he learned of plans to build the first of the germplasm repositories in Corvallis, Oregon. As a member of the committee, Royce would play a role in its planning, and would be present for its dedication.[186] He would eventually send samples of his most important cultivars and of significant wild strawberry clones for preservation there.

Life In Transition

In the years following his department chairmanship, Royce still had no lack of administrative duties. His work as department chairman had given him valuable experience which enabled him to participate in some of the various committees necessary to the functioning both of the department and of the university. The tightening of budgets continued to be a dominant theme as these committees met, and after one meeting on cost cutting in university orchards, Royce observed, "It's one of those circular arguments where the only way is to eliminate jobs and fire people. Attitudes vary concerning this, but the support of it seems to depend upon 'whose ox is being gored;' you favor it if it affects someone else, and oppose it if you are to be hurt in any way. I can see a lot of things that should be done and can't, because it would involve dismissal for people who cannot be fired. It is well that this is so in a way, because most older ones would be fired if it were done, a pretty cruel way to do business."[187]

One of Royce's committee assignments was the campus-wide Educational Policy Committee, which helped determine standards for college admission, as well as the organization of the departments. This presented a different series of challenges, but the tight university budget always loomed in the background "We had the Educational Policy Committee meeting, going over a rather large agenda for the most part talking about things we can't do too much about, including admissions criteria. That can be very confusing, because one person can take a given set of data and 'prove' that GPA (grade point average) is better than SAT scores or a combination of the two, while another can 'prove' just as convincingly that the opposite is true using the same data. We also dealt with what appears to be a half open threat to the existence of the classics department, largely because it is small. It had a familiar ring of when I became chairman and was forced to swing the budget cut hatchet. No one enjoys that role, least of all those being 'cut.'"[188]

In addition to budgetary matters, these committees had to reckon with the social and political issues which the student protest movement of the '60s and '70s had brought to the fore. As a response to the protest movement, student representatives had now been placed on some of the committees, a move Royce looked upon as misguided, as the students seemed oblivious to the actual purpose for the committees, and to the complex functions of the university. After one Educational Policy Committee meeting, he wrote, "Again we were busy with the extraneous nit-picking nonsense dragged in by the students on the committee, a series of non-issues that seem to have no end and serve only to give them some sort of feeling of power. Their lack of understanding and paranoia about being done in or ignored has to be seen to be believed. Truly the inmates have taken over the asylum, in a manner of speaking."[189] After a later meeting, he added, "I do get tired of listening to a lot of students as they use the floor to air vague 'social and political concerns' which they for some peculiar reason think originates with them. 'Oh I was like that when a lad?' How foolish much of their rhetoric is when I listen to it now, and yet they are so earnest with all."[190] One of the committee meetings was actually held while a student protest was taking place outside the administration building. "The chancellor was not there because of pickets at Mrak Hall over the current rage: Anti support of businesses that do business with South Africa[191], agricultural mechanization, and the Bakke discrimination case[192] now before the Supreme Court. The pickets were few and the enthusiasm flagging. When we got out of

424

the meeting at 5:00 p.m. they were gone with all their signs and junk."[193]

One of these issues which did affect Royce's work was the matter of agricultural mechanization, which was opposed by a newly-emerging farmworker protest movement, on the grounds that mechanization of the agricultural industry could cost farm workers their jobs. Royce was unsympathetic with this view—not only was he one of a generation of scientists who had revolutionized American agriculture, in large measure through mechanization, but he also had lived and worked in countries of Latin America where most of the California farm workers had originated. By this time he had also become somewhat weary of the protest mentality which pervaded university culture, and increasingly, the society at large. As an agricultural expert, Royce was invited to attend a meeting of the university regents in Los Angeles where the issue was being addressed. "The subject was agricultural mechanization research. I spoke out in favor of continuation, naturally. I felt all right about it, but being next to last after five hours, I was definitely less than in an advantageous position to speak. We had already listened to Cesar Chavez[194] and Tom Hayden[195] both delivering 'self-serving' half truths along with an orchestrated gang of students giving empty headed gullible support as 'cheerleaders' (or the equivalent). They regard the role of the University as a grand welfare agency charged with creating an unlikely world to serve them and the refugees from Mexico who wouldn't get the time of day under similar circumstances if they were across the border. I couldn't believe what I was hearing at times. They have nothing to do with reason, only the cause of the moment."[196] Such themes would continue to emerge as Royce worked on these committees.

The structure of the University of California was that of a statewide university system divided into individual campuses such as Davis, which really existed not as separate institutions, but as part of a united whole. In 1977, Royce was appointed to a statewide committee charged with determining educational policy for the entire university system. The committee met monthly, alternating between Los Angeles and San Francisco, and Royce added this travel to his already busy schedule, describing the meetings in succinct notes in his journal, starting with his first meeting in September of 1977. "Up early, caught Western Airlines to Los Angeles and attended the first meeting of the University-wide Committee on Educational Policy (UCEP). It was quite interesting, lasting from 10:00 to 4:00 p.m. This is a very rough 'merry-go-round' I am on and it sort of runs me. I don't really run it."[197] In this committee the matter of preferential admission for minority students had to be addressed, and Royce's opinions had not changed, though he believed in basic principles of fairness. "We spend much of our time struggling with various aspects of discrimination past, and the reverse discrimination of the present. I find one to be as odious as the other. It isn't even equally unfair, never."[198] After a later meeting addressing a similar topic, he wrote with more moderation, "We approved a final version of our evaluation of the affirmative action report. I think that we treated that important issue fairly. Not much will result from this exercise in futility anyway. We are fishing in 'pools' with few 'fish' of the desired types (women and minorities) in most disciplines."[199]

Royce quickly became an important voice on the committee, and his congenial nature, measured wisdom and sense of fairness made his opinions much sought after. After less than a year on the committee his role was expanded. "I received a not too welcome call from Alex McCalla, state committee on committees, asking if I would take the vice chairmanship for statewide UCEP for next year, with the understanding that I would chair it the following year. With some reservation I accepted—foolishly, I suppose."[200] Despite his initial reluctance, this chairmanship would become a major and memorable assignment for Royce, one in which he would have a far-reaching impact on the university system. His first meeting as chairman was a challenging one, dealing with a contentious issue forced upon the faculty by the university administration. "I convened the meeting at 10:10 a.m. with almost a full complement of members, the tardy ones arrived within a half hour. We made good progress in the morning, completing most of our necessary agenda. At noon we had an excellent lunch of Maui Maui fish from Hawaii. There the good times ended. The administration 'dog and pony' show was concerned with a proposed preposterous individual detailed record keeping scheme to satisfy the federals. I cannot believe that sane rational persons would devise such a costly, useless scheme and try to impose it on a great institution as the University of California. Anyway, we listened with patience until about 3:15 p.m. There was some counterattack without making it personal. I then terminated the meeting. We left at 3:30 and reached home a short time after 5:15 with no difficulty."[201]

At one of these meetings, a matter came up which was to have a significant impact on Royce's own life and career. "I heard one thing of interest to me. Retirement age may be changed to 70 years beginning about 1982. That would include me, and I could go on until about 1989, another ten years, something I would have to consider. I might do that, depending upon my health."[202] The opportunity for delayed retirement was one that appealed to Royce. His personal circumstances had delayed his entry into a profession that he had found fulfilling and satisfying, and in which he had been able to make significant contributions to his field. When the change of policy came into effect some years later Royce would be quick to take advantage of it, extending his professional career an additional five years.

Although Royce enjoyed his committee assignments and felt a sense of accomplishment as he performed them, he realized that they were taking their toll on other aspects of his profession, particularly his scientific work. The problem was not one of research, for Royce had continued to relentlessly pursue his investigations in every aspect of the strawberry program; rather, it was in finding the time to prepare his work for publication. At one meeting with his colleagues from across the country, he remarked, "We are delinquent in not having published. We have more data than anyone else."[203] When he at last completed the first paper for the *American Journal of Botany* on the wild strawberry work he had done with Jim Hancock, he commented, "That is good for me. I am breaking the ice on writing again. I have yet to catch up from being chairman."[204] A few months later, he wrote, "I spent the day working on papers, without much success, although I do have two important papers shaping up well. If I could settle down to it there could be at least a dozen, all relatively valuable."[205] In truth, despite his organizational shortcomings, Royce was at the height of a very productive career. His success in strawberry breeding was unquestioned, and in concert with Victor Voth, he was also supervising a remarkably complex series of ongoing experiments in growing techniques and in plant diseases. He was continuing and expanding his investigations of the wild strawberry populations, and defining their interrelationships. He was leading the way in exploring new laboratory techniques and applying them to every aspect of his work. He was highly esteemed as a teacher, not only in his university courses, but in gatherings of strawberrymen throughout the state. And he was playing an increasingly prominent role in the decisionmaking bodies of the university.[206]

Royce's extensive work with strawberries occasionally caught the public eye, and he was not infrequently interviewed by reporters, generally about the astonishing success of the California strawberry program. He appeared on local television programs, and once was featured on a nationally syndicated radio broadcast,[207] but most often he appeared on the pages of local newspapers, usually with attractive photographs of strawberries, often accompanied by Royce's winning smile. In May of 1977, when the strawberry harvest was at its height on the Central Coast, Royce was contacted and interviewed by a staff writer from the Sacramento Bee named Steve Duscha, who asked the usual questions about his work and the prospects for strawberry breeding. But when the article appeared in a Saturday issue of the paper, it turned out that the friendly questioning had been a deception, and the article which resulted proved a diatribe against the California strawberry program, based largely on information from an activist involved in the farm worker protest movement.[208] Royce wrote, "I thought it was the usual report about the industry and its importance to California. However, he had set up an attack on the industry based upon a background situation he had not even implied to me—a real non-issue involving the leader of the 'Mexican freeloader' groups in the business in the Central Coast and elsewhere, all bankrolled by the government." The article, which featured an uncharacteristically dour photograph of Royce, gave harsh treatment to the strawberries he had developed. "Advertising writers still describe the strawberry as sumptuous, succulent, sweet and juicy – the fresh fruit of spring. But critics say today's berries 'taste like potatoes' and bounce like rubber balls. And the strawberry of the future is expected to be even sturdier, 'a tough mother' as one grower put it."[209]

The article made something of a stir among Royce's friends and associates, some of whom handed him copies. Royce took a dismissive attitude in his journal. "The article was biased, slanted, and built around half-truths and nonsense. Pearl was incensed by it. I merely said, 'let it ride.' I wouldn't let her send a letter in on it, nor would I. It would be counter-productive. It made the front page of the Bee with my picture. I also said as did Harry Truman, 'If you can't take the heat, get out of the kitchen,' and as someone else said, 'I don't care what you say about me, but get the name right.' This they did. End of story."[210] There is no question that Royce felt the article was unfair and was not to be taken seriously, but the charge of poor flavor in commercial berries was especially hard to take, and surely made an impact on Royce. He knew that flavor in commercial berries had much more to do with the stage of harvest and the state of ripeness than with the berry varieties themselves, but in the upcoming generation of berries he would pay special attention to flavor as a factor, and his next releases would include some of the most flavorful berries of his career.

Meanwhile, life went on in the Bringhurst family. When Margaret, then a junior in High School, went out on her first date, Royce wrote good-naturedly, "She was home promptly at 12 as per instructions. I don't know whether her carriage would have changed to a pumpkin or not."[211] Although they had some happy times together, as Margaret neared high school graduation she seemed as distant as ever to Royce, and her relationship with her parents grew increasingly tense. Of the older children, only Ann remained in Davis, where she had recently given birth to her first child, a girl named Alisa. Starting in March of 1977, Pearl had begun caring for the baby during the day, as Ann went back to work to support her husband in graduate school. This was a significant undertaking for Pearl, for with the last of her own children shortly to leave home, she had looked forward to accompanying Royce on some of his many travels, but once again family responsibilities took precedence for her, and she would spend most of the next ten years caring for Ann and Ray's children on weekdays. The house would once again ring with the voices of young children, and Royce and Pearl would both become very fond of these grandchildren as they grew. John, meanwhile, nearing completion of his mission in Guatemala, had learned a Mayan dialect in addition to Spanish as part of his work, and had extended his mission an additional two months. He was now writing home about transferring from UC Davis to Brigham Young University when he returned home in order to participate in Church translation there. This left Royce with ambivalent feelings, since he had always regarded BYU as an academically inferior institution. As the time approached for John's mission to end, Royce and Pearl had received no word of his travel plans, and particularly Pearl had begun to worry.

Thursday, March 2, 1978 would become a day Royce would never forget. He had arisen in the early hours of the morning to attend another of his monthly committee meetings. "I boarded a Western Air Lines 7:00 a.m. flight for Los Angeles, where I attended the University-wide Committee on Educational Policy (UCEP) meeting. I then hastened over to the air terminal to see if I could catch an earlier plane, and did that, since the 3:30 flight was delayed until 5:10. As I walked about seeking a place to sit down, I was really taken by surprise. Miracle of miracles, there stood John. He saw me first and called to me, and was booked on the same plane for Sacramento. This is all the more unusual because we had not heard when or how he was coming." John, returning from his mission in Guatemala, had been unable to contact his parents about travel plans because of a communications strike in that country, which had blocked the mail as well as telephone and telegraph service. Having been unexpectedly united in the gate of the terminal, Royce and John sat next to each other on the plane to Sacramento. "He has changed little in appearance, perhaps a bit bigger, otherwise he is fine. We surprised Pearl as we came home. I walked in and asked if she had heard from John, then when she said no, I had him walk in and then she saw him. This has been one of the best days ever. We called our family to let them know."[212]

Although John, fresh from his mission, regarded this chance meeting in the Los Angeles Airport as nothing short of miraculous, Royce, in his naturalistic frame of mind, calculated that there had been a good chance all along that something like this would happen. With some reflection, however, the event began to seem more and more impressive to him, and the following Sunday being Fast Sunday, he

429

took occasion to stand up in testimony meeting and tell of it. "I recounted the peculiar circumstances of our meeting in the LA airport last Thursday. The more I think about it, the more unusual it appears. Anyway, it is something to pass on to posterity and likely it will be amplified in the retelling, perhaps almost beyond recognition. Things often are."[213]

John had matured as a missionary, and his presence in the house was a pleasant interlude for Royce and Pearl, and for Margaret as well, for John got along well with his younger sister, and was able to act as something of a buffer between her and her parents. He returned to his energetic piano playing habits from before his mission, which filled the house with music and kept Royce in good humor. John was not destined to stay long, however, for after less than two months he again left for Guatemala to participate in a linguistics project, and would thereafter depart for study at Brigham Young University, where Margaret also had been accepted as a student, which pleased Royce, even though he secretly still had his doubts about the university as an educational institution.[214]

Before John and Margaret departed for college, Royce experienced another premonition of tragedy. "When I awake in the night feeling strange, it often portends something sad, and this day was no exception. I went to the plots then over to Akiyoshi's. Upon returning to the plot I was met with a note telling me to call Gloria on an emergency matter. I ran right over for my call and learned that my sister June was killed in an automobile accident near Elko, Nevada early this morning. It is sad; she has had such an unhappy life the last dozen years or so. She was just ten years younger than I am. I remember her best as a little girl who loved horses. I never really knew her after that. She went through her teens during the war, and I was away. She was quite a stranger to me when I returned. She is to be buried alongside our brother Smith. It is ironic. In her way, she followed him, and now they are the only members of the family who have died. I do wish that I had seen her the last time I was in Utah, but wishing won't change the fact that I did not."[215] Once again Royce traveled to Salt Lake City for the funeral. "I gave the family prayer and could hardly do it because I choked up so. It was a very sad occasion. Truly, it is not given to man to tread the earth for long. I don't think that is important, however. The quality of life does matter, and for much of hers it was lacking. She was only 49, and was really very promising but her career was spotty. Her greatest achievement was nursing which she did well, but human frailty tainted that too. She was laid alongside of our Brother Smith, not far from Dad in the Elysean Garden cemetery. It was a pleasant day, with a freshening breeze. I wish I could say something worthy to the occasion, but I feel drained, empty and sad. Smith's ex-wife Peggy was there. She looks about the same. Her life has been no bed of roses. I'm not sure that we have helped with it either. Kindness could have changed much for Smith and for June. 'Then there were five.'"[216]

A few weeks later, Royce departed for a brief trip to Italy, which serves as a fitting conclusion to this chapter of his life's story. This travel was at the invitation of the owners of a private nursery in northern Italy who hoped to benefit from Royce's expertise in the various aspects of strawberry production. Royce, in turn, hoped to learn some lessons from the Italian nurserymen, who had made major strides in propagating strawberries through tissue culture. The nursery was situated in the Po River Valley, near the northern Italian city of Ferrara, were Royce had flown some of his most dangerous missions back in 1944. The trip brought many reflections of the past and present which Royce recorded in his journal, and which provide a glimpse into his soul as he contemplated his past and reflected upon his future.

"Old memories surged up as we crossed the old bridge at Ferrara, our target during World War II at least twice when I was along. Then we passed by Faenza, where I came the closest to being killed in my B-25 days. I still remember the details of how it felt to be caught in direct gunfire. I am happy to have remained among the living. How very soon those who died have been forgotten. It is almost as if they had not lived and their candle were never lit. No one really cares any more about what happened then to individuals, and particularly those like Frank Denstad of whose death I learned when I visited Foggia at that time, and again I thought of him, but only as a shadow who crossed my own feeble light for but a brief moment and then was gone, never to return. His wife remarried, and I doubt that he is now more than a shadow to her or their one daughter. I was interested in the many castles we passed en route, mostly on hilltops still guarding the ghosts of those who built them, as if it really matters now.

"I even thought more strongly on the brevity of man's existence as I observed the miles upon miles of aged but ageless olive trees that grace the hills and plains near Bari. Some are surely older than the American Revolution, and from their appearance, will continue to be there when the present generations of men are forgotten, as they most surely shall be by the teeming masses sure to follow. "I am, I am, I am, please don't forget me while I yet live," is perhaps the best we can hope for with some few exceptions. The olive trees bear the scars of time with a calm dignity and beauty that only can grace such as a tree, for few men exhibit similar traits. Age and ugliness as well as utter uselessness and hopelessness are our unhappy lot, try as we might to make it be otherwise; we don't really regret the passing of

431

most of our aged, but rather, without being obvious about it, we contrive in our secret ways to 'hurry them on' and have done with it. The space, facilities, and even the air they breathe is always needed by someone else more worthy and more deserving for their little moment. Clear the way, yield to those to whom the torch is passed. Die quietly without making a fuss about it. Even a true belief in life after death changes this but very little, unfortunately. We still urge them on to the 'pie in the sky' and begrudge them the 'crust of dry bread' that they need while yet they linger. The tragedy of all this is that they but show the way of all flesh to us, and each generation in its way must face the inevitable rejection and abandonment, with the hope, but not necessarily the expectation, that it might be mixed with a modicum of mercy and kindness. Slave on, you 'drudge,' and make it last while you can. The night is already shading the gate."[217]

Illustrations for Chapter 10

Bringhurst family at home on Luis Pereira Street in Santiago, Chile: Royce, Pearl, Ann and Margaret, with John on balcony.

Royce (left) as counselor in branch presidency, Nuñoa Branch.

Royce in a wild area of Chile with his yellow field journal, which he carried to record notes related to the various aspects of his work.

Victor Voth and Royce in an experimental strawberry planting.

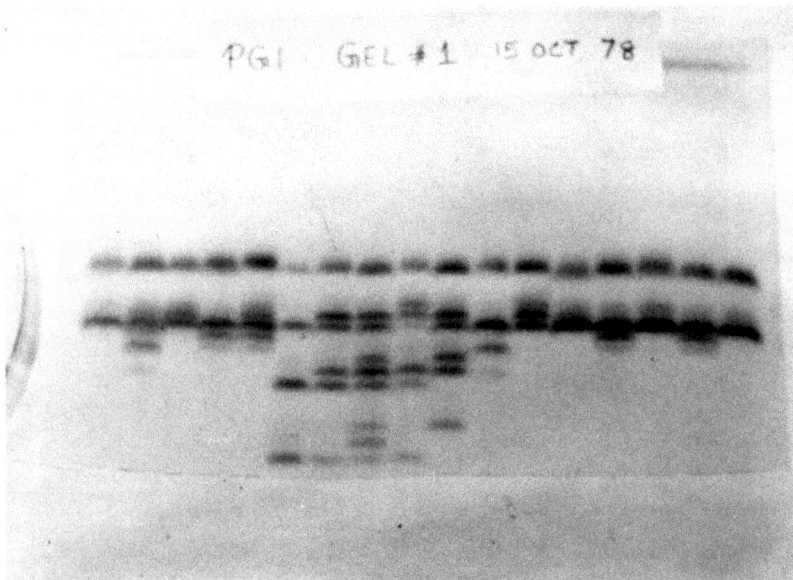

An electrophoresis gel, comparing different samples of strawberries. The band pattern of a specific sample differed according to the species or variety of strawberry, and could help to identify a specific variety.

Patent disclosure for strawberry variety Aiko.

435

Royce (left) appearing in an operetta with the Davis Comic Opera Company.[218]

Pearl Davidson Bringhurst on graduation day, 1977.

Davis: The Later Years

Chapter 11

The Empty Nest

In 1978, Royce was entering the last decade of a professional career which, if once delayed by war, would shortly be lengthened by the new university policy which allowed him to defer retirement until age 70. Royce had of course aged, but in a sense the entire country had aged with him. The Vietnam war had at last run its course, but the student protest movement it fostered during the 1960s and 70s had left the nation more divided and weakened, both at home and abroad, than it had ever been since before the Second World War. A sitting president had resigned in disgrace for the first time in the nation's history, and the succeeding administrations were plagued by diminished public confidence and ineffectual leadership.[1] The country was contending with economic malaise and political disunion at home, while abroad the "Cold War" against the communist states continued unabated, and ominous religious and political conflicts were emerging in the Middle East and elsewhere.

The departure of John and Margaret Bringhurst for Utah at the end of August, 1978 left Royce and Pearl presiding over an empty house, except for the presence of Ann's young daughter Alisa, whom Pearl cared for on weekdays while Ann worked. Shortly after Royce's return from Italy in the fall of 1978, he and Pearl drove to Utah to visit their extended family and their two youngest children who had just started school. Arriving in Salt Lake City, they traveled to nearby Draper to visit Royce's mother. "She did not know us except in a very general way, but she is happy and seems to like living there. I believe this might have been different had she continued to do the things she used to do. Age has taken its toll."[2] Still as regretful of the situation as ever, Royce had yet to reconcile himself with his mother's ongoing mental decline.

Christmas of 1978 brought much of the extended family back to Davis to visit, particularly since two days after Christmas day Royce celebrated his sixtieth birthday. Royce's sister Naida came from Oakland with her husband Ernie to help the rest of the family honor the occasion, and Royce for his part was grateful for his robust health. On his birthday he wrote, "I ran in the morning, a matter of principle. At 60 I can still get around as well as many much younger than I am. I have a lot to be grateful for, especially Pearl. She is such an excellent woman. How fortunate I am to have her as my wife. I should even let her know that."[3] For the Christmas holiday John and Margaret had made the long drive together from Brigham Young University, and John, back from a mission for less than a year, was accompanied on the trip by a young woman named Betty Farnsworth, whom he had recently met at his work there. Two days after his arrival, he drove the sixty miles to Placerville where Betty lived, and brought her back home to meet his parents, a move which Royce regarded with guarded optimism. "She seems to be very nice. I suppose that this was inevitable, and should be welcome to us as it is. We just have to get used to the idea."[4] A few days later Royce observed with a touch of resignation, "He obviously wants to get married. That is for him to decide, being of age. I just want him to be aware of the consequences. You can do certain things, and when one decides to do one, it usually precludes doing another. I can't argue strongly against him choosing to marry. When Pearl and I took that step, my star certainly didn't look all that bright, and she contributed very positively to everything worthwhile that has happened since. The world has changed, but not that much."[5]

As the school year wore on, Royce corresponded with John, recommending strongly against a hasty marriage, but by March the couple called to announce their engagement, and a wedding was planned for that May.[6] On May 4, 1979, the family again made the journey to the Oakland Temple, where John and Betty were married in a "sealing" ceremony, the fifth of the six children to marry. A reception was held the same day in Placerville, and an open house the day after in the Bringhurst home in Davis. The couple had opted not to have a wedding cake, choosing instead to serve a large platter of strawberries to their guests, which Royce provided with obvious satisfaction. Royce had largely overcome his earlier reservations about the marriage, and after the open house he observed, "He and Betty appear to be so happy, and that is infectious."⁷ Shortly after their marriage, John set off with Betty to Guatemala to work on a translation project, and returning a couple of months later, they stopped by Davis only briefly to pay Royce and Pearl a visit before heading back to Utah. Their departure as a married couple left Royce thoughtful. "John has now left home. We no longer are involved except as observers and counselors, and that is as it should be. I remember leaving home too—so many years ago and you cannot return, ever."⁸

A Turning Point

That summer, on July 30, 1979, Royce flew to Columbus, Ohio to attend the annual meetings of the American Society for Horticultural Science. These yearly meetings were for him a valuable event, at which he was able to compare notes and share information with other colleagues in the field. On the last day of the conference Royce presented a paper on the newly released California strawberry cultivars, and that same evening he attended a meeting of the American Pomological Society, where an awards ceremony was held. During the ceremony Royce was surprised to be called up to receive the Wilder Medal, the society's top honor, given to individuals or groups who had rendered outstanding contributions to the field of fruit horticulture, particularly the development of new varieties of fruit which had been widely adopted. The award was granted yearly to one or more individuals, and Royce was the sole recipient for 1979, in recognition of his now widely acclaimed breeding work with strawberries.⁹

This was a landmark event in Royce's professional career. The Wilder Medal was by far the most important award he had ever received, and amounted to a nationwide recognition of the significance of his work. With it, Royce was literally at the top of his field. Not only was he managing a large and highly successful breeding program with its associated array of horticultural research, he was also still engaged in pathfinding work on natural strawberry populations, and had become a world expert in strawberry genetics and evolution. He had led the way in applying new laboratory techniques to the identification and study of both wild and cultivated strawberries, and his professional advice was being sought out by experts in the field from around the world. He had also taken a leadership role in the governance of the university, and enjoyed the respect and esteem of his colleagues. In spite of his own misgivings, Royce really was on top of the world professionally, but that world was about to fall apart.

On September 28, 1979, less than two months after receiving the prestigious medal, Royce wrote, "This turned out to be a very sad day for me and I am literally heartsick as I write. I received a call from the editor of *California Agriculture*, asking about doing a color bit on our

439

new strawberries for the November-December issue, so naturally I called Victor Voth as I usually do about such things. He had been very cool toward me the last two days in Redding, but I thought it was because he had been ill and is under medication. Anyway, he merely growled that he didn't care what I did about it. 'Do anything you want,' etc. and said he was through with the whole bunch at Davis. Up until then, he didn't direct it toward me particularly. I asked a few questions, and at first he said he didn't want to talk about it, but then he leveled in on me and really told me off in very harsh terms, saying that I had been taking advantage of him for years, and relegating him to second place or worse always, and so on. This was of course very painful for me to listen to, but I did without responding in kind. I attempted to pacify the situation, but he then simply said I was being a hypocrite.

"It turned out that the anger was sparked by at least two things. First, whatever Noel Sommer [the Pomology Department chairman] told him was completely the wrong thing for the time being, and I really don't know any of the details, nor do I think that I will pursue the matter. It really isn't any of my business, although it does affect me greatly. The second thing was the Wilder Medal. He really roared about that, and made me wish that I'd never heard about that, let alone received it. I do agree with him, it should have been a joint award given to both of us. Evidently he must believe that I connived to get it, and my attempt to tell him that I had nothing to do with the awarding of it brought on more vituperation, and made me wish I had said nothing. I really am quite an innocent victim in that regard, but I couldn't convince him so I stopped trying. Anyway, Victor as good as told me to get lost as far as he is concerned, although he will continue with the University.

"I must in candor point out that I must have given some cause for this, although I have not done so consciously or deliberately. I am resolved to see this through and reestablish a good working relationship. I am resolved, further, to do everything in my power to see that all credit due to him, and more, is given. I personally think the University and the industry have been more than good to both of us. I also realize he doesn't really understand what my responsibilities are, nor does he really understand what it means to be a full functioning faculty member of a major university. I am fully aware of my weaknesses, and I suppose much of the hurt I feel inside right now is due to having them thrown in my face by someone I hold in high regard. This has truly been one of the most rotten days of my sixty plus years. I hope that I deserved better than what I received today, but I shall try to make it up nevertheless as if all he said were really true."[10]

The episode with Victor Voth served as something of a wakeup call for Royce, not just in his professional associations, but in other areas of his life as well. That Sunday at church, his pride still shattered, Royce rose to speak during the bearing of testimonies at fast meeting. "I had something to say, although I had not planned to do so. It was partly out of the turmoil and heartache of the past few days, though I did not make reference to that, nor shall I in any public or private meeting. This must be ironed out fairly with Victor. I can't imagine not squaring it away. I'm willing to go as far as necessary for reconciliation." At the same time, Royce was determined to address the flaws in his character which he knew had been bringing him down spiritually. "I have decided to commit myself to more righteous living, and a rededication to strict ethical behavior toward everyone. This is a big but necessary thing for me. The hypocrite within isn't easy to manage."[11]

With time the shock of the encounter began to pass, and Royce's outlook became more hopeful. "I am getting hold of myself once more so that I feel self confident and well now. I won't be beaten down by this, and I shall proceed forward and do things of my own choosing. In the meantime, I shall try to improve my attitude in every respect."[12] A week after the original episode, Royce once again had occasion to speak with Victor on the telephone, and was relieved to find that both were now willing to work beyond the fractured relationship. "I chatted with Victor this morning and nothing was said about the previous talk. I carried on as if nothing had happened and so did he. Perhaps that is the end of it, I hope so. I shall be most careful to see that I go the extra mile with him."[13]

This event, which so rocked Royce's professional life, seemed to have had a profound effect on his spiritual life as well. Largely gone from his journal were the cavalier put-downs of fellow Church members, and his comments about church meetings began to be less complaining and more positive. His attitudes about spiritual things in general grew less cynical and more thoughtful, and there appeared a sincere desire to refine those aspects of his character which he knew were holding him down. Coincidentally, just two weeks after Royce's fateful conversation with Victor, he delivered a lecture to students at the LDS Institute, which had recently been relocated to a picturesque remodeled house across the street from the University campus. Royce entitled his lecture "Academics and Faith," and he spoke with a new and thoughtful earnestness about his spiritual life, and what factors had caused him to remain faithful to the Church. The notes from this lecture offer a rare glimpse into his personal thoughts at this pivotal time in his life.

Royce began by quoting several scriptures which had served as guiding principles for him during his academic life: "With all thy getting, get understanding" (Proverbs 4:7), "The glory of God is intelligence" (Doctrine and Covenants 93:36), and "Whatsoever principle of intelligence we attain unto in this life, it will rise with us in the resurrection" (Doctrine and Covenants 130:18). He then told about his young life in the religiously insular rural Utah community of his boyhood, when the Church governed almost every aspect of life, and a few of the individuals in that community who had encouraged him on to greater achievement, and had inspired him to seek higher education at the agricultural college. He then spoke of his experience as the first member of his family to attend college, and of the six-year interruption of his schooling for a mission and for military service. Royce described his service in the war as a turning point, when his unquestioned faith in the Church and its teachings were first seriously challenged, and when the cynical side of his character began to emerge. At the same time, he said, "I should point out that in the military, I made my first Church commitment in the context of 'choose you this day whom ye will serve,' among real, compelling alternatives. I faced these issues on much of Church dogma, including such mundane things as the Word of Wisdom and the sexual code in the deceptive face of 'everyone is doing it,' and 'a war is on, what does it matter.' For the most part I chose wisely, as I see now in retrospect."

Royce then spoke of his own struggle to reconcile his belief in evolutionary theory with the doctrines of the Church, particularly when some members equated a belief in organic evolution with atheism. He noted, "This has led to some interesting (to me at least) intellectual activity on the part of many LDS scientists and scholars who are required to face the issue, among whom I have presumed to include myself. I could spend much time on the problem, and my own personal rationalization. First, that the Genesis creation makes biological sense if we accept it as a figurative rather than literal account. And second, it is not logical to speak of eternal progression (or might I say, 'evolution'), the cornerstone of Mormon doctrine (and a real mind-boggler if you put much thought into it) and not accept organic evolution, which is routinely examined in literally thousands of laboratories around the world at the experimental level—the real nuts and bolts of what we are, a simple earth-bound model of the loftier abstraction called 'eternal progression.' However, rationalizations of the sort referred to here are not what has really kept me in the Church. The intellectual approach is a crutch at best."

Royce then came to what he regarded as the key to his spiritual life: "I made a fundamental commitment when I was a graduate student coming to grips with the problem at the University of Wisconsin, that I have renewed regularly ever since. It is, that come what may (conflict, contradiction, or whatever), *I would remain an active mem-*

ber of this Church, attending my meetings and accepting responsibility. This has involved a continuing act of faith on my part, strong at times, weak at others, but always there. This faith (believing but not really knowing) has been the key thing in my church life. In my 61st year, I am an active member, and most other things have been secondary, most of the time. My actions and remarks (in the perception of others) have not always reflected this faith. However, the crucial point is that I have maintained my commitment, exercising faith in varying degrees during my entire academic life. As a result, my faith is strong now.

"Another positive result of this approach has come to me too. For the most part, I am tolerant of the beliefs and most actions of others, and do not insist that they see things or act as I do. I believe that I am more happy and tranquil than I was 25 years ago. The cynic within me has mellowed with time, and the Lord has blessed me with a more understanding heart. Do I understand better the things that puzzled me as a graduate student and young professional? Some blanks have perhaps been filled in, but my understanding is still most imperfect. However, I realize now that it doesn't matter. I have concluded that with enough faith you neither have to do all the things that you once thought that you must do, nor understand everything you once thought necessary." Royce concluded his lecture with a quote from Dag Hammarskjöld: "God does not die on the day when we cease to believe in a personal deity, but we die on the day that our lives cease to be illumined by the steady radiance, renewed daily, of a wonder, the source of which is beyond all reason."[14]

Overwork and Its Consequences

In the months that followed his fateful conversation with Victor, Royce pressed forward as vigorously as ever on his work both with the university and with the strawberry program. This continued to involve early risings with morning exercise, and late nights preparing papers and making notes, interspersed with long trips, usually by car, to the various strawberry areas. Royce had long realized that there was some danger in taking these long drives with so little sleep, and one day after making the nearly three-hour trip to Watsonville after a full day in the office, he observed, "I am tired. If I don't watch it, I'll be killed in that car some day, or maimed. I do get weary."[15]

Scarcely two weeks later, the feared accident finally happened. "A day I shall remember for a long time, and with no pleasure in one way, and a good deal of gratitude in another. I drove to San Jose and picked up Victor at 8:20. We stopped at the field station, and then drove over the hill to Santa Cruz, heading south on Highway 1. Just before reaching Soquel, after 41st street and before Bay, two cars go-

443

ing the same direction in front of a Standard Oil gasoline truck in front of us, collided. I could not see them. The truck stopped with air brakes, but we didn't, at least not fast enough. I had no choice but skid right into the back of the truck, which had veered to the right of the collision. My face struck the steering wheel just below the ridge and on my upper gums and lips. I was bleeding from both nostrils and the doors would not open. When we got out, we could see how fortunate we were that we had our seat belts buckled, and that no fire developed with all that gasoline.

"Victor went into shock and fainted.[16] He was not battered visibly, just bruised where the belt tightened on his ribs, I think. I was bloody from head to foot with my own blood. The highway patrol checked the accident. The two women driving the cars that collided were taken in stretchers. My yellow wagon was hauled away by a tow truck; I fear that it is beyond repair, fortunately not so us. We called Naturipe and were rescued and taken to Larry (director of the Watsonville station). Larry and I went to Ford's department store, where I bought a complete outfit for $80.00. I rescued some of the things, and then went to Aptos to attend the annual Naturipe growers meeting where Victor and I spoke. I didn't feel too much like doing it, because my nose was still bleeding." After the meeting, Royce caught a ride to Davis with a colleague who was heading to the strawberry nurseries in Redding. "The ride wasn't too comfortable for me, although the car was nice. I ate some dinner, then went over to Student Health for a physical check. Nothing is broken. I'll just be sore for a few days, and I am grateful for that. It felt good to get to bed, but I was uncomfortable and didn't sleep very well."[17]

The day after the crash Royce held doggedly to his exercise routine, in spite of the soreness. "We ran a couple of miles, because I do not want my legs to get out of shape. I felt all right. I have some sore muscles freshly showing from yesterday's accident. I also have whiplash injury to my neck. I spent part of the morning filling out injury report forms and the wreck report, got a replacement car, and prepared my lecture for tomorrow."[18] In truth, Royce was fortunate not to have suffered more serious injury. Although he went on with his work without interruption, he continued to have soreness for many days after the accident, which made it difficult for him to sleep, and he was left with a permanent numb spot in his upper gum, and with an altered sense of smell which lasted for years after the accident.

Meanwhile, he continued to struggle with the sheer volume of work to which he had obligated himself, above and beyond the strawberry program, which he always considered his highest priority. His run-in with Victor had given him a new sense of perspective, and perhaps an unaccustomed degree of courage, to resist the urge to take on new responsibilities, though he did this with a sense of ambivalence, for declining a request for help ran contrary to his nature. The month after the crucial phone conversation with Victor he recorded, "I made one decision today, saying 'no' to an add-on job that I really shouldn't take, but was tempted to accept just because I usually say yes whether I am prepared to follow up on it or not. Today, the issue was whether to take a partial administrative position with the Egypt program under Frank Child. A fraction of my time was to be purchased from the department. It was hard for me to say no, because Frank is a personal friend, but I did say no, and made it stick. I think that I shall try it next time someone asks if I feel, as I did this time, that I should not."[19]

Royce's work with university committees continued to occupy his attention, particularly the University of California Educational Policy Committee, of which he was now serving as chairman. This was an important assignment, but required frequent travel for statewide meetings and brought large administrative burdens. "It is almost incredible to see how much paper flows in a day's time. My desk is piled high, and there is no end."[20] At the same time, Royce felt the work he was doing as committee chairman was significant, and he found himself well suited to the position. After one meeting held at the Hilton Hotel in San Francisco, he recorded, "Never have I felt more calm or comfortable in that role, and things did go well. We finished one hour ahead of my designated time, and had worthwhile talks on many issues."[21] The members of this statewide committee appreciated Royce's tact and leadership skill, and their esteem for him was by no means diminished by the flats of freshly picked ripe strawberries which he frequently brought to their meetings.

In these committee assignments Royce continued to encounter the social issues which the Civil Rights movement had thrust to the forefront at that time in history. In these discussions Royce never dropped his objections to what he felt was the unjust practice of extending favoritism to any group, for any reason. After one statewide conference held in the beautiful coastal setting of Asilomar, he recorded, "The two developers of the discussion paper, a Mexican and a black, developed it all around minorities, whereas the title was 'Student Expectations.' I was one of several voices objecting to the perversion, but it did little good. I do agree that they need some special attention, but they are not the only people around with problems, nor will they be. They tend to demand without contributing to things, and to deal in vague ideas that really amount to a form of verbal blackmail." Royce's experience with the Japanese strawberry growers had given him new respect for what an oppressed minority group could accomplish, but his attempts to bring that up availed him little.

"When the Japanese are mentioned as a 'role model,' they dismiss that with a rather flip, 'Oh, but they have a tradition of education,' whatever that may mean. No racial group has suffered as much overt, even official discrimination and persecution as they have in recent times. I can't accept the flippant dismissal, because they are just as identifiable as a group. The difference, as I see it, is in the family structure, and willingness to work hard regardless of obstacles, mostly the latter. Traditions of 'let someone else do the work' and fight them by nailing them with a guilt complex for the sins of their ancestors, real or imagined, seems to be the chosen way. I tire of the eternal whining cry for special this and special that, all at public expense. They wouldn't get the time of day for that in Mexico or anywhere else."[22]

Despite the positive contribution Royce felt he was making on these committees, the day-to-day work they generated was at times overwhelming, and taxed his best efforts at organization. His time was stretched almost to the breaking point between his obligatory trips to oversee the various aspects of the strawberry project, his meticulous preparation for committee work and for classes, and his professional writing. There were days and weeks at a time when he felt as though the harder he worked, the more he fell behind. "There is always no end of things to be done. I am definitely overcommitted, and it is a question of what to add to the neglected list."[23]

All these obligations were weighing on him when, in early 1980, Royce had an unexpected spiritual prompting. He and Pearl had been approached about serving as temple workers, an assignment generally reserved for Church members who were at or nearing retirement, since it tended to be quite demanding and required a substantial weekly time commitment. Royce's extremely busy schedule seemed to put the possibility out of the question, but one evening after attending a broadcast of the priesthood session of General Conference, he wrote, "I was impressed that I probably should try to fill a temple assignment that is pending. I have so little time, however."[24] The following month, Royce and Pearl presented themselves for an interview at the Oakland Temple with the temple president, Richard B. Sonne. Royce recorded, "He wants us to serve as temple workers. Pearl is certainly worthy, but I question whether I am. However, I am inclined to make the effort to try to do it, including adjusting my attitude, where most of the problem is."[25]

In spite of his own misgivings, both about his own worthiness to participate in temple ordinances and about the limited time he had at his disposal, Royce approached the new assignment with an unaccustomed degree of faith, which seemed to be confirmed as he began the course of training which was required prior to starting the assignment. "I was set apart as a temple worker (Pearl was last week), and we attended class all afternoon, three hours. As time went on, it became clearer and clearer to me that I am doing the right thing. My power of learning seemed to increase. I can certainly master it all with the best if I live right and apply myself. I feel better about this than most anything I have done in the Church or elsewhere. We drove home very peacefully."[26]

After a period of weekday training which lasted several weeks, Royce and Pearl were assigned to work at the Oakland Temple on Saturdays, which was invariably the busiest day, and were given a morning shift which required them to arise before 4:00 a.m. every Saturday to make the drive from Davis to Oakland, not returning home until late afternoon or early evening. Their first regular Saturday assignment was on September 27, 1980, and Royce described the experience. "We were up at 4:00 a.m. and on our way to the temple at 4:30, arriving there about 5:45. We participated in most of the various things done there. I have in fact learned everything quite well, and make few mistakes now. All in all, it was a happy, rewarding experience. We left the temple by about 2:30, and reached home about 3:45. I do believe that we were able to accomplish about as much as I would have had I not gone to the temple. Anyway, I am going to attend as faithfully as I can. I am glad that we are in the program. I must improve my personal life to increase my worthiness. This certainly provides the incentive. 'How beautiful upon the mountains.'"[27] To Royce's surprise, the temple assignment did not seem to interfere with his many obligations, and it gave him a much-needed break from the pressures of everyday life. "While there, the outside world and its turmoil fades away."[28]

Royce's and Pearl would continue to serve in the temple for many years, and this proved to be a transformational experience for Royce. His faith in the Church and in things spiritual, which had been on the wane for many years, began once more to revive, for not only was he exposed frequently to genuine spiritual experiences while in the temple, but he also took seriously his responsibility to live a life worthy of being there, and undertook a sincere effort to reform both his actions and his thoughts. "I feel more at home all the time there. Again, this is having a very positive effect upon my personal life and attitude. Perhaps it does not show to others, but it is very clear to me. I simply feel better about myself in general, and perhaps I can be a better influence upon others as I change my attitude toward them. I feel none of the abrasiveness toward any of the people in the temple that sometimes I feel in my work or in the Church generally. That is good."[29]

As Royce and Pearl were still preparing for their assignment in the temple, the Sacramento Stake of the Church was again divided, creat-

447

ing a new stake centered in Davis. The stake president called to lead it was Peter Kenner, a dentist from Davis well known to the Bringhurst family, and before many weeks passed Royce was called in for an interview with the new stake president. "He wants me to be High Priest Group Leader. I know I shouldn't be, but I said yes."[30] The calling, which Royce had held years before, was another major imposition upon his time, and he still had apprehensions about his worthiness to serve, but he continued to stand by his commitment to accept and fulfill church assignments, and he began faithfully attending the requisite meetings and discharging his responsibilities.[31] At the same time, he made certain his leaders were aware of his work situation and of his temple assignment, and after six months of service he wrote, "I was released as High Priests Group Leader today. It grew out of my discussion with the bishop and stake president recently. I was sort of surprised, but not displeased. With my weekly temple commitment it is much more difficult to do the job that should be done. I frankly haven't the time."[32]

Meanwhile, life continued to move forward for Royce and Pearl's extended family. In February John's wife gave birth to their first child, and Royce and Pearl accepted an invitation to travel to Utah in March to witness the "blessing and naming" of the child, a daughter named Sarah. While in Utah, Royce once again stopped by the nursing home in Draper to see his mother. "How sad it is to see her in such a state of decline. She won't be with us much longer. She is frail and not herself. I almost weep with sadness. What a contrast to the lovely brand new child of John and Betty's."[33] That same month Florence called Royce and Pearl to advise them that she would be returning with Larry from Brazil by Christmas, and asked if they would be willing to sponsor the two children they had adopted, being unable to bear children of their own. To this Royce readily agreed.

That year Royce and Pearl's youngest daughter Margaret returned to stay at the house in Davis for a few months, having discontinued her studies at BYU. There was still a sharp divide between Margaret and her parents, and relations in the house were unhappy and tense. In the spring, Margaret announced her intention to marry Jack Dobbins, a man ten years her senior who had a son from a previous marriage. Jack was not a Church member, and unlike the spouses of Royce and Pearl's other children, was not college educated, but when they met to discuss wedding plans Royce found Jack likeable, and much easier for him to talk to than Margaret, with whom he still had a difficult relationship. Arrangements were made, and Royce called each of the other children to invite them to the wedding. The marriage was held on July 3, 1980, and Royce wrote, "Margaret's wedding day—we spent the entire morning working on preparations for the wedding, and continued into the afternoon. I even made the ice cream. Everything looks nice. At about 7:00 p.m. Margaret was married to John David (Jack) Dobbins in a simple ceremony in our backyard under the crepe myrtle tree, by Ernest Westover. This is Jack's second marriage, according to what we know. I hope it works out. Only a few members of the family and their close personal friends were present. It was satisfying and pleasing in a way."[34] Despite Royce's apprehensions, Margaret's marriage would prove to be an enduring one, and she and Jack would raise a family in nearby Woodland.

In his work, Royce now found that age was beginning to take its toll, and he was frequently frustrated about how little he was able to accomplish. "I worked in my office all day on various papers. I really get little done some days. I didn't realize how much of my drive I could lose. I used to do better. Now, I simply do not. I feel the tragedy of it all, but feel incapable of change."[35] At the same time, in the midst of his stacks of chores, Royce's scientific work continued to hold genuine fascination for him, and gave him moments of encouragement. "I worked in the office all day long, with only phone calls to break it up. In the evening I worked on papers that I am preparing. I can get quite a bit done sometimes that way, and that was so today. My fatigue just melts away when I am busy with something of real interest to me."[36]

At the same time, Royce could not escape a sense of inadequacy, particularly in the presence of his colleagues, and he perhaps considered the fact that with his professional life drawing to a close, there were academic benchmarks he had never achieved in the highly stratified world of the university. At a Christmas gathering at the university just a few weeks later, he wrote, "I still feel somewhat a stranger, or someone just passing through in the presence of some of my colleagues. It is difficult to overcome the idea that you and your life is of little worth as measured against those of your peers, if you have grown up feeling inferior to most in the ways that count, although you know within that you have the capacity to excel and lead if only things had worked out better. I still get that feeling, although by standards usually applied I would likely be judged a success by most. Truly 'we do think others are happier, more successful, etc. than we are,' and even worse, we think it obvious to even the casual observer. I have pushed, but often not in the right way. I am fortunate and appreciate the good fortune that has come my way, however."[37]

As 1981 dawned, Royce resolved to do something about the work habits which continued to dog him. "On New Years Eve, Pearl and I ran first, then had a bite to eat, and I went back to attack the mess in my 'private disaster area,' also called 'my office,' and this time I had prevailed before the day was out, except for a few odds and ends. My desk is clean and organized. My table is clean, and once more useful as a working surface, and my window shelf area is also clean. I had made up my mind to go into the new year with a clean slate as far as my office is concerned, and for once I have the possibility of keeping it that way. I am resolved right now, but we shall see how that works out. I have failed before, but as I say, 'this time it will be different.' I do not feel the need to say much about personal behavior. Since I have been working in the temple, I have changed a lot of little things for the better, and I plan to continue on others. I think that I truly am a better man than I was a year ago, but still have plenty of problems."[38]

Work Abroad

One of the factors which prevented Royce from being as efficient as he hoped at this time in his career was the frequent invitation by colleagues or organizations to make extended trips to various parts of the world as a consultant to strawberry industries in different stages of development. In 1981 he made three such trips, the first to Egypt, which he had visited under similar circumstances a decade before. Royce had recently declined a request to take charge of the entire UC Davis agricultural program in Egypt, but when offered an invitation to take a brief assignment as a consultant there, he jumped at the chance.[39] Arriving in March, Royce was impressed by the changes he saw since his previous visit. Much of his time was spent in the area of a city called Ismailia on the Suez Canal, which had been off limits during his previous visit, but which had since become a major strawberry growing area. Ismailia was quiet compared to the noisy streets of Cairo, and Royce was pleased to see that the technical know-how he had imparted on his previous visit appeared to have been put to good use, for the industry was thriving, and making good use of California varieties.

Royce also witnessed first-hand how the historical events had affected the area. "En route up the Suez, we saw the place of the major Israeli crossing where they came over between two Egyptian armies and outflanked both. Vestiges of the rubbish of war still show. They need no more of that. The canal was full of ships moving in an orderly way through, a source of great national pride and income. One good thing about it is the U. S. money, men and equipment that did most of the dredging to clear the waterway and put it back in business after years of closure from the 1967 war. Once in a while we have done a few things right. We drove back through the desert to Cairo (the noisy) after a pleasant interlude at a beach site on Great Bitter Lake. They are trying with varying success to make the desert 'blossom as a rose' in parts of this area. Where there is water (wells or canals) there is life, and where there is not, sand stretches as far as the eye can see."[40] Royce visited Italy during his return trip, and observed there similar improvement in strawberry culture since his previous visit.

Three months later, in July, 1981, Royce accepted an invitation to visit the world-renowned strawberry taxonomist Günter Staudt in Germany, and accompanied by Pearl, he traveled to Freiburg, an old city which had undergone considerable rebuilding after bombing during the War. The day after their arrival, Staudt took Royce on a tour of wild strawberry colony sites in the region, as Royce had done for him back in California. The next afternoon, Royce joined Günter in his office, where the two went over materials of mutual interest. "We discussed a huge manuscript he has done on the taxonomy of the American strawberry species. It is interesting, worthwhile, and should be published, but in English, not German as written, otherwise it will not get the attention that it should. I agreed to look into how and where it should be published, and I am willing to assist with and attempt to arrange for translation to English. The paper certainly clarifies a number of issues, and makes a lot more sense to me at both the octoploid and diploid level."[41] On their way home, Royce and Pearl stopped by Scotland and England, to visit Pearl's relatives and to observe work there with strawberries and other fruits.

Royce's final international trip for the year would be another first for him. In October of 1981, he was contacted by a nurseryman in New Zealand, inviting him to make a two-week visit, and Royce immediately agreed. "I have always wanted to go there, since my father talked of it all his life. That was where he went on his mission."[42] Touring both islands, Royce once again found good use being made of the California varieties with some local adaptations, though he was able to suggest many improvements to refine their growing practices based on his experience in California. "They need two-row beds and summer planting here. They are now equivalent to the early Shasta era in Central Coast California—poor winter planting, followed by a second year, for a crop of rather poor quality."[43] On the South Island Royce visited a strawberry breeder and other plant breeders, and was treated to a helicopter ride to Mount Cook, on whose flanks he found strawberries which had escaped cultivation. On the North Island, he was able to visit the Church's New Zealand Temple before returning to the United States. Royce considered his visit to New Zealand a fruitful one, and it was destined to lead to significant improvements in the strawberry industry there. The host company had hoped the visit would lead to licensing of UC strawberry patents for use in New Zealand, and that hope was eventually realized, but not as they would

451

have wished, for when the licensing was approved by the university, the agreement was made with a competing nurseryman. [44]

Strawberry Program—The Next Generation

By the early 1980s, Royce and Victor had made considerable headway on yet another generation of strawberries, in a sequence of planting and field testing which had by now become a systematic pattern, resulting in a predictable release of new varieties approximately every four years. Each cycle began with Royce carefully selecting appropriate parents from among the many named varieties and advanced cultivars, a process which was varied from year to year according to a predetermined schedule. Controlled pollinations were made by hand from the selected parents, and when the resulting fruit matured, the seeds were prepared and planted. The many thousands of seedlings which resulted were potted, and when they had grown sufficiently they were transplanted, either to the Wolfskill Experimental Orchard in Winters, or to the South Coast Field Station in Santa Ana, where the plants were grown to maturity. The plants and fruit were then systematically evaluated to determine which of the crosses had the most promise. From among these, the best plants were selected for continued evaluation, and the rest were eliminated.[45]

The selected plants were propagated, or "increased," by means of natural runners, usually in commercial nurseries, a process which produced enough plants of each selection to begin experimental plantings. This required Royce and Victor to travel to the nurseries near Redding, in the far north of the state, first to plant out systematically the large number of selections, and later to oversee the machine harvest of the resulting plants and perform the trimming and packaging necessary to prepare them for transplanting. This was a laborious process which took days, and had to be performed by hand.[46] The carefully packaged and labeled bundles of plants then had to be transported in an air-conditioned car to the experimental plots in Watsonville on the Central Coast and to Santa Ana in the far south. There separate plantings were made[47] to test how the same selections would perform in the different locations and under different conditions, such as the winter and summer planting systems, and whether they were best propagated in high elevation or low elevation nurseries.[48] In each case, the experimental plots had to be planned out, planted, mapped and recorded. When the plants came into fruiting, the fruit from the individual plots had to be harvested by hand, and careful records were kept of the timing and volume of harvest, as well as the quality, size, and firmness of the fruit from each selection. This information was used to designate which clones would be tested as advanced selections, and eventually to determine whether a selection

452

would be considered for release as a named variety, saved as a possible parent for further crosses, or discarded.[49] At the same time, testing of selections for resistance to plant diseases such as *Verticillium* wilt continued to be carried out, particularly at the research station in San Jose.[50]

This was a system of enormous complexity, with many experimental components which overlapped one another but had to be individually tracked. Royce and Victor each had permanent staff, which in Royce's case included Larry Fulton, who maintained the plots at Watsonville, and Gus Macias, who did the same for the Wolfskill Experimental Orchard in Winters. Temporary workers were hired as needed, but much of the physical labor, and virtually all the analysis of the data generated, was done by Royce and Victor themselves, and the distances involved meant that at certain times of year they were traveling almost constantly, often for days at a time, covering large segments of the state. Royce and Victor also were also obliged to attend meetings at regular intervals with the California Strawberry Advisory Board which sponsored their work, and with the other organizations which participated in it, as well as with the growers themselves, who depended on their expertise and recommendations. In Royce's case, this work came in addition to his teaching duties, his administrative work, and his other responsibilities as a faculty member at the university, not to mention the professional meetings and international travel which fell to him.

At this time Royce also faced the task of rebuilding his broken relationship with Victor Voth, a delicate and emotionally taxing process, but an essential one considering the long hours and sometimes days at a time they had to travel and work together. Royce was constantly aware of this, and it was a frequent topic in his journal. For example, after a meeting of the Strawberry Advisory Board in which the researchers presented a large budget proposal for the coming year, he commented, "Victor tends to perceive this as his doing, and that I just ride his coat tails, but this really isn't true. He writes up next to nothing, and still assumes that all of this would just happen because he is 'Victor,' with or without me. I cannot accept that, but have avoided most direct confrontation with him, and hope to continue doing so. Our destinies are so closely tied that they cannot be separated now."[51] The relationship between the two researchers would continue to be strained, and periodic flare-ups occurred. After a drive together to Redding in 1981 Royce wrote, "Victor unloaded another of his grievances which he tends to harbor for occasions such as this. This had to do with senior authorships of papers. I consented to a reasonable change, which tends to compromise my interests, not that his case is not without merit, because in all candor it is. It is just that I would feel better about sharing senior authorship of the breed-ing reports and patents if he were ever truly the author of any of them. I write all of them in the final analysis, and have always put his name first on cultural things. However, I prefer to compromise and have peace. It really doesn't matter that much to me, because his con-tributions are at least equal, and in many cases more than mine."[52]

By Spring of 1982, work on the advanced strawberry selections had proceeded to the point that there was a push from the industry for an early release of the next generation of new varieties to allow their propagation for commercial use, a matter which held some ur-gency for the growers, who evidently pressed the issue through. After a meeting of the Strawberry Advisory Board research committee, Royce wrote: "The topic was the release of the new varieties: C3, C5, C6, C24, CN12, and CN 18. This means that they go to the nurseries in time for spring planting. We shall see how this works. Our hand has been forced once more."[53]

A month later, Royce learned the Victor Voth had been hospital-ized for chest pains, and would be required to undergo coronary by-pass surgery.[54] This was a blow for Royce, for in addition to the feel-ings and fears he obviously held for a lifelong colleague, there was the added awareness of the recent problems which had recently darkened their relationship. Royce called Victor's wife Virginia repeatedly to check on his well-being, and was relieved to learn that things had gone according to plan, and that Victor was recovering as hoped. Vic-tor's illness meant that Royce had to handle some of the tasks that usually fell to him, which included evaluation of the year's crop of seedlings which had been planted in the South Coast Field Station. A couple of weeks after Victor's surgery Royce flew to Santa Ana and spent most of the day working in the seedlings. Later in the day he went to see his colleague. "I visited with Victor in the afternoon, and it was very gratifying to find him looking so well and optimistic. His color is good, and he should recover fully. Meeting him affected me emotionally."[55] Victor eventually made an uneventful recovery, and after a few months was back to work as before. In spite of their seri-ous differences, Royce and Victor had at last succeeded in establish-ing a new groundwork for their relationship, which though still cordi-al, would always be more strained and less cooperative than it had been in years past.[56]

Even before Victor's bypass surgery, another serious challenge had come up in Royce's part of the strawberry program. The Sev-enth-Day Adventist group which leased the Watsonville property where Royce's experimental plots were located had decided to change the use of their land, and requested that the strawberry plots be moved to a portion of the property which was far less satisfactory.

This prompted a search for a new location for the experimental plantings. Royce considered a number of possibilities, including a parcel of land at the UC Santa Cruz campus, and a USDA field station where he had maintained plots years before. Eventually he learned of a property in the Watsonville area owned by a Gertberg family, which appeared not just acceptable, but actually superior to the existing location, and Royce arranged to transfer the experimental station to this new property in time for the summer planting in 1982.[57]

With the six new varieties already being propagated for commercial release, Royce and Victor consulted on names for the new varieties, and Royce set to work on the laborious task of writing the disclosure statements[58] now required for each of them. In late September of 1982 he wrote, "I did in fact finish the work on the disclosures, and mailed them in today. It corresponds to almost the day when Douglas, Pajaro, Vista, Brighton, Aptos and Hecker went in in 1978. These are Chandler = C24, Parker = C3, Santana = C5, Tustin = C6, Fern = CN12, and Selva = CN18. There is always an element of real satisfaction in reaching this point. These are the 'tracks' that count when an impact is left from which people benefit. I anticipate that the big commercial 'winners' will be CN18 (Selva), C24 (Chandler), and C3 (Parker) in about that order, with CN12 (Fern) the next in order, followed by C5 (Santana) and a poor 6th in Tustin. We shall see. It is now out of our hands, or at least, shortly will be. There is also a strong element of relief in release too. We no longer have to 'prove' anything for these. They either find their place or they fall by the wayside. We now turn to the next generations, probably the last for Voth and me."[59] As before, the names of these varieties held special significance for Royce or Victor, and three of them were named in honor of deceased acquaintances. "Chandler" was named for Dr. William Henry Chandler, former dean of the UC Davis Pomology Department who had recommended Royce for the position of strawberry breeder in 1953, while "Parker" was named after Bob Parker, an energetic and respected nurseryman from Redding who had died while the varieties were being evaluated.[60] "Fern" was named after Fern Miller, the wife of Royce's old Utah State classmate Harry Miller, who had died suddenly just a few months before, a sentimental gesture by Royce.[61]

Less that two weeks later, Royce was working on the final harvests at the old experimental field, when he came to a decision regarding an additional variety, which he recorded in his journal. "I stopped at Kuni Shinta's to look at C95. I now say 'release' for sure. I must think of a name. It is almost certain to succeed as a replacement for much of the Aiko. The color is a bit light, but the size is good." The next day he added, "Today, I decided that C95 should be named. I think that the name should be a Central Coast one—why not 'Soquel?'

That is where Victor and I had the terrible accident. Why not commemorate that? It was a lesson in vulnerability."[62] A couple of days later, Royce drove to Winters to get leaf samples of C95 for electrophoresis classification, which was now considered a necessary part of the patent process.[63]

Royce got the paperwork on "Soquel" done promptly enough that it was processed with the other six varieties. In November, Royce attended a meeting with plant pathologists in Berkeley to iron out some technical problems with the release,[64] and in December he completed the final step. "I went to the Christmas party in the academic senate office, then to the Academic Affairs office to have my signatures on the new patents notarized. They are: Day-neutrals: 'Fern' and 'Selva' and short-day types: 'Chandler,' 'Parker,' 'Santana,' 'Soquel,' and 'Tustin.'"[65] The patents were forwarded to Victor for his signature, and a few days after Christmas the old trouble between the two resurfaced. Royce wrote with some exasperation, "I spent the morning in my office where I was unjustly chewed out by Victor on the phone over the way the names were on the patent disclosures. I called Diepenbrock to change them. Victor can be most unfair and unreasonable. He has a persecution complex. It really ruined my day. I was not privy to the way this was handled. I did send in the forms correctly ordered. Victor chooses to deceive himself, and believe otherwise. Maybe he will come around on this, but I am sure that he won't say anything to me."[66] When the official patents were submitted, Royce made certain Victor's name appeared first on four of the seven patents, including Chandler and Parker, the two short-day varieties that Royce and Victor considered most likely to succeed commercially.[67]

The simultaneous official release in 1983 of seven varieties from the University of California strawberry program was an unprecedented feat, which would never again be duplicated. Of these seven, a number would play important roles commercially in the coming years. With Fern and Selva, the day-neutral strawberry had at last come into its own, and with its abundance of firm fruit and prolonged growing season, Selva would be the first day-neutral to play a truly major role in the industry. Of the short-day varieties, Parker would become an important and widely-used cultivar, but Chandler, with its rich, red color, juicy texture and abundant yields, would eclipse all the other varieties. It would dominate the California industry for years, and its outstanding flavor would silence critics who complained of tasteless berries from California. Chandler, like some of its finest predecessors, would gain international use and acclaim, and in 1989 it would earn Royce and Victor the Outstanding Fruit Cultivar Award from the American Society for Horticultural Science.

Isoenzyme Studies

As these landmark strawberry varieties were being developed, Royce continued to refine his use of electrophoresis of various isoenzyme types to characterize strawberry species. Most of this work was done in collaboration with one of his graduate students, S. Arulsekar, whose graduate work was centered on this technique. There were two main areas of application for this work with isoenzymes (or isozymes, as they were often called). The first of these was the study of the wild strawberry colonies, both diploid and octoploid, in order to shed light on the genetics and evolution of the various species by observing their isoenzyme patterns. The second application was to use the isoenzyme patterns as a means of identifying specific varieties of cultivated strawberries, a purely practical use of the technique, but one which offered many possibilities.

This technique was of particular fascination to Royce in the later years of his career, and the more he worked with it, the more it seemed to him to be the next logical step in advancing the field of strawberry genetics. His journal entries reflected the satisfaction he felt as these studies were performed on plant material he had selected. "The gels on the material gathered yesterday were run. Few surprises, but some interesting things fell into place so that we understand a bit more of the genetics."[68] Shortly after he added, "We really had good luck with today's gels (test crosses with diploids, with excellent fits to expected ratios). Of such are all things composed, even man."[69] In the case of the wild strawberries, this study involved careful collections of leaf material from a sampling of plants in the various wild colonies, which was often performed on the side during trips to the Central Coast to work in the experimental plots. "I picked up Sekar at 6:15 and we drove to Watsonville, then went on to Point Sur where we collected a transect of *Fragaria chiloensis*, some *F. vesca*, *Potentilla glandulosa* and *P. anserina*, along with my usual bit of poison oak. We want to run the enzyme systems on them."[70]

The first publication to emerge from this work was a study of the diploid *Fragaria vesca*, which Royce and Sekar had collected extensively and systematically at two sites near the coast,[71] where from previous study Royce knew there were colonies of *F. vesca* with complex isoenzyme patterns. He and Sekar collected leaf samples from every colony they could find at both sites, and because these diploid strawberries had just one pair of chromosomes, by studying the banding patterns of a single enzyme, phosphoglucoisomerase (PGI), they were able to infer quite a bit about the genetics of the species. When the colonies came into fruiting, they also collected fruit from the colonies and planted out the seeds, to study the patterns of inheritance of the isoenzymes in the daughter plants. They discovered that nearly all of them had resulted from self-pollination, suggesting that this was an important reproductive characteristic of *Fragaria vesca* as a species.

Royce also had an extensive collection of plant samples of *F. vesca* which he had brought back from Italy, and when he and Sekar ran electrophoresis on these, they found that all had a simple pattern identical to all other specimens of European origin, but differing from all the American specimens, as would have been predicted. Royce had, in addition, a few samples of other diploid strawberry species from Europe and Asia, including some *Fragaria viridis* which had been previously given to him by Günter Staudt, and he and Sekar ran electrophoresis on these plants as well, and reported the results, which were published in 1981.[72] On his return from his trip to Germany shortly after the study was published, Royce performed electrophoresis on samples of a hybrid he had collected in the wild while visiting with Günter Staudt, and was gratified to find his conclusions confirmed. "We ran the *viridis-vesca* hybrid material, and did indeed verify the presence of what we have considered strictly viridis bands (very slow PGI). It is nice to verify something only held in theory up to that time. The story is quite clear: *viridis* must outcross to a considerable extent, and *vesca* only selfs. This would cause erosion of the viridis gene pool, making it scarce except where it might have some yet to be determined competitive advantage."[73]

In addition to their study of wild diploid strawberries, Royce and Sekar also performed a study of isoenzyme inheritance in octoploid strawberries, and for this purpose they used both commercial cultivars from the California program and samples of wild *Fragaria chiloensis* which they had collected on their excursions. Because the octoploid strawberry species each had four pairs of chromosome sets, the possible combinations of inheritance patterns were nearly endless, such that it had been almost impossible to tease out the inheritance patterns by studying the traits alone. However, the isoenzyme patterns were a much more specific tool, and offered the possibility of examining inheritance in much greater detail. When they examined their results, they seemed to confirm the idea that the octoploid strawberries had originated genetically from diploid species, and contained four diploid-like series of chromosomes which reproduced concurrently, making it a "diploidized octoploid." The results of this study were published in 1981.[74]

Published at the same time was an additional study involving University of California cultivars, which though much simpler, was to gain far more attention because of its practical applications. As plant patents came into widespread use, it became important to be able to make precise identification of the characteristics which distinguished a specific variety from others, in order that it qualify as an "invention" under patent law. This identification was traditionally done through descriptions and measurements of the plants and their behavior, but such measurements could vary depending on the growing conditions, and were often quite subjective, and prone to errors in human judgment. Very early in his work with isoenzyme electrophoresis, Royce had realized that this technique held promise to objectively distinguish one variety from another on the basis of its isoenzyme pattern on electrophoresis. Royce and Sekar selected twenty-two of the California varieties, and ran electrophoresis on all of them. By observing the PGI patterns of each, these varieties could be divided into six distinct groups of cultivars, which could be distinguished from one another by their banding patterns. Hence, varieties from one group could not be mistaken for varieties from another if this technique was applied, and in a couple of cases, a single variety could be definitively identified by its unique PGI pattern.

To better refine the process, Royce and Sekar added a second enzyme, leucine aminopeptidase (LAP), then a third, phosphoglucomutase (PGM), to better distinguish between cultivars. By using two of the three systems, they found that more clear differentiation could be made, and by using all three, twelve of the twenty-two cultivars could be individually distinguished from all the others, and the other ten nearly so. They reasoned that if still other enzyme systems were identified and used, it might be possible to assign a unique series of isoenzyme patterns to each California variety, which would act as a kind of "fingerprint" to the variety. By adding this information to the variety descriptions, identification became much more reliable, and the patenting process more secure. The resulting paper, which bore the title, "Electrophoretic Characterization of Strawberry Cultivars," bore the name of Royce as the chief author, and of Sekar, Jim Hancock, and Victor Voth as co-authors. It generated widespread interest among plant breeders, for it at last opened the possibility of identifying individual cultivars by molecular means, rather than relying solely on physical characteristics. For this paper, Royce and his co-authors in 1982 became the recipients of the Joseph H. Gourley Award, a prestigious prize given annually by the *American Fruit Grower* magazine for the year's best paper on general fruit horticulture.[75]

There were other practical applications to this technique as well. Even before the paper was published, for example, Royce had been able to use isoenzyme patterns to distinguish a promising advanced selection from other plants which had been inadvertently mixed in with it in the experimental plots.[76] Some years later, when growers purchased large lots of Chandler plants from a nursery, then found them to be of an inferior variety, Royce was consulted, and found that the offending plants had an electrophoresis profile identical to that of Tufts—this did not prove that they were Tufts, of course, since different plants could have an identical pattern of isoenzymes, but it proved definitely that the plants were not Chandler as advertised.[77] Royce and Sekar later prepared a comprehensive review of all the isoenzyme work which they had performed in the various *Fragaria* species, which was published in a collection of articles on the use of isozymes in plant genetics and breeding. In it they summarized the genetic significance of their work and suggested areas for future study.[78]

Plant Breeding Theory

In addition to his work with Isoenzymes, Royce was by now widely recognized for his special expertise in the theory and techniques of plant breeding. The extraordinary commercial success of the very complex California strawberry program, coupled with the scientific work it generated, put Royce in a prominent position among his fellow plant breeders, and around 1980 he was approached by a colleague from Purdue University named Jules Janick, to contribute a chapter on fruit breeding strategy to a book on fruit breeding. The chapter, published in 1983, gives an excellent overview of Royce's approach to his craft as a plant breeder in the later years of his career. He undertook the task systematically, outlining first the long history of fruit breeding, then defining the four basic steps which lead to the domestication and improvement of fruits originally found in the wild. Step one was to identify superior fruiting plants, either from the wild or from chance seedlings; step two was to propagate these selections in an agricultural setting; step three was to develop cultural practices which enhanced the performance of the selections, and finally, step four was to breed hybrids from among the best fruiting plants, then to select the superior offspring to use agriculturally, and to serve as the parents of the next generation, a process that could continue on indefinitely.

Royce used three examples of cultivated fruits to illustrate these steps in the breeding process. The first of these was the avocado, which had been grown as an agricultural crop in the United States for less than a hundred years, and which he regarded as still being at step 3, since in his estimation no genuine breeding work had been completed on the fruit. Instead, chance seedlings were still being used as the principal varieties, and almost all advancements had been made in cultural practices rather than breeding. His second example was the blueberry, which had been treated as a wild berry until the 1920s, when the first bred varieties were released, and the fruit officially domesticated. Since that time steady progress had been attained in variety improvement, and breeding efforts were actively underway. Royce's third example was the strawberry, which had already been domesticated for centuries, but in which dramatic advances had taken place by application of sound breeding practice. Always at the heart of this practice was "recurrent mass selection," by which Royce meant breeding the best available parents, from among whose progeny the best offspring were selected for use as the parents of the following generation.

These elements of successful fruit breeding were so basic and so simply stated, that Royce devoted the remainder of the chapter to the various pitfalls which often prevented breeding programs from achieving their full potential. The first of these pitfalls was insufficient financial support. Royce observed, "While monetary support is not everything and problems are not necessarily solved just because money is spent on them, little will be accomplished without adequate, long-term financial support." He emphasized that the financial investment had to be a long-term commitment that limited pressure for a quick "payoff," which tended to lead to the frequent release of lackluster cultivars at the expense of more gradual but genuine achievement of genetic advances in the breeding population. Royce recognized the value of the relationship that existed between his program and the strawberry industry. He considered the partnership between plant breeders and their respective industries to be essential to success, so much so that he held that, "as a general principle, breeding programs that fail to generate support from the industries they were created to serve should be questioned and, in many cases, terminated."

The second pitfall Royce identified was that of a faulty strategy, which he defined rather simply as "anything that doesn't work," though he clearly was referring to practices which violated basic principles of plant breeding. He pointed out that fruit breeding often involved large plants with long life spans, so early decisions tended to have weighty consequences. "Fruit breeders cannot afford to waste a single cycle since, quite definitely, time is not on their side. With the financial outlay required, not a single square meter of space should be committed to new seedlings or performance testing unless it is justified by the contributions that the consequences of committing that space will make to the program involved." He pointed out that bad decisions tended to set in motion a series of undesirable results, and seldom was the decision made to pull out the plants and start again. Royce used once more as an example the avocado industry, in which the breeding program had been allowed to continue for years despite lack of progress, based on what he regarded as a faulty strategy.[79]

Royce's third pitfall was that of a misdirected emphasis, by which

461

he meant focusing efforts so strongly on secondary breeding objectives that the primary one failed to be achieved. He defined secondary breeding objectives as those which, while desirable, are not essential to the success of the cultivars developed. These were generally objectives which had to do with the vegetative part of the plant rather than the quantity and quality of the fruit, and he used as a prime example the practice of breeding chiefly for disease resistance. While Royce considered a resistance program to be an extremely important part of every breeding effort, he recognized that if resistance were used as the principal screen in a breeding program, it was very unlikely that varieties with the rare combinations of genes that resulted in excellent fruit would ever be selected, and he pointed out that even when programs were successful in breeding disease-resistant plants, it was often only a matter of time before the pathogen or pest developed the ability to circumvent the resistance.

The fourth pitfall was under the category of faulty procedures, of which Royce listed several. One of these was "failure to measure things properly," and Royce emphasized the need to use well established scales to measure traits whenever possible, rather than relying only on subjective descriptions. He also pointed out that it was essential to have a breeding population large enough to have a statistical chance of developing superior varieties, and he advocated using randomized small samplings of each progeny for detailed analysis to select the best progeny groups, then using traditional methods of observation among those groups, a technique which enabled the breeder to screen a much larger number of seedlings than would otherwise be possible.

A second faulty procedure was "failure to start at the most advanced level possible," which meant failing to use as parents the best cultivars already available, starting instead with more primitive stock. A similar problem was "failure to move through the generations," which generally meant continuing to use a same parent over and over again, rather than making use of its improved offspring, a practice which was destined to result in little progress. Royce also listed "failure to eliminate the losers" as a roadblock to success, pointing out that if a breeder hung on to poor performers, the breeding stock would quickly grow too large to be manageable or to make reasonable progress, a process Royce referred to as "stagnation." His solution was simple. "As 50 new selections enter inventory, 50 old ones must be eliminated. The alternative will ultimately lead to a lack of productivity and loss of financial support."

Royce additionally warned against "failure in breadth," by which he meant that it was not only possible but necessary to have multiple breeding programs underway at the same time, particularly since virtually all fruiting crops took multiple years to progress from seedling to selection. The final pitfall Royce mentioned was that of faulty exploitation of improved cultivars. He felt it was the breeder's responsibility, not just to develop new and superior varieties, but to see that the necessary field testing was performed to determine how best to grow the varieties, and to communicate that information to the growers as the new varieties were released. This was, of course, an area in which he and Victor had excelled, and without which he knew that even the best of their varieties could not have approached their potential.

In concluding his chapter, Royce reflected on the future of fruit breeding, which he saw as bright, with much good work yet to be done. He wrote, "No successful fruit breeder has ever deluded himself into thinking that he has bred a perfect cultivar not amenable to improvement. No cultivar of any fruit species lacks defective traits—each is in need of improvement. Moreover, what may be a near perfect cultivar today, may not be so tomorrow because the conception of what is desirable may change." While correctly predicting that the emerging new field of genetic engineering might shortly offer the possibility of transferring genes with desirable traits directly from one plant to another through gene splicing, Royce still felt that ultimately, the principal strategy for good breeding would continued to be the straightforward, "pick the winners" approach which had been in use for generation after generation. He spoke out against any who supposed that the best gains in fruit breeding had already been achieved, and stated what had probably been the motivating factor in his and Victor Voth's work: "One of the intangible elements of good breeding strategy is eternal optimism. The future in fruit breeding belongs only to those who seize the opportunity, believing that they will succeed."[80]

In another publication, Royce commented extensively on the use of hybridization in plant breeding, which had been regular part of his own breeding program for years. Since the domesticated strawberry was an octoploid, the easiest hybrids were those created by crossing standard varieties with either the coastal strawberry *Fragaria chiloensis*, or with the meadow strawberry *Fragaria virginiana*, both of which were octoploids and crossed freely with domesticated strawberries. Royce pointed out that it was the use of hybridization with wild *F. chiloensis* which had likely provided the California strawberries with their disease resistance and their ability to grow vigorously during the winter, and of course it was hybridization with *F. virginiana glauca* which had brought the day-neutral gene into the California program, and into other breeding programs as well.

It was more challenging to transfer genes from strawberry species with other chromosome numbers, such as the diploid *Fragaria vesca*. Royce had explored three possible ways of transferring genes from the diploid *F. vesca* to the octoploid commercial strawberry varieties. The first of these was to create a decaploid intermediate (i.e. ten sets of chromosomes), which could be done in a number of ways. The second approach to transferring genes from the diploid to octoploid strawberries was by making use of 9-ploid plants, similar to the one Royce had discovered in the wild at Wright's beach. Royce found that by hybridizing pentaploids with octoploid plants, 9-ploids could be generated in large numbers, since most of the offspring of the pentaploids had unreduced gametes. These 9-ploids tended to be very fertile, and could be crossed readily with octoploids, transferring the desired diploid genes into standard octoploid plants. "The strategy is simply to select for the target gene or combination of genes from the diploid species and then, through backcrossing, incorporate those genes into the derivatives, and finally, by eliminating undesirable chromosomes, arriving back at the octoploid level." A third approach, which had been pioneered by another investigator, involved doubling the chromosomes in a tetraploid to produce an octoploid.

Royce had employed most of these techniques in his attempts to transfer genes, chiefly from diploid *Fragaria vesca*, but also from a variety of other strawberry species. He noted that successful crosses with plants of the closely-related *Potentilla* genus had also been successfully carried out, opening the possibility for transfer of genes from those plants into commercial strawberry varieties. Royce correctly pointed out that only a tiny fraction of the potential germplasm for strawberries had been put to use in modern cultivars, and he declared, "as needs arise, almost unlimited possibilities exist for improving the performance and fruit of cultivated strawberries through conventional means, or by incorporating desirable genes from other *Fragaria* species and the genus *Potentilla*."[81]

In the final years of Royce's career, pioneering work in recombinant DNA had opened up a new field of genetics which would come to be known as genetic engineering, a radically different approach to genetic improvement. Rather than the conventional, time-honored breeding techniques used by Royce and other plant breeders, this new technology involved isolating a desired gene, sometimes from an entirely different species, and inserting it into the genetic material of a target plant or animal. Similar techniques had already been used to create bacteria with the ability to produce human insulin, for example. Some scientists looked at these emerging techniques as a great advance, while others raised concern about the possible unintended consequences. Royce acknowledged that genetic engineering might some day play an important role in plant breeding, but he expressed reserve when an opportunity actually presented itself. "I had a discussion with a blue sky salesman on genetic engineering, not really believing much of anything he said. He knows little of biology, and is long on hype. His name is Goldfab, and the company is Calgene. We shall see how that 'exciting' enterprising organization fares. I want no part of it as I see things now."[82]

In part, Royce must have realized that these newer genetic techniques, if they came into their own, would become the work of future generations of researchers. His plate was already full, and he was beginning to feel the effects of age as he approached retirement. It was perhaps in relation to this same technology that he wrote a few months later, "I spent the day at Sacramento listening and talking in a gene resources meeting on strawberries. I'm not sure that I want much to do with them, but we shall see about that. They seem to be more interested in kingdom building than anything else. Probably part of my problem is that I feel threatened to a certain extent. It is getting harder to deliver as I get older. I no longer have the energy I used to, and never enough time."[83]

Though at times it seemed that the new technological advances would render Royce's generation of plant breeders obsolete, there were still some indicators that in this brave new world of rapid advance in genetics, there was still an important role for the time-honored methods of the traditional plant breeder. In the Summer of 1982, UC Davis sponsored a Genetic Engineering Symposium, for which Royce was preparing a large exhibit on *Fragaria* genetics. Royce registered for the symposium, and attended the opening sessions. "Noteworthy was Professor Symmonds' remarks (from Edinburgh, Scotland). They served to remind of the real world of genetics and plant breeding, through which any product must pass no matter how artfully crafted the engineering accomplishments might be."[84] Royce observed attentively the various sessions of the symposium, and the next day he wrote, "Surely there is no Plant Breeding Revolution yet, as Symmonds said there would not be. Almost, I am convinced of the value of all this, but it is still downstream."[85]

By the time Royce was approached by a representative of another genetic engineering firm some years later, he had had more time to consider the possibilities, and was confident that if such work were done with strawberries, it was many years in the future, and he felt certain that success would not be achieved without basic breeding experience such as he had acquired over decades of work. "They know absolutely nothing about berries, and want to start at the top, genetically engineering them. I predict failure and more failure until they at least learn something."[86]

The Germplasm Repository Program

In July of 1980, while attending a symposium in British Columbia, Royce was informed that he had been selected as a member of the governing committee for a new germplasm center which was under construction in Corvallis, Oregon. The National Plant Germplasm System, newly established under the U. S. Department of Agriculture, had been given the task of preserving, in field specimens and in tissue culture, the valuable and irreplaceable fruit varieties developed in the various breeding programs across the country, as well as significant wild plant specimens of potential value. In past years such collections had largely been gathered and maintained by individual plant breeders, mostly at universities, and the plants were often lost when the scientists retired, changed their research focus, or encountered funding shortfalls. Royce, who had made his own substantial collection of strawberry materials, had been a strong advocate of this type of variety preservation.

The Corvallis, Oregon site had been chosen as the first of the new repositories in part because of the influence of Wilson Foote, the associate director of the Oregon State University Agricultural Experiment Station located there, where a large collection of fruiting trees and plants already existed.[87] Royce had been a classmate of Foote as an undergraduate, and the first meeting of the committee held that fall, before the repository was completed, brought back memories. "We went on campus and met all morning in our first session, just talk. Wilson Foote was there. In 1937, he and I were two thin young men majoring in Agronomy at Utah State. He ran right on through, and I went on a mission, then into the military. He has had a good career as a cereal breeder at Oregon State, married out of the Church, and is inactive. I of course am active, although probably more of a hypocrite than he is. He is now the dean, etc. I am what I am, no more, no less."[88]

The dedication of the Corvallis Germplasm Repository took place six months later, in April, 1981. Royce, returning home from his second agricultural mission to Egypt, made a stop in Oregon en route to attend the dedicatory proceedings. After an exhausting series of flights and delays, he arrived in Portland after midnight, and did not get into his hotel until after 1:00 in the morning. The next day he wrote, "I woke up early since I am off schedule still, and so I ran up the road and back for about half an hour. I then drove to Corvallis and visited with Dick Converse, then had lunch with him, Whitey Lawrence, and others. In the afternoon, the Germplasm Center dedication was completed. It rained for much of the time." A meeting of the committee was scheduled for the following morning, by which time Royce, exhausted from lack of sleep and suffering from jet lag,

466

had developed a cold and had lost his voice almost completely. "At 8:00 a.m. our germplasm meeting started and went on all day. I have felt better in my life, but there was some solid accomplishment. I have to get with it on the strawberries, and soon. There is much to do."[89]

In 1981, the USDA took steps to establish a second repository at Davis, making use of the existing Wolfskill Experimental Orchard in Winters, which Royce had used since his arrival in Davis for planting and propagating strawberries. Royce participated in the meetings for the new center, which involved the UC Davis Pomology Department, and which got off to a rocky start. "I drove to the Sacramento Airport to pick up Victor, and we then drove to Sacramento to attend the germplasm meeting. It dragged on and on, and the two people involved said less and less. It was intended as a pitch for funding, but it went nowhere. I was glad to be able to leave, having an excuse for leaving at 4:00 p.m."[90] A few days later the department met to consider the proposal. Royce recorded, "I attended a pomology staff meeting until about 10:00 a.m., discussing the improvements to be put in at WEO at the expense of the USDA to make it a germplasm center. They are weasling on things now."[91]

The relationship of the pomology department and the staff for the new center were strained, with department members feeling that the center had plans at cross purposes with their own work. Near the end of 1981 Royce had become frustrated with the project. "I talked to the germ plasm people again, and told them off in somewhat blunt terms. Even they are catching on to the fact that we know that they are just trying to use us for their own ends, and that we aren't really going to cooperate with them much more, if at all."[92] In spite of these early problems, the Davis Germplasm Repository was eventually completed, and was dedicated on April 4, 1984.[93] It would play an essential role in the repository system, becoming the center for preservation of larger temperate fruit plants, such as pit fruits, nut trees, olives, grapes, pomegranates, persimmons, and kiwifruits.

Meanwhile, Royce continued to attend meetings and work with the committee at the Northwest Germplasm center in Corvallis, Oregon, preparing plants from his own strawberry collection for preservation there. Finally, in December of 1984 he prepared a shipment of plants to send to the repository. This included fourteen of the named California varieties (excluding those under patent), as well as more than thirty selections from the California breeding stock, ranging from parents of named varieties to key hybrids with plants from wild colonies. Royce also included almost forty strawberry clones originally from the wild, including the seedling of the *Fragaria virginiana glauca* plant from Utah from which all the day-neutrals were descended, as well as the natural hybrids he had discovered along the Pacific coast. Together with an additional ten specimen plants given to him from various sources, these added up to exactly one hundred of the most valuable plants from Royce's collection, which he hoped in this way would be preserved for future generations.[94] These became part of the initial core of what would become a large collection of *Fragaria* clones kept at the center.

Change and Perspective

The 1980s proved to be a time of considerable personal growth and refinement for Royce, even though he was already in his sixties and rapidly approaching the customary age of retirement. His problems with Victor had given him the incentive, and his work in the temple the means and motivation, to examine both his faith and his personal conduct, including his private thoughts and actions, which though unobserved, were often recorded or inferred in his journal. Though he continued to exercise faithfully and run in the early mornings, and as a result still enjoyed the health and stamina of much younger men, he was beginning to feel the effects of his age, and knew that his most productive years were rapidly passing. His attitudes and reflections by this time had become more positive and hopeful, and colored much less by cynicism, and more by quiet reflection than in the past.

An example was Royce's account of a Sunday testimony meeting at church, at which a visitor to the ward had caught his interest. "Last Sunday a Brother McComber came to the ward, and after a ward member had given a very doleful report on her family, describing her father as a 'terminal case,' this brother from Idaho, here for a training session on cannery operation and over 70 years old, said as a preface to other interesting remarks in the same vein, and with complete optimism, that 'we are all terminal cases,' and that it is only a matter of when and how. He then died effectively in his sleep here in Davis on Tuesday morning, and I am sure that he died smiling, at least figuratively. I was really touched by this. There is a time for all things, and death lurks nearby ready to move in when the time is right. This doesn't mean that I believe that we are destined to die at a certain time and place, but rather that we are certain to die, and that we may as well be at peace, and accept it whenever it touches us with good grace."[95]

The very next day, Royce and Pearl had their own scrape with mortality while traveling near Watsonville, a day prior to their weekly assignment at the temple. "I came near to killing us both, almost making my words of yesterday prophetic; the type of self-fulfilling prophecy one can do without. As we came over Highway 17 on our way back on rain-slick streets, I headed into a downhill right turn on a steep area going too fast. I suppose that I was thinking I was in the heavier UC car with a longer wheelbase which I usually drive, and I found myself going straight into the concrete median barrier. When I turned the wheel, the car hydroplaned, then overcompensated. It spun out to the right and after banging the curbing, came to rest completely reversed in direction. I got out, expecting great damage, and found that we had lost a hub cap and nothing else. When I think what might have happened, I shudder. As it is said, 'someone up there must like us.' We certainly are not ready to 'go' as I reflect on the incident, but death did brush our cheeks, and surely the 'destroying angel has passed us by and did not slay us.' I like to think that our attempt at more righteous living to be worthy of the temple where we go tomorrow, might have something to do with it. I don't fear death, but I do indeed prefer the alternative for the present, and that most emphatically! Strangely, we were both calm during and after the incident. I trust that this reflects a mature view of the whole matter. We did have our belts buckled and that likely made a difference. The postscript is that we went right to sleep at the motel."[96]

In fact, Royce's experience as a temple worker was already exerting a significant effect on his life and outlook. A couple of weeks after this incident he wrote, "I do enjoy temple work and I hope that I am overcoming weaknesses as a partial result. I believe that I am, but I have a long way to go."[97] Not only did his temple assignment give Royce a motivation to refine his personal attitudes and behavior, it also provided a succession of genuinely spiritual experiences which bolstered his faith. One of these was the sealing of his daughter Florence and her husband Larry with their two children in January of 1981. "At 10:00 a.m. their children, Daniel and Elizabeth, were sealed to them by Brother Jenks. I could not help but think of how different the lives of those two lovely little children would have been had they not been adopted by Florence and Larry. How sad that more can't be so helped."[98] The day after Royce's exhausting return from his 1981 trip to Egypt, he had a similar experience, involving a temple co-worker, who performed an ordinance for a deceased relative. "I was tired in the temple but still enjoyed it. I helped Brother Allen with the initiatory for his late son, John. It was a moving, thrilling experience."[99] It was perhaps this event which inspired Royce to undertake similar work for his own deceased family members, for the following month he wrote, "This was an important family day for us. I did the initiatory and endowment work for my brother John Smith Bringhurst, now dead for a number of years, and Pearl did the same for my sister June. I felt good about it, and feel that it should have been done."[100]

The temple calling gave Royce not just a respite from the often overwhelming rigors of his professional work, but also an important sense of purpose in an area of his life which he had long neglected. His ability with Spanish enabled him to be of special service when Spanish-speaking Church members came to the temple,[101] and likely awakened spiritual feelings from his past service as a Spanish-speaking missionary. The work in the temple mostly involved the uniquely Latter-day Saint practice of performing religious ordinances for people who were dead, and in his journal Royce began to contemplate the lives of those deceased individuals for whom he had served as proxy in temple ordinances. "I went to the session for one 'William Barley,' who had his little hour in the sun over a century ago. Truly 'it is not given to man to tread the grass for long.' Was he old? Did he have a family? Was he large or small? Was he a gentle person? Did he seek after what I did for him this day? These questions have no answers."[102]

At the same time, Royce was aware of the striking difference which existed between the powerful but largely symbolic and faith-based experience of temple worship, and the analytical and evidence-based world of biological science which was the basis of his professional career. The contrast seemed especially pronounced one week when he had attended a lecture on the origin and evolution of humans and apes not long before attending the temple. He wrote, "How out of context the above is with the simple temple story—worlds apart, but somehow it must be reconciled—I cannot live forever partly in two worlds apart from each other. Where will this division in my psyche lead to? We may understand sometime certainly, that cannot be reconciled now."[103] Royce recognized that his experiences with spiritual things were at some level no less authentic than his intellectual and scientific insights, though they were less readily explained. One Sunday after a testimony meeting in which he had contributed his own testimony, he wrote, "I went to fast meeting where I decided to have my say. I sometimes wonder at what I say, but as I say it, it is true for me."[104]

As Royce mastered his duties at the temple he was in due course recommended for a position of higher responsibility. "I was told that my name had gone in to be made an 'officiator.' I suppose I want that, but I also question my worthiness, as I do in many things. Do others have the same doubts, and mixed feelings as I do? Maybe I'm not as unusual in matters such as this as I have thought at times."[105] These persistent feelings of unworthiness that Royce expressed again and again in his journal stemmed in part from a long series of private indulgences in which he had engaged in times long past, during spiritual low points in his life. The recollection of these had for years cast a shadow over his spiritual life, and caused him to feel he was playing the role of hypocrite whenever he followed his spiritual impulses, a feeling that was intensified in the more sacred setting of temple worship. Indeed, it was probably this, more than any conflict he perceived between his scientific discipline and the teachings of his religion, which had made Royce feel at times like an outcast in the Church, despite his unwavering commitment to remain an active and faithful member. One Sunday after a lesson on Church discipline, he reflected, "I suppose if the truth were all known about me, I would have been a fit subject for excommunication at times. Sometimes, I think I am little better than a louse."[106]

These matters from years long past remained hidden in the background until the summer of 1982, when Royce's stake president was made aware of some of them through the disclosure of a family member who had been affected by them, and Royce was summoned for an interview. He recorded, "About 4:00 p.m. President Kenner called. I was asked to see him which I did at 5:30. It was voices from the past for me. I didn't enjoy talking about it after almost thirty years. Consequently, I did not sleep well." The following day Royce was permitted to attend to his assignment in the temple as before, and that evening he had another interview with the stake president. "Upon returning I did some yard work, then went to see Pete Kenner, settling the problem, I hope for all time."[107] Although it was a relief for Royce to at last be able to unburden himself of a dark aspect of his past, in truth the matter had not been completely resolved, and it would still weigh on him in the coming years, as he continued in his efforts to refine his character.

Concurrent with these concerns was the condition of Royce's mother, who had long been in decline. In May of 1981 Royce had written, "Mother is no better, and isn't likely to be here much longer. How sad her situation is, but life must go on. I sincerely hope that I never reach such a state."[108] His mother's mental decline represented Royce's greatest fear for his own life, even amid his attempts at self improvement. "I wish to be better than I am, and perhaps might even succeed if I live long enough. I certainly hope to keep my mind clear and active until I die. That is a must with me. Anything less is not acceptable. I don't fear death, but I do fear loss of my mental faculties. Somehow, I am convinced that it need not happen."[109]

One evening in September Royce sensed a need to call his brother in Utah. "At night I felt that I should talk to George. This turned out to be wise. Mother is evidently in a coma, and may die at any time now. My premonition may have been correct again."[110] As it turned out, his mother's slow decline dragged on an additional four months, until at last she died quietly in January of 1982, at the age of 96. On learning of it, Royce wrote, "How my heart pains when I think of what much of her last 10 years have been, and particularly the last two; not knowing much of anything, and not knowing any of us. Had she lived until the 2nd of February, it would have been twenty years since my father died."[111] Royce and Pearl flew to Utah, where they were met by his brother George, and were driven to the funeral home for a viewing. Royce, who had opposed in principle even holding a viewing, wrote, "In death mother looked quite well thanks to the people who prepared her. At least I dropped my objections (officially) to the 'viewing,' although there would have been none had I chosen. We met old friends and relatives. It was all in all a very happy occasion. When death is appropriate, there is no sadness, only nostalgia."[112]

The funeral was on the following day, and Royce was lost in reflection. "We met more friends of years long gone from almost a different world. I conducted the service. George gave the family prayer, Dean gave a synopsis of Mother's life, and Garnett Player Spoke. I had the closing words, 'Come, peaceful Death, come blest repose, come, into freedom Oh lead me, for of this world I am weary; come soon and set me free; my eyelids gently close, come peaceful death.' I thought it appropriate. Dean read from several of Dad's letters while he was on his mission. We then rode to the cemetery in the limousine of the mortuary. The sun came out like a benediction upon the snowy ground into which her mortal remains were placed alongside of Dad's grave. She of course was not in her wasted body. It was discarded as an instar only, and the real person moves on to better things. At least, I like to believe that is so. The vanity of life is of little consequence at the end. Nothing visible goes with us, and it matters not if one is rich or poor, lowly or mighty. We all pass through the same gate, and must bow down to get through."[113]

Some months after his mother's funeral, Royce faced a serious difficulty in his workplace when a dispute erupted between a university staff member named K———, whose hiring Royce had supported, and Royce's staff research associate, Hamid Ahmadi, on whom he relied heavily. The dispute, which had to do with the use of lab materials, caused something of a furor in the department, and Royce, after attempting unsuccessfully to stay aloof from the matter, found himself in the position of having to defend his staff research associate against what he felt certain were unjust allegations. At one point the incident led to physical contact between the two antagonists, for which police were summoned, and eventually the matter threatened the standing of some of the staff and graduate students. The entire episode left a bad taste in Royce's mouth, and when it had dragged on for more than half a year, he wrote, "Sometimes I think I should quit and have done with my work. I have lost some of the enthusiasm I have always had before, at least on certain days. I don't really command a very 'tight ship,' and some things, particularly the K——— business, are downright unpleasant. I would just as soon make peace with him, but there does not seem to be any easy way to do that, at least at the moment. Much of the difficulty still lies with Hamid, though in fairness to him, he does many useful things that we could ill afford to do without."[114]

Another worry for Royce was the disposition of his share of money from the strawberry royalties, which were beginning to add up to significant amounts, and which still made him uncomfortable. "I spent most of the first three hours of the day working on income taxes. We will be owing a substantial amount. We then went to the bank to deposit my royalty checks that I am somewhat loath to accept, because I really question the rightness of receiving it, since I am paid well already." As these annual patent royalties began to roll in, Royce followed his resolve to give half of his share as a contribution to the university, and placed as much as was allowed into a retirement account, but was at a loss as to what to do with the remainder. In addition to his donations to UC Davis, he began making yearly contributions to the universities where he had received his undergraduate training,[115] and in the wake of his near-accidents with Pearl, he determined that year to buy a new car, paying cash for an Oldsmobile Ciera, which he hoped would be safer than his older vehicle. "This will be our 40th wedding anniversary gift to one another. We do need a reliable car if we are to continue at the temple, and this seemed to be the wise thing to do at this time. It is really comfortable."[116] On the day of their anniversary, Royce took Pearl out for a fine prime rib dinner, and observed, "When we were married, what we spent on this meal would have fed us for a week or so." Royce and Pearl had been blessed with a long and enduring marriage, and although their relationship had had its turbulent moments, Royce sincerely valued the marriage, and his fondness for Pearl was never in question, though he was not very demonstrative generally about his feelings. "Ours has been a very good marriage. So many have had much less. I have received more than I have given, and what I have gotten from her has been the best. Sometime I should tell her, but I think that she understands."[117]

The passage of time had assumed greater importance for Royce as he realized the limitations of what he could accomplish with the years that remained to him. When Noel Sommer, who had succeeded him as chairman of the Pomology Department, completed his own term in that position, Royce reflected, "It brought back memories of the freedom I felt some six years ago when I stepped down. It is too bad that some of the resolve I felt then as far as the future was concerned was not converted into action. Truly we reap what we sow, and seldom realize more than a fraction of our true potentiality. I am too much involved with things of little consequence, among my numerous faults and failings. Tomorrow and tomorrow and tomorrow and—oblivion? Who knows. I must confess that some days, I certainly do not."[118] Despite their good health and vigor, it was becoming more and more difficult for Royce and Pearl to avoid noticing how age was taking its toll on their friends and acquaintances. After a dinner gathering, Royce observed, "We are all getting older, and some of us are a bit in decline. I suppose there is no cure for aging other than to attempt to face it with reasonable grace."[119]

Life was proceeding forward for their children as well. In April of 1981 the entire extended family had gathered in Davis for the first time since Ann's wedding in 1975, and no fewer than thirteen grandchildren were present, and that number would continue to grow in the ensuing years.[120] That fall, Marla's husband John had weathered a serious illness for which he had been hospitalized, and the same year Royce's son John had applied to medical school, and was accepted to begin study in 1982 at UC San Diego. Royce, who had encouraged his application, had promised to bear the cost of his training. Florence and her husband Larry meanwhile had moved to Mission Viejo, not far from San Diego, and Royce visited them as often as he could during his many trips to Southern California. That year for Christmas Royce and Pearl traveled to Las Vegas to spend the holiday with their oldest daughter Jean, with whom they had had very little contact over the years, and Royce reflected on the passing generations, and on his own struggles of faith. "We spent the evening talking and getting acquainted once more with their children. We really haven't seen them very much. Life is passing us by, and even some of our grandchildren are strangers. As we get old and pass away, I fear that we won't be much of a memory to them; only shadows that never quite came into clear focus. I hope that it is true that we do not really die in the sense of obliteration. I like what Shelley said in 'Adonais':

Peace, peace, he is not dead, he doth not sleep
He hath awakened from the dream of life.

"May that be true? This is both my question and desire."[121]

As Royce approached what a few years before would have been the mandatory age for retirement, his life became busier than ever. His work in the strawberry program went on as vigorously as before, and resulted in endless requests for help from many quarters. Additionally, Royce continued his university duties, teaching his usual classes and shouldering a large share of administrative obligations which did not seem to diminish with time. The pace was relentless, and after one busy day at the office at which he had handled endless paperwork and received visitors and phone consultations, Royce observed, "There are so many things to take care of, and so little time. My life should be getting simpler as I get older, but it is getting more complicated."[122]

One of the important duties which Royce continued to pursue during these later years was his work on the various academic committees to which he was assigned, both at the department and at the university level. Royce felt some reluctance at accepting these committee assignments, which he felt took him away from the areas of his professional life in which he could make more lasting contributions,[123] but he continued to feel a compulsion, dating from his earliest years, to shoulder his share of the burden, and his long experience and conscientious approach to administration made him a desirable candidate for this type of work. In May of 1983 he was asked to serve on the Academic Personnel Committee of the university, which was charged with making decisions on the standing and advancement of faculty members, as well as non-faculty staff, whose cases were considered in a companion committee called the Joint Personnel Committee. This was an area which was new to Royce, and he attended his first weekly meeting of the committee in September of 1983. "We passed judgment on a number of items with reasonable dispatch; appointments, appeals, and letters of explanation. It was quite interesting."[124]

Royce quickly came to understand both the functions of the committee and the sensitive nature of its work, and he spent many hours reviewing the records of faculty colleagues and staff members, in order to discharge what he considered a ponderous duty. "I spent the morning getting prepared for the meeting of the CAP Committee. In afternoon we held the meeting, dealing with the future of many people, a very serious responsibility."[125] Although diligent and fair-minded, Royce was never completely comfortably with the role of passing judgment on his colleagues. "I went to Joint Personnel Committee meeting (JPL) of the college, where we judged the fate of several people quite literally as to advancement. Playing 'God' is not my favorite role. In the afternoon we did the same thing for CAP (Campus Committee on Academic Personnel), about thirty cases here. Again we had to act the role as assigned. I am always a bit uneasy, but it has made me to know how fortunate I have been in my career here."[126] Royce's work on the committee was particularly challenging when decisions had to be made about colleagues with whom he had worked closely. After one such meeting he wrote, "It involved some that I know well. It is hard to be negative to such, but one must be fair."[127] When Victor Voth's name came up for advancement,

475

Royce wisely declined to participate at all in the decision making, and a separate committee was assigned to make the decisions in his case.[128]

Despite the delicate matters it addressed, Royce enjoyed his work on this committee, which was essential to the functioning of the university. "It is always a bit interesting. We have to be able to differentiate between the truly worthy and the blatantly greedy. Some really work the system better than others, and are willing to walk right up over the bodies of their 'friends and colleagues'."[129] In general Royce felt good about his participation, and did his best to see that the work was done justly. "In the CAP meeting, we had some sad cases along with some good. I felt that right and fairness prevailed in most of it. I was happy to be more helpful. I am more effective as I know more."[130] When his two-year term approached its completion, Royce considered signing on for another term, but in the end felt it wisdom to leave the task to others. "My feelings are ambivalent. I keep wondering why I was chosen in the first place. I really don't merit the confidence placed in me, although I have tried to be diligent in carrying my share of the load."[131]

Another important issue which came up at the same time was the matter of a successor to take command of the strawberry program at the time of Royce's own retirement, and in July of 1983 a meeting was held to address the issue. "We went to the faculty club for a meeting of the Research Committee of Strawberry Advisory Board, the dean, the chairman of the department and us, to hammer out a commitment regarding a future replacement of Voth and me. It was agreed that we should submit papers with a proposal to overlap by four to five years my retirement, financed by SAB and patent funds. We shall see how that goes. It was sort of like being present at your own funeral. I see the need of it, however."[132] The enormous complexity of the strawberry program explained why the extraordinary step was being taken to hire an incoming professor years before the outgoing one retired. A week later the proposal was agreed upon with the department chairman, and later that month it was provisionally approved in a meeting of the entire department.

The recruitment and hiring process was to involve both the university and members of the Strawberry Advisory Board, who wanted their own input on the successor. Two months later Royce met with the dean to discuss the university's position on the hiring. "They will authorize as long as it is understood that the person selected has no guarantee of the ladder position when available, just a shot at it. The industry may or may not understand this full well. We shall see. It is difficult to understand how the UC system operates in such matters unless you are a part of it. Anyway, I shan't lose any sleep over it."[133]

It would take an additional year before the details were hammered out and approved, and in January, 1985, official permission was granted for the Pomology Department to begin recruiting for the position.[134] The matter continued to be discussed by industry representatives. "We had a research meeting with the Strawberry Advisory Board. The question is what type of research structure is to be developed for continuing. It was agreed that there should be one regular faculty position as now, to be advertised for (mine), and two specialist types: one for SCFS and one for the Central Coast. We shall see. Part of the funding will have to come out of extension."[135] In spite of these proposals, the university at first agreed only to the hiring of a geneticist, and later consented to a single agronomist to take the place of Victor, though that hiring would be delayed some years.

By October of 1985, the department had received applications for five candidates for Royce's position, and as Royce reviewed the paperwork outlining the academic qualifications of the five applicants, he ranked one named Douglas V. Shaw a "poor fifth" compared to the others, for although trained at UC Davis, Shaw had worked for a private forestry firm for some years since obtaining his doctorate, and lacked the academic qualifications of the other applicants. However, in the ensuing weeks as each applicant came to interview and make a presentation to the department, Shaw proved to be by far the best speaker and the most experienced and mature of the group, and Royce was immediately convinced that he would be the one hired as his replacement. Royce, though hardly enthusiastic about handing over the reins, found this acceptable, though noting the academic records of the candidates which he had reviewed, he observed, "He won't survive if he doesn't publish more than he has up to now."[136] On November 8, 1985, a special department meeting was held to select Royce's successor, and Douglas Shaw was chosen as expected. He made a visit for orientation at the end of 1985, and Royce wrote, "I met with Doug Shaw at about 9:00 a.m. as per previous arrangement, and spent the entire day with him, showing and explaining about as much as I could to him about the project and letting him explain as best he could what plans he might have formed up in his own mind for working during the several years of overlap. I still am concerned about the type of science he is prepared to do for the sake of his own survival. I took him home for lunch. He is very personable, and I have no problem dealing with him."[137] Over the next four years Royce and Douglas Shaw would work closely together while Royce completed his work in the strawberry program.

In the meantime, Royce continued to receive invitations to travel abroad to give presentations and provide consulting work in various areas of the world. In the summer of 1983 he traveled with Pearl to

Hawaii at the invitation of a man who had offered to help him find wild colonies of *Fragaria chiloensis* which were reputed to grow there. This held a special interest for Royce, for although the diploid *Fragaria vesca* was widely distributed throughout the northern hemisphere, the octoploid *F. chiloensis* was limited to narrow areas on the Pacific coastline of North and South America, and there had been a few reports of the species growing wild in Hawaii as well. Royce was disappointed to find none of the octoploid berries he sought; rather, all the wild strawberries shown to him were common *vesca* colonies, though he did find a few wild blackberry and blueberry relatives which were of interest to him. The following year Royce returned to Hawaii at the invitation of an agricultural research investment company which had hoped to establish a strawberry industry there. Royce had provided them with plants of some of the earlier varieties the previous winter, and he wrote, "I talked quite a while to the person chiefly responsible for the actual growing. He does not know anything about strawberries. What they did with plants that they got from me is a complete disaster, since they let them all mat in, and mat they did. It can be used for a bit of information, however. I found that the mother plants were all flowering with as many as seven flower stalks per plant, all smothered of course by swarms of runner plants."[138] Royce prepared a talk the following day on how strawberries might be made to work there, and later in the day, on the slopes of a volcano, he at last found the colonies of native *Fragaria chiloensis* which he had sought, and took some runner tips to propagate back in California.

That same year, 1984, Royce was invited to Ecuador at the invitation of a man named Eduardo Crespo, who was involved in a strawberry growing operation and had sought Royce's advice. Landing in Quito, Royce was transported to lower elevation where the strawberry plantings were. Here once again he found the California varieties in use, though the patented varieties had arrived there illegally. "They said Aliso, but it turned out that Tufts is the most planted variety. The plants came from Israel, and they apparently changed the name to cover their tracks. I don't understand why, because they did not do the same with Douglas, which the Israelis also sent. I don't think too highly of their ethics in this situation." Royce examined a number of fields, and spent a day giving a brief course in strawberry production, though he was dubious about the prospects for the success of the enterprise. "The folks are nice here, but what they really want isn't that easy to give. Everything has to be fitted into their context. They are of course interested in off-season U. S. markets, and also on processed berries, which aren't really worth all that much.".[139]

In May of 1985 Royce traveled to southern Spain as an adviser to

478

a booming strawberry industry which had developed around the Atlantic coast city of Huelva, near the Portuguese border. The climate in Huelva was similar to that in California, and Royce found California varieties in use. "It is essentially all Douglas, and it does look good to me, probably about as good as any coming out of Southern California right now."[140] Having prepared to visit and speak as a consultant, Royce was taken aback by the warm welcome he was given. "I am received here as something 'special,' because their entire business is built around our varieties, and almost without doubt would not exist without them. This is rather sobering to think about."[141] While in Huelva, Royce gave a presentation for the growers, and in a ceremony before he left, he was presented a four-volume set on the history of Huelva. The same night he attended a final dinner, for which the entertainment was authentic Flamenco singing, which he found a little overpowering.

The pace of these trips picked up as the years went by. Later in 1985 Royce traveled with Pearl to Scandinavia for meetings and tours of facilities there, stopping once again in Scotland and England on the way home. He made a consulting trip to Italy in 1986, and another in 1987.[142] In December of 1987, he made a brief trip to Mexico to consult with strawberrymen there, where he had made his first foreign consulting visit back in 1963.[143] In 1988 as retirement approached, he would take trips to no less than five overseas countries. In March he traveled to Jamaica to advise on the prospects for strawberry growing there, which he did not consider very hopeful. In April he traveled with Pearl to Nottingham, England on an invitation to present a lecture there, and in May the two traveled to Italy for an international strawberry symposium at Cesena, at which Royce spoke. Finally, in November Royce and Pearl departed for the most memorable trip, first to attend a congress on strawberry varieties in Huelva, Spain, and then for an extended trip to Egypt, where Royce found that the strawberry program had continued to make impressive gains. He and Pearl took a side trip to upper Egypt and back on a cruise ship on the Nile, an unforgettable experience for both.

Meanwhile, as Royce aged he continued his pattern of daily running and exercise, which had begun back in 1976. "My health and longevity depend upon this. If I am not killed or maimed in an automobile accident, I think that I should live long, provided I continue to work on it. I know that my legs and heart are reasonably good, else I wouldn't be able to run and not be excessively weary."[144] In spite of his daily running habit, Royce had for years continued to be a few pounds overweight, mostly because of his eating habits, which were aggravated by his frequent stops at good restaurants when he was on the road. "It's hard to fight that. I enjoy my food."[145] However, as he reached age 65, he at last succeeded in reaching what he regarded as an ideal weight. "I am down to about 155 lbs., well within my target area, feeling better and, I assume, looking better. This time, I believe I am stable, with a real change in eating habits. Among other things, no butter, no salad dressing, and no soda pop. It is about as easy to eat rightly as wrongly, once the mind is set."[146]

Although the morning runs were a refreshing and unifying part of Royce and Pearl's daily routine, and were certainly beneficial to their health, they also held their hazards. One spring morning Royce slipped on wet cement after a rain and landed on an outstretched hand, fracturing a bone and requiring a heavy plaster cast. Royce wrote unhappily, "I shall have my style badly cramped for the next six weeks or so. It could be much worse."[147] Only a few days later, Royce and Pearl, while caring for Margaret's young daughter Angela, took their usual run while the child was asleep in the house, and came home to find the girl missing, failing to observe the note Margaret had left when she had picked her up. "Pearl took off in a panic, running toward the parking lot, and stubbed her feet on bad paving, then literally dived head first into the blacktop, striking her head above the right eye, raising a huge goose egg. She also hit her nose, chin, and several other less obvious places. She called the police, and then we started to drive toward Woodland. At least three officers responded. In the meantime, Betty saw the note and about the same time one of the police came back to look around, and was informed. He pursued us, and all was well. Thus we experienced the full circle of emotions; first, the agony of a child lost, the extreme anxiety of the search, and then the joy of child found. Pearl's eyes went completely black, and almost closed tightly. We had a lot of explaining to do, particularly to those who did not see me with the cast on my hand last week. Frankly, Pearl looks for all the world as if I had worked her over most cruelly with my fists or a blunt instrument. I am amazed that her injuries are not much worse."[148] In spite of the injuries the running continued, and Royce recorded two days later, "Pearl and I ran. We are a handsome pair, what with her black eye and bruised jaw and my broken hand. It would appear as if we had one terrible fight, both losing."[149]

Although Royce seldom took time off work even on holidays, he tended to observe them privately in his thoughts and in his journal, commenting on their meaning. Even Saint Patrick's Day usually was generally commemorated with the phrase, "Up the Irish!", which he had probably learned from his friend McNally during the war. Royce was particularly mindful of those holidays which reminded him of the "great adventure" of World War II, which continued to cast its shadow over his life after four decades. On Independence Day in 1983, he wrote, "I have to think about those who died that freedom might prevail here. I might have, but did not. I was ready for death in my way, but it never touched me, and I must conclude that either I was fortunate or that my time hadn't come. Death passed close enough to me that I can appreciate being alive and well in this blessed land. I am pleased that I at one time quite willingly placed my life on the line. I am also pleased that it was not taken. I don't believe that I can claim true bravery in all candor, but I made the offer and faced the worst that came my way with some degree of courage when I was young and had my whole life yet to come, or at least that portion which counts."[150]

Royce almost never made mention of his wartime experience, but he frequently encountered reminders which caused him to reflect on it. One spring he hosted a graduate student named Ferdinand Schmitz from Bonn, Germany, who on the night of his arrival had dinner with Royce and Pearl, and spent the night as a guest in their home. Royce observed, "His father may have even shot at me in World War II, where as a 17 year old, he was a flak battery soldier at Bolzano, Italy, where he was later captured. It is a small world."[151] Though not one to participate in veteran commemorations, Royce attended one year a reunion of his old Air Corps unit in Las Vegas, Nevada, which only served to remind him how much the world had changed. "We went to the 57th Bomb Wing hospitality room, and truly you can't go home again. I did meet a number who were more or less familiar. But the world that we knew has long since faded into memory, and some of the things we think we remember may not even be true. The greatest adventure of my young years is now a bit vague, and talking does not bring it back."[152]

If the memories were vague, the effects of the war still lingered in Royce's life. His bitter cynicism and explosive temper had at last mellowed, but four decades later the war continued to haunt his dreams, and from time to time he found himself lashing out physically in his sleep. Although Royce had never struck Pearl when conscious, even in anger, in 1981 he wrote, "I should mention that I struck Pearl on the head in my sleep. This is the third time, and I have no memory of this event whatsoever."[153] As the years passed, Royce became more prone to injure himself than Pearl during these tortured dreams. A couple of years later, he wrote, "During the night as best I can recall it, I had a dream where I was threatened. To escape, I threw myself out of the bed, landing on my head on the floor, temporarily stunned, calling to Pearl. Fortunately, neither Pearl nor I was hurt. We went back to sleep."[154] Later that same year, he recorded, "Last night I had a nightmare and in trying to escape, smashed my nose against the headboard. My nose bled."[155] The most serious incident took place a

481

couple of years before retirement. "At 1:30 a.m. I had another war-time nightmare and bailed out of bed, striking my ear on the sharp corner of the nightstand. I woke up bleeding badly, and semi-stunned. We bandaged it up and cleaned up most of the blood. I took some aspirin and went back to bed to sleep fitfully. At 9:00 a.m. I went to the medical center where Dr. Sobeck took care of me. They bandaged it up without stitches. It looks as if I had been beaten."[156] Having to go to work with a bulky bandage about his head, Royce told associates and family members only that he had injured himself falling out of bed.

At church, Royce continued to find himself in demand as a teacher, this time in his priesthood quorum, an assignment he accepted with some ambivalence,[157] though like all his other Church assignments, he put his full effort into his preparation, and his lessons were perceptive and thought-provoking. One Sunday he wrote, "I spent all day on my priesthood lesson on forgiveness. I think that I can relate to this as a subject. It went quite well, because I forced them to think about what it really means; among other things, that you cannot afford the luxury of enemies. It is completely out of harmony with Christian principles. Most everyone falls short here, including myself, and this is where it is at. Repentance is closely allied, and justice must have its way too, but unless we forgive, we are not doing the most important thing: loving our neighbor."[158] Another week he recorded, "I spent the morning preparing my priesthood lesson on resurrection, which went quite well. Mostly, I try to get them to think about it. As a rule, we don't really do that. It is simply easier to mouth the words and go on without realizing that the concept is not simple, and we don't really understand any better than those who contemplated what they perceived as the risen Christ, and had not believed until then."[159] Sometimes the lessons presented a particular challenge, owing to Royce's personal spiritual limitations. "I spent the evening at work on my lesson on prayer for tomorrow morning. It isn't easy for me to do. This isn't my strongest area, and it is not right to teach something that you don't practice as well as you should. At least I think that I gained something from the preparations. I am getting older and less prone to be contrary in some ways."[160]

Although Royce had experienced a genuine resurgence of faith and was sincere in his efforts to reform his actions and attitudes, he still struggled and at times harbored lingering misgivings, particularly as he pondered his accomplishments, his choices of the past, and his spiritual standing. One day after a visit from his boyhood friend and faithful Church member Art Wallace, Royce wrote resignedly, "I am still not overly excited about how my life has gone so far. I fall so far short of my potentiality. Perhaps there is a merciful God who will forgive everything and make all things right. I was at one time much more certain of that than I now am. We should never grow up, too many things come into clear focus that are best viewed slightly out of focus, and other things once clear, clean and beautiful are now faded and obscure. Will I ever be able to return to the simple life that once appeared to be so real and understandable? I fear not, and perhaps it is just as well."[161]

However, little by little clarity was returning. It was in his work at the Oakland Temple that Royce found his greatest spiritual strength, and the greatest motivation for self improvement. He realized that his unexpected temple calling had been a life-changing experience,[162] and despite the seeming importunity of the assignment, he continued to value it, and felt determined to do whatever was necessary to keep it a part of his life. This was sometimes still a mental challenge, particularly when Royce's rational views as a research scientist seemed to clash with the simpler and more traditional world of some of the other temple workers.[163] "Sometimes, I wonder if I should continue working there. At times I feel really at home, but all too often I have the uneasy thought that I am a 'stranger in a strange land,' and I ask myself what I am doing and why."[164] At the same time, Royce knew that there was an authentic spiritual power to the temple experience, and that however it might differ from his life as a scientist, and however great the challenge of reconciling each to the other, there was room for both in his life. Once he was even allowed the opportunity to express his scientific views to fellow temple workers, which he did honestly but judiciously. "At the last hour, we talked and I was asked to say something about biology and living things, so I did. It appeared to be received well enough, although some of them are very conservative. I toned it down to fit the clientele, but still told them my true feelings for the most part. I just did not elaborate more than is necessary."[165]

Given the sacred nature of temple work, Royce still questioned his basic spiritual fitness to work there, particularly when he was asked to take on additional responsibilities. "I am now officially an assistant supervisor. I certainly did not seek this, and feel inadequate and somewhat unworthy. Within myself, I feel that every one I work with is more genuinely dedicated to the work and living a better life than I am."[166] The spirit of the temple was undeniable, and though it defied any physical explanation or description, it was often commented upon by Royce. "The day was exceptional in many respects. There was a particularly nice feeling that most everyone seemed to sense. We were very busy but all seemed cheerful and moved. One of the workers told me that a weeping woman told him that a voice spoke to her saying, 'I accept this work, all of it.' I have heard of similar things and have no reason to question them. I suppose that it is my lack of something or other that it never happens to me, and I sort of feel that it never will, although, as I noted above, I felt something (presence or otherwise) singularly extraordinary today."[167] When asked to extend their term of temple service, Royce and Pearl unhesitatingly agreed to do so. "We were interviewed today, and gave our consent to continue indefinitely. There are some things that we can consent to most casually. We have done this for six or seven years, and we consented to it once more, a seventh of our time. There is much satisfaction in this. We have what amounts to our closest friends there, and they are exceptional people."[168]

As his spirituality increased concurrently with his temple service, Royce felt less troubled by the seeming conflicts between the religious and scientific aspects of his life, though at the same time, he remained completely convinced of the basic truth of evolutionary theory in particular. These two aspects of his life seemed worlds apart, and in his mind he periodically made his own attempts at reconciling them, a task that sometimes amounted to an almost titanic effort which would burst forth from his pen into his journal. Early in his temple service he had written, "I still believe with Dobzhansky[169] that nothing in biology makes any sense except in terms of evolution. Indeed, nothing in this world makes any sense without it, including religion. Perhaps I am not as candid (or arrogant?) with the subject as I once was, and everyone in the Church thinks I have changed on it, but not so!! Religion, too, must accommodate the concept, including our Church. Sooner or later, the truth of it will prevail in its elegant beauty. It makes man a nobler, not baser creature, for he 'thinks' and therefore is. He 'acts,' and is not just acted upon. He is a creator in his own right, a God above all living things. Immortality is within him, and to a certain extent, of him. His will is done upon earth, and he defies death even as he dies and returns to the dust of which he is made. He loves, cares about, and shelters his fellows. He is his brother's keeper. He is both friend and foe. His perfect side can only be seen imperfectly. Something of him will always continue after the structure is destroyed. He wrestles with the Almighty and justifies his position of 'a little lower than the angels.' He is both blest and cursed with self-awareness. He can do no other than magnify his humanity. When he falls there is an empty space in the forest in which he stood. His intellect towers above the best that lesser animals have to offer. He is both mortal and immortal, and even as he dies, is made alive again. The best of him leaves an echo which will never die and the hollow corridors of time will ring with it. He stands alone, and yet is never quite the solitary one that he often imagines he is. The 'windows' of his mind are worlds in their own right, and how beautiful are his creations. He is both the builder and the destroyer, and needs no special hand to bring the things of beauty about. He is both sin and forgiveness, the unrighteous and the redeemed. He has touched the fountain of youth, drunk of its waters, and lives forever, and yet has fallen into the pit of his own digging and pulled the covering soil over himself. He wears the crown, the jester's bells, and the rags of the despised. He defiles the very ground he walks upon, and yet knows he should remove his shoes because the place where he stands is holy. He is woman as well as man, and neither can survive more than a generation without the other. He kills the thing that he loves, and yet he gave it life. I fear that I have said too much on this already. The believer within me says, 'believe and all will be well.' The cynic within me says, 'it doesn't matter, life has no meaning beyond the little hour given to each.' I bless and curse those who have trusted me. I betray all and yet am faithful to the end. I love life and yet have failed to understand what it is about. I know nothing, and yet I see it all. I am a paradox, a puzzle, a mystery, an enigma, and yet within me there is something good and pure and easily understood by all. Enough is enough is enough, and I say no more this day."[170]

The Strawberry Program—The Final Push

After the landmark release of the seven new varieties at the beginning of 1983, the work on the strawberry program continued as vigorously as ever. Evaluation was already underway for the next generation, and there was much testing yet to be done on the varieties just released. Victor Voth, working at the South Coast Field Station, concentrated on the short-day varieties, testing out different planting schemes and different methods of fertilizer placement, while Royce, working at Watsonville, focused on the day-neutral varieties, working out how best to handle the planting and harvest in order to optimize yield.[171] These efforts were already beginning to result in wider use of the cultivars, and the day-neutral releases Fern and especially Selva had become highly successful, creating at last a major place in the industry for the day-neutral strawberry, as Royce had predicted.

Royce's life continued to revolve around the complex agricultural cycle of the strawberries, which with the day-neutrals now involved multiple planting dates and harvests that occupied much of the year, besides the management of seedling populations and the propagation of promising selections. The various aspects of the process took place simultaneously in different parts of the state, from the coastal fields in Southern California to the nurseries near the Oregon border to the north, causing Royce to remark, "I still feel as if I'm on a treadmill with no control over its movements."[172] In addition to setting up experiments and collecting data, Royce frequently received visitors from

485

other states and from abroad, and he was often sought out by strawberry growers for advice. On the new experimental property in Watsonville, he oversaw the construction of a new house for the grading of fruit,[173] and continued to explore better ways to gather and record the large amounts of plant data.[174] The routine was demanding and at times exhausting, but Royce still savored his work in the fields.[175]

Royce worked also with the fruit processing industry, and at the peak of the 1984 harvest season provided samples of the new strawberry varieties to representatives of a processing company for testing. A month later, Royce was requested to meet with the company. Accompanied by Victor, Royce first gave a talk to company representatives. "We then took care of them at the plots on a field trip, exhibiting fruit jam, preserves, and IQF[176] berries from the varieties. Uniformly, they like Chandler, fresh and processed."[177] Indeed, the short-day variety Chandler was beginning to stand out among growers as well, not just for its high yields but for its quality, especially its exceptional taste. That same month Royce visited the plantings of Kuni Shinta, and saw the variety being harvested on a commercial scale. "Chandler was large and beautiful, and the flavor excellent. That is a good berry."[178] Chandler was quickly coming to dominate the market, supplemented by Parker and by the day-neutrals Selva and Fern, which together provided fruit nearly year-round. Chandler also continued to be the favorite variety for the processed market, and some years later another processing company took the unprecedented step of making their own plantings to assure an adequate supply. "They are apparently serious about growing much of the fruit that they use for jam. Chandler is their choice."[179]

Meanwhile, work was well underway on the next generation of strawberry selections. The key to continued progress in the strawberry program was the evaluation of each generation of new seedlings, and Royce was insistent on using sound scientific methods of evaluation, which he knew gave the best chance of selecting good varieties, though he and Victor did not always see eye to eye on this methodology. One day after visiting the plantings at the South Coast Field Station, he wrote, "The seedlings are beautiful, and we spent time in there, although the method Victor uses is not good. Too much bias enters in, and too many good ones are never really looked at. His judgment, though usually good, can be bad too when he takes a certain attitude."[180] Royce continued from time to time to clash with Victor, who in their frequent presentations together had fallen into a pattern of exceeding his time limit at Royce's expense.[181] Royce raised no voice of criticism except in his journal, and he took special precautions to keep from antagonizing his lifelong colleague any further. When interviewed by a writer doing a special report on strawberry research, he wrote, "I made sure that equal treatment was given to Victor Voth. I certainly don't want another outburst from him."[182]

In 1984, Royce and Victor faced a new concern when an outspoken representative of the industry took measures that threatened to cut off Advisory Board funding for the strawberry program. Royce wrote, "I am concerned about the budget for next year. If we don't get it, our problems would be most grave. Herb Baum is really after us at the moment. I feel somewhat in between a 'rock and a hard place' in dealing with the issues."[183] The following day he added, "I was not invited to a university meeting about us, instigated by Herb Baum, much of it not complementary and much of it unfair. We shall see how that comes out in due time. I think the industry support is much more solid than would appear from listening to Herb. We should have our say before this goes much further. Herb really is not playing fair, nor is he understanding what is involved. It's their money, though."[184] The following week he remarked, "I worked on our request for Advisory Board funding for 1985. There will only be four more that I have to worry about. This request is an increase of about 10% (in line with inflation), up to about $280,000. That is a lot of money no matter how one looks at it."[185] The necessary funding was in due course approved, but the haggling by the board, which evidently dragged on to varying degrees for months, was so burdensome to Royce that two months later it caused him to remark, "I am tempted to retire and have done with all of it."[186] In taking stock of his own life and accomplishments, Royce could not escape a feeling of having fallen short of his possibilities. One day after a meeting with a new generation of farm advisers, he wrote, "Sometimes I think I just ought to quit the whole thing, and face up to the fact that part of life has passed me by without my having even touched it, and it will never again be within my reach. I could have done so much better."[187]

The hiring of Douglas Shaw 1986 to eventually take over Royce's position as geneticist in the strawberry program naturally brought mixed feelings, and when the two began working together, Royce's initial reactions to his successor were tentative, particularly when he learned of his color blindness. "I don't know how he is going to work out. He really isn't that well prepared to go on his own, and to top it off, he is red-green color blind. I wonder how he sees red berries. Well, we shall see. He is pleasant to talk to, and seems eager to start."[188] In spite of his misgivings, Royce found Shaw personable and easy to work with, and he began to involve him in the various aspects of the strawberry program. When a warm winter brought on an early harvest in the Watsonville plots, Royce observed, "With Shaw already on board my life should be simpler. I'll have to think about how to make it so, to the advantage of all concerned."[189] With four years left before retirement, Royce set to work training the young geneticist with the techniques specific to strawberries. "I went to Winters, taking Doug Shaw with me to show him how to emasculate and pollinate, basic skills. I did some pollinations that I want done at the same time."[190]

This mentoring partnership was not without its rough spots, particularly when early in their work together Doug Shaw once failed to follow the rigorous scientific procedures which Royce considered essential to obtain unbiased data. "I was seriously upset with Shaw when I found that he did not set the harvest plots up the way I had carefully set up, a randomized complete block design. He completely underestimates me, and that better not happen again or I won't be giving him the support he needs if he is going to survive. Part of it may have been my fault, but he should have caught on."[191] The support to which Royce referred was in part his assistance in performing and publishing scientific research, which was an essential prerequisite to Shaw's maintaining a position on the university faculty. In the end it appears that Royce did in fact provide that support, and that Douglas Shaw for his part had become well aware of the academic requirements for his position, for he immediately embarked on research based on the California breeding populations, using data from crosses which were made in 1985, just before his arrival, and Royce collaborated on the research, which resulted in two scientific papers published before Royce's retirement, one a study of the heritability of traits contributing to fruit quality in strawberries,[192] and other other on a plant disease known as Leaf Spot.[193]

Though the overlapping appointments of Royce and Douglas Shaw promised some continuity in the strawberry breeding program, the Pomology Department stalled when it came to the agronomy position presently occupied by Victor Voth. After a protracted meeting of the department in 1986, Royce wrote, "Most of the discussion was on the strawberry position(s), primarily Voth. The collective 'unwisdom' of the Department of Pomology was never at a higher level. Those in a position to know the least talked the most, as usual. They are capable of solving anyone's problems other than their own, or at least so they imagine."[194] Eventually the pomology department would relent and approve one agronomist to fill Victor's position, but this would not take place until years after Royce's retirement. Royce and Doug Shaw, however, continued to plan for the future of the strawberry program, including an expansion at Watsonville. "Shaw and I went to the Strawberry Advisory Board where we had lunch, then discussed an expanded program at the Research Facility, effectively increasing the operation by about 50%."[195] While they worked together as their respective positions demanded, the relationship between Royce and his successor was by nature an uneasy one. For Royce, Shaw represented the end of a professional career to which he had invested tremendous personal effort, and which he doubtless would have continued had not university policy mandated his retirement. Emotionally unprepared to relinquish his life's work, Royce doggedly continued to assert his leadership of the strawberry program, and Doug Shaw, well aware of his predecessor's professional stature and his ambivalence about his mandatory retirement, to the extent possible went about his own work and let Royce go about his.[196]

Meanwhile, Royce and Victor were making plans for the next generation of strawberry releases. Their efforts had focused more fully on the day-neutral varieties, and they had at least two of these well evaluated and ready for release, as well as one excellent short-day cultivar, which was thought to have some advantages over its predecessors. One of the day-neutral varieties, designated CN17, had been successfully tested over an extended period, and Royce had considered releasing it years before.[197] However, the plant patent system had by now grown very complex, and the release of plant varieties now required cooperation between multiple offices, including the patent office and their attorneys as well as the university administration and the industry's board, and it took Royce considerable effort to surmount the necessary hurdles. In April of 1986 he wrote, "I spent about two hours on the phone trying to arrange the release of our new strawberries through the UC bureaucracy. Things used to be simpler. Now everyone wants to have their say. Victor didn't help, by having 'told everyone off,' as he loves to do, thinking that will clear the way." The next day he added hopefully, "I spent a good bit of the morning on the phone on the business of release too. I think that all the wheels are moving now."[198] A few days later Royce had to smooth the ruffled feathers of the industry. "We went to the CSAB office to meet with Ed Kurtz and Dave Riggs for almost two hours, talking about Doug's budget and about their wounded feelings regarding the release of three berry varieties. They are getting more difficult, and seem to want to try to run everything."[199]

By the following month the release had reached a snag with the university administration, which Royce by now accepted with resignation. "Victor called at night, telling me that our attempt to release 3 strawberries was thwarted. It's all right with me. CN17 is the only one that goes, since the others have virus problems. I suppose that I don't really care. C43 and CN75 were rejected."[200] The process was held up for most of a year before the complicated requirements were met on the remaining varieties, and by then the industry was pressing for information about their release. "At 10:00 we went to a meeting of the

'Pomology' committee of the Strawberry Advisory Board. It was a long drawn-out affair with very tedious discussions of things that seemed a bit out of focus to me. The only thing I did was to assure them that C43 & CN17 would definitely be released. I stick my neck out on this, because the approval of the dean may be questionable. I will not consent to submitting the papers until release is assured. To me it makes no sense."[201] By this time an additional day-neutral variety had been added to the release list, bringing the total to four varieties, only one of which was short-day. Later that month Royce started on the necessary paperwork, which was completed in June of 1987. "I finished the writeup for he disclosure on CN 17 ('Muir'). It has been quite difficult to get it done, but that is pretty well behind me now. I shall do C43 ('Oso') next, and then do CN27 and CN75. That will be all for the present."[202]

Of the three day-neutral varieties, CN17, the most promising, was given the name "Muir" after the famous California naturalist; the others were called "Mrak" after a former UC Davis Chancellor who had just died,[203] and "Yolo," after the county where UC Davis was located. Of these three varieties, only Muir would become important commercially, though less so than its sister variety Selva. The single short-day release was called "Oso Grande," a name probably proposed by Royce. The first part of the name commemorated Oso Flaco Lake, the most southerly point in California at which Royce had found native *Fragaria chiloensis* colonies, and the second part referred to the fruit's exceptionally large fruit, bigger than Chandler and even than Douglas, the largest-fruited California cultivar then in use. Although Chandler would continue to play the dominant role in the industry through the coming decade, Oso Grande, or "Oso," as it was generally called, would eventually come to be one of the most widely planted cultivars coming out of the program.

With Royce and Victor's retirement now imminent, they began making yearly releases of day-neutral varieties as soon as field testing seemed to justify it. In 1988 they released a day-neutral variety which had been selected at the South Coast Field Station, giving it the name "Irvine," after the ranch on which the field station was situated. In 1989 they released two additional day-neutral varieties, which they called "Seascape" and "Capitola," after two towns located near the research station at Watsonville. Of these three, "Seascape," with its heavy yields and deeply colored and highly flavorful fruit, was destined to become a commercial success in the coming years, and would be the principal day-neutral variety in use after Royce's retirement.

At the same time, Royce continued to carefully plan crosses which he hoped would lead to further improvements in the strawberry cultivars, a process over which he had begun to feel an element of possessiveness. "I went to WEO to pollinate. I'm having good luck with that. I decided to do much of it myself, since others take credit when I just tell them to do so, and I get it done better too."[204] Royce realized that as he neared the end of his career, much of the selection and evaluation process would be left in the hands of others. "I am literally sowing that which I shall not reap, but do so because I feel so compelled."[205] He also spent hours out in the field making initial observations on some of the seedlings which had already grown. "I went to Winters first thing, and started with note taking on the seedlings (about 8,000) at the crack of dawn, and finished about 11:15. It was cool and nice, and the fruit was very good; perhaps the best populations that we have ever grown. Some of it is huge. I then gathered the last group of crosses from the greenhouse; a very large group. At least I shall be going out with a great flourish, giving Shaw the impression that all is easy.[206] Royce determined once again which of these seedlings would be propagated to become part of the breeding stock for the university program. Ever mindful of the importance of this step, Royce wrote, "I hope that my judgment is good. That I cannot be sure of is the only thing of which I am sure."[207] Royce's judgment as a plant breeder would vindicate itself to the very end. In 1988, the year before his official retirement, a cross was made between Douglas and another advanced cultivar, from which Victor Voth would select a short-day progeny that would undergo accelerated testing, to be officially released in 1992 as "Camarosa." With fruit larger, firmer, and earlier than Chandler, Camarosa would take its place as the premier California cultivar, a position it would continue to hold for many years.

In 1989, his official year of retirement, Royce drafted a final review of California strawberry cultivars for publication, in which he summarized the advances made in the strawberry industry, with dramatic increases in yields over the time he and Victor Voth had worked together, about half the increase attributed to improved cultural practices, led chiefly by Victor, and the other half to improved strawberry varieties, mainly under the direction of Royce. These had included a long succession of highly successful varieties, each better than the last, from Fresno, to Tioga, to Tufts, to Aiko, to Douglas and Pajaro, to Chandler and Selva, and beyond.[208] These had impacted the strawberry industry not just in California, but in other areas of the United States and in nations around the world, and had made Royce the recipient of numerous professional awards and honors recognizing the impact of his work. But ultimately, it was the work in the fields, with its attendant and overlapping cycles of pollination, seed germination, planting, selection, propagation, and harvest that were at the heart of Royce's professional life, and that brought him the greatest satisfaction. He frequently commented on his good fortune in being able to make a living doing the things he loved the most—he was in effect living the life he had only dreamed about when as a youthful farmer he had watched the work of F. V. Owen in his father' beet fields. After an evening in the experiment station in Watsonville he wrote, "I went to the strawberry planting and stayed until dark. Surely, they are my life and ever have been. Other than Pearl and the family, the berries come first. I justify that by noting that I 'should be about a good work.'"[209]

Final Scholarly Work

As his retirement neared, Royce had actively continued his investigations into the evolution and genetics of the strawberry, which resulted in a last series of published articles at the beginning of the 1990s. The first of these, authored solely by Royce, was an analytical summary of all the work he had done over the years with the wild strawberry colonies along the Pacific coast. Entitled "Cytogenetics and Evolution in American *Fragaria*," the paper began with the statement that the modern cultivated strawberry, known scientifically as *Fragaria* x *ananassa*, represented the culmination of strawberry evolution, resulting as it had from human intervention in the evolutionary process which had taken place in the gardens of Europe during the 18th century. Royce's objective in the paper was to demonstrate, through his study of the natural strawberry hybrids he had discovered along the Pacific Coast, the evolutionary forces and processes which had given rise to the modern octoploid strawberry, and which were likely to continue their course if the wild strawberry colonies along the coast were allowed to survive.

Royce began by proposing a change in the known genomic structure of the octoploid strawberry. He had last recommended, in his 1967 paper with Don Senanayake,[210] that the eight sets of chromosomes in octoploid strawberries be categorized as AAA'A'BBBB, a designation which had been widely accepted by the scientific community. Royce now proposed that the genome structure be modified to AAA'A'BBB'B', based on subsequent published evidence by himself and others that the octoploid strawberries were highly diploidized, meaning that each of the four pairs of chromosome sets seemed to reproduce separately and independently of each other, and were not interchangeable.

Royce described the many colonies of pentaploid hybrids which had resulted from natural crosses between the octoploid *Fragaria chiloensis* and the diploid *F. vesca* which he had discovered along the Pacific coast, as well as the single instances of a hexaploid and an enneaploid (9-ploid) colony, both resulting from unreduced gametes in

492

one of the parents. Royce pointed out that the octoploid *F. chiloensis* was almost certainly the female parent of all these hybrids for two reasons: first, offspring of the female octoploids tended to segregate evenly by gender as was the case with the natural hybrids, and second, the diploid *F. vesca* strawberries had a strong tendency to self-pollinate, making them an unlikely source of outside crosses.

Royce pointed out that a decaploid (containing ten sets of chromosomes) was the next higher balanced chromosome level after the octoploids then in existence, and it was likely that natural decaploid colonies either already existed or had existed in the past. He pointed out that when seeds from the fruit of the natural hexaploid hybrid had been germinated, more than 60% of the offspring had been decaploids, and although the natural pentaploid hybrids were less fertile, when they did bear fruit, more than 80% of the offspring had unreduced gametes, and these often resulted in decaploids. Hence, it was but a small evolutionary step for decaploids, containing chromosomes from both *F. chiloensis* and *F. vesca*, to become established and begin reproducing among themselves,[211] in effect creating a new fertile species of strawberries, and it was not unlikely that such a step had already taken place in the wild.

Of special interest was the single enneaploid (9x) colony which Royce had discovered at Wright's Beach, which had probably resulted from a combination of a naturally reduced gamete from diploid *F. vesca* with an unreduced gamete from an octoploid *F. chiloensis*.[212] This colony was shown to be as fully fertile as the natural octoploids, and when pollinated by an octoploid (a highly likely occurrence in the wild), the offspring which resulted generally included a few octoploid plants, a case of "introgression," or reversion from an abnormal number of chromosome sets back to a normal one. In these cases, a full set of chromosomes from the diploid *Fragaria vesca* could readily be transferred into an octoploid species without the need of stabilizing a population at a different chromosome level. Among other things, this held great promise in transferring desired traits from natural diploids into octoploid strawberry breeding populations.

In conclusion, Royce noted that the processes of polyploidization and introgression which he had observed had always been present in strawberries, and not only explained the variety of species and hybrids that had been observed in the wild, but also held promise for future evolution, and for plant breeding. However, he pointed out that two of the three best hybrid sites he had discovered had already fallen victims to coastline development, and the others were at high risk. Royce had gathered and preserved the best specimens from these sites at the clonal Germplasm Repository in Corvallis, Oregon, so they were not entirely lost, and he concluded his paper by suggesting,

"An attempt should be made to reestablish some of the natural hybrids in as natural a situation as possible so that a study of the ongoing processes might be continued. However, it will not be the same as the undisturbed colonies, and steps should be taken to preserve certain of the colonies as intact as possible, particularly those at Point Sur."[213]

Royce also collaborated in the publication of an entire series of genetic studies with his staff research assistant Hamid Ahmadi, an outspoken Iranian with a jovial sense of humor who had become a favorite with Royce and Pearl's grown children. The first of these studies was on the inheritance of photoperiodism, meaning the effect of day length on the flowering and fruiting of strawberries. This was of particular importance to Royce, since one of his principal contributions as a plant breeder had been the introduction of the day-neutral trait into the California breeding lines. Royce had long believed that the trait in octoploid berries acted like a simple autosomal dominant gene, while other authorities had proposed that multiple genes were involved, though Royce, with long experience in breeding day-neutrals, doubted that they had used adequate criteria for determining day-neutrality in the offspring. Based on data from the breeding populations and from carefully designed crosses, Royce and Hamid set out to define the precise inheritance pattern of this trait, in both octoploid and diploid strawberries, taking care to point out the three known sources of a day-neutral trait in octoploid strawberry breeding lines: one from 19th century France, one from 19th century America, and the *F. virginiana glauca* gene that Royce had encountered in the wild in 1954, which had served as a parent for all California day-neutral cultivars.

To test the heritability of the day-neutral trait in diploid *Fragaria vesca*, Royce and Hamid began with a day-neutral strain of woodland strawberry from Europe, which had been known and cultivated for centuries. When this was crossed with a short-day European strain, all of the offspring were short-day, but among the second-generation offspring, one in four were day-neutral, consistent with a simple recessive gene governing inheritance of the trait. However, when the European day-neutral was crossed with a New World variety of *F. vesca* from California, only one in 64 of the second generation were day-neutral, which improved to one in eight when the first generation was back-crossed to the original day-neutral strain. Mathematically, this meant that not just one but three separate recessive genes appeared to govern the day-neutral trait in the diploid *Fragaria vesca*.

The inheritance of the day-neutral trait among the octoploid cultivars proved to be much simpler, though Royce and Hamid first had to establish a series of strict criteria for designating a plant as day-

494

neutral, to avoid confounding factors which could lead to faulty conclusions as had been drawn in the past. When those criteria were applied, they found that whenever a day-neutral strawberry was hybridized with a short-day variety, exactly half of the offspring were day-neutral. When one day-neutral was bred with another, three-quarters of the offspring were day-neutral. Both of these findings were consistent with the day-neutral trait being determined by a single dominant gene, for which the California day-neutral cultivars were heterozygous (meaning, that they had one set of chromosomes with the day-neutral gene, and one set without). To further prove their hypothesis, Royce and Hamid bred multiple generations of selfed day-neutrals, in order to establish a homozygous day-neutral line (meaning a line of plants with both sets of chromosomes containing the day-neutral trait, rather than just one).When they crossed the resulting offspring with short-day varieties, they found that all the progeny were day-neutral, which definitively confirmed the simple dominant inheritance of the trait. They also found that this dominant day-neutral gene was passed to the offspring when hybrids were made between standard cultivars and wild octoploid *Fragaria chiloensis*, the diploid *F. vesca*, and even hybrids with the related genus *Potentilla*. Royce and Hamid published these results in 1990.[214]

A second study conducted by Royce and Hamid explored in detail the genetics of sex expression in strawberries. It was well known among botanists that nearly all diploid species of strawberry were hermaphroditic (containing both genders in the same flower); the one exception happened to be one subspecies of *Fragaria vesca* found in California, of which some plants were female and others hermaphroditic. It was likewise known that most octoploid strawberries were dioecious, meaning that individual plants produced either male or female flowers, with only occasional plants containing both sexes. The one exception to this pattern was the commercial strawberry, which had been bred to be completely hermaphroditic.

Although many of the details of sex expression in the strawberry were known, in most cases the genetics and inheritance patterns were still unclear, and Royce and Hamid undertook an extensive series of plant crosses to define the genetics of sex expression. For diploid plants, they used hermaphroditic plants from Europe and California, as well as one of the California female diploids which Royce had found at Hecker Pass, and they also included several diploid species from the closely related genus *Potentilla*. For octoploids, they used a collection of male and female *F. chiloensis* clones collected along the coast, as well as hermaphroditic commercial cultivars, both short day and day-neutral, and a couple specimens of wild hermaphrodites. They also experimented with a wide variety of natural and synthetic hybrids between the species, for which they determined sexual expression in the parents and offspring.

As they conducted these experiments, clear patterns of sex expression began to emerge. Among diploids, crosses of hermaphrodites produced only hermaphrodite offspring, whereas a cross between a female and a hermaphrodite produced half female and half hermaphroditic offspring, showing that the female gene, though rare, was dominant over the hermaphrodite one. In the octoploid species, which had male as well as hermaphroditic and female plants, it was found that the gene for hermaphroditism was dominant over the gene for male sex, while the gene for female sex was dominant over both the male and the hermaphrodite gene. The dominance of the female gene was manifest even in complex decaploid hybrids where one female gene was matched against three hermaphrodite genes, and could not be altered by the application of plant hormones. Natural tetraploid and hexaploid species, by contrast, were noted to have a different system of inheritance, with males being dominant over hermaphrodites, and females dominant over both hermaphrodites and males. This significant study, which at last definitively clarified the genetics of sex expression in the various strawberry species, was published in the *American Journal of Botany*.[215]

A third and final genetic paper published by Royce and Hamid Ahmadi was a study of strawberry breeding at the decaploid level, which meant using as breeding stock plants with ten chromosome sets rather than the standard eight. This study, which was something of a culmination of Royce's work with interspecies hybrids and polyploid plants, had several objectives: first, to determine if it was possible to obtain fertile decaploid clones with enough desirable traits to form the basis for a breeding program; second, to compare the efficiency of different breeding procedures to establish decaploid lines, and third, to describe and evaluate the attributes of selected decaploid clones. The use of decaploids was of particular interest, because it offered the possibility of combining the genetic material from different species of strawberries, and particularly to join the best qualities of diploid and octoploid strawberries in the same fruit.

Royce and Hamid tried out a variety of methods in an attempt to develop decaploid plants, but the best by far was to start with pentaploid hybrid crosses of octoploids and diploids, either natural or manmade, and double the chromosome number from five to ten sets using colchicine. They also had some success with crosses between a standard octoploid variety and the natural hexaploid which Royce had discovered on the coast.[216] They were able to develop a number of day-neutral decaploid lines by hybridizing standard day-neutral varieties with various diploid species and doubling the chromosome num-

ber, and the resulting selections were often aromatic and good flavored, with textures ranging from soft to firm, and they also showed
a range of resistance to various plant diseases, suggesting that these
traits could be selectively improved by further breeding. Even in this
infant stage of the plant breeding process, some of the best selections
had fruit size and yields which were not far inferior to the commercial
varieties then in use. Royce and Hamid calculated that by applying the
time-honored method of recurrent mass selection, it would be possible to quickly develop superior cultivars with excellent fruit, disease
resistance, and adaptation to commercial growing conditions. This
study at last seemed to bring within practical reach the dream expressed by Henry A. Wallace some thirty years before, of incorporating the fragrance and flavor of the small European strawberries into
standard commercial varieties. The study was published in 1992, two
years after Royce's retirement.[217]

Looking Back and Looking Forward

As Royce neared the end of his professional career, there were
frequent events which gave him occasion to reflect back on earlier
episodes of his life. One of these was a lecture by a plant breeder
from a private company, that took Royce back to his early days at
Davis. "I went to a lecture on evolution at night: Brown from Pioneer
Seeds. He knew my benefactor Henry A. Wallace very well. I talked
to him for a time after."[218] Another was when Royce attended agricultural meetings at the University of Wisconsin, which he had not seen
since he attended graduate school there. "For the afternoon, I tried to
recapture the past by taking a walking tour of the campus, after being
away for 35 plus years, a sentimental journey into the past, proving
only once more that you cannot go home ever again. There are only
faint to mute echoes of the past remaining. The lake (Mendota) is still
there, frozen over, Bascom Hall with the seated Lincoln brooding
upon the sins of his generation in the same bronze casting; the
agronomy building is still there (now Horticulture). I could even see
the window of the room where I spent all those hours of trying to do
some science. The bell tower and the observatory are about the same.
Only the ghosts of those whom I knew are left, and they only in fading memories. I called Vera Smith (WK's widow). I would like to
have gone to see her. WK died at Thanksgiving time last year, making
it one year too late already."[219]

Not long after, a series of international horticultural meetings was
hosted at the UC Davis campus, which kept Royce busy, but also
gave him the opportunity to meet many of his former graduate students, including Daud Khan, Facundo Barrientos, Khalifa Okasha,
and Jim Hancock. Even a field day at the South Coast Field Station,

held on a clear day in Santa Ana, was enough to bring back a flood of memories of former years. "We were up early, a beautiful perfect day in Southern California, almost as I remember it from my first visit many years ago. I was young then, and yet to be changed by the many things that have subsequently changed my life almost completely from missionary to soldier to student and finally professor and worker with strawberries."[220]

As Royce's retirement approached, his and Pearl's children were experiencing a combination of opportunities and heartache in their own lives. That year Royce's son John graduated from Medical school, and Royce, who had supported him financially while obtaining his degree, attended the ceremony with Pearl. But at about the same time, Royce learned that his oldest daughter Jean had experienced serious marital difficulties, which would eventually lead to a divorce. Royce, who since returning from the war had never had a good relationship with Jean, could not help feeling partially responsible for her unhappiness.[221] The situation cast something of a shadow over the marriage of Royce and Pearl's first grandchild that same year. "Jean and Bruce's daughter Gina was married to Brian Nelson in the St. George Temple at 9:15 a.m. What should have been a happy occasion was tempered by the unhappy pending breakup of Jean and Bruce's marriage."[222] By that time their fourth daughter Ann's husband Ray, having completed a law degree and a doctorate in economics, had been offered a faculty position at the University of Tennessee, and her family made ready to leave. Ann and Ray by now had three children to whom Royce and Pearl felt especially close, since for a full ten years Pearl had cared for the children while Ann had worked to put Ray through school. In the spring of that year they left to pursue their new opportunity, and on Mother's Day that year Royce observed, "We had dinner for the first time with just the two of us. Being alone has its advantages. I told Pearl that she had a 10 years to life sentence of grandmothering, but got off with only 10 years for good behavior. Life is all right. Health makes it worthwhile."[223]

Royce and Pearl had continued to exercise daily, and both enjoyed robust health, aside from a minor hand tremor which Royce had developed in recent years, and despite occasional mishaps, such as one he described in 1989 after doing some tree trimming in the back yard. "I lucked out, having about a fifteen-foot fall from our redwood, lacerating my left upper thigh rather badly. I was most fortunate to walk away. I could have been injured terribly."[224] That summer, with Ann's children gone, Royce and Pearl had the opportunity to acquire a new pet when Larry Fulton, who ran the Watsonville field station, bred a litter of energetic puppies of a valuable Australian breed. "Larry brought in the puppies and asked me to choose one. I did, choosing a little silver colored female. They are Australian shepherds, and cute as can be. I'll take ours home next week."[225] They called the puppy Sally, and began to include her in their morning walks. Royce, ever a lover of animals, became very attached to the new pet, and indulging in his own peculiar brand of dog training, soon had Sally perching at his command atop the peak of her dog house.

Royce and Pearl continued their regular service at the temple, and they went faithfully to church services, which Royce now found much more satisfying than in the past. "Some days I almost enjoy it as much as some claim that they do all the time. This is sure to be interpreted wrongly by some who might read it. My loyalty to the Church is in fact very firm indeed."[226] By this time Royce had solidified his spiritual convictions, and was once again serving in the capacity of High Priest Group Leader. In accordance with his lifelong commitment, he continued to conscientiously discharge his duties, which included participation in leadership meetings, encouraging temple attendance, and organizing and leading the high priests in performing monthly visits to the members known as home teaching. He also held periodic activities in his home for the members of his quorum. "We had over 60 people come, and everyone seemed to enjoy it. It was delightfully cool all evening. Pearl made the ice cream (strawberry of course). I sliced the berries for it, and I gave everyone berries to take home with them. We had the tables and chairs back by 9:00 p.m. and the last guests leaving. All in all a successful affair, I thought."[227]

Royce's chief difficulty at church now came from his diminished hearing, which made it increasingly hard for him to follow the proceedings. One Fast Sunday after coming home from church he wrote, "We ran first thing, then went to church meetings and experienced the usual ups and downs in testimony meeting. I hear very little there at times." Although his hearing had begun to diminish noticeably, Royce's chief fear of aging was not so much his sense of hearing as the possibility of losing his mental faculties, and his fears were still largely governed by his experience with his mother, who had spent her last days in a nursing home. The same day he remarked, "I wonder how things will be when we really age. I could not stand the 'warehouse' treatment so common these days. I'd much rather be dead."[228]

Meanwhile Royce continued his work at the university, which during his career had expanded from a minor agricultural college into a major academic institution with colleges in a wide array of disciplines. He attended meetings on agronomy and genetics both at home and abroad, and accepted occasional invitations to present papers. One of the most important of these came in 1986, when he was invited back to the East Malling Research Station in Kent, England to present the prestigious Amos Memorial Lecture, in which he outlined the history and future prospects of the California strawberry program.[229] He also continued his work on the Germplasm Repository committee, making periodic visits to Corvallis to attend to business there. He entertained guest experts from all over the world, who came to observe his breeding program. As his retirement approached, he decided to donate his substantial collection of scientific journals for use in universities of mainland China.[230] In 1987, he received a group of Chinese specialists at his experimental station in Watsonville. "They brought me some plants, and were interested in what we are doing and the varieties. Of course, I had to tell them that I could not give them plants. I am putting them in contact with the patent office. I doubt if they have ever seen berries quite like what we have here."[231] A week later, he wrote, "I went to my office and spent the entire day packing my professional journals for shipment to China (*Genetics*, *Journal of Heredity*, *American Journal of Botany*, *American Journal of Horticultural Science*, *HortScience*, *Phytopathology*, *Plant Diseases*, and *American Naturalist*). It took sixteen large boxes to hold them all."[232]

As his mandatory retirement approached, Royce began to consider what type of projects he might undertake in the future. "I have to give serious thought to how much or how little I shall be doing after my retirement. There are some things that I feel I should do, provided I am not in the way. We shall see about that in good time. How time does go by, and I seem to get older but not wiser."[233] Meanwhile, he began wrapping up his professional duties at the university. He gave the last of his popular Small Fruits lectures at the university, and administered his last final. In the spring of 1989, a strawberry symposium was conducted at UC Davis in honor of his imminent retirement—"Quite an affair, not by my choice."[234] That May his friend Günter Staudt and his wife Annliese visited the Bringhursts in Davis, and Günter remained for six weeks as Royce accompanied him on field trips to visit wild colonies, and assisted him in preparing his landmark paper on strawberry taxonomy for publication in English.[235] The paper, entitled *Systematics and Geographic Distribution of the American Strawberry Species*, was later published by the University of California Press, and became the definitive work on Strawberry taxonomy in the Americas. In it, Günter Staudt officially classified the coastal hybrid colonies Royce had discovered as a separate hybrid species, which he designated *Fragaria* x *bringhurstii*.[236] In the summer, Royce and Pearl took another trip to England, and Royce for the first time was able to visit the ancestral village of Bringhurst, in the countryside north of London.

In connection with his retirement, between 1989 and 1990 Royce was to receive a flood of awards and recognitions, which were added to the Wilder Medal which he had received in 1979, and the Gourley award in 1982. Along with Victor Voth, in 1989 he became the recipient of the Outstanding Fruit Cultivar Award for the strawberry variety "Chandler" (later, with Victor Voth and Douglas Shaw, he would again win the award in 2002 for the variety "Camarosa."[237]) In 1989 Royce was elected a fellow of the American Association for the Advancement of Science and was awarded the Outstanding Researcher Award by the American Society of Horticultural Science for his extensive research on the strawberry,[238] and in 1990 he would receive the Milo Gibson Award[239] from the North American Fruit Explorers. In addition, he received a host of other recognitions. In 1989 at the biennial meeting of the California nurserymen's organization in Redding, he was recognized and presented with a plaque listing the many strawberry varieties which he and Victor had released. In November of 1989 Royce and Victor were invited to a special commemoration dinner at the Japanese-American National Museum at Los Angeles, where each was given a plaque and vase honoring their contributions to the strawberry industry of which the Japanese had been such an integral part; this was accompanied by certificates of commendation from the city and county of Los Angeles, and from the California State Senate. In 1990, Royce joined a handful of distinguished faculty in being presented the Award of Distinction from UC Davis for his academic and professional achievements. Finally, in 1991, the California Strawberry Advisory Board held a dinner in honor of the retirement of both Royce and Victor Voth, and there they were each presented with commendations from the State Department of Food and Agriculture, the Congressional Award from a member of the United States Congress, and a letter of commendation from the governor of California. Plaques and certificates from around the world began to appear about the Bringhurst home on walls and shelves to join the strawberry-themed dinner plates and decor that the family had accumulated over the years.

As Royce finished up his professional duties, the world around him was also being changed by momentous events. In October 1989, the great "Loma Prieta" earthquake struck the coast mountain range of California, nearly destroying the town of Watsonville as well as a large portion of Santa Cruz, and toppling a section of the Oakland Bay Bridge and a length of elevated freeway in Oakland. But by far the most important changes were those which were happening abroad, especially for Royce, who had lived his adult life in the aftermath of the Second World War. "To me, the most important world event during the period was the passing of the Soviet Union and associated events. I never thought that I would live to see this hap-

501

pen—what a difference this will make in the long run. Germany once again one, and the USSR many. Poland and all the rest of the client states are free. The Berlin Wall is gone. What a happy world this might be. Only China, North Korea, Cuba and Vietnam remain somewhat the same, but islands of 'captivity' in a free world won't survive as such without profound change. With the exception of those above plus a few backward African states, the Church now has missionaries in virtually all the world, and is continuing to expand."[240] The fall of the "Iron Curtain" marked a pivotal turning point in the history of the world, even as Royce faced the transition to the final chapter of his life.

Illustrations for Chapter 11

Royce examining experimental plants in a UC Davis greenhouse.

Central Coast experimental station in Watsonville, California. Experimental varieties underwent field testing at this site.

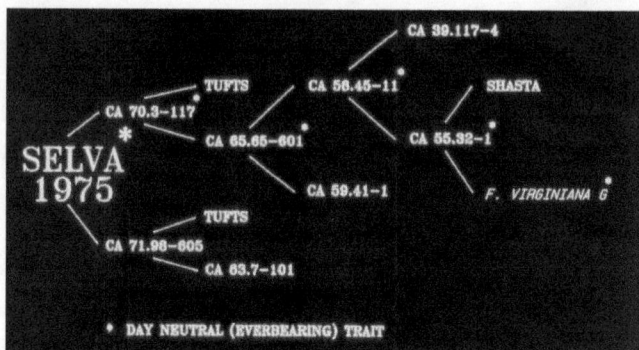

Pedigree for variety Selva, showing its descent from a wild strawberry collected by Royce in 1953. Selva was the first commercially successful day-neutral variety.

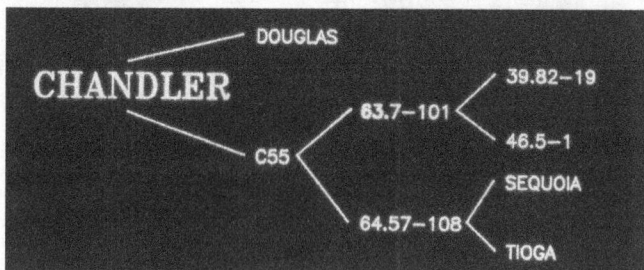

Pedigree for variety Chandler, released with Selva in 1982.

Fruit of the Chandler variety. Known for its superb flavor, Chandler received the Outstanding Fruit Cultivar Award from the ASHS in 1989.

Royce examining wild strawberry plants in the field.

Northwest Plant Germplasm Repository in Corvallis, Oregon. Designed to preserve valuable plant material both wild and cultivated, this was the first facility of its kind. Royce was on the planning committee.

505

Royce in the laboratory.

Royce in his office at Wickson Hall, shortly before retirement. Note mud on boots, from work in the fields the same day.

Royce and Pearl with family and descendants in Davis, April, 1981.

UC Davis Pomology Department in 1988, shortly before Royce's retirement. Royce is 4th from left on the 2nd row. His successor, Douglas Shaw, is 3rd from right on the front row.

Royce receiving one of many awards, in this case the faculty Award of Distinction from UC Davis, in 1990.

With colleague Victor Voth at joint retirement commemoration in Monterey, 1990.

Retirement

Chapter 12

Royce S. Bringhurst celebrated his 70th birthday on December 27, 1988, and at the end of the 1988-1989 academic year he took mandatory retirement as required by the University of California, having served 39 years in the university system, 36 of those years at Davis. At the graduation ceremonies in June of 1989, Royce was invited by university authorities to sing the National Anthem at the UC Davis commencement exercises, and his rich tenor voice, soaring above the full concert band that accompanied him, brought a burst of spontaneous applause from the graduates and the audience which Royce found very gratifying. A few days later, Royce left with Pearl for a *Rubus* symposium in Great Britain, and while there, the two again visited the ancestral village of Bringhurst with its ancient stone church, and they made certain to attend the London Temple prior to their return home.

As a retired professor, Royce was relieved of teaching duties and of burdensome committee assignments, and at last had unaccustomed free time to catch up on his writing. Douglas Shaw had taken over responsibility for the strawberry breeding program, although Victor Voth, younger than Royce and still a year away from retirement, continued his work of selecting and testing in the South Coast Field Station using the plant crosses Royce had previously made, while Doug Shaw assumed the same work at the Watsonville and Wolfskill experimental stations, and began making the first crosses of his own. Royce, for his part, remained as involved in the strawberry program

509

as he could, assisting in the digging of nursery plants in October of 1981, and writing up the disclosure paperwork for the new day-neutral varieties Seascape (CN49) and Capitola (CN93), which was filed in January of 1990.[1] As circumstances permitted, he continued to attend field days and meetings of the California Strawberry Advisory Board and other grower groups, though increasingly he did this as an observer only.[2]

Outside the California strawberry program, Royce continued to be a sought-after teacher and consultant. He continued to teach an annual strawberry breeding class to the students of a colleague named Leroy Baker from Chico State University, stopping only after 1991 when Dr. Baker also retired. He lectured, as he had in the past, to groups of students from Brigham Young University, and from Ricks College (later BYU Idaho), both run by the Church. He continued to meet with members of the Germplasm Repository committee in Corvallis, Oregon, and found most of the strawberry plants he had sent there still thriving. In July of 1991 he presented a paper at a conference in Wisconsin, and he and Pearl made a nostalgic visit to Baraboo and Badger Village.[3] A year later he traveled to St. Louis, Missouri as a consultant for a major company which was considering genetic engineering of strawberries.[4]

Royce also continued to travel frequently to various countries, both as a speaker and as a consultant for the strawberry industries abroad. In July of 1989 he traveled to Thailand to visit horticultural experts there,[5] and in March of 1990 he and Pearl made a three-week journey to Argentina, where they crisscrossed the country, looking at extensive strawberry plantings, though he trip was marred when Pearl's purse was stolen in Buenos Aires on Easter Sunday, and with it most of Royce's notes from the trip.[6] In July, 1990 Royce returned with Pearl to New Zealand, where he gave a talk on strawberry breeding and the patent process, and while there they made a side trip to Australia, where Royce spoke with Australian strawberry growers. During this trip they saw a wide array of the wild animal life for which New Zealand and Australia are famous, and Royce described this as "one of the wonderful trips of my life."[7] The following year, in 1991, Royce was invited to present a paper at a strawberry meeting in France, and he travelled there with Pearl, also checking out strawberry nurseries in Segovia, Spain, where they found the new California day-neutral varieties Seascape and Capitola in fruit.[8] In November of 1991 as spring came to South America Royce returned once more to Chile with Pearl, traveling to the south where he was able to find wild strawberries in full flower. Once again, a trip to Latin America was spoiled by theft. "We returned to Santiago, only to be robbed. I lost my best slides (not replaceable), so I had to give my talks without them. It went well, but I could weep at the loss of those slides."[9]

In September of 1990 Royce attended the retirement gala for Victor Voth, which was held at the Disneyland park in Anaheim,[10] and the following spring he attended the series of seminars presented by candidates for Victor's position. The seminar on April 9, 1991, was given by Kirk D. Larson, a plant physiologist, who was subsequently hired for the job, and began work at the South Coast Field Station in partnership with Douglas Shaw.[11] In June, 1991 the California Strawberry Advisory Board held a special retirement gala for Royce and Victor jointly at a beautiful hotel in Monterey at which many guests were present, including most members of Royce's own family. Of this event Royce wrote, "A flattering affair at Monterey. Hundreds were there—the passing of an era." Weary of the long string of professional recognitions and still a little self-conscious about appearances in public, Royce added, "Enough is enough is too much."[12] Many of Royce's postwar generation of colleagues were also reaching retirement age, and Royce found himself not infrequently attending, and sometimes speaking, at similar commemorations for them. When in 1992 a new alumni center was dedicated on the UC Davis campus, Royce and Pearl were invited to the dedication ceremony, and Royce wrote, "One of the gardens is named for us because of the substantial contributions we have made."[13]

On May 11, 1992, Royce and Pearl made the long drive from Davis to Salt Lake City to lend some help to Pearl's sister Helen, who had fallen and broken her hip not long before. That month marked the fiftieth anniversary of Royce and Pearl's hasty wartime wedding, and when the anniversary arrived, Royce wrote, "We went to the Salt Lake Temple to celebrate and remember our 50th (golden) wedding anniversary. From there we went to Park City, checking in at a very good hotel. This was nice, since we had no celebration when we were married, as I went to war shortly after."[14]

Royce and Pearl returned to California on Monday, May 18, 1992, and a few days later they received a phone call from Richard Brimhall, who represented the Ezra Taft Benson Institute,[15] a nonprofit humanitarian organization based at Brigham Young University. The Benson Institute specialized in elevating the standard of living in developing countries by improving small-scale agriculture, and Brimhall requested that Royce and Pearl send in a resume so that they could be considered for a humanitarian service mission to Chile on behalf of the institute. This request came as a complete surprise to Royce and Pearl, who had scheduled many activities for that year, but on May 29 they sent in an affirmative response, and began preparing for a mission. They met with an attorney to put their now complicated financial affairs in order, and called upon John's wife Betty to take care of their finances while they were gone. In July they traveled to Provo, Utah to receive training for their upcoming mission, which was to take place in a city called San Felipe.[16] Shortly after, Royce and Pearl departed for a previously planned trip to Hawaii, and upon their return Royce attended the funeral of a colleague, which proved to be a reflection of the times. "I went to a memorial service for Richard Snow, who died of AIDS last winter. It was a rather odd service, but I'm glad we went. I have always respected him."[17]

That August Royce and Pearl planned a family reunion in Davis, which would serve both as a celebration of their 50th anniversary and as a send-off for their upcoming mission to Chile. On August 12, 1992, the guests began to arrive, starting with Jean and her family from Las Vegas, and Florence and Larry Nielsen, who stopped in Davis on their move from southern California to Cedarville in the far north, where Florence had accepted a teaching position. John and Marla Vaughn brought their family from Oroville, and Ann arrived from Tennessee. John and Betty by now lived in nearby Woodland, where they had moved when John completed his medical training. Margaret was already living in Woodland with her husband Jack, and both their families came, as well as some of Pearls' and Royce's relatives from Utah. On Saturday, August 15, Royce and Pearl rented a city park for a grand picnic with games, and that Sunday the entire family attended Sacrament meeting, which served as a missionary farewell for Royce and Pearl. The church service was effectively turned over to the Bringhurst family, and was described by Royce: "All of our children talked (Jean, Florence, Marla, Ann, John and Margaret); then we all sang, including George and others of the family, a version of 'How Firm a Foundation' not so well known in the Church.[18] I chose that, and I must say it was well done. John led it, and Sarah (John's daughter) played the piano—almost 30 participated and it sounded as good as most ward choirs. Collectively, we have good voices. Pearl and I then talked, and we finished on time. My brother George gave the opening prayer, and Betty (John's wife) the closing. It was a good meeting. I read from my father's mission letters. The place was packed to overflowing, and we were pleased to see our 'temple family' represented. How good it was to see them, and we go away feeling good."[19] After the meeting the entire family retired to the Bringhurst home for an open house, and that week Royce and Pearl did their final packing.

Mission to Chile
On the early morning of August 22, 1992, Royce and Pearl began the series of flights to Chile, departing from Miami for the last overnight flight just ahead of a major Florida hurricane.[20] Arriving in San-

tiago the morning of Sunday, August 23, they were met at the airport by a Church employee named Lito Magnere,[21] who drove them about an hour northward to the city of San Felipe which would be their home for the next eighteen months. San Felipe was a small, picturesque city nestled at the base of the great Andes Mountain range which divides Chile from Argentina, almost within the shadow of the famed Mount Aconcagua, the tallest peak in the Americas. Royce and Pearl were to work in cooperation with a recently formed private college called Universidad de Aconcagua,[22] where Royce would teach and do experimental work in cooperation with the dean of Agronomy there, a professor named Victor Gamboa, while Pearl taught English classes to a group of students. Royce and Pearl were accommodated in a hotel during their first week while arrangements for housing were being made, but that Saturday they were moved into a more permanent apartment, and on Sunday they walked to church in San Felipe for the first time through a heavy rain, and found that their services were immediately in demand. "I blessed the Sacrament for the first time in years, and gave a talk at a baptismal service," Royce wrote, adding, "I am having difficulty hearing and understanding."[23] Both Royce and Pearl would struggle with the language initially, Pearl because she had never mastered Spanish, and Royce because of his partial deafness. Three weeks after their arrival Royce wrote, "Pearl is doing very well with the language and by the time we return, I think that she will have gained some self-confidence, although she is fairly fearless and does not feel wounded when she makes a mistake or is not understood. I have my problems too, mostly hearing and understanding, since I can say most anything I want and have it understood reasonably well."[24]

At the beginning of September their formal teaching duties began, and Royce recorded, "At 10:00 a.m., or rather 10:20, Pearl had her first 'English as a Second Language' class. It went well, and should improve over time. Pearl will be teaching more than she thought, I less, it now appears."[25] Two days later Royce taught his first class. "Yesterday and most of today I spent preparing my lecture for today. My strawberry class was at 4:00 p.m. to 5:15. It wasn't the greatest. There were seven students present. This should improve." Still getting his bearings, Royce wrote, "I don't feel comfortable with what we are doing (at least with what I am doing). We have not received our books and papers yet, so it is not easy to function. My language perception is improving, but I still have a long way to go."[26] Royce had trouble readjusting to the particular style of Spanish spoken in Chile. "The problem is the very rapid delivery. My eyes aren't that good, to say nothing of my ears."[27] As for the much-needed books and papers, these had inadvertently been sent by overland mail, and

513

would take months to arrive, so Royce had to begin his work without them.

It took another week before Royce began to feel that he was getting his feet on the ground, and began to gain confidence with the language. A couple of weeks after their arrival he wrote, "I worked in the *sala de profesores*, and while there had a good talk with one of the professors of Genetic Engineering. I had no trouble with her at all. Most of the time it was almost as if she were speaking English. The main principle involved is confidence. If they think you can communicate, you can. It is the first real boost I have had since getting here. After I talked with her I managed to grab a few minutes with the Dean of *Agronomía*, Victor Gamboa, about some small time and size research with students. We talked about polyethylene cover experiments (solarization and poly mulch). I think I am now getting to the point where I can generate some enthusiasm, particularly if I can include some strawberry plants in my program. It should work here, particularly if I can get some day-neutral plants ('Selva' or even 'Fern,' and perhaps some 'Seascape')."[28]

Royce's principal assignment in Chile was to work with soybeans, which served an important role in the Benson Institute's small-scale agricultural scheme, mainly to provide feed for livestock. He was to perform experiments to determine the optimal conditions for the growth of the legume, and he received more details of this a month after his arrival in San Felipe. "The soya planting scheme calls for two varieties (Hamilton and Fremont), eight times of planting, two special treatments, five locations, and a whole flock of planting dates." That same day he noted, "I gave my second strawberry lecture today, to about twelve students. It went fine—I picked up a few words to be used with their proper meaning, for example, *fotoperíodo* for photoperiod, *parcela* for plot, *hilera* for raised bed, etc. I worked three shifts and was busy."[29] A few days later Royce began the planting for the first of the soybean experiments, which were carried out on experimental plots at a place called Almendral.

Royce's first look at the soybean planting a couple weeks later was a little discouraging. "I met Victor Gamboa, and went out to see how the soybeans we planted two weeks ago are doing. The stand isn't the greatest, and now we see the real problem, nut grass and bermuda in the whole field. All is sort of well, but I have my doubts about many things."[30] It was only when he undertook some writing about the proposed objectives of the project that Royce began to feel better about the work he had begun. "I finally completed a couple of articles concerning Benson and small scale farming. I wrote in English, and they will translate. It did me some good to think a bit about why we are here, and to put it in writing. There is nothing that they do that cannot be improved upon, and we should be able to improve the quality of life for everyone we have anything to do with." Noting that important obstacles remained, Royce added, "'Easier said than done,' especially when communication problems are involved. We still have not received the things that were supposed to come by boat. I wonder sometimes why this had to happen. We have been seriously handicapped by not having things that we counted on being here."[31]

Because of Royce's special expertise with strawberries, he was also assigned to give a series of lectures throughout the country on strawberry production, and several weeks later he and Pearl made a trip southward to begin this process, which led to the renewal of several old acquaintances, including that of Vilma Villagrán, the pioneer in Chilean strawberry agronomy with whom Royce had worked during his year-long visit to Chile many years before. The trip took him through the picturesque regions of southern Chile, and Royce reflected, "This is the end of our second month in Chile. I can see how we can really have the time fly by as we get more involved in things that are really happening. As I look out of our hotel window toward the river, I can't help but think of how beautiful it is. The first Spaniards must have thought it a 'Garden of Eden'."[32]

Royce meanwhile continued to make slow progress on the language, in which he was hampered in part by his diminished hearing, and in part by the peculiarities of the Spanish spoken in the region. Although Royce had lived in Chile before, his association with the Benson Institute required a command of vocabulary still unfamiliar to him, and this sometimes led to embarrassing situations, as when he attempted to cash some checks from the university at a local bank, but did not know simple banking terminology. It was only after much stumbling that it occurred to him to ask if any of the bank employees spoke English, and he was quickly and courteously attended in his native tongue. Of this episode, Royce observed, "Humility should be my strong suit because I have plenty to be humble about. I still have my lessons to learn. I would give a great deal to be able to hear and understand as I should."[33] When it came to strawberries, however, Royce was in his element, and had little difficulty with the language, as when he delivered one of his strawberry lectures a couple of weeks later. "We rode the bus to Santiago where I gave a talk at the Central University of Santiago. It went very well. There were well-thought-out, appropriate questions. I understood the questions and comments and thanked them for it. It was a real fun time for me, because I knew the subject matter, and was fluent with the language. The best I have been in Chile, I believe."[34]

Shortly after this lecture, Royce and Pearl attended a workshop with representatives of the university and of the Benson Institute, at which they discussed the expectations for their work in Chile. Although technically missionaries for the Church, and under the leadership of the Chile Viña del Mar Mission, Royce and Pearl were required to be especially cautious about their relationship with the local Church and its activities, so as not to jeopardize the purely humanitarian work which they had been sent to perform. Nevertheless, they had ample contact with the Church on Sundays, and Royce was especially impressed with the quality of the missionaries in the area, and the large proportion of native Spanish-speaking missionaries, in contrast to their previous experience in Chile. After a zone meeting of missionaries which he and Pearl attended, he wrote, "It is of interest to note that of the 100 or so missionaries, fully half were Latin American (mostly Chilean). Nearly every U.S. missionary in the zone has a native speaking companion. I do believe this portends great things for the future of the Church here. Within ten years there should be an effective native base for Church leadership in place and moving, that will transform the Church here in a very positive way. I am impressed with the quality."[35]

At the same time, Royce had reservations about the way missionary work was being conducted in Chile at the time, perhaps recalling the extreme caution and extensive preparation which had been required to perform a baptism during his own mission nearly half a century before. In San Felipe baptismal services were held nearly weekly, and after one of these Royce observed, "There are still more baptisms than conversions. They move them too fast, still trying to get first visit pledges to baptize. The missionaries should spend as much time with baptized members to get them moving into the Church as they do with seeking those willing to be baptized."[36] Initially Royce and Pearl were not given callings in the local branch, probably to avoid any appearance of proselyting activity, but eventually Pearl was asked to direct the music, and Royce to serve as an instructor in genealogy, where he had to learn to reckon with the Spanish system of surnames. "I started on the genealogy class, and it went quite well. The only thing we have not decided is how to put the names down. It has to almost be as we do it in English, or there will be ample confusion, more than now."[37]

Less agreeable for Royce and Pearl were their dealings with the Chilean Bureaucracy when it came time to renew their visas. After three months had elapsed Royce wrote, "Our visas have now run out, and we have no *carnet*, and thus technically are in the country illegally. Otherwise all is well, we think."[38] Their passports arrived with the necessary visas not long after, but the *carnet*, an identification card required for legal residency in Chile, involved a much more exacting process, which would require multiple trips to Santiago to meet requirements which seemed to change with every new visit to a government office. A full six months after their arrival in Chile, Royce's identification card finally arrived in San Felipe,[39] but Pearl's was still delayed owing to a petty detail involving her surname from their previous stay in Chile decades before, which multiple inquiries at sundry offices failed to resolve. Nearing the end of his patience, Royce wrote, "The case is obvious, except to the most obtuse reasoning by a genuine, dedicated bureaucrat."[40] It would take additional trips to the capital, a series of letters from the American consulate, and no fewer than four sets of fingerprints before the issue was finally resolved, but not until Royce and Pearl had been in the country for most of a year.

As spring came to San Felipe the weather warmed, and Royce and Pearl began to acquire some of the items which they had felt essential to life there. "We now have a refrigerator and it works, and I expect a car in my hands before the week is out. It is a Lada (Russian). I am always a bit suspicious about what is offered as a good deal here. We shall see."[41] The car came as hoped, and proved to be a welcome addition, as it gave Royce and Pearl a new degree of flexibility in their travels. "It runs well, and I complain not. This can change the way I view things. I'm feeling much better generally."[42] The springtime strawberry crop in Chile had by that time come into full production, and Royce began to acquire samples of California-bred fruit as he had back in Davis, to take to his class. By the end of the year the mission appeared to be looking up. Pearl was well-established in her English classes, and Royce was completing his work as instructor in what had proven to be a very successful course in small fruit cultivation, and had finished planting out most of the year's soybean experiments while making plans for a similar series of strawberry plantings the following winter. With Christmas came the hot days of summer, and Royce and Pearl felt a growing sense of optimism about their mission, though they were shortly to pass through one of their first major misadventures in the country.

On New Year's Eve Royce and Pearl were invited as guests to Lito Magnere's home in Viña del Mar, from which they planned a trip to the south of Chile to meet with their friend and strawberry expert Vilma Villagrán to obtain strawberry plants for the winter planting. They departed on January 3 with Lito's son Gastón, but as they traversed a steep downhill section on a series of tunnels near Santiago they experienced a near-disaster. "We had driven into the middle of the second tunnel (Lo Prado) when the car stopped, and would not start again. Between fifteen and thirty minutes later a 'rescue' *camillón* with three people in it pulled in front and towed us through the tunnel to safety on the Santiago side. You can appreciate what a terrifying experience that was. It could have resulted in a horrible accident."[43] Their rescuers managed to get the car working again, but when it stalled once more on a busy thoroughfare in front of the central railway station in Santiago, they aborted the journey and returned to San Felipe by bus. Royce wrote, "We were very fortunate not to have had a serious accident, and felt that the Lord was looking out for us, and we count our blessings."[44]

The stall in the pitch dark tunnel proved to be one of the most frightening and memorable of Royce and Pearl's mission experiences, but it was certainly not the last of their misfortunes. A few days later they returned to pick up the repaired car and resume their trip southward, and near Temuco Royce attempted to locate plants of a native *Fragaria chiloensis* strawberry which was still under cultivation there. Arriving at the southern city of Osorno, Royce and Pearl took a side trip which, though starting quite pleasantly, proved to be another near-disaster for Royce. "We drove to the Parque Nacional Puyehue, where we visited a waterfall (Salto del Indio) and a huge tree there. We left on horseback for about a six-hour trip through the forest, with wild bamboo and Chilean native hardwoods everywhere. We crossed the river on horseback going and coming, where we really got wet on the return. My horse was a beautiful ball-faced sorrel named 'Papocho,' well-trained, gentle and lively. On our trip back, just before crossing the river I was looking to the left when I should have been looking ahead. As my horse went under a fallen tree trunk, I had not ducked low enough, and my right cheek, forehead, and eye struck the tree, really banging up my head. It was fortunate that I did not lose an eye. I have a black eye, and some quite severe cuts below my right eye. I feel fortunate that it was not worse. When we got back to the hotel, an *enfermera* was called who dressed the cut and scraped area, and treated to reduce the swelling. It is quite painful, but I'll settle for that, since it could have been so much worse. How difficult it would be for me to lose an eye at my age. I see poorly enough as it is."[45]

Returning from this trip to San Felipe, Royce quickly discovered that the problems with the car were even worse than he had imagined. "It turns out that we drove all the way south to Osorno and back, and over to Viña and back without a spare tire. I have no idea how this happened, but will not say anything to Pearl or to Lito. Someone was looking out for us. The spare tire was in the back, but flat, and the *cámara* (tube) was absolutely unusable. I have no idea either why the people who did the repairs in Santiago did not say anything. I suppose that I should count our blessings and accept the rest. I will buy another tire. I don't see how we can do without."[46] Though the car had been provided for Royce's use and he was responsible for it, other university personnel had been making frequent use of the

518

vehicle, which not only made it unavailable for days at a time, but also presented a safety hazard for Royce and Pearl, since the other users felt no responsibility for its maintenance. Royce met with university officials to restrict use of the car, but even so, a month later when one of the front tires began to leak, it was discovered that the tire had a half dozen nails and a badly damaged sidewall, and Royce made immediate arrangements to purchase new tires. Royce sought for official approval for the repair, but wrote, "I intend to do the repair even if I have to pay for it myself. I think it is a matter of personal hazard that must not be overlooked. I would never drive with those tires at home. Why should I here? This gives me a double reason for never loaning this car out. Such is the life on a 'service' mission. I think that we are being looked after in many ways. The right front tire could have (even should have) blown out so easily."[47]

Not long after his near-accident in the tunnel, Royce one day received an unexpected phone call from the college where he had earned his bachelor's degree nearly five decades before. "I received a pleasant surprise today from President George H. Emert of Utah State University, informing me that I have been selected to receive an honorary doctorate degree at Utah State on Saturday, June 5, 1993 at the 100th annual commencement. We have decided to accept it and go, one way or another."[48] Royce and Pearl had already planned a trip home in March to prepare their income taxes, and they quickly made arrangements to postpone their trip until late May, to allow Royce to participate in the graduation. This was a singular honor for Royce, who had fond memories of his years at the Agricultural College, and he wrote a letter of acceptance, indicating that he and Pearl would bear the cost of their travel, and requesting that only family members be notified of the event. In the ensuing months Royce and Pearl began making preparations for the trip.

Shortly after, Royce received a letter from Douglas Shaw, indicating that another generation of strawberry varieties was ready for release. Since these had resulted from crosses Royce had made while still in charge of the program, his signatures would be required, and it was hoped that in March he would be present to help prepare the necessary patent paperwork. Royce responded that he would be not be making the trip in March as he had planned, and he instead arranged to receive by mail the disclosure forms for six new strawberry varieties, the last group of now famous California strawberry patents which would include his name as one of the "inventors."[49]

Meanwhile, Royce continued making his monthly experimental soybean plantings with a professor at the university named José Godoy, and the two concurrently made preparations for similar experiments with strawberries. "At 4:00 p.m. it was cool, and Godoy and I

519

planted the last crop of *Soya*. Last evening and this morning I mapped out the strawberry planting, probably to be winter planted in early May or late April. The first thing is to solar treat the ground. I trust that it is not too late for that. We need at least nine weeks." the following day Royce wrote, "Godoy and I finished most of the planning for strawberry planting: three varieties (Chandler, Selva and Fern), five treatments: Solar treatment, methyl bromide fumigation, clear poly, grey poly, and black. He had some good suggestions, such as furrow mulching with bamboo. We bought the polyethylene for solar and methyl bromide fumigation, a 27 meter by 1 meter tube, at a cost of 5,500 pesos. It is heavy, and should hold together."[50] A few days later Royce and Godoy began the solar treatment, which consisted of applying sheets of clear polyethylene to unplanted and irrigated soil, in an attempt to eliminate both weeds and soil pathogens by heating the soil with sunlight until it was sterilized. That same week Royce traveled to a berry farm at a place called Polpaico, where he at last met up with his friend Vilma Villagrán, and received her assurance that she could provide the strawberry plants necessary for the winter planting.[51]

Although strawberries were Royce's chief interest, by this time he had gathered enough data on the soybeans to begin the statistical analysis on the experiments he had been sent to perform. "The *Soya* project is proving to be more interesting than I thought; our performance estimates are looking better. The solar treatments are proving interesting. I have data to work with on strawberries too, so I am not feeling as underemployed as I might have felt. I need to get on with some writing. I didn't want to get going with it too soon, but now it approaches procrastination."[52] As the harvest of the various soybean plantings got underway, Pearl began accompanying Royce to record the yield data, first by counting the pods on live plants, and later by harvesting the pods and counting them. After compiling the data Royce wrote, "I worked on *Soya* data in the office all day, finally making a comparison of the destructive with the non-destructive method of taking data, separated by two weeks. Pearl and I did both, and the two sets are remarkably similar. This pleases me, because I finally have something to talk about here besides strawberries. This is just the first planting of four that were made at monthly intervals starting 16 October 1993. I'll make up a report of this and we shall follow up on it. I'm working with José Godoy on this, and he is great. I'm generating a bit of enthusiasm now. At least I didn't let anything drop by the wayside."[53] In a letter home, he added, "I am pleased to pronounce this high priority item with 'Benson' a success, and to find skills I haven't used for a while still available for recall. I think this is the trick to keeping the old brain in working order."[54]

As he directed the soybean research, Royce became aware that most of the literature on the crop was in English, and he sought to rectify the situation by translating a few materials into Spanish, despite his own insecurities with the language. "I spent most of the day in my office translating for the most part, things published on soybeans in English, but not available in Spanish. I quite enjoy doing it, and I learn a lot about *soya* and quite a bit of Spanish I am shaky on in the bargain."[55] Royce proved to have more ability as a translator than he expected, and as time went on he was called upon to translate other kinds of scholarly materials as well. One day in the fall he wrote, "In the evening, a couple from upstairs came by to ask if I would translate a paper on oceanic studies with radar. I told them I would try." The next day he added, "I worked most of the day on the translation, finishing about 6:00 p.m. It proved to be more interesting and easier than I had thought. I felt good about it, and guess that I do have some ability in this area."[56] When fall college classes came back into session in March, Royce began again to teach his small fruits course, and in addition to his field work and talks on strawberry production, he was also asked to give a series of lectures in his areas of expertise. For the first time since his retirement he began to feel genuine concern about overextending himself.

That April Royce received a letter from his brother George, informing him of the death of Steve Mackay, his old friend from Murray, who had been in ill health since suffering a stroke the year before. The news left Royce contemplative of yet another piece of his past that had vanished forever. "There is something wonderful about the 'best friend' of your youth. We camped, fished and hiked together, and I have always treasured the memory of those days we spent together. How lovely it is to recall the memory of the days when we were young, and thought those days would last forever."[57] Shortly thereafter Royce and Pearl received another letter from home with more welcome news. "Today we received a wonderful letter from Marla in which she told us she had gone through the temple for her own endowment on last Wednesday, 7 April. Some things work out well. All of them have now been to the temple. We are very happy about this."[58] To have had all their children attend the temple was particularly meaningful for Royce and Pearl, for whom temple work had become a cherished part of their life.

As they looked forward to returning home in late May to attend the Utah State graduation in June, Royce began making final preparations for the winter strawberry plantings, confirming his arrangements with Vilma Villagrán for the necessary nursery plants. The strawberry experiments called for a variety of soil treatments to determine the optimum conditions for growing. Part of the plots had already been solarized for planting, and now Royce and José Godoy turned to the task of fumigation. "Today we put the methyl bromide under the row designated for that purpose. The temperatures still hit about 90 degrees F. I was tired from my activities—I guess that I'm getting old."[59] As the day for departure approached, Royce took stock of the experiments he had underway. "We haven't yet gotten the plants, but are ready to go when we do. The weeds are doing well too, too well. The *Soya* is just about threshed out. The project is probably set for another year at least. I have finished with the analysis of the first solar data. We averaged about 8° C. gain in temperature with the clear poly on the warm (north here) side of the bed. I think I have been of some help." He added, "We will have been here half our scheduled time as of May 23. Things are moving more rapidly now."[60] The long-awaited strawberry plants would not arrive until after Royce and Pearl had departed, and Royce left careful instructions to have them planted according to the experimental scheme which he and Godoy had worked out.

As the day of their departure neared, Royce and Pearl made ready to move into different living quarters, as a larger apartment had been found for them which was much closer to the university. The move was scheduled for just three days before they were scheduled to leave, and coincided with the arrival of visiting professors from California. Royce and Pearl spent the day packing their belongings, and it was not until the move was underway that they discovered that the new quarters were not as expected. "Pearl was frantic because the place was such a mess and it seemed so hopeless, and the folks from the university that came to move us could not get the van to move us until after 5:00 p.m. I actually had a tough time keeping things moving and choosing where to put them as we took the four loads over. To top it off, the first of the visiting U.C. Santa Cruz professors, a sociologist, had no towels, and we couldn't help because our alternate set was at the laundry. We finally ate late in a messy filthy kitchen, vienna sausages and scrambled eggs. They were great, since we were hungry."[61] As Royce made final arrangements for the all-essential strawberry plantings and taught his last small fruits class, Pearl continued to labor on the house, which by the time they left Royce would declare "not ship-shape, but sanitary." The two packed their suitcases, and made ready to depart the following day.

Honorary Degree at Utah State University

When they set off for their trip to the United States on Thursday, May 27, 1993, Royce and Pearl were destined for another misadventure which would once again pit them against a relentless Chilean bureaucracy, but would ultimately restore their faith in human kindness.

Royce recorded the events. "Victor Gamboa picked us up at 4:00 p.m. and drove us to the Santiago Airport, where he saw us leave for the International Police door, and then bid us farewell and left to return to San Felipe. Then the trouble started, when we were denied permission to board the aircraft because we had no *salvo conductos*, as required for people with temporary residencies, as in our case. We were stuck and absolutely beside ourselves, not knowing what to do. Pearl fussed and shouted, making herself known to all present.[62] The agents were actually very kind and most considerate. As things calmed, we accepted what was an accomplished fact, no flight for us tonight.

"The airline (American) arranged our tickets for tomorrow and put us up in a hotel, where we came into contact with the person who was to be our 'savior' for tomorrow, Nelson Hermosillo-Silva, an Australian-Chilean Christian gentleman. He spent 3-1/2 hours with us, starting at 8:00 a.m., and before 11:30 we had the required document signed, sealed and delivered: a '*salvo-conducto*.' We spent the rest of the day at the hotel. However, after leading us around and getting all the necessary documents including the Chile version of the IRS clearance and paying the taxi fare, Mr. Hermosillo (may he be blest, our hero for the day) would not accept anything for his services or the taxi. How wonderful. I would not have believed that generous, unselfish services of this type are given freely in our lifetime, but they are. Never underestimate the inherent goodness of mankind. I shall always believe that he was sent as an answer to our prayers. We spent the afternoon at the hotel, then went early to pass the gates and go aboard. How happy we were."[63]

By the time Royce and Pearl arrived in California Pearl had come down with a severe cold and fever, and felt miserable. She and Royce were greeted in Davis by Florence, Marla, John, and Margaret and their respective families, and a few days later they made their way to Utah, where they stayed in the home of relatives. Arriving in Logan the next day, they checked in at the university and met up with old friends. "The first people we met were our very dear friends of many years, Arthur and Elna Wallace. Arthur and I started first grade together at Plymouth School in Taylorsville. I would never have gone to UCLA to work in 1950, were it not for Arthur. We have always kept track of one another."[64] Art Wallace was in Logan celebrating the 50th anniversary of his own graduation, his schooling not having been interrupted as Royce's had been. Royce saw many old colleagues and friends as he attended a reception at Utah State and toured the agricultural building.

The graduation ceremonies were held on Saturday, June 5th, and Royce kept a record of the proceedings. "This was the big day for me.

We went to an early celebrity breakfast, and it was good, although I was not terribly hungry. After that we dressed in the Old Main, and later joined the procession in a march to the large field house, where I sat on the stand at the side of President Emert, a very handsome and talented man. Jake Garn gave the address to the graduates. He was the Utah congressman who flew on one of the space shuttle missions, an active member of the Church. He also received an honorary doctorate. I was attended by Clark Ballard, my old classmate at Utah State, and received an honorary doctorate in Agricultural Science. I felt very honored, and was happy to accept for our strawberry work. Enough 'honor' has certainly fallen upon me for that. I think that this is the last of this sort of thing. I can't help but remember my humble beginnings, and still feel somewhat unworthy to sit in the 'seats of the mighty.' I think that I might prefer to be the 'doorkeeper in the house of the Lord.' Well, what is done is done, and who am I to judge such matters.

"We then joined in the picnic, and there all of our living brothers and sisters save Naida were present, plus part of our children, John and Margaret and families (the rest were all tied up with teaching activities—for most it was the last week of school). In the afternoon, we went to the special activity for the College of Agriculture graduates. I was asked to speak there; of that I only say that I am by no means a very competent public speaker. I always come away thinking about what I should have said but did not. Anyway, it was good to meet old friends and classmates once again, and feel the thrill of the academic life of my youth once more. I have always loved Logan."[65]

On Sunday Royce and Pearl went to the Salt Lake Tabernacle to attend the weekly program of the Mormon Tabernacle Choir, and visited with relatives in Salt Lake before making the long drive back to California, where the family continued to gather. Larry Fulton, hearing that Royce was in town, dropped by from Watsonville with three crates of strawberries, which added to the celebration. That Sunday, on the eve of their return to Chile, Royce wrote, "We went to church in our ward and quite enjoyed it, although to come home in mid-mission isn't the wisest thing to do. I'm glad we went only once to the ward. I don't enjoy explaining why we were there over and over again."[66] Royce and Pearl were packed and ready to go by the following morning.

Completing the Mission

On June 14, 1993, Ann and Ray drove Royce and Pearl to the airport to board the first of their flights back to Chile, where they arrived the following day. "After a bad night of virtually no sleep, we arrived in Santiago about 8:30 a.m. where we were met by José Godoy, who was waiting for us. He is a sad man, having lost his wife on the 29th of May, right after we left. There is nothing to say that would assuage his grief. We had no trouble with entry, and found the house fairly clean."[67] Winter cold had set in at San Felipe, and there was frost on the car windows the following day when Royce, Pearl, and José Godoy departed for a Benson Institute meeting on the soybean project in Viña del Mar.

A couple of days after their return, Royce went to Almendral and looked over the strawberry plantings which were done while he and Pearl were in the United States. "I would have done it somewhat differently, but their judgment is probably good as mine."[68] Still awaiting planting were the day-neutral plants, which required a later planting date, and here Royce discovered another problem. "We went at 2:00 p.m. to plant the last rows of 'Selva' at Almendral. We found that the plants have sprouted and that many stems around the crowns are rotting badly. We had to go through the whole thing and peel off the rot from around the crowns. It took us until almost dark to clean them up and plant half of the remaining 6 rows in the poly planting. The others have to wait until Monday."[69]

As winter progressed it came time to apply the different mulch treatments which had been planned for the strawberry plots, which had been planted with three types of soil treatments: solarization, fumigation and no treatment (control). Groups of plants with each of these treatments were now subdivided to receive different types of mulching, which included clear polyethylene, smoke-grey polyethylene, and control plots with no polyethylene. Royce began to take careful temperature measurements in the test and control plots, and before long began to note differences in performance based on the various treatments. "We have quite a weed differential, much less to hardly any in the fumigated and the solar than in the control plots with no treatment, and only clear poly. It is particularly so in the solar. The smoke-grey works out fine for weed control, and it apparently blocks light about as well as white/black or black does. I don't think the plants grow any better or as well as no poly. The most interesting thing is the solar treatment. 9 weeks in Dec-Jan-Feb-March appears to be adequate. We certainly reached some high temperatures."[70]

In addition to Royce's agricultural work, Royce and Pearl continued to participate in the local branch of the Church, and on a cold winter morning in July they attended a conference which proved to be an important event for the members there. "On July 4 the mission president and Elder Mickelsen from the First Quorum of the Seventy came to San Felipe and created the Los Andes Stake, with five wards and two branches. This reflects the near explosive growth of the

Church here. There are baptisms almost every Sunday after Sacrament meeting. There have been as many as seven convert baptisms per week. Some are really quality people, ready to move into leadership positions with minimal training. Many others will require a lot of help before they get moving, and there is more attrition than there should be. Part of that is due to the travel problems associated with active membership."[71] Reflecting on his and Pearl's own role in the local Church, Royce continued, "Our regular church role here fits in quite well. We add a touch of stability where it is really of value. Mostly we are there on time, regularly, ready to extend a helping hand where it is needed, particularly to the new members who are quite lost when they first attempt to be 'active' members. It isn't easy for most of them."[72]

In August Royce and Pearl had completed twelve months of missionary service, and shortly after, they traveled south again to Temuco, where they looked over some of the Benson Institute projects in that area, including a strawberry nursery. While there, Royce had the opportunity to review his program with the nationwide director. Later that month, as Lito Magnere prepared to travel to Utah to report to the Institute, Royce met with him again. "We went over the things I brought for him to use in his presentation on solar treatment. I'll be writing this up for publication in the Benson Institute Review. I have more than enough for a fairly detailed preliminary report."[73] Evidently the presentation went well, for some time after Lito's return, Royce wrote, "I received a phone call from Lito asking about a visit for next Tuesday with other Benson folks. Apparently the word has gotten out about the initial success of our little solar experiment."[74]

Meanwhile in San Felipe Royce continued with his usual activities, which included the small fruits class, which he now felt very comfortable teaching. "I prepared for class in the morning and all went well with the presentation, probably as good as I have ever done with a class up to date. I presented the material on day-neutrals using slides, and they are good, speaking to the point without too much redundancy."[75] At the same time he continued collecting data on the strawberries, and in this new setting he investigated some aspects of the plantings which had not been part of his work in California. "I went into the field with Pearl and we measured every plant. I should have done this sort of thing out at Watsonville, and it would now be part of the literature, but I did not. It wasn't hard to do here, and the temperature data are good."[76] As the experiments progressed, they generated increased interest at the university, particularly when Royce took his small fruits class and the dean of Agronomy to observe the plantings on the same day. "Twenty students showed up for my class today, and went to Almendral where we reviewed the status of the strawberry planting experiment. So far have observed runners on Fern only, which received two extra weeks of cold storage, since it was harvested first. I then took Gamboa out, and I know he has never seen anything like this. He seems to be very pleased with what I have done so far. Aside from being planted too late, it has gone well, a little bit of everything. The solar and fumigation weed control systems have worked very well. I have counts and measurements and we are having them do the same. Plant size differences are very striking, and it will be interesting when some of the differences reverse."[77] By the end of the September the plots were just beginning to set fruit.

With spring came the strawberry harvest, and Royce began making weekly measurements of the fruit produced on the various experimental plots. The ripe strawberries presented a special problem when Royce took his small fruits class to the field again to see the experiments. "They are rather difficult to handle out there, because there are too many. They walk on the beds and pick the fruit even though I tell them not to time after time."[78] When he returned later with the class, he made certain that the harvest data had already been collected. "I broke free early enough to meet with Godoy and harvest the berries. It was the largest crop so far on those 5 rows with 1200 plants. We took off more than 8 kilos this time. Heaviest was on Chandler, then Pajaro, last Fern. The fruit size was best yet too. In the afternoon I took the class out and they worked on weeds."[79] Returning from the plots, Royce compiled the data which had been gathered. "I worked in my office most of the day crunching down data I have collected. Some of it looks pretty good. Even the pitifully meagre bit of harvest data is interesting. I have good data on runners, plant size, weeds and even brix (sugar) readings."[80]

As the months passed, Royce and Pearl became concerned about the renewal of their visas and identification cards, which had caused so many problems the year before. Contacting the mission headquarters, they were assured that the mission secretary would attend to the renewals, and on November 5, they set out with the necessary paperwork for the mission headquarters in Viña del Mar to obtain their visa extensions. This led to yet another adventure, recounted by Royce in his journal: "We left almost as scheduled, a few minutes after 8:00 a.m. The car was a bit 'choppy,' and that developed into a classic case of 'hesitations.' Anyway, as we were passing the city of Quillota, the situation became acute, and the car stopped and would not start again. All in all we were reasonably calm, and I could not help but feel very strongly that all would be made well. Sure enough, within about half an hour a gentleman from Quillota who was traveling the same way as we were perceived correctly that we were stranded and in need of aid. This 'good Samaritan' traveling with wife and four-year-old child, crowded us into his *camioneta* and took us to a garage in Quillota where our needs were taken care of by the owner. He towed us to his repair shop. He found that the problem was very bad plugs and an even worse rotor of the distributor. He replaced both of these, and set the time, etc. The mechanic was excellent and his charges very modest; the equivalent of $50.00 U.S. dollars. He assured us that henceforth the car should run well, and it did, to make a long story short. We completed our trip to Viña del Mar mission home and back. The car runs better that it ever has. I can't but conclude that Providence has once more intervened in our behalf on our mission."[81]

Although Royce and Pearl had once again been rescued through the kindness of strangers, their car trouble was far from over. Just three days later during a trip south it failed again, and Royce and José Godoy had to return to San Felipe by bus. Pearl had not accompanied Royce on this trip, since she had assumed new teaching responsibilities at the Chilean-North American Institute of Culture. The car was in the shop all the next day, and finally two days later Royce wrote, "Happy day, the car was returned to us today. I feel much better now that the car may work for a good while yet. However, I would <u>never</u> buy that car. It excels in nothing."[82]

One Sunday while attending church services, Royce had an unexpected visitor. "I was in church up in front doing the Sacrament when a familiar back view appeared and it turned out to be Dick Converse from Corvallis, Oregon. He had taken the bus up to see us on the way through to home." Just two days later Royce heard from another colleague. "The mail brought a very nice letter from Jim Hancock, who is in Talca for the next 5-6 months. This should afford us ample opportunity to meet with him and to include Dick Converse. This turned out to be one of our better days here."[83] Royce had worked with Dick Converse as part of the Germplasm Repository committee in Oregon, and he had kept in touch with Jim Hancock since he had done graduate work with Royce at U.C. Davis in the 1970s. The opportunity for these three to get together came that November, when meetings were called at a university in Talca, a city a good distance beyond Santiago to the south, and Royce drove there with Pearl and José Godoy. "The meeting at Talca was intended to get all the 'Gringos' in Chile working on small fruits together in an attempt to inform and coordinate. My former student, Jim Hancock (there with his wife Ann) was the co-leader. After our meeting we drove down to Cauquenes to look at the *F. chiloensis* there under the care of J. Arturo Lavian Acevedo at the substation there. We were able to authenticate the presence of the thick-leafed cultivated types, white and red fruited, among others. I want to get some of these for the Corvallis Collection."[84] The *Fragaria chiloensis* plants which Royce spoke of were of special interest to him and his colleagues, for they were not the now standard strawberries used in their breeding programs (known as *Fragaria x ananassa*), but rather one of their early ancestors, the native species which had been bred and cultivated by native Chileans for centuries, and which were still found in cultivation in some remote areas.

As November gave way to December the days in San Felipe became hot, and on one particular Friday Royce spent the entire day in his office at the university to escape the heat. He returned to his office after dinner, and this led to still another misadventure. "This turned out to be one of those days to remember. When I went into my office at 6:00 p.m. I had no idea they would be locking the building right away. This had to do with election day tomorrow. Anyway, when I went to leave the facilities at about 6:15 p.m., I found myself locked in with no way to get help. I called loudly at the hall door and no one heard. I was able to get outside, but was closed in by a three-meter wall. There I put the two Spanish chairs and climbed to the top of the wall, and held myself up far enough to see where I was, but no one was inside except two dogs who barked loudly. Someone heard my calls for help, and eventually hunted up a person who could get the keys and open the two doors, and freed me from my predicament. Pearl, unbeknownst to her, directed them to where someone with the keys could be encountered. I know now why the U.S. has laws against having doors that cannot be opened from the inside on all public buildings. I got free about one hour after my ordeal started. All I have to show for it is a pair of scraped arms and some stiff legs and arms. All is well, and I could have fallen so very easily with broken this and that. Can you picture a 75-year-old (lacking about 10 days), scaling a 9-foot plus wall? I hope never to repeat that experience or anything even remotely like it."[85]

As Christmas approached, Royce completed the strawberry harvest in the experimental plots with Pearl's help, and set to work analyzing the harvest data. With summer vacation drawing near, he also taught the last of his small fruits classes, and administered the final examination. "Fortunately everyone was present, so I don't have that to contend with, only some of the most wide-open cheating I have ever observed. I don't think that it made a lot of difference; in fact, most were hurt more than helped. There are 21 in the class. I went right to work, and by bedtime, I had half the corrections made. The exam is better than I thought it was."[86]

Shortly after completing the small fruits class, Royce received a letter from his friend Günter Staudt of Germany, asking him to investigate a rare colony of wild strawberries in the far south. "He wants me to go to Chiloé to look at the Chepu *F. chiloensis*. I would like very much to do so, if I can generate a bit of courage in driving the Lada, or getting a rental car that I feel good about, and if I can get Godoy to go with me. Most of our problems have had to do with the Lada. A series of road failures has been enough to make me hesitate more than a little. It has been very dangerous."[87] The next week as if on cue, the car once again broke down. "In the afternoon I took the car to COPEC to get gas, and had difficulty starting it, and more keeping it going. I drove over to Torconal Auto (where it came from) to try to get it repaired as soon as I realized it is impossible to keep it going. What a pain. This makes me very discouraged. Spare me from another 'Lada' forever. After returning to my office, Godoy came looking for me. We decided that if possible, we would drive to the south (Chiloé) next week."[88] It was not until Christmas eve that the repairs on the car were finished, and even then the problems continued. Royce wrote, "I did finally get the car back at noon today, but the speedometer wasn't working until Godoy took it over to have it fixed, and the starter was intermittent when tried, but I didn't do anything about it since it was too late."[89]

The Christmas season was a time of reflection for Royce, as it meant that the service mission was nearly over, and it was necessary to start looking to the future. Royce had always enjoyed good health, but now signs of aging were beginning to appear. "I must start thinking of what I can do when we return. I wonder how long I shall live. Surely Pearl will live longer than I do. My hand is as steady as it gets these days right now. That is a good sign. My hands have had a distinct tremor most of the time recently. It is nice to have them steady again, if only for the time being. I can't complain. I only hope that when the ghost of death calls for me, the voice will be gentle and my response one of welcome."[90] The tremor in his hands, particularly on the right, had been present since before his retirement, but it had become more accentuated during his mission to the point that he had begun to limit his writing, and he found that he could operate better with a pencil than a pen.

Two days after Christmas was Royce's birthday, and his trip southward to investigate both wild and cultivated *Fragaria chiloensis* strawberries began. "My 75th birthday, and Pearl and I left San Felipe with José Godoy at 8:00 a.m. heading for Concepción, where we spent the night with Hugo Barrales Pizarro, Department of Botany at the University of Concepción, and his wife. We didn't give them much warning, but the evening was pleasant, and we enjoyed a good meal, although their dog was generous with his fleas and I am well bitten. We were late getting to bed (1:00 a.m.). The next morning we left early and drove first to Punén where we went up high to 'Manzanal Alto' where I have been several times before, but this is the first time I have been there when the fruit crop (white-pink Chilean) has been on full. I hope that my photos come out well—fruit, plants and flowers. We picked up a kilo of fruit, paying for it. There is a mix in the plants with about 2-5% pink-red type. We have observed the mix before. The white-pinks are the best of the cultivated F. chiloensis, with a rich aromatic odor that most people like. I would like to run a backcross program to *Fragaria chiloensis* to improve them, but retain most of the good 'stuff'. We left them and drove all the way to Puerto Montt where we spent the night in a very nice Cabana. It was late and we didn't get to eat till midnight."[91]

The following day they arrived at their destination, though not without some difficulty. "We left the cabin early in the morning and caught the ferry over to Ancud, Chiloé, where we had breakfast and then drove into Chepu, about 20 kilometers in, where we got some help from a native, then walked in to the beach. It was very steep, impossible to drive. I thought that by going north as advised, we would easily find the berries. Not so. We had to stop at a river, not really feasible to cross by foot, and by that time Pearl's knee was giving her painful trouble, so we turned around and climbed the steep cliff back to where we had left the car. Godoy wisely walked back ahead of us, where he arranged for two horses and the boys to guide us in to where the berries were, one 11 and one 16 years old. It turned out that with Pearl we had actually walked to near the berries, just around this cliff after crossing the river on horseback, which was relatively easy. The berries aren't in great shape, but are surviving. They are all hermaphrodite with thick glossy leaves similar to those at Cape Mendocino and Oso Flaco in California, except they are all outspreading compared with the California appressed. Also these had 5-8 petals compared with the normal five in California. The fruit size is relatively large, and certainly selection could likely do the rest. They have a rich aroma, and also are firmer than the white-pinks I mentioned. (Incidentally, the reds mixed with the whites at Purén are also firmer than the whites.) I collected fruit, leaves and runner plants for inclusion in the materials from Chile. These are of particular value, and are in some ways unique. Of one thing I am certain—these plants are native there and were not moved there by man. This was a very successful trip, thanks to the boys and horses. As I have noted before, it is handy to be able to ride. I enjoy horses."[92] The long drive home proved relatively uneventful. "We did not even have a car breakdown. The 'Lada' chugged along in a respectable manner, and we only had one flat tire, which the car had the decency to present us with right in front of a repair place on Ruta 5, at Rancagua. I had to buy a tire, but we even got a good deal on that. This trip will always be one of my happy memories."[93]

After their return from Chiloé, Pearl went back to teaching English, while Royce spent his days analyzing the data they had collected from the strawberry experiments. The car meanwhile remained a constant worry, for not only did it have endless mechanical problems, it also continued to be borrowed for other uses and not returned as promised, which often left Royce without expected transportation. When once more the car was borrowed and did not reappear after four days, with little time in the country remaining, Royce decided to relinquish the car for the remainder of the mission. As his time in San Felipe drew to a close, Royce hurried to complete his analysis of the data he had collected on the strawberry plantings, and did some final consulting with colleagues whom he had seen while in Chile. He and Pearl took one last bus trip to see Vilma Villagrán, who had asked Royce to help identify a problem which had developed in some of the strawberries. Shortly after, he and Pearl were visited by the Hancocks, and Royce provided Jim Hancock with some materials for an upcoming professional talk.[94]

During the final weeks of their mission Royce and Pearl spoke in Sacrament meeting, and as summer vacation began in February, they made one last trip to the southern city of Temuco for a nationwide youth conference for members of the Church, including a trip to the shore of Lake Villarrica, which brought back to Royce a recollection of his first visits to that country. "The view of Volcán Villarrica was really beautiful. It was puffing out smoke on a very regular basis, and there were no clouds, just the same as when I first saw it over 25 years ago." Royce had been asked to give a talk to the young people, and on the final day of the conference, he wrote, "I had breakfast, then went to conference at the chapel, within walking distance of the hotel. I was requested to talk to them. I even bought a tie for the occasion. I kept my remarks to the personal level, starting with the chain of events that has followed my call to be a Spanish-American missionary in 1939."[95]

Back in San Felipe, Royce continued to compile and analyze the data from research he had headed, and his urgency to get the work done was coupled with uncertainty about its true value. "I spent most of the day in my office finishing the last table for the solar write-up. There is a total of eight tables now. If I can crank up José Godoy, we should be able to finish this and the write-up. This is more easily said than done, because José is busy with the grape harvest right now, and who knows when that will be done. It won't be long before that will grind to a halt with everything else for us. I really don't know how this operation measures up. It is difficult to judge. It could have been quite different had I had a car in my hands all the time that was reliable. However, that too may have been my responsibility."[96]

During the last week Royce and Pearl had their last visits with colleagues and Church members. "The Relief Society had a farewell get-together for Pearl to which I was also invited. These folks have been good to us, and we shall miss them. We can learn a great deal from them concerning reasonable expectations. We always ask too much, particularly in areas of no great consequence."[97] The last Sunday, Royce and Pearl were asked to give short farewell talks in church, and they took pictures with the members. They had one last dinner with their university colleagues, and on February 23, 1994, did their final packing and headed for the airport. Recalling their previous experience attempting to leave the country, Royce observed, "We boarded the plane on schedule. We cannot re-enter Chile for six months. Three months is normal—the restrictions come because we have not yet received our new visas. Ours is in process after the old one ran out on December 3, 1993. It is incredible the way they drag everything out."[98] Royce and Pearl made the overnight series of flights back to Sacramento, where they were met at the airport by John and Betty and their family. Pearl, who had been suffering from a broken tooth, wasted no time in getting to the dentist and having it repaired that same day.[99]

Royce and Pearl's mission to Chile was a landmark event in their lives. It had subjected them at an advancing age to challenges and circumstances which would have daunted much younger individuals, and which helped to put their own lives into perspective. For Pearl, it afforded the opportunity to at last gain mastery of the Spanish language which had eluded her on her first stay in Chile. For Royce, it presented a final opportunity to perform both agricultural research and botanical field investigation in a meaningful way, and its conclusion in that sense brought an end to his work as a research scientist. It brought a similar end to his journal, for with increasingly tremulous hands, he would not write in it again beyond the first few days home. For both of them, the mission presented an exercise in service to mankind, and an affirmation of the value of that service.[100]

Life After the Mission

Back in Davis at last, Royce and Pearl set to work reconnecting themselves to life at home. The morning after their arrival they went on campus to attend their exercise class, but found that its numbers had been diminished by the poor health of some members and the death of others. Their morning exercise routine continued where they had left off, except that their morning run had turned into a brisk walk. That Sunday they attended Church in Davis and greeted their old friends in the ward, then attended the Sacrament meeting of the

Spanish Branch in Woodland, where both had been invited to speak by their son John, who was then serving as branch president. Of this, Royce wrote, "I gave one of my worst talks ever. I just did not have it today. Pearl's talk was OK. We also sang in between, and that was good."[101]

Royce was naturally interested in the progress of the University of California strawberry program, and the following week he had the opportunity to meet with his successor, Douglas Shaw. His long absence in Chile had been good for their sometimes tense relationship—it had brought a sense of finality to Royce's retirement from the program, and it had allowed Shaw an interval to carry it forward on his own terms. Their meeting was an amicable one. "I visited with Doug Shaw in my office, where he brought me up to date on what the new varieties are doing. Camarosa is doing well in Southern California, and has a chance of competing with Chandler there. It was nice of Shaw to do this. I think that what differences we had are being set aside."[102]

That Saturday Royce and Pearl drove to the Oakland Temple, where after attending a couple of endowment sessions they met with a member of the temple presidency, who called them and set them apart as temple ordinance workers for Saturdays, the same position they had held before their mission.[103] The following week they resumed their weekly early morning trips to the temple, a practice they would continue for many years, with the additional benefit that Pearl was now also able to help with the ordinances in Spanish. The temple assignment continued to be a spiritually fortifying experience for Royce, who had by now overcome most of his former misgivings about spiritual things. His work in the temple and the mission in Chile had solidified his faith, and had largely enabled him to reconcile it with his scientific convictions, though he acknowledged that there were some for whom this was still a challenge. Writing home from his mission about a dying colleague who had lost his faith in an afterlife, Royce had remarked, "I don't quite understand how so many people in my professional area end up that way. I may not think about things the way many of our Church people do, but I have a really strong belief in the life hereafter and have no fear of death because I can view it as part of a living process, and when we accepted earthly life, we accepted death with one being just as acceptable and natural as the other."[104]

Two Sundays after their arrival home a stake conference was held, and Royce and Pearl were invited to speak briefly to the large congregation about their mission experience. A couple of days later they attended an interview with the stake president which was a requirement for all returned missionaries. "At 8:30 p.m. we met with Stake

President Sheldon Smith to report on our mission. It was a very good experience. We discussed all aspects of it, good and not so good, and Pearl had the opportunity to talk and ask questions about something that has long bothered her. That went well too. I hope that she can put that aside now. He gave her about the same advice that I have, 'forgive and forget.' This has troubled her a great deal, and I should have been of more help to her on it."[105] Pearl's difficulty had to do with a misdeed by another Church member during their early days in Davis which she had recently learned had been unjustly attributed to her, a situation over which she felt great indignation, and which in the coming years would occasionally provoke outbursts of anger in Pearl, to which Royce responded as a calming influence.[106] Royce and Pearl were extended callings as ward missionaries on their return to Davis, and they began to attend local missionary meetings and to invite the full-time missionaries over for dinner regularly, a practice they would continue for years, even after their church calling changed.

A few days after their interview with the stake president, Royce took Pearl on a drive to the strawberry fields at Watsonville, where Royce was at last able to cast a critical eye over the new strawberry releases. "We visited Kuni's place first. He has planted virtually all Seascape. The best experimental plantings are 'Cuesta' (day-neutral) and 'Camarosa' (also very good in Southern California). 'Laguna' and 'Carlsbad' look fair, but are too light-colored. 'Sunset' is too soft, and 'Anaheim' doesn't have much going for it. We went out to lunch with Kuni (the host, as usual), John Jarse, and Bob from Lassen Canyon, as well as Larry Fulton. We then went back to the plots, and picked up our dog Sally who has been with Larry since we left for Chile."[107] Sally, the energetic "Australian Shepherd"[108] which Royce had acquired as a puppy from Larry, would accompany Royce and Pearl on their morning walks for many years, and kept company with an elegant long-haired grey cat named Blue. As he aged, Royce, who had kept pets since his boyhood, would continue to cherish these animals and to enjoy their singular personalities.

Though by now well removed from the strawberry program, Royce continued making visits to the various growers, nurseries and field stations, where he was received as a welcome guest. Larry Fulton continued to provide crates of experimental strawberries during harvest season which Royce and Pearl would distribute to friends and neighbors as in times past, but Royce's role on these visits was now mostly an honorary one. He made appearances at the annual field days at the experimental stations, and at these events he had occasional contact with Victor Voth, who like Royce found it difficult to stay away from the strawberries, though their work was largely finished, and the industry was moving forward without them. Attending with Pearl an event sponsored by packing company in Fresno, Royce found that he no longer knew most of the younger generation of growers there.[109]

Though he had little to do with the California strawberry program in an official capacity, Royce continued to be in demand as a speaker and consultant. While still on the mission to Chile, he had been invited to an international symposium on strawberries to be held in Portugal, and in April, 1994, two months after his return to Davis, he traveled there with Pearl and presented a talk entitled, "Origin and Characteristics of University of California Strawberry Varieties."[110] In February, 1995 he attended strawberry meetings in Orlando, Florida, where he again met up with his former student Jim Hancock, and with strawberry colleagues he had met in Portugal, Argentina, and New Zealand. In April of the same year, he was invited by a radio station in Santa Cruz to be interviewed on the air about the strawberry industry, and in May Royce and Pearl drove up to Corvallis, Oregon, to look over the strawberry plants in the Germplasm Repository and to visit with the Converses. In October he and Pearl were again invited to Chile, where over a span of ten days Royce visited strawberry plantings, consulted with his old friend Vilma Villagrán, and presented a paper to Chilean strawberry experts. Before returning home, he and Pearl spent a Sunday in San Felipe, attending church and visiting with old friends and associates. [111]

By this time Royce had become increasingly frustrated by a persistent tremor, which chiefly affected his right hand. This had been present since before his retirement, when it had been little more than a minor nuisance, but in Chile it had progressed to the point of making it difficult to write, though it was still only intermittent and had remained manageably subtle. As the years passed it had become severe and almost constant, and began more and more to interfere with basic activities such as eating and drinking. Royce's son John, now a physician, had long been convinced that Royce was suffering from Parkinson's disease,[112] and at last in early 1996 Royce was referred by his own doctor to a neurologist who confirmed the diagnosis and started Royce on one of the standard medications for the disease. He would continue to take medications intermittently for the disorder for the remainder of his life, reckoning with both the effects of the disease and the side effects of the medications used to treat it. As his disease progressed, scientific evidence began to mount that exposure to agricultural pesticides increased the risk of Parkinson's disease, and Royce became convinced that this was likely the case with him. Recalling his years in the fields as a boy, a student, and a professor, he acknowledged resignedly, "We practically bathed in the stuff."[113]

Royce remained as active as he could, in spite of the Parkinson's

disease. He continued to travel to the strawberry fields and experimental stations, and attended industry and university events whenever
the opportunity arose. He made trips with Pearl to the coast, hiking
to view the familiar colonies of wild strawberries which he had now
been observing for decades. In the summer of 1995 he took a back-
packing trip to the high sierras with John and his two oldest sons,
where they fished for trout in the high alpine lakes, a reminder of
similar trips in years past, and in September of 1997 he and Pearl accompanied John and Betty and their children to the north coast for a
camping trip, which they enjoyed immensely, though it would prove
to be their last.[114] Royce and Pearl also continued to make the yearly
drive each summer to Salt Lake City to attend family reunions—for
Pearl's large extended family these were well-planned events, but for
Royce's smaller and now diminishing family they amounted to a
spontaneous dinner engagement whenever Royce and Pearl were in
town. They traveled weekly to the Oakland Temple to serve as ordinance workers, and Royce frequently officiated at the temple sessions,
which with his long experience he was now able to perform expertly,
despite his tremor which was now obvious to all. Royce and Pearl
during this time also received a new church assignment, taking a shift
weekly at the Family History Center which was a part of a newly-
constructed stake center in Woodland.[115] This assignment was well
suited to their physical limitations, and they continued to serve in it
faithfully for years.

At the same time, Royce's advancing age and persistent tremor
meant that his ability to participate in work of a professional nature
was rapidly drawing to a close. In March of 1997 he was invited once
more to travel alone to Egypt, where he was to speak and observe the
development of the strawberry industry there, but he lacked his former vigor, and the two-week trip left him exhausted. In July that
same year, Royce and Pearl drove to Utah for horticultural meetings
which were scheduled in Salt Lake City for July 24, which coincided
with Utah's Pioneer Day, and which that year happened to mark the
150th anniversary of the arrival of the Latter-day Saint pioneers in the
Salt Lake Valley. Scheduled to speak that afternoon, Royce missed his
session due to a miscommunication, and had to be rescheduled for an
evening session the same day. While statewide Pioneer Day celebra-
tions were underway outside, Royce, perhaps appropriately to the
occasion, related to his audience the circumstances of his discovery of
the day-neutral strawberry, a landmark event for the California strawberry breeding program, which had taken place in Brighton, Utah,
only a few miles distant from where he was speaking. Afterward
Royce and Pearl once again visited with Royce's former student Jim
Hancock, who had already become a respected fruit breeder in his

own right, and with Gene Galletta, a long-time friend and colleague who ran a highly acclaimed USDA strawberry breeding program in Beltsville, Maryland. The following day Royce and Pearl traveled to Logan, where they attended the fifty-year reunion of Royce's graduating class.[116]

In late 1997, two of Royce and Pearl's grandsons received simultaneous calls to serve as full-time missionaries for the Church, a significant occasion for Royce, whose missionary service in his youth had been one of the key events of his life. On November 8 he and Pearl flew to Las Vegas to hear the farewell talk of Jean's son Jonathan Anderson, who had been called on a mission to northern California, not far from where Royce and Pearl lived. The following week, on Sunday, November 16, Royce and Pearl drove to Oroville to hear the farewell talk of Marla's son Samuel Vaughn, who was departing to serve a mission in northern Italy, the same region over which Royce had flown during the war. After dinner with the family in Oroville, Royce and Pearl bade farewell and made an early start for Davis so they could get there before dark.

Their trip home from Oroville proved to be a harrowing adventure, and something of a turning point in Royce's life. Royce and Pearl had recently purchased a new car, and had decided on a Cadillac which, though more modest in appearance than most cars of that famous make, was still equipped with a broad array of dashboard controls which were unfamiliar to Royce. As they left Oroville to return to Davis, Royce made a wrong turn and headed northward, and continued on despite Pearl's vehement protests that they were headed in the wrong direction. Royce continued undeterred up into the mountains, and it was not until night was falling and they found themselves amid lakeside campgrounds where signs warned of the need for chains or snow tires ahead, that Royce was finally persuaded to turn the car around and head down the mountain. The trip home was made at night in the pouring rain in the still unfamiliar car, and as he drove southward toward Davis Royce inadvertently activated the high-beam headlights, and amid the jumble of controls on the new Cadillac was unable to find how to lower them, to the consternation of oncoming motorists who repeatedly flashed their own headlights at Royce and Pearl as they plodded on through the driving rain. At length they arrived safely home to the great relief of Pearl, but the episode had left both of them shaken.[117]

The next day Royce was still feeling abashed and perplexed about the experience, particularly his inability to solve the headlight problem the night before, when John stopped by with his son Benjamin to borrow some scrap lumber for a Boy Scout project. Royce, still feeling unsettled about his adventure the previous day, insisted that the

Cadillac be moved from the garage to the driveway, and that John and Benjamin work from a ladder, although they could reach the wood easily enough simply by standing on the sturdy workbench on one side of the garage. To humor Royce, John obediently set up the ladder, and from this new vantage point he discovered a series of cardboard boxes which had been stored in the loft and forgotten years before. Lowering the boxes, he discovered that they contained letters written by Royce and Pearl in years long past, items which had long been presumed lost. One last wooden box contained journals Royce had kept in his boyhood, and during his mission and military service. The discovery of these missing records brought an immediate end to Royce's bewilderment, as he realized that it was his misadventure with the car which had led unwittingly to the recovery of these treasures from the past. In the coming days he would spend countless hours pouring over the journals, and after decades of silence he at last became willing to open up to his family about his war experiences. Pearl for her part began typing his war journal entries into a newly acquired computer.

By this time, Royce's physical decline was becoming apparent to all, despite his efforts to adapt to it. Royce continued to drive despite his handicap, and he and Pearl developed a kind of eccentric driving partnership in which Royce would continue behind the wheel, while Pearl, who had superior vision but was more nervous about driving, would direct him where to go, warn him of upcoming signals and obstacles, and scold him whenever the process led to a missed turn or a traffic hazard. Although they continued their weekly drives to the temple, in January of 1988 they had their scheduled arrival time changed so that they could leave home later—at 5:00 a.m. rather than the customary 3:00 a.m. Royce increasingly complained of poor vision, and in May of that year he underwent surgery to correct the problem, which resulted in some improvement, but it soon became apparent that his vision was still not what it had once been, and other problems had begun to develop. Royce started having episodes of confusion, and in July of 1998 Pearl was approached by other workers at the temple who were concerned about his diminishing performance there. Royce and Pearl at last made the decision to seek a release from their temple assignment, and they received their official release as ordinance workers in the Oakland Temple on July 27, 1998, having served there faithfully for a full eighteen years, excepting the eighteen month leave of absence for their mission in Chile.

The Final Years

The rapid physical decline which led to Royce's release from his temple calling reversed itself in large measure when his neurologist

539

discontinued his Parkinson's medication the following week, but the disease itself continued to run its inexorable course. Combined with the usual problems attendant to age, it would begin over the ensuing years to deprive Royce little by little of the vitality and the basic physical tools which had made him the man he was, and had enabled him to excel in his field. His hearing, already damaged in the war, would eventually worsen to the point that ordinary conversation and even amplified voices became hard to discern. His once sharp vision would continue gradually to fade until he found even simple reading difficult. His hand tremor had long made writing out of the question, and now his formerly brisk gait had slowed to a stiff shuffle. His face, once vibrant with wisdom and wit, would begin increasingly to assume the dull, expressionless stare typical of those with the disease, to which the usual array of human emotion could only be applied momentarily, and with great effort. Even with his intellectual powers still largely intact, in the grip of his illness Royce would gradually begin to fade into a kind of sensory fog which left him unable to see, hear, read, or act, but only to sit and think for long interminable hours, until toward the end of his life even the power of reason would at length give way and begin to fade.

These changes, however, were so gradual as to be almost imperceptible from day to day, and life continued forward as before. In October of 1998 Royce ordained his grandson William Dobbins (Margaret's oldest son) to the Aaronic Priesthood, for which he required the assistance of his son John; a month later, he ordained John himself as a high priest in the Melchizedek Priesthood, again requiring assistance, this time from the stake president.[118] The next day, November 2, 1998, he and Pearl departed for a tour of the Holy Land with a group of Latter-day Saints, composed chiefly of friends from their work at the Oakland Temple. They spent two weeks traveling from Jerusalem to Galilee to the Dead Sea, visiting Biblical and historical sites as they went. In Jerusalem they were allowed inside the Dome of the Rock, where Pearl was required to dress in a long robe and wear a head covering. Royce participated in a swim in the Dead Sea, where he was so buoyant that he did not have the strength to sink his feet to the bottom when it came time to leave, and required assistance to exit the water. Before returning home, the two crossed over into Jordan, where they visited the ancient city of Petra and the modern city of Amman.[119] That December, two days after Christmas Royce commemorated his eightieth birthday, and on December 29, 1998, the extended family gathered in the home of John in Woodland for a joint celebration for both Royce and Pearl, whose birthdays fell within a few days of each other.

As time wore on, travel became more difficult for Royce. Twice in succession he fell in airports while attempting to negotiate escalators with his luggage, and though he and Pearl began using alternative routes to avoid the escalators, the experience left Royce nervous about flying.[120] He and Pearl increasingly had difficulty managing their complex financial affairs, and as a precaution gave their son John increased authority to act on their behalf. At the same time, both felt an obligation to be generous with the money Royce's work had brought them. Royce continued to make substantial contributions to the two universities where he had been trained, to Brigham Young University, and particularly to UC Davis, to which he occasionally made extra payments. He and Pearl agreed to support one of their grandchildren on a mission and another at college, and they provided help to their adult children and their families as financial need arose.[121] They also decided to begin making joint contributions to a variety of different charities, a practice which within a short time resulted in a veritable flood of mailed requests for donations, to which in future years it would fall to Pearl to respond. Pearl, who had experienced no small amount of material want in her early life, felt an obligation to donate to these charitable organizations whenever she received their requests by mail, and continued to do so even when mentally and emotionally exhausted by the effort. As Royce's condition worsened, Pearl increasingly had to take responsibility for financial affairs, a need which became especially obvious after an incident in which Royce, in a moment of confusion, wrote a check for tithing which was for a much greater sum than he had intended or could afford.[122] Although Pearl managed the finances from that time forward, Royce continued to carry the money and make payment for purchases.

Despite the steady decline in his energy and health, Royce continued to make occasional appearances at professional events. In January of 1999, he was invited to the annual Small Fruits Conference in Oregon, along with some of the other early authorities in the field, including Victor Voth. Accompanied by a UC Davis plant pathologist named Doug Gubler, a member of the Davis Ward who with Royce's encouragement had done professional work on strawberry diseases, Royce made the flight to Oregon and delivered a talk at this conference.[123] He continued to make his way to the Watsonville plots, sometimes riding with others and sometimes driving himself with the vigorous assistance of Pearl. Yearning to see the wild strawberry colonies, he made a drive with Pearl to the north coast "just to see if he could do it," allowing Pearl to drive part of the way home.[124] In June of 1999, Royce received a group of Chilean visitors who had been brought by Doug Gubler to discuss strawberries with him, and in November of that year, he and Pearl drove to Watsonville to attend an event for the retirement of Herb Baum, one of the most outspoken leaders of the strawberry industry, and there, a full ten years after his retirement, Royce once again had the opportunity to meet with his old colleague Victor Voth, though by then it was challenging for him and Pearl even to get to and from the event. In 2001, Herb Baum contacted Royce for information for a book he was writing on the strawberry industry,[125] and the next week Royce and Pearl again made the drive to Watsonville for the retirement of Larry Fulton, the director of the Watsonville field station. Once again they had difficulty making the trip, and Royce suffering a minor collision en route.[126]

As Royce's health continued to decline, he and Pearl tried to maintain their usual activities, though of necessity these became more limited. They continued to man their Tuesday evening shifts in the Family History Center in Woodland, and they began meeting regularly with a dinner group composed of older members of the Davis Ward, people who in spite of occasional old differences had grown dear to them through long association in good times and bad. They also continued their daily exercise routine, taking their daily walk with the dog Sally, but Royce was walking more slowly now, and often would cut the walks short, particularly when construction work on the university campus made portions of their original walking route impassible. His driving grew worse, and was limited now to trips around town or to the Family History Center, with occasional drives to Lake Berryessa in the hills behind Winters, near where the Wolfskill Experimental Orchard was located. By now Royce's difficulty with driving was becoming more obvious, and in May of 2001 he underwent cataract surgery of his right eye in an effort to improve his fading vision, but there was little real improvement, and in August of that year Royce was finally persuaded to give up driving entirely and leave it to Pearl.[127]

The inability to drive came as a difficult blow to Royce. He was painfully aware of his own physical and mental decline, and of his increasing dependence after a life of activity and vigor. He had already had to discontinue his work in the temple which had brought such meaning and purpose into his life, and to give up driving as well seemed to be to relinquish his last measure of independence. Habitually of a cheerful temperament and never prone to depression, Royce nonetheless became deeply discouraged at this time by his incapacities, though his outward attitude remained resigned and uncomplaining. This was, after all, a time of happy anticipation of important family events, for two of Marla's children were to be wed in temples that summer, and John's two oldest children had just received mission calls, and were to begin their service in October.[128] These occasions were important ones for Royce, but the bustle of preparation which surrounded them accentuated his own inability to help, and left him feeling alone and useless.[129]

On August 17, 2001, Royce and Pearl departed on an early flight to Salt Lake City for a week-long trip which was scheduled to coincide with one of the two temple marriages, but there were other events planned as well. They first attended a reunion of the Davidson family, which at last had begun to diminish in numbers as Pearl's oldest siblings reached the end of their long lives, and they went to church on Sunday with Pearl's sister Helen. On Monday, August 20, they traveled to Brigham Young University in Provo, where they had been invited to meet with Don Norton, a BYU professor of English who was engaged in making a collection of oral histories of Latter-day Saint veterans of World War II, who were by then a rapidly vanishing generation.[130] Royce and Pearl spent several hours with Dr. Norton discussing Royce's war experiences, which Royce still found so emotionally taxing after nearly sixty years that he wept openly.[131] Pearl had brought Royce's war diary which was copied for the collection, and she later sent in a typed copy. Four days after the interview Royce and Pearl were taken to the Salt Lake Temple, where they were present for the sealing of their granddaughter Katie Vaughn to her husband Cameron Richie.

Back in California the significant family events continued to unfold. On Saturday, September 8, 2001 Royce and Pearl rode with John and Betty to the Oakland Temple, where their grandchildren Sarah and Benjamin Bringhurst were to receive their respective endowments in anticipation of their upcoming missionary service. This day proved an especially bright spot for Royce and Pearl at a difficult time, for their community of temple friends had heard of the occasion, and gave them a warm reception. There were other happy events to look forward to, for a family gathering was to be held the following Saturday at Marla and John's house in Oroville in honor of the two weddings, and most of the family would be there, including Ann, who planned to fly in from Washington to be present.

That Tuesday, on September 11, 2001, Royce and Pearl took their usual early morning walk with Sally, but on arriving home they turned on the television and were stunned by the news that airliners loaded with passengers had slammed into the twin towers of the World Trade Center in New York and the Pentagon building in Washington. With the rest of the country they watched aghast as the massive skyscrapers collapsed on national television, killing thousands. The occasional hijacking of commercial aircraft by zealots or cranks had long been a fact of life, but their use as a weapon of destruction was a new and unthinkable development, and brought upon the entire nation a sense of shock and horror not felt since the Japanese attack on Pearl

Harbor some sixty years before. All commercial flights were immediately cancelled, bringing an unfamiliar quiet to the usually busy skyways, and in Davis dark clouds rolled in overhead, adding to the day's gloom.

This historic turn of events cast its shadow over the much anticipated family events. The grounded flights forced Ann to cancel her plans to visit, and John found himself stranded in Salt Lake City, where he had traveled for another translation project, and it was only with some difficulty that he found a ride home in the back of a crowded van with the family of Cameron Richie, Marla's new son-in-law. That Saturday Royce and Pearl traveled to Marla and John's home in Oroville, where a wedding celebration was held with those of the family who could still attend, and the following day they attended church services in the Woodland chapel where Benjamin and Sarah were scheduled to give their farewell talks before leaving on their missions. By the instruction of Church leaders, the meeting was converted into a memorial service for the week's events, and the two young departing missionaries gave much more subdued talks than planned, on topics appropriate to the occasion. Royce and Pearl were present three weeks later when the two were set apart as missionaries.

In the ensuing weeks and months Royce continued to struggle with his health. He underwent cataract surgery in his left eye, in another effort to improve his failing vision. He continued to perform chores around house and in the yard, but he had become unsteady on his feet, and suffered occasional falls. In the spring of 2002 his mental functioning seemed to worsen rapidly, and at times he experienced episodes of mental confusion, while at other times he seemed to sense other people in the room besides Pearl, occasionally even conversing with them. Aware through his interaction with Pearl that it was impossible for them really to be present, Royce began good-naturedly to refer to these apparitions as "ghosts."[132]

On the morning of April 17, 2002, Royce awoke lucidly enough, but with a fixed determination that he needed to make another large donation to UC Davis, so Pearl drove him to his office for that purpose and wrote out the check for the donation, in accordance with his wishes. That night and the following morning he experienced rapid heart beating and chest pains, and after consulting with their son John, Royce was taken to the hospital in Woodland, where he was admitted with heart disease. He was transferred the following day to another hospital in Sacramento, and on the afternoon of April 19, 2002, he underwent a stenting procedure to open blocked coronary arteries. The procedure went without incident, but during the night Royce, still strongly affected by the anesthesia, awakened as nurses entered during the night to perform their routine duties. Finding him-

self in the dark in a strange environment, he became completely diso-
riented and began to strike out at the nursing staff, and at John and
Pearl who, having remained to spend the night with him, jumped to
their assistance. For Royce it must have seemed like another of his
old war dreams, except that this time there was no awakening—these
were real people present and speaking to him, real hands holding him
down, and real objects painfully piercing his skin, and the familiar
faces of John and Pearl intermingled with those of strangers as he
fought back with remarkable physical strength.

Royce was at length subdued and placed in restraints to prevent
him from harming himself, but at some time in the course of this
miserable night he became convinced that he had died, and supposed
that he had entered into some hellish purgatory where he was subjected to an array of carefully inflicted pains, and deprived of his dignity or his power to act. As day dawned the next morning Royce
calmed and became more cooperative, but he had received so much
medication to quell his outbursts during the night that his discharge
home was delayed until late in the afternoon, and even after arriving
in the familiar confines of his home he remained in a kind of daze,
continuing to believe that he had died the previous day. The next day
was a Sunday, and things were little changed when John came to
check on him after his own church meetings. It was not until John sat
down on an impulse and began to play hymns on the piano that
Royce at last began to return to reality, being brought back to life, as
it were, by the music, for as John continued to play, Royce began fee-
bly to sing, and gradually his singing gained strength until, as his rich
tenor voice began to fill the house, he came to his full senses, and was
at last persuaded that he was still in the land of the living.[133]

The following month Royce and Pearl celebrated their 60th wedding anniversary, and all their children and many grandchildren gathered at their home for a family celebration. Royce had by then fully
recovered from his procedure, and appeared to have been considerably benefited by it, for there was a noticeable improvement in his
physical stamina, and especially in his mental functioning. It was perhaps this new clear-mindedness, or perhaps the memory of having
thought himself dead and subjected to a punishing afterlife, coupled
with the awareness that real death could not be far off, which
prompted Royce to reflect once more upon the secret indulgences of
his past which had impeded his spiritual growth and repeatedly
caused him to feel unworthy of his callings in the Church and in the
Temple. At last that summer he opened up to Pearl about the matter,
and two days later both of them met with the stake president where
Royce unburdened his conscience to him. The president received the
information with unexpected kindness, and gently counseled Royce as

to the actions he should take to make amends. Though this was an issue unknown outside the family and only to a very few within it, to them and to Royce it was of utmost importance, bringing at least some kind of a conclusion to a problem which had troubled him for many years, and leaving him with a genuinely clear conscience for the first time in decades.[134]

In September of 2002, Royce received a telephone call from a Mexican geneticist named Pedro Dávalos,[135] who had visited Royce in California some fifteen years before to consult with him about strawberry breeding and culture.[136] Dávalos, the head of a strawberry breeding program in the state of Guanajuato, Mexico, invited Royce to appear and speak the following month at a strawberry conference to be held there, in the same area where Royce had done his first foreign consulting nearly four decades before. Pearl was terrified at the prospect of making the trip, but was placated when their son John, who spoke fluent Spanish, agreed to go with them. For his part Royce was excited at the prospect, and he began going through professional slides which he had not touched in years, in order to prepare a presentation. On October 23, 2002 the three set out for the Oakland airport to catch an overnight flight to Mexico, and arriving the following morning, Royce, Pearl and John were taken to their hotel in the city of Irapuato, the center of the strawberry industry in the region known as *El Bajío*. Later the same day they were taken by car to tour the very old Spanish colonial city of Guanajuato, which Royce had explored on foot during his early trip back in 1963.

The strawberry conference took place on the morning of October 25, and after a brief introduction Royce rose as first speaker to give the keynote address. On the topic of strawberries he was as lucid as ever, and using his slides as a guide, he outlined his work with the California strawberry program in fluid and cogent Spanish, much as he had in prior years, except that the Parkinson's disease had now so weakened his voice that it began to fatigue as he spoke, growing softer and softer, until by the time he concluded some 45 minutes later it had been reduced to a mere whisper. When he had finished speaking Royce was presented with a plaque for his lifetime achievements, then sat down to rest, and to listen with interest to the remainder of the conference. During their visit Pedro Dávalos drove them out to his experimental strawberry plots, explaining on the way that although the California varieties were being widely grown in the region, he was attempting to hybridize local strawberries with the standard cultivars, in an effort to develop berries better adapted to the region, a practice that Royce was convinced had been the foundation of the successful program in California. The trip to Mexico would be Royce's last foreign travel, and his talk in Spanish at the Irapuato conference would

546

prove to be his last professional presentation.[137]

Royce continued to slow down physically, and it became more of a chore for him and Pearl to perform basic domestic tasks. In February of 2003 their dog Sally, a constant companion on their morning walks for fifteen years, at last grew weak and died, a blow to Royce, who had a fond relationship with the unruly hound. He and Pearl continued to take their morning walks, though Royce kept a much slower pace now. Both looked forward to the frequent visits from family members and friends, whom they would often invite to their favorite eating spots in Davis such as the "Silver Dragon" for Chinese food, or "The Symposium" for fine Greek fare (including Royce's favorite calamari). They still made occasional trips to the Temple, but now they rode with members of the family rather than attempting the drive themselves. Royce also received occasional visits from his professional colleagues. In April of 2003 he was visited by Vilma Villagrán, his old friend and associate from Chile, who brought him a copy of a book she had just written on Strawberry culture.[138] A few days later he invited his old graduate student Hamid Ahmadi for dinner, and Hamid continued to pick up Royce periodically and take him to university functions.

Royce had expressed a desire to make one last trip to Salt Lake City, an idea which Pearl resisted, but once again John agreed to assist them, and in May of 2003 they made the long trip overland at a leisurely pace, stopping in Nevada on the way. They stayed in a fine hotel on Temple Square, visited the Church museum there, then drove to see some of the places where they had spent their early years, including Pearl's old home.[139] The visit brought back many recollections to Royce and Pearl of their young life in the Salt Lake Valley. They enjoyed a dinner with Pearl's surviving brothers and sisters, and the next day had lunch with Royce's brother George and his wife Laverne before starting back for California. The two brothers had always remained close, and that day they had what would prove to be their last talk together, for scarcely a month later word came that George had suffered a debilitating stroke from which he would never recover.[140]

As time passed Royce became more feeble, and he and Pearl increasingly struggled to maintain what had been once been a simple routine. Travel became difficult, and in November of 2003 Royce and Pearl at last gave up their work in the Family History Center, as Pearl no longer felt safe driving at night. That December John's daughter Sarah was scheduled to be married to Tony Familia in the San Diego Temple on Royce's birthday, two days after Christmas, and Pearl and Royce had planned to attend, but as the day approached Pearl grew anxious about Royce's ability to travel, and finally asked John to cancel their tickets. Royce and Pearl attended Sarah and Tony's reception in the chapel in Woodland the day after Christmas, and as the family left for the marriage in San Diego, Royce spent his 85th birthday quietly at home with Pearl, inviting the local sister missionaries to stay and eat with them when they happened to stop by close to dinner time.[141] In their old age the two had become an endearing couple, and the fondness they had for each other was obvious to all. The following week, on the first Fast Sunday of January, 2004, both of them rose to bear their testimonies, each expressing gratitude for their long life together in such tender terms that many ward members approached them afterward to thank them for their remarks.[142]

On a Sunday the following month, Pearl was returning from a ward choir practice when she was startled to find Royce seated on a chair near the kitchen door with a blood-soaked towel wrapped about his head. He explained that in attempting to sit down he had missed the chair and fallen, striking his head on a cupboard. John was again summoned, and drove his father to the emergency room where he spent the evening cleaning and suturing his scalp wound. The wound healed quickly enough, but the fall and head injury had taken their toll on Royce. He began again to suffer bouts of serious confusion, and became more unsteady on his feet. The "ghosts" he had seen before his heart procedure returned, and he began to experience double vision. His welfare became a constant worry for Pearl, and matters became much worse when one day in April Pearl became distracted by Royce while driving and suffered an automobile collision at an intersection, resulting in major damage to the car. Pearl was emotionally shaken by the accident, and remained beside herself with anxiety for weeks after, and she and Royce were left without transportation for an extended period while they waited for the car to be repaired. At this time of need, once more it was John's wife Betty who came to their aid, driving regularly to Davis to look in on them and taking them to do their weekly grocery shopping.

As Royce's condition continued slowly to deteriorate, it became a full-time task for Pearl just to attend to his needs. She and Royce continued to take their early morning walks, but now Royce's movements were so affected by advancing Parkinson's disease that his walking had been reduced to a quick forward-leaning shuffle which was sometimes so rapid that Pearl had difficulty keeping up. Royce also began to have sudden lapses of consciousness, which added to the worry. By the fall of 2004, the demands of Royce's care had become so overwhelming for Pearl that a home service agency was engaged, which sent attendants daily during working hours to provide companionship and assist in performing housework and attending to Royce. With this timely aid a great burden was lifted from Pearl, and life returned to a more manageable routine. She and Royce continued to attend church weekly, and to go out to shop and to attend to events around town, and John and Margaret and their families often came on Monday nights for a family home evening together.

In December of 2004 Pearl decided that a new car was needed, and went with Royce, John, and the attendant on duty to a local Honda dealership, where to everyone's surprise, Royce, who had always preferred conservative cars with subdued colors, selected a bright red Accord. Not long before Christmas Pearl's brother Daniel died, and with John's assistance, Royce and Pearl flew to Utah to attend the funeral, returning the same day. They had a pleasant enough visit with Pearl's family, but it was a difficult day for Royce, and he would never travel again.[143] On Christmas Eve Royce and Pearl had dinner at John and Betty's house, and they spent Christmas Day in 2004 pleasantly at home, where they were visited by all the families of their children who lived nearby. December 27, 2004 marked Royce's 86th birthday, and in a determined show of independence, he succeeded in slipping out the front door unnoticed after breakfast and going for a walk by himself in the rain. Later that day he and Pearl celebrated his birthday with John and Margaret and their families at a local restaurant.[144]

As 2005 dawned, Royce continued to suffer episodes of weakness and confusion. He increasingly had difficulty walking and seemed in constant danger of falling, yet when John and Pearl purchased him an excellent wheeled walker and later a wheelchair, Royce doggedly refused to use either. On May 14 Royce and Pearl celebrated their 63rd wedding anniversary, and once again their children gathered and shared a meal with them, but when Royce went to a routine check-up just a few days later his physician, observing his obvious decline, discontinued all his medications and suggested that Pearl consider Hospice care.[145] Unwilling to face the possibility that Royce was that near to death, Pearl declined the service, continuing instead to shoulder the overwhelming burden of Royce's care as his physical strength and powers of reason slowly slipped away. Finally one day in June Royce complained of chest pain, and a panicky caretaker called emergency services and had Royce transported by ambulance to the hospital, where a painful and fruitless series of tests were undertaken. After this episode Pearl was at last persuaded to accept Hospice care, and workers arrived a couple of days later, providing an adjustable hospital bed and other materials for Royce's physical comfort. Unable to walk safely and unwilling to use a wheelchair, Royce became bedbound.

Long before, in the years following the decline and death of his own mother, Royce had repeatedly expressed his feeling that he would rather die than suffer the loss of his mental faculties and his independence. He felt persuaded that with sufficient mental and physical discipline he would be able to avoid that eventuality in his own life, but he had not reckoned with the slow, relentless ravages of Parkinson's disease. Now his worst fears had been realized, and at some level Royce was cognizant that he had become so physically and intellectually disabled that he was completely dependent for the most basic necessities of life. Above all else, he had feared being consigned like his mother to a nursing facility for "custodial care," and when his son John after concluding his medical training had expressed that he would never be able to repay Royce for having supported him through medical school, Royce had responded with a quiet forcefulness tinged with bitterness, "Just don't let them put us in a rest home." Now, years later, the bountiful income Royce had earned through his strawberry work enabled that request to be fully honored—a new team of in-home caretakers were engaged to provide care around the clock, in addition to the frequent visits of the Hospice professionals. In effect, the nursing care was brought to him, and Royce was at least granted the blessing of remaining in his own home with Pearl by his side.

For Pearl this became a long and lonely vigil as Royce continued, almost imperceptibly, to worsen from day to day. She washed him, shaved him, prepared food for him and fed him, and talked to him, and when he became too weak to talk she sat by his side and read scriptures to him. John and Margaret and their families came frequently to visit or to hold family home evenings[146] with Royce and Pearl. During one of their home evenings the family sang together over Royce's bed the old familiar hymn "Love at Home," and were surprised to hear his now frail voice rising to join them in the chorus. Once again it was music which brought Royce back into the land of the living, if only momentarily, and from that time forward they sang with him frequently. One beautiful warm autumn afternoon John lifted Royce's now stiffened frame from his bed into a wheelchair and took him for a walk around the street under a cloudless blue sky. On their return home John asked Royce how he felt, and Royce, after weeks of bed confinement and scarcely able to speak, summoned enough strength to declare, "Wonderful! Marvelous!" Other family members came to visit, and Ann's children, who had spent much of their early years in Royce and Pearl's home, were deeply saddened to see their grandfather so weak and ill. There were other visitors too. Doug Gubler, the plant pathologist who had worked with Royce and now bishop of his ward, brought over a crate of strawberries, and Royce's colleague Kay Ryugo and his wife brought fruit from the experimental orchard in Winters. On Sundays, ward member came to administer the Sacrament to Royce and Pearl.

As Royce's vitality faded away, word came of other deaths in the family. Royce's brother George died in April, and in early October his older sister Rhea also died, leaving Royce as the last surviving member of his family. By November it became clear that Royce, too, was nearing the end of his life, and the family began to visit to pay their last respects. Thanksgiving day was a subdued affair, for as the family arrived to prepare and eat the traditional feast Royce lay in his bedroom, only intermittently conscious and completely unable to eat. That Saturday, November 26th, John phoned family members to let them know the end was near, and that evening, assisted by Marla's son Sam, he gave Royce a final anointing and blessing. The following day after his church duties John returned to the house to spend the night, and that evening Margaret and Jack came over to have a song and a prayer with Royce. Halfway through the night, in the early hours of Monday, November 28, 2005, Pearl and John were awakened by the attendant and hurried to Royce's bedside, where they found him breathing only intermittently. Standing on either side of the bed, Pearl took Royce's right hand while John took his left, noting that his formerly strong pulse was now weak and irregular. Although Royce was near death and had been scarcely arousable for days, in the darkness he appeared fully awake with his eyes open, gazing at Pearl, or rather at a place just beyond her. As they watched, Royce's breathing faded, until at last he took a couple of quick breaths, closed his eyes as though turning off the lights, then the fading pulse ceased and grew still. Royce S. Bringhurst, farm boy, student, missionary, soldier, scholar, professor, and world-renowned expert on the strawberry, was dead, one month short of his 87th birthday.

Royce's funeral was held on December 1, 2005, and although through an error of the funeral home the notice failed to appear in the newspaper, the chapel was filled to overflowing with relatives, Church members, and professional colleagues from the university and the strawberry industry. The funeral service was conducted by Doug Gubler, Royce's friend, associate, and bishop, and all of Royce's six children participated in the program. Margaret began by singing a musical version of a poem about death by Robert Louis Stevenson which Royce had loved and had sung himself,[147] and then each of Royce's daughters spoke in turn, beginning with Ann, who during her eulogy invited Herb Baum as an industry spokesman to offer some reflections on Royce's life and influence. After each of the other sisters had spoken in turn, the entire family rose to sing a final musical number,[148] and then John, acting partly as a Church leader, delivered a combined tribute and reflections on immortality, and Bishop Gubler closed the meeting with a few concluding remarks. Royce was buried that day near a flower bed in the peaceful little Davis cemetery at the east end of town, a place he had always been fond of. A tombstone later appeared over his grave; on it was engraved his name, the dates of his birth and death, an acknowledgement of his service in World War II, and at the bottom, the image of a strawberry.[149]

Illustrations for Chapter 12

Pearl and Royce at a University of Aconcagua building in San Felipe, Chile, where they served as agricultural missionaries with the Benson Institute.

Experimental strawberry plots in San Felipe. This was Royce's last formal scientific work.

Royce in his final journey as an international consultant, Egypt, 1997.

Royce and Pearl during visit to Irapuato,
Guanajuato, Mexico, 2002.

Pearl and Royce as workers in the Oakland Temple.

50th wedding anniversary portrait.

Postscript

By the time of his death in 2005, Royce's contributions to both the science and breeding of strawberries had become almost legendary. His partnership with Victor Voth had transformed California's struggling strawberry industry into a powerhouse that dominated strawberry production in the United States. The strawberry varieties Royce and Victor had developed have been used throughout the world, and almost every variety released by any major strawberry breeding program has included UC plants in its parentage. The day-neutral lines which Royce introduced have now been fully integrated into the California planting systems, having an impact which has revolutionized the industry.

The California strawberry breeding program which Royce and Victor had developed into such a potent force was carried vigorously forward after their retirement by Douglas Shaw and Kirk Larson, whose work expanded even more as the place of the day-neutral cultivars became better defined. The long string of releases of excellent cultivars has continued under their leadership, and has resulted in increasingly larger, better flavored, and more efficiently harvested fruit. The day-neutral lines which Royce established have led to many outstanding varieties which have become a major hallmark of the California program, and cultural methods have continued to be improved and refined as well. However, the academic and commercial world in which Shaw and Larson have had to operate is a far different one from that which existed in the early days when Royce began his work. Where once Royce exchanged both information and plant materials freely with other investigators, now legal constraints hold sway and patent attorneys jealously guard the university germplasm from outside use. The breeding process itself has become an increasingly expensive and high-stakes operation, and the relationship of the California program and the industry shifted in 1999 when the university at last allowed funding for the strawberry program to come directly from patent revenue (something Royce had advocated from the beginning), breaking the program's dependence on industry funding.

More recently, in 2012 Douglas Shaw and Kirk Larson, feeling the changing university environment increasingly incompatible with their strawberry work, announced their plans to leave the university and continue their work in a private setting in cooperation with a consortium of growers, a move reminiscent of the departure of Thomas and Goldsmith nearly seventy years before. The announcement unleashed a furor in the industry, the university, the press, and the state legislature. In 2015 the University of California announced the hiring of professor Steve Knapp, a geneticist with expertise in genomics, and a staff research associate named Glenn Cole, an experienced plant breeder, to rebuild the California breeding program using existing breeding stock. (In a curious turn of history, Glenn Cole was hired from the Pioneer Seed Company, originally founded by Henry A. Wallace, Royce's old mentor and benefactor.) Douglas Shaw and Kirk Larson meanwhile have continued their own breeding program under a private organization called California Berry Cultivars, conducting their field research in the same Watsonville plot that Royce had leased for the university years before. At the same time, the private Driscoll program formed by Thomas and Goldsmith continues its breeding work in competition with both. These changes coincide with the final phase out of the methyl bromide fumigation system which has been the mainstay of disease control for over half a century, leaving the industry scrambling for alternatives. With these new developments, the ultimate future of the UC Davis strawberry program is uncertain, though its profound contribution to the world of plant breeding remains unquestioned.

Although Royce left behind a strong and vibrant breeding program, not all his other efforts as a scientist were to have as enduring an influence. His work with interspecies hybrids, including the decaploid work he had performed with Hamid Ahmadi, was not continued after his retirement, and the interspecific breeding populations they created were discarded by the university. Although Royce's use of isoenzyme electrophoresis to identify specific varieties and to trace heredity had been pathfinding in its time, that technique was quickly replaced as advancing technology allowed the widespread use of much more specific DNA markers. The tissue culture propagation technique which Royce had considered so promising never came into wide use. Not only was it extremely expensive for a crop which self-propagated with such ease, it also tended to produce mutant off-types, a problem which was never satisfactorily resolved. In the end the use of tissue culture has remained limited to the production of virus-free plant stock, much as it was in Royce's day. Finally, despite Royce's keen interest in the genetic diversity of wild strawberry germplasm, and his conviction that its inclusion in breeding programs could provide a valuable infusion of essential new genetic traits, to date there are almost no modern instances of a wild trait having made an important contribution to the betterment of strawberry cultivars.

The single notable exception is the day-neutral trait, which Royce had introduced from wild berries he had found on a family vacation in Utah.

The sheer variety of these investigations was an indication of the breadth and vitality of Royce's scientific pursuits. Where the strawberry was concerned, Royce was a Renaissance Man—not only did he become the world's most successful breeder of strawberries of his time, he also made significant and lasting contributions in the fields of agronomy, plant pathology, genetics, taxonomy, and evolutionary botany. His name was known and his advice sought by strawberry experts from around the world. He was an excellent teacher, and a thoughtful and insightful speaker. His modest dignity and kind demeanor were well known to both strawberrymen and to colleagues, and he shared the respect of all. His impact continues to be felt in the universities to which he trained and worked, and UC Davis, Utah State University, and the University of Wisconsin all have scholarship programs in his name, developed from his many financial contributions over the years.

In his private life, Royce was a man of great intellectual and spiritual depth and integrity. In spite of his personal struggles, to the end of his life he remained faithful both to his scientific convictions and to his spiritual commitments as a member of the Church of Jesus Christ of Latter-day Saints. His religious opinions, though sometimes unconventional, were widely respected, and among his temple colleagues he was beloved and revered. He was generous in his regard for people of other faiths, and he earned their respect in return. His personal thought and choices were reflected not just in his own life, but in the lives of his descendants. All of his children remained faithful members of the Church, though not all of their spouses are members. While Royce was the first of his own family to receive a higher education, all of his children attended college, and a number of his descendants hold advanced degrees. Various of his descendants have followed his precedent of serving full-time missions for the Church, and two of his grandsons rendered service in the military as well. Music, particularly classical music which he so loved, has continued to enrich the lives of many of his posterity to this day.

But perhaps one of the most lasting legacies Royce left his children was his example of unflinching and lifelong devotion to his wife Pearl. Though their relationship had sometimes been a turbulent one, their declining years reinforced the companionship and affection between them. Pearl lovingly and selflessly attended to Royce's needs and comfort during the waning years of his life, and Royce in turn had provided amply for Pearl's subsequent care and comfort during the remainder of her life. As she underwent her own slow decline years later she was enabled to remain in the comfort of her own home. Like Royce, she was the last surviving member of her family, and when seven years after Royce she too breathed her last in the early morning hours in March of 2012, it was surrounded by members of the family. Pearl is now buried by Royce's side in the Davis Cemetery.

Maps

Italian Campaign, 1944

Southern Europe in 1944. The 489th Bomb Squadron was based
at Alesan, Corsica. Royce flew against targets in Italy, France,
and Yugoslavia.

California Strawberry Areas, 1953-1990

Map of California, showing wild and cultivated strawberry areas. The
strawberry program spanned the length of the state.

Notes and References

CHAPTER 1:

[1] Joseph Granville Leach, L.L.B., *History of the Bringhurst Family, with notes on the Clarkson, De Peyster, and Boude Families*, 1901, p. 17-25.

[2] Among other things, the Bringhursts of Pennsylvania were credited with having made a fine carriage for George Washington. Royce's relating of this story is confirmed in Pearl Bringhurst's journal, from a church meeting near the end of Royce's life. "Several of us told of interesting incidents in the lives of their ancestors. Royce told of his great-great-grandfather building a carriage for President Washington." (PDB journal, September 2000 to June 2001, entry for March 17, 2001.)

[3] Luzon Bringhurst Glines of Toquerville, Utah, as quoted in Church News, Deseret News, August 27, 2011, p. 16.

[4] Beitler is sometimes spelled "Beidler" in family records.

[5] Now Florence, Nebraska. The extraordinary circumstances of their stay in Winter Quarters was recounted by Luzon Bringhurst Gines, Church News, Deseret News, August 27, 2011, p. 16.

[6] Eliza R. Snow dedicated a poem of friendship to Eleanor Bringhurst in 1848. See *The Personal Writings of Eliza Roxcy Snow*, Ed. Maureen Ursenbach Beecher, University of Utah Press, 1995, p. 221-222.

[7] See *Bringhurst's, Those Who Came West: Samuel, William & Descendants*, compiled by Gayle Bringhurst Young, 2003, p. 3. Samuel Bringhurst was mentioned in an official Church history as having "welded an axle tree which was the first blacksmith work done with stone coal in the territory" (B. H. Roberts, *Comprehensive History of the Church*, vol. 3, chapter xci, footnote 27).

[8] Taped interview, September 1997, content condensed and edited. It is noteworthy that a man named Samuel Bringhurst served as foreman of the jury that convicted George Reynolds of polygamy on April 21, 1875. (*Millennial Star*, Vol. XXXVII, May 10, 1875, as cited in *Utah: The Storied Domain*, J. Cecil Alter, The American Historical Society, Inc. Chicago and New York, 1932, p. 421.) Reynolds, then serving as secretary to the First Presidency, had been asked by Brigham Young to submit to arrest and stand trial to test the constitutionality of the 1862 Morrill Anti-Bigamy Act, and his conviction was later upheld by the U.S. Supreme Court, which opened the door for the widespread incarceration of Latter-day Saint polygamists. At the time of this trial, Samuel Bringhurst was 62 years old, and according to the same source (J. Cecil Alter, op. cit. p. 211) he had moved to West Jordan in 1874, which lends plausibility to the family tradition, recounted by Royce, that the move was motivated by a disagreement with leaders over polygamy. A respected Latter-day Saint known to oppose or at least to refuse to practice polygamy would have been considered ideally qualified to serve on a jury at a time when Latter-day Saints were frequently excluded from jury service on such controversial cases. Samuel Bringhurst had a son also named Samuel, but as he was only 25 years old at the time, it is unlikely he would have been selected a jury foreman. (My thanks to Newell Bringhurst for bringing this infor-

mation and reference to my attention.)

[9] This was not the case with other family members. Samuel's brother William, six years his junior, helped found a temporary Church settlement in Las Vegas, Nevada (referred to for a time as Bringhurst Fort), before settling in Springville, Utah, where he became a community leader and bishop. William had at least two wives. Samuel's eldest son, William Augustus Bringhurst, also had several wives and numerous descendants. Of him, Royce remarked, "He was a pioneer in his own right, because he was born in Pennsylvania. He ended up in Toquerville, and whenever you hear anybody ask, are you related to the Bringhursts in Toquerville, you can say, yes indeed, they would be descendants of my great grandfather's oldest brother." (Taped interview with RSB, September 1997, content edited.)

[10] A twin sister, Mary Elizabeth Bringhurst, has a birthday listed the following day; hence, they were probably born near midnight. They were born while the family was still in Salt Lake City.

[11] Taped interview, September 1997, content edited.

[12] Samuel Brannan was an early Latter-day Saint leader from New York who, as the main body of the Church prepared in Illinois for their epic trek to the Intermountain West in 1846, led a group of Saints in a sea journey around South America to the west coast aboard the sailing ship Brooklyn. Arriving in California in the midst of the revolt from Mexico, Brannon established an important Church settlement in the town of Yerba Buena (now San Francisco). He tried unsuccessfully to persuade Brigham Young to bring the pioneers to the west coast and settle in what is now California, which he knew to be a far more hospitable region. Brannan played an important role in the history of California, including making public the discovery of gold in the Sierra Nevada mountains, which led to the 1849 gold rush and rapid California statehood. He was never reconciled with Brigham Young, and eventually left the Church. A successful businessman and purportedly the first millionaire in California, Brannon eventually lost all his financial holdings and died penniless.

[13] Letter, RSB to PD, November 8, 1941. Royce would, of course, later settle with his own family in California, but would always refer to Utah as "back home."

[14] See Bringhurst's, Those Who Came West: Samuel, William & Descendants, compiled by Gayle Bringhurst Young, 2003, p. 249.

[15] Enoch Bartlett Tripp was a shoemaker by trade. Jessie Eddings Smith was the fourth of five wives.

[16] Although a registered Democrat for most of his adult life, Royce was fairly independent in his political views. While in the military he evidently voted against Franklin D. Roosevelt, the most prominent 20th century Democrat politician, because he held him partly responsible for the Second World War. During his professional years the academic community tended to strongly favor the Democratic party, and this may have influenced Royce's thinking, particularly when Ronald Reagan, the Republican governor of California, pursued a course which most faculty considered contrary to the interests of the university. However, when Reagan later ran for president, Royce voted for him in preference to what he considered to be an inferior Democratic

candidate. "I decided with considerable reluctance to vote for Ronald Reagan, something I thought I would never do, but times do change." (Davis Journal, volume 5, November 4, 1980.)

[17] See *Bringhurst's, Those Who Came West: Samuel, William & Descendants,* compiled by Gayle Bringhurst Young, 2003, p. 249.

[18] Taped interview, September 1997. Royce described his grandfather in a letter sent from Corsica during combat service in World War II. "I remember how I always looked forward to visiting my grandparents on either side. I was quite small when Grandfather Bringhurst died but I remember him well. He was always so kind to us. He was six feet two and a bit slender. He was a good man. I don't remember Grandmother so well. I was very young when she died." (Letter, RSB to PDB, Wednesday, August 16, 1944, content condensed.)

[19] This story is told in detail in *William A. Bringhurst: From Devout Latter-day Saint to Condemned California Killer—A Personal Confrontation with the Past,* an unpublished typed manuscript by Newell G. Bringhurst, Royce's nephew who, though never active in the Church as an adult, became a respected scholar of Church history. A copy of this account was given to me at Royce's funeral by Newell, with whom I have also discussed the story. Bill Bringhurst's criminal actions and his conviction and execution must have had a profound effect upon his quiet older brother Jack. Royce was five years old at the time of the execution in 1924, but it is not clear how much he knew of the events, which were not spoken of within the family; his younger brother George, who himself was very close to Royce, learned about them only indirectly from a cousin. It is Newell's opinion that it may have been the widely-known criminal behavior of this family relative which caused many of the Bringhursts, including Royce and his brother George, Newell's father, to always be "anxiously engaged" in some useful undertaking (as explained by Newell on p. 39-40 of his manuscript).

[20] Ironically, though himself an upstanding member of the community and faithful Church member, Royce's father, John Tripp Bringhurst had some kind of disagreement with Samuel Enoch Bringhurst, probably relating to a family property, and held him in low esteem, a fact mentioned to me by one of my sisters and confirmed to me by Newell Bringhurst. This feeling evidently was not shared by Royce, who commented upon hearing of his uncle's call as a mission president while serving in the military overseas, "I was happy to hear about Uncle Sam. I think he will make a very fine mission president." (Letter, RSB to Bringhurst family, Saturday, September 9, 1944.)

[21] The "Uncle Sam" spoken of was Samuel Bringhurst (1850-1936), who married Sarah Elizabeth Orr ("Aunt Sarah", 1835-1958) in 1888. An older brother to John Beitler Bringhurst, he is not to be confused with Samuel Enoch Bringhurst, Royce's uncle in the succeeding generation. "Uncle Louie" was Louis Bringhurst (1856-1939), John Beitler Bringhurst's youngest brother.

[22] Davis Journal, volume 11, July 24, 1993. This remembrance was part of a Pioneer Day reflection written by Royce well into his retirement, when he was serving as a service missionary in San Felipe, Chile.

[23] See Beret N. Smith, *A Biography of George Fred Smith, Compiled by his Grand-*

son, 2011, p. 25.

24 At the time of the marriage, Florence's father, George Fred Smith, was serving a mission to England while his wife and children were at home. (Beret N. Smith, *A Biography of George Fred Smith, Compiled by his Grandson*, 2011, p. 51.) George Fred Smith had an excellent singing voice, and as Royce's father was also a singer, Royce probably inherited musical ability from both sides of the family.

25 Details from the preceding two paragraphs are taken from a typed manuscript, "Golden Wedding, John T. and Florence Smith Bringhurst, Married June 13, 1907", which was either distributed at the anniversary celebration or was typed up from a recording made there. The term "Seventy" refers to a priesthood office in the Church, patterned after a Biblical reference from the New Testament (Luke 10:1).

26 George Bringhurst, interview by Newell Bringhurst on June 4, 1998, p. 14-16, content condensed.

27 Royce S. Bringhurst, talk notes, approximately 1979, content condensed, rearranged, and edited. The original concluded, "even at the time of his death he had been outside chatting with a neighbor as was his custom, and had been fatally stricken as he returned home, falling across the path under a tall pine tree he had planted some 20 years before."

28 Davis Journal, volume 6, January 9, 1982, content revised and condensed. This account was given on the occasion of his mother's death.

29 Royce's mother, Florence Elizabeth Smith Bringhurst, recalled to Royce a quarter of a century later the circumstances of his birth, which held some irony at the time, as he was then in military training. "You were born just after the close of the last war and at the time we were having such a lot of flu and many deaths. Everyone should have been so happy on account of the war coming to a sudden end, but there was so much sickness and death that people didn't know what to expect next. And then when you were born we were so happy to have another boy, and I used to lay in bed and look at you and feel so thankful to think you would never have to go to war, because they used to say they would have things fixed up so we would have no more wars. And here it is your twenty-fifth birthday and you are in a training camp and we are at war again." (Letter, Florence Smith Bringhurst to RSB, December 25-27, 1943, content edited and condensed.)

30 Taped interview, September 1997, content condensed and edited.

31 George Bringhurst, interview by Newell Bringhurst on June 4, 1998, p. 22-23, content condensed and edited.

32 Taped interview, Tape 2, 1997, content edited.

33 George Bringhurst, interview by Newell Bringhurst on June 4, 1998, p. 1, content condensed.

34 Letter, Florence Smith Bringhurst to RSB, July 8, 1940.

35 George S. Bringhurst, interview by Newell Bringhurst, June 4, 1998, p. 2-3 and 16, content condensed and edited. The context of this segment was George's distaste for farming.

36 Taped interview, September 1997.

37 Letter, RSB to Bringhurst family, 27 July, 1943. In the same letter, written while in military training in South Dakota, Royce reminisced about life on

the old farm: "I can't think of a single part of the old place now without it reminding me of something that I treasure as a happy memory. That goes back to where we used to herd cows in the upper field when the big ditch was still there. That was when we had the old cow named "Coal." We used to go up there with Rhea and take a bottle of milk. Also, I remember when George and I used to fish in the ditch with bent wire for hooks."

[38] Taped interview, September 1997, content condensed and edited. In the original interview Royce mentioned seating fourteen young people on the horse, which is implausible; when he recounted the same event at his brother Smith's funeral at a much younger age, the number he recalled was five, which is more likely and is used here. It is probable that fourteen is the number of young people the horse pulled in the sled.

[39] A brooder house is a heated structure for housing hatchling poultry.

[40] This narration was given at a Sunday School class for young adults in the Davis Ward on July 27, 1975, and as the author was a member of the class, it remained a vivid memory. The story quoted was found written on index cards prepared for the class, and is mostly taking verbatim, with slight adjustments according to the author's recollection of how the story was told. The climax of the tale as he told it was his father's wish to have bought the man a new pair of shoes, which brought emotion to Royce's voice.

[41] Taped interview, September 1997, content condensed.

[42] Taped interview, September 1997, in which Royce was describing photographs taken in his youth. Content edited.

[43] This is a lake accessed from Salt Lake via the Big Cottonwood Canyon.

[44] From taped interview, September 1997, supplemented with direct quotes from the "Golden Wedding Anniversary" document, content edited.

[45] In continuation, the narration mentions a notable coincidence: "When your mother and I decided to get married, when we were clearing up the various papers we had to sign and testify to, it turned out that Pearl and I were baptized the same day, same place. Her stake was there same as my stake." (Taped interview, Tape 2, 1997, content edited.)

[46] Royce's brother George gave details in a 1998 narration: "Of course, we all went to school and we had to travel about a mile and one-half from home, but we did have the luxury of the bus to ride, and one of the bus stops was on Redwood Road right in front of our house. So all we had to do was make sure we were out there in time to catch the bus as it would come by and stop for us, and take us about a mile and one-half north to what was the old Plymouth School which fronted on Redwood Road and 4800 South. Our education from first grade through junior high, from first to ninth grade, took place down at Plymouth School." (Interview with George S. Bringhurst, June 4, 1998, Taped by Newell Bringhurst, p. 3, content condensed.)

[47] Granite High School, established in 1906, had the distinction of being the location for the first LDS Seminary program in 1912. Granite High remained in operation for over a century, closing as a public school in 2009.

[48] Five year diary, February 16, 1937; the "prosecuting attorney" for this court was none other than James E. Faust (see Granite High School Yearbook, 1937 "Granitian", p. 15).

[49] Five year diary, entry for February 3, 1937.

[50] This was a maternal uncle, Glen Smith, with whom Royce would renew his acquaintance at the funeral of an aunt in 1976. Of their meeting, Royce wrote, "I was caused to reflect back on the time when Glen visited us when I was a young boy on the farm. How I envied him, 'a Forest Ranger.' I do believe that it was then that I decided to get a college education someday. I even thought I might be a Forest Ranger – but there was a mission and a war that had to be completed before I could complete the education process. I have no regrets about any of my choices in that regard." (Davis journal, volume 1, January 3, 1976, content slightly condensed.)

[51] Letter, RSB to PDB, Sunday, August 6, 1944. At the time he wrote this letter he was flying combat missions out of Corsica, and had been in the military nearly two years. He concluded the paragraph by saying: "I'm afraid the old cords are pretty rusty by now but I think I could pick up again."

[52] Taped interview, Tape 2, 1997, content edited and condensed. Meade and Royce apparently enjoyed a close friendship. Royce included in his 1997 narration the sad conclusion to the story: "He was really talented, but he...anyway, he was killed in France. He got struck with shrapnel right to the forehead; I believe it was shrapnel, or rifle or machine-gun burst, the first time he was in action. I guess it was in the invasion. He was a sergeant. You'd never believe he was a military man. Meade, Meade Steadman...."

[53] Five year diary, January 8, 1937, also Granite H.S. Yearbook, 1937 "Granitian", p. 78. At the same competition Royce's best friend, Steve Mackay, took first in crops judging. This was also the event at which they drank. It is perhaps the drinking episode which precluded individual mention of Steve and Royce in the yearbook, though the competition and the prizes were mentioned.

[54] Seminary refers to a class in Church doctrine which was (and still is) permitted to be taught on a site near the school, during the regular school schedule.

[55] Five year diary, entries for January 8 and 9, 1937. Pearl's recollection was that it was the adviser himself who supplied the drinks. "Hillam. He's the one that was their Ag. teacher, that gave them...got some liquor for them once when they went to judge something back in Utah State, and Royce said he drank some of it, and I just told him if he ever did it again, he didn't need to bother to call me up." (Pearl D. Bringhurst, recording, March 1, 2009.)

[56] Five year diary, January 11, 1937.

[57] Five year diary, August 4, 1937. Royce's older brother Smith developed a drinking problem, and in a letter to Pearl the following year from college, Royce wrote, "I don't know whether you saw Smith the other night at the dance. I didn't. He had been drinking again. Dad saw him, he felt quite bad about it. He was asking me about it. I'm glad mother didn't see him. He's twenty-one now so I guess it is his own business; no one else can do anything about it. Nearly every time I go somewhere I can see just one more good reason why you didn't want me to drink." (Letter, RSB to PD, February 28, 1938.)

[58] Five year diary, January 10, 1937. His church attendance tapered off somewhat when farm duties became heavy, and again later when he was

given heavy work loads during college classwork.

[59] Five year diary, January 10, 1937.

[60] Most of the diary entries about this were subsequently scribbled over by Royce, and as he obviously did not want them to pass to posterity, I have refrained from attempting to decipher them.

[61] Pearl recorded her own account of this incident: "Mr. Hatch, our high school principal sent a messenger to my Seminary class one day toward the end of the school year and called me into his office. I went & he talked to me about my relationship with Royce. He said I probably allowed Royce privileges that I would not give to another boy—I couldn't believe my ears. He was accusing me of having sex with Royce. Shy little creature that I was I could hardly answer at all. I only shook my head. That anyone would even think such a thing of me was devastating to me...I got the feeling he still believed I had given myself to Royce because next period he called Royce in and openly accused him of it. Royce of course told him we had done nothing." Pearl later suspected a vindictive friend on the student council of having planted the rumor. (Pearl's black history journal. See also Pearl D. Bringhurst *My Life and My Poetry*, p. 40.)

[62] Five year diary, February 23, 24, 25 and March 2, 1937. A year after the incident, on February 23, 1938, Royce wrote in a P.S. of a letter to Pearl: "Did you notice in your diary a year ago today, the day Mr. Hatch called us in the office. I still don't think that was necessary. I thot then that he had some right to call us in but I don't now. He thot he was doing the right and only thing to do I guess" (Letter, RSB to PD, February 23, 1938.)

[63] Though the story of Royce's skipping commencement exercises was evidently untrue, Royce's third daughter Marla later verified that the family stopped by the school en route to California, presumably around 1950 after earning his PhD, and she remembers waiting in the car while his father entered the school to talk to his old principal.

[64] Hyrum Smith was the brother of Joseph Smith, the founding prophet of the Church of Jesus Christ of Latter-day Saints. Joseph F. Smith was Hyrum Smith's son who later became president of the Church, and Joseph Fielding Smith was a grandson, who likewise led the Church as president.

[65] David O. McKay was a Latter-day Saint apostle who served as president of the Church during much of the time Royce lived in Davis, and who was still familiar to his audience at this time (1979). Emma Ray McKay was Royce's high school speech teacher.

[66] Notes from "Academics and Faith" talk given by Royce at the Davis Institute on October 12, 1979, repeated on June 2, 1988 and March 8, 1991, content edited.

[67] Five year diary, January 12, 1937.

[68] Five year diary, January 26, 1937.

[69] Five year diary, April 14 and April 20, 1937.

[70] Five year diary, on respective dates in 1937.

[71] Five year diary, May 7 and May 11, 1937, content condensed.

[72] Five year diary, May 14, 1937.

[73] Members of the quartet besides Royce were Meade Steadman, the musician whom Royce greatly admired and who would sing at his missionary

farewell, James E. ("Jim") Faust, who would later become an apostle and a member of the First Presidency of the Church, and Loren Ferre. They sang "Jesus, Lover of My Soul" (Granite L.D.S. Seminary commencement program, Sunday, May 16, 1937).

[74] Royce's Senior yearbook, the 1937 "Granitian", remained among his belongings, bearing the signatures of dozens of well-wishers, an occasional prankster, words of encouragement from some faculty members, and an entire page by his High School sweetheart Pearl, all written in youthful hand in flowing ink (before the days of the ball-point pen) or in pencil, interspersed with the black-and-white images of classmates, many of whom died in the war. Pearl's book is similar.

[75] The basis of the tradition that Royce declined to participate in the commencement is probably the fact that his name did not appear in the printed commencement program, perhaps having been removed when his qualifications to graduate were still in doubt. The principal was held responsible. On a copy of the program saved among Royce and Pearl's papers, Pearl had written indignantly, "His Holiness 'Principal Hatch' had Royce Bringhurst deleted from the program."

[76] John Payne taught the Sociology class attended by both Pearl and Royce. Granite High School yearbook 1937 "Granitian", p. 11.

[77] Five year diary, May 15-May 21, 1937. The watch was of very fine quality, and Royce wore it for years. Seven years later, while in bomber crew training in Columbia, South Carolina in World War II, the crystal mounting screws worked loose, and Royce lost the crystal and damaged one of the hands of the watch. Heartsick, he sent the watch back to his wife and requested that she look into having it repaired. (See letter, RSB to PDB, January 7, 1944.)

[78] First two photos from Pearl's family album; last two from the Granite H.S. Yearbook, 1936 "Granitian".

[79] Granite H.S. Yearbook 1937 "Granitian", p. 29, p. 83, p. 74.

[80] Granite H.S. Yearbook, 1937 "Granitian", p. 78.

[81] Granite H.S. Yearbook, 1937 "Granitian" p. 28, p. 105.

CHAPTER 2:

[1] Now known as Utah State University. USAC was a Land Grant institution, meaning that it was established under a program, developed during the American Civil War, under which federal lands and funds were provided to establish colleges in various states for the purpose of providing higher education to the "industrial class," meaning those not belonging to the "professional classes" which included clergy, medicine, and law. These colleges also were charged with providing expertise and support to agriculture and industries in their respective states, a charge which was to profoundly influence Royce's later life. The Utah State Agricultural College was established by an act of the state legislature in 1888, though it did not open in Logan, Utah, until some years later. Its name was later changed to its present one to better reflect its function as a liberal arts university. For a history of the institution see Ricks, Joel E., "The Origin of the Utah State Agricultural College" (1953), *Joel Ricks Collection*, Paper 8, posted online by **digitalcommons.usu-**

.edu.

[2] LaMont, known by all as "Mont", became prominent in the Church and in politics. In the 1960s he was called as president of the Toronto, Canada mission, and he also served a term as Secretary of State for the state of Utah, and later ran for governor, losing by a narrow margin. He and Helen raised their family in a home on a cul-de-sac called Country Club Circle, southwest of the city center, and for many years they housed Pearl's mother, Jane Davidson, in a basement apartment. Theirs was usually the home where Royce's family stayed in later years, during visits to the Salt Lake area from California.

[3] Saltair was an amusement center and ballroom which had been erected on the shore of the Great Salt Lake, not far from downtown Salt Lake City. It was a popular destination for many years, and lavish events were held there. Subsequently rebuilt, it has largely fallen into decay.

[4] Five year diary, entry for June 11, 1937. The information for this paragraph is drawn from entries in the diary from May 22 to June 13, 1937.

[5] Five year diary, June 25, 1937.

[6] Five year diary, June 3, 1937.

[7] Five year diary, May 25, 1937, content edited. A hootch pot was a kind of trap used to capture codling moths, whose larvae spoil the mature fruit.

[8] Five year diary, June 7, 1937.

[9] Five year diary, June 8, 1937.

[10] Five year diary, May 29, 1937.

[11] Five year diary, April 17, May 13, May 29, June 12, June 24, September 1 and 2, 1937.

[12] Five year diary, June 27, 1937.

[13] Five year diary, July 6, 7 and 8 1937, content condensed.

[14] Five year diary, July 14, 1937.

[15] Five year diary, July 17, 1937. Royce subsequently purchased a $1000 life insurance policy from Mutual Life, though he did not mention his motivation for doing so, nor who he named as beneficiary. (Five year diary, July 19, 1937.)

[16] See Five year diary, July 24, 1937. Pioneer Day is July 24, the anniversary of the arrival of Brigham Young into the Salt Lake Valley with the first pioneer settlers in 1847. This is still a major state holiday, with public observances which in Utah rival those on Independence Day.

[17] Five year diary, August 2, 1937. The entry stated: "Doc. Owen was out. Weve got nematode in beets. He told me to look up a Keller when I get to Logan. Forestry bldg."

[18] In an interview following his retirement, Royce recalled the influence of Dr. Owen in his life:

> RSB: One of the reasons I went into breeding in the first place was because I knew Dr. Owen. F. V. Owen was his name. And he was involved in a very important project with sugar beets, developing monogerm sugar beets, sort of like the day neutral strawberry, ones that would handle the way they wanted them to handle, because they were very expensive to grow, and hard to mechanize.
>
> Q: And he did some research on your farm, didn't he?

RSB: On our farm, yes. My father used to rent a piece of land to him every year. And I used to look at Dr. Owen, and I said, that's for me. Q: And he rented the land and did some research, and he inspired you into going into breeding work, did he? Is that right? RSB: Well...they actually offered me a job on sugar beets when I finished my graduate work. I was sorely tempted, because I could think back, of Dr. Owen, I used to think, what a wonderful way you'd travel about, and do all these things, and do whatever you want every day of the week. Farming in those days, from my point of view, was pretty hard work.
(Interview of RSB by Anton Kofranic, UCD Emeriti Association, 1996, content condensed.)

Dr. Owen made major advances in beet breeding, including the discovery of cytoplasmically inherited male sterility which led to major advances in hybrid development. (See R. J. Hecker and R. H. Helmerick, "Sugar-beet breeding in the United States," chapter 2 of *Progress in Plant Breeding 1*, ed. G. E. Russell, 1985, p. 39. For a contemporaneous description of Dr. Owen's work and objectives, see Eubanks Carsner and F. V. Owen, "Saving our Sugar Beets," from USDA publication *The Yearbook in Agriculture 1943-1947: Science in Farming*, 1947, p. 357-362.)

[19] Five year diary, August 18, 1937, trip details are described in diary from August 14 to August 18.
[20] Five year diary, September 4, 1937.
[21] Five year diary, September 17, 18, 19, and 22, 1937.
[22] Five year diary, September 23 and 24, 1937.
[23] Only Steve Mackay was named by Royce in his recollections; Sam Oliver, named frequently in Royce's journal but rarely with a last name, is identified presumptively because, besides being a member of Royce's judging team, he was the only member of the graduating class with the name "Sam" (Granite H. S. Yearbook, 1937 "Granitian, p. 76, p. 37). Dave Miller (pictured in the Granite H. S. Yearbook, 1937 "Granitian" on p. 35), occasionally named before college began and much more frequently after, was the only one of Royce's original roommates who continued to live with him after the first couple quarters of college, as confirmed by his inscription in Royce's freshman yearbook: "It looks like we...were the only members of the old gang to stick it out to the end of this year" (Royce's 1938 "Buzzer", page entitled "Campus Views"). Emil Stenstrom was referred to in Royce's writings only as "Swede", but in his yearbook autograph in their Junior year, he signed his nickname "Swede" beneath his actual name (Granite H. S. Yearbook, 1936 "Granitian", Royce's copy, inside back cover), revealing his identity. The Scandinavian name explains the nickname.
[24] Five year diary, September 23-27 (quotes from September 23 and 24), 1937.)
[25] Five year diary, September 28, 1937.
[26] Five year diary, September 29, 1937, content condensed.
[27] Reserve Officer Training Corps, a program of uniformed military training for college students.
[28] Letter, RSB to PD, October 21, 1937.

[29] Five year diary, October 4, 1937, content condensed.

[30] Five year diary, October 9, 1937.

[31] Five year diary, November 6, 1937, content condensed.

[32] Five year diary, November 14 and 15, 1937, content condensed and edited.

[33] Five year diary, November 25, 1937.

[34] Five year diary, December 15, 1937.

[35] Five year diary, December 6, 1937.

[36] Five year diary, January 1, 1938.

[37] Courses included Botany, Bacteriology, Agricultural Economics, and Military Science, and the cost of registration was $22. (Five year diary, January 3, 4 and 5, 1938.)

[38] Five year diary, January 12, 1938.

[39] Letter, RSB to PD, February 28.1938. The same day the wrote in his journal: "4-H club school starts today. Resolved right now that I do a good job of being leader this year." (Five year diary, February 28, 1938.)

[40] Five year diary, February 12 and February 26, 1938.

[41] Only a handful of these letters remain among Royce's and Pearl's papers. The surviving letters contain quite a bit of detail about Royce's activities, and are quoted in this chapter.

[42] Letter, RSB to PD, March 3, 1938. In his five year diary, dated the same day, Royce recorded: "Got a letter from Pearl & answered it. It was very important to me and to her to I think." Pearl's letter to Royce was lost or destroyed.

[43] Five year diary, February "29", second space (1938). This was written on the blank space left in his diary for February 29 (1938 being a non-leap year).

[44] Royce wrote to Pearl, "Sam, Steve, and Swede are thinking about quitting. I think they probably will. If they do I think Dave and I will move down with the Gold's, and let this house go, because we couldn't afford to keep this house going just the two of us." (Letter, RSB to PD, March 6, 1938, content edited.) In the same letter he explained that Sherm was an upper-classman assigned to correct papers for an Ag. Econ class Royce had taken.

[45] Letter, RSB to PD, dated March 11, 1938, though this segment was written the following morning, on March 12, when the letter was postmarked.

[46] Five year diary, March 12, 1938.

[47] Five Year Diary, March 14-19, 1938, content edited.

[48] Letter, RSB to PD, misdated March 14, 1938, but obviously written (and postmarked) a week later on March 21, 1938. His journal for March 20 reads: "Got up early. Steve, Sam & I left for Logan at 7 got there at nine. They got their things & left. I cleaned the house then went over to Sherms. We ate & went to a show." (Five year diary, March 21, 1938.)

[49] None of the letters Royce received from his mother while in Logan survive, but he retained numerous letters she sent him while on his mission, and they bespeak a close and loving relationship. Royce's mother held him in high esteem, and in her first letter to him when he departed on his mission she wrote: "Royce I just want to say to you I am so proud of you and if every missionary went out with as clean a record as you did there wouldn't be so many of them go wrong when they return. You have never caused me any

worry so far as your conduct was concerned." (Florence Smith Bringhurst to RSB, November 24, 1939.) Unlike his mother, who wrote him faithfully, Royce's father only wrote him twice while on his mission.

50 Five year diary, December 10, 1938.

51 Five year diary, April 9, 1938.

52 Five year diary, October 30, 1937 and May 29, 1938. Harry Miller was later an associate of Royce's at UC Davis, and for years served as a member of the Davis City Council. After a priesthood meeting in Davis some 40 years later taught by Miller, Royce wrote, "Harry Miller asked that members of the class respond by relating experiences we had in our life that were of particular interest. Harry told of going as Danforth Scholar when he was at Logan. I knew him then, and related how his life touched mine then. I looked up to him, a freshman newly off the farm looking toward an 'experienced' & wise upperclassman." (Davis Journal #2, July 3, 1977, content condensed.) Harry and his wife later lived in Davis, California in a house with a large park-like yard on South Campus Way. After the death of Harry's wife, Royce named one of his successful strawberry selections after her—"Fern."

53 Five year diary, March 8, May 20, June 1, September 8, 1939. Yet another friend was Ralph Willes, an old classmate from Granite High who, like Royce, was attending college at Utah State. Once, when Ralph was eating dinner at their house, Royce and his roommates pulled a prank which Royce recalled with particular relish. The event was later retold by Pearl: "Royce did a lot of cooking, because he'd taken cooking classes in high school. And so when they went up to Utah State, and were working, Ralph Willes was doing the dishes one night when there were about four of them there, and Royce would dry them, and they'd put them over where the dirty dishes were, and he'd wash them again, and after it went on for a long, long time, he finally realized he was washing too many dishes, and he caught on that that's what they were doing" (PDB, taped interview, March 3, 2009, content edited). Upon discovering the prank, Ralph was quoted by Royce as looking up and with a sheepish smile exclaiming, "You dirty guys!"

54 Five year diary, April 2, 1938.

55 RSB, taped interview, 1997.

56 Five year diary, May 14, 15, and 29, 1938.

57 Five year diary, June 2 and 3, 1938.

58 Five year diary, June 5, 1938.

59 Five year diary, June 6, 1938.

60 Five year diary, September 2 and 9, 1938.

61 Five year diary, June 12, 1938.

62 Five year diary, July 31, 1938.

63 Five year diary, September 18, 1938.

64 Five year diary, August 28, 1938. Royce's mother held a party for Smith and his new wife Peggy a few days later (Five year diary, September 1, 1938). Smith's way of life would be far different from Royce's, and though there was no enmity between them, their paths in life quickly diverged, and their relationship drifted apart. While in the military, Royce wrote of Smith in a letter to Pearl, "He hasn't changed much but I guess maybe I have because when I came home before I went into the army he seemed almost like a

stranger to me. I don't know, but it always seemed to me that he got off on the wrong foot about the time he hit high school. Untill that time our lives were more or less the same. He started to pal around with boys who wern't the best company for him and I guess he wasn't the best company for them either. Anyway, among the bunch of them they go to raising merry 'you know whats' and that was that. I started along on a minor scale about the same way but never quite felt good about it and with a few new influences, number one being you, I cut it out. I guess I should be glad that Smith never got into any real trouble. I think he has settled down a lot these last several years though I doubt if any radical changes will take place in his life for quite some time. They have their life to live just the same as you and I and they are entitled to make it what they want." (Letter, RSB to PDB, Thursday, July 6, 1944, written from Corsica during combat service.)

[65] Five year diary, September 23, 24 and 25, 1938.

[66] Five year diary, October 9, 1938.

[67] Five year diary, October 23 and 28, 1938.

[68] Letter, RSB to PD, October 17, 1938.

[69] Five year diary, November 9, 1938.

[70] Five year diary, December 13, 1938.

[71] Among those present was Royce's uncle Ray Smith, and a humorous event happened in relation to him that remained a favorite memory, and many years later was recounted with much laughter by both Royce and Pearl in perfect tandem: (Pearl:) "Royce invited me out to go to Thanksgiving dinner at his mother's place, and the whole Smith family came to Thanksgiving dinner. And I came, and I walked in and said, 'Well, hello Ray, how are you?' And he said, 'I'm fine, but who are you?' And I said, 'Oh, come on, you're my Sunday School teacher.' And he said, 'No, I'm not.' And I said, 'OK, if you want to play that game, I'll play it too.' This was on Thanksgiving, on Thursday. So Sunday Royce comes to Sunday School with me, and he walks into our Sunday School class, and he says, 'Well, Uncle Ray, what are you doing here?' And he says, 'I'm not your Uncle Ray!'" (Royce:) "Same thing, repeated." (Pearl:) "Two men, the same name," (Royce:) "both bald, like my mother's brothers," (Pearl:) "the spitting image of one another!" (Royce:) "And here was the living image of my Uncle Ray! With the same name!" (Pearl:) "I just couldn't believe it. It wasn't until then, when he came up and said, 'I'm not your Uncle Ray,' then I realized, that guy out there really wasn't my Sunday School teacher." (Royce and Pearl Bringhurst, taped interview, October 20, 1997, content condensed.)

Royce's brother George also mentioned Thanksgiving dinner at the old Smith home, which appears to have been a yearly event: "The nicest thing that I remember about my grandparents are my maternal ones, because we used to go up to the Smith's, we used to go up there quite a bit. Thanksgiving was always spent at their place. The whole family, all of their kids, all of my aunts, uncles, and cousins, all of them anywhere in the area, why it was Thanksgiving and Grandma and Grandpa Smith's." (George S. Bringhurst, interview by Newell Bringhurst, June 4, 1998, p. 30.) The Smith house was located on the north side of Vine street, a short distance west of the place where Royce's parents later built their new home. It still stands at the time of

this writing, and can be distinguished by a section of the front which pro-
trudes in a partial hexagon.
[72] Five year diary, October 30, 1938, content condensed.
[73] Five year diary, entries for October 23, November 13 and 20, December 4
and 11, 1938, content condensed.
[74] Five year diary, December 27, 1938.
[75] Five year diary, January 23, 1939: "Back to School again The usual proced-
er. What a speller I am I don't know why I don't get after myself & make me
take more pains with it."
[76] Five year diary, February 23, 1939.
[77] Five year diary, January 8, 1939.
[78] Five year diary, January 11, 1939.
[79] Five year diary, March 8-11, 1939. Royce never mentioned the name of the
opera, but details are found on p. 56 of the USAC yearbook, the 1939
"Buzzer."
[80] Royce never did master typing, and after the hiatus of his mission and
military service, apparently never went back to it—all his professional work
appears to have been either handwritten or dictated to a transcriptionist.
[81] Five year diary, April 7 and 8, 1939.
[82] The arrest and fine are recorded in the Five year diary, April 29 and 30,
1939. Pearl and Royce later described the incident in a taped interview: "It
was up State Street. She lived on 7th East and 35th South, and I drove up
Redwood, into Murray, and then drove up to 33rd South. Before I got to
33rd South I was going a pretty good clip. It wasn't all that fast, but it was
enough. They picked me up, and when I got up to her place, her brother was
going with me, and her brother wasn't quite ready." Pearl added, "Dan and
Mary. And we waited for Mary. He didn't have to hurry." (taped interview,
1997, tape 2, content edited).

During Royce's mission his mother sent him a newspaper clipping with a
photograph, which identified the arresting officer as state highway patrolman
George Pazell. In the margins, Royce's mother had written: "I thought this
picture would bring back happy memories. By the way I think he picked up
your youngest brother not so long ago, so he is still on the job." (Letter,
Florence Smith Bringhurst to RSB, March 3, 1941.) Royce's brother George
confirmed this in an interview by his son Newell many years later:
"(George:) I got a ticket for running a red light. The first ticket I ever got in
my life and it was by a county sheriff; he's dead now, but George Pazell. I
ran the red light at 27th South and State Street, I believe it was. I just wasn't
thinking, and George Pazell picked me up and he sent me to my Uncle Art.
(Newell:) What was his reaction when you appeared before him? (George:)
Oh, just a five-dollar fine. (Newell:) He didn't show you any favoritism, huh?
(George:) No. He said, 'I have to charge you something.' My uncle Art was
also a veteran of World War I. He served in France in World War I. I don't
know how he became a "J.P." I think that was a political job." (George
Bringhurst, interview by Newell Bringhurst on 4 June, 1998, p. 60-61, content condensed.)
[83] For Latter-day Saints, a statement of personal conviction.
[84] Quoted from a three-page, carefully handwritten paper, "Why I believe

Joseph Smith is a Prophet of God", dated May 7, 1939, and preserved among the papers found in Royce's green binder. The assignment evidently was to write an outline of a testimony, and at the conclusion Royce wrote apologetically in pencil "This isn't so much an outline but I hope it fills the assignment." The faculty mentor, presumably the Brother Sessions mentioned by Royce in his journal, wrote in response: "Very good – The Lord bless you in the development that must come to you." Joseph Smith (1805-1844) was the 19th century visionary who was the founding prophet of the Church of Jesus Christ of Latter-day Saints; Royce's testimony was a statement of belief in the fundamental Latter-day Saint teaching that the ancient Christian gospel had been restored in modern times through the prophet Joseph Smith.

[85] The Church practice is to designate one Sunday each month, generally the first Sunday, as "Fast Sunday". For approximately 24 hours faithful members refrain from food or drink, and in place of the usual Sacrament services (at which individual members are assigned in advance to give spiritual messages or "talks"), a "Testimony Meeting" is held, at which members of the congregation are free to come forward and give brief unrehearsed expressions of conviction, or "testimonies". The act of fasting is believed to bring spiritual strength, though there is also a practical side to the fasting—members are encouraged to donate an amount at least equivalent to the cost of the meals which have been skipped, and these moneys, known as "fast offerings", are placed in a fund to assist poor and needy members of the congregation.

[86] Five year diary, May 6 and 7, 1939, content condensed. It is possible that this was Royce's first experience "bearing his testimony" in public, as preparing a written testimony in advance was a departure from the usual practice at such meetings. Though Royce came to be a superb extemporaneous speaker, even in his later years he would often write out key points of his talks which he wished to be expressed in a certain way.

[87] Five year diary, May 10 and 11, 1939. Royce came to regard all learning as a spiritual pursuit. He loved and often quoted a statement from the revelations of Joseph Smith: "The glory of God is intelligence" (Doctrine and Covenants 93:36) and a later excerpt from a Joseph Smith discourse: "Whatever principle of intelligence we attain unto in this life, it will rise with us in the resurrection. And if a person gains more knowledge and intelligence in this life through his diligence and obedience than another, he will have so much the advantage in the world to come" (Doctrine and Covenants, 130:18-19).

[88] Five year diary, May 21, 22 and 23, 1939.

[89] Five year diary, May 24 and 27, 1939.

[90] Five year diary for the dates indicated. The total expenses are summed in the final pages of the diary, categorized by incoming cash rather than outgoing expenses. Presumably he spent everything that came in.

[91] Five year diary, June 5 to June 10, 1939.

[92] Five year diary, June 14 to June 16.

[93] Five year diary, June 20 and 22, 1939.

[94] Five year diary. On June 26, 1939, Royce recorded earning $33.45 for his first ten days; two weeks later, on July 10, he was paid $39.29, and his third

wage payment, received July 25 a week after he had been laid off, was for $40.50.

[95] Five year diary, July 16, 1939. Royce's account was, "Had her out to dinner – I made her ride the horse etc. She seemed to have a good time." Royce seemed to regard the ability to ride a horse as a basic life skill, like riding a bicycle or swimming. Royce taught the author, his son, the fundamentals of horseback riding in just a few minutes, his chief instruction being, "always let the horse know who's boss." Though by no means an accomplished equestrian, the author has felt confident on horseback ever since.

[96] Royce's journal entry for the day read, "Hot er than all Get out today – I got my patriarchical blessing from Bish. Diamond tonite. I hope it may be an inspiration to me." (Five year journal, July 7, 1939.) A patriarchal blessing is a formal blessing of a prophetic nature which a member of the Church of Jesus Christ of Latter-day Saints may receive at the hand of an ordained patriarch (this is a specific office in the priesthood, similar to priest, elder, or bishop.) This blessing is recorded in writing, and becomes a sort of personal revelation or scripture to help guide the life of the recipient, and a copy is retained by the Church. Thomas Dimond was a neighbor farmer and former bishop, then serving as patriarch.

[97] Five year journal, Monday, August 7, 1939. Presumably this occurred during an interview with his bishop, probably Samuel S. Smith, from whom he had sought counsel during his senior year of high school (see Chapter 1).

[98] Five year journal, August 31, September 1, 2, and 3, 1939, content condensed and edited.

[99] The following spring Royce's mother wrote to him on his mission: "June is doing fine in school. You were here just long enough to help her out all she needed." (Florence Smith Bringhurst to RSB, April 7, 1940.)

[100] All information from five year diary, 1939. Reflections on the beginning of World War II are found in entries on August 31 to September 3, starting work at the "Section" on September 6, and layoff from the railroad on September 30, 1939.

[101] According to Royce's five year diary, Mont returned from his mission on July 17. On October 17 he wrote: "Jennette came home today she looks fine." She spoke at church on Sunday, October 22. Ralph Willes' farewell is mentioned in the diary on Sunday, October 15, 1939. Royce's college friend Norm Johnson was called on a mission to the Eastern States mission, entering the Salt Lake City mission home just after Royce left it to begin his own mission. Just prior to this he stopped by to visit Pearl at her work on December 2, 1939, which she recounted to Royce by letter (Letter, PD to RSB, December 3, 1939.) Royce made no mention of Sam Oliver in the diary after September 1938, but enclosed in the diary is a newspaper clipping of his missionary farewell prior to his departure to the Northern States mission. The date of the farewell was listed as Thursday, January 18, 1940, and speaking at his farewell was Leroy W. Hillam, their old Crops Judging coach from Granite High School.

[102] Five year diary, October 8, 1939.

[103] Five year diary, October 8, 1939: "Monday – Disked all morning – Went up to be interviewed by Dr. Jos. F. Merril in the afternoon "O.K." Got Pearl

& went after my suit etc. had supper at her place." Joseph Francis Merrill had been ordained an apostle in 1931, and was the first native of Utah to receive a doctorate, hence the reference to "Dr. Merrill."
[104] Five-year diary, October 12, 1939, content condensed.
[105] Five-year diary, October 27, 1939. Recounting this episode at his brother Smith's funeral, Royce called the horse "Babe," but the correct name was recorded in his journal.

Many years later, Royce gave a much more detailed account: "My brother Smith had a strawberry roan horse that he had trained. We always used to go on a fishing trip, but somebody always had to stay home to take care of the animals, and I usually had that luck, so I would tend the animals when our family went. Well, it seems that just as it was time for the family to go on that picnic up to the lakes, before they went up, the horses were horsing around, as it were, in the corral, and one of them kicked the other one, and he went into the fence, and a big splinter of wood from the wooden fence went right into his rump side, and he got infected immediately. There was no such thing as antibiotics and that sort of business in those days. The horse looked like he was a goner. So they told me, 'Well, we're going on our trip. If the horse gets down and can't get up, you'll have to take care of him.' So the veterinarian did what he could, and then he left, and the family left, and I was left with the horse. So one day, about three days after they had gone, I went out and there was the poor old horse lying on his side, obviously unable to get up. I had been given instructions by my father to shoot the horse if he didn't get better, and if he looked like he was suffering. Uncle Art had let my father have a French rifle that he got in World War I, when he was serving with the expeditionary forces over there. I went in, the only thing I could find in the house was that rifle, it was a French 7 mm—La Belle was the kind of rifle it was. So I got it and loaded it up, and went out there, and walked up to the horse, and literally shut my eyes and fired. The first time I just shot into the wall. But I took one look at that horse, and he seemed to know that I was trying to help him, so the next time I didn't miss. I hit him right in the temple, and the horse dropped dead. That was the saddest...that one was hard to handle. That horse...well...." The story died away in the telling. (Taped interview, September 1997, content condensed.) The existence of a World War I French rifle known as La Belle was verified.
[106] First mission journal, page 1, content condensed.
[107] First mission journal page 1 & 2.
[108] Five year diary, November 12, 1939.
[109] First mission journal page 2.
[110] The picture of "Old Main" is from the USAC yearbook, the 1938 "Buzzer" on the introductory "Faculty" page. The pictures of Royce's roommates are yearbook photographs from the Granite High School yearbooks—Steve Mackay's is from the 1936 "Granitian", and the photos of "Swede" Stenstrom, Sam Oliver, and Dave Miller are from the 1937 "Granitian". Royce's Five year diary contained two newspaper clippings announcing missionary farewells, one of Sam Oliver and the other his own.
[111] The first and third photos are from the USAC yearbook, the 1939 "Buzzer," p. 145 and p. 252; the one showing Royce in military uniform is a detail

of a group photo of 4-H Club members for that year. The photo of Norman Johnson is from a family photo album organized by Pearl, and on the reverse of the print Royce had written, "Norm Johnson & I at the A C Winter Quarter 1938 (February?)," and in pencil below, "Norm was killed in France WWII."

[112] The left photograph is a detail of a larger one which Royce would show repeatedkly over the years, commenting on the wistfulness he saw in his father's face, reflecting on the fact that he considered his own opportunities to be an extension of his father's aspirations. The right photograph was taken at the Smith family property located on Vine Street in the neighborhood called Cottonwood, south of Salt Lake City. Gatherings of Royce's extended family were often held there.

CHAPTER 3:

[1] The mission home was at that time located at 31 North State Street, and contained dormitories for newly called missionaries. Training was rudimentary by current standards, and was mostly performed by General Authorities of the Church. The training lasted only a few days, and no language instruction was given.

[2] First mission journal, p. 3, November 16, 1939. Latter-day Saint temples are separate structures very distinct from ordinary meetinghouses, specially set apart and dedicated as holy places for the performance of the most sacred ordinances of the religion. To be permitted to enter the Temple and participate in these ordinances, one must be a member in good standing of the Church of Jesus Christ of Latter-day Saints, and have passed through an interview process to establish moral worthiness, which includes adherence to Church principles as well as to high standards of personal conduct. The temple "endowment" spoken of is a series of highly symbolic rituals and ceremonies representing the individual's progression toward the presence of God. In the process, the recipient makes solemn covenants to follow specific principles of the gospel.

[3] First missionary journal, p. 3-4, November 18, 1939 (quoted) and Nov. 21, 1939.

[4] First mission journal, p. 4, November 22, 1939. Evidently most missionaries of that time served as Elders, but those from Royce's stake were ordained Seventies because Royce's stake president believed that this was more in keeping with Church doctrine.

The ordinance of "setting apart" is very similar to that of "ordaining", and both are performed by means of the laying on of hands on the head of the individual, a prayer being pronounced by a priesthood bearer holding authority and being authorized to do so. The term "ordaining" refers to the conferring of an office in the priesthood, such as that of priest or elder; that of "setting apart" refers to granting the individual a specific calling, such as that of missionary.

[5] Royce's first impressions of Los Angeles are worthy of note. "Oh, it is beautiful down here, feels like spring. Everything is pretty and green, flowers, palm trees, shrubs – Any way, I wish you could see it. I don't think

much of Los Angeles, too funny the way its put together, streets running every which way and every one in a terrific hurry – I could get lost easily." (Letter RSB to PD November 25, 1939.)

[6] Letter, RSB to PD, misdated November 17, 1939, postmarked November 27.

[7] First mission journal, p. 5, November 27, 1939, content condensed.

[8] First mission journal, p. 6 & 7, November 28 and 29, 1939.

[9] Latter-day Saint missionaries are virtually always assigned a companion with whom to labor. Companion assignments change from time to time, as determined by the mission president.

[10] First mission journal, p. 8, December 4, 1939, content condensed and edited.

[11] First mission journal, p. 13, December 17, 1939, and p. 35, December 30, 1939, content condensed.

[12] First mission journal, entries for December 10, 1939 (p. 10-11), December 17, 1939 (p. 13) and December 24, 1939 (p. 16), content edited and condensed.

[13] First mission journal, p. 18-19, December 31, 1939 and January 1, 1940.

[14] "Moved today from the tourist camp to the home of Mrs. Cronkrite, a very nice lady. I like the new place. It is so much cleaner, neater & free from cockroaches, anyway, it is really lots better. She's not a Mormon but I wish she were, she would make a good one I bet." (First mission journal, p. 19, January 4, 1940.) There may have been no formal address, as in a list of residences at the beginning of the journal Royce listed first "Tourist camp at Fabens Texas" and subsequently "Home of Mrs. Cronkrite Fabens, Texas." The mailing address was simply "P.O. Box 1, Fabens, Texas". Royce appeared in the 1940 U.S. Census as a lodger in the home of Myrtle Cronkrite, together with Wilmer Porter May and Myrtle's two sons. She was listed as a school teacher. (U.S. 1940 Census, State of Texas, County of El Paso, unincorporated town of Fabens located in Justice Precinct 5, S.D. 16 E.D. 71-20, sheet 1A. Royce appears on line 4.)

[15] First mission journal, p. 43, March 25,1940, content condensed and edited.

[16] First missionary journal, p. 25, January 24, 1940; also p. 30, February 8 and 9, content condensed and portions translated from Spanish. In the February 9 entry, Royce added, "Met the usual group of Catholics. Were about as welcome to the Catholics up there as the plague but we arn't giving up yet. If we havn't done anything on that side of town but this, we have made a few friends."

[17] To "administer" in this setting is to perform the priesthood ordinance of anointing and blessing on the sick, referred to biblically in James 5:14-15. It is generally performed by two men, one of whom anoints the sick person with olive oil previously consecrated to that purpose, and the second of whom seals the anointing and gives a blessing. This appears to be the first such experience recorded by Royce.

[18] Royce kept these typewritten talks in a small six-ring binder together with notes from conferences and quotes that he found useful in his missionary work; eventually he added popular songs in Spanish. His first brief typewritten talk was dated December 3, just after his arrival in Fabens, but was not

given at that time; it contained numerous notations for the unfamiliar Spanish pronunciation. Subsequent talks, dated January 14 and February 11, show increasing facility with the language, but obviously required the help of his senior companion.

[19] First mission journal, p. 31, February 11, 1940.

[20] First mission journal, p. 24, January 18, 1940 and January 21, 1940, content condensed.

[21] First mission journal, p. 39, March 10, 1940.

[22] First mission journal, p. 41, March 17, 1940.

[23] First mission journal, p. 44, March 27, 1940.

[24] First mission journal, p. 51, April 19, 1940.

[25] First mission journal, p. 32, February 13, 1940, content condensed and edited.

[26] First mission journal, p. 36, February 28, 1940.

[27] First mission journal, p. 37, March 3, 1940.

[28] First mission journal, p. 38, March 4, 1940, content edited and condensed. Describing the caverns to Pearl a few days later Royce wrote, "By a large pillar of Rock in there they call "The Rock of Ages" every one sits down (200 in our party) then they turn the lights out. It is so black you can almost feel it (750 feet underground). Then they sing 'Rock of Ages' down the caverns (a quartet). Next they start turning the lights on 1/2 mile away and gradually bring them toward you. It just thrills you all over. You can imagine any shape or form that you like in the things under there." (Letter, RSB to PD, March 9, 1940.)

[29] First mission journal, p. 41, March 17, 1940.

[30] First mission journal, p. 42-43, March 24, 1940.

[31] Taped recollections, September 1997, Tape 2.

[32] First mission journal, p. 61, May 24, 1940.

[33] Letter, RSB to PD, June 7, 1940. The marriages spoken of were temple sealings, in accord with the Latter-day Saint doctrine of eternal family relationships. Taking place exclusively in temples, sealings are ordinances which link husbands to wives and parents to children in a relationship that is believed to endure beyond death into the eternities. On that day in his journal, Royce wrote, "The pleurisy pains have left me. This whole day was wonderful. First, we spent all morning in a session thru the Mesa Temple. It is the first time I have ever witnessed sealings. I am grateful that I had the priviledge. In the afternoon we had our missionary report and testimony meeting. A wonderful spirit was present from start to finish." (First mission journal, p. 62, May 28, 1940.)

[34] First mission journal, p. 63-64, May 30-June 8, 1940. Royce appears to have avoided any mention of his serious illness in his letters home. His mother, who constantly fretted about his health, made no mention of it in her letters to him, and although he described the Mesa conference events in detail to Pearl, the most he wrote of his illness was, "We were not even able to come back here untill Wednesday because of certain unavoidable reasons." (Letter, RSB to PD, June 7, 1940.)

[35] Taped interview, September 1997, Tape 2.

[36] First mission journal, p. 64, June 5, 1940 and p. 65, June 9, 1940.

[37] First mission journal, p. 65, June 11, 1940.

[38] This refers to an instruction given to Joseph Smith, the founding prophet of the Church of Jesus Christ of Latter-day Saints, in 1833, which is widely known by Church members as the "Word of Wisdom" (see Doctrine and Covenants, section 89). In this revelation, Church members are counseled to abstain from alcohol, tobacco, coffee and tea. This later came to be regarded as a basic commandment for members of the Church, and a requirement to enter the temple or hold positions of responsibility. In a society in which drinking, smoking, and the use of coffee or tea not only was acceptable but in many cases considered a social necessity, it was this practice perhaps more than any other which set the Latter-day Saints apart as a people at that time.

[39] First mission journal, p. 71, June 30, 1940.

[40] Letter, RSB to PD, July 27, 1940. It was ironically the corrupt "patronage" system of Mexico which probably spared the missionaries more serious problems. By gaining the favor of a local authority, the rules were superseded.

[41] First mission journal, p. 78, July 21, 1940, content condensed.

[42] Letter, RSB to PB, July 27, 1940, content condensed.

[43] Senior companion; LDS missionaries generally work in companionships of two, with one missionary designated as "senior" companion with principal responsibility for the work in the area, and for the training of the junior companion. This involved a significant increase in responsibility.

[44] First mission journal, page 79 and 80, July 23 and 24, 1940. The "Lamanites" are a reference to peoples mentioned in the Book of Mormon, a religious record of ancient inhabitants of the Americas which is considered scripture to members of the Church of Jesus Christ of Latter-day Saints. The book tells of two nations, the Nephites, who at the end of the record were annihilated, and the Lamanites, presumed to be among the ancestors of the American Indians, to whom promises of ultimate redemption are extended in Book of Mormon prophecies. Because Mexicans and other Latin Americans have a strong admixture of Native American heritage, they are believed to be heirs of these promises, hence the interest of the missionaries in this concept.

[45] Quotes are from first mission journal, p. 80, July 26, 1940 and letters from RSB to PD, July 27 and August 3, 1940.

[46] Letter, RSB to PD, July 27, 1940, content condensed.

[47] First mission journal, p. 81, July 28, 1940.

[48] First mission journal, p. 80, July 26 and 27, 1940, and p. 85 and 86, August 11, 1940.

[49] First mission journal p. 83, August 3, 1940.

[50] Royce had described this in a letter to Pearl after a missionary meeting. "He was almost miraculously saved from death by faith and prayer at his opperation. The doctor said he had done all he could and said he could not possibly live. His appendics had broken and the poison was already in his body. The doctor said nothing remained but prayer. They administered to him and today he is just about well again." (Letter, RSB to PD, March 9, 1940.)

[51] First mission journal, p. 84, August 7, 1940.

[52] Genealogy is an important aspect of Latter-day Saint religion because of the belief in vicarious work (baptisms and other ordinances) on behalf of deceased persons, who in the afterlife may receive the blessings of the gospel if they choose to accept them. Church members search out their ancestors in order to provide these ordinances.

[53] Primary is an organization of the Church established in the 1800s specifically for the religious instruction of children.

[54] Letter, RSB to PD, August 12, 1940, content condensed and edited.

[55] First mission journal, p. 86-87, August 13, 14 and 15, 1940, content condensed. Word of the flood made news in the Salt Lake City area, as on August 19 Pearl wrote, "We read in the paper about a flood at Tucson. I do hope you were all right, although I'm sure you were." (Letter, PD to RSB, August 19, 1940.)

[56] First mission journal, p. 88, August 18, 1940.

[57] First mission journal, p. 88, August 20, 1940.

[58] First mission journal, p. 90, August 24 & 25, 1940.

[59] Near the end of a letter in August, Pearl wrote, "I hope you have to hold a street meeting or two. That's where Christ taught." (Letter, PD to RSB, August 7, 1940.) Royce had replied, "About your street meetings, about the only place they hold them in this mission is in Texas *al sur*. I do hope I get the opertunity to attend some sometime. It would be just about imposible to have them here." (Letter, RSB to PD, August 12, 1940.)

[60] First mission journal, August 31, 1940.

[61] First mission journal, p. 94-95, September 4, 1940, content condensed and edited.

[62] First mission journal, p. 98, September 13, 1940, content condensed.

[63] Letter, RSB to PD, September 7, 1940.

[64] First mission journal, p. 109-110, October 11, 12 and 13, 1940.

[65] Letter, RSB to PD, September 21, 1940.

[66] First mission journal, p. 111, October 16, 1940.

[67] Mission journal, p. 124, November 18, 1940.

[68] First mission journal, p. 113, October 21 & 22, 1940, and p. 114, October 23, 1940.

[69] First mission journal, p. 149, January 15, 1941.

[70] George Albert Smith would become president of the Church in 1945 at the death of President Heber J. Grant.

[71] Taped interview, September 1997, Tape 2. The time spent with the apostle was probably less than a week. Although Royce made no mention in his journal about interpreting for Elder Smith, the fact is also recorded in Pearl's handwriting on the reverse of the photograph of Royce with Elder Smith in this chapter: "Royce served as interpreter for Elder Geo. Albert Smith who was an apostle at the time".

[72] First mission journal, p. 120-122, November 9, 10 and 11.

[73] First mission journal, p. 125, November 21, 1940.

[74] On December 13, 1940, Royce wrote, "Today, I found out that my Dad and Elder Rowley's Dad were in the New Zealand Mission at the same time. Quite a coincidence to find their sons both together on a mission about thirty years later." (First mission journal, p. 135, December 13, 1940.)

[75] Letter, RSB to PD, January 1, 1941. In this passage Royce repeatedly alludes to a quote of Ralph Waldo Emerson, often used by (and attributed to) Heber J. Grant, who was then president of the Church: "That which we persist in doing becomes easier to do, not that the nature of the thing has changed, but that our ability to do has increased."

[76] First mission journal, p. 152-152, January 23 and 24, 1941, content condensed. The first reference to this member includes his name, but since in all subsequent entries Royce had either scribbled out the surname or had lined in a blank, and had simply neglected to find and scribble out the first, I have respected his wishes by leaving the member anonymous.

[77] Taped interview, September 1997.

[78] First mission journal, p. 170, March 7 and 8, 1941.

[79] The "Women's Relief Society" was founded by Joseph Smith in the 1840s in Nauvoo, Illinois as a service organization for the women of the Church. It was initially headed by Joseph Smith's wife, Emma Hale Smith, with a select membership which included many of the women of the area. After the Church's expulsion from Illinois the organization was revived in Utah by Brigham Young under the direction of Eliza R. Snow, and has been an important cultural and religious feature of the Church ever since.

[80] First mission journal, p. 174, March 16, 1941, and p. 175, March 18, 1941, content edited.

[81] First mission journal, p. 199, May 14, 1941.

[82] First mission journal, p. 193, May 2, 1941.

[83] First mission journal, p. 199 and 200, May 15 and 16, 1941.

[84] First mission journal, p. 197 and 198, May 12, 1941.

[85] Letter, RSB to PD, May 3, 1941.

[86] Letter, RSB to PD, June 16, 1941.

[87] First mission journal, p. 195, May 6, 1941, content edited and condensed.

[88] First mission journal, p. 197, May 10, 1941.

[89] First mission journal, p. 195, May 6, 1941. The chapel cleanup was recorded on p. 195-197, May 5 to May 10, 1941.

[90] First mission journal, p. 206, May 28, 1941. Royce added somewhat somberly, "I can hear guns from the fort rumbling tonite as we have the last few nights. May they never get any closer, but I fear they shall."

[91] First mission journal, p. 208, June 1, 1941, content condensed.

[92] First mission journal, p. 218, June 20, 1941, content condensed. In more recent years, as in the very early days of the Church, it has been the practice of missionaries to baptize investigators after a relatively short time when their understanding of Church doctrine is still relatively limited, and count on the fellowship of Church members to fill in the gaps. This was certainly not the case in Royce's time, and the Guzmán family, who clearly desired membership and in our era might have been baptized within a few weeks or even less, took nearly a year to become Church members.

[93] Letter, RSB to PD, July 11, 1941, content edited. Royce mentioned in his journal that this was the second Primary he had formed on his mission, the first being in Tucson (first mission journal, p. 229, July 10, 1941).

[94] Letter, RSB to PD, July 11, 1941, content edited and condensed. Royce would indeed run the family farm at least briefly, but the same month his

mother remarked on development underway in the west side of the valley, and also on the pressures of farming, which would eventually combine to bring an end to the family farm. "We were over to the farm the other day. The grain is ready to cut, also the second crop of hay so far. Smith has been working every day, but the smelter is going to shut down for a month or six weeks the first part of August. They are a little afraid about getting some of their men back if they give them a layoff as lots of them will try to get work in the ammunition plant that is going to be built just west of Redwood and Twenty-first South, and then there is that new airport just west off the Bingham Highway, so between the two there is lots of work going on. It is going to be a boom for the west side as they are going to need homes for a lot of people and if Smith can't work and run the farm I don't think we will have any trouble selling it or part of it, at least forty acres is too much to try and run and work too. Of course we won't do anything about it until after you come home and we see how things work out. Maybe the farm can keep you out of the draft and we don't have to sell it. But with things pulling that way it looks like the farm will be worth something and your Dad just can't do anything with it now. We are going over and spray the orchard again the last of the week. Dad has been doing the spraying. I think we will have apples enough to pay the taxes if we get any price for them." (Letter, Florence Smith Bringhurst to RSB, July 30, 1941.)

[95] First mission journal, p. 225, July 4, 1941.

[96] First mission journal, p. 250, August 15, 1941.

[97] First mission journal, p. 253-254, August 21, 1941.

[98] First mission journal, p. 255, August 25, 1941.

[99] First mission journal, p. 256, August 26, 1941.

[100] Within two weeks Elder Buchmiller would return to the area seeking work. At this time with the Great Depression nearing its end but still a factor, Los Angeles presented a rapidly growing area with a desirable climate and thriving industries, and a number of ex-missionaries were attracted back to the area. Royce himself would return to Los Angeles a decade later, upon completion of his postgraduate studies.

[101] Letter, RSB to PD, August 31, 1941.

[102] First year journal, p. 269, September 23, 1941, and p. 270, September 26, 1941.

[103] First mission journal, p. 263, September 11, 1941.

[104] First mission journal, p. 264, September 12, 1941.

[105] First mission journal, p. 271, Sunday, September 28, 1941, content condensed.

[106] First mission journal, p. 275, October 5, 1941. Regarding the Guzmáns, Royce had recorded in the margin of his journal shortly after his arrival in Los Angeles, "Elders Terry and Sampson ran into the Bapt. minister down at Guzman's. Apparently it done some good. He lit into them and *Hna.* Guzman lit into him" (first mission journal, p. 257, written in the margin above entries for August 28, 29, 30, and 31).

[107] Letter, RSB to PD, October 6, 1941, content edited.

[108] Letter, RSB to PD, October 18, 1941. Aimee Semple McPherson established a religious denomination, known as the Foursquare Church, in which

she combined fundamentalist teachings with a flair for technology and celebrity. She was famed for faith healings, and though she actively maintained important charities in addition to her preaching, details of her personal life, including multiple marriages, lawsuits, and a widely-publicized kidnapping, were often in the news.

[109] Letter, RSB to PD, October 25, 1941, content condensed. Royce described the details in his journal as follows: "We got back at the house about two P.M. and a girl came over and asked us to come over and hold funeral services for her mother at three P.M. She had been shot by her husband and then he killed himself. Their name was 'Molina' – Margarita was the girl's name."

[110] First mission journal, p. 287, October 29, 1941, content condensed and edited.

[111] Taped interview, Tape 2, 1997, content edited.

[112] First mission journal, p. 293, November 11, 1941.

[113] First mission journal, p. 295, November 13 and 14, 1941.

[114] First mission journal, p. 295-297, November 15 and 16, 1941, content condensed and edited. A sad postscript to this story is that Elder Richard R. Lyman was excommunicated from the Church for violation of the law of chastity scarcely two years later. See chapter 5.

[115] Letter, RSB to PD, December 1, 1941.

[116] Second mission journal, page 5, November 27, 1941.

[117] Second mission journal, p. 8 and 9, December 5 and 6, 1941.

[118] Taped interview, September 1997, tape 2, content edited. In the same narration, Pearl told of her experience the same day. "I was at church. And on my way home, Walt Vance came riding his bicycle up the street, and he says, 'Pearl, we're at war!' And I said, 'You're crazy, Walt.' And I walked home, and walked in the house, and my mother was sobbing, listening to the radio, and I thought, 'There goes my life. It's over.' I just thought...I knew what was going to happen. I knew he was going to war. Well, what could you do? It was going to be a lonely life."

[119] Second mission journal, p. 9-11, December 7 and December 8, 1941. In his journal Royce grossly underestimated the losses in the Pearl Harbor attack, perhaps relying on news reports of the day: "The Japs got one and possibly two of our battleships; by all indications their attack was very highly successful."

[120] Letter, RSB to PD, December 9, 1941, content condensed.

[121] Letter, RSB to Bringhurst family, December 17, 1941, content condensed.

[122] Royce described conditions to his parents shortly after the attack. "Perhaps the only change noticeable is that everyone seems to be a great deal relieved knowing at least in part what course we shall pursue now. War has been a part of this world for a long time. It appears to me like peoples' minds were pretty well prepared for this. There hasn't been any hysteria or violence of any kind here. I pass thru the Japanese section almost every day. They call it "little Tokyo." The only thing different about it is the large number of police officers that there are there now. Down in the city itself they have antiaircraft guns set up in different parts ready for action. Other than that and soldiers' barracks in the parks it isn't much different. Japanese

planes have been reported off the coast and over California. There must be a aircraft carrier or two out in the ocean. Our planes have been racing by during the last hour I have spent here at the house. Los Angeles hasn't been blacked out as yet but a number of coast cities have and radio stations down here have been silenced except for every half hour or so when they are allowed to summarize the news. This is to prevent them from being a guide to enemy aircraft they say, and also to prevent interference with military communication." (Letter, RSB to Bringhurst family, December 9, 1941.)

[123] Second mission journal, p. 11-12, December 10, 1941.

[124] Second mission journal, p. 12, December 11, 1941, content edited.

[125] Baptisms in the Church of Jesus Christ of Latter-day Saints are never performed before the age of 8, which is regarded as the "age of accountability", when a child has acquired sufficient maturity to begin to be responsible for his or her actions. The fact that this child was baptized on her eighth birthday indicates that she was not considered to be a "convert baptism", but rather the baptism of a child in a member family, or a "child of record."

[126] Second mission journal, p. 13-14, December 14, 1941, contents condensed.

[127] Second mission journal, p. 19-20, December 24 and 25, 1941, condensed.

[128] Letter, RSB to PD, December 26, 1941, content condensed.

[129] Letter, RSB to Bringhurst family, January 4, 1942.

[130] The *Improvement Era* was an official periodical published by the Church from 1897 until 1970, when it was replaced by the current magazines, the *Ensign*, for adults, and the *New Era*, for youth.

[131] Letter, RSB to PD, January 26, 1942.

[132] Second mission journal, p. 50, February 8, 1942.

[133] Second mission journal, p. 52, February 11, 1942, and p. 54, February 12, 1942, content condensed.

[134] Letter, RSB to PD, February 22, 1942.

[135] Second mission journal, p. 63, February 24, 1942, and p. 63-64, February 25, 1942. The shelling incident, on February 23, 1942, was an actual attack by a Japanese submarine, which surfaced and fired shells at a fuel depot near the coast at Santa Barbara, causing minimal damage. Although this proved to be an isolated incident never to be repeated during the war, there was no way for the nervous inhabitants of the seacoast cities to know the location, intentions, or capabilities of the enemy. The anti-aircraft fire over Los Angeles the following night was probably a result of heightened nerves in the wake of the shelling. Although Royce recorded in his journal a few days later (p. 65, February 26, 1942) that it had been confirmed that there were about 15 enemy planes over Los Angeles, in the end it was concluded that an anti-aircraft battery had simply opened fire on a stray weather balloon, and that others fired at the tracers left by the initial anti-aircraft bursts. In retrospect, it is almost impossible that Japanese planes could have been in the area, and of course no bombs struck the city. The event was later jokingly referred to as "The Great Los Angeles Air Raid of 1942."

[136] Norman Johnson was stationed at Thunderbird Field in Arizona, in pilot training, having completed a mission of his own. (Letter, RSB to PD, March 17, 1942.) He wrote a return letter to Royce and to his cousin Elder Johnson

who worked in Royce's district. Royce retained this letter, dated April 5, 1942. Both he and Elder Johnson were interested in the Air Corps, and Norm described his pilot training, including the entrance qualifications. Among other things, he confirmed that whereas only single men were enlisted when he entered, the wartime demand had altered the qualifications to include married men, and he recommended they both consider marriage first, to "have the temple work done" prior to entering combat.

[137] Second mission journal, p. 27, January 9, 1942. Royce would never in his later years have used a disparaging ethnic term like this, and his comment, meant to be humorous, must be taken in the context of the times.

[138] Second mission journal, p. 55, February 13, 1942.

[139] Second mission journal, p. 61-62, February 22, 1942, content edited.

[140] Second mission journal, p. 70-71, March 4, 1942.

[141] Second mission journal, p. 81, March 13, 1942.

[142] Second mission journal, p. 86-87, March 19, 1942, content condensed.

[143] Second mission journal, p. 112, April 17, 1942, content condensed.

[144] Letter, RSB to Bringhurst family, January 24, 1942.

[145] Second mission journal, quotation from p. 74, March 7, 1942, and 75-76, March 8, 1942. President Haymore also spoke at length with the Muñoz family with whom the missionaries lived, and Royce hoped that he would be baptized the following week. In the end, the Muñoz's were never baptized, though Royce recalled visiting them a decade later when the family moved to Los Angeles. "They were old. But they greeted me like I was a long lost brother. Never joined the Church. But they didn't think they had to." (Taped interview, 1997, Tape 2.) Royce wrote his Spanish translation of "Unfold Ye Portals" in his second mission journal on p. 37, January 25, 1942.

[146] Second mission journal, p. 94-95, March 27, 1942.

[147] Letter, RSB to PD, March 17, 1942.

[148] Second mission journal, p. 82, March 15, 1942. On the last regular Sunday of Royce's mission, April 19, 1942, he wrote "It's the best Sunday School I have ever been to down there. The room was right full. There was perfect order and really a fine spirit present. It will be fine to remember as the last regular Sunday School I ever attend in L.A." (p. 114, April 19, 1942).

[149] Letter, RSB to PD, March 20, 1942. Royce had kept among his personal treasures a letter from W. L. Frost (Bill), dated April 13, 1936, when he was still a junior in high school, and Bill had joined the Navy.

[150] Letter, RSB to PD, April 15, 1942. This was Royce's final letter to Pearl; by this time he had learned of the change of his release date.

[151] Letter, RSB to PD, March .

[152] Second mission journal, p. 103-104, April 5, 1942. The reference to a "flaming torch and a trumpet" was an allusion to the Biblical account of Israel's victory over the Midianites against great odds, recounted in Judges 7:19-23.

[153] Second mission journal, p. 108-109, April 12, 1942.

[154] Letter, RSB to PD, April 15, 1942.

[155] Second mission journal, p. 117, April 23, 1942, and p. 118, April 24, 1942. The Spanish term *unción* means "anointing", and refers to the practice of "administering to the sick," by anointing with consecrated olive oil.

156 Second mission journal, p. 119, April 25, 1942, and p. 120-121, April 26, 1942.
157 Letter, RSB to PD, March 17, 1942.
158 More than fifty years later, writing to his son who had served a mission of his own, Royce wrote, "My life was changed and eventually focused because of my Spanish American Mission call. I would not be where I am today; I would not have had the career I have enjoyed professionally if someone hadn't decided to send me to that particular mission, and if the president of the mission had not sent me to California (San Diego and Los Angeles) to spend my last 1-1/2 years. The path was sort of set from that time on, and I may have looked back some, but not very much. It was 'California all the way' from that time on." (Letter, RSB to John & Betty Bringhurst, August 30, 1993, content edited.)
159 *Doctrine and Covenants* 88:118.

CHAPTER 4:
1 Second mission journal, p. 122-124, April 28-April 30, 1942. In his taped interview (Tape 2, 1997), Royce recalled: "I took off, and went up to Oakland, to my sister Naida and her husband Ernie Buchan, and I had a very nice visit with them. I remember they bought me a new pair of shoes. I needed a new pair of shoes. I needed a new suit too, because the suit I had actually had a hole in the seat. I had to be very careful with that."
2 Second mission journal, p. 124, May 2, 1942.
3 According to Pearl, their initial plan was to marry on Saturday, but Pearl's mother, who had a superstitious streak, thought it a bad idea to marry on a Saturday, so they settled on a Thursday, which presumably was a convenient time for both families. One of Pearl's daughters recalled her saying that the 13th was ruled out by Pearl's mother as well, that being an unlucky number.
4 Letter, RSB to PD, April 15, 1942, content condensed. This letter was written twelve days before the official end to Royce's mission.
5 Taped interview, September 1997, Tape 2. The conversation continued: Royce: "They gave you a little bonus if you were married, and it seemed that it would take forever before you'd start giving enough to even help her. But it worked out. She went out...this mother of hers, made this possible too, because she lived with her all the time I was serving in the military." Pearl: "And Jeanette and Burnell supplied us with vegetables. His mother and dad had me out many a time for dinner, and other things when he was gone, and that's when I became really closely acquainted with Grandma Bringhurst." Royce: "Well, we made everything come true, of course. The argument against getting married is, the first thing you know she'll be pregnant..." Pearl: "And I was." Royce: ...and they were absolutely correct. So here we are with one child, that was Jean..." Pearl: "And he saw her for, what, nine days."
6 After a year of marriage Royce recalled the circumstances surrounding the engagement ring. "I always picked the funniest places to do things. I'll never forget coming from town on my way to home, hurrying into your place and finding you in the kitchen and giving you the rings without any formalities at

589

all. Maybe I should have waited untill that nite but I just couldn't and I think you were glad I gave them to you as I did because it was a bit different. I liked to see them on your hand." (Letter, RSB to PDB, Friday, July 30, 1943.)

[7] The requirements for a temple recommend included both a belief in the basic doctrines of the Church and faithfulness to its precepts, as well as compliance with the commandments and practices associated with faithful Church membership, such as basic honesty, moral cleanliness, and keeping of the Word of Wisdom, including abstinence from alcohol, tobacco, coffee and tea.

[8] This detail, which had been mentioned by Pearl in relation to their marriage, was confirmed by her during her final illness, a few days before her death in April of 2013. This was a matter of about which Royce felt quite self-conscious. Throughout their marriage Royce would make certain Pearl was well dressed, and he frequently picked clothes for her.

[9] See Pearl D. Bringhurst, *My Life and My Poetry*, 2012, p. 46.

[10] Royce's father found it increasingly difficult to manage the farm, as he worked full-time at the copper smelter. Royce's younger brother George was now married and had no interest in the farm. His older brother Smith had helped to manage the farm the previous year. It is not clear what he was doing at this time, but it may have been that he had problems of his own. In a final postcard to his parents the day his mission was completed, Royce had written cryptically, "Sorry to hear about Smith." (Postcard, RSB to Bringhurst family, April 29, 1942.)

The months after his marriage were a blank spot in Royce's writings. The last entry in his mission journal was dated May 12, 1942, two days before the marriage. Most of the information for the remainder of this chapter comes from later recollections of Royce and Pearl, as related either in subsequent letters or to their children in their older years, though some of the details are drawn from letters written by Pearl to her mother, Jane Davidson, as she left Salt Lake to visit two of her children in other states.

[11] The following year Royce made mention of this arrangement in a letter to Pearl: "Remember how we lived off your earnings the first weeks of our married life? You'll never know how much it made me appreciate you for what you are even though it did hurt my pride quite a bit to think of my wife keeping us. I was glad when you quit work even though we at the time hardly knew what to do or what we would do for money. That's a part of a man I guess. I never want you to work that way again. It made it so we were together and yet apart someway. It was so wonderful when we did start being together all the time. I guess that was our Honeymoon until the duration. I hardly think we could have been happier even had we had a so called Honeymoon of the usual type. I'll always be grateful for the help the folks gave us at that time because as never before in my life I needed help. I wish I could pay back everything they have done and given to me. I guess I appreciate that now more than anything. Faults yes, but its worth it to know you have people like that backing you up." (Letter, RSB to PDB, Monday September 6, 1943.)

[12] Letter, PDB to Jane Sutherland Davidson, November 1, 1942.

[13] RSB and PDB, recorded recollection, Tape 2, 1997, content condensed. In November, Pearl wrote in a letter to her mother: "Our little goat isn't so little anymore, and it's a regular pal to Helen's old white rooster." (Letter, PDB to Jean Sutherland Davidson, November 1-2, 1942.) Shortly after Royce's departure for military service, the white rooster probably became a meal for Pearl; Royce wrote after receiving the first of Pearl's letters, "Glad the rooster was good. I'd like to have tried some of it." (Letter, RSB to PDB, November 30, 1942.) The slaughtering and consuming of a domestic animal, even one that had become a pet, was a necessary aspect of farm life which Royce and Pearl always took in stride, though it would be much harder for their children to do so.

In 1944, while in military service in Africa, Royce saw a herd of goats, which brought back happy recollections of this time. "It kind of reminded me of the wedding present you didn't appreciate at first. I'll never forget the day I went over and got the bleating rascal. It must have cost the boys no little bit to send her all that way. I really got attached to the beast before we left and felt a bit badly about it just as you did when we had to leave home." (Letter, RSB to PDB, Wednesday, April 26, 1944.)

[14] RSB, recorded recollection, Tape 2, 1997. In a letter to Pearl during military training a year later, Royce recalled, "Remember how it was a year ago now? We were out on the farm together. You were if I remember correctly still working and trying to be a wife too. It wasn't very long after this that we found out that there would probably be three of us instead of two. I don't remember just how I did feel when we first found out. I guess I was a bit worried and afraid, but I believe I knew it was going to happen all right...I used to think about leaving you when I should have to go to war and I guess I always did think one way about it even before we were married, and that was that I wanted to leave you that way, with a baby in the making. Really the only objectional thing that ever entered my mind and it really did bother me plenty was finance. I was so afraid I would have to leave you that way with no money in the bank and no money on hand." (Letter, RSB to PDB, June 22, 1943, content condensed.)

[15] Letter, PDB to Jane Sutherland Davidson, July 28, 1942, content condensed. At that time Pearl's mother was in San Antonio, Texas, with Pearl's sister Laura who was due to have a baby.

[16] Letter, PDB to Jane Sutherland Davidson, August 19, 1942. This was actually the second of three letters, dated August 17, August 19, and a postscript dated August 20, 1942, which were sent to Pearl's mother in San Antonio, Texas.

[17] Letter, PDB to Jane Sutherland Davidson, July 28, 1942. Royce also later described the hailstorm: "Do you remember the stormy day when our tomatoe crop and the grain were beat to pieces, also the damage was done to the apple crop. We watched hail beat down hard till it covered the ground. I believe that I just barely got back to the house before the hail started." (Letter, RSB to PDB, Saturday, February 19, 1944.)

Evidently after Royce entered the military, the cannery attempted to hold him financially liable for the loss of the tomato crop. He expressed his feelings to Pearl: "As for those tomatoes, we lost all the work we put in on them and the factory loses the plants. That's the way it figures up to me. If they don't like it, they know what they can do. Call it the scotch in us or what you will. It's bad enough to work for no pay without paying and working and then recieving nothing." (Letter, RSB to PDB, December 10, 1942.)

[18] Letter, PDB to Jane Sutherland Davidson, August 19, 1942.

[19] Letter, PDB to Jane Sutherland Davidson, August 18 and August 20, 1942.

[20] Letters, PDB to Jane Sutherland Davidson, August 19, 1942.

[21] Letter, PDB to Jane Sutherland Davidson, November 1 and 2, 1942. Content rearranged and condensed.

[22] Royce S. Bringhurst, recorded recollection, September 1997, Tape 2, content edited and condensed. Jay later served in the Korean War, which was followed by mental illness and death by suicide in the early 1970s. Royce recorded his reactions in his journal years later, when he happened to see Jay's son at his sister Dean's house. "It was as if Jay were back from the dead (he was Dean's oldest boy, a delayed victim of the Korean War, where he was a photographer). I remember Jay most for the time death nearly touched him but passed him by on the farm in 1942. I often wondered why his life was preserved only to have the aftermath of tragedy in Korea destroy him. I think I see the reason in his son." (Davis Journal, volume 7, August 15, 1983.)

[23] Letter, RSB to PDB, January 22, 1943. Shortly after entering military service, Royce wrote to his family, "So your having trouble with the old truck again. I believe or rather thot that I had had enough with that thing to last for quite some time, but I guess not. I'm just glad that it didn't ever quit on me while I was up at the market with a load of apples. That would have been most embarrassing to say the least. I still think I had more good luck with the old thing than bad." (Letter, RSB to Bringhurst family, January 5, 1943.)

[24] Letter, PDB to Jane Sutherland Davidson, August 19, 1942.

[25] Pearl D. Bringhurst, Recorded interview, Tape 2, September 1997.

[26] Letter, RSB to PDB, Wednesday, February (labeled February 8), 1944.

[27] Letter, PDB to Jane Sutherland Davidson, November 1, 1942, content edited. Pearl's narration continued: "About $100.00 went for expenses of buying, boxes, baskets, paying pickers and paying for market stall. I went to market with Royce one morning. It was really something. We've kept track of every sale though and it has really been quite an experience to help keep track of something like that. Royce thinks I'm a pretty good bookkeeper."

[28] Evidently there was a government incentive for sugar beet production, which Royce made mention of in a letter home from the military the following year: "Now about the money you mentioned which is coming from the beet crop, really I had forgotten all about it. It's that government subsidy payment isn't it? The one payment from the company itself was made in December. I'll write to Dad about it. If we keep it all of course you will pay tithing on it and put it in the bank." (Letter, RSB to PDB, March 5, 1943.)

[29] Letter, RSB to PDB, June 22, 1943. Shortly before shipping overseas, Royce wrote longingly of the apple orchard. "I'll never forget the orchard. I loved that place. It always seemed more mine than anything around. I trimmed it year after year. I irrigated it. I built dikes in it and scattered alfalfa

seed to stop the washing away of the soil. I sprayed it and tried to be kind to it if you can be kind to trees. Anyway, it seemed that the old orchard kind of wanted to do something really nice for us so it produced the best crop of apples we ever had for you and me and sort of made it so I could leave home without worrying too much about taking care of you and the baby that was then on its way." (Letter, RSB to PDB, Sunday, March 19, 1944.)
[30] Letter, PDB to Jane Sutherland Davidson, Nov. 1-2, 1942. Pearl's mother was at this time staying with her son, Daniel Davidson, and his wife Mary, whom Royce had visited in San Bernardino, California during his mission.
[31] The purchaser was a man named Kiel, who Royce recalled "was referred to as 'the German' by boys my age. He was a hard working immigrant, cabinet builder and general carpenter as well as farmer. They bought our old farm in 1942 (a good buy in retrospect). My mother always regretted that sale but forgot that the 'far down' property they bought in 'Cottonwood' with that money was just as good a buy for them. 'Bill' Kiel was a good man, and I regret any negative comment I may have had for him as a boy. I believe that it was simply 'spin off' from WWI attitude toward Germans. You cannot escape the prejudice of your age and contemporaries (elders and peers, particularly the former)." (Davis Journal #3, September 24, 1978.) Bill Kiel evidently married a daughter of Royce's "Uncle Sam," the brother of his grandfather, so the property was still held by distant relatives after Royce's parents sold it.
[32] Letter, RSB to PDB, Thursday, August 17, 1944, content condensed.
[33] Pearl D. Bringhurst, recorded narration, Tape 2, September, 1997, content edited. Royce described the same event in a letter from aviation school: "It's been over four months now since I last saw you as I told you 'goodby' and you went one way to get in the car and I the other way to get in the train. I saw you cry and felt my own throat tighten up and choke. It wasn't easy to tell you 'goodby' that day. Time has gone by rapidly in a way, but the time just can't be hastened enough when we are to meet again."

The specific date of Royce's departure is inferred, since his orders required him to report to the Air Force Classification Center in Santa Ana on November 21, 1942, a date Royce later confirmed in his journal and in mail home. Presumably this was an overnight train which left the 20th and arrived at its destination the 21st, but depending on the timing of the train, the departure could have been either November 19 or November 20.
[34] This began another long separation for the newlywed couple, during which they would once again communicate only by mail, with an occasional phone call. Before a year was out, Royce would reflect in a letter to Pearl, "I like to think of you and of home and dream of the time when I can return and maybe not live the old life, but start a brand new one that will be better, finer and fuller. You see I have never lived with you yet when I knew that we were really together to stay. Ever since we first knew and loved each other we had to bridge accross years when we knew we were to be apart in order to reach in our thinking the time when we should be together. We havn't yet been able with untroubled thoughts to say that our uninterrupted life together has begun. I guess these interruptions have made us to grow a bit more and have made me a whole lot wiser. I like to think that they have helped us

and will help us to understand each other much better. I know it has made me appreciate the time we do manage to claim together. I don't know when we will at last be able to be together for always. Really, I don't even know whether or not our whole lives will be like this. We don't know what the future holds in store for us. I don't believe I want to know untill we live thru it. The savor would be gone from life if you couldn't look forward and wonder what tomorrow will bring...I do believe things will turn out all right. You stand by me always. I honestly don't know what I would do or what I'd be were you not such a vital part of my life, even though you are distant from me." (Letter, RSB to PDB, August 9, 1943.)

CHAPTER 5:
[1] Letter, RSB to PDB, November 24 & 25, 1942.
[2] On the orders from Fort Douglas, Utah, dated November 14, 1942, in the alphabetical listing of recruits Royce S. Bringhurst was listed 26th and Frank I. Denstad was listed 50th.
[3] Letter, RSB to Bringhurst family, December 6, 1942.
[4] Letter, RSB to PDB, November 24, 1942.
[5] Letter, RSB to PDB, December 4, 1942. In his next letter, Royce wrote more optimistically, "Don't worry about me. Honestly I don't. I felt down in the dumps for a few days but I didn't for one minute believe that I would wash out. You know I have never failed a single thing in my life and really I don't intend to begin now." (Letter, RSB to PDB, December 7, 1942.)
[6] Letter, RSB to PDB, December 20, 1942. Royce described the meeting similarly to his parents: "Today we held our first church service. We got one school house here. There were only twelve of us there and we had to sit on the floor but we had the sacrament and enjoyed a service that did everyone a whole lot of good. I really feel good about it." (Letter, RSB to Bringhurst family, December 20, 1942.)
[7] Letter, RSB to PDB, December 25, 1942, content condensed.
[8] At a time when long calls were a luxury, the two talked a full thirteen minutes, which cost them $4.60 counting tax (letter, RSB to PDB, December 28, 1942).
[9] Letter, RSB to PDB, December 28, 1942, content condensed and edited.
[10] Letter, RSB to PDB, January 3, 1943, content condensed. Royce left the base in company of Frank Denstad and Frank Van Limberg, and they arranged for a room to spend the night, then they separated so Royce could visit the members of his branch. Of the sleeping accommodations, Royce wrote, "The bed was terrible, not even worth the dollar it cost me. I will appreciate my old army cot when I get back to the air base." He added, "I am very sorry that I could not get to church this morning. Next to my wife I miss that more than anything." (First War Journal, Sunday, January 3, 1943.)
[11] Letter, RSB to PDB, January 5, 1943, content condensed and edited. Another important change in 1943 is that Royce began once again to keep a daily journal which had been sent him by Pearl and arrived just after the beginning of the year. He kept a daily record with some lapses until late October when, as the training schedule became more intense, he stopped writ-

ing altogether.

[12] Letter, RSB to PDB, January 5, 1943.

[13] Letter, RSB to PDB, January 25, 1943.

[14] Letter, RSB to PDB, January 22, 1943.

[15] Reid Ellsworth was transferred to Mather Field in Sacramento where he completed navigation training four months later, in May 1943. He immediately thereafter volunteered for combat duty, and was assigned to a B-17 squadron overseas without the benefit of crew training, and without being assigned to a formal crew.

[16] First war diary, February 17, 1943.

[17] Letter, RSB to PDB, January 18, 1943. The "W.P.A." was the Works Progress Administration, one of the major federal initiatives of Franklin Delano Roosevelt's "New Deal," which greatly expanded the function and scope of the federal government. At the height of the Great Depression of the 1930s, the W.P.A. employed literally millions of workers, engaging them in various projects ranging from public arts to infrastructure. The program was criticized for giving its participants incentives to work unproductively, and this was the basis for Royce's comment.

[18] First war diary, January 8, 1943.

[19] First war diary, January 16, 1943.

[20] Letter, RSB to PDB, February 3, 1943. In the same letter, Royce mentioned another feature of Santa Ana: "There are quite a few sport celebrities here as athletic instructors. Joe Dimagio the Yankee baseball player, and Fred Perry, the world champ tennis player are both here as instructors on our field. They are just enlisted men."

[21] Letter, RSB to PDB, February 2, 1943.

[22] Letter, RSB to PDB, February 2, 1943. Royce continued, "Now, one thing, I havn't seen a single one of the Mormon boys mixed up in a card game. Even the ones who don't care about anything else arn't doing it. They are different than the common run of the mill even though they drink and smoke."

[23] Letter, RSB to PDB, February 27, 1943. The context was the completion of pre-flight training. "It's certainly quiet right now in this old barn. There are only five of us in here instead of the usual gang. All the boys have gone to town for a big time. I hope they all make it back on time and intact. Some of them will really cut loose since we have finished academic work and get out of quarantine all at once. They really hit the L.A. night spots and go thru lots of money etc. etc. They will talk about it all next week. They are good fellows though and it is good to be with them. On the average their caliber is lots higher than that of the boys who come in here from the regular army."

[24] Letter, RSB to PDB, January 15, 1943.

[25] Letter, RSB to PDB, January 27, 1943.

[26] Letter, RSB to PDB, February 27, 1943, content condensed. After her birth, Royce took to referring to Jean as "the papoose."

[27] Letter, RSB to PDB, March 3, 1943. Royce learned of the assignment to Santa Maria after a member of the squadron caught a glimpse of the orders while on an assignment; the men were not officially informed until five days later, shortly before departure.

[28] Letter, RSB to PDB, Tuesday, March 9, 1943.
[29] First war journal, Monday, March 8, 1943.
[30] First war journal, Wednesday, March 10, 1943.
[31] Letter, RSB to PDB, Thursday, March 11, 1943, content condensed.
[32] Letter, RSB to PDB, Thursday, March 11, 1943, content greatly condensed. Royce gave more detail about the school's origins in a later letter: "It seems that this Mr. Hancock, who owns the field (school), part of Santa Maria, and the valley surrounding, is a philanthropist. As a multimillionaire he can afford to be one. They say he doesn't make anything on the cadets he trains here, just pays the help and keeps things up. He adds a dollar to the dollar the government appropriates for each cadet every day for food in order that we might have the best of everything good to eat. He is a world war pilot himself, plays in a symphony orchestra, and does no end of other things. I havn't seen him yet, but the cadets say he is usually arround in a greasy old pair of coveralls. I'll probably meet him soon." (Letter, RSB to PDB, March 14, 1943.)
[33] Letter, RSB to PDB, Friday, March 12, 1943, content condensed.
[34] First war journal, Friday, March 12, 1943.
[35] Letter, RSB to PDB, Saturday, March 13, 1943.
[36] Letter, RSB to PDB, Friday, March 12, 1943, content condensed.
[37] Letter, RSB to PDB, Saturday, March 13, 1943, content condensed. Between these two passages Royce added more detail: "Some of the fellows arn't feeling so good about it because they got sick at it. When you throw up in a plane you get a shower and some of the boys really came down messed up. We had to laugh about it."
[38] First war journal, Sunday, March 14, 1943.
[39] Letter, RSB to PDB, second letter dated Sunday, March 14, 1943.
[40] First war journal, Monday, March 15, 1943 and Tuesday, March 16, 1943.
[41] Letter, RSB to PDB, Tuesday, March 16, 1943. Royce drew on the letter the trajectory of a stalling plane to illustrate.
[42] First war diary, Thursday, March 18, 1943.
[43] Letter, RSB to PDB, Sunday, March 21, 1943, content condensed.
[44] First war journal, Monday, March 22, 1943.
[45] Letter, RSB to PDB, Monday, March 22, 1943. Five days later, after being informed he would be assigned to a "check ride" for lack of progress, Royce wrote, "There is one rather ironical thing in connection with we five who were with the same instructor. Of the five, there were two of us who did not use tobacco or alcohol. We are both up for check rides and the other three are going on doing all right. Of course it doesn't shake my belief in the Word of Wisdom or the value of living it, but it kind of gripes me a bit. I'll never, never smoke, use tea or coffee or drink, you can be assured of that. I believe all those things become more repulsive to me by the hour. That is one of the big things that my mission did do for me, even though I didn't go on it to reform." (Letter, RSB to PDB, Monday, March 29, 1943.)
[46] Letter, RSB to PDB, Thursday, March 25, 1943, content condensed.
[47] Letter, RSB to PDB, Sunday, March 28, 1943, content condensed.
[48] First war journal, Tuesday, March 30, 1943 and Wednesday, March 31, 1943, content edited.

[49] Letter, RSB to PDB, Thursday, April 1, 1943. Although Royce did not pass his check flight, another cadet under exactly the same circumstances was able to pass and return to flying, and over nine months later on a rainy day in Columbia, South Carolina, Royce would run into him again. "Tonight, as we walked toward the barracks, we encountered a couple of pilots whom we (McNally and I) saluted. It was raining hard and as we hurried by I almost failed to look them over. As I turned my head to do so one of them did likewise and sure enough it was one of the fellows whom I was with in cadets. His name is Hall. We were together at Santa Ana and Santa Maria. He graduated at the same time and at the same school as Frank. We had a happy little reunion right there untill we decided to move on before we got soaked. I'll be seeing more of him. He was up for elimination checks the same time I was, only he got by the Army ride, stayed and made it. He just arrived here. I still expect to see more of them. He said that Frank is probably over seas by now." (Letter, RSB to PDB, Friday, January 14, 1944.)
[50] Letter, RSB to PDB, Sunday, April 4, 1943.
[51] First war journal, Thursday, April 1, 1943.
[52] Letter, RSB to Bringhurst family, Friday, April 2, 1943.
[53] Letter, RSB to PDB, Friday, April 2, 1943, content greatly condensed. Royce would later visit Santa Maria frequently in connection with his work with strawberries, and would even use the grounds of the old airfield as an exercise area for his morning runs. More than thirty years later he still felt the disappointment of his failure there, and after a strawberry meeting there he wrote, "Santa Maria holds some bitter memories for me. I was washed out of pilot training there in 1943. I knew after I left and went to Fresno to start again in "radio operator" training (which I completed successfully), that I was too proud or loyal to say that I needed another chance. As I recall, I loved flying & felt comfortable doing it. I was completely surprised when I was told of the fateful 'check ride.' Even today, I tell myself that I should have made it & would have very successfully. This was my first real failure in life. I still so regard it." (Davis Journal #3, December 6, 1977.)
[54] Letter, RSB to Bringhurst family, Friday, April 2, 1943, content condensed.
[55] The Mormon Tabernacle Choir conducted weekly radio broadcasts of a program "Music and the Spoken Word," originating with radio station KSL in Salt Lake City. For Royce this was a real taste of home. He listened to the broadcasts most of his adult life.
[56] The blessing of a baby refers to a priesthood ordinance performed in the Church of Jesus Christ of Latter-day Saints, in which an infant child is given a name and a blessing by priesthood authority, which results in an official Church membership record being created for the child. This bears no resemblance to infant baptism which, though performed in many Christian churches, is contrary to Latter-day Saint teachings. The Church only permits baptism after the age of eight, and then only after an interview assures that the person is aware of the implications of baptism, and is in agreement.
[57] Letter, RSB to PDB, Sunday, April 4, 1943.
[58] First war journal, Monday, April 5, 1943.
[59] Letter, RSB to PDB, Monday, April 5 (misdated April 6), 1943. Royce also

made quick financial calculations to assure Pearl would be cared for adequately. "We will probably get more money than we have before. They will take twenty-five dollars from my base pay and then the government matches that and then I believe that they give about 12 dollars for each child." He concluded the letter, "I'll work hard Honey and try to make something out of myself for you. Pray for me always. Sometimes, I wonder if I'm being punished or what. It just seems that everything comes the hard way, and then I don't get a lot of the things I try so hard to get. I know that I have you and no one could ever change that."

[60] First war journal, Wednesday, April 7, 1943.

[61] Letter, RSB to PDB, Friday, April 9, 1943, content condensed.

[62] Letter, RSB to PDB, first letter dated Friday, April 16, 1943.

[63] Letter, RSB to PDB, Sunday, April 18, 1943.

[64] An article in the August, 2006 edition of *Chicago* magazine describes the history of the hotel, which was completed just prior to the Great Depression and brought financial and personal ruin to its founders, the Stevens family, two of whom lost their lives in the debacle. The third, Ernest Stevens, was initially convicted of fraud in the wake of the hotel's failure, but was later acquitted, and managed to rebuild his life managing a much smaller hotel. His son, John Paul Stevens, became a justice on the U.S. Supreme Court.

[65] Letter, RSB to PDB, April 21, 1943. The move to Chicago also marked a change in pay grade. In the same letter Royce wrote, " I suppose you notice Pfc. in front of my name in place of Pvt. I am now a Private First Class. That entitles me to one stripe on my sleeve and I'll get four dollars more per month. That will be just about enough to pay my insurance. We all recieved that raise in grade when we arrived here. There's one thing about this, I've taken my first step up anyway. I'll try to keep climbing up and up. I've got a long way to go anyway."

[66] Letter, RSB to Bringhurst Family, Friday, April 23, 1943.

[67] First war journal, Thursday, April 22 and Saturday, April 24, 1943, content condensed.

[68] Letter, RSB to PDB, Friday, May 7, 1943. Royce eventually chose not to pursue the air cadet option, which would have greatly extended his military training.

[69] First war journal, Thursday, May 13 and Friday, May 14, 1943, content condensed; also letters, PDB to Jane Davidson, May 15 and May 17, 1943. Royce recalled their meeting months later in a letter from Sioux Falls: "Every time I think of you and your picture comes to my mind, the same scene always returns. I see you walking toward me from the train in Chicago with our baby in your arms and a half afraid look on your face. I'll never forget how you brightened up when you saw me and how good it felt when I held you close to me after so long." (Letter, RSB to PDB, Wednesday, August 11, 1943.)

[70] Letter, RSB to Bringhurst family, May 24, 1943.

[71] Letter, PDB to Jane Davidson, May 15, 1943. While in Chicago, Pearl described a civil defense exercise in a letter home. "We had an air raid in Chicago yesterday. It lasted 2 hours and 150 planes participated, dropping paper bombs all over the city. We had the regular air raid drill sirens and

everything. It was really something too. They really take things seriously here." (Letter, PDB to Jane Davidson, May 24, 1943.)

[72] First war journal, segments dated Thursday, June 3, and Saturday, June 5, 1943, though the entries for June 3-5 appear to have been reconstructed from memory after Pearl and Jean left.

[73] Letter, RSB to PDB, Sunday, June 13, 1943.

[74] Letter, RSB to PDB, June 9, 1943, content condensed.

[75] First war journal, Saturday, June 12 and Sunday, June 13, 1943.

[76] Letter, RSB to PDB, Monday, June 14 (misdated June 12), 1943.

[77] First war journal, Monday, June 14, 1943.

[78] Letter, RSB to PDB, Tuesday, June 17, 1943, content condensed.

[79] Letter, RSB to PDB, Sunday, June 20, 1943, segments quoted.

[80] Letter, RSB to Bringhurst family, Friday, August 20, 1943.

[81] Letter, RSB to PDB, Monday, July 19, 1943, condensed and edited.

[82] Letter, RSB to Bringhurst family, Monday, July 19, 1943, content condensed.

[83] First war journal, Tuesday, August 3, 1943; letter, RSB to PDB, August 4, 1943.

[84] "They are making the course here a lot stiffer and have made the requirements in code harder. They have let us know that it is quality rather than numbers they want in their operators. They are also washing out some of the fellows. Don't worry about me though, I won't be one of them. I am ahead in code and up to par in mechanics anyway. In code for twenty words a minute you now have to write four minutes with a maximum of seven mistakes. That's hard to do. Eighty words at that speed about wears you out to get it down on paper. I've already passed that so I don't have to worry. It's my credited speed. However, I probably will have to go back and work on it a while longer." (Letter, RSB to PDB, Sunday, August 1, 1943; see also letter, RSB to PDB, Friday, August 13, 1943.)

[85] Letter, RSB to PDB, Thursday, August 26, 1943, content condensed.

[86] Letter, RSB to PDB, Wednesday, September 9, 1943.

[87] Letter, RSB to Bringhurst family, Tuesday, September 14, 1943, content condensed.

[88] Letter, RSB to PDB, Monday, September 13, 1943.

[89] Letter, RSB to PDB, Wednesday, September 15, 1943.

[90] Letter, RSB to PDB, Wednesday, September 15, 1943, content condensed.

[91] RSB, First war journal, Monday, September 27, 1943.

[92] Letter, RSB to PDB, Tuesday, October 12 (labeled October 13), 1943.

[93] Letter, RSB to PDB, Friday, September 24, 1943.

[94] Excerpts from letters, RSB to PDB, September 24, 25 and 26. Of the tear gas he wrote, "The only thing that was bothered by it particularly was a big fat dog (the original 'chow hound') which lives over here. He began to cry a bit and wondered what in the world the silly humans were trying to do. The wind will clear the gas out of your eyes in a hurry as soon as you get away from the contaminated area." (Letter, RSB to PDB, Friday, September 24, 1943.) Royce had no way of knowing that tear gas (chloropicrin) would play an important role in his professional life, as it was employed in the fumigation of strawberry fields.

[95] Letter, RSB to PDB, Sunday, September 26, 1943.

[96] Army Air Force Technical Training Course. (RSB first war journal, Tuesday, September 28, 1943.)

[97] The term "Jack Mormon" originated during the intense anti-Mormon persecution around Nauvoo, Illinois around the time of Joseph Smith's assassination. Originally it was used disparagingly by antagonists of the Latter-day Saints as a term for non-Mormon citizens who were sympathetic to plight of the Mormons, some of whom were expelled from the state along with the Latter-day Saints. By Royce's time it had evolved into a term used by Latter-day Saints to describe someone who, though officially a member of the Church, was living a life at obvious variance with its teachings.

[98] Letter, RSB to PDB, Thursday, September 30, 1943, content condensed.

[99] Letter, Frank Denstad to RSB, August 8, 1943.

[100] Letter, Reid Ellsworth to RSB, September 22, 1943. This was the last letter Royce would receive from Reid, as shortly after beginning combat duty he was shot down over Italy.

[101] Letter, RSB to PDB, October 6, 1943, content condensed.

[102] Letter, RSB to PDB, October 8, 1943.

[103] First war journal, Friday, October 8, 1943.

[104] Letter, RSB to PDB, Saturday, October 9, 1943, content condensed.

[105] Letter, RSB to PDB, Tuesday, October 12 (labeled October 13), 1943, content condensed. Royce added, "You recover completely in about a half minute, and there are no after effects at all."

[106] Letter, RSB to PDB, Wednesday, October 13, 1943. Royce's assessment of the survival rate of battle casualties in World War II relative to previous wars was correct.

[107] Letters, RSB to PDB, both dated Saturday, October 16, 1943, content condensed and edited.

[108] This was George Watkins, the younger of two brothers who had served together as missionary companions under Royce's leadership. The custom among Latter-day Saints is to address each other by the title "brother" or "sister." In accord with this custom the younger Elder Watkins was referred to by the missionaries as "*Hermanito*" (little brother) Watkins, which Royce in his letter abbreviated to "*Hnito*."

[109] Letter, RSB to PDB, Sunday, October 17, 1943, content condensed.

[110] Letter, RSB to PDB, Monday, October 18, 1943, content condensed. Note that the following day, Tuesday, October 19, 1943, was the last day Royce wrote in his First 1943 war journal, indicative of the difficult schedule he was required to maintain in gunnery training.

[111] Letter, RSB to PDB, Thursday, October 21, 1943. Gunnery training changed shortly after Royce's term in Harlingen. "Between the time when I graduated from gunnery school and when I hit combat, the whole theory of aerial gunnery changed. The stuff they taught us about firing at attacking aircraft was for the most part false. Gunners who had been getting hits, had been unconciously disregarding what they had been taught. All of that is changed now. They teach you the correct theory in the gunnery schools now. It's really quite simple, and I was taught such a complicated system in schools." (Letter, RSB to PDB, Thursday, September 14, 1944.)

[112] Letter, RSB to PDB, Monday, October 25, 1943, content condensed.

[113] Letter, RSB to PDB, Tuesday, October 26 (labeled October 27), 1943, content condensed.

[114] Letter, RSB to PDB, Wednesday, October 27, 1943.

[115] Letter, RSB to PDB, Saturday, November 6, 1943.

[116] Letter, RSB to PDB, Saturday, November 6, 1943, content condensed. To his parents Royce wrote, "Of all things the cartridges we fired were all made at the small arms plant at Salt Lake. They are those big ones that you have which George brought out to you. They really make a noise when they go off. It will take a little time to get used to their sound. They go off at the rate of about eight hundred per minute." (Letter, RSB to Bringhurst family, Sunday, November 7, 1943.)

[117] Letter, RSB to PDB, Wednesday, November 10, 1943, content condensed.

[118] Letter, RSB to Bringhurst family, Sunday, November 14, 1943, content condensed and edited.

[119] Letter, RSB to PDB, Tuesday, November 9, 1943. November 11, 1943, happened to be the 25th anniversary of Armistice Day, which commemorated the conclusion of the "war to end all wars," and the irony was not lost on Royce. "This armistice day is sort of a jest at what happened on this day some 25 years ago. We met a fellow working in the mess hall today who was in the last war and rather bitterly recollected how happy he was along with thousands of boys some twenty-five years ago today when the 'war to end all wars' drew to a close." (Letter, RSB to PDB, Thursday, November 11, 1943, content condensed.)

[120] Letter, RSB to PDB, Monday, November 15, 1943, content condensed.

[121] Letter, RSB to PDB, Tuesday, November 16, 1943, content slightly condensed.

[122] Letter, RSB to PDB, Thursday, November 18, 1943, content condensed. Royce was aware that some of his success was due to good fortune as much as skill. "A couple of times, I might have done lots better had I had better equipment. Then again I wouldn't have done so well had I had to use the equipment other fellows did on some of my best shooting."

[123] Letter, RSB to PDB, Saturday, November 20, 1943, content condensed.

[124] It was evidently on this day that Royce first learned of the excommunication of Richard R. Lyman, the apostle he had met and admired on his mission, who had performed his temple marriage just the year before. He was excommunicated that month for "violations of the law of chastity," which apparently involved an unlawful polygamous relationship. Royce characteristically was unwilling to write anything uncomplimentary of a Church leader, only stating in a letter to Pearl, "One thing we talked about wasn't very pleasant, in fact it was a shock. I'll talk it over with you when I get home. You have probably heard about it too. I'd rather not write anything about it right now. It really cut me to the core." (Letter, RSB to PDB, Sunday, November 21, 1943.)

[125] Letter, RSB to PDB, Tuesday, November 23, 1943, content condensed.

[126] Letter, RSB to PDB, Thursday, November 25, 1943.

[127] Letter, RSB to PDB, Thursday, November 25, 1943.

128 Letter, RSB to PDB, Sunday, November 28, 1943.

129 "Barry, I have spoke of him very often, washed out on air firing. He didn't have enough hits, I'll tell you about it after. It kind of took a bit of the joy out of it for me." (Letter, RSB to PDB, Sunday, November 28, 1943.) Robert Barry was reassigned to a troop carrier command in Missouri, and was eventually trained as a radioman on a glider tow ship. Such ships did not carry weapons, so gunnery was not required. He remained at a lower pay grade, and Royce felt his friend had been given something of a raw deal.

130 Letter, RSB to PDB, Tuesday, January 25, 1944; see also letter of Tuesday, December 21, 1943.

131 Letter, RSB to PDB, Saturday, December 18, 1943.

132 Letter, RSB to PDB, Wednesday, December 22, 1943.

133 Letter, RSB to PDB, Tuesday, December 21, 1943.

134 Letter, RSB to Bringhurst family, Tuesday, December 21, 1943.

135 Letter, RSB to PDB, Tuesday, December 21, 1943, content condensed. Royce later considered that his assignment to a B-25 crew was probably the reason he survived the war. Speaking of a friend who had been shot down and killed, he said "He was on a B-24. I trained part of the time on a B-24, and could have had the same fate he had, because they shot those things down by the dozens sometimes, B-24's and B-17's. They flew at about 20,000 feet or higher, maybe as high as 30,000 feet. I flew around 8,000 feet, eight to ten, maybe as high as twelve." (Taped interview, September 1997, tape 2.)

136 Letter, RSB to PDB, Wednesday, December 22, 1943, content condensed. Royce was uncertain how secret the cannon-bearing B-25H still was, and he asked Pearl not to discuss it with others.

137 Letter, RSB to PDB, Friday, December 24, 1943, content greatly condensed.

138 Letter, RSB to PDB, Friday, October 27, 1944, content condensed. This was written while Royce was at his combat post the following year, recalling the occasion.

139 Letter, RSB to PDB, Saturday, December 25, 1943, content condensed. In the same letter, Royce mentioned a kind Christmas gesture: "Santa Clause didn't quite forget us here. One of the boys who went down town last nite & we don't know which one bought us each the nicest little bag of marbles. We found them placed on our bunks when we came back from dinner today. We had a pretty good laugh out of it anyway. It's a little too cold to get down and play on the floor or we would have most probably have had a few games going."

140 Letter, RSB to PDB, Wednesday, December 29, 1943, content condensed.

141 Letter, RSB to PDB, Thursday, December 29, 1943, content condensed.

142 Letter, RSB to PDB, Saturday, December 31, 1943, content condensed and edited.

143 Letter, RSB to PDB, Sunday, January 1, 1944. In continuation, he described the next day's aftermath of Andy's New Year celebration: "He tossed all his dinner up this afternoon when he was flying."

144 Letter, RSB to PDB, Saturday, January 31, 1943.

[145] Letter, RSB to PDB, Thursday, December 29, 1943.

[146] Second war journal, entry for January 1, 1944; also letter, RSB to PDB, Saturday, January 1, 1944. The journal was actually purchased on January 3, so the first entries were written from memory.

Royce never forgot the feel of the recoil from nose-mounted cannon, and many decades later he could still describe the sensation. "You'd fire that thing, and the plane would just literally stop. 'Whoomp!'" (Taped interview, RSB & PDB, tape 2, 1997.)

[147] Letter, RSB to PDB, Thursday, January 6, 1944.

[148] Letter, RSB to PDB, Sunday, January 9, 1944, slightly condensed. Since missing airmen who survived most generally either returned to base or were reported captured shortly after, Royce had assumed the worst. As it turned out, Reid was still alive. While serving as a navigator aboard a B-17 based in Tunisia, his plane had been disabled and left formation, and was attacked and set afire by German fighters. Reid had parachuted successfully from the plane before it exploded, and landing in a plowed field, he was immediately met and hidden by sympathetic Italians. At the time Royce learned of his 'missing in action' status, Reid had evaded capture for nearly two months, and was being shuttled from house to house in rural Italy. On January 24, 1944 he was betrayed to German authorities by a Fascist sympathizer, and spent the remainder of the war in a German prisoner-of-war camp. His remarkable story is chronicled in a book, *The Reid F. Ellsworth Story: An Account of War and Divine Interposition*, self-published in 1977. Found among Royce's journals after his death was a copy of this book with a handwritten inscription by the author: "Dear Royce, I send this book to you with my regards and esteem to the one whom I consider to have been my best missionary companion, notwithstanding that we were together for only a rather brief period. My hopes and prayers are that your health problems will not be too devastating to you. Again – With Love and Best Wishes, Reid."

[149] Letter, RSB to PDB, Sunday, January 9, 1944, content condensed.

[150] Letter, RSB to PDB, Tuesday, January 17, 1944, content condensed.

[151] Letter, RSB to PDB, Tuesday, January 11, 1944, content condensed. The situation likely had a spiritual effect on Royce as well. The next day he wrote, "Sometimes of late it has almost felt like I was a stranger in Church. I guess it has been all this going to meetings where I knew no one. I hope I have been doing right, Honey. I'm trying to do my best to live as I should and remember the teaching I have received. This is different than anything else I have ever experienced. At times I am almost separated from the Church. I hope it isn't hurting me, Darling. I don't think it is. If I can't continue to take it this way I'm not much good am I?" (Letter, RSB to PDB, Wednesday, January 12, 1944.)

[152] Letter, RSB to PDB, Saturday, January 22, 1944.

[153] Letter, RSB to PDB, Tuesday, February 1, 1944, content condensed.

[154] Undated letter, RSB to PDB, sent February 2, 1944, content edited.

[155] Letter, RSB to PDB, Friday, February 18, 1944, content condensed.

[156] Letter, RSB to Bringhurst family, Sunday, February 20, 1944.

[157] Letter, RSB to PDB, Monday, February 21, 1944, content condensed.

[158] Letter, RSB to PDB, Wednesday, February 23, 1944, content condensed.

603

[159] Second war journal, page 8, February 26.
[160] Letter, RSB to PDB, Saturday, February 26, 1944, content condensed.
[161] Letter, RSB to PDB, Wednesday, March 1, 1944 and Saturday, March 11, 1944, content condensed.
[162] Letter, RSB to PDB, Monday, March 6, 1944, content condensed.
[163] Letter, RSB to PDB, Sunday, March 5, 1944.
[164] Letter, RSB to PDB, Tuesday, March 21, 1944, content condensed.
[165] Letter, RSB to PDB, Monday, March 13, 1944.
[166] Letter, RSB to PDB, Friday, March 17, 1944, content condensed.
[167] Letter, RSB to PDB, Monday, March 20, 1944, content condensed.
[168] Letter, RSB to PDB, Monday, March 27, 1944.
[169] Letter, RSB to PDB, Wednesday, March 29, 1944, content condensed.
[170] Letter, RSB to PDB, Tuesday, March 21, 1944.
[171] Letter, RSB to PDB, Sunday, April 2, 1944.
[172] Letter, RSB to PDB, Wednesday (mislabeled Thursday), April 5, 1944.
[173] Letter, RSB to PDB, Thursday, April 6, 1944.
[174] Letter, RSB to PDB, Saturday, April 8, 1944, content condensed.
[175] Letter, RSB to PDB, Sunday, April 9, 1944. Old acquaintances that Royce met at church included a soldier named Durham, with whom Royce had had contact in Sioux Falls, and who had also served in the Spanish-American mission and was then serving as a radio mechanic at the base; a son of President Haymore, Royce's mission president, who was an enlisted man in an antiaircraft artillery unit stationed there; and an officer named Johnson in the same unit, who had been his bacteriology lab instructor at the agricultural college, and whose wife he also knew. All these, like Royce, had put their lives on hold and were playing individual roles in the war effort.
[176] Letter, RSB to PDB, Monday, April 10, 1944.
[177] Letter, RSB to PDB, Tuesday, April 11, 1944.
[178] Letter, RSB to PDB, Tuesday, April 11, 1944. Royce was responding to a letter Pearl had written on April 6.
[179] Letter, RSB to PDB, Friday, April 14, 1944.
[180] Letter, RSB to PDB, Sunday, August 20, 1944, content condensed.

CHAPTER 6:
[1] Borinquen Airfield is on the northwest corner of the island, near the coast. The flight took 5 hours and 25 minutes.
[2] Second war journal, entries for April 15, 16, 17 and 18, 1944, p. 12-16, some entries condensed. Royce noted the flying time by each entry: From Puerto Rico toward British Guiana and back to Trinidad, 5 hours 22 minutes; from Trinidad to British Guiana, 2 hours 35 minutes; from British Guiana to Belem, Brazil, 5 hours 6 minutes, and from Belem to Natal, Brazil, 5 hours 55 minutes.
[3] Second war journal, entries for April 22, 23 and 24, 1944, pages 16-19, content condensed. Royce again entered the flying time for each leg of the journey: From Natal, Brazil to Ascension, 7 hours 28 minutes; from Ascension to Liberia, 5 hours 45 minutes; from Liberia to Dakar (Senegal), 4 hours 25 minutes.

⁴ Letter, RSB to PDB, Wednesday, April 26, 1944, content condensed.
⁵ Second war journal, entries for April 28 and 29, 1944, p. 19-21, content condensed. The trip from Dakar to Marrakech involved 7 hours 45 minutes flying time, with a stop in Atar for a fuel check. From Marrakech to Algiers took 4 hours 17 minutes.

Royce described to Pearl the stay in Algiers. "I went to town several times with Andy and the boys. It's quite a city. We didn't see all of it because there are certain Arab districts that are out of bounds. I went to a couple of American shows there. The speaking is in English then they flash the French words on for the French part of the audience. We didn't eat much there because it is unsafe to eat off an army base over here. It's too easy to pick up intestinal diseases etc. We talked to quite a few boys from the British Army. They are all nice fellows though it is kind of hard to understand them at first. They don't seem to have as much trouble understanding us as we do understanding them. I have met quite a number of French soldiers too since this is French country." (Letter, RSB to PDB, Thursday, May 4, 1944.)
⁶ Second war journal, May 3, 1944, p. 21-22.
⁷ Letter, RSB to PDB, Thursday, May 4, 1944.
⁸ Letter, RSB to PDB, Friday, May 12, 1944.
⁹ Letter, RSB to PDB, Friday, May 12, 1944.
¹⁰ Second war journal, May 14, 1944, p. 23, content condensed. The flight from the coast of North Africa took three hours.
¹¹ Second war journal, May 14, 1944, p. 24-26, content condensed.
¹² Letter, RSB to PDB, Thursday, May 18 (misdated May 8), 1944.
¹³ Second war journal, May 14, 1944, p. 26-27, content condensed.
¹⁴ Royce later explained this to Pearl: "Now about the crew business; When green men come over here they first are split up and flown with experienced crews. That is good sense and as it should be. Later, each man is assigned to a definate crew. I was fortunate to be placed on my old pilot's crew. He flew as co-pilot for a time after he came over. When he was given a crew I was on it. Harper was assigned to the same crew. Wendell is working on the line as a mechanic and Robinson was placed on another crew because they needed gunners before Lt. Brassfield had his crew formed here." (Letter, RSB to PDB, Thursday, June 29, 1944.)
¹⁵ Second war journal, May 24, 1944, p. 28-29, content edited.
¹⁶ Second war journal, May 25, 1944, p. 29.
¹⁷ Second war journal, May 27, 1944, p. 30-31, content condensed and edited. Royce added, as he often did, an expression of admiration about the country he saw: "Some of the most beautiful mountains and valleys in the world are there. They really caught my eye." Royce recorded a flight time of 2 hours and 30 minutes.
¹⁸ Letter, RSB to PDB, Thursday, June 22, 1944, content condensed.
¹⁹ Second war journal, June 5, 1944, p. 35-36, content condensed. The name of the place bombed was Narni, and the flight took two hours. This was a significant mission, and Royce, together with all the crew members, was awarded the Air Medal for his participation, to which he would later receive multiple oak leaf clusters. Of the aircraft mentioned, B-26s were medium bombers like the B-25s, but somewhat faster; P-47s were one-seat fighter-

bombers, and Spitfires were small, oval-winged British fighters which had earned fame defending the British homeland in the Battle of Britain. The Spitfires were assigned to escort the bombers and defend them against attacking enemy fighters. Royce mentioned these in the margin: "About 30 fighters did approach but the Spits lit into them. Two Spits went down but the Germans got away. (I didn't see it.)"

[20] Letter, RSB to PDB, Friday, June 2 (misdated May 2), 1944.

[21] Letter, RSB to PDB, May 25, 1944, content condensed. See also second war journal, May 25, 1944, p. 29-30.

[22] Letter, RSB to PDB, May 29, 1944, content condensed.

[23] Letter, RSB to PDB, Monday, June 5, 1944.

[24] Letter, RSB to PDB, Saturday, July 15, 1944.

[25] Letters, RSB to PDB, Sunday, July 23 and Sunday, August 27, 1944.

[26] This passage was composed from the following letters: RSB to PDB, Friday, June 2 (misdated May 2), 1944, Tuesday, June 6, 1944, and Tuesday, June 13, 1944.

[27] Letter, RSB to PDB, Thursday, June 29, 1944. A month later the men took a raft to the beach to practice procedures in case of an emergency ditching. "We had a large rubber raft down there and had quite a time. We were practicing the handling of the raft in the water just in case. They turn over easily and you have to be able to right them O.K. It's no trick as long as you know how to swim. We spent most of the time fighting to see who stayed on the thing. I do pretty well in the water so I get along all right." (Letter, RSB to PDB, Sunday, July 30, 1944.) A couple of days later he wrote of Robbie, "He has learned to handle himself pretty well in the water now. I told you that he couldn't swim a stroke when we arrived here and was terrified in the water. That's all about past now. He can keep himself up and has lost most of his fear of the water." (Letter, RSB to PDB, Monday, July 31, 1944.)

[28] Letter, RSB to PDB, Tuesday, May 30, 1944, content condensed. The show Royce saw that particular day had been one entitled *Jack London*.

[29] Letter, RSB to PDB, Wednesday, August 9, 1944.

[30] Letter, RSB to PDB, Thursday, October 26, 1944, content condensed.

[31] Vmail, RSB to PDB, Sunday, June 4, 1944.

[32] Second war journal, June 6, 1944, p. 37, and letter, RSB to PDB, Tuesday, June 6, 1944, content condensed. Having begun to keep a journal in 1976, he observed the anniversary of the attack: "D-day Europe 1944—I remember, I remember flying over Italy that day & hearing of the invasion first from Germany & then England. I remember but dimly now. It was another world, another time. Was I really there, or is it but a dream? Perhaps a little of both, since memory is selective." (Davis Journal #1, 6 June, 1976.) After his retirement, Royce corroborated his account of picking up the radio account of the invasion, first from a German propaganda broadcast. "We picked up information, and the Germans said that there was an attack taking place, they were being repulsed and driven into the sea. And the British word came through with more modesty, saying 'Everything is going according to plan.'" (Videotaped interview of RSB by Anton Kofranic, UCD Emeriti Association, 1996.)

[33] Letter, RSB to PDB, Wednesday, June 7, 1944. Photos of the plane are found in the 489th Bomb Squadron "Yearbook" on p. 50 (which shows a crew without Royce but including the Pilot, Brassfield, the Bombardier, Harper, and the tail gunner Robinson), and on p. 100 (which shows the nose art, including the stenciled names of Brassfield and Harper, as well as the ground crew chief, Edward F. Bedell).

Royce mentioned some of the other aircraft in a later letter to his family. "You've asked me time and time again the name of the bomber I fly in. I'm sorry not to have made it clear that I fly in nearly every ship we have. Today, it was 'Black Jack.' Several days ago it was a ship with 'Wabbit Twacks' on one side and 'Dorothea' on the other. I don't know why it has two names. Another is 'Lady Luck' and some others are, 'That's All Brother,' 'C Ration,' 'Knockout,' 'Mission Completed' and 'Queen Mary.' There are others too. Most of them have pictures of girls painted on them in various stages of undress. The ship I was on before I started flying as photographer has the worst name on the place I think though the picture is cute. The picture is a baby and a little dog and the name is 'Snot Nose.' I don't know who thought it up. We get a chuckle out of lots of the names and figures on them." (Letter, RSB to Bringhurst family, Saturday, September 30, 1944.)

[34] Second war journal, June 8, 1944, p. 38.

[35] Second war journal, June 13, 1944, p. 39.

[36] Second war journal, June 15, 1944, p. 40-41, content condensed.

[37] Second war journal, June 15, 1944, p. 41. News of Andy's wounding probably came in a letter from McNally, which Royce received the same day. He later learned further details from McNally, who had corresponded with Andy's mother. "He's coming around all right according to the information I received. They had to put a plate in his legg above the knee. Of course I don't know exactly how bad his wounds are but I think the war is over for him. His mother said he sounded blue in his letters and expressed the wish that Mac and I write to him." (Letter, RSB to PDB, Friday, July 7, 1944.)

[38] Elba, a small German-held Island lying between Corsica and the Italian mainland, was invaded by French forces on June 17, and the last surviving Germans left the island on June 20, the day before Royce's 13th mission.

[39] Second war journal, June 21, 1944, p. 42-43.

[40] Second war journal, June 22, 1944. p. 43-44.

[41] See Second war journal, June 23, 1944, p. 44.

[42] Letter, RSB to PDB, June 22, 1944.

[43] Letters, RSB to PDB, Wednesday, June 14, 1944 and RSB to Bringhurst family, Sunday, June 19, 1944. Royce gave an example some days later: "We received our P.X. rations one day early. Mine are already gone. Terrible, am I not? All week long we kind of crave something like that and then when it does come I'm not happy until it's all under the belt. Today, it was five small candy bars. I wish I didn't have to speak of them in the past tense, but we have to face the facts, they were good while they lasted. One of them was a small Hershey almond bar. I naturally saved that till last since it's my favorite. For a while, I thought I might even have the will power to save it till tomorrow but I've always been taught not to put off till tomorrow, so down it went. I'll think about it the rest of the week then do exactly the same

again."
[44] The "bomb line" in this setting describes the line beyond which there was no danger of hitting friendly troops if bombs were released; this approximately corresponded to the enemy's front lines.
[45] Second war journal, June 30, 1944, p. 45-47, content condensed.
[46] Second war journal, July 1, 1944, p. 47-48, content condensed and rearranged.
[47] Second war journal, July 3, 1944, p. 49-51. Royce would many years later make peacetime visits to Ferrara as a strawberry specialist.
[48] Letter, RSB to PDB, Saturday, July 1, 1944.
[49] Letter, RSB to PDB, Sunday, July 16, 1944.
[50] Letter, RSB to PDB, Tuesday, July 4, 1944, content condensed and edited.
[51] Composite of two letters, RSB to PDB, Saturday, July 1, 1944, and RSB to Bringhurst family, Wednesday, July 5, 1944.
[52] Letter, RSB to PDB, Sunday, August 13, 1944, content slightly condensed.
[53] Letter, RSB to Bringhurst family, Saturday, October 21, 1944, content condensed.
[54] See Second war journal, July 12, 1944, p. 53-54.
[55] Second war journal, July 13, 1944, p. 54-55, content condensed.
[56] The "F. W. 190s" spoken of were "Focke-Wulf FW 190s", a radial-engined single-seater German fighter. The Me. 109 was the Messerschmitt BF 109, the most famous and distinguished German fighter plane of the war.
[57] The "I.P." was the Initial Point, or the predetermined point where the flight of bombers began their bombing run. At that point the steering of the entire box was taken over by the lead bombardier, and generally the ships were unable to do evasive maneuvers until after the bomb run. They were especially vulnerable to anti-aircraft fire during the bomb run.
[58] Second war journal, July 14, 1944, p. 56-58, content edited and condensed. The mission took 3 hours and 25 minutes, and Royce flew with a pilot named Montgomery in the #1 position.
[59] Second war journal, July 15, 1944, p. 59.
[60] "V-mail" was designed as a cheaper patriotic alternative to regular mail. A letter was written on a single sheet, and after military censoring was photographed, then developed and printed on a small photographic sheet on arrival at its destination. The system lent itself only to brief messages.
[61] Letter, RSB to PDB, Monday, July 17, 1944, content condensed.
[62] Letter, RSB to PDB, Sunday, July 9, 1944, content condensed.
[63] Letter, RSB to PDB, Tuesday, July 18, 1944, content condensed. According to Benjamin Angland, another of Royce's friends from that time, Royce occasionally had to "straighten out" Eikhoff on account of his "wild behavior." (Telephone conversation with Benjamin Angland, July 13, 2010.)
[64] Segment compiled from letters, RSB to PDB, Sunday, July 9, 16, 23, and 30, 1944, content edited and condensed.
[65] Letter, RSB to PDB, Thursday, July 13, 1944, content condensed.
[66] Letter, RSB to PDB, Tuesday, July 25, 1944. Norm was killed flying a P-51 fighter over France.
[67] Letter, RSB to PDB, Thursday, July 27, 1944. Frank Van Limburg had

finally made his way through radio school and was assigned as a radio operator and gunner on a B-24 bomber when he went down over Europe.
[68] Letter, RSB to PDB, Thursday, August 3, 1944.
[69] Letter, RSB to PDB, Tuesday, July 18, 1944.
[70] Letter, RSB to PDB, Wednesday, July 19, 1944, content condensed.
[71] Second war journal, July 26, 1944, p. 61-62. The mission took 3 hours and 10 minutes.
[72] Letter, RSB to PDB, Wednesday, July 26, 1944.
[73] Second war journal, July 27 and July 30, 1944, p. 63. Entries from the squadron's tactical journal documented the alert: "July 27-29...At noon the Group (and all the other Groups and forces on the island of Corsica) was alerted for a possible invasion by the enemy from the air and the sea. Higher Allied Intelligence has acquired certain knowledge which indicates the imminence of a large-scale attack from Southern France or from Northern Italy. July 30-31—Since the 27th we have been on stand-down. The danger of an invasion of the island still remains. Fighter planes race back and forth over the island day and night. We continue to carry guns and gas masks and remain alerted for any emergency." (Tactical Diary from 489th Bomb Squadron "Yearbook" p. 207.)
[74] Second war journal, August 2 and 3, 1944, p. 64 & 65.
[75] The first day of stand-down was due to a new policy with the 340th Bomb Group, in which only three of the four squadrons comprising the group would fly on any given day, while the fourth would rest on a rotating basis; the second was due to inclement weather.
[76] Second war journal, August 6, 1944, p. 66-67.
[77] Second war journal, August 7 & 8, 1944, p. 67-69.
[78] Second war journal, August 9, 1944, p. 69.
[79] Second war journal, August 14, 1944, p. 70-71.
[80] In the margin Royce wrote the names of the crewmen: "Pilot Thomas, Co-pilot Swanson, Bombardier England, Gunner _____, Radio Buchanan, Tail gunner Williamson." Royce had flown in the downed plane, 9-P with Thomas and Swanson as pilot and co-pilot just two days before; he had flown missions with all the men named except the radioman. He later included in the margin the fate of the crew: "All these men except Williamson, who was killed and _____, who was captured evaded the enemy and made their way back to our lines. They went down near Avignon, got back to the squadron Sept. 4th. Williamson left a wife and child." (Second war journal, margin of the entry for August 15, 1944, p. 72.) The unnamed gunner was probably Luther S. Craver, whose name is listed among the deceased or missing in action from the 489th Bomb Squadron, with the date as 15 August 1944, and the location as "D-day," southern France (489th Bomb Squadron "Yearbook," p. 7.) Royce had flown with him on mission #13.
[81] Second war journal, August 15, 1944, p. 71-73. The "D-day" mission was a long one, taking 4 hours and 20 minutes. Royce flew it in an aircraft designated 9-F to which he had never before been assigned, though he flew with most of his old training crew, including pilot Brassfield (who flew as co-pilot), bombardier Harper, and Royce's tent-mate Robinson as tail gunner. See Second war journal, p. 146, mission 33.

[82] Second war journal, August 18, 1944, p. 74, content slightly condensed.
[83] Second war journal, August 23, 1944, p. 76-77.
[84] Letter, RSB to PDB, Wednesday, August 23, 1944.
[85] Second war journal, August 19, 1944, p. 75-76.
[86] Letter, RSB to Bringhurst family, Friday, August 25, 1944, content condensed.
[87] Letter, RSB to PDB, Thursday, August 24, 1944, content condensed.
[88] Letter, RSB to PDB, Wednesday, July 26, 1944.
[89] Letter, RSB to PDB, Sunday, August 27, 1944, content condensed.
[90] Second war journal, August 27, 1944, p. 78. The mission must have made life easier for the invading forces, as Royce received a fifth cluster to his air medal for his participation.
[91] Letter, RSB to PDB, Monday, August 28, 1944, content greatly condensed.
[92] Letter, RSB to PDB, Tuesday, September 5, 1944, content condensed and edited.
[93] Letter, RSB to PDB, Thursday, September 7, 1944.
[94] Letter, RSB to PDB, Monday, September 4, 1944, content condensed.
[95] Letter, RSB to PDB, Tuesday, September 5, 1944. Royce continued, "We don't indiscriminately bomb civilians as the Germans so many times have done. We always go after purely military objectives but sometimes bombs go wild and people suffer." Royce was quite shaken by the interaction with the woman who had been wounded by Allied bombs, and years later made specific mention of it to one of his daughters.
[96] Second war journal, September 9, 1944, p. 80.
[97] Letter, RSB to PDB, Saturday, September 9, 1944.
[98] Second war journal, September 12, 1944, p. 81. "Prop wash" refers to the disturbed air left in the wake of the spinning propellers of an aircraft ahead, which can be treacherous when one ship lands closely behind another.
[99] Letter, RSB to PDB, Saturday September 16, 1944, content condensed.
[100] Letter, RSB to PDB, Tuesday, September 19, 1944.
[101] Second war journal, September 22, 1944, p. 84-84. The mission took four hours.
[102] Letter, RSB to PDB, Friday, September 22, 1944. Royce continued, "If anyone ever asks me how many times I have been frightened over here I'm going to have a definate answer as long as I remember how many missions I have had. I think anyone who says differently about it isn't being honest. The stuff they put up there after you is the real thing. This is war though and I suppose we have it much easier than many do."
[103] Letter, RSB to PDB, Tuesday, September 19, 1944.
[104] Compiled from two letters, RSB to Bringhurst family, Sunday, September 24, 1944, and RSB to PDB, Saturday, September 23, 1944.
[105] Letter, RSB to Bringhurst family, Sunday, September 24, 1944.
[106] Letter, RSB to PDB, Friday, September 29, 1944, content condensed. The original text actually gave the number of years since a Christmas at home as five, but Royce corrected himself in a later letter—it would actually total six, three as a missionary, then three as a soldier.
[107] Letter, RSB to PDB, Monday, October 2, 1944, content condensed.

[108] Second war journal, October 1, 1944, p. 86. The first mission, to destroy the gas plant at Piacenza, earned Royce the seventh cluster to his Air Medal.
[109] Second war journal, October 3, 1944, p. 87-88. Royce made two marginal notes, evidently of information which came to him after the journal entry. On page 87: "Miskov had a piece of flak go through his flak suit and part way through a bible he had in his breast pocket." Miskov was a gunner in Royce's squadron. On page 88: "2 ships of the 321st Grp. were seen to go down." The 321st Bombardment Group was the nearby B-25 group of which Royce's friend McNally was a part. Royce's squadron (the 489th) was part of the 340th Bombardment Group.
[110] Letter, RSB to PDB, Tuesday, October 3, 1944, content condensed.
[111] Second war journal, October 4, 1944, p. 88. Royce described the new position in the aircraft to Pearl. "You wondered how it feels to ride in different parts of the ship. It's more or less the same while in flight but you do notice a difference taking off and landing. I rather prefer that up front because you are able to see better where you are going and you don't notice the shock of landing so much. You don't take off or land up in the nose of the ship. It's a bad place to be in a crash. I always stand in back of the copilot. In an emergency, I could brace my back against his to take the shock." (Letter, RSB to PDB, Sunday, October 29, 1944.)
[112] Letter, RSB to Bringhurst family, Sunday, October 8, 1944, content condensed.
[113] Letter, RSB to Bringhurst family, Saturday, October 21, 1944.
[114] Letter, RSB to PDB, Wednesday, October 4, 1944, content condensed. Normally bombardiers were commissioned officers like Royce's crewmate Harper, and the shortage of commissioned bombardiers explained why Harper had completed his missions ahead of Royce and the others of the crew, as the bombardiers had been needed to fly on nearly every mission rather than rotate through them as the other crew members did.
[115] Letters, RSB to PDB, Friday, October 6 and Tuesday, October 10, 1944, content condensed and edited. Another reason the bombardiers had to have some facility with navigation was explained in a later letter. "Part of it, is to always know where we are so I can pinpoint any observations any of the crew make." (Letter, RSB to PDB, Tuesday, November 7, 1944.) Observations the crew had made during the flight were logged and reported during the debriefing held after each mission. Their location was important, since information acquired in this way was used assess the combat situation and to plan future military actions.
[116] Letter, RSB to PDB, Saturday, October 7, 1944.
[117] Letter, RSB to PDB, Monday, October 9, 1944, content slightly condensed.
[118] Letter, RSB to PDB, Saturday, October 14, 1944, content condensed.
[119] Letter, RSB to PDB, Friday, October 27, 1944.
[120] Second war journal, October 6 and October 11, 1944, p. 89.
[121] Letter, RSB to PDB, Wednesday, October 11, 1944.
[122] Second war journal, October 15, 1944, p. 91.
[123] Telephone conversation with Ben Angland, July 13, 2010. Ben Angland called the author to get in touch with Royce who was already deceased. Ac-

cording to Angland, Royce was "very much involved" in the development of the radio bomb release, and was the only radio operator who was so involved.

[124] Letter, RSB to PDB, Wednesday, October 18, 1944, content condensed.

[125] Second war journal, October 18, 1944, p. 92. Royce described the scenery to Pearl. "You would have held your breath with awe if you could have seen the ocean of clouds with tops of snow-covered mountains poking through them. I was nice and warm and feeling swell."

[126] Second war journal, October 19, 1944, p. 92-93. The mission took 3 hours and 15 minutes.

[127] Second war journal, October 20, 1944, p. 93-94, content condensed.

[128] Letter, RSB to PDB, Friday, October 20, 1944, content condensed.

[129] Letter, RSB to PDB, Friday, October 27, 1944.

[130] Letter, RSB to PDB, Monday, October 16, 1944, content condensed.

[131] Letter, RSB to PDB, Tuesday, October 17, 1944, content condensed.

[132] Letter, RSB to PDB, Sunday, October 22, 1944. Royce added, "There was one fellow there from Pocatello, Idaho who asked me if I knew Reid Ellsworth when he learned that I was in the same mission. He knew that Reid is a German prisoner."

[133] Letter, RSB to PDB, Monday, October 23, 1944, content condensed.

[134] V-mail, RSB to PDB, July 8, 1944.

[135] Letter, RSB to PDB, Thursday, July 27, 1944.

[136] Letter, RSB to PDB, Sunday, September 10, 1944.

[137] Letter, RSB to PDB, Wednesday, October 25, 1944, content condensed.

[138] Letter, RSB to PDB, Sunday, October 29, 1944, content condensed.

[139] Letter, RSB to PDB, Sunday, November 12, 1944.

[140] Letter, RSB to PDB, Wednesday, November 1, 1944.

[141] Second war journal, November 4, 1944, p. 95-96. In the 489th Squadron "yearbook," all these men are listed among the killed or missing in action except the radio operator Harris.

[142] Second war journal, November 6, 1944, p. 97-98, content condensed. Both these missions were long ones, the Padua mission taking 3 hours and 35 minutes, and the one at Brenner Pass, 3 hours and 10 minutes. As an aside to his account of the mission, Royce recorded a mysterious disappearance which was never explained. "One ship of the 486th went down in the water as we were going through clouds. None of them have been found. No one saw them go down. It's just missing." The missing plane and its crew were never heard from again.

[143] Second war journal, November 7, 1944, p. 98-99.

[144] Letter, RSB to PDB, Monday, November 6, 1944, content condensed. Royce had begun to record not just his missions but his total combat flying time in his journal, and in continuation he wrote, "I now have about 165 hours combat time. I'd like you to check with Marie on just how much time and how many missions Frank now has. He hasn't written to me but once."

[145] Letter, RSB to PDB, Tuesday, November 28, 1944.

[146] Letter, RSB to PDB, Tuesday, November 7, 1944.

[147] Letter, RSB to PDB, Wednesday, November 8, 1944, content condensed.

[148] Second war journal, November 10, 1944, p. 99-100.

[149] Letter, RSB to PDB, Friday, November 10, 1944.

[150] Second war journal, November 11, 1944, p. 100-101.

[151] Second war journal, November 16, 1942, p. 102-103.

[152] Recorded, September, 1997, Tape 2, content edited.

[153] Asked in a Priesthood class more than three decades later to relate an experience of particular interest, Royce recounted, "I related the Faenza Italy mission experience—telling of how my life was probably saved by obeying the prompting that came to me and moving out of the line of shrapnel that cut through the plane. I still have the .50 caliber cartridge that exploded in my ammo box before my face. From that time on I seemed to know that I would survive the war. I should write that up in detail sometime. There is a brief account in my war diary. I may have referred to it in a letter too. I still remember that incident very lucidly." (Davis Journal #2, July 3, 1977.) Royce never did make the written account as he intended, and the account here was reconstructed from journals, letters, and from memory of his occasional retellings.

[154] Letter, RSB to PDB, Friday, November 17, 1944, content condensed. Bill Mauldin was a cartoonist who won a Pulitzer Prize for his telling depiction of average foot soldiers involved in the Italian campaign. A few days before, Royce had sent home another Bill Mauldin cartoon depicting a woman standing in the ruins of her destroyed home, staring accusingly at a passing infantryman who replies over his shoulder in the caption, "Don't look at me, lady, I didn't do it" (Bill Mauldin, "Up Front..." from *Stars and Stripes*, Tuesday, November 7, 1944). The presumption for Royce was that her home had been destroyed by aerial bombing, and he wrote, "I guess it isn't funny, just tragically true." (Letter, RSB to PDB, Tuesday, November 7, 1944.)

[155] Second war journal, November 17 and 18, 1944, p. 104.

[156] Letter, RSB to PDB, Sunday, November 19, 1944.

[157] Letter, RSB to Bringhurst family, Monday, November 20, 1944.

[158] Letter, RSB to PDB, Wednesday, November 22, 1944.

[159] Letter, RSB to PDB, Wednesday, November 22, 1944.

[160] Letter, RSB to PDB, Wednesday, November 22, 1944, content condensed and rearranged for clarity.

[161] Letter, RSB to Bringhurst family, Thursday, November 23, 1944,

[162] Letter, RSB to PDB, Sunday, November 26, 1944, content condensed.

[163] Letter, RSB to PDB, Saturday, November 25, 1944, content rearranged.

[164] Letter, RSB to PDB, Monday, November 27, 1944. Royce continued with news about Harper, the bombardier on the original Brassfield crew. "Harper should be leaving for home soon unless something drastic happens and I don't think it will. Did I ever tell you that he had a 'purple heart?' He had a piece hit his hand on his last mission. It was only a minor wound and is all healed up now."

[165] Letter, RSB to PDB, Wednesday, November 29, 1944, content condensed.

[166] Second war journal, December 1, 1944, p. 105. The squadron tactical diary reported the mission as follows: "1 December, 1944—We put six airplanes into the air and sent them to Villavernia to bomb a bridge just outside of town. The bomb pattern, instead of covering the bridge, was laid directly

across the town. The inhabitants probably paid dearly in destruction of property and in loss of life as a result of the town's fateful proximity to the bridge." (489th Squadron "Yearbook", p. 212.)

[167] V-mail, RSB to PDB, Friday, December 1, 1944.

[168] This quotation was a composite from a letter, RSB to PDB on Monday, December 4, 1944, and Second war journal, December 2, 1944, p. 106. Frank's death personified for Royce the disillusionment he increasingly felt about the war. Nearly two decades later, he wrote, "War, and the hopeless, dreary waiting. War and the concern for those near and dear. War and its waste, its sorrow and devastation. It seems as but a dream now. I'm sometimes not even certain it really happened. The skills of war making so carefully learned, and now forgotten; the faces of the young men who shared the common lot and who came to be more than brothers—all but a dream, a desperate dream that perhaps wasn't even real; the slogans, the words of warning to a world gone mad.

"One thing I remember most vividly is the face of a blue-eyed, honey blond-haired boy with a look of joy almost always on his face and no reason to die. I believe that in my memory or dream Frank Denstad personifies the many that I knew and know no more. I see his face and yet it isn't real either because he isn't here. With the legions of young men, he is but a ghost of a thought that was but is no more. These are the things that crowd my thoughts as I reflect on an event you faced alone 21 years ago because of a war we did not want and yet did want." (Letter, RSB to PDB, March 12, 1963.)

[169] Letter, RSB to PDB, Monday, December 4, 1944, content condensed. More than fifty years later Royce and Pearl told of this incident, and Pearl related that Royce had entered pilot training with five other Latter-day Saints, and he was the only one to survive the war. Royce's own recollections were tinged with sadness. "When I came home from overseas, I went up to see Guy Anderson in the hospital up at Brigham City, and then, right after I came home, I was entitled to a rest and recuperation, and they sent me to California, and she could go with me. Right after I got home, I received a phone call, and it was Frank Denstad's wife calling, to ask if I would (pause) offer the prayer at Frank's funeral. See, they'd just gotten information for sure that he was dead. He'd been dead about a year and a half, he was shot down over the south of France, and all of them were killed. He was reported missing, but a Catholic priest took the trouble to dig up their graves, look up their dog tags, get the names on them, and identify all the bodies, so that's why they were able to get his body. There was a pretty good chance that the body really was his body, that was shipped back to Salt Lake. And that was my closest associate quite dead (sad chuckle). Yeah, we saw, up in Salt Lake, their names, you know, it's like those marble statues, there's his name, and when he was killed, and where. There were others.... His wife remarried. (Pearl:) His daughter that was born while he was in training... (Royce:) ...the same age as Jean... (Pearl:) ...was married, and we went to her wedding. (Royce:) Yeah, that's right. (sad chuckle) Dead, dead, dead. They didn't make much fuss over one more dead soldier. There was plenty of them around." (Recorded interview, September 1997, tape 2, content condensed.)

Not long after Royce found out about Frank he did in fact receive a letter from Pearl reporting that he was missing in action, though this had been in response to Royce's suspicions that something had happened to him, and his specific request that she not hide bad news from him.

[170] Letter, RSB to PDB, Monday, December 4, 1944, content condensed.

[171] Letter, RSB to PDB, Monday, December 4, 1944, content condensed.

[172] Taped interview, September, 1997, tape 2.

[173] Letter, RSB to PDB, Monday, December 4, 1944.

[174] Letter, RSB to PDB, Saturday, December 9, 1944. Pearl had sent a picture of Guy Anderson which she had cut out of the newspaper.

[175] Letter, RSB to PDB, Wednesday, December 6, 1944.

[176] Composite from letter, RSB to PDB, Wednesday, December 6, 1944, and Second war journal, December 6, p. 108.

[177] Second war journal, December 6, 1944, p. 108.

[178] Second war journal, December 8, 1944, p. 109.

[179] Letter, RSB to PDB, Saturday, December 9, 1944, content condensed.

[180] Letter, RSB to PDB, Sunday, December 10, 1944.

[181] Letter, RSB to PDB, Monday, December 11, 1944. Besides filling out his flying record, Royce requested letters of recommendation from senior officers for his post-combat assignment. One, by a Captain Scofield, a bombing officer with the squadron, was particularly complementary.

"This letter has reference to T/Sgt. Royce S. Bringhurst, radio-gunner, who has just completed a tour of 65 combat missions with this squadron. When the squadron experienced a serious shortage of bombardiers last September, T/Sgt. Bringhurst volunteered for the job. Since then he has flown 16 combat missions as wing bombardier, always carrying out his duties with proficiency and reliability. He flew against some of the most heavily defended targets in Italy and Southern France, respected by all his crew members for his coolness and clear thinking. To supplement the practical knowledge of bombing and navigation which he gained in the air, he attended several hours of navigation classes given by the navigation officer and learned to operate the Norden bombsight.

"In addition to his thorough knowledge of the B-25 airplane, T/Sgt. Bringhurst has the reputation of being one of the best soldiers in the squadron. I wish to put my whole-hearted approval behind his plans to enter bombardier training or whatever course he does choose to pursue upon his return to the States. His instructors will find that he catches on quickly and applies himself to his work with diligence. He is one of the finest examples of officer material with whom I have come in contact." (Letter, Captain Francis W. Scofield to Commanding Officer, Redistribution Center.)

[182] Letter, RSB to PDB, Tuesday, December 12, 1944.

[183] Letter, RSB to PDB, Saturday, December 16, 1944. Royce continued, "I guess that every generation of parents say that at one time or another and yet, the same thing goes on. Boys are born, grow up to be men and then find themselves mixed up in a war they had no hand in starting. I guess we would all grow pretty bitter if we didn't nurture the fond hope that this mad cycle is going to come to an end." Royce then quoted a poem of A. E. Housman, which he had clipped and carried in his journal:

"Here lie we dead, because we did not choose
To live, and shame the land from which we sprung.
Life is perhaps no great thing to lose,
But young men think it is, and we were young."
    Regarding Meade Steadman, at the anniversary of the start of the war nearly fifty years later he would write, "Pearl Harbor Day. Yes, I remember and in my heart give honor to those who failed to return from the great adventure of our lives, a just war. I have recently had my attention called to the Steadman family. Meade was my close friend and he died in France just after the invasion. I have a hard time placing him in the infantry as a platoon sergeant. He was such a gentle person." (Davis Journal, volume 11, December 7, 1993.)

[184] Vmail, RSB to PDB, Thursday, December 21, 1944. The remainder of this letter showed that Royce's sense of humor was little changed. "We have a new fellow in our tent now. He's a radio gunner from Massachusetts who just came in the outfit. His name is Beckwith. That makes six of us counting our star boarder, the toad. I don't think I told you about him. We don't know exactly when he moved in but he has been here with us a long time and seems quite happy. We rarely if ever see him though he lets us know he is still here by croaking every evening. We first discovered him under some boards we were going to use. Robbie didn't like his being inside so he took a little stick and paddled him out with it. He took his time making little hops. Anyway, about a half hour later I opened the door to throw out some water and the rascal hopped back in and headed for a hole in the floor. We didn't have the heart to throw the bum out so I guess he's here to stay. In fact, we're quite attached to him now."

[185] Vmail, RSB to PDB, Monday, December 25, 1944.

[186] Vmail, RSB to PDB, Tuesday, December 26, 1944.

[187] Vmail, RSB to Bringhurst family, Wednesday, December 27, 1944.

[188] Vmail, RSB to PDB, Wednesday, December 27, 1944.

[189] Most of these details were received in a telephone conversation with Benjamin Angland on July 18, 2010. Although Angland was not mentioned in Royce's journals or surviving letters from overseas, together with his wife he was associated with Royce and Pearl after their return to the U.S. and were mentioned in their letters at that time.

[190] Taped conversation, September 1997, tape 2, content edited and condensed.

[191] Letter, RSB to Bringhurst family, Friday, March 15, 1945; see also letter, PDB to Daniel Davidson, March 8, 1945.

[192] Letter, RSB to PDB, Tuesday, March 20, 1945.

[193] Letter, RSB to Bringhurst Family, March 15, 1945.

[194] Letter, RSB to Bringhurst family, Friday, March 15, 1945. These quotes are from the same letter, written after Royce's arrival in Laredo.

[195] Letter, RSB to Bringhurst family, Sunday, March 18, 1945.

[196] Letter, RSB to PDB, Monday, March 19, 1945, content condensed.

[197] All quotations from letter, RSB to Bringhurst family, Sunday, March 18, 1945, content edited.

[198] Letter, RSB to Bringhurst family, Sunday, April 8, 1945.

[199] Letter, RSB to Bringhurst family, Sunday, April 1, 1945.

[200] Letter, RSB to Bringhurst family, Saturday, April 21, 1945.

[201] All these quotes are from a letter, RSB to Bringhurst family, Thursday, April 26, 1945.

[202] Letter, RSB to Bringhurst family, Monday, April 30, 1945.

[203] Letter, RSB to Bringhurst family, Tuesday, May 1, 1945.

[204] Letter, RSB to Bringhurst family, Sunday, June 3 (labeled June 2), 1945, content condensed.

[205] Letter, RSB to Bringhurst family, Sunday, June 10, 1945. The following month when Royce had McNally take pictures of his family, almost all the pictures were of Jean.

[206] Letter, RSB to Bringhurst family, Saturday, June 16, 1945, content condensed.

[207] Letter, RSB to Bringhurst family, Tuesday, July 3, 1945, content condensed.

[208] Letter, RSB to Bringhurst family, Sunday, July 15, 1945.

[209] Letter, RSB to Bringhurst family, Sunday, July 15, 1945, content condensed.

[210] Letter, RSB to Bringhurst family, Monday, July 30, 1945.

[211] Letter, RSB to Bringhurst family, August 22, 1945, content condensed.

[212] More than thirty years later, Royce would reflect in his journal, "How fortunate are they who sleep as well as I did before my military years." (Davis Journal #2, June 23, 1977.)

[213] Author's personal recollection. It was not until 1962, some eighteen years after his war experience, that Pearl remarked that fireworks no longer made Royce jump in alarm (Letter, PDB to Jane Davidson, September 11, 1962), and it was not until five years later, in 1967, that she could report that he was finally sleeping as well as he did before the war (Letter, PDB to Jane Davidson, December 3, 1967).

[214] See Veterans Administration "Award of Disability Compensation or Pension" dated December 1, 1945 file # C-512 97 86; also letter, RSB to James E. Higgs, Asst. Chief, Registrar Division, Veteran's Administration Hospital, December 24, 1958.

[215] On Pearl Harbor day in 1980, Royce wrote, "39 years ago today on that dim Sunday in another world, we entered WWII, courtesy of the Japanese and our bumbling government. How the world was irrevocably changed by that war can only be thought about. One can never know what might have been. We must remember and remember and remember! to forestall a global conflict ever happening again. The one bright spot is that Japan & the USA, the great enemies of that day, are now interdependent allies, dedicated to the same peaceful objectives." (Davis Journal, number 5, Sunday, December 7, 1980)

A year later, he wrote, "Forty years ago the drums of war started to beat in earnest for this nation. The 'great experience' of all our lives, that changed our lives and the world forever. Lest we forget the terribleness of it all, I think that we should reflect on it and pass it on to the generations yet unborn. War is the unthinkable ultimate of evil." (Davis Journal, volume 6, December 7, 1981.)

For Christmas of 1989, more than 44 years after the war ended, Royce gave his son John and his wife a book containing a compilation of excerpts from great stories about the war. On the inside cover he wrote, "Of all our many lives World War II was the great adventure. It is important that we remember."
[216] Photo from 489th Squadron "yearbook."

CHAPTER 7:
[1] The legislation, officially known as the Serviceman's Readjustment Act of 1944 but known ever after as the "G.I. Bill of Rights" or simply the "G.I. Bill," was signed into law on June 22, 1944.
[2] Letter, RSB to PDB, Sunday, September 14, 1944, content condensed. Like Pearl, Royce's mother reacted favorably to the idea of college, and Royce's description of her reaction also gives an indication of his mother's influence on his life. "She told me how happy it made her to know I wanted to finish my schooling when the war is over. Mother has always expected quite a bit from me and I don't know why. I never liked to think that she thought of me as her favorite son. In fact she did write me once and tell me how she felt about her boys. She said she loved one just as much as the other but that I was the only one who seemed inclined to be what she desired her son to be. I'm not patting myself on the back because I know me a lot better than she does and I can't ever call myself good when I can see my faults staring me in the face. She has always worked so hard and she has made the world about her good. It's made me always desire to be the son she thinks I am." (Letter, RSB to PDB, Thursday, October 19, 1944.)
[3] The job with the electric company was included on a handwritten outline used by Royce to prepare his application for a faculty position at UCLA in 1950. He indicated he worked for the company from September to December, 1945.
[4] There are few family records from this time, but when Pearl wrote to her brother Daniel in November and December of 1945, they were living with her mother, and when she began writing to her mother in January of 1946, she made allusion to having lived with her. Both Pearl's letters to her mother and Florence's blessing certificate placed them in the Springview Ward of the Grant Stake, which included the area of the valley where Pearl's mother lived (see letter, PDB to Jane Davidson, March 4, 1946). When Royce began receiving his G.I. Bill stipend, a portion was sent to Pearl's mother, probably to compensate her for their having stayed with her.
[5] See Florence Bringhurst Certificate of Blessing, PDB's Book of Remembrance.
[6] See Ricks, Joel E., "The Origin of the Utah State Agricultural College" (1953). *Joel Rick Collection*, Paper 8, posted on DigitalCommons@USU.
[7] The increase in rent may have been due in part to the fact that the "G.I. Bill" provided a housing allowance for veteran students. This influx of federal moneys likely contributed to an inflation of housing prices proportional to the allowance. There was also the simple factor of supply and demand. The influx of new students surely made housing a scarce commodity.

⁸ This agency was the Office of Price Administration, established as a wartime emergency measure with authority to ration goods and cap prices of certain items, including rent.

⁹ The information in this segment is drawn largely from letters from PDB to Jane Davidson, dated January 19, January 26, February 5, February 10, February 14, March 4, March 10, and March 17, 1946.

¹⁰ See letter, PDB to Daniel & Mary Davidson, April 11, 1946.

¹¹ On April 6 Pearl wrote, "Royce is really running himself ragged with all his work. I hope this extra work doesn't drag his school grades down. He's right in the top with his studies so far.... He had to go up to school Saturday until 1:00 p.m. I guess he will be working every Saturday now." (Letter, PDB to Jane Davidson, April 10, 1946.)

¹² Toward the end of Summer quarter Pearl wrote, "Royce has finished 2 quarters with straight A's. If he passed this last quarter with straight A's he will have a scholarship award. I think he might have made it. He surely does study hard enough. It will probably help him out a great deal when it comes to getting a job after school is out." (Letter, PDB to Jane Davidson, July 22, 1946.) It does not appear at this time that the possibility of graduate school had come up between them.

¹³ Delmar C. Tingey was a member of the Utah State Agricultural Faculty, and had been an important figure in plant genetics, using wheat hybridization to overcome disease problems in winter wheat. (See Delmar C. Tingey, "The Bunt Problem in Relation to Winter Wheat Breeding," ninth annual USAC faculty research lecture, 1950, posted on USU Digital Commons.) Royce mentioned his job with Dr. Tingey in his resume outline for his UCLA faculty application in 1950.

¹⁴ Of the influence of Dr. Tingey in his life, Royce later remarked, "Del Tingey was a very wonderful person to work with, because he liked young people. He liked to see you learn to do something, and do it well. And I've always prided myself on having followed a lot of the things that he did." (Videotaped interview of RSB by Anton Kofranic, UCD Emeriti Association, 1996, content condensed.)

¹⁵ After his retirement, Royce explained some of the reasons he preferred Wisconsin. "I applied to three institutions, and was accepted at all of them, but I chose Wisconsin. The word Wisconsin always grabbed me a little bit. It was cold back there. I was also offered a job at the same time in Minnesota, colder. Both of them were fine. I really didn't want to work on cereal crops, and that's what I would have had to work on at Minnesota. I worked on agronomic crops." (Videotaped interview of RSB by Anton Kofranic, UCD Emeriti Association, 1996, content condensed.)

¹⁶ The G.I. Bill allotment was for $120 per month living expenses, in addition to payment of tuition. According to Royce's official paperwork, his major subject was Agronomy and Genetics, under the direction of W. K. Smith and D. C. Cooper, and his minor subject was Botany, under C. L. Huskins. All three of these men were prominent in the study of crop plant genetics, Smith with sweet clover, Cooper with corn, and Huskins with oats and wheat.

¹⁷ This is the name of the foundation as recorded in Royce's doctoral thesis.

Notes and References

The coumarin in sweet clover, under the influence of fungus mold, breaks down into dicoumarol, which acts as an anticoagulant by counteracting the effects of Vitamin K. This had been established as the cause of the so-called "bleeding disease" in cattle, and is the reason Dr. Smith was attempting to breed coumarin-free strains of sweet clover. These discoveries led to some practical uses as well, including the development and release of the anticoagulant warfarin (named after the Wisconsin Alumni Research Foundation which funded Royce's research). Warfarin in high doses was already being used as a commercial rat poison at the time Royce arrived in Wisconsin, and in the following years highly controlled low doses would come into use as an anticoagulant for use in humans with various clotting disorders. Warfarin served as the dominant medical anticoagulant for more than seven decades, and at the time of this writing is still in wide use, though at last is beginning to be replaced by newer compounds with different modes of action.

[18] The address was Unit 84-C, Badger, Wisconsin. The apartment was in a section of Badger Village known as South Badger.

[19] See *Badger Village Mini-movie Script*, a PDF transcript of a brief movie documentary of the village compiled from University of Wisconsin library archives, found at *archives.library.wisc.edu*.

[20] Letter, PDB to Jane Davidson, August 12, 1947.

[21] Details from letter, PDB to Jane Davidson, August 12, 1947.

[22] Davis Journal #1, August 8, 1976, content edited. Royce's account continued: "It wasn't well received, to put it mildly, and she won't ever forget it. I meant no offence, but it didn't sell. We paid the hospital & doctor with military pay received for all the leave time I didn't get while in the service, & had enough left to buy a washing machine. How good it was not to go in debt for that. Those were happy days that I failed to enjoy as much as I should have, and didn't really let the small family we had then enjoy much either."

[23] This was an oft-related episode in family folklore.

[24] Pearl remarked in a letter to her mother, "Our little Sunday School is going right along but I am afraid if Royce & I ever moved from the village there would be a fade out here. We are starting a Relief Society. We will have our first meeting at our house tomorrow night. I'm afraid two of the girls aren't going to help out much but as long as the other two will come along & give me some help I'll be glad to try." (Letter, PDB to Jane Davidson, January 6, 1948 (labeled 1946), also February 8 and February 29, 1948.)

[25] Others were involved in Royce's research, including Dr. D. C. Cooper, who appears on the acknowledgment page of Royce's thesis as having contributed advice and guidance in his cytological study. There were also other faculty involved in his training, one of which was a relative of Royce, and paid him a visit many years later, as Royce was making his way through paperwork in his Davis office. "I was almost through with it when who should walk in but D. C. Smith, my old professor from Wisconsin (plant breeding). He is also my cousin on my mother's side. It was good to talk to him." (Davis Journal, volume 7, January 27, 1983.

[26] Videotaped interview of RSB by Anton Kofranic, UCD Emeriti Association, 1996, content condensed.

[27] This episode is recounted from memory. It was told with some humor,

and I was unable to locate a recording which included it.

[28] Florence Bringhurst Nielsen, who provided this detail, added, "The neighbors all predicted that we would die from eating them, but as we continued to flourish, some of them began to gather puffballs as well. They were delicious!

[29] Letter, RSB to Bringhurst family, June 21, 1948.

[30] Royce's daughter Florence recalled that Royce learned and frequently sang the Wisconsin fight song, "On Wisconsin," to his children.

[31] Letter, PDB to Jane Davidson, September 1, 1948.

[32] Letter, PDB to Jane Davidson, November 21, 1948.

[33] Letter, PDB to Jane Davidson, January 23, 1949, content condensed.

[34] Letter, PDB to Jane Davidson, December 27, 1948. Florence and Marla, on learning of this incident, remarked that it was unusual that Florence had been given gum, as that was a forbidden item in the household for as long as they could remember. Royce had always disliked chewing gum, and a complete prohibition may have resulted from this incident.

[35] Some of this discussion was recounted by Pearl in a letter to her mother (Letter, PDB to Jane Davidson, January 9, 1949).

[36] Letters, PDB to Jane Davidson, March 13 and March 20, 1949. Florence recalled that to call the doctor, Royce had to walk to a phone located across the same field where the family had gathered puffballs. Penicillin was a truly revolutionary medication which rendered previously deadly infections readily curable. It was still only available as an injected medication at that time, as an effective oral formulation had yet to be developed.

Also in Wisconsin, Florence evidently contracted a mild case of polio, which was not discovered until later when she was noted to lack normal dexterity in her legs. Hence, each of the girls in turn became seriously ill during the stay in Wisconsin.

[37] Jean's account of the incident was as follows: "I hated the black Buick. When we went to see the fireworks, I was sitting in the back seat doorway of the Buick. We were in our nightgowns. It was a Roman candle that was shot off by university students that went wild. I remember watching it come across the field right at me. It hit me in the arm, caught my nightgown on fire and burned the inside of the Buick. Dad rolled me in the grass but I didn't have much nightgown left when he was done. I didn't like sitting of the driver's side of the Buick because I could see the burn marks and that horrid memory would come back. I always sat on the right side because I could look out the window and pretend nothing had ever happened." (Jean B. Anderson, Recollections #3, 2010.)

[38] In a letter to F. V. Owen a few months later, Royce referred to the busy pace of his graduate studies, and remarked, "Thank you again for the time you spent with me during my vacation. It was very enjoyable." (Letter, RSB to F. V. Owen, January 27, 1950.) In a later letter after Royce was established in Los Angeles, Dr. Owen mentioned his continued work on John Tripp Bringhurst's property. "I have not seen your folks for some time but I hope to be checking on the fall plantings in the near future and expect to be seeing them. It is surely nice of your father to grow a little plot for us on his place and it is a pleasure to get around to see him. I only regret that I can't

make it more frequent." (Letter, F. V. Owen to RSB, February 2, 1951.)

[39] Letter, PDB to Jane Davidson, May 7, 1950, content condensed.

[40] Letter, PDB to Jane Davidson, May 14, 1950.

[41] It does so by negating the effect of Vitamin K, a vitamin essential to many steps in the coagulation process. Even now, administration of Vitamin K is the principal antidote for Warfarin toxicity.

[42] All the information from this section is condensed from Bringhurst, Royce S., *Genetic Analysis of Chlorophyll Deficiency in Melilotus alba × M. dentata with Some Observations of Meiotic Irregularities*, Doctoral Thesis, U. of Wisconsin, 1950.

[43] All the initial crosses were made using the defective hybrids as the female plant. Although this assured that the offspring were truly hybrids, the fertility rate was so low as to make this a very difficult method. The process developed by Royce involved using the normal plant as the female parent, with pollen from the defective plant. Normally this was considered an imprecise method, because the normal plants were very prone to self-pollination, and the offspring were difficult to distinguish in many cases from hybrids. By developing a scale of physical characteristics which distinguished true hybrids from self-pollinated offspring, Royce was able to overcome this difficulty and produce hybrids in large numbers.

[44] Dr. W. K. Smith died in December of 1985, when Royce himself was not far from retirement. Royce and Pearl had contacted his widow, Vera, and the following March she sent them a letter of thanks in which she enclosed a memorial resolution from the University of Wisconsin, honoring Emeritus Professor Smith and summarizing his professional contributions. It is largely from this memorial that information about William K. Smith's work with sweet clover was taken.

[45] Pearl gave a description of the oral examination (Letter, PDB to Jane Davidson, June 2 (labeled May 2), 1950.)

[46] This episode, alluded to by Pearl years later, was thought to be no more than family folklore, until Marla verified that she remembered waiting in the car while her father went to talk to the principal. She was not yet three at the time.

[47] This photograph was taken in 1958, in the Salt Lake area.

CHAPTER 8:

[1] Letter, J. C. Ripperton to RSB, January 4, 1949; the reply was dated January 10, 1949.

[2] Letter, RSB to Dr. D. W. Thorne, October 18, 1949.

[3] Art, who had been on Royce's crop judging team at Granite High School and had been a member of Royce and Pearl's Seminary class, had remained a friend to both. His field at UCLA was plant nutrition and plant physiology.

[4] Letters from S. H. Cameron to RSB are dated October 27, 1949, November 14, 1949, December 10, 1949, and February 25, 1950. A letter from Art Wallace was sent February 8, and another informal handwritten letter was evidently included with Dr. Cameron's letter on February 25, 1950. Royce's replies to Dr. Cameron are dated November 2, 1949, November 27, 1949,

December 23, 1949, and February 15, 1950.

[5] The position at Utah State was filled by a colleague named DeVere McAllister. As Royce approached retirement, he wrote, "I replied to a letter from DeVere McAllister, who took the job that was offered me at Utah State. I wonder what my life would have been had I taken it. One can never know the answer to that." (Davis Journal, volume 7, February 9, 1983.)

[6] Dr. Cameron sent Royce a telegram in March indicating that he recommended his appointment at the higher salary which Royce, at Art Wallace's prompting, had requested. The official acceptance letter from S. H. Cameron to Royce was dated May 8, 1950, though by then Royce had already been given verbal notification of his appointment. His reply of acceptance to S. H. Cameron was dated May 11, 1950. It is worthy of note that Royce's appointment had to be submitted through Dr. Claude B. Hutchison, dean of the College of Agriculture at U. C. Berkeley, since in 1950 UCLA was still technically a subsidiary of the original University of California located at Berkeley. The Los Angeles campus would not achieve co-equal status with the Berkeley campus until the following year; hence, faculty appointments in 1950 still had to be approved through Berkeley. This explains much of the delay in Royce's final approval.

[7] Letter, RSB to F. V. Owen, January 27, 1950. Royce referred to the machine as a "Swedish seeder." He received a reply from Dr. Owen on February 3 with pictures of the seeder, which he referred to as a beet seed drill, made in Hilleshög, Sweden. (Letter, F. V. Owen to RSB, February 3, 1950.)

[8] Letter, R. D. Rands to RSB, May 22, 1950. The reply is dated May 27, 1950.

[9] The facilities were described in one of the letters from Dr. Cameron. (Letter, S. H. Cameron to RSB, December 12, 1949.)

[10] This experimental plot, central to Royce's work, is described in C. A. Schroeder, "Recollections of Avocado History at U.C.L.A.," *California Avocado Society 1992 Yearbook*, 76:77-83. The author, Art Schroeder, was well known to Royce and helped him and Pearl move in to the home they later had built in Reseda. Dr. Schroeder described the plot as "possibly the best ever assembled for teaching and research of subtropical fruits according to visitors from many institutions and experiment stations throughout the world." The planting of this experimental orchard actually preceded the opening of the UCLA campus, and the College of Agriculture was its pioneer college. As the university grew, the College of Agriculture would be closed and shifted to U.C. Riverside, and the extraordinary orchard would succumb to the growing needs of the UCLA campus.

[11] Letter, S. H. Cameron to RSB, October 27, 1949.

[12] Jean wrote of the trip, "We did not drive to Los Angeles in the Buick. We rode the train. Dad drove the Buick out.... I recall we were awakened in the night during that trip. I recall looking back and seeing train cars lying on their side. I don't know if part of our train came off the tracks or another train. I just remembered being frightened." (Jean B. Anderson, recollections #3, 2012.) Pearl had also recounted the derailment, and recalled being awakened at night to running of footsteps and anxious voices, and looking out and seeing the car adjacent to theirs derailed.

[13] Letter, PDB to Jane Davidson, dated July 20, July 31, and August 6, 1950.

[14] Pearl described the incident: "Sunday a jet bomber flew low over our units and it sounded like our roof was torn off. I would hate to be bombed by one of them. It left us all shaking. It sounded like a very loud clap of thunder only it lasted a few minutes. Royce said it sounded to him like a terrible explosion." (Letter, PDB to Jane Davidson, October 10, 1950.) Jet aircraft were still relatively new at the time, and it is probable that the family had never heard one at close quarters. The concern about bombing resulted from the heightened tensions between the U. S. and China, and the fear that the two countries would go to war.

[15] Letter, RSB to Bringhurst family, December 3, 1950, content slightly condensed.

[16] Letter, PDB to Jane Davidson, August 27, 1950.

[17] The Church maintained, and continues to maintain, an extensive welfare system to provide for poor members. This included a series of canneries manned on a volunteer basis by local Church members. Pearl described the experience to her mother: "Last week I went over to the church cannery and canned two cases of peaches for us as well as peaches for the welfare. It was really a night of canning. We started canning at 6:00 p.m. and finished the next morning at 2:30 a.m. At 11:00 p.m. when everyone was tired anyway, a whole new truckload of peaches came in and we had to can them or they would spoil. The only thing that kept me going was thinking of the fruit situation at home and hoping some of this might find its way to you. It was just Beverly Hills Ward doing it and only about 25 or 30 people showed up. Some of those ladies were fancy ladies too with lots of money but they dug right in and worked with the best of them the full time without a complaint. It truly amazed me and made me feel this truly was the gospel in action. Sunday I asked one of them how she felt after it and she said fine, that it didn't hurt her a lot. We really worked hard. I was worn out the next day. We had men along that really had a hard job too." (Letter, PDB to Jane Davidson, September 6, 1950.)

[18] The chapel was located on the northeast corner of the temple lot, and the groundbreaking took place on Saturday, January 13, 1951, under the direction of LeGrand Richards, then a member of the presiding bishopric but later an apostle. Pearl wrote, "Last Saturday morning Royce went over to the church site to help work on our chapel. He swung a pick all day and has had sore muscles in his back from it. He's been away from that kind of work for a long time." (Letter, PDB to Jane Davidson, February 5, 1951.)

Plans for construction of the Los Angeles Temple had been announced in 1937 when the Church acquired the property. Construction on the temple did begin in 1951, though the completion and dedication did not occur until 1956, when Royce and Pearl had already moved from the Los Angeles area. The temple property was within a few blocks of their apartment near UCLA.

[19] This quake occurred when they were dining at the home of Aunt Neit, a relative of Royce on his mother's side.

[20] The loan guarantee taken advantage of by Royce and Pearl was a part of the "G.I. Bill of Rights," which had also funded Royce's post-war college education. The move was probably also motivated by loud all-night parties held several times a week by the upstairs neighbors. (Letter, PDB to Jane

Davidson, February 5, 1951.)
²¹ Pearl in her later years was a capable driver, but in Los Angeles her inexperience evidently showed. Her daughter Florence later recalled the fear she and her sisters would feel when their mother piloted the car. She recalls the three of them turned with their faces to the back seat crying "Don't let mommy drive!" (Personal recollection, Florence Bringhurst Nielsen, March 2015.)
²² C. Arthur Schroeder at that time was an Associate Professor of Subtropical Horticulture, and was an important name in the Avocado field. He made many trips to Mexico and Central America in search of promising varieties of avocado and other exotic fruits, for use by California growers. (See "Awards of Honor," *California Avocado Society 1953-54 Yearbook*, vol. 38 p. 11.)
²³ See *California Avocado Association Annual Report*, 1:5-6, 1915. The organization in its first meeting also acted to determine what name the industry would use to refer to the fruit. All agreed that the name "alligator pear," then in common usage, would be a poor choice, but some who had studied the fruit in Latin America favored the Spanish word *ahuacate*, derived from the Nahuatl name *ahuacatl*. The association agreed on the term "avocado," since it had become an accepted American word for the fruit, was easily pronounced by English speakers, and because it was already in use by the U. S. Department of Agriculture.
²⁴ Wilson Popenoe, who became a researcher for the United Fruit Company, did much of the pioneer work for the industry in California, and had introduced most of the major California varieties in the early years (see C. A. Schroeder, "Recollections of Avocado History at UCLA," *California Avocado Society 1992 Yearbook*, 76:77-83.) This pioneering work was a family enterprise, and at the first meeting of the California Avocado Association in 1915, Wilson's father, F. O. Popenoe, had advocated strongly for the careful selection of superior varieties, rather than a hodge-podge of often inferior varieties which then existed, and which proved costly to grow. (F. O. Popenoe, "Varieties of the Avocado," *California Avocado Association Annual Report*, 1915, 1:44-69.)
²⁵ The Hass avocado, which continues to dominate the industry as of this writing, is an example of the random ways in which these varieties appeared. A postman named Rudolph Hass decided to plant a grove of avocados, and ordered seedlings on which to graft them. Few of the grafts took, but one of the seedlings grew and produced abundant fruit which the Hass children preferred to any others. The variety proved to bear extraordinarily well, and was eventually patented and marketed in 1935. Despite the fact that the fruit had a different color and texture from standard varieties, it had excellent flavor and produced more consistently, which brought a great boost to the avocado industry. It is still the dominant avocado variety not just in California but in much of the world. (See "The Hass Mother Tree," *California Avocado Society 1973-74 Yearbook*, 57:16-17.)
²⁶ Dr. Lammerts summarized the first years of his research in W. E. Lammerts, "Progress Report on Avocado Breeding," *California Avocado Society 1942 Yearbook*, 27:36-41.

27 Walter E. Lammerts, "The Avocado Breeding Project," *California Avocado Society 1945 Yearbook* 30:74-80. Dr. Lammerts had joined the staff of a private firm, where he planned to concentrate on roses, camellias, and ornamental plants.

28 In the initial letter Royce received from Dr. Cameron advising him of the position, he was told, "The Experiment Station part of the job will involve the continuation and expansion of a breeding program with some of our subtropical fruits – a program that was started about ten years ago by Dr. W. E. Lammerts and discontinued, except for the maintenance of some of his progeny plantings, when he resigned about three years ago." (Letter, S. H. Cameron to RSB, October 27, 1979.)

29 Letter, S. H. Cameron to RSB, December 12, 1949. The letter continued, "The breeding work with the other fruits, so far as its practical aspects are concerned, involves merely a search for better commercial varieties. In the case of the cherimoya we should like a variety that does not require hand pollination. So far as I am aware, none of these fruits present any pollination difficulties comparable to that of the avocado. As I implied in one of my earlier letters, we are also interested in developing varieties of some of the temperate zone fruits that will succeed reasonably well in southern California. Some of our other subtropical fruits, such as the loquat and guava, offer interesting opportunities for genetical studies and variety improvement. We prefer not to delimit the field of work for the man who may accept this position. Except for the avocado, we are not under any industry pressure to work on the above mentioned fruits. We can work on them or not, as we choose."

30 This detail was recalled by one of Royce's daughters.

31 In 1997, long after retirement, Royce described the process: "You had to get them so that they were blooming at the same time. And then we brought in the honey bees, and we put those inside of the chambers, and they moved about in there, and pollinated, and it worked very effectively. I developed the greenhouses that we used for that purpose. I got stung pretty good. They were mean rascals. We're talking about plain old honey bees. And those rascals, we'd put them in the greenhouse, and they'd get on you and sting the devil out of you. I was very sensitive to them. But I used to move them at night. Then I didn't get stung much." (Taped interview, September 1997, tape 2.)

32 See R. S. Bringhurst, "Influence of Glasshouse Conditions on Flower Behavior of Hass and Anaheim Avocados," *California Avocado Society 1951 Yearbook*, 36:164-168. Based on his first series of observations, he determined that the best time to hand pollinate was early afternoon.

33 This is inferred from a listing of seedlings and crosses sent to Royce in 1954 by J. Nicholas Thille, an avocado grower who was willing to dedicate an orchard to experimental varieties. Many of the trees bore numbers indicating the year the cross was made, such as "51-11," "52-48" and "5339." There were a handful from the early 1940s, such as "43-47," "44078," and "45001." Presumably these had been planted by Dr. Lammerts when he was in charge of the program. One of the two crosses originally made by Royce and later named as a variety by others was known as "UCLA 51007." Royce

would later use a similar system to keep track of strawberry cultivars.

[34] Avocados were considered to be divided into "races" with certain characteristics. Most commercial varieties were considered "Guatemalan" varieties, and had features typical of those seen in American markets. Mexicola was of the "Mexican" type, which had a smaller fruit with a smooth, thin skin which was often edible. A third type was the "Caribbean" type, which was of little importance in California because it bore inferior fruit and was poorly adapted to the climate. Some of this class are grown in Florida.

[35] These stages were described by Royce in Bringhurst, R.S., "Influence of Glasshouse Conditions on Flower Behavior of Hass and Anaheim avocados," *California Avocado Society 1951 Yearbook*, 36: 164-168. This phenomenon had first been described in detail 28 years earlier in Stout, A. B., "A Study in Cross-pollination of Avocados in Southern California," *California Avocado Association Annual Report 1922-1923*, 8:29-45.

[36] J. W. Lesley and R. S. Bringhurst, "Environmental Conditions Affecting Pollination of Avocados," *California Avocado Society 1951 Yearbook*, 36:169-173.

[37] R. S. Bringhurst "Influence of Glasshouse Conditions on Flower Behavior of Hass and Anaheim Avocados," *California Avocado Society 1951 Yearbook*, 36:164-168.

[38] One neighbor family whose friendship Royce and Pearl especially valued were Whitey and Ann Kutcher, to whom Royce paid a visit many years later, during one of his many trips to attend to strawberry business. "I drove to Reseda and stopped in to see Whitey Kutcher. They were really our best friends when we lived there (New York origin Jews). I have always been fond of them and whenever I drive through I stop to see them when I have any time at all. I am always somewhat envious of Jews. I think I could have enjoyed being one." (Davis Journal #3, February 16, 1978, content condensed.)

[39]Letter, PDB to Jane Davidson, May 10, 1951.

[40] Letter, PDB to Jane Davidson, June 5, 1951. The sacrament service on Fast Sunday has no assigned speakers; instead, members are allowed to come forward to "bear their testimonies," meaning to give a brief but often heartfelt expression of their religious convictions. As a matter of custom, certain priesthood ordinances such as the blessing and naming of infants, and the confirmation of individuals who have been baptized (which is not permitted until at least eight years of age) are performed in this meeting prior to the bearing of testimonies. In the Reseda Ward, there were so many infants and children of baptism age that these ceremonies occupied the entire meeting, leaving no time for the customary testimonies. The numerical growth of the ward was mentioned in a later letter (Letter, PDB to Jane Davidson, Thursday, September 27, 1951.)

[41] See explanation and footnote in Chapter 3. Royce had been ordained a Seventy at the commencement of his mission, unlike present practice, and hence for many years he would be a member of a priesthood quorum whose principal focus was missionary work. At this time the Sunday School was a more autonomous organization than is now the case, as were priesthood quorums; hence it was that Royce held a calling as a Sunday School teacher

in his ward while also performing Seventies assignments for his stake.

[42] Letters, PDB to Jane Davidson, July 1 and July 23, 1951. The letter of June 23 indicates that a photo of the extended Davidson family was taken during this trip.

[43] Letters, PDB to JD, September 27, 1951 and October 10, 1951.

[44] Letter, PDB to JD, November 30, 1951. Seven months later Pearl wrote, "Royce fell on his sore foot day before yesterday and thinks he might have knocked something back in place because he says his foot feels better. It has been bothering him all along. He is going to keep putting on his elastic bandage to see if it will get completely well." (Letter, PDB to Jane Davidson, June 26, 1952.) From a medical standpoint, I do not have a good explanation for an additional injury causing improvement. It is probable, given the duration of the pain, that he had suffered a fracture or a significant joint injury. He eventually did recover fully, and was not bothered by the foot for the remainder of his life.

[45] Letter, PDB to Jane Davidson, January 19, 1952. Pearl gave an excellent description of the flood. In the following letter she wrote, "A bridge on Reseda Blvd 3 blocks from us buckled from torrents in the flood canal. I used to laugh at that deep canal last summer with a little puddle of water in it. I am not laughing any more though. I am just very thankful it is there. Friday when Royce & I went over to the bridge I took one look at the water in the canal. I don't think a human being could possibly survive if they fell in. It was boiling & churning and moving more swiftly than I have ever seen water move." (Letter, PDB to Jane Davidson, January 27, 1952.)

[46] Marla and Jean recalled this event as follows:
Marla: "We had a flood, and our street flooded, and Dad was out on a rowboat, and a neighbor. I don't know if they were going to see how bad the flood was, or see if people needed help, or just lookie-looing...
Jean: And Dad was young and dumb.
Marla: ...They were out rowing a boat out in our street. Our house didn't flood though. We were up above. The street did, I think the canal overflowed, and he's out there in a boat...
Jean: "Having a good time. Then they got caught, and the current got faster, and all of a sudden they recognized they were headed for the canal, and they were going to be in the ocean very soon. And as I recall, it was a truck of some sort that literally drove in front of them, parked, and helped them out. And two very humble men came back." (Taped Recording, Jean B. Anderson and Marla B. Vaughn, January 22, 2011, content condensed.)

[47] Pearl wrote, "The lowest the University would take was $925. Royce said there's sort of an agreement among the U. men not to go high on them. Royce figured the next logical bid was $950, so he bid $951. He was $1.00 higher than the next highest bidder. So we got the car." (Letter, PDB to Jane Davidson, May 26, 1952.) Royce's father may have assisted with the downpayment on the vehicle (letter, PDB to Jane Davidson, February 6, 1952.)

[48] Letters, PDB to Jane Davidson, April 13, 1952 and July 8, 1952.

[49] Letter, PDB to Jane Davidson, July 30, 1952. The three girls remembered the earthquake distinctly. Florence remembered being shaken awake by her mother, while Jean recalled awakening with the surreal feeling that she was

being rocked in a cradle. Marla recalled being sent to bed, while Florence remembered staying in their parents' bed, both in accord with the account in Pearl's letter. (Jean B. Anderson, Florence B. Nielsen, and Marla B. Vaughn, Taped Conversation, January 22, 2011.)

[50] Meiosis is the process by which sex cells, in this case pollen and ova, are produced. Normal plant and animal cells have two complete sets of chromosomes, one from each parent. When the plant or animal is to reproduce, the sex cells undergo meiosis, which is similar to mitosis (regular cell division), except that only one of each pair of chromosomes is retained, allowing the offspring plant to have a single set of chromosomes from each parent.

[51] Royce published this information in 1954. See R. S. Bringhurst, "Interspecific Hybridization in and Chromosome Numbers in *Persea*, reprinted from *Proceedings of the American Society for Horticultural Science*, vol. 63, p. 239-242, 1954.

[52] The biological effect of colchicine is to inhibit the function of microtubules, which play a role in a number of cellular functions, and are essential for the migration of chromosomes during mitosis or meiosis. In modern medicine, colchicine is still used as an effective treatment for acute gout attacks.

[53] See R. S. Bringhurst, "Breeding Tetraploid Avocados," reprinted from *Proceedings of the American Society for Horticultural Science*, vol. 67, 1956. The article was prepared for publication in 1955, two years after Royce left U.C.L.A.

[54] Letter, RSB to Dr. D. C. Cooper, September 6, 1950. Royce was writing to ask how to obtain dissecting needles such as he had used in Wisconsin, as they were not available in Los Angeles.

[55] See letters, RSB to W. K. Smith, September 19, 1950 and October 10, 1950, and W. K. Smith to RSB, October 2, 1950. Dr. Smith expressed concern that Royce might be duplicating some breeding work being performed by another graduate student, but Royce assured him that this was not the case. He wrote a later letter to W. K. Smith reporting on his use of the seed, and asking him if he required an additional supply of ch3ch3 plant seed, as Royce had had to propagate some for the catalase experiments. In the same letter, Royce indicated that he was preparing a manuscript on his work with *Melilotus* for publication, though I have found no indication that this was ever published.

[56] See letters, RSB to R. W. Woodward, October 9, October 19, and October 30, 1950, and R. W. Woodward to RSB, October 15, 1950. Dr. Appleman determined that the defective barley plants were low in catalase, as were most albino lethal plants.

[57] *California Avocado Society 1952 Yearbook* 37:210-214.

[58] This was written in Royce's hand in a brochure of the meeting, and had been saved in a folder entitled "Memoirs." It was chiefly this folder which brought to my attention the fact that Royce had intended to write his life's story, and which induced me to write this history. The account ended with the words, "As I drove home that evening, sick at heart and unhappy, I resolved to leave UCLA and seek a job elsewhere. It was in this mood that I looked at a sugar beet job with USDA and then went after the strawberry job at Davis when it was offered the following year."

[59] William Henry Chandler (1878-1970) was a distinguished pomologist originally from Missouri, who as Pomology Department chairman at Davis had worked with Dr. Tufts who was the current chairman. Dr. Chandler had transferred to UCLA to serve as assistant dean of the college of agriculture, and was in retirement when Royce arrived. He had learned of the vacancy in the strawberry program from Dr. Tufts, and they both evidently considered Royce to be a good candidate for the position (see Interview of Royce S. Bringhurst, UCD Emeriti Association, 1996.) Royce would eventually name one of his most successful strawberry cultivars "Chandler" in his honor.

[60] Royce spoke highly of Dr. Cameron at the time of his death, which he learned of while serving as a committee member at the Corvallis Germplasm Repository in 1986. "Ron Cameron was present this afternoon, and he informed me that Dr. Sidney Cameron died in November at about 90 years of age. I take note of this because he is the one that hired me at Los Angeles in 1950. I owe a great deal to him, certainly my life would have been very different if ... and you can never really probe what it would have led to. I pause in silent tribute to one more good man who has influenced my life. We never stand alone." (Davis Journal, volume 9, January 8, 1987.)

[61] Some who knew Royce might question that he had a serious fear of crowds, but there is good reason to suppose this was so. In early 1956, Royce needed to schedule a trip to the Southern California strawberry growing areas at about the time the Los Angeles Temple was being dedicated. In a letter to her Mother, Pearl wrote, "I tried to get him to schedule his trip so he could go to the temple dedication, but he still has that terrible dread of crowds. I wonder if he will ever get over it." (Letter, PDB to Jane Davidson, February 20, 1953.) If such a fear of crowds did exist, in later life Royce learned either to control it or to hide it well.

[62] All these details were outlined in a letter from Pearl to her mother (Letter, PDB to Jane Davidson, May 31, 1953).

[63] Letter, PDB to Jane Davidson, July 29, 1953.

[64] Royce's daughter Jean recalled this occasion. "I remember Dad driving us to where we could see the whole San Fernando Valley. He stopped the car and had us get out. He said, "Take a good look. This will soon be gone." We saw nothing but orange groves as far as our eye could see. Sure enough, it was soon nothing but houses. I thought that part of Sacrament meeting was reading lots of names of new people to our ward. It wasn't until we moved to Davis that I discovered this was unique to the San Fernando Valley." (Jean B. Anderson, Recollections #3, 2010.)

[65] Nearly three decades later, Royce returned to the old home during one of his many automobile journeys for the strawberry program. "I made a nostalgic journey to LeMay Street & Kitridge St. to look at our home. It needs paint and yard care, more than the last time I saw it, when it was newly painted and everything was well cared for.... How very tenuous is our hold on anything that we hold close in life. As we grow older, it is gone with the passing days and yesterdays are but a misty dream that perhaps didn't even happen except in our minds. (Davis Journal, volume 5, June 25, 1981.)

[66] Royce regarded Dr. Peterson as a friend, and forty years later ran into him at a conference in Ames, Iowa. "In particular, I spent time with Peter Peter-

son (old friend of UCLA – me, and Riverside – him days). He has quite an interesting pitch on gene rescue in maize. I understand part of it. I think his optimism a bit high, but it is evidently good work." (Davis Journal, volume 7, March 8, 1983.)

[67] Royce, almost certainly referring to this work with the avocado, later wrote, "After experiencing initial difficulties with pollination control in a certain tree fruit crop, the breeder decided that it was not worthwhile to try to make the hybrids that obviously should be made and determined to grow nothing but open pollinated seedlings of several of the leading cultivars. The breeder assumed that the desired types of hybrids would occur naturally and ignored the fact that the cultivars in question had been selected because they were highly self fruitful when planted in solid blocks. Furthermore the fact that the segregating progenies from these cultivars (selfed or intercrossed) did not yield the phenotypes specifically sought and needed desperately by the industry was ignored, inasmuch as thousands of open pollinated seedlings of the principal parents involved had already been grown and observed through the fruiting stage. The total waste of resources was immense, since juvenility in the seedlings lasted from five to as long as 15 years and the tree size was relatively large. The grower organization supporting the activity realized virtually nothing from their investment. The information generated from it was of little scientific value and the project never moved from square one in the real game that might have resulted from the employment of appropriate breeding strategy, starting with the difficult hybridizations that could and should have been made in the first place." (R. S. Bringhurst, "Breeding Strategy," Chapter 10 of James N. Moore and Jules Janick, ed., *Methods in Fruit Breeding*, 1983, p. 149-150.)

[68] Information from the *Avocado Variety Database*, an on-line resource listing about a thousand cultivars.

CHAPTER 9:

[1] Warren Porter Tufts served as chairman of the Department of Pomology (fruit science) for many years, and was involved in Royce's hiring. He made major contributions in tree fruit horticulture. He retired in 1958.

[2] Letter, PDB to Jane Davidson, September 11, 1953.

[3] George Ledyard Stebbins (1906-2000) was one of the first scientists to bring together emerging discoveries in the field of genetics with Darwinian theories of natural selection. His book *Variation and Evolution in Plants*, published in 1950, was considered one of the key publications of the 20th century in the field of evolutionary biology. Stebbins had written and taught extensively on the role of hybridization and polyploidy in the evolution of plants, which became an important focus of Royce's work with wild strawberries. Dr. Stebbins was considered a brilliant teacher, and Royce valued his friendship. A biological preserve near Davis still bears his name. The house he rented to the Bringhurst family was located at 653 Miller Drive in Davis.

[4] Brighton is now best known as a world-class ski resort, the construction of which was distressing to Pearl, as she felt it had destroyed the place's natural beauty. This trip was memorable to the older girls in the family because of

an incident in which Royce, hidden in the shrubbery beside the trail, growled like a bear, causing Jean and Florence to flee down a hillside trail in panic, with Royce running along behind trying to reassure them.

5 This species, an octoploid which was widely distributed throughout both the Rocky Mountains and the Sierra Nevada of California, was later reclassified as a subspecies of *Fragaria virginiana* which was common in the eastern United States, and would be referred to as *Fragaria virginiana-glauca*.

6 The high-altitude strawberries carried the gene responsible for the day-neutral strawberry. The timing of the trip was almost surely July of 1954, since Ann was still a baby in arms, as recalled by the older sisters.

7 The French intelligence agent, Frèzier, unaware that these strawberries were dioecious (had separate male and female plants), probably brought back only female plants of *F. chiloensis*, and though they had produced well in Chile, lacking a consistent pollinator they bore little fruit in Europe. Eventually the plants began to be interplanted with octoploid *F. virginiana* plants, and by cross-pollination they were rendered productive. See Gunter Staudt, "The Origin and History of the Large-Fruited Garden Strawberry Fragaria x Ananassa Duch." *Der Zuchter* 31,5, 1961 (from a typescript copy in RSB's collection p. 4-5); Staudt concluded that the first European specimens of *Fragaria virginiana* had been imported to France from Canada (same source, p. 2-3).

8 See "The Strawberry from Chile," chapter 4 of George M. Darrow, *The Strawberry: History, Breeding and Physiology*, 1966, p. 24-39; also Gunter Staudt, "The Origin and History of the Large-Fruited Garden Strawberry Fragaria x Ananassa Duch." *Der Zuchter* 31,5, 1961 (from a typescript copy in RSB's collection, p. 7-9). The species name *ananassa* means "pineapple," and these berries were often referred to as pine strawberries. The first truly scientific studies of the strawberry were performed by another Frenchman, Antoine Nicolas Duchesne, who died in the early 1800s.

9 See "Albert Etter: Fruit Breeder" by R. Fishman, *Fruit Varieties Journal*, vol. 41 #1, p. 40-46, 1987. According to this source, Etter's breeding materials were donated to the University of California breeding program. George M. Darrow, in a letter to Royce in 1962, gave some details of Etter's life, and a recollection of visiting the family ranch in Humboldt County (Letter, George M. Darrow to RSB, February 8, 1962, found in alphabetized collection of letters from 1960-62, RSB Collection). Darrow speculated as to why Etter ended his highly successful strawberry breeding program. "It seemed to me probable that Etter got in virus sometime in the early twenties and that he did not know why his selections did not hold up. I think he stopped breeding about 1926 for that reason. Possibly he might have had other diseases or some insects that damaged him." Darrow summated his information on Etter in his landmark book on strawberries in 1966. Near the conclusion of Etter's life sketch, he wrote, "While ill during 1938, he lost most of his strawberry material and only a few of his varieties and selections are known to still exist. His genetic material now exists in E-80 and E-121 and in varieties of other breeders derived in part from some of his varieties." (George M. Darrow, *The Strawberry: History, Breeding and Physiology*, 1966, p. 183-187.) This seems to contradict the later claim that Etter's material ended up in the

UC breeding program.

Of note, a grandnephew of Albert Etter named Michael Etter contacted Royce in 1982 requesting plants of a variety named Ettersburg 121, one of Etter's favorite cultivars, which he wished to propagate for sentimental reasons on the family property. Royce replied, "We shall be most pleased to provide you with several plants of Ettersburg 121. I shall have them propagated very soon and they should be ready to pick up in several weeks. I am most pleased that you are taking interest in the accomplishments of your great uncle. His work is appreciated and known by strawberry breeders worldwide." (See letter, Michael Etter to RSB, postmarked May 12, 1982, and Royce's response, dated May 17, 1982.)

Of Etter's contributions to his later breeding work, Royce wrote, "I personally believe that the most important genes responsible for adaptation to California growing conditions come directly from *F. chiloensis* via the 'Lassen' variety. I commented upon that recently (see Breeding Octoploid Strawberries, Iowa Stat J. Res. 58). Etter was indeed the original and in many respects, the most important California strawberry breeder. I have visited the place where he worked (Ettersburg) and recently his nephew contacted me seeking information about him." (Letter, RSB to Dr. R. W. Becking, October 9, 1984.)

[10] Royce described Earl V. Goldsmith in response to an inquiry by Henry A. Wallace: "Yes, I knew Goldsmith. He was a very hard working, interesting man. I think that his contribution to strawberry breeding was at least as great as Etter's. He was very much interested in developing the 'ideal' strawberry, and his concept of that ideal berry was very good. He gave me the impression of being a veritable encyclopedia of knowledge relative to strawberries, and this was true. Selection numbers and their positive and negative breeding characteristics came from him like the flow of proverbs from Cervantes' Sancho Panza. I tried to take careful notes at first but by the second visit I realized that his memory was not entirely reliable, as contradictions were fairly frequent. I concluded that it was unfortunate that many of the old University breeding records were never put into written form." (Letter, RSB to Henry A. Wallace, October 29, 1962.)

[11] See "Strawberry Breeding of North American Agricultural Experiment Stations," chapter 15 of George M. Darrow, *The Strawberry: History, Breeding and Physiology*, 1966. The section on California (p. 227-234) written by Royce Bringhurst, was relied upon for much of this summary. Thomas and Goldsmith concentrated much of their later breeding effort on so-called everbearing strawberries which extended the growing season, and gave Driscoll a proprietary advantage in the Central Coast area. The departure of Royce and Victor's predecessors from the university and their subsequent use of university breeding stock for private research was later called into question by some industry leaders (see Herbert Baum, *Quest for the Perfect Strawberry: a case study of the California Strawberry Commission and the Strawberry Industry*, p. 161, 2005); curiously, at the time of this writing a similar drama is playing out with the departure of Royce and Victor's successors, Douglas Shaw and Kirk Larson, from UC Davis to form a private strawberry firm, which resulted in a law suit against the university by the California Strawberry Commission,

alleging privatization of public university materials.

After his arrival in Davis, Royce met at least twice with Earl V. Goldsmith, who freely shared with Royce much of his own valuable knowledge and insights into strawberry breeding. Goldsmith died suddenly shortly thereafter, so their relationship was cut short almost as soon as it began. Royce's relationship with Harold E. Thomas was much more enduring. Although Thomas's Driscoll organization was technically in competition with the University program, in actual fact he and Royce became friends and professional associates, referring visitors to each other and assisting and consulting with one another (see letters, RSB to Harold E. Thomas, October 12, 1960, and December 21, 1961, also letter, Harold E. Thomas to RSB, August 13, 1962, and response, RSB to Harold E. Thomas, August 15, 1962.)
[12] These varieties, Campbell and Cupertino, were evidently released at the behest of commercial growers who had been involved in their testing; they were described by their developers in not very glowing terms. (Richard E. Baker and Victor Voth, *Breeding and Testing Strawberry Varieties*, California Agricultural Experiment Station Bulletin 714, October, 1949, p. 12-14.)
[13] See "George M. Darrow, *The Strawberry: History, Breeding and Physiology*, 1966 (the section on California is on p. 227-228.) Royce recognized the often-unappreciated importance Dr. Baker had had in the strawberry program. Asked to provide a reference for Baker in 1960, he wrote, "When I assumed responsibility for the project, I found that a comprehensive, state-wide program had already been organized and was in operation. The program was designed to yield fundamental information relative to the genetic and cytogenic structure of small-fruits plants (particularly Fragaria) and also to analyze certain associated physiological relationships. In addition to this, an integral part of the program included provisions to solve the important practical problems of the industries.

"Significant progress was made by Dr. Baker before he left the University as evidenced by the publications that resulted. Furthermore, the progress we have been able to make since 1953 has in large part been based upon the work he initiated." (Letter, RSB re: Dr. Richard E. Baker, January 4, 1960.)
[14] The Wolfskill Experimental Orchard was located on property adjacent to Putah Creek, originally owned by the William Wolfskill family, a group of enterprising pioneer immigrants from Kentucky who received the property as a land grant from the Mexican government in the early 1800s. (See Iris Higbie Wilson, *William Wolfskill 1798-1866: Frontier Trapper to California Ranchero*, 1965, p. 142. A plastic-bound photocopy of portions of this book is located in the RSB papers.) According to a brochure published a few years before Royce's hiring, the Wolfskill Experimental Orchard was used as an isolation nursery, since it was geographically distant from any commercial or home-garden plantings of strawberries which could be infected with virus. (See Richard E. Baker and Victor Voth, *Breeding and Testing Strawberry Varieties*, California Agricultural Experiment Station Bulletin 714, October 1949, p. 3.)
[15] Interview of Royce S. Bringhurst, UCD Emeriti Association, 1996. Royce later named one of his important cultivars "Tufts" in honor of the department chairman who had hired him.

[16] Another important figure in these early years was David Van Hook, who served as staff research assistant in the program at Davis.

[17] Interview of Royce S. Bringhurst, UCD Emeriti Association, 1996.

[18] See "Commitment to Excellence," an informational videotape produced by the California Strawberry Advisory Board in 1992.

[19] This method of seed extraction was referred to by Royce as the "blender-extraction, flotation-separation method" (RSB and V. Voth, "Effect of Stratification on Strawberry Seed Germination," *American Society for Horticultural Science*, vol. 70, p. 144. Darrow gave a description and illustration of this method. (George M. Darrow, *The Strawberry: History, Breeding, and Physiology*, 1966, plate 14-1.) However, the author, and Royce's other children, recall Royce smashing fruit with a flat blade onto absorbent paper, and later collecting the seeds. This process may have been used with smaller berries, or for a specific purpose. Generally the seeds required some kind of processing, such as scarification or application of potent acid, in order to germinate. Royce also confirmed that stratification (storage for a period of time at cold temperatures) greatly increased the germination rate.

[20] Letter, PDB to Jane Davidson, July 12, 1955.

[21] Letters, PDB to Jane Davidson, December 25, 1955 and January 8 and 12, 1956.

[22] Letter, PDB to Jane Davidson, January 23, 1956.

[23] Both these accounts were retold by Jean Anderson, Florence Nielsen, and Marla Vaughn, recorded January 22, 2011.

[24] The button proved, ironically, to be a campaign button for the American Red Cross. The name of the physician, Dr. Tillotson, was recalled by Marla in a recording on January 22, 2011. Royce and Pearl were unaware he was a Latter-day Saint until Royce recognized him years later at a Church meeting in Sacramento, and introduced him to the son whose life he saved. The Davis physician was Dr. Leo Cronan, a general practitioner who remained the family physician for years. The best account of this event was found in PDB journal, 1953-1961, entry dated February 14, 1959.

[25] DDT (dichlorodiphenyltrichloroethane) is a highly effective insecticide which was used worldwide for malaria control, and later in agricultural settings. The chemical breaks down very slowly, however, and was later shown to seriously reduce populations of certain carnivorous birds. Its banning in the 70s and 80s was accompanied by a resurgence of malaria in many areas of the world. It was decades later that insecticides were implicated as a contributing cause of Parkinson's Disease, which Royce eventually developed.

[26] Letter, PDB to Jane Davidson, January 29, 1957.

[27] These details about the house are found in letters from PDB to Jane Davidson, February 13 and May 8, 1957.

[28] See letter, PDB to Jane Davidson, May 7, 1959.

[29] Royce's solo appearances with the orchestra proved to be something of a painful ordeal, as Royce was petrified by stage fright, particularly when he was required to take a soloist's position in front of the orchestra, and found himself unable to hear the orchestra behind him. On one such occasion he actually walked off the stage during the performance, and Alex Gould, turning around from the conductor's stand, delivered the solo himself (recollec-

tion of Florence Bringhurst Nielsen, January, 2016). Royce eventually withdrew from the Sacramento Symphony Chorus after his year-long stay in Chile, but he continued to sing vocally and in choirs as he got the chance, and each Saturday when he was home he would listen to the weekly New York Metropolitan Opera broadcasts. His musical abilities continued to be hampered by his poor hearing, which, coupled with the powerful volume of his own voice, made it difficult for him to hear and follow accompaniments.

[30] Letter, PDB to Jane Davidson, June 22, 1958.

[31] The accident happened on April 9, 1959, as recorded in Royce's date book: "John's leg was fractured today. It could have been worse." See also letters, PDB to Jane Davidson, May 7 and June 21, 1959. John, the author, remembers the placing of the cast is one of his earliest recollections.

[32] Florence's injury happened on Friday, June 24, 1960, and according to Royce's appointment book, she was hospitalized from June 27 to July 8, and underwent the procedure to reduce her fracture on July 1. She was not taken out of the cast until November 14, and was still required to use crutches until December 10, and was still not completely healed months later. (Letter, Florence Bringhurst to Jane Davidson, June 28, 1960; also letters, PDB to Jane Davidson, July 6, November 9, and December 19, 1960.) Florence in the worst of times always retained a happy disposition—her mother called her "sunny-natured," a description her siblings could verify.

[33] On November 18, 1960, Royce and other committee members received a memo from UC Davis Chancellor Emil M. Mrak, requesting that they inform him if any of the scholarships contained discriminatory clauses to prevent their presentation to any minority groups, presaging the Equal Rights movement which would gain momentum during the 1960s. Royce admired the chancellor, and would later name one of his strawberry varieties "Mrak" in his honor.

[34] Dr. Dillon S. Brown, then the Pomology Department chairman, made a detailed description of the construction of the new building:

"In the proposed plan for the new structure, windowless laboratories were arranged to the inside so as to utilize a central utility core, while offices with windows were located along the outside wall. To minimize staff objections to the plan, which were based mostly on the possibility of claustrophobic feelings being a problem with users of the inside laboratories, the architects included windows on the outside walls of the laboratories and the inside walls of the offices. These windows, which provided some 'line of sight' from the laboratories across the hallway separating them from the offices and thence through the offices to the out-of-doors, were intended to give more of an 'open' feeling to the laboratories....

"Construction of the Horticulture-Viticulture Building, later named Wickson Hall, began in late 1956 and was completed in early 1959. It was the first office-laboratory building on the Davis campus to be constructed with provision for air-conditioning. It was built with three stories housing offices, laboratories and classrooms on the north half and only one story, primarily space devoted to controlled atmosphere (temperature) storage and related postharvest research facilities on the south side....

"After it was occupied, no one complained of claustrophobia in the inside laboratories. In fact, the windows which had been provided to give a more 'open' feeling were soon covered with posters, paper or curtains, both in the laboratories and the offices, probably to provide privacy from passersby in the hallways into which laboratories and offices opened." Dillon S. Brown, *The Department of Pomology: University of California, Davis—The Contemporary Years, 1945-1981*, p. 98-99.

[35] Pearl would have one subsequent pregnancy, which ended in miscarriage.

[36] According to Pearl's account, Margaret was to be the last birth attended by Dr. Cronan, the family doctor (PDB journal 1953-1961, entry for March 17, 1961.

[37] Quoted from an undated talk, given around 1978.

[38] Letter, PDB to Jane Davidson, February 8, 1962.

[39] R. S. Bringhurst and Victor Voth, "Effect of Stratification on Strawberry Seed Germination," *American Society for Horticultural Science*, vol. 70, p. 144-149, 1956. As a matter of mutual agreement, both Royce and Victor's names appeared as authors on the majority of their publications, indicating that both were collaborators in the study. The fact that Royce's name came first indicates that he was the principal investigator in this study; on other studies, particularly of growing practices, Victor's name or that of another investigator often came first. The original hand-recorded data for this study were kept in a yellow field journal labeled "Seed Germination" from 1953-55, and were recorded in Royce's hand (RSB Collection).

[40] See George F. Waldo, Royce S. Bringhurst, and Victor Voth, *Commercial Strawberry Growing in the Pacific Coast States*, USDA Bulletin #2236, 1971, p. 13.

[41] The term "mulch" refers to any material applied around a crop to cover the surface of the ground. The mulch in this case consisted of sheets of polyethylene plastic rather than organic material.

[42] See Victor Voth and R. S. Bringhurst, "Polyethylene Over Strawberries," *California Agriculture*, May, 1959, p. 5 & 14. Victor and Royce, while investigating the use of plastic mulch, also experimented with various degrees of plant pruning, a common industry practice, and found that it actually hampered yield under some conditions. See Victor Voth and R. S. Bringhurst, "Pruning and Polyethylene Mulching of Summer-Planted Strawberries in Southern California, *Proceedings of the American Society for Horticultural Science*, vol. 78, 1961, p. 275-280. Clear polyethylene was the only material which advanced the harvest, but it did not control weeds, which were able to grow under the plastic. In areas where an early harvest was not an objective, black polyethylene could provide excellent weed control, along with the other advantages of growing on plastic. Its only drawback was that it could heat up in very warm weather, damaging the fruit.

[43] Royce and Victor, continuing their experiments with polyethylene mulch, were able to show that in winter plantings, applying the plastic as early as possible in the winter months dramatically increased the yield (Victor Voth and R. S. Bringhurst, "Early Mulched Strawberries," *California Agriculture*, February, 1962).

It is interesting to note that although Victor Voth appears as the primary author on this paper and likely performed experimental work and data collection, it appears that Royce proposed and wrote the paper, and performed statistical analysis on the data. He submitted a draft to Victor for review, and evidently had it reviewed subsequently by the department chairman before submitting it for publication. (Letters, RSB to Victor Voth, December 1, 1961, and Victor Voth to RSB, December 4, 1961.) These letters do suggest the nature of Royce and Victor's collaboration on these joint publications. It is certainly the case that both were fully involved and in agreement on the essential elements of their numerous joint publications. The letters also make clear that in addition to publishing such data, they used them in field presentations to the growers themselves.

[44] See Victor Voth and R. S. Bringhurst, "Fruiting and Vegetative Response of Lassen Strawberries in Southern California as Influenced by Nursery Source, Time of Planting, and Plant Chilling History," *Proceedings of the American Society for Horticultural Science*, vol. 72, 1958, p. 186-197.

[45] Limited data from these 1954 experiments are found in a yellow field journal labeled "Virus, Heat Inactivation" (RSB Collection). A unique approach was to attempt to eliminate virus by growing the plants in a naturally hot environment, for which a planting was made in the Imperial Valley, an inland area near the Mexican border known for its sweltering hot summers. This research was done in cooperation with an entomologist from Berkeley, and the approach was a first in agronomy, though not very practical in application. (See N. W. Frazier, V. Voth, and R. S. Bringhurst, "Inactivation of Two Strawberry Viruses in Plants Grown in a Natural High-Temperature Environment," *Phytopathology*, Vol. 55 No. 11, November 1965, p. 1203-1205.)

[46] See the yellow field journal labeled "Virus Index 1960, 1961, 1965," particularly the second page of the initial segment entitled "1960 virus indexing" (RSB Collection). A later yellow journal entitled "Runner Tip Culture Virus Inactivation 1964" outlines the specific technique used at that time to culture virus-free plants from runner tips, together with some data on the outcome. This was recorded in an unknown hand, but was titled in Royce's handwriting, and was evidently kept by him for reference (RSB Collection).

[47] See R. S. Bringhurst and Victor Voth, "Strawberry Virus Transmission by Grafting Excised Leaves," *Plant Disease Reporter*, Vol. 40, No. 7, July 15, 1956, p. 596-600. In a letter to her mother, Pearl wrote, "One day I went into his office just as his dept. head was leaving. After he left Royce told me that they had named him as the one who had done the most important piece of scientific research in the dept. He got it for a leaf graft test for virus that he worked out on his strawberries. Now everyone in the dept. is using it and it does save time." (Letter, PDB to Jane Davidson, August 7, 1956.)

[48] Fumigation was, of course, not without its dangers, as both agents were toxic to humans. Methyl bromide, a very simple chemical, is produced naturally in large amounts by a number of organisms, but it was synthesized chemically for agricultural use. Its effectiveness as a soil fumigant made it essential to the early success of the California strawberry industry, but it later came under increased scrutiny as the environmental effects of agricultural chemicals began to be addressed. Besides its extreme toxicity to humans and animals, it was much later shown to be an ozone-depleting chemical, and starting in 1989 its use began to be phased out, though strawberries were

initially exempted from the phase-out because of the essential role it has played in pest control. Chloropicrin, a form of tear gas, is included for control of soil fungus. Though less dangerous, it also has obvious properties toxic to humans, and has been subjected to similar scrutiny.

At the time of this writing, methyl bromide has recently been phased out completely as an agricultural chemical, and the use of chloropicrin has been greatly restricted. Satisfactory alternatives have yet to be found, and this remains an important challenge to the strawberry industry at the present time.

[49] In 1971, Royce recorded in his field journal an outline of the history of the California Strawberry Advisory Board as given by Malcolm Douglas, who served for a time as its president. According to the outline, the board was established in June of 1955, and was funded with a $50,000 loan. Its research budget had risen from $5,000 in 1957 to $92,000 in 1970. (Yellow field journal labeled "May 1970 thru Feb. 1972," entry for September 16, 1971. The occasion was a biennial strawberry meeting in Redding.

[50] See R. S. Bringhurst and Victor Voth, "Solana Strawberry," *California Agriculture*, January, 1958, p. 13. The fruit of "Solana" was not only far better flavored than "Lassen," it was also firmer and stored much longer. The foliage was unusual in that the leaves often had four or five leaflets rather than the usual three.

[51] The Lancaster site was later eliminated, and the Torrey Pines operations were later moved to Santa Ana.

[52] Royce and Victor published detailed descriptions of these new varieties, together with growing recommendations. (R. S. Bringhurst and Victor Voth, "Fresno, Torrey, Wiltguard for California growing areas," *California Agriculture*, 1961, p. 11-12.) The initial intent was to release four varieties; the fourth variety may have been the one later named Tioga.

[53] Royce proposed the naming of "Fresno" in a letter to Victor: "Yesterday, I saw Fred Hirasuna in Fresno and discussed the new varieties with him. They fruited 53.9-10 in a summer planting at Fresno last year in comparison with Lassen and four other selections and picked 53.9-10 as superior to Lassen in every replicate. The field had a lot of salt injury and 53.9-10 stood up well. This is the second performance report on 53.9-10 at Fresno and both have been the same, 'the best in the field'. I suggest we name it Fresno. Will you buy that name? Let me know your reaction." (Letter, RSB to Victor Voth, October 14, 1960.)

[54] R.S. Bringhurst, Stephen Wilhelm, and Victor Voth, "Pathogen Variability and Breeding Verticillium Wilt Resistant Strawberries," *Phytopathology*, Vol. 51, no. 11, November, 1961, p. 786-794.

[55] See R. S. Bringhurst, P. E. Hansche, and Victor Voth, "Inheritance of *Verticillium* Wilt Resistance and the Correlation of Resistance with Performance Traits of the Strawberry," *American Society for Horticultural Science*, vol. 92, 1968, p. 369-375.

Royce explained this problem in a summary of California breeding work. "In the California program there is interest in a possible negative relationship between Verticillium wilt resistance and desirable performance traits. It has been noted that, with intensive selection for desirable traits, most of the selected clones are susceptible, even though both parents may be resistant.

Genes conditioning wilt resistance may be linked with genes that condition undesirable performance traits." (George M. Darrow, *The Strawberry: History, Breeding and Physiology*, 1966, p. 229. The section on the California breeding program was written by Royce.)

[56] Royce was not unique among scientists in pursuing this line of investigation, but it provided him with valuable experience which would later enable him to make truly unique contributions.

[57] At that time Royce referred to these plants found in the high-altitude areas of the Rocky Mountains and Sierra Nevada as *Fragaria ovalis*; they were subsequently reclassified as *Fragaria virginiana-glauca*, and Royce immediately adopted that term (see letter, RSB to Henry A. Wallace, September 11, 1962).

[58] Letter, RSB to Henry A. Wallace, March 3, 1961.

[59] See R. S. Bringhurst and Victor Voth, "Larger Strawberries through Plant Breeding," *California Agriculture*, February, 1960, p. 8.

[60] These appear to have been "Mark Sensing Cards," similar to those then used for standardized tests, in which data entry was made by using special pencils to mark spaces on the card. Royce sent an explanation of the details to Victor. (Letter, RSB to Victor Voth, September 15, 1960.)

[61] These were cards designed to be read by early computers. The elongated cards were "punched" with perforations which recorded data for processing. In effect, this was a form of data storage which, while primitive by later standards, was effective in its time, and far superior to handwritten tables.

[62] In his section on the California strawberry breeding program, written for Darrow's landmark book on the strawberry, Royce wrote, "Computer technology can play an important role in large scale breeding work. This view is supported by various factors, some of which cannot yet be fully realized in terms of their potentialities. First, the computer can be programmed to reduce into a comprehensible form, at minimal cost, the large quantities of data which breeders amass each year. In the California program, field data are summarized, the standard deviations are calculated for the various fruit traits that are measured, and 'performance,' a value which is weighed heavily by yield but considers appearance, fruit size, and firmness as well, is calculated. Data can be recorded on cards in the field to reduce the cost of obtaining analyzable data. At the end of the harvest season (four to seven months), the values can be obtained for seasonal summarization by machine." (George M. Darrow, *The Strawberry: History, Breeding and Physiology*, 1966, p. 229.)

[63] Letter, RSB to Victor Voth, September 15, 1960.

[64] R. S. Bringhurst and Victor Voth, "Tioga: A New California Strawberry," *California Agriculture*, April, 1964, p. 12-13.

[65] See letters, G. Staudt to RSB dated January 28, February 22, October 25, and December 5, 1960, and those from RSB to G. Staudt dated February 5, March 16, and November 2, 1960. Regarding the reclassification of *Fragaria ovalis* as *Fragaria virginiana* ssp. *glauca*, Royce wrote, "I certainly agree with you that the mountain strawberries we have referred to could more accurately be referred to as ssp. glauca. This is very descriptive of the types that we commonly encounter in the mountains here. The leaves almost invariably have the grayish-blue cast suggestive of the name." (Letter, RSB to G. Staudt,

November 2, 1960.)

Royce actually tried to arrange for Staudt to be offered a visiting professorship at UC Davis, but in the end could not obtain funding. (See letters from RSB to G. Staudt, February 5 and March 16, 1960.)

[66] See R. S. Bringhurst, Stephen Wilhelm, and Victor Voth, "Verticillium Wilt Resistance in Natural Populations of Fragaria chiloensis in California," *Phytopathology*, vol. 56, no. 2, February, 1966, p. 219-232.

[67] Letter, Henry A. Wallace to RSB, January 11, 1961.

[68] Letter, RSB to Henry A. Wallace, March 3, 1961.

[69] Royce seems to have regarded Henry A. Wallace more as an agricultural figure than a political one, and in a later interview referred to him only as "the Secretary of Agriculture under Roosevelt." Royce held him in high regard. "He was the founder of the Pioneer Seed Company, the first hybrid seed company. He was the first one that was really successful. He received a lot of bad [publicity], but this man was a wonderful man. He was a humble agriculturalist of the first order."

After Wallace died, Royce was contacted by archivists caring for Henry A. Wallace's papers. "They sent me a letter, and said, 'We have on file numerous letters that you have written to Mr. Wallace, and there is not a copy of a single letter that he wrote to you. And yet we know he did, because he was responding to your letters.' Well, it turned out, he used to either write them long hand, or he typed them out himself, and he always scribbled a few little notes on them. And here I had this whole gang of letters, I had all of them. And you can imagine the library that was interested in accumulating his papers. They were delighted to get these, because I donated them to them, and they were a substantial part of the letters, because it gave them a different view of this man." (Videotaped interview of RSB by Anton Kofranic, UCD Emeriti Association, 1996, content condensed. and edited.)

The letters between Royce and Henry A. Wallace now exist in two collections. Those from 1960 to the beginning of 1963 are in the Royce S. Bringhurst Collection, while later letters, evidently provided by Royce, are now in the Henry A. Wallace Papers, located in the University of Iowa Libraries, Iowa City, Iowa. It is clear from the existing letters that others are missing, including all of Royce's letters after 1962. The existing letters have been obtained in microfilm, and copies of all the letters have been reproduced, and provided in electronic copy to the Wallace collection and in both printout and electronic copy to the Bringhurst collection. Unfortunately, letters Wallace had received from Royce were not found among the Wallace papers, and are presumed lost, except for a few of which Royce had retained copies.

[70] By the time Wallace wrote his introductory chapter for George M. Darrow's landmark book on the strawberry, he had been forced to conclude, "I am sure that the very practical and very successful strawberry growers of California will never fool with either of the two wild European strawberries." (George M. Darrow, *The Strawberry: History, Breeding, and Physiology*, 1966, p. 9-10.)

[71] George M. Darrow evidently visited Royce to exchange information in November of 1963. A letter written by Wallace to Darrow that month begins with the words, "I trust you had a swell time with your family and are

now enjoying yourself with Dr. Bringhurst. You will get slants from him which no one else can give." (Letter, Henry A. Wallace to George M. Darrow, November 29, 1963, University of Iowa Library microfilm 18270, Reel 54, frame 446.) Darrow's book appeared in print in 1966, the year after Henry A. Wallace's death. The introductory chapter was written by Wallace, and the dedicatory page read, "To the memory of Henry A. Wallace, who conceived this book and nursed it to completion; a distinguished scientist, farsighted statesman and sincere humanitarian." (George M. Darrow, *The Strawberry: History, Breeding, and Physiology*, 1966.)

[72] Letter, RSB to Henry A. Wallace, December 1, 1961, content condensed.

[73] In later years he would combine these excursions with family vacations, and his children would recall with a mix of fondness and terror the long drives along Highway 1, between a steep hillside on the one hand and a sheer cliff down to the pounding surf on the other, Royce at the wheel seemingly heedless of the winding road as he scanned the cliffs for strawberries.

[74] In his first year at Davis Royce had experimented with crosses of octoploid and diploid strawberries, and had used similar measurements to confirm the hybrid status of the offspring. (See yellow field journal entitled "Species Hybrids Strawberry," entries from December, 1953.) Royce had developed and used a similar set of criteria during his graduate studies in Wisconsin, to tell in uncertain cases if the progeny of a normal clover was a hybrid instead of a self-pollinated plant (See Bringhurst, Royce S., *Genetic Analysis of Chlorophyll Deficiency in Melilotus alba × M. dentata with Some Observations of Meiotic Irregularities*, Doctoral Thesis, U. of Wisconsin, 1950, p. 64-65.)

[75] See yellow field journal, dated October 1961 through July 1962. Royce summarized his findings as follows: "In summary re this colony: It occupies an area between the 2 culverts (east of the road) about half way between the two clumps of cypress just above a deep gulch running west of road to the sea—the area it occupies is badly eroded and lies just below colonies of normal male & female chiloensis – normal female chilo. are growing with *F. vesca* in & by the low shrubs bordering the hybrid colony (east of it and uphill from it.)

"Apparently, the hybrid originated some years ago (appears to be a single female clone) and it has been able to colonize the eroded area ahead of normal chiloensis. it appears to produce stolons prolifically—this is in line with our experimental experience with this hybrid in GH [greenhouse]."

The appearance of hybrid colonies on both sides of the road meant that the hybrid had occurred before the road had been graded. Royce later contacted the Highway Department in Sacramento, and learned that the road had been graded and constructed between 1928 and 1933, meaning that the hybrid colony had already been well established three decades previously. (Same journal, last recorded page.)

[76] In fact, Royce later discovered that these pentaploid colonies had slight fertility, and would later advance understanding of strawberry evolution by studying their progeny (see R. S. Bringhurst and Tarlock Gill, "Origin of Fragaria Polyploids. II. Unreduced and Doubled-Unreduced Gametes," *American Journal of Botany*, vol. 57 no. 8, 1970, p. 969-976.

[77] The name of Royce's guest was Vivian Lee, a science reporter for *Life* magazine who was then a student at Stanford, and who was conducting research to assist George M. Darrow in the writing of his book on strawberries. Lee eventually wrote chapters three to five of that book, which dealt with the history of the strawberry. Vivian Lee, though interested in the *Fragaria chiloensis* and *vesca* colonies pointed out to her by Royce, was evidently unaware of his discovery of the hybrid colony. The chromosome number of the second colony had not yet been determined when Royce wrote Wallace. (Yellow field journal, October 1961 to July 1962, see entry for June 13, 1962; also letter, RSB to Henry A. Wallace, June 22, 1962, and letters, Vivian Lee to RSB, May 21 and June 6, 1962, and RSB to Vivian Lee, May 28, 1962. See also George M. Darrow, *The Strawberry: History, Breeding, and Physiology*, 1966, p. x, where mention is made of contributions to the book both by Vivian Lee and by Royce S. Bringhurst.)

[78] Daud Ahmad Khan was Royce's first PhD student, and worked with him from 1960 to 1964. Among Royce's papers is a yellow field journal kept by Daud. His work was with hybrid crosses between natural tetraploid strawberries and colchicine-induced tetraploids, and the foliar and floral abnormalities that resulted. See Daud Ahmad Khan and R. S. Bringhurst, "Foliar and Floral Abnormalities in the Strawberry Hybrids," *Pakistan Journal of Agricultural Sciences*, vol. 2 no. 1, March 1965, p. 66-70.

Additionally, Royce worked with Daud on demonstrating biochemical relationships among the different strawberry species, and in 1964 Royce presented a paper at a meeting of the American Botanical Society (showing Daud Khan and Victor Voth as co-authors) entitled "Biochemical Relationships Among Fragaria Species." Of this, Royce's department chairman wrote, "It was a significant paper because it confirmed phytogenetic relationships among the strawberry species through use of chromatographic separation of phenolic extracts of foliar tissue." (Dillon S. Brown, *The Department of Pomology, University of California, Davis: The Contemporary Years, 1945-1981*, p. 200.) This work was similar in its approach to work Royce would do more than a decade later with another graduate student, using electrophoresis gels.

Many years later, after Royce's retirement, Daud wrote him for information, hoping to establish a strawberry growing system in Islamabad, Pakistan (letter, Daud Ahmad Khan, PhD to RSB, June 28, 1994).

[79] Letter, RSB to Henry A. Wallace, October 29, 1962, content slightly condensed.

[80] R. S. Bringhurst and Daud Ahmad Khan, "Natural Pentaploid Fragaria chiloensis-F. vesca Hybrids in Coastal California and their Significance in Polyploid Fragaria Evolution," *American Journal of Botany*, vol. 50, no. 7, August 1963, p. 658-661. Regarding Wallace's request to keep his financial support anonymous, see the letter from Henry A. Wallace to RSB dated November 4, 1962.

[81] Henry A. Wallace developed Amyotrophic Lateral Sclerosis (or ALS, commonly referred to as "Lou Gehrig's Disease") and died in November, 1965, at the age of 77.

[82] The Bringhurst family Christmas newsletter for 1964, speaking of Royce,

says, "He will take a six-month sabbatical beginning in March and will spend his time studying the wild strawberries of the California coast, with a possible trip or two to more distant sites, including South America." (Bringhurst family newsletter, included in letter from PDB to Jane Davidson, December 7, 1964).

[83] The hexaploid colony, discovered on April 8, 1965 at Pacifica, was clearly different from the other hybrid clones, and was given the accession number 1327. (Yellow field journal labeled March 1965-April 28, 1965, entry for 4/8/65.) He described it as follows: "Designated Pac. Hybrid (4) – should look over area again – top of leaves very hairy – One thicker leaf observed – definitely different from #3. Vesca ♂ & ♀ chilo both of same rugose type as ♀ collected growing with it in same clump – fairly new colony - sparse.... Note – #(4) is only one that possibly has higher chr number, leaves are particularly thick." He noted that this collection was made on a rainy day. On June 16, 1966, he described a similar acquisition: "...Near this is an obviously fertile hybrid type – doesn't appear to be thick leafed enough to be 9x but everything is so dry – hard to tell." He later wrote "Prob. 6x" between the lines. This plant was designated Acquisition 1615. On a facing page, he drew a map of the Pacifica site, indicating the location of the various clones, or "acquisitions" which he had collected there. (Yellow field journal labeled Sept 1965 to Jan 1967, entry for June 16, 1966.)

[84] The enneaploid (9-ploid), found on April 22, 1965 near a road at Wrights Beach, was designated with accession number 1364. Royce jotted a map of the area, noting its location, and wrote "This very thick leafed hybrid – could be decaploid??" He later added, "Note proved to have 63 chrs" (yellow field journal labeled March 1965-April 28, 1965, entry for 4/22/65).

[85] Royce noted that of all the hybrids he had collected, about 10% involved unreduced gametes, a very high percentage considering how rarely unreduced gametes occur, much less result in successful pollination. He concluded that the reason for their high frequency had mostly to do with a survival advantage they inherited from their combined traits, and he noted that both the hexaploid and the enneaploid colony showed aggressive growth. (See R. S. Bringhurst and Y. D. A. Senanayake, "The Evolutionary Significance of Natural Fragaria chiloensis x F. vesca Hybrids Resulting from Unreduced Gametes," *American Journal of Botany*, vol. 53, no. 10, November-December 1966, p. 1004.) Royce pointed out that the hexaploid must have resulted from a combination of a normally reduced gamete of Fragaria chiloensis (4 chromosome sets) with an unreduced gamete of F. *vesca* (2 chromosome sets), while the enneaploid must have resulted from the opposite situation, an normally reduced gamete of Fragaria vesca (1 chromosome set) with an unreduced gamete of F. *chiloensis* (8 chromosome sets).

[86] R. S. Bringhurst and Y. D. A. Senanayake, "The Evolutionary Significance of Natural Fragaria chiloensis x F. vesca Hybrids Resulting from Unreduced Gametes," *American Journal of Botany*, vol. 53, no. 10, November-December 1966, p. 1000-1006. Royce had collected and preserved individual plants of each of these natural hybrids, giving each of them an "Acquisition number," by which they would ever after be known. In 1966 he planted clones of many of these acquisitions in the back yard of his home on Mulberry Lane,

making a careful map of the plantings in his field journal (see Yellow field journal labeled Sept. 1965 to Jan 1967, entry for "April 1966," located between April 20 and May 4, 1966). The plants were a small but attractive addition to the landscape, and other family members were completely ignorant of their significance.

[87] R. S. Bringhurst and Y. D. A. Senanayake, "Origin of Fragaria Polyploids. I. Cytological Analysis," *American Journal of Botany*, vol. 54 no. 2, 1967, p. 221-228. Until this publication, the octoploid strawberries were assumed to have acquired their chromosomes from three separate sources, and the genomic formula was characterized as AABBBBCC. By comparing chromosome behavior of hybrids made with *F. vesca* with those made with the strawberry relative *Potentilla glandulosa*, Royce was able to show that the "C" chromosome was really a subset of the "A" chromosome, also found in the diploid *Fragaria vesca*, and he proposed altering the genetic formula to AAA'A'BBBB. This paper was part one of a two-part series; the second part would be published in 1970 (R. S. Bringhurst and Tarlock Gill, "Origin of Fragaria Polyploids. II. Unreduced and Doubled-unreduced Gametes," *American Journal of Botany*, vol. 57 no. 8, 1970, p. 969-976).

[88] R. S. Bringhurst and Tarlock Gill, "Origin of Fragaria Polyploids. II. Unreduced and Double-Unreduced Gametes," *American Journal of Botany*, vol. 57 no. 8, 1970, p. 969-979. In most cases, Royce had simply collected the fruits from the natural hybrids in the wild. In the case of the Wright's Beach 9-ploid, known as "Acquisition 1364," when he checked the colony for fruit on April 20, 1966, he discovered that it had set no fruit, and that there appeared to be no bee activity in the area, which was essential to pollination. Accordingly, he took the process into his own hands, pollinating 35 flowers with pollen from nearby *F. chiloensis* flowers on April 20. When he checked the plants for fruit set on May 10, he pollinated an additional 21 flowers, and on May 26 he collected the first of the fruit (from dead stalks which had ripened prematurely), and pollinated an additional eight flowers, discovering three new hybrids at the same time. On each of these dates he marked the individual pollinated flowers with a distinct color of marker for each date. On June 13, 1966, he returned to the site to assess fruit set, and found that nearly all his hand pollinations had set fruit, and that one additional fruit had set from natural pollination. He wrote in conclusion, "General lack of set of 9x [is] due to lack of pollination @ site. Obvious that it pollinates readily when conditions are right and insects active." (See Yellow field journal labeled Sept. 1965 to Jan 1967, pages marked April 20, May 10, May 26 and June 13, 1967, each marked with a paper clip in the journal.) Although Royce likely planted out the seeds from both the natural and hand-pollinated fruit immediately, he did not complete the study until after his return from Chile in 1968.

[89] *Systemic and Geographic Distribution of the American Strawberry Species: Taxonomic Studies in the Genus*, Günter Staudt, University of California Press, 1999, p. 135-138. Although these hybrids do not reproduce sexually in the wild to any great degree, they do self-propagate by aggressive runnering, hence the designation of a separate hybrid species. It is also referred to as *Fragaria x bringhurstii*, just as the modern commercial strawberry, also of hybrid origin,

is frequently referred to as *Fragaria x ananassa*.

[90] Now Zimbabwe.

[91] Perhaps for this reason, during the nine weeks he was in Mexico, Royce would keep a daily journal, similar to that which he had kept while in the mission field in Texas, Arizona, and California. This journal, the first he had kept since the war, was thorough and reflective, and serves as something of a window into the soul of a man now at the forefront of his field and at the height of his intellectual powers.

[92] Mexico journal, entry for February 1, 1963.

[93] Mexico journal, entry for Sunday, February 3, 1963, greatly condensed. The quote was from a Biblical passage, and is found in Matthew 25:36-40.

[94] Mexico journal, entry dated Tuesday, March 5, 1963.

[95] These were Robert K. Flake and his wife, both of whom had served in the Spanish-American Mission with Royce, and had completed their missions ahead of him. They were in Mexico working on chapel construction.

[96] Mexico journal, entry for Sunday, March 24, 1963, content condensed.

[97] Mexico journal, entry for Monday, March 25, 1963.

[98] Royce used the briefcase in his work until it began to show wear, and for the remainder of his life it housed his collection of vocal music.

[99] Letter, PDB to Jane Davidson, April 28, 1963. The infection was probably acquired from Pearl, who had had "blood poisoning" (lymphangitis) in the right hand while Royce was in Mexico (see letter, PDB to Jane Davidson, March 24, 1963). This was probably a Staphylococcal infection.

[100] See letters, PDB to Jane Davidson, July 31 and August 12, 1963. The trip was cut short by a day when Royce developed an abscess on his nose, a continuation of the skin infections he and Pearl had experienced.

[101] See letter, PDB to Jane Davidson, August 12, 1963.

[102] Letter, PDB to Jane Davidson, March 30, 1964. The meeting house was dedicated by Elder S. Dilworth Young, one of the seven Presidents of the Seventy, who presided at a conference of the Sacramento Stake that same day. The author, then a boy of eight, recalls feeling awestruck by the completed chapel.

[103] Jean's boyfriend was Jewish. He graduated in 1964, after which he had a military obligation, and he pressured her to marry him prior to his departure. She declined, and broke off the engagement in January of 1965. (Letter, PDB to Jane Davidson, January 8, 1965, erroneously dated 1964.)

[104] Royce regretted the poor relationship he had with Jean. On her birthday in 1977, he would write, "Today was Jean's birthday. I always have a pang of regret when I think upon Jean for a number of reasons, some known only to her. Not the least of course is the fact that I never really knew her as a baby or little girl. I came home from the war a stranger to her, and could not understand why she was so frightened of me. Little help for that now. The years of my youth are fled and time has healed most of the little wounds left from those turbulent unsettled years of conflict and uncertainty. Pearl of course was constant, but I have given pain to her too." (Davis Journal #2, March 12, 1977.)

[105] Letters, PDB to Jan Davidson, June 3 and August 9, 1964.

[106] Letter, PDB to Jane Davidson, December 31, 1964. The oratorio was put

on by the stake, with the assistance of some members of the Sacramento Symphony Chorus. Royce sang the challenging tenor solos at the beginning of the piece, including "Comfort ye, my people," "The voice of one crying in the wilderness," and "Every valley shall be exalted."

[107] See letter, RSB to PDB, May 21, 1965, sent from Buenos Aires.

[108] Royce's work on this trip was hampered by the malfunction of a new camera, which left him unable to take photographs for much of the trip, a major objective of his going there. He made do with borrowed cameras, and collected and preserved leaf samples and seeds from wild berries. (See letters, PDB to Jane Davidson dated March 2 and March 10, 1965; also letters from RSB to PDB dated March 1, March 2, March 3, March 4, March 8, March 13, and March 14, 1965.) One of his hosts on the trip was S. C. Harland, a prominent British geneticist who had retired to a house on the Peruvian coast, and who had an interest in strawberries. (Letter, RSB to PDB, March 14, 1965.)

[109] Royce studied strawberry populations as far north as British Columbia (Bringhurst family newsletter, 1965).

[110] Letter, PDB to Jane Davidson, August 1, 1965. Another event significant for Royce that year was the injury of his brother Smith, who was thrown from a horse, sustaining major brain damage. (Letter, PDB to Jane Davidson, April 7, 1965.) Smith, who had lifelong problems with alcohol, had suffered the mishap while intoxicated. Though he lived for years after the accident, he never recovered normal mental function, and the author recalls that after a rare visit to Smith, Royce remarked sadly that he was no more than a shell of his former self.

[111] Letter, PDB to Jane Davidson, September 2, 1965, content condensed.

The author remembers vividly the sight of a bear rearing on its hind feet with paws outstretched, and what sounded like the howls of creatures of all types echoing through the forest. These were California brown bears, the grizzly having been extinct in the state for many years.

[112] By this time the '57 Chevy had been traded in on a Plymouth Fury II station wagon.

[113] At that time, the office of high priest was regarded as the highest of the three offices of the Melchizedek Priesthood ordinarily conferred upon male Church members by ordination. The three offices were that of elder, Seventy, and high priest. Royce had been ordained an elder as a youth in Murray, and was shortly thereafter ordained a Seventy as he began his mission. Having served as a Seventy for over twenty-five years, Royce was ordained a high priest in 1966 by Franklin Dewey Richards, then an Assistant to the Quorum of the Twelve Apostles.

[114] The new stake center, a modern building which featured nearby athletic fields for church competitions, would only be a used for a few years before the stake was divided, and the Davis members returned to the old stake center which was located next to a park on 51st and Dover Streets in Sacramento. The members several times participated in the building of a new stake center, only to have the stake divide and return to the old—this became something of a running joke among the members, who in general were supportive of the process and happy about the growth that made it necessary.

---

[115] "Bracero" is a Spanish term for manual laborer; the Bracero program was a guest laborer program established during World War II to allow Mexican nationals to travel to the United States to provide farm labor, of which there was a shortage owing to the huge manpower drain for the war effort. The program was extended after the war because of the perceived labor need, but was finally discontinued in 1964.

[116] Letter, PDB to Jane Davidson, September 26, 1965. The labor shortage had a significant impact on the labor-intensive strawberry industry. The lack of affordable labor led quickly to a huge upsurge in illegal immigration of Mexican nationals which continues to this day.

[117] Pearl was struck broadside from the left by a rapidly moving car as she entered an unmarked intersection, and was incensed when the policeman, a friend of the other driver, issued her a citation. She contested the citation in court armed with charts and diagrams, and succeeded in having the charges dismissed. (See letters, PDB to Jane Davidson, dated February 5, February 14, and February 22, 1966.)

[118] Letter, PDB to Jane Davidson, April 23, 1967.

[119] Letters, PDB to Jane Davidson, January 24, 1967 and March 13, 1967.

[120] Letter, PDB to Jane Davidson, May 16, 1967. For this "silver" anniversary, Royce and the children purchased Pearl a silver tray.

[121] An unconventional couple, Florence and Larry opted for a large multi-layered carrot cake as their wedding cake.

[122] Photograph from *California Agriculture*, December 1955, p. 6.

CHAPTER 10:

[1] Letter, PDB to Jane Davidson, September 24, 1967.

[2] The house and its surroundings are described in the letter from Pearl to Jane Davidson, September 24, 1967, and was well remembered by the author, who was privileged to occupy the bedroom with the balcony.

[3] According to one source, "The basic goals of the project were to strengthen the University of Chile as an agent of progressive reform in South America and to create a mutual learning experience for faculty and students of both universities" (Ann F. Scheuring, *Abundant Harvest: The History of the University of California, Davis*, 2001, p. 228.) This description tends to give a political ring to the program objectives, an irony since political turmoil in 1973 caused a shut down of all social science aspects of the program. In the end, most of its accomplishments were in the agricultural and academic fields.

[4] Pearl reported, "Royce has been giving lectures to the avocado people here, and they want him to do some teaching so they may give him a temporary appointment to the University of Chile in place of a man who is going away for a year." (Letter, PDB to Jane Davidson, November 12, 1967.)

[5] See Vilma Villagrán Díaz, *El Cultivo de la Frutilla*, 2002, p. 9-10. This was a manual on strawberry production, written by Vilma Villagrán for the Chilean industry. It acknowledges Royce's part in the development of the Chilean strawberry industry, though some details are not entirely accurate; for example, the book implied that Royce's activities took place in 1969, when it is more likely that in that year the results of his work of 1967-68 began to be

observed.

[6] A summary of the Convenio program's work noted that insufficient training and poorly equipped facilities, coupled with political unrest, had greatly hampered the program's activities in these first years, however by the time of its conclusion, "the Convenio's Agricultural program was credited with installing and equipping Chile's most important plant pathology laboratory, establishing a strawberry culture program, identifying common fungus/virus diseases in Chilean fruit trees, and collecting thousands of Chilean insect and nematode specimens." (Ann F. Scheuring, *Abundant Harvest: The History of the University of California, Davis*, 2001, p. 228.)

[7] These results were reported in a strawberry breeding symposium in Scotland three years after Royce's return from Chile. (See R. S. Bringhurst, "Contemporary Chilean *Fragaria chiloensis* and Frèzier's of 1712," from *Symposium on Strawberry Breeding*, The Fruit Group of Eucarpia, in association with the Scottish Horticultural Research Institute, July 24-28, 1971, abstracts, p. 22.)

[8] See Letter, PDB to Jane Davidson, November 26, 1967. In the same letter, Pearl recounted that Ann had begun to interest her friends in the Book of Mormon, and had given out multiple copies. After John received the Aaronic Priesthood he went on visits with the elders, and the missionaries were frequent dinner guests. Most of the local Church members were fairly recent converts, and the family, far from their usual circle of acquaintance, played the role of strengthening the faith of others.

[9] Pearl concluded a rather romanticized account in a letter to her mother, "As we left Chan Chan and I looked back over the lake to it I thought of Brigadoon & Shangri La—those places of peace that don't really exist & which you only visit once and then are lost until one day you want to go back enough to turn again into the mist. This was our Christmas." (Letter, PDB to Jane Davidson, December 27, 1967.) The author's recollections are similar.

[10] See letter, PDB to Jane Davidson, December 30, 1967.

[11] Typically most organizations in the Church of Jesus Christ of Latter-day Saints are headed by a president assisted by two counselors (designated first and second counselor), who form a presidency. In a ward of the Church, the position of president is occupied by a bishop, who, with his two counselors, are referred to as a bishopric. Since the Nuñoa Branch was not yet organized as a ward, it had a branch presidency rather than a bishopric.

[12] It was thought by the family that John, who became sick in February, had acquired the illness from an unpasteurized ice cream cone which a neighbor boy had purchased for him on his birthday in January, and which he felt obligated to eat. The rest of the family remained well, aside from minor intestinal ailments, but as a precaution they stopped eating lettuce, which they had previously eaten after treating it with iodine solution.

[13] The family was taken aback when Isabel, after serving the children fish steaks, brought to Pearl and Royce the fish heads, which were considered a delicacy. Royce explained that the family was not accustomed to eating the heads, and offered them to Isabel, who accepted them with relish.

[14] This was Salvador Allende, a Marxist physician and politician whose election in 1970 was followed by radical social reforms, and a realignment of the

649

country, at the height of the Cold War, toward the Communist Bloc. Allende's breakup of large land holdings and nationalization of major industries were followed by economic collapse, and his defiance of the Chilean congress and courts led to a constitutional crisis. He was deposed in a violent military coup in 1973, during which he took his own life. This led to a long period of military dictatorship, bringing to an end Chile's long tradition of democratic rule, which had been almost unique in Latin America.

[15] PDB journal 1968-1969, entry for June 21, 1968.

[16] Pearl, who had made the choice not to drive while in Chile because of the aggressive driving style of the Chileans, was somewhat unfamiliar with the controls of the car, which probably led to the accident. The author was seated on the front of the car when it began to move, and remembers jumping off just as the car shot backward. Panicked at the sight of blood, he ran back to the chapel to inform his father and the branch president, both of whom ran to attend to the injured woman. The incident was related by Pearl in a letter to her mother (letter, PDB to Jane Davidson, June 9, 1968); also in Pearl's journal (PDB journal 1968-1969, entry for June 3, 1968).

[17] Letter, PDB to Jane Davidson, June 22, 1968.

[18] See letter, PDB to Jane Davidson, September 16, 1969. A lover of history, Newell strongly encouraged the author in the writing of this biography.

[19] This detail was barely recalled by the author, and the specific circumstances were best recalled by his sisters Ann and Margaret. Royce met Pat years later as a married woman, while he served as a worker in the Oakland Temple. On one occasion he wrote, "I saw Pat Galleger, who lived with us as an extra daughter back in the early 1970s. She is happily married, and is fine and really young-looking, particularly when you realize that she has 6 children. I like to believe that we influenced her life for good." (Davis Journal, volume 8, September 8, 1984.)

[20] Backcrossing is a technique used in plant breeding, whereby a new trait is introduced into an elite line of plants by crossing one of the plants with a parent containing the desired trait, then repeatedly crossing the offspring with plants from advanced strains, in an effort to return the offspring line to the very high quality of the commercial strain, while still retaining the desired trait. The trait in this case was the gene for day-neutral flowering behavior.

[21] See R. S. Bringhurst and Victor Voth, "Twenty Years of Strawberry Breeding: Pomology 1972-73 SNB Report #8." Pearl commented on Royce's excitement in a letter to her mother: "Royce right now is very happy over a new strawberry he developed some time back. It is a cross between a wild berry he picked up at Brighton and one of the California varieties. The thing that seems to tickle him most is the fact that it came from home. He picked it up over ten years ago when we were up at Brighton once. It has taken this long to develop the berry. It is an ever-bearer when planted on the coast. He isn't going to release it for a couple of years he says when he finds out what it will do. He is as happy as a little kid with a new toy." (Letter, PDB to Jane Davidson, January 9, 1969.) Royce initially referred to these plants as everbearers, but changed the terminology to "day-neutral" when it had become clear that they behaved differently from plants which had previously been designated as everbearing.

[22] The calyx is the small circle of leaves at the stem end of the strawberry.
[23] This description of the state of the breeding program and its objectives are gleaned from a talk prepared by Royce for the Eucarpia Meeting on Strawberry Breeding, held at Auchincruive, Scotland, in July 1971, entitled, "Breeding Strawberries for Arid Mediterranean Environments."
[24] R. S. Bringhurst and Victor Voth, "Twenty years of strawberry breeding," Pomology 1972-73 SNB Report #8.
[25] See R. S. Bringhurst and Tarlock Gill, "Origin of Fragaria Polyploids II: Unreduced and Double-Unreduced Gametes," *American Journal of Botany*, vol. 17 no. 8, 1970, p. 969-978. The specific details of this research were mentioned in Chapter 9 of this book.
[26] This was a symposium on strawberry breeding, sponsored by the Fruit Group of Eucarpia, in association with the Scottish Horticultural Research Institute, held in Auchincruive, Scotland, on July 24-28, 1971. Royce's description of wild Chilean strawberries is found in the abstracts from the symposium, p. 22. Royce also presented a paper on Verticillium wilt resistance in wild California strawberry populations (p. 14 of the abstracts) and on strawberry breeding for arid Mediterranean environments (typescript).
[27] See handwritten notes from the R. S. Bringhurst collection, which can be dated from 1971, as they reference a "fundamental paper" on breeding work with strawberry polyploids, written "20 years earlier" (1951) by D. H. Scott. In these notes, evidently for a presentation given that year, he mentioned five different ways of producing decaploid strawberries, and also discussed gene transfer by introgression of enneaploids (9-ploid) to the octoploid level. In 1976, he recorded his efforts to create and use hexaploid plants, making use of similar methods: "I ... brought in a few 10x plants to cross with the P. glandulosa and other diploids (Fragaria). I want more 6x plants and that may be the best way to generate them. If it works I should have 10x plants from them bearing only one dose of P. glandulosa. I of course can get the same thing directly by crossing the 10x hybrids (amphiploids of F. chiloensis-P. glandulosa) with the same involving F. vesca. So much for science. I do want to get a lot of things done before I get much older." (Davis Journal, volume 1, April 22, 1976.)
[28] A. N. Kishaba, Victor Voth, A. F. Howland, R. S. Bringhurst, and H. H. Toba, "Twospotted Spider Mite Resistance in California Strawberries," *Journal of Economic Entomology*, vol. 65 no. 1, February, 1972, p. 117-119. Although Royce was the fourth author in this study, an entry in his field journal for April 1969-April 1970 indicated that he was actively involved in the research. In 1976, he reported in his journal planting out an additional experiment dealing with twospotted spider mites (Davis Journal, volume 1, November 24, 1976).
[29] The name "United Arab Republic" had originated with a union of Syria and Egypt under a single government, which had been established under the leadership of Egyptian president Gamal Abdel Nasser. Although the union of the two countries lasted only a short time, Egypt continued to retain "United Arab Republic," abbreviated U.A.R., as its official title until after Nasser's death.
[30] Royce wrote home nearly daily, leaving a vivid record of his activities and

thought during this pivotal time.

[31] Letter, RSB to PDB, September 1, 1970, content condensed.

[32] A week after the hijackings, Pearl wrote her mother, "Royce was supposed to leave for Egypt last Monday September 7th. It was quite a jolt to wake up to the news that a plane was blown up at Cairo. A TWA [Trans World Airlines] at that, which was what he was taking. I still haven't received word that he has arrived safely but I look for a letter today." (Letter, PDB to Jane Davidson, September 14, 1970.) Pearl evidently worried enough to contact government officials, for the next day Royce wrote, "I received a wire today from Wash. D. C. Saying that you had not heard from me. I presume that this meant since my arrival in Egypt. I wrote several times from Rome and this is letter #7 from Cairo. You should have received all the letters from Rome earlier and since your letters are getting to me regularly after about 7 to 10 days enroute, you should be getting mine by now in regular order." (Letter, RSB to PDB, Tuesday, September 15, 1970.)

[33] Letter, RSB to PDB, Tuesday, September 8, 1970.

[34] As always, Royce delighted in describing the animals. "I rode on, of all things, a camel, and stayed on it throughout. Camel riding is a little rough. You have to assume western saddle stance and sort of roll with the animal. I could have walked faster, and been more comfortable in the process, but the novelty of it was very interesting. Several times the camel turned that great head around and gazed at me with soulful eyes. I guess it thought that it didn't make much sense, but a living is a living." (Letter, RSB to PDB, Friday, September 11, 1970, content condensed.)

[35] Letter, RSB to PDB, Sunday, September 20, 1970. The narration continues, "It is a world apart from that which we know. If one wished to look for it, there is plenty to deplore, on the other hand if one wishes to look for it there is also plenty to admire. Stately beauty is present in much of the city. Even some of the recently built apartment buildings are designed and positioned in such a way that it is pleasing in an aesthetic way. Many similar structures in Davis are much less attractive."

[36] These lectures are outlined on a page of the yellow field journal labeled "May 1970 thru Feb. 1972."

[37] Letter, RSB to PDB, Friday, September 25, 1970. The trip gave Royce a completely new perspective on the situation in the Middle East, which in the United States tended to be viewed from the standpoint of America's close ally Israel. Somewhat later, he reflected with considerable foresight, "Of course, the problems over here are really out of this world, political, economic, social. I feel a great sympathy for the people here. They are good and they mean well, but I am certainly not in harmony with many of the official attitudes. However, it would do some of the 'let's fight and get it over with' gang a great deal of good to look at the world from the other fellow's back yard. War, humiliation, and constant dependency on other nations is an evil combination that is very difficult to stomach, and regardless of the beginning or background of the 1967 war, the fact is that Israel still parks on a huge chunk of Egypt, and has a military force that probably outclasses the combined forces of all possible foes here. As if that were not enough, we are forced (willingly) into a position of being the really bad guy behind the bad

guy (in the Arabs' eyes) by accepting the proposition of giving arms (selling them) to Israel at a time when a U.S. originated peace initiative is on the table. You can imagine how popular that idea is over here. I believe that they will negotiate if Israel will. There is definitely a second side to this." (Letter, RSB to PDB, Wednesday, October 7, 1970.)

[38] After the momentous events of September had passed, Royce commented on the state-run press: "A controlled press is a very interesting phenomenon in a way. It is so deadly boring in scope that no one really takes it seriously. The real news is that which circulates in pieces here and there, some right, some wrong. As a result the whole objective behind the controlled press is frustrated. It isn't that people know very much more, it is just that they have little if any confidence in what the 'house organ' plays to them as news." (Letter, RSB to PDB, October 5, 1970.) Royce periodically came upon other reminders that he was living in a controlled society, as when he wrote Pearl, "If a friend has you in for a visit, he must report that visit to the police. How about that?" (Letter, RSB to PDB, Sunday, September 27, 1970.)

[39] Letter, RSB to PDB, Monday, September 28, 1970, content condensed.

[40] Letter, RSB to PDB, Tuesday, September 29, 1970, content condensed.

[41] Letter, RSB to PDB, October 1, 1970 (mistakenly labeled November 1), content condensed.

[42] Letter, RSB to PDB, Wednesday, October 21, 1970.

[43] Letter, RSB to PDB, Monday, October 12, 1970.

[44] Letter, RSB to PDB, Wednesday, October 14, 1970.

[45] According to Dillon Brown's history of the department, the dean at that time should have been Chester O. McCorkle, who had replaced James H. Meyer as Dean of Agriculture when he became chancellor of the university in place of Emil Mrak, who had retired. According to the same history Dean McCorkle was replaced by Alex F. McCalla in 1971, but the letter from the chancellor, as related by Pearl, indicated that the nomination was made by Dean McCalla, suggesting that he already occupied the office in late 1970. (See Dillon S. Brown, *The Department of Pomology, University of California, Davis: The Contemporary Years*, p. 253, also letter, PDB-RSB, September 13, 1970.)

[46] Quoted in letter, PDB-RSB, September 13, 1970.

[47] The author recalls being present for this event, and the date is verified by Royce's ordination certificate, signed by Joseph Fielding Smith, who was then president of the Church.

[48] Dillon S. Brown, *The Department of Pomology, University of California, Davis: The Contemporary Years*, p. 260-265.

[49] The author recalls Royce relating this anecdote. The university's stringent Affirmative Action policies eventually brought it to the forefront of national attention when the Supreme Court ruled, in the 1978 Bakke decision, that the university's minority quota system for admissions was unconstitutional.

[50] Letter, PDB to Jane Davidson, June 21, 1972.

[51] Now known as Hepatitis B. This disease was known as "serum hepatitis" since transmission had been linked to the injection of human blood products, specifically serum used in the preparation of vaccines. The other major form of hepatitis, known as "infectious hepatitis," was transmitted through contaminated water and foodstuffs, and was later designated Hepatitis A.

⁵² The visit was recalled by the author; the events are recorded in letter, PDB to JD, December 28, 1970. Hospitalization is no longer routinely required for hepatitis.

⁵³ The author attended one of the shows, and was awed by the performance. The male lead was Bob Cello, a prominent figure in the Veterinary Medicine department who later founded the Davis Comic Opera Company. Eva Child, who played one of the female leads, had performed in USO shows during World War II, and both were exceptional performers. The barbershop quartet, in which Royce sang the lead tenor, was a comic success.

⁵⁴ G. Staudt, "Flavour components of strawberry species investigated by gas chromatography," abstracts from Symposium on Strawberry Breeding, The Fruit Group of Eucarpia, in association with The Scottish Horticultural Research Institute, July 24-28, 1971, p. 27-28. The final comment was recorded in Royce's yellow field journal labeled "May 1970 thru Feb. 1972," which also contains a detailed listing of the twelve rolls of slides taken during this trip.

⁵⁵ Entomology is the study of insects.

⁵⁶ See Royce's notes entitled "Asilomar Conference (Genetics) 1 July 74," in his yellow field journal labeled "July 1974 to April 1975." Royce's notes included detailed accounts of the UCLA and UC Berkeley approach to the issue.

⁵⁷ "College: Who Needs It," *UC Davis Spectator*, vol. 7 no. 7, June-July 1976, p. 6. Royce's was one of a number of faculty responses to an article by the same name which had appeared in the April 26, 1976 edition of *Newsweek* magazine, questioning the value of a higher education at a time when job opportunities were slim among graduates. Royce had a low opinion of the medical profession, in part relating to his experience with the pre-med students. "I think the majority of our MD's are good at pocketbook diagnosis only. They guess at most of the rest, give you some pills, a few grunts and then send a large bill for 'services rendered.' The insurance associated with it isn't much better in my estimation. I could be wrong on much of this but right now I don't think so, and the new generation of them waiting in the wings are worse—many med & pre-med students are unethical, predatory rascals with only one objective—make it big, fast, & then get in on that 'country-club' living, starting at the top." (Davis Journal, volume 1, January 22, 1976.)

Royce's remarks prompted some positive responses among medical school faculty. In response either to his article or the opinions he expressed, one wrote, "Your comments on professional schools, particularly medicine, are right to the point—and have needed to be said for a long time. I have long suggested that many students involved in sports and other extracurricular activities and carrying a B average would make better Doctors." (Memo, Ivan J. Thomason, Professor of Hematology to RSB, July 9, 1979.)

⁵⁸ A few years later, when Royce's son John actually applied for medical school, the medical schools themselves had begun to address the problem by giving increased consideration to a person's breadth of education and experience, rather than merely his or her grade point average. This began to alter the incentives for preprofessional students, and it was almost certainly this

change of emphasis which allowed John to gain admission.

[59] Regarding his work as department chairman, Royce later said, "The one thing that came about and has lasted even now, was the good association that we have with the Extension Service. I always felt that they should be an integral part of the department, and we set that up that way. We talked about it, we got together with them, held a bunch of meetings, and before we got through with it, they began to feel like they belonged to the department. And I think they should feel that way, because there are a lot of them, and they're out there where we wish we were, in many instances. And they must be treated with respect. It is still continuing, and it is somewhat of a model on campus among the agricultural groups, for what can be done. I don't claim credit for it. The one that really bored that thing home was the late Ed Maxie, and others." (Videotaped interview of RSB by Anton Kofranic, UCD Emeriti Association, 1996, content condensed.)

[60] Davis Journal, volume 3, December 31, 1977. This was after watching a New Year's Eve performance of *La Traviata*.

[61] Davis Journal, September 20, 1976. On a later occasion he wrote, "I watched the opera 'Vanessa' by Samuel Barber, text of Gian Carlo Menatti. It was excellent. I am still the frustrated non-achieving opera singer. I still think that it might have been. I know I have the voice; training, ear and discipline is lacking." (Davis Journal, January 31, 1979.)

[62] The 1970 election was a close one, with Allende gaining a slight edge over the nearest of two other candidates. Since he did not win by a majority vote, the winner had to be decided by the Chilean congress, which upheld the democratic process by ratifying Allende's election, though the congress would later be complicit in the military coup which deposed him.

[63] Allende's death was officially reported as suicide, and though the report was contested by supporters, later investigations confirmed that it was probably factual.

[64] The military government eventually came under the leadership of Augusto Pinochet, Allende's appointee as chief of the army who, by some accounts, was a reluctant participant in the coup, but who eventually emerged as the clear national leader. Pinochet reversed many of the controversial actions taken by Allende, and re-privatized industries which had been nationalized, resulting in a strong economic recovery. His often autocratic rule, which generally receives particularly harsh judgment from historians, continued until 1988.

[65] The author recalls this detail being reported by Royce on his return, though he made no mention of it in his letters.

[66] Letter, RSB to PDB, November 19, 1993, content edited.

[67] Letter, RSB to PDB, November 19, 1973. The quote contains two separate excerpts.

[68] Letter, RSB to PDB, November 25, 1973. It is possible that this overnight trip was the one taken to Mount Campana at the conclusion of Royce's stay. There remain several photographs of Stebbins examining plants in Chile.

[69] Letter, RSB to PDB, November 21, 1973, content condensed. In the same letter, Royce described some of his activities. "Yesterday (Tuesday) I had conferences with various people all day long including the Pro Rector, an

interesting gentleman, and Luciano Campos (the dean). You remember the latter. Today I was in a class all morning with students interested in strawberries. I spent the entire afternoon talking with two of the professors, helping them to lay out some experiments. Tomorrow, I go to Melipilla."

[70] Spanish short for "*golpe de estado*," a translation of the French "*coup d'etat*," referring to the government takeover.

[71] Letter, RSB to PDB, November 21, 1973. The account continued, "She told me that they are a stake now and Nuñoa is a ward with a bishop. The school is going very well now and growing, and she says that the Church is the greatest thing in her life. The two little girls are growing up and of course they did not remember me at all. I'll see them at Church. I found out the meeting schedule and may be able to go next Sunday after all."

[72] Letter, RSB to PDB, 24 November, 1973.

[73] This detail was not included in Royce's letters home, but the author recalls him recounting it after his return. Royce was probably using discretion in what he mentioned in his letters home, perhaps to avoid alarming Pearl, though it may also have been that he considered the possibility that his mail was being monitored by Chilean authorities, especially after he had been detained.

[74] Letter, RSB to PDB, November 25, 1973, content greatly condensed.

[75] Letter, RSB to PDB, November 26, 1973.

[76] Royce's date book for Saturday, December 1, 1973 contains a notation: "German brown, 6-1/2 kg = 14 #." The author infers the rest.

[77] Nearly five years later, a colleague from Chile visited Royce and commented on conditions there. "Miguel Lagaraga (Chile) was waiting in my office when I came in. I spent half the afternoon discussing berries and the situation in Chile. He gave a best possible case appraisal, indicating that everything is going well, inflation under control and the economic situation getting better all the time with democracy just around the corner. This is quite the opposite to that which comes from our popular press. The truth probably is somewhere in between. I do know this: When I was there last (1973), things were much improved over my previous time there (1968). One cannot expect Chile to operate as this country does." (Davis Journal, volume 3, August 1, 1978.)

[78] Yellow field journal labeled "July 1974 to April 1975." The entry containing this name designation was in a series of notes dated July 19, 1974.

[79] Davis Journal, volume 1, October 13, 1976, content edited.

[80] Pearl wrote in 1967, "Royce had asked a couple of years ago that his strawberry varieties be patented. The university turned a deaf ear on him. Just last week they called him in before the University patent officer & discovered that if they had patented his varieties and collected 50¢ on each thousand plants sold they would have collected about $50,000.00 a year on each variety. If they had been patented in the beginning, by now the University would have collected a quarter of a million dollars. I think the patents are going through now." (Letter, PDB to Jane Davidson, February 27, 1967, content condensed.) Royce made mention of this meeting in his field journal, and later quantified the potential patent earnings and the proposed royalty scheme as follows: "30¢/1000 grower charge, 20¢/1000 nurserymen,

estimate 150,000,000 plants/yr." (See Yellow field journal labeled Jan 1967 to May, entry between February 15 and February 25, 1967, and entry between March 15 and March 17, 1967.)

[81] Pearl wrote to her mother, "They are now in the process of patenting his berry varieties. Under the law he will get 40%. He is fighting it, I think he is allergic to money. I see his point though. Some of the fellows are not working on anything that can be patented and jealousies arise. Well, we'll see how that comes out. We could use a little more money here to help our house look a little more like a home." (Letter, PDB to Jane Davidson, January 9, 1969.)

[82] The details of this are a little unclear. At the biennial strawberry meeting held in Redding in September, 1971, a presenter named McOwens described the history of the university patent process, and Royce's notes include the phrase, "Sequoia may be sued, but hopes some kind of solution." (Yellow field journal labeled "May 1970 thru Feb. 1972," entry for September 16, 1971, under heading "McOwens.") Hamid Ahmadi was later adamant that Tufts was the first patented UC variety (interview, February, 2016), so it is likely that any patent procedures on Sequoia were simply deemed invalid, and Tufts therefore became the first legitimately patented variety.

[83] Yellow field journal labeled "Aug. 1975 to Mar 1976," initial page dated August 21, 1975.

[84] This listing is found in the same yellow field journal, between entries for October 17 and October 29, 1975. As an example of the new system, the short-day cultivar 71.101-611 (part of the 1971 batch of breedings) was designated C45, and when released, was given the name Pajaro. The day-neutral cultivar 70.3-108, bred in 1970, was designated CN8, and was released as the variety Brighton.

[85] "Went to Watsonville to examine strawberries shipped to Chicago from Florida then moved to San Francisco – had been sent as 'tufts' & 'tioga' – they were that as far as I could tell. The 'Tufts' fruit was flared out at the end (wider than @ base) – Ca fruit is even – these look very much like some 'Tufts' seedlings. In order to categorically say w/o some nagging doubt that it is 'Tufts' and none other, would have to look at the fruit on the plants. This will be done about the weekend of Feb. 16. Took pictures as evidence – general attitude now seems to be one of proceeding with licensing out of state nurserymen or orderly control. Is generally conceded that what they are trying to do right now with the patent as an instrument is not legally defensible – since is aimed at monopoly – expressly forbidden under pl. patent act – that is trade restriction." (Yellow field journal labeled "Aug. 1975-Mar 1976," entry for January 22, 1976.)

[86] See Davis Journal, volume 1, January 22, 1976. Of the two crates of berries from Florida which he examined, Royce declared with apparent satisfaction, "They looked pretty good."

[87] Davis Journal, volume 1, February 15, 1976.

[88] Of a meeting with the Strawberry Advisory Board, Royce wrote, "They agreed to be 'tough' & to a certain extent unreasonable about the patent, and passed the ball back to the regents. Chief issue now is Miss Opalkas' demand for 7% of the gross plant sales royalty, rather than flat per 1000. This would

discriminate against winter plant users, since they pay almost twice as much for plants yielding half as much now." (Davis Journal, volume 1, June 15, 1976.)

[89] Davis Journal, volume 1, September 20, 1976, content slightly condensed.
[90] Davis Journal, volume 2, January 13, 1977. The name of the defendant in the case was Ronnie Fulwood, a Florida nurseryman.
[91] Davis Journal, volume 2, March 31, 1977, content edited. The decision of the patent board was recorded on March 8, 1976. Royce was involved in a limited way on some of the negotiations regarding foreign licensing. On May 6 he wrote, "I went in to the Advisory Board H.Q. to meet with Joe Tomasello, Malcolm Douglas, Don Scott & Zanzi from Italy, where I participated in discussions relative to a license for Italians to grow Tufts for sale. I advise only on technical matters and stay out of the money end of it. Agreements in principle were reached, and the details will have to be worked out to mutual satisfaction of the Board, State Dept. of Ag. and the UC. It will involve lump sum payment plus a royalty arrangement involving an advance thereafter related to sales." (Davis Journal, volume 2, May 6, 1977.)
[92] Of one such instance Royce wrote, "Finished getting 'interrogatories' off to Lawyer Diepenbrock after overcoming my general frustration and exasperations. The legal profession still amazes me, most lawyers are basically predatory rascals interested mostly in lining their pockets & secondarily in seeing justice served (again, my biased opinion). It takes one to know one, and maybe I am that at heart too – who knows?" (Davis Journal, volume 3, November 28, 1977.)
[93] See Davis Journal, volume 3, entries for January 27 and 30, and February 1 and 2, 1978. In the February 2 entry, Royce acknowledged, "Hopefully the suit will be settled. I see no reason not to, although our case is less secure on all counts than I previously had thought. We have little to win by going to court, if anything." The matter was brought before the patent committee of the Strawberry Advisory Board on March 23, 1978, and Royce wrote, "We discussed settlement of the Florida (Fulwood) dispute with reasonable attitude exhibited by all. I think that it can be resolved to the mutual advantage of all concerned. We also pretty well resolved the problem of distribution of plants for testing, patent (disclosure) filing and 'release.' I am staying by my guns here. Victor and I agree on the procedure in principle but not on certain unimportant details on naming."
[94] Davis Journal, volume 3, August 2 and 3, 1978.
[95] The expert was a colleague named Gene Galletta, who ran a USDA strawberry breeding program at Beltsville, Maryland (see Davis Journal, volume 3, August 4, 1978).
[96] Davis Journal, volume 3, December 11, 1978. According to a later account by Royce's successor Douglas Shaw, the existing licensee on the Tufts patent was the California Strawberry Advisory Board, and the suit against the Florida nurserymen was in part an attempt to maintain exclusive use of the variety for California growers. According to Shaw, the Florida growers issued a counter suit for anti-trust violations, and it was this which induced the Advisory Board to turn the strawberry patent process over to the university. The California growers, who contributed to Advisory Board funding for the U.C.

strawberry program, were given a preferential status which included a discount on patent royalties. (Douglas V. Shaw, telephone interview, January 21, 1916.)
[97] Davis Journal, volume 3, December 15, 1978. Those who met with Royce were Burt Rae and Roger Ditzel.
[98] Davis Journal, volume 2, February 3, 1977.
[99] Davis Journal, volume 2, March 3, 1977. Even as he neared retirement, Royce's opinion of the patent system had not changed. "Met with plant patent auditor this day, giving my opinion about plant patents in general, and how the UC program is run, not too favorable. I think that much of the money is wasted, and that we get too much of it, among other things." (Davis Journal, volume 9, August 28, 1985.)
[100] That this would eventually be the case was apparent by 1979, when Royce had a meeting with Roger Ditzel, the University of California patent administrator. "We talked mostly about control procedures and transfer of title following release, but also talked of the money that will be coming in, since the sum will be rather great next year. I must do something about it. In time, this could amount to almost as much as my salary." (Davis Journal 4, May 18, 1979.) Royce found the semi-commercial aspect of the patent process unsavory. At around the same time, he wrote, "Did have a visit with a New Zealander named _____. I gave him an encouraging impression of possibility of them getting our patented plants. As a rep. of the University, I am expected to promote business, but sometimes am loath to do it." (Davis Journal 4, June 22, 1979.)
[101] After Margaret and John attended Brigham Young University, Royce began making contributions to that institution as well.
[102] Account related by Marla B. Vaughn in 2015.
[103] In Royce's 1973 date book, Friday, October 24 bears the notation, "Smith died," with an asterisk leading to a further notation: "I dreamed of his death in great detail last nite—except I had it mixed with Dad."
[104] Royce's talk read in part, "I recall the warm, close relationship we had as children—he was my older brother, then strong, self reliant, of a naturally happy disposition, and proud as a young boy to do a man's work, driving the team, mowing, plowing, and so on. He was born too late perhaps to exploit one of his many talents to the fullest. He was a gifted horse trainer, and from the time he was old enough to handle the job physically, he shoed all the horses, and it was done right. He was a born trail rider, and many of his best hours were spent mounted upon a horse." Royce then quoted the words of one of his favorite musical works, Johannes Brahms' *Requiem*: "'Lord, make me to know the measure of my days on earth, and consider my frailty, that I must perish.' The tragic foreboding in these words serve to remind us of the brevity of man's life, at best brief for all, and doubly brief for one whose productive life was cut short in the prime middle years by an accident, ironically involving a horse." Funeral talk notes, 1975, content condensed and edited.
[105] The "administering" referred to here is an anointing with consecrated olive oil followed by a blessing, explained in a previous chapter. The author recalls feeling almost startled by the immediate relief he experienced after the

blessing. His only remaining symptom was a hoarse voice, which given his usual talkative nature, was probably regarded by the family as a bonus blessing.

[106] Florence and Larry departed on May 21, 1976 (Davis Journal, volume 1, May 20, 1976). Since both they and John, on his mission in Guatemala, were abroad at the same time, Royce and Pearl developed a weekly practice of letter-writing. Royce, who had received almost no letters from his father during his own mission, wrote to John faithfully once a week, while Pearl sent a weekly letter to Florence and Larry.

[107] Davis Journal, volume 1, February 23, 1976, content condensed. This was written in his journal eight months after completing his term as chairman.

[108] Davis Journal 1, June 22, 1976.

[109] Davis Journal, volume 1, January 9, 1976, content condensed.

[110] Davis Journal, volume 3, July 28, 1978. The Biblical quotation is found in Isaiah 40:31, and is repeated in the Latter-day Saint revelation on physical health known as the Word of Wisdom, found in Doctrine and Covenants 89:20.

[111] During his class in 1978 he described his haul of fruit from the South Coast Field Station: "Returning to the SCFS, I collected avocados, citron, cherimoyas & macadamia nuts for class. The cherimoyas were particularly good. The fruit was beautiful." (Davis Journal, volume 3, February 17, 1978.) At the conclusion of the class for that same year, he wrote, "The last day of class for me this quarter. I spent the morning doing paper work less than effectively, and getting ready for my last lecture. It went well. One girl brought a gift of strawberries and when they clapped vigorously as I finished, I told them in jest that there would be a final exam anyway. I left them with a good feeling. I almost wish I didn't have to grade them." (Davis Journal, volume 3, March 15, 1978.)

[112] Davis Journal, volume 1, January 8, 1976.

[113] Davis Journal, volume 1, January 6, 1976. Not long after, Royce wrote, "You must sacrifice something for any worthwhile activity, and I believe that I have failed to do that in many instances. Oh, I am busy enough, but much of it isn't all that well organized. I seem to avoid certain things very effectively when I make up my mind to do so. The reverse is also true when I deem it wise to really go for something. Much of what I do however is dictated by circumstances & expediency rather than by well thought-out design or even strong desire. I keep telling myself never to promise anything I don't intend to immediately pursue or that does not really interest me. Unfortunately I tell myself all this and then do not accept my own advice." (Davis Journal, volume 1, March 10, 1976.)

[114] Davis Journal, volume 1, March 4 and 5, 1976.

[115] The phrase "You can't go home again" was the title of a book by American novelist Thomas Wolfe which Royce had recently read, and which had left a profound impression on him. He would cite the phrase again and again in his journal.

[116] Davis Journal, volume 1, April 7, 1976, content greatly condensed. After a visit a few months later, when his mother had been moved to a new facility, Royce wrote, "We drove out, picked up George & went to Draper to see

Mother. The new place is better than the old in almost every respect. There is no change in Mother. I am not certain that she even knows us, but she talks as if she does most of the time. She looks just fine and is pleasant and nice. I guess this is the best for her." (Davis Journal, volume 1, August 15, 1976.)

[117] Davis Journal, volume 4, December 18, 1979.

[118] Davis Journal, volume 3, January 15, 1979. Even before his mother was placed in a nursing home, Royce had remarked, "I do not really fear death, I just don't want to be a burden of any kind before I leave." (Davis Journal, volume 1, February 23, 1976.)

[119] Davis Journal, volume 1, June 19, 1976.

[120] Davis Journal, volume 1, May 14, 1976.

[121] This episode was witnessed by the author, who happened to attend the Sacrament meeting of the University Ward on that day. After the member had expressed his doubts, a number of strong and forceful testimonies were given, and at the conclusion of the meeting Royce rose and called for understanding and kindness, stating that his only regret was that his son might have attended that day with the thought of sharing his own testimony, and been unable to do so because of the events during the meeting. He said this with a gentleness of spirit that the author recalls to this day. A "testimony" in this setting means a public expression of one's convictions.

[122] On one Sunday he observed, "My students, including Margaret, likely don't realize how much I talk to myself as I talk to them." (Davis Journal, volume 2, Sunday, September 25, 1977.)

[123] This was a teaching of Joseph Smith, the Latter-day Saint prophet, found in Doctrine and Covenants 130:6—"It is impossible for a man to be saved in ignorance."

[124] As bishop of the University Ward, Royce had written an "editorial" in a two-page LDS Institute newsletter, in which he had made an appeal on this point. Entitled "First Things First," it read: "For a long time I have been disturbed by the unfortunate fact that many of us in our relationships with others, are preoccupied with matters over which we have little or no control, either now or in the foreseeable future. Much of it involves a cumbersome form of neo-fundamentalism in matters ecclesiastical, 'out of joint' with reality. Some of it involves aggressive political orientation, left or right, often confused with religion. All of it involves a form of offensive selfishness characterized by covert and overt efforts to coerce others into our own thought channels whether it is reasonable to them or not. Impelled by our misguided zeal, we seek to prescribe and proscribe patterns of intellectual action for friends and colleagues that may be inherently repugnant to those we hope to influence. Let us not forget that our principal responsibility as Christians is to emulate the Master, and that this means rigorous self discipline, rendering unselfish service to our fellows, scrupulous honesty and granting others the freedom to choose, just as we claim the privilege for ourself –FIRST THINGS FIRST." ("Davis Discovery" newsletter, LDS Institute of Religion, February 1971. I am indebted to Newell Bringhurst for providing me a copy of this newsletter and bringing the quote to my attention.)

[125] Notes for a talk entitled "Evolution and LDS Membership," given January 27, 1974, content condensed and slightly edited. For years Royce was invited yearly by the Institute director, Sheldon Dahl, to give this presentation, which Sheldon considered very valuable. (Personal conversation with Sheldon Dahl, October 25, 2015.)

[126] Davis Journal, volume 1, May 15, 1976, contents condensed and edited.

[127] Davis Journal, volume 1, February 22, 1976, content edited.

[128] Davis Journal, volume 1, August 18, 1976, content edited and condensed. This was written after a conversation which Royce had had with two BYU professors named Farmer and Jeffreys, while attending scientific meetings in Salt Lake. Jeffreys was editor of *Dialogue*, a scholarly journal concerned with topics related to the Latter-day Saint movement.

[129] Of a training session which he had held for teachers in Woodland, Royce wrote, "I talked on the theme that Joseph Smith envisioned the Church as a community of scholars. It could have been an excellent talk had I spent another 3 hours preparing. Actually it was pretty good, but much below what I consider acceptable in terms of what I can do." (Davis Journal, volume 4, December 6, 1979.)

[130] Davis Journal, volume 2, February 6, 1977. This passage ended with a New Testament quotation from the Book of Matthew, which in reference to God, states, "he maketh his sun to rise on the evil and on the good, and sendeth rain on the just and on the unjust" (Matthew 5:45).

[131] Davis Journal, volume 3, Sunday, February 4, 1979.

[132] Davis Journal, volume 3, November 26, 1978, content condensed. On another occasion, after being uninspired by a missionary homecoming, he wrote, "My main problem is lack of enthusiasm for some things. I guess I'm getting old or lazy or both. Nothing thrills me much." (Davis Journal, volume 3, November 13, 1977.)

[133] Davis Journal, volume 3, January 1, 1978, content condensed.

[134] Davis Journal, volume 1, December 12, 1976. The speakers were a family of the name Winters.

[135] Davis Journal, volume 2, March 8, 1977.

[136] Davis Journal, volume 1, February 19, 1976, content condensed.

[137] Davis Journal, volume 4, June 27, 1979.

[138] Davis Journal, volume 1, November 19, 1976.

[139] Davis Journal, volume 2, July 1, 1977.

[140] Davis Journal, volume 2, August 2, 1977.

[141] Davis Journal, volume 3, December 31, 1977. This passage was preceded by the following: "We stayed until late talking of 'dear hearts & gentle people' from yesteryear, again emphasizing the fragile memories we have and the fleeting nature of our brief day in the sun; too soon old and all is gone, just as youth is fled from us now. There is an ironic twist to the roles we try to play with pomp and the assurance that we are something special."

[142] The author, while attending the young adult class taught by Royce in 1975, came to realize that Royce frequently used his lessons to share personal feelings with his son which he could not bring himself to express in private, but was able to discuss more freely in the more detached setting of a class.

[143] Davis Journal, volume 2, July 3, 1977. Royce seemed to feel this sense of isolation particularly during the Christmas season. "This is a lonely season for me (us). We have lots of 'general' friends but no specific ones. We are never asked to go with someone & we never ask anyone to go anywhere with us. I guess I am a loner at heart." (Davis Journal, volume 3, December 20, 1977.)

[144] At a church meeting in which Royce sat on the stand, he remarked, "I didn't get a whole lot out of the meeting. Part of the problem is the same difficulty I had when I was bishop of the University Ward. I could never hear most of the female speakers when I was seated behind them." (Davis Journal, volume 1, January 18, 1976.) At a reception before a University Senate meeting a couple years later, he expressed a similar problem. "The reception was the usual 'talk-talk' which I cannot hear too well and am always a bit uncomfortable with that sort of thing, even when cocktails are not served." (Davis Journal, volume 3, May 23, 1978.)

[145] Davis Journal, volume 4, June 1, 1979.

[146] After a Protestant funeral for the widow of a professional associate, he wrote, "Went to Lybby Maxie's funeral today and heard the same funeral sermon from the same Community Church minister (Presbyterian) as always. Only the name changes. It is really cold and impersonal with nothing about the person's life & times. It was almost as if she had never lived – oblivion before the casket is closed. I wish we could make a brief personal history of each passing person rather than support the florist trade. The flowers wither and fade so soon but the written record could be made to last forever and at such modest cost. Our church really teaches this, but few practice it as far as I can see." (Davis Journal, volume 1, May 11, 1976, content condensed.)

After attending a Jewish wedding ceremony, he wrote, "We went to the Rappaport wedding at the Jewish fellowship in the afternoon and were treated to a most interesting ceremony. The Rabbi explained or translated everything he did and we were amazed at how much of it was similar to our own marriage ritual & tradition in the temple, complete with the rib symbolism. Our people could learn something from them about many things." (Davis Journal, volume 1, June 20, 1976, content condensed.)

His reaction to receiving anti-Mormon flyers in the mail was even more telling. "We got our annual anti-Mormon literature from Wash. DC address (printed by hand, plain envelope). This is the 2nd year of this. More tracts this time. They concentrate on former Mormons, converts, ex-missionaries, I was raised a faithful Mormon, etc. Even if you accept their arguments which I find a bit silly, what they offer as a substitute is pretty dreary stuff. I could never go for that." (Davis Journal, volume 1, January 19, 1976.)

[147] Davis Journal, volume 1, July 24, 1976.

[148] Davis Journal, volume 1, October 3, 1976, content condensed. The Church did in fact later reduce the number of conferences.

[149] Davis Journal, volume 3, June 9, 1978, content condensed and edited. The matter of "revelation" here alluded to is often difficult for a non-believer in Church theology to understand. The founding principle of the Church of Jesus Christ of Latter-day Saints is the concept of continued revelation and the belief in living prophets. The practice of withholding the

priesthood to black Church members had been established and subsequently reviewed by past Church presidents, who had concluded that it was "the will of the Lord" for their time. A belief in continuing revelation through a living prophet mandated that such a policy could only be overturned by a subsequent revelation, similarly reflecting God's will. Hence, this was not a simple policy matter, but a doctrinal one, which Church presidents (who clearly believed in the same foundational principle of continuing revelation) did not feel they had the prerogative of changing without a manifest revelation. This explains why the practice continued long after most Church leaders doubtless would have preferred to change it. Only when Spencer W. Kimball, as president of the Church and "prophet, seer, and revelator," felt he had received such a revelation was the policy overturned, and the decision was immediately and unanimously ratified in all leading councils of the Church.

[150] Davis Journal, volume 4, January 13, 1980.

[151] Davis Journal, volume 4, April 13, 1980. Royce expressed his feelings more fully following a Priesthood meeting not long after. "We discussed Revelation. I believe in it in my way. I am not much for the trumpet call from 'on high' but can accept the small voice which prompts you when you are right, whether you be the lay member of modest circumstances within or the President. Certainly, one must recognize the wisdom of the various 'revelations' of President Kimball. We must overcome the conception that going to many meetings contributes to the practice of true religion. There has to be more to it than that, for that is indeed the personification of vanity." (Davis Journal 4, January 20, 1980, content condensed.)

[152] Davis Journal, volume 1, December 31, 1976, content condensed.

[153] Davis Journal, volume 1, January 20, 1976.

[154] Davis Journal, volume 2, May 25, 1977, content edited.

[155] An example of the efforts to achieve greater efficiency in the selection process is found in Royce's summary of the performance of the 1971 crop of seedlings during the 1974-75 winter and summer planting seasons. Comparing data from the Southern California Field Station (SCFS) in Santa Ana with that from the Watsonville station in the Central Coast, he noted that many of the new seedlings performed better than standard varieties in both stations. However, when there was a discrepancy between the two, Royce noted that a much higher percentage performed better only in SCFS than did in Watsonville, leading him to conclude, "We lose little if we run the preliminary screen at SCFS omitting Watsonville; much if we reverse the procedure." (Yellow field journal labeled, "April 1975 to Aug. 1975," Summary of performance of '71 seedlings, near end of journal.)

[156] Royce wrote in the Summer of 1979, "I am writing the release notices for the new strawberries, 'Douglas' = C51, 'Pajaro' = C45, 'Vista' = C38, 'Hecker' = CN7 & 'Brighton' = CN8. I'll be glad to see them on the move and out of our hair. They will make some money once they have been available for a year or so, not all, but some. We really can't be sure at this stage." (Davis Journal, volume 4, June 25, 1979.) Just over a month later, he wrote, "Spent the evening at home writing the disclosure on another strawberry (CN15) to be released as 'Aptos.' (Davis Journal 4, August 30, 1979.) On signing and sending in the release notice on the first five varieties, Royce predicted, "We

shall see how that goes. I think that Douglas & Pajaro will both be winners, and that Brighton & Hecker will enjoy modest success as the first day-neutrals of that type." (Davis Journal, volume 4, July 6, 1979.)
[157] Of these varieties, "Douglas" was named after Malcolm B. Douglas, a manager of the California Strawberry Advisory Board who had recently died; "Pajaro" was named for Pajaro Valley, an important strawberry-growing area in Santa Cruz County, and "Vista" was named for the city of Vista, in San Diego County.
[158] "Hecker" was named for Hecker Pass, a mountain pass in the same county through which Royce often passed, while "Brighton" was named for the Brighton Valley in the Big Cottonwood Canyon near Salt Lake, where Royce had discovered the wild strawberry with the day-neutral trait. The variety "Aptos" was named for the city of Aptos in Santa Cruz county.
[159] See Royce S. Bringhurst and Victor Voth, "Six new strawberry varieties released," *California Agriculture*, February, 1980.
[160] This information is taken from R. S. Bringhurst and Victor Voth, "California Strawberry Cultivars—Past, Present and Prospects," *Fruit Varieties Journal*, vol. 33 no. 2, 1979.
[161] Davis Journal, volume 1, December 30, 1976, content edited and condensed.
[162] Davis Journal, volume 4, April 28, 1979.
[163] Davis Journal, volume 4, June 16, 1979.
[164] Work with meristem culture was done in collaboration with Ruth Hilton (later Ruth H. Mullin) a research associate in the UC Berkeley Department of Plant Pathology, who had succeeded in adapting this methodology to strawberries in order to provide virus-free stock of the various California cultivars. Royce referred to her often in his field journals from this period. (See R. H. Mullin, S. H. Smith, N. W. Frazier, D. E. Schlegel, and S. R. McCall, "Meristem Culture Frees Strawberries of Mild Yellow Edge, Pallidosis, and Mottle Diseases," *Phytopathology*, vol. 64, November 1974, p. 1425-1429.)
[165] Davis Journal, volume 2, January 7, 1977, content edited. Royce learned about this technique at the Annual Small Fruits Disease Conference, held yearly in Oregon, where it was probably presented as a means of virus elimination.
[166] See Royce's yellow field journal labeled "July 1974 to April 1975," an entry dated February 19, 1975 at Santa Ana, where the 1976 budget was being discussed.
[167] Electrophoresis provides a means of separating and classifying large organic chemicals such as proteins, based on their electrical charges. In this method, an adequate sampling of plant material is liquefied, and a sample is placed at the edge of a column or sheet composed of a gel matrix. A uniform electrical charge is applied over the gel, causing the charged organic particles to migrate through the gel. Different substances migrate at different speeds, depending on their physical characteristics and their relative charges. After a specified time the electrical charge is removed, and a special stain is applied to the gel to reveal the location of a single target, such as a specific enzyme. A particular pattern of bands will appear from the staining, depend-

ing on the components of the enzyme. The banding pattern can be quite unique for individual plants and could vary between populations, and it was this characteristic that interested Royce.

[168] PGI was referred to in some papers as GPI, or glucosephosphate isomerase. (See S. Arulsekar and R. S. Bringhurst, *Isozymes in Plant Genetics and Breeding, Part B*, Elsevier Science Publishers B. V., Amsterdam, 1983, p. 391-400.)

[169] When this was first attempted with strawberry tissue, Royce wrote, "First good electrophoresis on varieties, and it looks good—we can distinguish varieties, particularly Tufts." (Davis Journal, volume 1, November 13, 1976.)

[170] One day during this period of investigation, Royce observed, "I made a trip to WEO to get leaves for electrophoresis study—a really interesting group. Sekar is running them (PGI). We have nice segregation in the populations currently under study." (Davis Journal, volume 2, October 26, 1977, content edited.) Royce also instructed Sekar at the microscope in the study of chromosomes, at which Royce had become something of an expert. "I haven't lost my touch there. I still can see & interpret as well as advise on how to do things. There is as much art as science in making good chromosome preps." (Davis Journal, volume 3, January 9, 1978.)

[171] An account of this trip is found in Royce's yellow field journal labeled "March 1972 to June 1974," in an entry dated August 25, 1973. The author recalls the German scientist being a guest in the Bringhurst home, though he was unaware of the significance of the visit, or of Staudt's professional stature.

[172] Royce recorded making collections from eighteen plants, including a number of new acquisitions, after which he recorded, "On all of these – leaves and flowers were collected for electrophoresis – on those so designated (with AC numbers) plants were also collected." (Yellow field journal labeled "July 1974 to April 1975," entry for April 3, 1975.)

[173] James F. Hancock was to follow in Royce's footsteps, and became an acclaimed plant breeder and researcher at Michigan State University. Specializing in strawberries and blueberries, he became a world authority, and authored a definitive book on the strawberry. Perhaps because of their parallel interests, when asked about graduate students who had become particularly prominent, Royce singled out Hancock. "He's a leader in his field now, does a lot of the things that he did with me. We worked a lot with the wild strawberries. We had a great time with those. I remember the first time I took him out where those things grow, he couldn't believe what he was seeing." (Videotaped interview of RSB by Anton Kofranic, UCD Emeriti Association, 1996.) In turn, Jim Hancock spoke of his former professor with great devotion. "Having been close to Royce Bringhurst as his student and friend, my admiration for him has only grown as the years go by. He has worked amazingly hard and has been extremely innovative. He is supportive of others and is a caring individual. Royce S. Bringhurst is a most impressive man and the strawberry industry has been fortunate to have him as their champion." (James F. Hancock, "Dedication: Royce S. Bringhurst, Pre-eminent Strawberry Breeder," in Jules Janick, ed. *Plant Breeding Reviews, Volume 9*, John Wiley & Sons, Inc. 1992, p. 4.)

[174] The published reports of this study indicate these excursions took place

from July 15 to August 18, 1975, whereas Royce's yellow journals described one extensive trip from August 4 to 6, 1975 to the coastal areas, and subsequent coast trips on September 30-October 1, a trip to the Sierras on October 7-8, and further trips to the coast on October 15 and October 29. (Yellow field journals labeled "April 1975 to Aug. 1975" and "Aug. 1975-Mar 1976.")

[175] The "tap-root" described here combined an elongated old crown which terminated into a single large woody root which extended deeply into the sand before branching, giving an appearance similar to that of a taproot which occurs on many plants, such as the carrot.

[176] These environmental variations in the octoploid strawberries were particularly important, since they showed that some of the species' adaptations were the result of natural selection in a specific environment, and were not simply the result of an increased variety of genes due to the higher number of chromosome sets. Royce and Jim had also included the size of the achenes (or "seeds"), which varied with different growing conditions, as one of the adaptations which may have favored survival in different environments. This was evidently called into question by other experts, and in response Royce and Jim performed experiments germinating seeds from different sites on soil samples from the various sites, to verify their hypothesis that variations in achene size were an adaptation to the environment from which the plants had come. They found instead that the ability of seeds to germinate on soil similar to that from which they had come was no more than could be expected by random chance. They published a report of this experiment, which concluded with the words, "We thank several anonymous reviewers who insisted on making us see the truth." (James F. Hancock and R. S. Bringhurst, "Plasticity in the Germination of California *Fragaria* Seeds," *Madroño*, vol. 26, no. 3, July 1979, p. 145-146.)

[177] See J. F. Hancock, Jr. and R. S. Bringhurst, "Sexual Dimorphism in the Strawberry *Fragaria chiloensis*," *Evolution*, vol. 34, no. 4, 1980, p. 762-768; J. F. Hancock, Jr. and R. S. Bringhurst, "Hermaphroditism in predominately dioecious populations of *Fragaria chiloensis* (L.) Duch.," *Bulletin of the Torrey Botanical Club*, vol. 106, no. 3, p. 229-231; and J. F. Hancock, Jr. and R. S. Bringhurst, "Evolution in California Populations of Diploid and Octoploid Fragaria (Rosaceae): A Comparison," *American Journal of Botany*, vol. 68, no. 1, 1981, p. 1-5. This article was sent for publication in May, 1979, and a revised version was accepted for publication in August of 1980. By that time Jim Hancock had become established in the Department of Horticulture at Michigan State University.

[178] Davis Journal, volume 2, May 30, 1977.

[179] Davis Journal, volume 3, November 1, 1978.

[180] "We drove to Pigeon Point and places between there and Pacifica, checking the wild strawberries. In general, they are well, but drought is hitting hard. The Moss Beach colony of hybrids is threatened but still there. I can't find a single hybrid colony at Pacifica. The bulldozers have won." (Davis Journal, volume 2, April 22, 1977.)

[181] Royce S. Bringhurst, James F. Hancock, and Victor Voth, "The beach strawberry, an important natural resource," *California Agriculture*, September

1977, p. 10, content condensed and edited.
[182] Davis Journal, volume 2, September 23, 1977.
[183] Davis Journal, volume 3, November 7, 1977, content edited. The next day Royce wrote, "Met with Andy Manus and Geo. Rackelman (State Parks & Rec) & Geo Dong (same) to discuss Año Nuevo matter. It went well." (Davis Journal, volume 3, November 8, 1977.) See also yellow journal labeled "Oct 1977-Nov 16, 1978," entry labeled "Nov 77;" also letter, Andy Manus, Area Marine Advisor to RSB, July 13, 1978, and related documents.
[184] Davis Journal, volume 3, December 15, 1977.
[185] In 1982, Royce wrote to the executive director of the California Coastal Commission, in which he emphasized the value of these colonies: "We have studied the wild strawberries of California for almost 30 years with emphasis on the California beach strawberry, *Fragaria chiloensis*. In my estimation, the California colonies of this species contain the most valuable extant strawberry germplasm in the world, particularly those found south of the mouth of the Russian River. The species is confined strictly to the narrow, relatively dense fog belt along the coast. Enclosed are copies of a few of our publications dealing with *F. chiloensis*.

"For some time I have been concerned about the survival of the species south of the Russian River in general and in certain important areas in particular. I am alarmed at what has been happening to the colony at Oso Flaco Lake which I have observed regularly since the first time I visited there in 1954. The increase in dune buggy [traffic] and similar activities there and in adjacent areas threaten to destroy the strawberries and most of the other natural flora at the site.

"I cannot speak for the other flora but the loss of these particular strawberries would be very important. This is the southernmost colony in California and by the way, nearer to the equator than any of the wild counterparts found in coastal Chile in the southern hemisphere. The nearest colony to Oso Flaco is at the mouth of the Big Sur River.

"We have used an excellent (relatively rare) hermaphroditic clone from Oso Flaco Lake in our breeding work with good results. Among other things, the wild plants from there are endowed with the ability to grow vigorously during the winter months, a particularly valuable characteristic for southern California strawberry areas. We have lost one valuable group of plants from the Pacifica area as a result of road and home building. Ironically, they were in the area where there were problems in January due to rain. I am concerned also about colonies at Point Sur, Año Nuevo and Franklin Dunes, Pigeon Point, Point Reyes and Bodega Bay to mention a few. I hope the plight of the endangered flora at Oso Flaco Lake will be considered in the use planning for the area." (Letter, RSB to Michael Fischer, February 12, 1982.)
[186] See yellow field journal labeled "Apr 1975 to July 1975," entry labeled 7/2/75; also yellow field journal labeled "Oct 1977-Nov 16, 1978," notes from ASHS meeting in Salt Lake City starting October 12, 1979 (fifth page of the notes). In February of 1979, Royce received a letter requesting native plant specimens of *Fragaria* species to begin a germplasm collection at Beltsville, Maryland. One of the signers of the letter was Arlin Draper, with

whom Royce would later work at the Germplasm Repository in Corvallis, Oregon (letter, A. D. Draper, J. L. Maas and J. R. McGrew to "colleagues," addressed to RSB, February 2, 1976.)

[187] Davis Journal, volume 1, January 13, 1976, content edited.

[188] Davis Journal, volume 1, February 25, 1976, content edited.

[189] Davis Journal, volume 2, February 15, 1977, content edited.

[190] Davis Journal, volume 2, June 14, 1977. The line "Oh, I was like that when a lad" originates with a Gilbert and Sullivan operetta (Trial by Jury) in which Royce had sung.

[191] This was part of a political movement to oppose the South African government by placing pressure on major institutions to divest themselves of economic investments that involved South Africa, in protest of that country's discriminatory *Apartheid* policy. The complex financial affiliations of investment funds made this a nearly impossible task initially, but in the 1980s the movement began to gain traction, and the University of California system eventually did divest itself of investments in South Africa, as did many other institutions.

[192] The Bakke case was a landmark reverse-discrimination case involving the University, brought by Alan Bakke, a medical school applicant who claimed he had been denied admission to UC Davis medical school when the school's Affirmative Action policies had allowed the admission of less qualified minority applicants. The case was before the U. S. Supreme court at the time of this protest, and was particularly contentious, since it threatened to overturn all "Affirmative Action" programs which gave preferential consideration to minority groups. The Supreme Court verdict, given in 1978, took a middle ground, upholding the concept of racial preference embodied by Affirmative Action, but finding that the quota system established by UC Davis was unconstitutional, and requiring that Bakke be admitted.

[193] Davis Journal, volume 2, May 27, 1977, content edited.

[194] Cesar Chavez was a Mexican American activist who helped to found the United Farm Workers union, which championed the rights of Mexican immigrant farm workers in California, organizing strikes and boycotts, and influencing legislation. During his lifetime his activities were controversial even in the Hispanic community, particularly when he opposed continued immigration of undocumented Mexican workers. Since his death he has become an iconic figure in the Mexican-American community, and in many liberal-leaning communities. The elementary school attended by the author in Davis (formerly West Davis Elementary School) is now named after him.

[195] Tom Hayden was a leader in the radical student movement in the 1960s, and one of the founding members of the SDS, or Students for a Democratic Society, a nationwide left-wing student organization. During the Vietnam conflict, he made heavily-publicized visits to North Vietnam at a time when the United States was engaged in military action against it, and was sympathetic to the Communist cause. He later became a political figure in California. His participation in this meeting was doubtless part of his left-wing activism, which Royce did not regard very highly..

[196] Davis Journal, volume 3, February 17, 1978, content edited.

[197] Davis Journal, volume 2, September 29, 1977, content condensed.

[198] Davis Journal, volume 3, April 20, 1978.

[199] Davis Journal, volume 3, May 31, 1978, content condensed.

[200] Davis Journal, volume 3, May 25, 1978. In 1993, Royce summarized his committee work as part of a response to a colleague from Utah State University, who had requested information about his professional and personal life, to include in a history of the USU Plant Science Department. "In the years following my chairmanship, my most important University assignment involved the Committee on Academic Policy of the Academic Senate. I served first as a member of the Davis Campus Committee, later as chairman of that, then as member, later as Vice Chairman and finally Chairman of the State-wide All University Committee on Academic Policy of the Senate. This is one of the most important committees of the Senate." (Letter, RSB to Devere R. McAllistar, February 22, 1983.)

[201] Davis Journal, volume 4, October 4, 1979, content condensed and edited.

[202] Davis Journal, volume 3, March 4, 1978.

[203] Davis Journal, volume 2, October 11, 1977. The occasion was a meeting of the American Society of Horticultural Science in Salt Lake City.

[204] Davis Journal, volume 3, June 6, 1978.

[205] Davis Journal, volume 3, December 18, 1978.

[206] Significantly, in 1975 Royce had been offered a position at the University of Massachusetts as dean of the College of Food and Natural Resources and Director of the Experimental Station and Director of the Extension Service, a position which would have been considered a marked advancement in academic status. Royce declined the offer. "I have thought the matter over carefully and have concluded that I would rather not do so. I have an interesting career here that I would prefer to continue with at the present." (Letter, RSB to Dr. William J. Bramlage, August 29, 1975; see also the attached letter from William J. Bramlage to RSB dated August 15, 1975.)

[207] This program was known as "Science Editor," a program produced by the UC system, and broadcast in cooperation with the CBS Radio Network. The segment on strawberries, written by Camille Parker and narrated by Mike Francisco, was aired on September 7, 1983, and featured Royce's voice describing some aspects of his breeding work.

[208] The activist's name was Steven Huffstutlar, who had been involved in forming groups of Mexican farm workers into cooperatives in Salinas and Watsonville, using public grant money, in an attempt to grow strawberries commercially. Royce had had contact with these cooperatives at some of the grower meetings, and held a low opinion of the entire enterprise, which attempted to put poorly educated farm workers, some of whom spoke little or no English, in charge of complex and demanding farming operations. By the time Huffstutlar and another author published a very positive report of these activities, most of the cooperatives had already failed. (See Refugio I. Rochin and Steven Huffstutlar, "California's low-income producer cooperatives," *California Agriculture*, March-April 1983, p. 21-23.)

[209] Steve Duscha, "Our Strawberries Get Razzberries," *Sacramento Bee*, Saturday, May 7, 1977, p. A1 and A8.

[210] Davis Journal, volume 2, May 7, 1977.

[211] Davis Journal, volume 1, August 21, 1976.

212 Davis Journal, volume 3, March 2, 1978, content condensed and edited. This event was memorable for John as well as Royce. Because of the communications strike, John was himself uncertain of the flight date until shortly before his departure. Not only could he not tell his parents, he had also spent his last American money paying an exit tax at the Guatemala City airport, and had no money to make a phone call. Pearl, meanwhile, was beside herself with worry, particularly since a plane from Latin America had crashed at the Los Angeles airport just a couple of days before.

213 Davis Journal, volume 3, March 5, 1978.

214 John, who attended both UC Davis and BYU as an undergraduate, had no doubt that BYU compared favorably at least as an undergraduate institution. Royce would eventually begin to make annual contributions to BYU, as he did to UC Davis and to the two universities where he had trained, Utah State and Wisconsin.

215 Davis Journal, volume 3, August 8, 1978, content condensed.

216 Davis Journal, volume 3, August 11, 1978, content condensed.

217 Davis Journal, volume 3, September 9, 1978, content condensed and edited.

218 The operetta shown was Ruddigore, and part of the scenery included a large framed painting of Royce in this costume. The painting was given to Royce on his retirement, and is still owned by the family.

CHAPTER 11

1 In August of 1974, President Richard M. Nixon had resigned in the wake of the Watergate scandal, and was replaced by Gerald Ford, who as Speaker of the House had occupied the vice presidency with the resignation of Spiro T. Agnew for corruption charges in 1963. Ford, a Republican, was defeated by Jimmy Carter, the Democratic candidate in the 1976 presidential election. Both of these headed weak and embattled administrations.

2 Davis Journal, volume 3, September 23, 1978, content edited.

3 Davis Journal, volume 3, December 27, 1978, content edited.

4 Davis Journal, volume 3, December 22, 1978.

5 Davis Journal, volume 3, December 31, 1978.

6 On March 4, Royce recorded, "I wrote to John today recommending strongly against hasty marriage. He will still do what he wants to, of course, but I think my advise is sound." (Davis Journal, volume 3, March 4, 1979.) Scarcely a week later, he wrote, "John called us tonight and talked marriage plans. He is looking toward a May wedding. It's all right with me. I just tried to point out where things are right now with him. He must be able to carry his own load now." (Davis Journal 3, March 11, 1979.)

7 Davis Journal, volume 4, May 5, 1979. John's marriage coincided with a nationwide fuel shortage which impacted Royce, with his many travels on strawberry business. A couple of weeks after the wedding Royce wrote, "I returned to Davis and went to gas the car up but found that the University is out of gas. I got some at a station in town after about a 20 minute wait." (Davis Journal, volume 4, May 18, 1979.)

8 Davis Journal, volume 4, July 16, 1979.

[9] Davis Journal, volume 4, August 3, 1979.

[10] Davis Journal, volume 4, September 28, 1979, content condensed and edited.

[11] Davis Journal, volume 4, Sunday, September 30, 1979, content edited.

[12] Davis Journal, volume 4, October 3, 1979, content condensed.

[13] Davis Journal, volume 4, October 5, 1979. A month later, Royce wrote of Victor, "Things seem to be cleared up between us. At least, he has treated me about like he used to. I have gone out of my way to be as nice as possible. Maybe that helped but it could be just time as well." (Davis Journal, volume 4, November 5, 1979.)

[14] Notes for lecture "Academics and Faith," first given October 12, 1979, content condensed and edited. The quote from Dag Hammarskjöld is from his book *Markings*, p. 56, and may have been used only in the later versions of the lecture. The talk was repeated on at least two subsequent occasions, on June 2, 1988, and on March 8, 1991, making use of the same notes with some alterations. On the day of this talk Royce recorded in his journal, "It was good that I wrote it out for the most part. It went well. I was honest and talked of my faith and the reason for always staying with the Church." (Davis Journal, volume 4, October 12, 1979.)

[15] Davis Journal, volume 4, January 9, 1980.

[16] Victor's "shock" was not caused by massive internal blood loss from injuries, as the term is generally used in a medical sense in the setting of a major accident. Rather, he almost certainly suffered a simple vasovagal episode, commonly referred to as a "fainting spell," presumably brought on by seeing Royce covered in blood. Such episodes are common in this setting.

[17] Davis Journal, volume 4, January 21, 1980, content condensed and edited. An entry is made in Royce's field journal, indicating the location of the accident in very shaky hand. (Yellow field journal labeled "Nov 1978-Apr 80," page between entries for October 9, 1979 and January 30, 1980.)

[18] Davis Journal, volume 4, January 22, 1980, content condensed. The following day Royce got a more detailed account of the accident. "Talked to Bob Kavet & Kuni Shinta today. They pointed out that the accident was caused by a 79 year old woman wanting to make a right turn from the left lane and stopping abruptly when she could not do so. The other woman plowed into her, driving both into the center guard rail. The gas truck swerved right to avoid hitting them as he braked, partially blocking the right lane, and he came to an abrupt stop parallel & snugly to the right of the wrecked cars. This left us with no place to go & traffic to our right. I am feeling a bit better today, but not great." (Davis Journal, volume 4, January 24, 1980.)

[19] Davis Journal, volume 4, November 19, 1979, content condensed and edited.

[20] Davis Journal, volume 4, November 30, 1979.

[21] Davis Journal, volume 4, February 7, 1980. Some of the meetings Royce chaired did not go as well as he would have hoped. After one such meeting in Los Angeles he wrote, "This wasn't one of my best; partly the agenda and partly my handling of it. I then went by taxi to UCLA where I attended the last session of a meeting on 'experiential learning.' I really would rather not

be there. Nostalgically I passed by old avocados that I knew well when they and I were young so many years ago. I wish I could return and do it all over again sometimes." (Davis Journal, volume 4, May 8, 1980, content condensed.)

22 Davis Journal, volume 4, February 11, 1980, content condensed and edited. Royce likewise continued to have little regard for the practice of including on the committees student representatives, whom he felt tended to favor a somewhat mindless protest mentality over a studied analysis of the issues, something of which he had long since grown weary. Of the same meeting, he recorded, "We listened to some more student outburst. It serves as an emotional outlet, I suppose. I keep hearing the same thing: 'Our way or we will rant, rave, and shout or weep.'"

23 Davis Journal, volume 4, April 1, 1980.

24 Davis Journal, volume 4, April 5, 1980. The "temple assignment" here alluded to was a calling to serve as a worker in the temple. Temples, as previously explained, are considered by Latter-day Saints to be the most sacred structures on earth, where ordinances (such as endowments and marriage sealings) are received by faithful members. The performance of these ordinances, or ceremonies, requires the assistance of a fair number of Church members to perform the various functions involved. Royce and Pearl were being asked to join the group of volunteers who assisted in performing these ordinances. A basic level of "worthiness," defined as a conviction of the basic tenets of the faith and compliance with its commandments, was required.

25 Davis Journal, volume 4, May 30, 1980, content condensed.

26 Davis Journal, volume 4, August 19, 1980, content edited. Because of the sacred nature of the temple ordinances, all the learning had to be done within temple walls, and notes could not be taken. After a later training session, Royce wrote, "Things went quite well. We do seem to be learning the things we must, letter perfect. There seems to be hope for the old brains yet, although I have wondered about it at times. I made my share of errors. It is amazing how one can learn in the temple with no homework permitted." (Davis Journal, volume 5, September 2, 1980.)

27 Davis Journal, volume 5, September 27, 1980. The quote was a reference to a passage from Isaiah in the Old Testament, "How beautiful upon the mountains are the feet of him that bringeth good tidings, that publisheth peace; that bringeth good tidings of good, that publisheth salvation; that saith unto Zion, Thy God reigneth!" (Isaiah 52:7.)

28 Davis Journal, volume 5, September 11, 1980.

29 Davis Journal, volume 5, October 28, 1980.

30 Davis Journal, volume 4, August 3, 1980.

31 Royce's concern with his calling was at least partly based on his worry that he would not be able to perform it in the way he felt it needed to be done. A month after being called he wrote, "Went over to a 7 a.m. meeting with the Bishop. I frankly didn't need the H. P. group leader's position. I don't handle that sort of thing to my satisfaction." (Davis Journal, volume 5, September 7, 1980.) At the same time, Royce held to his commitment to do the best job he could, and there was no question of holding back, or regarding it as

an honorific assignment. The following Sunday he added, "I worked on the home teaching assignments for the high priests—I'm trying to do that job right." (Davis Journal, volume 5, September 14, 1980.)

[32] Davis Journal, volume 5, December 14, 1980.

[33] Davis Journal, volume 4, March 29, 1980, content condensed.

[34] Davis Journal, volume 4, July 3, 1980, content condensed and edited.

[35] Davis Journal, volume 5, September 23, 1980.

[36] Davis Journal, volume 5, October 9, 1980.

[37] Davis Journal, volume 5, December 15, 1980.

[38] Davis Journal, volume 5, December 31, 1980, and January 1, 1981, content condensed and edited.

[39] Davis Journal, volume 5, October 23, 1980.

[40] Davis Journal, volume 5, April 2, 1981, content edited and condensed. Omitted for clarity's sake was the following comment: "Here, I found an interesting comment on the peace that should prevail. Painted on the man's adobe house were three symbols: the crescent of Muslims, the cross of Christians and the Star of David of the Jews, side by side in the same world as they should be. Peace is important to the Egyptians, because they have recently known the terrible consequences of war."

[41] Davis Journal, volume 5, July 22, 1981, account condensed and edited. Royce did make preparations for the translation of Staudt's manuscript, soliciting the help of member of the Davis ward who was a native German with technical expertise. "I am prepared to handle the translation and arrange for the publication of your manuscript on the Taxonomy of American Strawberries, including all costs associated therewith. Hans J. Rocke, Associate Librarian of the University of California, Davis Library (Librarian for Agricultural and Biological Sciences) has agreed to do the translation. He grew up in Germany and has perfect command of both languages. Moreover his area of competence is Agricultural and Biological Sciences. I have spoken with Jules Janick, Editor of the Journal of the American Society for Horticultural Sciences and he agreed to work with me in arranging for the manuscript to be published in the most appropriate place. Indeed, he is very enthusiastic about it. If you will send the material to me, we will take it from there. I would like to see this project underway as soon as possible." (Letter, RSB to Günter Staudt, December 22, 1981.) It would take over a decade to prepare the manuscript for submission in its final form (see letter, Günter Staudt to RSB, August 9, 1994), and then the approval process for publication by the University of California Press was so prolonged that in December, 1995 Staudt threatened to withdraw the manuscript and have it published in Germany (email communication, Günter Staudt to Rose Ann White and reply, December 20, 1995). The University of California Press finally did publish the manuscript in 1999. (Günter Staudt, *Systematics and Geographic Distribution of the American Strawberry Species: Taxonomic Studies in the Genus Fragaria (Rosaceae: Potentilleae)*, University of California Publications, Botany series Volume 81, 1999.)

[42] Davis Journal, volume 6, October 8, 1981, content edited.

[43] Davis Journal, volume 6, November 17 = 18 NZ, 1981.

[44] This decision was evidently made by the UCD patent office, independent

of Royce.

[45] Royce provided a detailed description of this process to an investigator in Norway who was hoping to organize a nationwide breeding system patterned after the California program. (Letter, RSB to Johannes Øydvin, April 10, 1985.)

[46] Royce's journal entries in the fall of 1980 provide a good example of what this entailed. "I left to pick up Victor at 8:45. We then drove straight to Redding, and then to MacArthur, having lunch in Burney. We then started to dig plants and continued until dark, getting all dug except nine plots from a total of 43. We then drove back to Burney and checked in for the night." The next day he wrote, "We reached the sheds at 8:00 a.m. and packed plants until all of the workers went home. Victor supervised the last of the digging, and I took care of the trimming and packing. We completed about two thirds of it, perhaps a bit more. We got back to Burney about 8:30." The third day the digging was at last completed. "Victor and I had breakfast, then went right to work packing and preparing plants for planting. We didn't quite finish by noon, but did by 1:45 after going over for a quick lunch. We then drove to Redding and picked up the plants we sent down yesterday. We then drove straight through to the airport in Sacramento, reaching there about 7:00 p.m. Victor and I had a very serious talk on the way down. I believe that most of what was perceived as a problem by him has now dissipated, at least I hope so. It isn't that he was without some justification." (Davis Journal, volume 5, October 29, 30, and 31, 1980.)

[47] As an example, the Watsonville winter planting for 1980 was described as follows: "I drove to WEO (Wolfskill Experimental Orchard) and picked up the rest of the plants scheduled to go in. I reached Watsonville by about 10:00 a.m. We then proceeded to plant 17 rows. These were the plants from MacArthur, dug last week. The planting conditions were just about perfect. I worked in the field until it was too dark (about 5:30), then went to the Star Motel." The next day planting continued. "Up and planting, or at least preparing to, by 7:00 a.m. Gus and Hamid didn't show until about 10:00 a.m., but we still finished by noon. I then delivered plants to Richard Ishibashi for experimental planting, and headed for home, reaching there before 6:00 p.m." (Davis Journal, volume 5, November 4 and 5, 1980, content condensed.)

[48] At about this time, Royce wrote an excellent and very clear explanation of the two systems.

"There are two cultural systems in California: 1. *Summer planting*—for this system, plants are harvested from low elevation (Sacramento-San Joaquin Valleys) nurseries, in late December to mid-February, trimmed of all leaves and stored in containers lined with thin polyethylene film at 28° to 30°F (-2.2° to 1°C) until they are set in fruit production fields from early August to mid-September depending upon the cultivar and the location. Summer planting is useful in all growing areas. There is a small 'crown crop' of fruit followed by runner production shortly after plantings are established. That fruit may or may not be harvested, runners are removed and the real 'first year' commercial crop begins about seven months after planting (late March at the earliest). Yields are very high and the fruit quality very good on first year harvest.

"2. *Winter planting*—for this system, plants should be harvested from high elevation (over 3000 ft = 900 m.) nurseries near McArthur in Shasta County from late October to about November 1, and planted shortly thereafter. This system is useful only in the relatively warm-winter coastal to semi-coastal areas. Plants should be fertilized in the planting slot, and clear polyethylene bed mulch should be applied shortly after planting, to raise the bed temperature and thus ensure that the plants grow continuously during the short days of winter, hence the name 'winter planting'. Production on winter plantings commences about three months after planting, February in the best areas. Consequently, winter plantings start fruiting at least one month earlier than summer plantings at a given area, although planted some two months later. Yield per plant on winter plantings is often only one-half that of summer plantings but since the plant size is much smaller, the yield differential may be reduced to almost nothing under favorable circumstances by increasing the plant density. Fruit quality and size on winter plantings is particularly good. The precise timing of plantings is mandated by cultivar and location for both systems.

"Under both systems, the soil is normally sprinkler irrigated before planting and during the establishment period, in order to facilitate ground preparation and planting, and to minimize salt accumulation. Thereafter, through the fruiting period (five to seven months usually) plantings are either drip irrigated (now the recommended practice and increasing greatly in popularity) or furrow irrigated." (From Undated Publications folder, RSB papers.)

[49] Royce put considerable worry into the selection process, realizing the importance of this step of the breeding cycle. "I must make the right decisions on the current group of advanced selections. It now appears to me that of those sent to high elevation last season for 1981 testing, only one of those not selected for the 1981 nursery possibly should have gone (a Douglas-Pajaro hybrid). We shall see how the yield comes out. There is always next year, for a few years yet, but they too will pass, and I will then have only yesterdays to contemplate, and someone else will stand where I have stood and walk where I walked. My fragile hold on the little hour that is mine will be lost." (Davis Journal, volume 5, May 28, 1981.)

[50] Although Royce continued his experiments on *Verticillium* resistance, he was not greatly impressed with their usefulness. After a presentation at a plant pathology meeting in Oregon in 1981 he wrote, "I spoke first on Verticillium Wilt reaction of California cultivars, then on the same on F. vesca & F. chiloensis. Both talks were rather bland, since our accomplishments have been rather modest in relation to the amount of time, effort and money we have put into it." (Davis Journal, volume 5, January 14, 1981.)

[51] Davis Journal, volume 5, September 25, 1980, content condensed. The amount was $225,000. As he prepared for the meeting in Davis after the amount was calculated, Royce had remarked, "I just cannot comfortably conceive of spending that much money, but surely shall if it comes." (Davis Journal, volume 5, September 25, 1980.) The increase in budget was probably only partly due to expansion of the strawberry program itself. It was also due in part to the dramatic rise of the inflation rate in the United States,

which at that time exceeded 10% per year.

[52] Davis Journal, volume 6, September 23, 1981, content condensed and edited. It is probable that the lack of senior authorship in publications had limited Victor's ability to earn advancement within the university system, as well as professional recognition for his pathfinding work. Royce recognized this inequity and tried to compensate for it as he could. When Victor's case for a professional acknowledgement came up for Royce's review, he wrote, "I had to go over the candidates for 'fellow' for the Hort. Society. I voted for most of them, including for Victor. In this case it was partly against my 'better' judgment to a certain extent, but I justify it on the basis that he has contributed more to true advancement of Horticulture in his own way than anyone else I know. He just hasn't done much for the society in the sense that most have, and I fear that a majority of the people voting will go against him for that reason. Perhaps mine is a vote of loyalty more than conviction in a manner of speaking, but I know his capability and accomplishments better than others do." (Davis Journal, volume 5, April 30, 1981.)

[53] Davis Journal, volume 6, March 17, 1982.

[54] "I called Virginia and learned that Victor remains in the hospital, and will remain there until he has open heart surgery (triple bypass). This is scheduled for Monday. My work will be cut out for me, that is for sure." (Davis Journal, volume 6, April 15, 1982.) At the time, Royce was having some chest pains of his own, and he recorded, "I ran and exercised, pain in the chest notwithstanding. I did this yesterday, and it is not wise I shall admit. The pain in the lower bronchial area is intense, particularly as I cough." (Davis Journal, volume 6, April 16, 1982.) However, Royce's problem was pulmonary rather than cardiac, and it resolved with a course of antibiotics.

[55] Davis Journal, volume 6, April 28, 1982.

[56] Colleagues would describe Royce and Victor's relationship in these later years as "competitive," as opposed to their very cooperative early relationship. According to Douglas V. Shaw, who came to the university some years later, there was no detectable animosity between the two, but he described their relationship as one of a "team of rivals," and acknowledged that he sensed this particularly on the part of Victor.

[57] A year later, the author had the opportunity to assist Royce with one of the experimental plantings on this site, and gave a brief description: "The plot was located on a swell of hill lined by windswept Monterey conifers, overlooking a then mist-filled valley with occasional hedged clumps of vegetation scattered among the fields." (John R. Bringhurst journal, September 1982 to October 1983, entry for September 21, 1983.)

[58] The "disclosure" referred to here was an official declaration to the University of California that an item was deemed ready for patenting. This had to be accompanied by information sufficient to accurately identify the patented item and justify the patent.

[59] Davis Journal, volume 6, September 24, 1982.

[60] Parker had been very supportive of Royce and Victor's efforts, and following his death Royce made repeated mention of him during trips to the nurseries. In 1980 he had written, "How changed it is with Bob Parker dead and gone. Things will never be the same here." (Davis Journal 4, May 26,

1980.)

[61] Of Fern Miller's death, Royce wrote, "How sad the day; Fern Miller died. She was stricken with a heart attack some weeks ago, but was at home and appeared to be recovering, but that was all cut short. She was born the same year as I was, only in Feb rather than December." (Davis Journal, volume 6, May 23, 1982.)

[62] Davis Journal, volume 6, October 4 and 5, 1982, content condensed and edited.

[63] "Drove over to WEO to get the leaves of C95. I got them back, classified them, and finished the writeup on C95 so that it can be released reasonably soon after the others. I finished that and also wrote a letter to Lowell Lewis re: plant breeder's protocol for release." (Davis Journal, volume 6, October 7, 1982.)

[64] "Picked up Hansche and Durzan, and went to Berkeley where we held a meeting on release protocol. Primarily a dispute over wording vis a vis plant pathologists and plant breeders. Meeting around the table, we came to a reasonably quick understanding and acceptable wording." (Davis Journal, volume 6, November 24, 1982.)

[65] Davis Journal, volume 6, December 16, 1982, content edited.

[66] Davis Journal, volume 6, December 29, 1982.

[67] Royce was named first only on the two day-neutral patents "Fern" and "Selva," and on the short-day variety "Soquel," which Royce had made the decision to release.

[68] Davis Journal, volume 5, October 7, 1980.

[69] Davis Journal, volume 5, October 13, 1980.

[70] Davis Journal, volume 5, November 28, 1980, content condensed and edited.

[71] These were Hecker Pass on highway 129, and a place called Glenwood, near Santa Cruz. Royce kept records of these collections in his field journal. (See Yellow field journal labeled "Nov 1978-Apr 80," entries dated April 11, 18 and 24, May 30, and June 11 and 20, 1979.)

[72] S. Arulsekar and Royce S. Bringhurst, "Genetic model for the enzyme marker PGI in diploid California *Fragaria vesca* L.," *The Journal of Heredity* 73:117-120, March/April 1981. Of specimens collected with Günter Staudt, Royce wrote, "The *Fragaria vesca* we collected while there proved to be what I anticipated it would be electrophoretically, fixed for almost all the enzyme systems in the same way as other European *F. vesca* examined previously and distinctly different from California *F. vesca*." (Letter, RSB to Günter Staudt, December 22, 1981.)

[73] Davis Journal, volume 6, December 23, 1981.

[74] S. Arulsekar, R. S. Bringhurst, and Victor Voth, "Inheritance of PGI and LAP Isozymes in Octoploid Cultivated Strawberries," *Journal of the American Society of Horticultural Science*, 106(5): September, 1981, p. 679-683. It is doubtful that Victor played a significant role in the study, but because it involved cultivars from the California breeding program, his name was included as a co-author.

[75] This award was given as part of the annual meetings of the American Society of Horticultural Science, held in August, 1982, near Des Moines, Iowa.

76 This occurred in November of 1980, and was recorded in four entries in Royce's journal: "I worked in my office all day except for going to WEO where I picked up the leaves from one of the selections having a mixture (a promising selection), hoping to separate it by the allozymes. I think it will work. I know part of the enzymes for the one I am after anyway, and I have two plants of it in the greenhouse." (Davis Journal, volume 5, October 2, 1980.) "I was able to distinguish the rogue material from the selection on either PGI or LAP, and that is nice. (Davis Journal, volume 5, October 3, 1980.)

77 The sequence of events is outlined in the yellow field journal labeled "from Jan 1982 to Dec 1985," entries dated March 29, April 26, and May 3, 1985.

78 See S. Arulsekar and R.S. Bringhurst, "Strawberry," published in S. D. Tanksley and T. J. Orton (Editors) *Isozymes in Plant Genetics and Breeding, Part B*, Amsterdam, 1983.

79 Royce did not specifically mention the California avocado program, but describes precisely the condition which existed there, in which the breeder, faced with the difficulties of avocado pollination, chose to use open pollinations rather than controlled crosses. Royce concluded, "The grower organization supporting the activity realized virtually nothing from their investment. The information generated from it was of little scientific value and the project never moved from square one in the real game that might have resulted from the employment of appropriate breeding strategy, starting with the difficult hybridizations that could and should have been made in the first place."

80 R. S. Bringhurst, "Breeding Strategy," chapter 10 of James N. Moore and Jules Janick (editors), *Methods in Fruit Breeding*, Purdue University Press, 1983, p. 147-153.

81 R. S. Bringhurst and Victor Voth, "Breeding Octoploid Strawberries," *Iowa State Journal of Research*, vol. 58, no. 4, May, 1984, p. 371-381. The segment on hybridization was taken in large measure from a previous article (see Royce S. Bringhurst and Victor Voth, "Hybridization in Strawberries," *California Agriculture*, August, 1982, p. 25).

82 Davis Journal, volume 5, May 1, 1981, content condensed. It should be noted that Calgene did in fact succeed as a company, made significant strides in genetic engineering, and was later purchased by Monsanto, a major genetic engineering firm. Royce's unwillingness to collaborate was probably a boon to the strawberry industry, given the subsequent public furor over "genetically modified organisms," often abbreviated as "GMOs."

Royce had not discounted a possible future role for genetic engineering. At the conclusion of his chapter on breeding strategy, he had stated, "Within the coming decades, it is likely that major breakthroughs will be realized in applying some of the techniques of genetic manipulation of propagated crops, such as the selective insertion of DNA carrying genes that condition desirable traits. This may very well change the strategy employed in transferring traits malleable to the procedure, particularly if it can be coupled to a shortening of the juvenility period as plants are regenerated." (R. S. Bringhurst, "Breeding Strategy," chapter 10 of James N. Moore and Jules Janick (editors), *Methods in Fruit Breeding*, Purdue University Press, 1983, p. 152.)

[83] Davis Journal, volume 6, January 21, 1982, content condensed.

[84] Davis Journal, volume 6, August 16, 1982, content condensed. That day Royce set up his strawberry exhibits, and prepared to speak to attendees. "It was gratifying to find a very positive response of the two groups that came through: species, interspecific, and intergeneric *Fragaria-Potentilla* hybrids. There was genuine interest and many questions asked."

[85] Davis Journal, volume 6, August 17, 1982, content condensed.

[86] Davis Journal, volume 7, February 24, 1984. The representative was a man named Bob Garcia from Plant Genetics, "a gene splicing outfit nearby."

[87] See Joseph Postman and Kim Hummer et. al., "Fruit and Nut Genebanks in the U.S. National Plant Germplasm System," *Hort Science* 41(5): p. 1188-1194, 2006, which describes the history of the system.

[88] Davis Journal, volume 5, October 15, 1980.

[89] Davis Journal, volume 5, April 15-16, 1981, content condensed and edited. Royce found another colleague at the repository who was an active member of the Church. "Arlin Draper and I went to and from the meeting together. He told me about the 19 temples to be built, the new wave, a definite step forward in my estimation."

[90] Davis Journal, volume 5, August 14, 1981, content condensed and edited.

[91] Davis Journal, volume 5, August 19, 1981, content edited.

[92] Davis Journal, volume 6, December 18, 1981.

[93] That day Royce wrote, "I had to stay in my office until noon, went home for dinner, then returned to my office and went out to the dedication of the Germplasm Center. It wasn't too bad. The talks were brief. The praise of the builders was overdone." (Davis Journal, volume 8, April 4, 1984.)

[94] On December 5, 1984, Royce wrote, "I went to WEO with Gus and dug the plants from the nursery, scheduled to go into the National Germplasm Repository." The next day, he added, "Harry Ingerstadt from Corvallis, Oregon came in at 8:30, and we went first to the field, then to the various greenhouses. We finished by noon, then had lunch at home. I then spent the afternoon writing descriptions of the items going to the repository. I didn't finish until 6:00 p.m. It turns out that there are exactly 100 items. We had figured on 75 to 100, and without counting, we hit the high number." (Davis Journal, volume 8, December 5 and 6, 1984, content edited.) A specific listing of clones and their descriptions, dated December 6, 1984 and written in Royce's hand, is included in a file entitled "Germ Plasm Repository" in the RSB collection.

[95] Davis Journal, volume 5, November 6, 1980, content edited and condensed.

[96] Davis Journal, volume 5, November 7, 1980, content condensed and edited.

[97] Davis Journal, volume 5, November 16, 1980.

[98] Davis Journal, volume 5, January 31, 1981.

[99] Davis Journal, volume 5, April 18, 1981.

[100] Davis Journal, volume 5, May 23, 1981.

[101] One such occasion, which had overtones of divine intervention, was recorded in Royce's journal: "We reached the temple before 6:30, and shortly

after my arrival, I had my scheduled marriages assignment changed to that of greeter. I didn't think much about that at the time, although I have at times remarked that 'I would rather be a doorkeeper in the House of the Lord than to sit in the seats of the mighty,' in a jocular sense. As it turned out, within an hour of my starting with that assignment a young couple from Guatemala who could speak no English arrived with another couple, the male of which was not a member. The first-mentioned couple want to be sealed. I was the only one of about 30+ men whom they might have picked for that assignment who spoke Spanish. To make things brief, I was able to make the arrangements so that they could have it done. Perhaps it was a coincidence, but I was certainly at the right place when my modest talent (with Spanish) was needed. Bro. McKnight who wasn't expected in today, came in about noon; as far as I know, one of two authorized and capable of sealing in Spanish. Brother Nielson and Fonseca also appeared. Consequently, everything was done for them, and they went through the last session where I was assistant officiator. It was a good experience, and I felt wanted and needed, as well as useful." (Davis Journal, volume 7, November 26, 1983.)

[102] Davis Journal, volume 6, September 19, 1981, content condensed.
[103] Davis Journal, volume 6, October 23, 1982.
[104] Davis Journal, volume 5, June 7, 1981.
[105] Davis Journal, volume 6, October 24, 1981.
[106] Davis Journal, volume 6, September 13-14, 1981, content condensed and edited.
[107] Davis Journal, volume 6, July 9-10, 1982, content condensed.
[108] Davis Journal, volume 5, May 11, 1981.
[109] Davis Journal, volume 5, June 21, 1981.
[110] Davis Journal, volume 6, September 13, 1981.
[111] Davis Journal, volume 6, January 9, 1982.
[112] Davis Journal, volume 6, January 12, 1982, content condensed and edited.
[113] Davis Journal, volume 6, January 13, 1982.
[114] Davis Journal, volume 6, December 20, 1982, content edited. Other details of the dispute are found in the entries for May 27 and 28, June 6 and 16, July 13, and August 4, 1982; also in Davis Journal, volume 7, May 26 and August 31, 1983.
[115] Royce began making equivalent contributions to Brigham Young University, where his youngest two children had attended. In 1984, he recorded, "I wrote out a whole series of donations to the various institutions that I help support, including UCD, Wisconsin, Utah State, and BYU. I give out more money now than I used to receive." (Davis Journal, volume 8, December 29, 1984.) This practice continued throughout his life.
[116] Davis Journal, volume 6, March 29, 1982, content condensed.
[117] Davis Journal, volume 6, May 14, 1982, content condensed.
[118] Davis Journal, volume 6, September 23, 1981.
[119] Davis Journal, volume 6, June 5, 1982.
[120] This coincided with Picnic Day, an annual open-house on the U.C. Davis campus, for which Royce customarily prepared an impressive exhibit of fresh strawberries. "I rose at 5:30 or perhaps earlier and, taking John with

me, I prepared the strawberry exhibit; this was better than usual. John did the printing. I had the various varieties plus appropriate selections. It took me several hours but I suppose it was worth it, at least it seemed so because many people observed it. We had our entire family present today for the first time since John went on his mission. We have 13 grandchildren. It was pleasant, although hectic." (Davis Journal, volume 5, April 25, 1981, content condensed. See also the family photo taken on this occasion, p. 507)

[121] Davis Journal, volume 6, December 23, 1982.

[122] Davis Journal, volume 8, January 18, 1985.

[123] For example, in 1983 Royce wrote, "Went to a meeting of the steering committee for the review of plant & animal genetics. This involves one more job that I haven't time nor in some ways the inclination to do; must practice saying 'no' with convincing conviction." (Davis Journal, volume 7, April 7, 1983.)

[124] Davis Journal, volume 7, September 28, 1983.

[125] Davis Journal, volume 7, October 27, 1983, content condensed.

[126] Davis Journal, volume 7, February 2, 1984, content condensed.

[127] Davis Journal, volume 8, May 17, 1984.

[128] See Davis Journal, volume 8, November 14 and 21, 1983, and Davis Journal, volume 9, April 5, 1985. Royce was dissatisfied with the system of advancement set up for non-academic staff members, perhaps in part due to his experience with Victor who, though an energetic researcher who made tremendous contributions to his field, still did not meet the qualifications for advancement. A couple of years later he wrote, "Put up with problems of advancement among the non-academic ranks. The system is bad, the judges are biased, and no one is happy with it." (Davis Journal, volume 10, June 18, 1987).

[129] Davis Journal, volume 8, January 24, 1985, content condensed.

[130] Davis Journal, volume 9, April 5, 1985, content condensed.

[131] Davis Journal, volume 8, February 22, 1985.

[132] Davis Journal, volume 7, July 1, 1983, content condensed. Four days later, Royce wrote, "I then came back to work again, dealing mostly with the problem of getting paperwork and thinking about my replacement into focus. I feel as if my funeral were imminent." (Davis Journal, volume 7, July 5, 1983.)

[133] Davis Journal, volume 7, September 13, 1983. The author was privileged to attend Royce's meeting with the California Strawberry Advisory Board a week later, and recorded his impressions of the event. "The meeting here has turned to mighty disputations over questions which arose over a bittersweet matter—that of finding a replacement for Dad to continue his work on the occasion of his retirement. They intended to help pay with the university the salary of a second professor as an apprentice to Dad during the last years of his career; now they are discussing funding a research person independent of the university, and the talk goes on. Dad made his presentation with the wisdom and eloquence that I have seen grow in him throughout my life, his short stature shuffling humbly to and fro as he spoke, which in a strange way endeared me to him as never before. He sits silently to my left now listening to the proceedings and I wonder vaguely what he is feeling

now. The respect the board has for him is obvious, and the room is full of old friends and comrades who, I suppose, owe much of their livelihood to things which Dad has done. ... His words to the Board, 'I accept my own mortality,' so very like him yet strange to hear to an outsider of a son sitting quietly behind, drew laughter at first from those present until he continued in a quietly earnest voice laced with a touch of emotion that he meant it seriously; in those moments an occasional glance came my way, and I wondered what they saw there through eyes that now so honor my father." (John R. Bringhurst journal, September 1982 to October 1983, entries for September 21 and 23, 1983.)

[134] Royce recorded in his journal the various steps of the process. "I attended pomology staff meeting at 8:30. The main item was to vote on whether an overlapping hiring should be made as offered to the strawberry industry by the dean. The vote was positive." (Davis Journal, volume 8, November 9, 1984.)

[135] Davis Journal, volume 9, March 29, 1985.

[136] Davis Journal, volume 9, October 28, 1985; see also entry for October 4, 1985.

[137] Davis Journal, volume 9, December 30, 1985.

[138] Davis Journal, volume 8, April 9, 1984.

[139] Davis Journal, volume 8, July 10, 1985, content edited.

[140] Davis Journal, volume 9, May 27, 1985.

[141] Davis Journal, volume 9, May 28, 1985.

[142] The first of these two trips was at the request of the Zanzi nursery, to investigate a claim of having received for propagation from a California nursery an off-variety rather than Chandler. (See Davis Journal, volume 9, April 4 to April 11, 1986). The matter later resulted in a lawsuit in which Royce gave evidence: "I drove to Berkeley where I met with Pino Zanzi and his lawyer, Bob Fissell and Roger Ditzel (Patent Office). This is a follow-up on my trip to Italy in the spring of 1986 over the Mislabeled Chandler mixup of Pevehouse Nursery. I signed an affidavit after making them rewrite it." (Davis Journal, volume 9, January 26, 1987.)

The second trip was as a consultant at the invitation of a nursery from southern Italy, and Royce went accompanied by Pearl. He found that the industry continued to depend completely on the California varieties, though he also found that breeding work was being conducted, using both the UC varieties and Driscoll varieties as parents. (Davis Journal, volume 10, April 25 to May 3, 1987.)

[143] See Davis Journal, volume 10, entry for December 6-7, 1987. This was probably at the behest of Pedro Antonio Dávalos, from Irapuato, Guanajuato, Mexico. Dávalos had visited Royce in 1985, spending the night in the Bringhurst home, and accompanying him to see his experimental plantings and to collect wild strawberries from the coast. (See Davis Journal, volume 9, November 24, 25 an 26, 1985.)

[144] Davis Journal, volume 7, December 19, 1983.

[145] Davis Journal, volume 7, July 2, 1983.

[146] Davis Journal, volume 8, April 16, 1984.

[147] Davis Journal, volume 8, June 6, 1984.

[148] Davis Journal, volume 8, June 17, 1984, content condensed and edited.
[149] Davis Journal, volume 8, June 19, 1984.
[150] Davis Journal, volume 7, July 4, 1983, content condensed and edited.
[151] Davis Journal, volume 8, March 6, 1985, passage edited.
[152] Davis Journal, volume 9, February 27, 1987, content condensed. The full entry included more details: "I did meet Col. Kaufman who was our commanding officer on Corsica, and also Col. Lynch who was a lead bombardier whom I flew with a number of times (same box, that is). They look similar to then, and both are in good health. Also met a Seegmiller from St. George Utah whom I knew that I met in Church on the island and at Foggia. There were a number of others who were more or less familiar."
[153] Davis Journal, volume 5, February 1, 1981.
[154] Davis Journal, volume 7, January 18, 1983.
[155] Davis Journal, volume 7, July 29, 1983.
[156] Davis Journal, volume 10, June 19, 1987, content condensed and edited. As a result of this episode, Royce and Pearl removed the nightstand from Royce's side of the bed. Royce had experienced similar nightmares on the road. (PDB Journal 1986-1988, entry marked Tuesday, January 12, 1988.)
[157] Royce's concerns mainly related to the official lesson manuals, which he felt sometimes gave shallow treatment to serious subjects. When his assignment became a permanent calling, he wrote, "I was officially installed as the only teacher. I have mixed feelings about this. Part of the lessons, I don't like. I have to modify them so that I have the stomach to teach them. I don't see why I find myself in this role so much, but I do." (Davis Journal, volume 7, October 9, 1983.)
[158] Davis Journal, volume 7, July 3, 1983. The next day, Royce added, "I should also note that in relation to forgiveness I told the story of my Uncle Lou (my grandfather's youngest brother) and of Scott (my uncle, his only son) and the lifelong enmity between them, the reason for which I likely never knew because I remember nothing about it now at all, except the tragic fact that it darkened most of their days, and even as death came, first to the father and then to the son, forgiveness was not offered by either or received. How sad that this should have happened. Life is so brief at best." (Davis Journal, volume 7, July 4, 1983.)
[159] Davis Journal, volume 7, October 16, 1983, content edited.
[160] Davis Journal, volume 7, March 3, 1984.
[161] Davis Journal, volume 9, June 18, 1985.
[162] Royce later learned of the origin of the assignment at the funeral of his former stake president in the Sacramento Stake, Homer N. Stephenson, who died in November of 1985. "President Sonne [the Oakland Temple president] cleared up what was a mystery to me. He told us that it was Homer who recommended us as temple workers. I was always certain that the bishop did not. I am certain that he never even thought of such things, and did not view us in that role at all. It is nice to know some things." (Davis Journal, volume 9, November 29, 1985.)
[163] One day after working in the temple, Royce wrote, "In the waiting period, there was some, to me, nonsensical fundamentalist discussion going back to the 18th and early 19th century. They displayed complete ignorance of mod-

ern geology, biology, and related hard science, still fighting windmills that go on turning no matter what they say or do. Nature does not conform to anyone's dogma, and at this point it does not matter at all what they say about it. It just makes them sound foolish, vain, and arrogant in a way. I chose not to comment. When a question was directed to me, I simply raised my hands and declined to comment as it would be futile. I am quite certain that were I to comment, some would consider me 'apostate,' and perhaps I am, if their narrow, simpleminded view really prevailed in the Church at the official level. I won't back down on this, just back off when it seems to be that I shall be absolutely put down by all, and there is no way to reason with them. I think institutionalized ignorance is the worst possible tyranny against civilization and intellectual progress. Not that I am a great intellectual. I certainly don't claim that, but I do know what I can or cannot accommodate with a clear conscience as far as belief in relation to the known world is concerned." (Davis Journal, volume 7, March 10, 1984.)

[164] Davis Journal, volume 8, March 23, 1985.

[165] Davis Journal, volume 8, September 15, 1984.

[166] Davis Journal, volume 9, February 1, 1986.

[167] Davis Journal, volume 9, January 24, 1987.

[168] Davis Journal, volume 9, February 21, 1987.

[169] Theodosius Dobzhansky, one of the key figures of the 20th century in the field of evolutionary genetics, had spent the last years of his life as an emeritus professor at UC Davis, and was known to Royce, who especially respected him as an acclaimed scientific figure who still retained his faith in God. Like Royce, he rejected as disingenuous the arguments of so-called creationists. The oft-quoted passage, which originally read, "Nothing in biology makes sense except in the light of evolution," was written while Dobzhansky was at Davis.

[170] Davis Journal, volume 7, February 7, 1984, content edited. This lengthy series of thoughts stemmed from Royce's faculty assignment to administer a qualifying exam to a graduate student. The portion quoted was preceded by the words, "I was responsible for 'evolution.' I wonder how some of the folks in the temple would regard me if they could understand my views on the subject."

[171] See the various issues of the CSAB "Strawberry News Bulletin" from 1983 and 1984.

[172] Davis Journal, volume 7, May 11, 1983.

[173] The house was constructed by one of the minimum-security prisoners who were often employed at the station. "We have the new house for grading fruit etc. well under construction. It is being built by a prisoner who is a carpenter-contractor. He only has a couple more weeks to serve, but we should finish by then. He is happy working, and we are getting the best, picture windows and all." (Davis Journal, volume 8, April 17, 1984.)

[174] Royce had made early use of a band-style tape recorder, and when they became available, he began making use of portable computerized recorders. "I ... spent an hour with Hamid working on programming our data recorders that I hope to use this next season at Watsonville." (Davis Journal, volume 8, January 25, 1985.)

[175] Royce also looked forward to retirement with trepidation. At the conclusion of the harvest in 1984, he wrote, "We finally have the printout on the Watsonville plantings for 83-84; only four more full harvest years remain for me, then out to pasture I go and begin the next life. It doesn't look that great. This is not an old folks' world at all. I am less able than when I was younger, and the game will end with a whimper." Davis Journal, volume 8, September 28, 1984.

[176] IQF stands for "individually quick frozen," a method of rapid freezing which results in good preservation and no fruit clumping.

[177] Davis Journal, volume 8, June 25, 1984, note context.

[178] Davis Journal, volume 8, June 12, 1984.

[179] Davis Journal, volume 10, March 10, 1987.

[180] Davis Journal, volume 7, April 27, 1983. The previous month, Royce had written, "Victor is not doing things the way they should be done, but no chance of getting him to change." (Davis Journal, volume 7, March 22, 1983.)

[181] An example was recorded by Royce in his journal: "Morning was an in-house meeting, with Voth talking twice as long as he should and with minimal discussion as a result. I decided to say less, but am losing patience. In the afternoon, the same thing (1 hour and 10 minutes of monologue with dozens of repetitious slides and no chance for audience uptake). I did mine in 25 minutes (I was supposed to have half an hour as was Victor. This was a grower's meeting, and I had arranged for us to be first so we could leave. I was quite upset at his utter inconsideration. He is oblivious to his audience, and it gets worse with time." (Davis Journal, volume 9, February 19, 1986.)

[182] Davis Journal, volume 9, May 9, 1986.

[183] Davis Journal, volume 8, September 6, 1984.

[184] Davis Journal, volume 8, September 7, 1984. Herb Baum ironically would later be invited to speak at Royce's funeral, having written a book on the industry which was complementary to Royce and Victor.

[185] Davis Journal, volume 8, September 7, 1984. Two weeks later, Royce wrote, "I went to the CSAB re: strawberry budget for 1985. Ours was upped to over 270,000 dollars. I suppose that it is justified. If they don't want us to continue, I guess they can cut it off, but we need lead time since we have started the whole process already. The total involved in all requests is over a half million dollars. I am sure that they won't fund it all." (Davis Journal, volume 8, September 21, 1984.)

[186] Davis Journal, volume 8, November 14, 1984.

[187] Davis Journal, volume 7, February 6, 1984.

[188] Davis Journal, volume 9, January 27, 1986. Douglas Shaw, in a later interview, acknowledged having a form of red-green colorblindness, though he felt that this gave him no difficulty in distinguishing the color of strawberries, and perhaps even enhanced it somewhat, as he seemed better able to judge the luster of the fruit than others. He would later pride himself on the appearance of the fruit he had released. He did acknowledge that at times he had difficulty distinguishing greens from browns, which sometimes gave him problems in assessing the health of the plants, but this would never be a significant factor during a long career in strawberry breeding. (Phone inter-

view with Douglas V. Shaw, January 21, 2016.)

[189] Davis Journal, volume 9, February 16, 1986.

[190] Davis Journal, volume 9, February 20, 1987.

[191] Davis Journal, volume 9, April 25, 1986, content condensed and edited.

[192] Douglas V. Shaw, Royce S. Bringhurst and Victor Voth, "Genetic Variation for Quality Traits in an Advanced-cycle Breeding Population of Strawberries," *Journal of the American Society of Horticultural Science*, vol. 112, no. 4, 1987, p. 699-702. This study was in some respects similar to work done in the 1960s with Royce by Paul E. Hansche, which had compared the heritability of yield, fruit size, fruit appearance, and firmness. (See P. E. Hansche, R. S. Bringhurst and V. Voth, "Estimates of Genetic and Environmental Parameters in the Strawberry," *Proceedings of the American Society for Horticultural Science*, volume 92, p. 338-345.)

[193] Douglas V. Shaw, Royce S. Bringhurst and Victor Voth, "Quantitative Genetic Variation for Resistance to Leaf Spot (*Ramularia tulasnei*) in California Strawberries," *Journal of the American Society of Horticultural Science*, vol. 113, no. 3, 1988, p. 451-456. Although the disease was common in California, it had generally been controlled by chemical fungicides, and no formal selection for resistance had been undertaken..

[194] Davis Journal, volume 9, March 20, 1986.

[195] Davis Journal, volume 10, May 26, 1987, content condensed.

[196] Phone interview with Douglas V. Shaw, January 21, 2016.

[197] "CN17 and 75.45-115 looked particularly good just now on summer planting, along with Selva. I think CN17 should be released and have done with it. We won't learn too much more, I fear, by waiting. I'll think about it." (Davis Journal, volume 8, March 22, 1984.) The variety 75.45-115 would later be designated CN75, and would be released as the "Yolo" variety.

[198] Davis Journal, volume 9, April 14 and 15, 1986.

[199] Davis Journal, volume 9, April 22, 1986.

[200] Davis Journal, volume 9, May 6, 1986.

[201] Davis Journal, volume 9, February 12, 1987, content condensed.

[202] Davis Journal, volume 9, February 20, 1987. The following month the disclosures were completed. "I went over to my office early in order to 'ramrod' through the typing (final) of the four disclosures, and did in fact get it done before the day was out, and got it ready to send over to the patent office." (Davis Journal, volume 10, March 18, 1987, content condensed.)

[203] Former chancellor Mrak died while the disclosure process was still underway, and Royce wrote, "Emil Mrak, former UCD Chancellor, died yesterday at age 85. He was always good to me. I think that one of the strawberries currently pending as varieties will now get his name." (Davis Journal, volume 10, April 10, 1987.)

[204] Davis Journal, volume 9, March 2, 1987, content condensed and edited.

[205] Davis Journal, volume 10, April 3, 1987.

[206] Davis Journal, volume 10, April 24, 1987.

[207] Davis Journal, volume 10, May 20, 1987.

[208] R. S. Bringhurst and Victor Voth, "California Strawberry Cultivars," *Fruit Varieties Journal*, vol. 43 no. 1, 1989, p. 12-19.

[209] Davis Journal, volume 10, April 19, 1987. The quote was probably a ref-

erence to a verse of LDS scripture: "Verily I say, men should be anxiously engaged in a good cause, and do many things of their own free will, and bring to pass much righteousness" (Doctrine and Covenants 58:27).

[210] Y. D. A. Senanayake and R. S. Bringhurst, "Origin of *Fragaria* polyploids, I. Cytological analysis," *American Journal of Botany*, volume 54, 1976, p. 221-228.

[211] Royce mentioned that the pentaploid "male" hybrids from Bodega Bay actually tended to be hermaphroditic and self-fruitful, and hence could generate decaploids through self-pollination. However, by the time of his paper, all the pentaploid colonies at Bodega Bay had been destroyed by development.

[212] Royce pointed out that the enneaploid may also had resulted as the partially unreduced offspring of one of the pentaploid hybrids, although no such hybrids had been found at the Wright's Beach site.

[213] Royce S. Bringhurst, "Cytogenetics and Evolution in American *Fragaria*," HortScience, vol. 25 number 8, August 1990, p. 879-881.

[214] Hamid Ahmadi, Royce S. Bringhurst, and Victor Voth, "Modes of Inheritance of Photoperiodism in *Fragaria*," *Journal of the American Society of Horticultural Science*, vol. 115, no. 1, 1990, p. 146-152. While collecting the data for these experiments, Royce expressed satisfaction at being able to verify scientifically what he had long believed. "I went to WEO at daybreak where I walked the seedlings, picking up a few good ones. I also took notes on day-neutrals, coming out with perfect Mendelian fits when calculating it as a dominant trait. I shall do it for most of the field." (Davis Journal, volume 10, June 23, 1987.)

[215] Hamid Ahmadi and Royce S. Bringhurst, "Genetics of Sex Expression in Fragaria Species," *American Journal of Botany* vol. 78 no. 4, 1991, p. 504-514. Also published in 1991 was a study in which both Royce and Hamid Ahmadi participated, on the possibilities of infecting strawberry roots with a type of bacteria called *Agrobacterium*, which had the capability of introducing foreign genes into infected plants. Diploid *Fragaria vesca* plants were selected because of their simpler genetic structure, and when roots were scored and inoculated, tumors were observed to occur, and transfer of genetic material was confirmed. It was suggested that the diploid *F. vesca* was suitable for molecular studies; the possibility was also raised of using *Agrobacterium* species to transfer useful traits into strawberries, such as herbicide, pest, or disease resistance. Once introduced into *F. vesca*, the genes could then be transferred into standard octoploid varieties by the means Royce and Hamid had outlined. (Sandra L. Uratsu, Hamid Ahmadi, Royce S. Bringhurst, and Abbaya M. Dandekar, "Relative Virulence of *Agrobacterium* Strains on Strawberry," *HortScience*, vol. 26 no. 2, 1991, p. 196-199.

[216] Attempts to use different ploidy numbers such as 12-ploids resulted in sterile plants and were unsuccessful, as were attempts to develop decaploid hybrids using the very pungent musk strawberry, a hexaploid species referred to as *Fragaria moschata*.

[217] Hamid Ahmadi and Royce S. Bringhurst, "Breeding Strawberries at the Decaploid Level," *Journal of the American Society of Horticultural Science*, vol. 117, no. 5, 1992, p. 856-862. Royce evidently expected widespread interest in his

research with decaploids, and he ordered hundreds of reprints of this article, which were found among his papers at his death. However, interest in the project disappeared after Royce's retirement, and the breeding population he and Hamid had established was later discarded. This does not appear to be an area of active investigation in breeding programs now. (Personal communication with Hamid Ahmadi, February 2016.)

[218] Davis Journal, volume 9, April 4, 1985.

[219] Davis Journal, volume 9, February 22, 1986, content condensed.

[220] "Davis Journal, volume 10, March 17, 1987.

[221] "We have a sad situation in our family now that we might have realized would happen, but have hoped that it wouldn't. Jean wants to leave Bruce, and I suppose divorce him. Apparently she is terribly unhappy and has been for a long time. What can we do? Precious little at the moment. We have determined to be supportive of whatever as long as it is rational. I feel some personal responsibility about the whole thing. It is sad." (Davis Journal, volume 10, May 29, 1987, content condensed.)

[222] Davis Journal, volume 10, August 7, 1987. Coincident with these events, Royce for several years stopped writing regularly in his journal, and his account of the marriage of Jean's daughter Gina ended with these words: "Perhaps that is why I stopped keeping records—there are things that I do not care to remember."

[223] Davis Journal, volume 10, May 10, 1987, content condensed.

[224] Davis Journal, volume 10, April 4, 1987.

[225] Davis Journal, volume 10, July 1, 1987.

[226] Davis Journal, volume 10, May 17, 1987. At the time of this writing Royce was home alone, Pearl having traveled to Martinez, California where John's wife had just given birth to a new child.

[227] Davis Journal, volume 10, June 4, 1987.

[228] Davis Journal, volume 10, June 7, 1987, content condensed.

[229] The title of Royce's lecture was "Breeding short-day and day-neutral strawberries in California." (See letter, RSB to Dr. J. E. Jackson, August 22, 1986.) This invitation was one of the most significant honors Royce received in connection with his work. He described the trip briefly in his journal. "Pearl and I went to England where we visited in East Malling, where I had been invited to give the Amos Memorial Lecture. It rained much of the time, but we were kept busy and entertained constantly. Our last day was spent in London doing shopping and sightseeing. We stayed in a West Malling hotel (The Swan) that is centuries old. I had to watch the mushrooms. They seem to serve them all the time." (Davis Journal, volume 9, November 16-23, 1986, content condensed.) Royce had an allergy for mushrooms, but loved their flavor.

[230] This appeared to have originated from Royce's connections in Oregon. "I have offered, or intend to offer to do some things and to donate my journals to China. They have need of them." (Davis Journal, volume 9, January 9, 1987.) "Sam Dietz visited me for a time. He will be arranging to take over my books and ship them to China. I am donating them." (Davis Journal, volume 10, April 23, 1987.)

[231] Davis Journal, volume 10, June 17, 1987.

[232] Davis Journal, volume 10, June 24, 1987, content edited.

[233] Davis Journal, volume 10, April 1, 1987.

[234] Davis Journal, volume 10, April 28, 1989.

[235] Although no longer formally making journal entries at that time, Royce recorded, "29 May 89: Günter Staudt & wife Annliese visited as home guests, for 6 wks for Günter, few days for his wife. Purpose is to complete paper (his most important on *Fragaria* systematics). 29-30: Bodega Bay to Pt. Sur with Staudt—A good trip." (Davis Journal, volume 10, May 29-30, 1989.)

[236] Günter Staudt, *Systematics and Geographic distribution of the American Strawberry Species: Taxonomic Studies in the Genus Fragaria (Rosaceae: Potentilleae)*, University of California Publications in Botany, volume 81, 1999. Since Staudt was defining a new species, he inserted a description in Latin, as follows: "*FRAGARIA x BRINGHURSTII* SPEC. NOV. Diagnosis: Hybridae probabiles inter *Fragaria Chiloensium* et *Fragaria vesca* subsp. *Californicam*. Plantae Plerumque steriles raro aliquantum fertiles; *F. chiloensium* simulantes sed quoad pubescentiam adaxialem foliorum, crassitiem foliorum, numeros dentium laminarum, crassitiem petiolorum atque folios luteo-virides minus nitidos aliquantum intermedie Chromosomatum numeri $2n = 35$, $2n = 42$, $2n = 63$." (p. 135).

[237] This award was presented annually by the Fruit Breeding Working Group of the American Society for Horticultural Science. See Gayle M. Volk, James W. Olmstead, Chad E. Finn, and Jules Janick, "The ASHS Outstanding Fruit Cultivar Award: A 25-year Retrospective," *HortScience* vol. 48 no. 1, January 2013, p. 4-12. Royce received a congratulatory letter from a member of the selection committee. "Dear Royce: Recently, the 'Chandler' strawberry was nominated for the Outstanding Fruit Cultivar Award given by the Fruit breeding Working Group of the ASHS. On behalf of the selection committee, I am pleased to inform you that 'Chandler' was chosen as a winner of this year's award which will be presented at the Fruit Breeding Working Group business meeting. I hope you can attend. 'Chandler' is the envy of many plant breeders and I would like to congratulate you on its well documented success." (Letter, Bruce Reisch to RSB, June 27, 1989.)

[238] See letter of congratulations from ASHS president Roy A. Larson to Royce S. Bringhurst, June 19, 1989.

[239] This curious award came in the form of a hand-lathed wooden plate and pedestal, on the base of which was inscribed by hand, "1990 Milo Gibson Award, to Royce Bringhurst, in recognition of his ability to solve the knotty problems of strawberries. From North American Fruit Explorers." The plate, which had large knots in the wood, sat in the Bringhurst kitchen for years, where it was used by Pearl as a cake plate. Most of the family remaining ignorant of its origin and the fact that it was an award until after Pearl's death, when it was examined more closely.

[240] Davis Journal, volume 10, notation before July 20, 1987.

CHAPTER 12:
[1] Davis Journal, volume 10, January 5, 1990.

2 Of one such meeting in 1991 (the Annual Strawberry Research Meeting at Watsonville) Royce wrote, "I no longer talk at these. They make a point of this (Shaw mostly). They lose more than I do." (Davis Journal, volume 10, February 20, 1991.) After a similar meeting a year later, he wrote, "Annual Wat. Strawberry Research Meeting held. A lot of reinventing of wheels these days. Between Shaw, Welch and even Voth, a lot of redoing takes place." (Davis Journal, volume 10, February 12, 1992.) In fact, it seems clear that Douglas Shaw was simply asserting his leadership of the strawberry program in the wake of Royce's retirement, just as Royce had done previous to it.

3 Davis Journal, volume 10, July 11-15, 1991.

4 "Went to Salt Lake, staying 2 days visiting family, then on to St. Louis Missouri for consultation with Monsanto Chem Co (Maude Kinkchee) re: genetic engineering of *Fragaria*." (Davis Journal, volume 10, June 20, 1992.) Monsanto was a company which had its start in agricultural chemicals. In the late 20th century, it became one of the pioneers in genetically modified agricultural plants. At the time of this writing, it remains a leader in that controversial field. Attempts at genetic modification of strawberries have never resulted in viable commercial varieties.

5 See yellow field journal labeled "1988-89," entry dated July 15, 1989.

6 "Went to Argentina covering much of that huge country, back and forth. They grow a lot of strawberries – favorite is 'Pajaro' – also grow some 'Heidi' which is called 'Kubo Seis.' It is kind of soft. We were robbed of Pearl's purse on Easter day (15 April) in Buenos Aires just before we returned to California. Unfortunately we lost most of my notes that I dictated to Pearl." (Davis Journal, volume 10, March 26-April 17, 1990.) Many details of the Argentina trip are found in the yellow field journal labeled "Apr 1990 – ," first several pages. In his record for April 16, 1990, Royce gives specific recommendation for the Argentine strawberry industry.

7 Davis Journal, volume 10, July 12, 18, 20, 23, 25 and August 4, 1990.

8 Davis Journal, volume 10, October 12, 22, 24, and 25, 1991.

9 Davis Journal, volume 10, November 19, 1991, content edited. The prior entry, dated November 13, 1991, reads, "Headed south with Fusas [Sudaki's] tech. We found lots of wild strawberries, all in flower, and for the first time was able to see the wonderful Aurocaria pines of Chile, Yanguay among other places. I brought back a few seeds. *Araucaria araucaria*. Saw cultivated *F. chiloensis* near Pehuen." Some notes from this trip are found in the yellow field journal labeled "Nov 1991 Chile," and after the first few pages Royce again included a series of specific recommendations for strawberry growers in Chile. These are written in a very shaky hand, possibly in a moving vehicle.

10 "Retirement Gala for Victor Voth at Disneyland. This was beyond belief in some aspects; for example, a nearly nude stripper stepping out of a cake. Perhaps this made it appropriate to be at Disneyland—Never-never world." (Davis Journal, volume 10, September 22, 1990.)

11 See Davis Journal, volume 10, April 9, 1991. Victor continued to work at the South Coast Field Station, probably to effect a smooth transition with Kirk Larson as Royce had with Douglas Shaw. Nearly a year later, Royce attended the field day at SCFS and noted, "Voth's plantings look good. He

still hangs on." (Davis Journal, volume 10, March 10, 1992.)

[12] Davis Journal, volume 10, June 22, 1991. Three months later Royce, who had been honored by the nurserymen two years before, attended a nursery-men's meeting at Redding. "This time they properly mentioned and honored Voth. I was mentioned in passing only. So much for that." (Davis Journal number 10, September 5, 1991.)

[13] Davis Journal, volume 10, April 10, 1992. The "contributions" spoken of were simply the portion of Royce's royalty shares which he had signed over to the university. These were eventually applied to a scholarship fund in Royce and Pearl's name.

[14] Davis Journal, volume 10, May 14, 1992.

[15] Officially known as the Ezra Taft Benson Agriculture and Food Institute, the organization was named for Ezra Taft Benson, a Latter-day Saint apostle who had served as U.S. Secretary of Agriculture under President Dwight D. Eisenhower, and later became president of the Church. Its objective was to improve the lot of families in poorer nations by applying modern agricultural methods to small farming.

[16] "Flew to Salt Lake, then on to Provo for Benson Institute training. Was not too clear in my case, and I wondered if they had the right person." (Davis Journal, volume 10, July 25, 1992.) Notes from these meetings are found in the yellow journal labeled "Nov 1991 Chile," in an entry dated July 27, 1992.

[17] Davis Journal, volume 10, August 9, 1992.

[18] This was a musical setting which Royce had first heard at Protestant services while in active military duty on Corsica.

[19] Davis Journal, volume 10, August 16, 1992, content edited.

[20] Royce and Pearl considered asking for a later start date, but were later grateful they had not. "Fortunately I did not press to leave later. Had we done so, we would have been hit and delayed by a tremendous hurricane that hit Miami right after, and left people stranded there for 3-4 days. It is 'Andrew,' the hurricane of the century." (Davis Journal, volume 10, August 22, 1992.)

[21] His full name was Eleazer F. Magnere (letter, RSB to John & Betty Bringhurst, September 20, 1992).

[22] The campus of the newly-formed university was evidently not very impressive, for after seeing other colleges in the same system Royce wrote, "I should note at this point that both the Universidad de la Frontera (Temuco) and the Universidad Austral (Valdivia) both look like universities, contrary to Aconcagua which doesn't look that way at all. I'm talking about the physical set-up, buildings, library classrooms and students too. The faculty at Aconcagua look fine. Too many don't even live in San Felipe. They are in Viña del Mar." (Davis Journal, volume 10, October 23, 1992.)

[23] Davis Journal, volume 10, August 30, 1992, content edited.

[24] Letter, RSB to John & Betty Bringhurst, September 14, 1992.

[25] Davis Journal, volume 10, September 2, 1992, content condensed. In a letter, Royce wrote, "Pearl is teaching 6 classes in English as a second language each week. I help her some. I'm teaching one class a week." (Letter, RSB to John & Betty Bringhurst, September 7, 1992.)

[26] Davis Journal, volume 10, September 4, 1992, content condensed.

[27] Davis Journal, volume 10, September 6, 1992.

[28] Davis Journal, volume 10, September 9, 1992, content condensed.

[29] Davis Journal, volume 10, September 25, 1992, content edited.

[30] Davis Journal, volume 10, October 28, 1992, content edited and condensed.

[31] Davis Journal, volume 10, October 19, 1992, content condensed. It was the author and his wife who had failed to send the books by air mail. Of this Royce wrote, "This is annoying to me, since they assured us that they would send air freight." (Davis Journal, volume 10, September 24, 1992.) When he finally received the box, he wrote, "I got the things from the post office about 5:30. They were good to get, but I thought that we had sent more. Two months plus is a long time to wait for something, and you sort of forget what was packed. We are glad for what we have anyway, and what could we do about it if we don't like it?" (Davis Journal, volume 10, October 30, 1992.)

[32] Davis Journal, volume 10, October 23, 1992.

[33] Davis Journal, volume 10, November 16, 1992, content condensed.

[34] Davis Journal, volume 10, November 30, 1992.

[35] Letter, RSB to John & Betty Bringhurst, September 20, 1992.

[36] Davis Journal, volume 11, April 11, 1993, content edited.

[37] Davis Journal, volume 10, December 6, 1992. The interest of Church members in genealogy has to do with the belief that the living can perform vicarious ordinance work such as baptisms and endowments in the Temple on behalf of deceased ancestors.

[38] Davis Journal, volume 10, November 25, 1992.

[39] "It is exactly 6 months since we came to Chile, and no thanks to many, today I received my carnet card. Pearl's was not here yet, probably because it was handled by a different office. My no. is 6-068364-6, and must be renewed by 2 Dec 1993, or at least the process started, and that is slow." (Davis Journal, volume 10, February 23, 1993.)

[40] Davis Journal, volume 10, March 24, 1993.

[41] Davis Journal, volume 10, October 27, 1992.

[42] Davis Journal, volume 10, November 10, 1992.

[43] Letter, RSB to John & Betty Bringhurst, January 4, 1993.

[44] Davis Journal, volume 10, January 3, 1993, content edited.

[45] Davis Journal, volume 10, January 9, 1993, content edited. The following day Royce added, "We had breakfast and the *enfermera* came by about 9:15 a.m. to dress my injury again. If I had had my way, I would not have called the lady, and might have had a very serious problem." (Davis Journal, volume 10, January 10, 1993.)

[46] Davis Journal, volume 10, January 14, 1993.

[47] Davis Journal, volume 10, February 4, 1993, content edited.

[48] Davis Journal, volume 10, January 14, 1993, content edited.

[49] Royce wrote, "I had a letter from Doug Shaw about the new varieties. Had to tell him we won't come in March. It may be just as well, and they should be doing that sort of thing now." (Davis Journal, volume 10, February 23, 1993.) Two of the six new varieties, selected by Douglas Shaw, were

called "Sunset" and "Cuesta," (Shaw later stated that "Sunset" had been selected by Royce prior to his retirement—Personal communication, January 21, 2016) while the other four, selected by Victor Voth, were named "Camarosa," "Anaheim," "Laguna" and "Carlsbad." Of these, the short-day variety Camarosa would come to play a dominant role, while Carlsbad and Cuesta would also see some success. For details of the release, see Anne Michaud, "Building a Better Berry: UC Research in Irvine, Davis Yields 6 New Plant Varieties," *Los Angeles Times*, June 20, 1993. (Although the article contains several inaccuracies, the variety names are correct.) Royce also during his mission received and processed paperwork to allow licensure of previously released varieties in other countries.

[50] Davis Journal, volume 10, January 21 and 22, 1993, content edited.

[51] Davis Journal, volume 10, January 23, 1993.

[52] Davis Journal, volume 10, February 19, 1993. Royce explained the significance of his soybean research in a letter home. "This crop has never been introduced into Chile. This was true in the USA before World War II. Now it is a major U.S. Crop, and U.S. leads the world in soya production. It is grown on a rotational basis throughout the middle west with corn, giving oils and proteins. It seems that *Soya* is a big thing with Benson Institute in Latin America, where along with other warm climate areas it has great potentiality for 'feeding the poor among us.' One month after we got here, we planted *Soya* here, and repeated it every month until January. The first and second planting are at the maturing level now. The *Soya* loves hot summer days. My contribution has been developing a simple method for estimating crop potentiality as it reaches mature levels. I have repeated the procedure 3 times with 3 of the 4 planting dates. Your mother and one Chilean professor (Godoy) have worked with me on this. Already it looks like a successful procedure. Anyway, it makes me feel good in the same way as strawberries do." (Letter, RSB to John & Betty Bringhurst, March 17, 1993.)

[53] Davis Journal, volume 10, March 3, 1993, content condensed.

[54] Letter, RSB to John & Betty Bringhurst, March 5, 1993.

[55] Davis Journal, volume 10, February 11, 1993, content edited.

[56] Davis Journal, volume 11, April 30 and May 1, 1993.

[57] Davis Journal, volume 11, April 22, 1993.

[58] Davis Journal, volume 11, April 19, 1993.

[59] Davis Journal, volume 11, May 18, 1993.

[60] Davis Journal, volume 11, May 22, 1993, content condensed and edited. The same day Royce recorded, "Pearl finally got her cédula—Happy day."

[61] Davis Journal, volume 11, May 24, 1993, content condensed and edited.

[62] Royce actually understated Pearl's reaction, if her own account is to be believed. "When we went through the international police they told us we could not leave. My temper exploded. I yelled, I threatened, I kicked walls and slammed my books on the floor. In fact I was an absolute stinker but I could see us waiting here to get a stupid document that no one had informed us that we needed. I could see Royce's honorary doctor's degree going down the drain. All the family disappointed. Fury overtook me and I turned loose like a wild woman, but we got attention. They had a wonderful calm agent that helped us. I turned sick. Royce shook like a leaf, but finally calmed me."

(PDB journal, January 31, 1993-October 7, 1993, entry for May 27, 1993). Pearl apologized to the airport agents the following day. Royce's shaking to which she referred was probably his Parkinsonian tremor, aggravated by the extreme anxiety of the situation.

[63] Davis Journal, volume 11, May 27-28, 1993, content edited and condensed.

[64] Davis Journal, volume 11, June 3, 1993.

[65] Davis Journal, volume 11, June 5, 1993, content condensed and edited. Royce continued, "After thanking the Dean and telling him that I did not want them to send me the money I paid out to get to Utah from Stgo Chile and back ($2600 +) but rather that it should go to the program of their choice in the Ag. college, we then drove back to Salt Lake via Tremonton."

[66] Davis Journal, volume 11, June 13, 1993, content condensed.

[67] Davis Journal, volume 11, June 15, 1993.

[68] Davis Journal, volume 11, June 17, 1993, content edited.

[69] Davis Journal, volume 11, June 26, 1993, content edited.

[70] Davis Journal, volume 11, August 5, 1993, content edited.

[71] Letter, RSB to John & Betty Bringhurst, August 9, 1993, content condensed. In his journal, Royce wrote, "Chile is the fastest growing Church area in the world, at a growth rate of about two stakes per year. This is certainly a high growth area. Everywhere in South America is good where missionaries are working unmolested, but Chile leads." (Davis Journal, volume 11, July 4, 1993, content edited.)

[72] Letter, RSB to John & Betty Bringhurst, August 9, 1993.

[73] Davis Journal, volume 11, September 28, 1993, content edited.

[74] Davis Journal, volume 11, November 30, 1993, content edited.

[75] Davis Journal, volume 11, September 8, 1993.

[76] Davis Journal, volume 11, September 9, 1993.

[77] Davis Journal, volume 11, September 15, 1993.

[78] Davis Journal, volume 11, October 27, 1993.

[79] Davis Journal, volume 11, November 17, 1993, content edited.

[80] Davis Journal, volume 11, October 28, 1993, content edited

[81] Davis Journal, volume 11, November 5, 1993, content condensed and edited.

[82] Davis Journal, volume 11, November 10, 1993, content condensed, and a crossed-out portion included in the narration.

[83] Davis Journal, volume 11, September 26 and 28, 1993, content edited.

[84] Davis Journal, volume 11, November 24 & 25, 1993, content edited and condensed.

[85] Davis Journal, volume 11, December 10, 1993, content condensed and edited. The next day Royce wrote, "Election day. All stores are closed. We spent the first 2 hours harvesting our berries (about 5-1/2 kilos). We gave most of them away, and won't be eating any. I don't trust the water." (Davis Journal, volume 11, December 11, 1993.) Royce the following day described the outcome of the election: "Eduardo Frei was elected president (effective in March) by a majority of some 58%. His father was president when we were in Chile in 1967-68, before Allende was elected." (Davis Journal, volume 11, December 12, 1993.)

[86] Davis Journal, volume 11, December 15, 1993.

[87] Davis Journal, volume 11, December 17, 1993, content edited. Günter Staudt's letter read in part, "In case you have time for a trip to the isle of Chiloé, I would like to propose the following: When I looked through the collection of Cameron, I found a *F. chil. f. patagonica* from coast of Chepu, on beach, on sand dunes (very thick stolons), which displayed the largest flower size of any *patagonica* investigated. The males had a diameter of 46.5 [mm] in average. Unfortunately no fruits on the male plants and no female plants from the same locality. It could be possible that the females have also bigger fruits than normal. Therefore my question: Are the *patagonica* plants (wild *chiloensis*) from Chepu having bigger fruits and are therefore the starting point for the origin of cultivated *F. chiloensis*? If you have time, I think it should be worthwhile to look at that place. See other accessions I have seen from Chile have the same small fruit size, typical for *pategonica*." (Letter, Günter Staudt to RSB, postmarked December 7, 1993.)

[88] Davis Journal, volume 11, December 21, 1993, content edited.

[89] Davis Journal, volume 11, December 24, 1993, content edited.

[90] Davis Journal, volume 11, December 21, 1993.

[91] Davis Journal, volume 11, December 27 and 28, 1993, content edited.

[92] Davis Journal, volume 11, December 29, 1993, content edited and condensed; see also letter, RSB to John & Betty Bringhurst, January 4, 1993.

[93] Letter, RSB to John & Betty Bringhurst, January 4, 1994. In this letter, Royce gave a detailed account of the search for berries at Chepu, Chiloé.

[94] "The Hancocks came in from Talca in the early afternoon yesterday. We had a very nice visit with them. I was able to provide a picture for a talk he plans to give in England, and a paper I wrote years ago that was on the same subject. They spent the night with us as house guests, and left for home (Talca) bus-train at 10:30. It was cool, and we were pleased to have them as house guests." (Davis Journal, volume 11, February 5, 1994.)

[95] Davis Journal, volume 11, February 11, 1994.

[96] Davis Journal, volume 11, February 14, 1994.

[97] Davis Journal, volume 11, February 15, 1994, content condensed.

[98] Davis Journal, volume 11, February 23, 1994, content edited and condensed.

[99] Davis Journal, volume 11, February 24, 1994.

[100] Royce quoted in a letter home a poem by famed Chilean poet Gabriela Mistral entitled, "The Pleasure of Service," which he copied in his journal:

"How sad the world would be if all within it were done,
if there were no rose garden to plant,
no enterprise to undertake.
"...Service is not only a task of inferior beings;
God who gives fruit and light serves.
He might well be called thus: "He who serves.
And he has his eyes upon our hands,
and he asks us each day:
"Did you render service today? "To whom?"

(English translation of poem by Gabriela Mistral entitled, "*El Placer de Servir*," found in Spanish in its entirety in Davis Journal, volume 11, inside back cover.)

[101] Davis Journal, volume 11, February 27, 1994. Royce's recounting of his and Pearl's talks was accurate. John recalled that of the two, Pearl seemed to have superior command of Spanish, which surprised him, since in earlier years Royce had always been fluent, and Pearl had struggled with the language.

[102] Davis Journal, volume 11, March 4, 1994, content edited. In spite of Royce's early misgivings, the strawberry program continued to do very well under Shaw's leadership, and resulted in a continued train of highly successful California cultivars. The variety Camarosa, selected in Southern California by Victor Voth shortly after Royce's retirement, quickly replaced Chandler as the premier short-day variety, and a variety called "Diamante," released in 1997, eventually supplanted Selva as the dominant day-neutral, having greater productivity and better flavor than its successor. Other very successful day-neutrals followed. "Albion," released in 2004, brought greatly improved flavor, more uniform fruit quality, and improved disease resistance over its predecessors, and continues to play a dominant role in the industry at the time of this writing. Day-neutrals "San Andreas" and "Monterey" were released in 2008. For a time the day-neutrals occupied a niche role in the Central Coast region, but with "Albion" and "San Andreas," they have come to play a major role in Southern California as well. The day-neutral varieties continue to be a major hallmark of the California breeding program.

[103] "We drove to the temple about 7:00 a.m., went on the 9:00 and 11:00 o'clock sessions, then met with Brother Daines of the temple presidency, and were accepted and set apart as ordinance workers for Saturdays. We start next week." (Davis Journal, volume 11, March 5, 1994.)

[104] Letter, RSB to John & Betty, October 10, 1993. Royce's statement reflects his conviction of the Latter-day Saint concept of pre-existence, the belief that all humankind lived with God as individual spirit beings prior to earth life, and that it was their choices in this pre-existence which enabled them to experience earth life.

[105] Davis Journal, volume 11, March 15, 1994, content edited.

[106] The issue had to do with a series of anonymous negative letters written in the distant past to leaders of the Davis Ward, which Pearl had recently learned had been attributed to her, and which she felt convinced had damaged her reputation in the ward. This incident had filled Pearl with a deep sense of indignation, and for years to come would affect her relationship even with members of the ward who had nothing to do with it. Eventually one of the older ward members wrote a letter to Pearl in which she confessed having written the letters, but it was only gradually that Pearl overcame her indignation, after a conscious struggle of many years. Four years later, in 1997, Pearl wrote, "Decided to put all that behind me and I will try to be friendly to all." (PDB Journal, July 1996-December 1997, entry for July 14, 1997.) As the parties aged, Pearl largely succeeded in laying her indignation aside and viewing the offending member as an old friend.

[107] Davis Journal, volume 11, March 16, 1994, content edited. This was Royce's last journal entry.

[108] The breed is more properly known as Australian Cattle Dog, also referred

to as a Queensland Blue Heeler, the term recalled by the author's wife Betty. Dogs of this breed are known to be intelligent, energetic, and faithful.
[109] Pearl's account of this event was as follows: "To Fresno with Royce to Sunnyside Packing Co. get together. Royce forgot how long it took to get there, and we were late. Victor leaving as we arrived again. Nice dinner, saw many old friends but the younger generation coming up, we didn't know the younger generation." (PDB journal, September 1994-December 1995, entry dated April 5, 1995.)
[110] A couple of months before the mission ended, Royce wrote, "I have been invited to Portugal (José Carreiro of Multiplanta, Cantanhede, Protugal) to participate in special strawberry symposium, Portugal Society of Horticulture Apr 28-29, 1994. I sent a fax to them today saying I accept." (Davis Journal, volume 11, December 6, 1993.)
[111] The trip to Chile was described in Pearl's journal (PDB journal, September 1994-December 1995, entries for October 19 to October 31, 1995).
[112] Parkinson's disease is a chronic neurodegenerative disorder which affects a region of the midbrain known as the substantia nigra. It most often begins with a characteristic tremor, but has varied manifestations, including muscular stiffness, difficulty initiating movements which results in a typical abnormal gait, imbalance, sleep disturbances, loss of facial expression, and eventually dementia. Royce had a very typical case, and over time developed all of these manifestations.
[113] Author's personal recollection.
[114] Royce enjoyed the fishing trip, though his tremor sufficiently damaged his dexterity that he had difficulty handling a fishing pole. It was on the camping trip in 1997 that John and Betty became aware of the problems Royce had developed in his driving, although Royce drove the entire distance. One of their children, who was in the vehicle with Royce and Pearl, described it as a harrowing experience.
[115] The Family History Center is a series of rooms open to the public, which contain records, materials, and computer resources to perform genealogical research. This was related to the Church's practice of performing vicarious work for the dead, though many of those in the community who make use of the center are not members of the Church.
[116] See PDB journal, July 1996-December 1997, entries dated July 24 and 25, 1997.
[117] See PDB Journal, July 1996-December 1997, entry dated November 16, 1997.
[118] This was Bill Marble, the son of a Davis Ward couple with whom Royce and Pearl had had close association for many years. Bill had just been made stake president when this ordination occurred.
[119] See PDB journal, December 1995 to September 1999, entry for November 2-16, 1998.
[120] The first of these episodes occurred on March 10, 1999, while Royce and Pearl were flying to Utah to attend the funeral of Mary Davidson, wife to Pearl's brother Dan; Royce injured his leg in the fall, and Pearl, who turned to help him, also fell. The second occurred on May 10, 1999 while Royce and Pearl were returning from Pullman, Washington, where they had attend-

ed the graduation of their granddaughter Alisa, daughter of Ann and Ray. (See PDB journal, December 1995 to September 1999, entries for March 10 and May 10, 1999, also PDB journal, September 2000 to June 2001, entry for November 12, 2000.)

[121] Royce and Pearl gave significant assistance to all six of their adult children and their families at some time in their lives, either financially or by providing childcare.

[122] This incident, which occurred in 2000, is recalled by the author, and was mentioned in Pearl's journal. Although tithing donations, being tax-deductible, were ordinarily considered nonrefundable, an exception was made in this very obvious instance, and the correct amount of tithing was eventually paid.

[123] Pearl made mention of the conference in her journal, and Dr. Gubler, who recalls the event, provided the details. It was decided that year to honor some of the founding members of the conference by inviting them to return and speak, and Royce and Victor were among those invited.

[124] See PDB journal, December 1995 to September 1999, entry for September 1-3, 1999.

[125] Herbert Baum's book, entitled *Quest for the Perfect Strawberry*, was published in 2005, the year of Royce's death. It described some of the history of the California industry from an industry perspective, and served as something of a tribute to Royce and Victor's contribution to the industry, though Pearl, ever jealous of Royce's position as head of the program, expressed dissatisfaction with the emphasis given to the role of Victor Voth. Herb Baum was CEO of a strawberry cooperative called Naturipe Berry Growers, and twice served as chairman of the California Strawberry Advisory Board.

[126] See PDB journal, September 2000 to June 2001, entry for January 24, 2001.

[127] These details are recounted in PDB journal, September 2000 to June 2001 and PDB journal, June 2001 to February 2003.

[128] Marla and John's son Samuel Vaughn was married to Rachel Morrissey in Oregon in June of 2001, and their daughter Katie Vaughn was married to Cameron Richie in Utah in August of 2001. The two missionaries were Sarah Bringhurst, called to Chile, and Benjamin Bringhurst, called to Mongolia; although the two siblings received their mission calls a month apart, they were both assigned to begin their mission the same day.

[129] The author recalls a particular day, probably at this time, when Royce, in a rare moment of candor, said to him quietly, "I feel so useless." It was an unforgettable and heart wrenching moment, and the author had no ready reply. Pearl shared in Royce's discouragement during this busy time. After an unannounced visit to John and Betty's house in early September she wrote, "We are too old for people to be bothered with us." (PDB journal, June 2001 to February 2003, entry for September 1, 2001.)

[130] Known as the "Saints at War" project, the collection is archived in the special collections section of the Harold B. Lee Library on the BYU campus.

[131] Pearl wrote, "We had an appointment with a Dr. Don Norton of BYU at 9:00 a.m. Instead we got there at 9:30 a.m. and were there until noon. He asked us about Royce's war experience. I chimed in on it." (PDB journal,

June 2001 to February 2003, entry for August 20, 2001.) Pearl later related to the author that Royce had cried during the interview as he relived his experiences. With his family he was always more reserved.

[132] See PDB journal, June 2001 to February 2003, entries for April 1, 5, and 15, 2002.

[133] Pearl's account of these events are found in PDB journal, June 2001 to February 2003, entries for April 18-21, 2002.

[134] See PDB journal, June 2001 to February 2003, entries for August 9, 11, and 22, 2002.

[135] Full name Dr. Pedro Antonio Dávalos González.

[136] See Davis journal, volume 10, June 30 and July 1, 1987. Dávalos also wrote Royce periodically seeking advice in his programs. (See letter, Pedro Dávalos to RSB dated February 1, 1985, July 4, 1992, and September 15, 1992.)

[137] PDB journal, June 2001 to February 2003, entries for October 23-27, 2002; see also JRB journal, December 2001 to January 2004, entries for October 23 and October 27, 2002.

[138] Vilma's book was entitled *El Cultivo de la Frutilla*, and it was published in Chile in 2002. Her dedicatory in the beginning reads, "*A mi maestro a quien debo mis conocimientos recopilados en este libro. Con mucho afecto, Vilma Villagrán, Abril 2003*" ("To my teacher to whom I owe the knowledge I have compiled in this book. With great affection, Vilma Villagrán, April 2003").

[139] Pearl's home had been moved from its previous location, and Royce's old home no longer stood.

[140] See PDB journal, February 2003 to January 2004, entries for May 5-10, 2003; also JRB journal, December 2001 to January 2004, entry for June 10, 2003. Although Royce's sister Rhea was still alive, Royce declined to visit her since she was suffering from advanced dementia, and he preferred to remember her as she had been in former years.

[141] On that day Pearl wrote, "Royce's 85th birthday. The day was fairly quiet. No one came to help us celebrate until our evening meal. I just fixed a sandwich. The lady missionaries came at 6:00 p.m. just when we were going to eat a bite. I asked if they had any place to go – they didn't, so I asked them to stay with us and eat with us. It made it nice to have company for Royce's birthday." (PDB journal, February 2003 to January 2004, entry for December 27, 2003.)

[142] See PDB journal, February 2003 to January 2004, entry for January 4, 2004.

[143] See PDB journal, January 2004 to January 2005, entry for December 20, 2004. Royce was very close to Daniel, having had associations with him both at Pearl's home and on his mission, when he stayed with Dan and his wife Mary while overseeing the missionary work in San Bernardino, California where Dan and Mary lived.

[144] See PDB journal, January 2004 to January 2005, entry for December 27, 2004.

[145] Hospice is a term used for a specialized form of nursing care designed for patients with terminal illnesses or who are in the last stages of life. It employs an approach which prioritizes the alleviation of symptoms, particularly

pain, over the treatment of the disease process, which is allowed to run its course. Hospice care is of particular value for family members of the patient, who often have never had to confront the dying process first-hand.

[146] Family Home Evening is a practice, promoted by the Church of Jesus Christ of Latter-day Saints, of holding a family gathering once a week, usually on Monday night, for companionship, spiritual instruction, and enjoyment. Often it involves a semiformal meeting with prayers and singing.

[147] This was Stevenson's "Requiem," which reads as follows:

Under the wide and starry sky
    Dig the grave and let me lie.
Glad did I live and gladly die,
    And I lay me down with a will.
This be the verse you grave for me;
    Here he lies where he longed to be,
Home is the sailor, home from sea,
    And the hunter home from the hill.

Margaret sang the poem to a tune of her own invention, with John accompanying on the piano. The tune was different from that which Royce had sung.

[148] This was "O Divine Redeemer" with music by Charles Gounod, and Royce's granddaughter Sarah Bringhurst accompanied on the piano.

[149] After Pearl's death the tombstone was replaced by a larger one which included Pearl's name, and the same strawberry image. The original tombstone was moved to Marla and John Vaughn's residence in Oroville.

Index

Christenson, (LDS line chief), 218
Chicago, Illinois, 140-144, 145-146, 178-179, 598-599 n. 64, n. 69, n. 71
China, 270, 284, 500, 502, 624 n. 14, 689 n. 230
chlorophyll deficiency, 268-274, 298
chloropicrin, 150, 158, 169, 330, 356, 599 n. 94, 638-639 n. 48
Church of Jesus Christ of Latter-day Saints, 2, 3, 4, 5, 12, 13-14, 16, 40, 207, 295, 347, 351, 362, 383-384, 392, 405-406, 410, 413-414, 415, 442-443, 558, 579 n. 1, 2, 4, 582 n. 44, 587 n. 125, 597 n. 56, 649 n. 11, 663-664 n. 149
Church welfare system, 262, 284, 295, 317, 352, 624 n. 17
citrus and subtropicals class, 402-403
Cittadella, Italy, 224
Civil Rights movement, 367, 445
Clayson, Elder, 91, 92
clover, see sweet clover
colchicine, 297, 334, 496, 629 n. 52, 643 n. 78
"Cold War," 437, 649-650 n. 14
Cole, Glen, 557
Colombia, 350
Columbia, South Carolina, 161-173, 180, 191, 206, 228, 233, 569 n. 77, 597 n. 49
Columbus, Ohio, 439
communism, 243, 284, 369, 437, 649-650 n. 14, 669 n. 195
computer analysis, 335, 640 n. 60-62, 685 n. 174
computer (navigator's), 214
Concepción, Chile, 309, 530
Constantine, Algeria, 184
Convenio program, 351, 360, 362, 363-365, 367, 648-649 n. 3, n. 6
Converse, Dick, 423, 466, 528, 536

Corle, Chester E., 221
Corvallis, Oregon, 423, 466, 467, 493, 500, 505, 510, 528-529, 536, 630 n. 60, 680 n. 94
Cooley, V. A. (chaplain), 218-219, 229, 232
Cooper, D. C., 619 n. 16, 620 n. 25
Cooper, James H., 199
Corbett, Ottis B., 91, 173
Corsica, 184-234, 376, 560
Cottonwood, Utah, 5, 72, 80, 108, 112, 267, 322, 593 n. 31
coumarin, 260, 262, 268-269, 274, 619-620 n. 17, 622 n. 41
Covert Training Unit, 149-150
Crane, Julian, 377
Craver, Luther S. 609 n. 80
creationists, 407, 685 n. 169
Cronan, Leo, MD, 635 n. 24, 637 n. 36
Cronkrite, Myrtle, 580 n. 14
"Cruz" strawberry, 394
CSAB, see California Strawberry Advisory Board
Cuba, 502
"Cuesta" strawberry, 535, 693-694 n. 49
"Cupertino" strawberry, 634 n. 12
curfew, 390, 392-393
Dahl, Sheldon, 662 n. 125
Dakar, French West Africa, 183, 604 n. 3, 605 n. 5
Dale, Elizabeth, 111
Danneberg, Robert A., 130, 133, 134
Darrow, George M., 339, 632 n. 8, 635 n. 19, 641-642 n. 70-71, 643 n. 77
Darwin, Charles, 393, 409, 631 n. 3
Dávalos, Pedro Antonio, 546, 683 n. 143, 700 n. 135-136
Davidson, Daniel, 45, 84, 236, 285, 575 n. 82, 700 n. 143
Davidson, Helen, see Toronto, Helen
Davidson, Jane, 21, 35, 37, 109-

www.ingramcontent.com/pod-product-compliance
Lightning Source LLC
Chambersburg PA
CBHW070646150426
42811CB00073B/1947